Shaping the Future of ICT

Shaping the Future of ICT

Trends in Information Technology, Communications Engineering, and Management

Edited by

Ibrahiem M. M. El Emary and Anna Brzozowska

CRC Press is an imprint of the
Taylor & Francis Group, an **informa** business

CRC Press
Taylor & Francis Group
6000 Broken Sound Parkway NW, Suite 300
Boca Raton, FL 33487-2742

CRC Press is an imprint of Taylor & Francis Group, an Informa business

Printed on acid-free paper

International Standard Book Number-13: 978-1-4987-8118-3 (Hardback)

Library of Congress Cataloging-in-Publication Data

Names: International Conference on Communications, Management, and
Information Technology (2016 : Calabria, Italy), author. | El Emary,
Ibrahiem, editor. | Brzozowska, Anna, editor.
Title: Shaping the future of ICT : trends in information technology,
communications engineering, and management / Ibrahiem El Emary and Anna Brzozowska.
Description: Boca Raton : CRC Press, [2018] | Selected and revised papers
from the International Conference on Communications, Management, and
Information Technology (ICCMIT'16) held April 26-29, 2016 in Calabria,
Italy sponsored by the Universal Society of Applied Research. | Includes
bibliographical references and index.
Identifiers: LCCN 2017001160| ISBN 9781498781183 (hardback : alk. paper) |
ISBN 9781498781190 (ebook)
Subjects: LCSH: Information technology--Congresses. | Computer
networks--Congresses. | Computer science--Congresses.
Classification: LCC T58.5 .I5662225 2016 | DDC 004--dc23
LC record available at https://lccn.loc.gov/2017001160

Visit the Taylor & Francis Web site at
http://www.taylorandfrancis.com

and the CRC Press Web site at
http://www.crcpress.com

Contents

Preface......ix
Editors......xiii
Contributors......xv

SECTION I INFORMATION TECHNOLOGY

Chapter 1 Computer Vision for Object Recognition and Tracking Based on Raspberry Pi......3

Ali A. Abed and Sara A. Rahman

Chapter 2 Modeling of the Teaching–Learning Process in E-Learning......15

Mustapha Bassiri, Mohamed Radid, and Said Belaaouad

Chapter 3 Solving the Problems of Linguistically Diverse First-Year University Students Using Digital Learning......41

Dace Ratniece and Sarma Cakula

Chapter 4 Moving Enterprise Architecture from Professional Certificates to Academic Credentials......49

Fekry Fouad

Chapter 5 Automatic Identification and Data Capture Techniques by Radio-Frequency Identification (RFID) Tags: Reader Authentication and Ethical Aspects......63

H. Saadi, R. Touhami, and M. C. E. Yagoub

Chapter 6 Artificial Intelligence in E-Learning......77

Hachem Alaoui Harouni, Elkaber Hachem, and Cherif Ziti

Chapter 7 Wireless Multimedia Sensor Networks: Cross-Layer Approach Protocols......93

Manal Abdullah and Ablah AlAmri

Chapter 8 Survey of E-Learning Standards......109

Manal Abdullah and Nashwa AbdelAziz Ali

Chapter 9 A Classification Perspective of Big Data Mining......123

Manal Abdullah and Nojod M. Alotaibi

Chapter 10 Streaming Data Classification in Clustered Wireless Sensor Networks 137

Manal Abdullah and Yassmeen Alghamdi

Chapter 11 Video Codecs Assessment over IPTV Using OPNET 153

Eman S. Sabry, Rabie A. Ramadan, M. H. Abd El-Azeem, and Hussien ElGouz

Chapter 12 Brain Storm Optimization (BSO) for Coverage and Connectivity Problems in Wireless Sensor Networks 165

Rabie A. Ramadan and A. Y. Khedr

Chapter 13 An Exploratory Evaluation of the Awareness of E-Government Services among Citizens in Saudi Arabia 177

S. Q. Al-Khalidi Al-Maliki

Chapter 14 An Internet of Things (IoT) Course for a Computer Science Graduate Program........ 187

Xing Liu and Orlando Baiocchi

Chapter 15 Modeling Guidelines of FreeRTOS in Event-B 195

Eman H. Alkhammash, Michael J. Butler, and Corina Cristea

Chapter 16 Generation of Use Case UML Diagram from User Requirement Specifications... 217

Wahiba Ben Abdessalem and Eman H. Alkhammash

SECTION II COMMUNICATION SYSTEMS

Chapter 17 Design and Simulation of Adaptive Cognitive Radio Based on Software-Defined Radio (SDR) Using Higher-Order Moments and Cumulants..................... 229

Ahmed Abdulridha Thabit and Hadi T. Ziboon

Chapter 18 Effect of Black Hole Attacks on Delay-Tolerant Networks..................................... 243

Alaa Hassan and Wafa Ahmed El Gali

Chapter 19 Proposed Adaptive Neural-Fuzzy Inference System (ANFIS) Identifier for M-ary Frequency Shift Keying (FSK) Signals with Low SNR.......................... 259

Hadi A. Hamed, Sattar B. Sadkhan, and Ashwaq Q. Hameed

Chapter 20 New Image Scrambling Algorithm Depending on Image Dimension Transformation Using Chaotic Flow Sequences................................... 269

Hadi T. Ziboon, Hikmat N. Abdullah, and Atheer J. Mansor

Chapter 21 Efficient Two-Stage Sensing Method for Improving Energy Consumption in Cognitive Radio Networks 281

Hikmat N. Abdullah and Hadeel S. Abed

Chapter 22 Current Status of Information Security Based on Hybrid Crypto and Stego Systems 297

Rana Saad Mohammed and Sattar B. Sadkhan

Chapter 23 The Status of Research Trends in Text Emotion Detection 309

Rusul Sattar B. Sadkhan and Sattar B. Sadkhan

Chapter 24 Reaction Automata Direct Graph (RADG) Design on Elliptic Curve Cryptography 321

Salah A. Albermany and Ali Hasan Alwan

Chapter 25 A Performance Comparison of Adaptive LMS, NLMS, RLS, and AFP Algorithms for Wireless Blind Channel Identification 333

Sami Hasan and Anas Fadhil

Chapter 26 Design of New Algorithms to Analyze RC4 Cipher Based on Its Biases 347

Ali M. Sagheer and Sura M. Searan

Chapter 27 Extracting Implicit Feedback from Users' GPS Tracks Dataset: A New Developed Method for Recommender Systems 363

Tawfiq A. Alasadi and Wadhah R. Baiee

SECTION III MANAGEMENT

Chapter 28 Management through Opportunities as an Unconventional Solution in the Theory of Strategic Management 377

Anna Brzozowska and Katarzyna Szymczyk

Chapter 29 Concept of Supply Chain Management in the Context of Shaping Public Value 389

Dagmara Bubel

Chapter 30 Current State of Information and Communication Provision of Ukraine and Spread of Information Technologies in Agricultural Sector 403

Antonina Kalinichenko and Oleksandr Chekhlatyi

Chapter 31 Selected Issues of Management of Green Logistics in Transport Sector 413

Marta Kadłubek

Chapter 32 HR Analytics as a Support of High- and Mid-Level Managers in Contemporary Enterprises in Eastern Poland 425

Monika Wawer and Piotr Muryjas

Chapter 33 Effects and Impact of Playing Computer Games on Gamers 439

W. Chmielarz and O. Szumski

Chapter 34 ICT Drivers of Intelligent Enterprises 455

Monika Łobaziewicz

Chapter 35 MVDR Beamformer Model for Array Response Vector Mismatch Reduction 467

Suhail Najm Shahab, Ayib Rosdi Zainun, Essa Ibrahim Essa, Nurul Hazlina Noordin, Izzeldin Ibrahim Mohamed, and Omar Khaldoon

Index 481

Preface

The current era is named the era of communication and information technology with its various and numerous applications in many areas of life, civilian and military, including transportation, satellites and aerospace, medicine, education, and other fields reflecting on all production and service sectors. There are also many jobs that will flourish thanks to information and communication, including technology, programming services, information services, digital services, repair satellites, maintenance and repair of computers, sea and ocean science, biotechnology, prosthetic technology, solar energy, web page designing, computer and information technology, satellites and satellite TV, mail armament, the digital economy and e-commerce, tourism, transportation, electronic devices, e-learning, robotics, medicine, e-libraries, and electronic media.

Information and communication technology in recent decades has witnessed rapid development, thereby increasing the capacity of communities and individuals to obtain and process information exponentially, as there are available to large segments of people new tools for learning and professional development. A computer has become the most important education technology used in formal education and informal education alike, and changed the nature of the information and its role and function. What was once limited is now unlimited and available in multiple formats, and after it had been a goal of education, communication technology has become a tool for capacity development.

As communication technology has expanded the sources of knowledge and doubled the websites available to obtain information, it has produced tools and new forms to gain knowledge and skills, building environments that can be considered a revolution in the field of education. The learner's site has changed and moved from a small environment to the wide open world and resulted in information technology contributing to the formation of a new global culture, beyond the characteristics of people and narrowing the differences between them.

However, developing countries face new challenges in the light of the concept of the "global village" and the rising of knowledge and information with its importance in achieving development. It is no secret that developing countries with the needs of progress and reform should identify challenges and study them, and require meeting these needs in their educational systems. So, knowledge and skills have become critical to achieving economic growth and the development of sound leadership and resource development. There is no doubt that technology can play an active role in the work of institutions and productivity across all stages of development and reform.

All this raises important questions in the field of teaching and learning, such as: What are the kinds of skills and learning styles that may be needed by the young people in different countries, so that they can develop their knowledge and skills continually in the rapidly changing world? How can educational systems take advantage of technological development to improve productivity? These questions apply to all stages of education, in general, and higher education, in particular. Higher education is an active factor in social and economic development, because of the field of research, dissemination of knowledge and skills development, and production of leaders with transformational and developmental impact. Higher education faces many challenges, including providing equal opportunities and the quality of education, and meeting the needs of growing research and development. Also developing countries face obstacles affecting the development of capabilities, such as the production and dissemination of knowledge, or, in the era of globalization and information, dissemination of a single language. Thus, the ability to access the information level of proficiency in the language should be associated with the development process in the developing countries, which highlights the issue of unequal access to information and ways to overcome this barrier, which restrain the development process.

Based on the previous facts, in pursuit of keeping up with what's new and human development in various educational, industrial, and commercial fields, this book offers a range of outstanding

scientific chapters in the fields of computers, communications, and administration that discuss, decompose, and illustrate many of the theories and algorithms as well as computer applications that benefit all paths of life. This book is divided into three sections.

Section I is dedicated to information technology and discusses subjects like

- Modeling of the teaching–learning process in e-learning
- Solving the problems of linguistically diverse students using digital learning
- Enterprise architecture moving from professional certificates to academic credentials
- Automatic identification and data capture techniques by RFID tags applied to reader authentication
- Artificial intelligence in e-learning
- Cross-layer approach protocols for wireless multimedia sensor networks
- E-learning standards
- Classification perspective of big data mining
- Streaming data classification in clustered wireless sensor networks
- Video Codec's assessment over IPTV using OPNET
- Brainstorming algorithm for coverage and connectivity problems in wireless sensor networks
- E-government services among citizens in Saudi Arabia
- MVDR beamformer model for array response vector mismatch reduction
- An IoT course for a computer science graduate program
- Modeling guidelines of FreeRTOS in Event-B
- Generation of use case UML diagram from user requirement specifications

Section II is devoted to communication that focuses on the advanced subjects including

- Design and simulation of adaptive cognitive radio based on SDR using higher-order moments
- The effect of black hole attacks on delay-tolerant networks
- Proposed adaptive neural-fuzzy inference system (ANFIS) identifier for M-Ary FSK signals with low SNR
- New image scrambling algorithm for image dimension transformation using chaotic flow sequences
- Efficient two-stage sensing method for improving energy consumption in cognitive radio networks
- Information security based on hybrid crypto and stego systems
- Text emotion detection
- RADG design on elliptic curve cryptography
- A performance comparison of adaptive LMS, NLMS, RLS, and AFP algorithms for wireless blind channel identification
- Design of new algorithms to analyze RC4 cipher based on its biases
- Extracting implicit feedbacks from users' GPS tracks

Section III covers management science where it discusses the following issues:

- Management through opportunities as an unconventional solution in the theory of strategic management
- Concept of supply chain management in the context of shaping public value
- Information and communication in Ukraine and spread of information technologies in the agricultural sector
- Management of green logistics in transport

- ICT drivers of intelligent enterprises
- HR analytics as a support for managers in enterprises in Eastern Poland
- Effects and impact of playing computer games on gamers

Finally, the Editorial Board of this book consider that the publication of this book is an important step on the road to building the culture and awareness of information and communication technology issues in the modern era, where scientific advances in every direction and every field leads us to the need to interact and take advantage of the fields of communications and information technology as well as management.

Editors

Ibrahiem M. M. El Emary, PhD, is a professor of computer science and engineering at King Abdulaziz University, Jeddah, Kingdom of Saudi Arabia. His research interests cover various analytic and discrete event simulation techniques, performance evaluation of communication networks, application of intelligent techniques in managing computer communication networks, and performing comparative studies between various policies and strategies of routing, congestion control, and subnetting of computer communication networks. He has published more than 200 articles in various refereed international journals and conferences covering the subjects of computer networks, artificial intelligence, expert systems, software agents, information retrieval, e-learning, case-based reasoning, image processing and pattern recognition, wireless sensor networks, cloud computing, and robotic engineering. Some of these articles were published in ISI journals with Impact Factor. He has published seven chapters in three international books (published by Springer Verlag, IGI, and NOVA Science), and was the author of two books edited by the international publisher Lambert Academic Publishing and another book with Taylor & Francis, which was selected as the book of the year in 2013 by the ACM Digital Library. He has participated in more than 30 international conferences as a keynote speaker and potential speaker. He is editor-in-chief or editor of more than 10 international refereed journals in computer science and engineering. He achieved the title of highly cited scholar from IGI in 2012. He has been included in Marquis *Who's Who in the World*, 2013 and 2015 editions.

Anna Brzozowska is associate professor in economics, in the discipline of management studies. Currently, she works in Department of Business Informatics, Faculty of Management, Czestochowa University of Technology, Poland. She is the author of over 50 articles in national and international journals and 18 papers at national and international conferences. She is also the author and co-author of 9 monographs, 20 chapters in a monograph, and 1 book. She has actively participated in over 20 conferences, mainly international, as well as in few research projects. She is the editor of *EUREKA Journal: Social and Humanities*, and the editor of selected editions of Research Bulletin of the Technical University of Częstochowa. She is also a scientific committee member of economics, *Management and Sustainability Scientific Journal.*

Research activities of Anna Brzozowska are focused on enterprise management processes and the integration chain, its logistics aspects and institutional environment, influencing management in the light of integration activities that oscillate around the use and management of the European Union projects. These issues develop systematically by conducting research in the Silesian Voivodeship, participating in thematic groups and research and development teams as well as through cooperation with business and local and regional authorities. Her main research and study is the broadly defined concept of management in agribusiness and the European Union integration processes; IT in business; managing EU projects and farms through absorption of the Rural Area Development Programme funds; organization of management, logistics, and transport processes; managing the transport process; systematic interpretation of logistic relations of the economic situation of transport sector operators; analysis of key transport management processes; identifying integration factors of shipping industry operators; defining the key concepts of transport facilities; analysis of transport, goods storage, completion, and dispatching processes; use of transport management methods and techniques; role of transport in the economy of business operators.

Contributors

M. H. Abd El-Azeem
College of Engineering and Technology
Arab Academy for Science
Technology and Maritime Transport
Cairo, Egypt

Hikmat N. Abdullah
College of Information Engineering
Al-Nahrain University
Baghdad, Iraq

Manal Abdullah
Faculty of Computing and Information Technology
King Abdulaziz University
Jeddah, Saudi Arabia

Hadeel S. Abed
College of Information Engineering
Al-Nahrain University
Baghdad, Iraq

Ali A. Abed
Department of Computer Engineering
University of Basra
Basra, Iraq

S. Q. Al-Khalidi Al-Maliki
Department of Management Information Systems
King Khalid University
Abha, Saudi Arabia

Ablah AlAmri
Jeddah Community College
King Abdulaziz University
Jeddah, Saudi Arabia

Tawfiq A. Alasadi
Software Department
University of Babylon
Babylon, Iraq

Salah A. Albermany
Department of Computer Science
University of Kufa
Kufa, Iraq

Yassmeen Alghamdi
Department of Computer Science
King Abdulaziz University
Jeddah, Saudi Arabia

Nashwa AbdelAziz Ali
Arab Academy for Science
Technology and Maritime Transport
Cairo, Egypt

Eman H. Alkhammash
Department of Computer Science
University of Taif
Taif, Saudi Arabia

Nojod M. Alotaibi
Department of Computer Science
King Abdulaziz University
Jeddah, Saudi Arabia

Ali Hasan Alwan
Department of Computer Science
University of Kufa
Kufa, Iraq

Wadhah R. Baiee
Software Department
University of Babylon
Babylon, Iraq

Orlando Baiocchi
Institute of Technology
University of Washington–Tacoma
Tacoma, Washington

Mustapha Bassiri
Department of Chemistry
and
Department of Technical Science of Physical Activities
Hassan II University of Casablanca
Casablanca, Morroco

Said Belaaouad
Department of Chemistry
Hassan II University of Casablanca
Casablanca, Morroco

Wahiba Ben Abdessalem
RIADI laboratory
University of Manouba
Manouba, Tunisia

and

Department of Computer Science
University of Taif
Taif, Saudi Arabia

and

Department of Computer Science
High Institute of Management of Tunis
University of Tunis
Tunis, Tunisia

Anna Brzozowska
Faculty of Management
Częstochowa University of Technology
Częstochowa, Poland

Dagmara Bubel
Main Library
Częstochowa University of Technology
Częstochowa, Poland

Michael J. Butler
Department of Electronics and Computer Science
University of Southampton
Southampton, UK

Sarma Cakula
Information Technology Department
Vidzeme University of Applied Science
Valmiera, Latvia

Oleksandr Chekhlatyi
Department of Economic Cybernetics and Information Technologies
Poltava State Agrarian Academy
Poltava, Ukraine

W. Chmielarz
Faculty of Management
University of Warsaw
Warsaw, Poland

Corina Cristea
Department of Electronics and Computer Science
University of Southampton
Southampton, UK

Hussien ElGouz
College of Engineering and Technology
Arab Academy for Science
Technology and Maritime Transport
Cairo, Egypt

Essa Ibrahim Essa
College of Computer Science and Mathematics
Tikrit University
Tikrit, Iraq

Anas Fadhil
College of Information Engineering
Al-Nahrain University
Baghdad, Iraq

Fekry Fouad
Information Science Department
King Abdulaziz University
Jeddah, Saudi Arabia

Wafa Ahmed El Gali
Computer and Internet Center
College of Dentistry
University of Kirkuk
Kirkuk, Iraq

Elkaber Hachem
Mathematics and Computer Department
Moulay Ismail University
Meknes, Morocco

Hadi A. Hamed
Al-Mussaib Technical College
Al-Furat Al-Awsat Technical University
Babylon, Iraq

Ashwaq Q. Hameed
Electrical Engineering Department
University of Technology
Bagdad, Iraq

Hachem Alaoui Harouni
Mathematics and Computer Department
Moulay Ismail University
Meknes, Morocco

Sami Hasan
College of Information Engineering
Al-Nahrain University
Baghdad, Iraq

Alaa Hassan
Computer and Internet Center
College of Dentistry
University of Kirkuk
Kirkuk, Iraq

Marta Kadłubek
Faculty of Management
Częstochowa University of Technology
Częstochowa, Poland

Antonina Kalinichenko
Department of Process Engineering
University of Opole
Opole, Poland

Omar Khaldoon
School of Computer and Communication Engineering
Universiti Malaysia Perlis
Arau, Malaysia

A. Y. Khedr
Systems and Computers Department
Alazhar University
Cairo, Egypt

and

Department of Computer Science and Software Engineering
University of Hail
Hail, Saudi Arabia

Xing Liu
Institute of Technology
University of Washington–Tacoma
Tacoma, Washington

Monika Łobaziewicz
Faculty of Management
Politechnika Lubelska
Lublin, Poland

Atheer J. Mansor
College of Electric and Electronic Engineering
University of Technology
Baghdad, Iraq

Izzeldin Ibrahim Mohamed
Faculty of Electrical and Electronics Engineering
Universiti Malaysia Pahang
Pahang, Malaysia

Rana Saad Mohammed
Education College
Mustansiriyah University
Baghdad, Iraq

Piotr Muryjas
Institute of Computer Science
Lublin University of Technology
Lublin, Poland

Nurul Hazlina Noordin
Faculty of Electrical and Electronics Engineering
Universiti Malaysia Pahang
Pahang, Malaysia

Mohamed Radid
Department of Chemistry
Hassan II University of Casablanca
Casablanca, Morroco

Sara A. Rahman
Department of Electrical Engineering
University of Basra
Basra, Iraq

Rabie A. Ramadan
Computer Engineering Department
Cairo University
Giza, Egypt

and

Department of Computer Science and Software Engineering
University of Hail
Hail, Saudi Arabia

Dace Ratniece
Distance Education Center
Riga Technical University
Riga, Latvia

and

University of Liepaja
Liepaja, Latvia

H. Saadi
Faculty of Electronics and Informatics
University of Science and Technology Houari Boumediene
Bab Ezzouar, Algeria

Eman S. Sabry
Department of Electronics and Electrical Communication
El Shorouk Academy
Cairo, Egypt

Rusul Sattar B. Sadkhan
Computer Engineering College
Razi University
Kermanshah, Iran

Sattar B. Sadkhan
IT College
Babylon University
Babylon, Iraq

Ali M. Sagheer
Department of Computer Science
University of Anbar
Anbar, Iraq

Sura M. Searan
Department of Computer Science
University of Anbar
Anbar, Iraq

Suhail Najm Shahab
Faculty of Electrical and Electronics Engineering
Universiti Malaysia Pahang
Pahang, Malaysia

and

Electrical Department
Al-Hawija Technical Institute
Northern Technical University
Baghdad, Iraq

O. Szumski
Faculty of Management
University of Warsaw
Warsaw, Poland

Katarzyna Szymczyk
Faculty of Management
Częstochowa University of Technology
Częstochowa, Poland

Ahmed Abdulridha Thabit
Department of Communications and Computers
Al-Rafidain University
Baghdad, Iraq

R. Touhami
Faculty of Electronics and Informatics
University of Science and Technology Houari Boumediene
Bab Ezzouar, Algeria

Monika Wawer
Institute of Economics and Management
John Paul II Catholic University of Lublin
Lublin, Poland

M. C. E. Yagoub
School of Electrical Engineering and Computer Science
University of Ottawa
Ottawa, Ontario, Canada

Ayib Rosdi Zainun
Faculty of Electrical and Electronics Engineering
Universiti Malaysia Pahang
Pahang, Malaysia

Hadi T. Ziboon
Department of Electrical and Electronic Engineering
University of Technology
Baghdad, Iraq

Cherif Ziti
Mathematics and Computer Department
Moulay Ismail University
Meknes, Morocco

Section I

Information Technology

1 Computer Vision for Object Recognition and Tracking Based on Raspberry Pi

Ali A. Abed and Sara A. Rahman

CONTENTS

1.1 Introduction 3
1.2 Object Tracking Using CamShift Algorithm 4
1.2.1 Mean-Shift Algorithm 4
1.2.2 CamShift Algorithm 4
1.2.3 Account for Search Window Size 5
1.3 Object Tracking Using Color Detection 6
1.4 Histogram Equalization 8
1.5 System Requirement 9
1.5.1 Raspberry Pi 9
1.5.2 Camera Module 9
1.6 Results 10
1.7 Conclusion and Future Works 13
References 14

1.1 INTRODUCTION

Vision-based systems have become part of everyday life; therefore this work will be in the field of artificial vision based on image processing suitable for many applications such as mobile robots navigation. Computer vision is a type of processing that inputs images producing output that could be a set of characteristics or parameters related to images. Its application in robotics, surveillance, monitoring, and security systems makes it very important and widespread worldwide. The work starts with creating a model to make useful decisions about real physical objects and scenes. A camera mimics and uses real-time digital videos for object recognition and tracking [1].

Object tracking is the main task in the field of computer vision. It has many applications in traffic control, human–computer interaction, digital forensics, gesture recognition, augmented reality, and visual surveillance [2]. An efficient tracking algorithm will lead to the best performance of higher-level vision tasks, such as automated monitors and human–computer interaction. Among the various tracking and recognition algorithms, CamShift tracking and color detection algorithms will be adopted in this chapter. CamShift is primarily intended to perform efficient head and face tracking in a perceptual user interface. It is based on an adaptation of mean shift that, given a probability density image, finds the mean (mode) of the distribution by iterating in the direction of maximum increase in probability density [3]. The aim of the color detection is to identify the category pixel color in a given image. Color detection had already gained the attention of researchers for the possibility of robust and efficient human body detection.

Zhao et al. [4] proposed a method for tracking objects with their size and shapes that change with time, on the basis of a group of mean-shift and affine structure. The results showed the object's tracking capability in during scale change and partial blockage. Emami [5] submitted an effective color-based CamShift algorithm for target tracking.

Altun et al. [6] suggested a method for efficient color detection in RGB space in hierarchical structure of neural networks. The results show that the proposed hierarchical structure of neural networks is best on traditional neural network classifier in color detection. Zhang et al. [7] used the color cooccurrence histogram (CH) for recognizing objects in images. The results show that the algorithm works in spite of confusing background clutter and moderate amounts of blockage and object praise. Although the color detection and CamShift algorithms have many advantages, they do not work in the dark and this makes it difficult to continue tracking during the lack of light, so in this chapter, we introduce an algorithm that makes these algorithms operate normally even in darkness.

In this chapter, a new method for recognition and tracking by using CamShift and color detection is proposed and compared. All the presented algorithms were programmed with Python programming language supported by OpenCV libraries, and executed with a credit card–size computer board called Raspberry Pi with attached external camera.

1.2 OBJECT TRACKING USING CAMSHIFT ALGORITHM

1.2.1 Mean-Shift Algorithm

Mean shift is a nonparametric feature-space analysis technique for locating the maxima of a density function. Cluster analysis or clustering is the task of grouping a set of objects in such a way that objects in the same group (called a cluster) are more similar (in some sense or another) to each other than to those in other groups (clusters). It is a main task of exploratory data mining. The mean-shift tracker provides accurate identification of the location and it is computationally possible. The form that is used widespread for target representation is color histograms, due to its independence from scaling and for the purposes of rotation and its durability to partial blockage. To keep track of the target using the mean-shift algorithm, it repeats the following steps [7]:

1. Choose a search window size and the initial site of the search window.
2. Account for the mean site in the search window.
3. Determine the center of the search window at the mean site computed in step 2.
4. Repeat steps 2 and 3 until rapprochement (or until the mean location moves less than a predefined threshold).

1.2.2 CamShift Algorithm

CamShift refers to the continuously adaptive mean shift algorithm. A probability distribution image of the color required in the video relay is created by using the CamShift algorithm. The CamShift algorithm first establishes a model of the desired hue using a color histogram and then uses the hue saturation value (HSV) color system, which is compatible with the projecting standard BGR color space along its director diagonal from white to black. Color distributions derived from video image sequences change with time, therefore the mean shift algorithm must be modified to adapt dynamically to the probability distribution and its relay.

The CamShift algorithm is dependent on an adaptation of mean shift algorithm. It is calculated as follows [7]:

1. Select the initial location of the search window.
2. Mean shift as described earlier (one or many repetitions); store the zeroth moment.
3. Adjust the search window size equal to a function of the zeroth moment located in step 2.
4. Repeat steps 2 and 3 until conversion (average site moves less than a predefined threshold).

1.2.3 Account for Search Window Size

In CamShift, to calculate the search window follows these steps [8]:

1. Compute the zero-order moment:

$$m_{00} = \sum_x \sum_y I(x, y) \tag{1.1}$$

2. Calculate the first-order moment:

$$m_{10} = \sum_x \sum_y xI(x, y) \tag{1.2}$$

$$m_{01} = \sum_x \sum_y yI(x, y) \tag{1.3}$$

3. Calculate the centroid position of the search window:

$$x_c = \frac{m_{10}}{m_{00}} \tag{1.4}$$

$$y_c = \frac{m_{01}}{m_{00}} \tag{1.5}$$

4. Then set the initial window size s and the relationship:

$$s = 2 * \sqrt{\frac{m_{00}}{256}} \tag{1.6}$$

Taking into account the symmetry, calculate the results to the nearest odd number. By calculating the two-order image, moments can be tracked in the target direction:

$$m_{20} = \sum_x \sum_y x^2 I(x, y) \tag{1.7}$$

$$m_{02} = \sum_x \sum_y y^2 I(x, y) \tag{1.8}$$

$$m_{11} = \sum_x \sum_y xyI(x, y) \tag{1.9}$$

Obtain the length and width of the search window as follows:

Window direction:

$$\theta = \frac{\arctan(b / a - c)}{2} \tag{1.10}$$

where:

$$a = \frac{m_{20}}{m_{00}} - x_c^2 \tag{1.11}$$

$$b = 2\left(\frac{m_{11}}{m_{00}} - x_c y_c\right) \tag{1.12}$$

$$c = \frac{m_{02}}{m_{00}} - y_c^2 \tag{1.13}$$

1.3 OBJECT TRACKING USING COLOR DETECTION

Color is used to identify and isolate objects. The RGB values of every pixel in the frame is read and turned into the HSL color space.

The flowchart of the CamShift algorithm is shown in Figure 1.1 [9]. The HSL color space is selected because that chromatic information is independent from the lighting situations. Hue identifies basic color, saturation decides the intensity of color, and lighting depends on lighting condition.

Because each color has its own range of H values, the program compares the H values of each pixel with a predefined group of H values of the relegation zone. If it is within 10%, the pixel is marked as being a part of relegation zone. When the group of H values is correctly chosen, the landing spot is identified more accurately. After passing all the pixels for one frame, the centroid of all the marked pixels is calculated. Conversion of the BGR color space to HSV color space is performed using the following Equation 1.14 [10]:

$$h = \begin{cases} 0, \ldots\ldots\ldots \text{if } \max = \min, \\ \left(60° * \dfrac{g-b}{\max - \min} + 360°\right), \ldots\ldots \text{if } \max = r, \\ 60° * \dfrac{b-r}{\max - \min} + 120°, \ldots\ldots \text{if } \max = g, \\ 60° * \dfrac{r-g}{\max - \min} + 240°, \ldots\ldots \text{if } \max = b, \end{cases} \tag{1.14}$$

where r (red), g (green), b (blue) $\in [0,1]$ are the coordinates of the BGR color space, and max and mini compatible to the greatest and least of r, g, and b, respectively. The hue angle $h \in [0,360]$ for HSV color space.

The value of h is normalized to lie between 0 and 180 to fit into an 8 bit gray scale image (0–255), and $h = 0$ is used when max = min, although hue has no geometric meaning for gray. The s and v values for IISV color space are determined as follows:

$$s = \begin{cases} 0, \ldots\ldots \text{if } \max = 0, \\ \dfrac{\max - \min}{\max} = 1 - \dfrac{\min}{\max}, \ldots\ldots \text{otherwise} \end{cases} \tag{1.15}$$

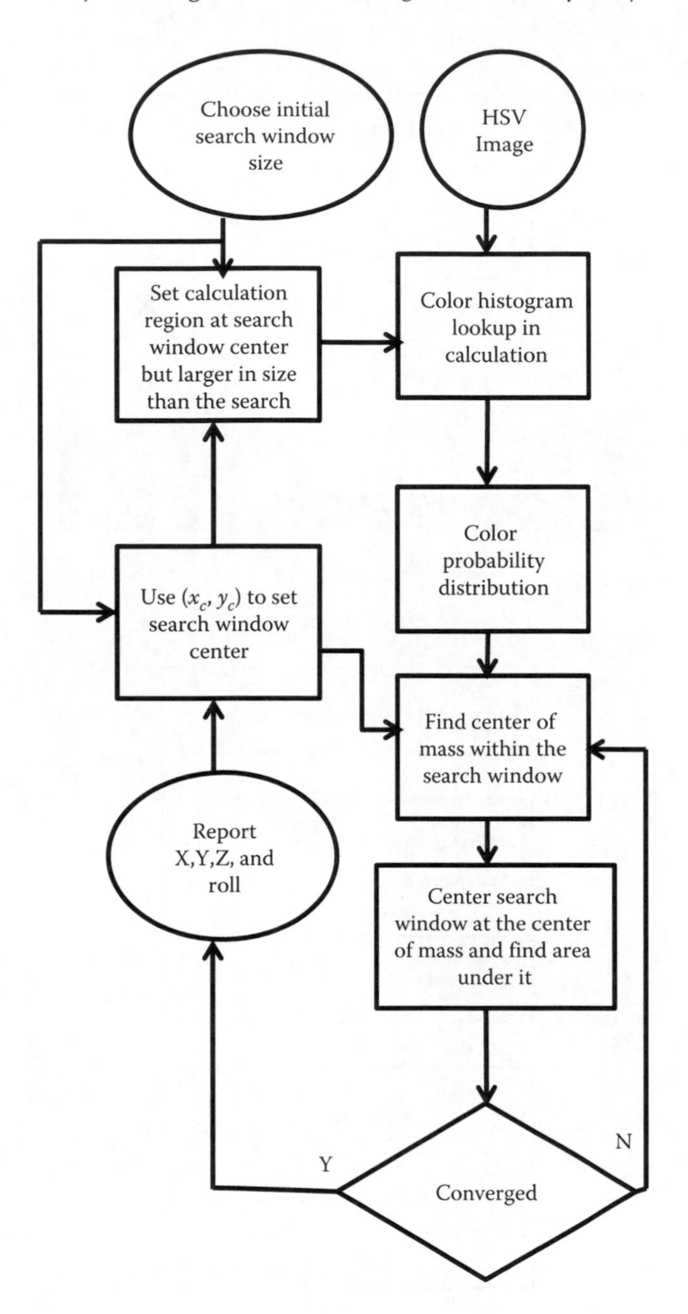

FIGURE 1.1 Flow chart of CamShift tracking algorithm.

The *v* or value channel represents the gray scale portion of the image. The threshold for the hue value of the image is scheduled on the basis of a color-mounted marker color. Using the threshold value, segmentation between the color required and other colors is performed. The resulting image is a binary image with white that indicates the desired color region and black is assumed to be the noisy region.

Visible colors located between 400 nm (violet) and 700 nm (red) on the electromagnetic spectrum are as shown in Figure 1.2 [11].

The block diagram for object tracking by color detection is shown in Figure 1.3 [12].

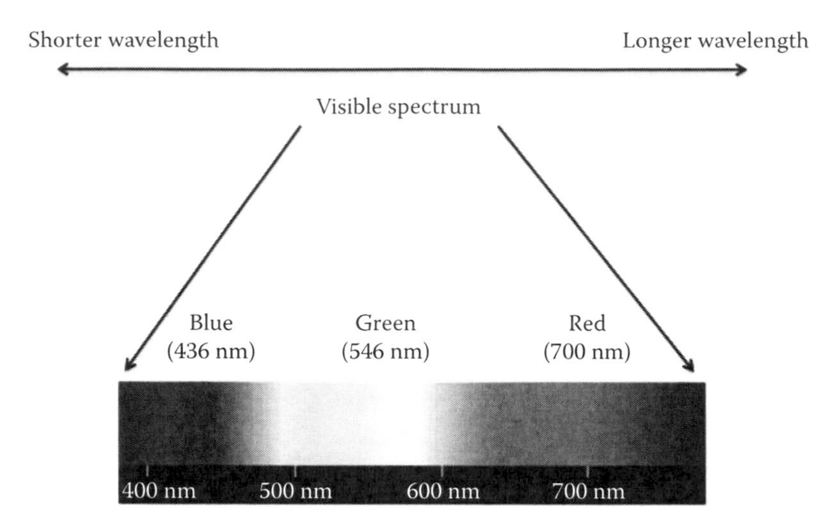

FIGURE 1.2 Visible spectrum.

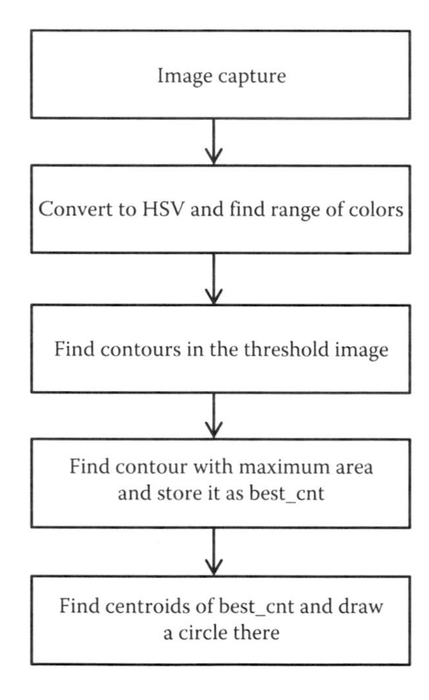

FIGURE 1.3 Flow chart for object tracking by color detection.

1.4 HISTOGRAM EQUALIZATION

Histogram equalization is a nonlinear technique for adjusting the contrast of an image using its histogram. It is the process that remaps the pixels of the image [13]. It increases the brightness of a gray scale image, which is different from the mean brightness of the original image. To following describes how to calculate the histogram to any image:

- The acquired image is converted into HSV (hue, saturation, and value).
- The image is then decomposed into two parts by using exposure threshold.

FIGURE 1.4 Raspberry Pi model B.

- The two parts are then equalized independently. Overenhancement is also controlled in this method by using the clipping threshold.

For measuring the performance of the enhanced image, entropy and contrast are calculated. Image enhancement is the process of improving the visual quality of an image so that the results are more suitable than the original.

1.5 SYSTEM REQUIREMENT

The hardware system consists of the following parts:

1. The Raspberry Pi Model B to run object recognition and tracking programs
2. An image capturing camera

1.5.1 Raspberry Pi

The Raspberry Pi Model B shown in Figure 1.4 is a low cost, credit card–size computer that plugs into a computer monitor or TV, and uses a standard keyboard and mouse. The Raspberry Pi 2 Model B now features more than 1 GB of RAM memory. The operating system kernel has been upgraded to take full advantage of the latest ARM Cortex-A7 technology and is available with the new version 1.4 of NOOBS software. The software NOOBS is known as New Out-Of-Box Software [14].

1.5.2 Camera Module

The camera module for Raspberry Pi shown in Figure 1.5 can be used for taking high-definition video, in addition to still photographs.

It is designed to contact to camera serial interface (CSI) of the Raspberry Pi. The camera module fits with all models and versions of Raspberry Pi. The camera module is very common in homeland security applications, and in wildlife and other locations. The camera module can work in programs by executing the following instructions [15]:

```
from time import sleep
from picamera import PiCamera
camera = PiCamera()
```

FIGURE 1.5 Raspberry Pi camera module.

1.6 RESULTS

Our algorithm is applied with python code using OpenCV library. CAMShift and color detection algorithms are used to track various objects based on color, in environments with different illumination. The experimental results show that the color detection gave the same results in two different conditions, while the CAMShift algorithm in natural lighting and very low lighting it is not gave the same results. It is used in this chapter with the Raspberry Pi Model B that is shown in Figure 1.4, mouse, keyboard, HDMI cable, and Raspberry Pi Camera Module shown in Figure 1.5. It is used during the tracking of the objects shown in Figure 1.6, where it has the blue color.

The results of color detection in natural lighting are shown in Figure 1.7. The emergence of on-screen objects is identified by their color. The algorithm gave the same results in different intensity lighting, as shown in Figure 1.8. In addition to using the color detection algorithm, the traditional CamShift algorithm was used for comparison, where it is noticed as shown in Figure 1.9 how the tracking process works in natural light.

In the case of CamShift algorithm, it is noticed that the results vary depending on the intensity of light as shown in Figure 1.10. The loss of the tracking process in the dark makes this algorithm inefficient, and therefore, it is found that the tracking process based on color detection more suitable.

Following the results of algorithms, it was noticed some of the tracked images may be less obvious due to some external factors, for example, the intensity of external lighting that negatively affects the tracking process. These results were conducted on the histogram equalization. After conducting this technical process, more results became visible and this ensures the success of the tracking process. The histogram equalization process is shown in Figure 1.11a,b.

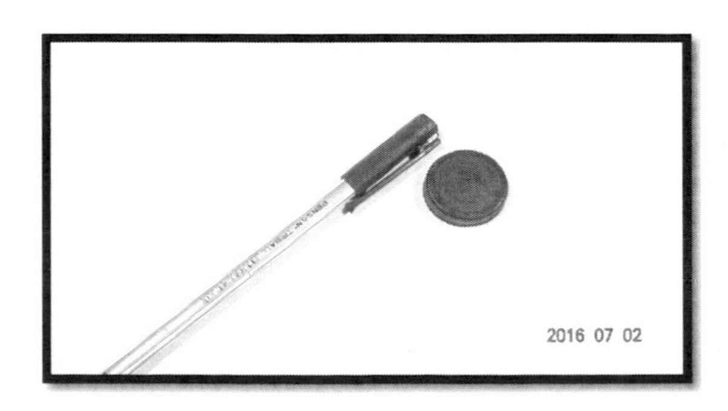

FIGURE 1.6 Examples of objects that are used to track.

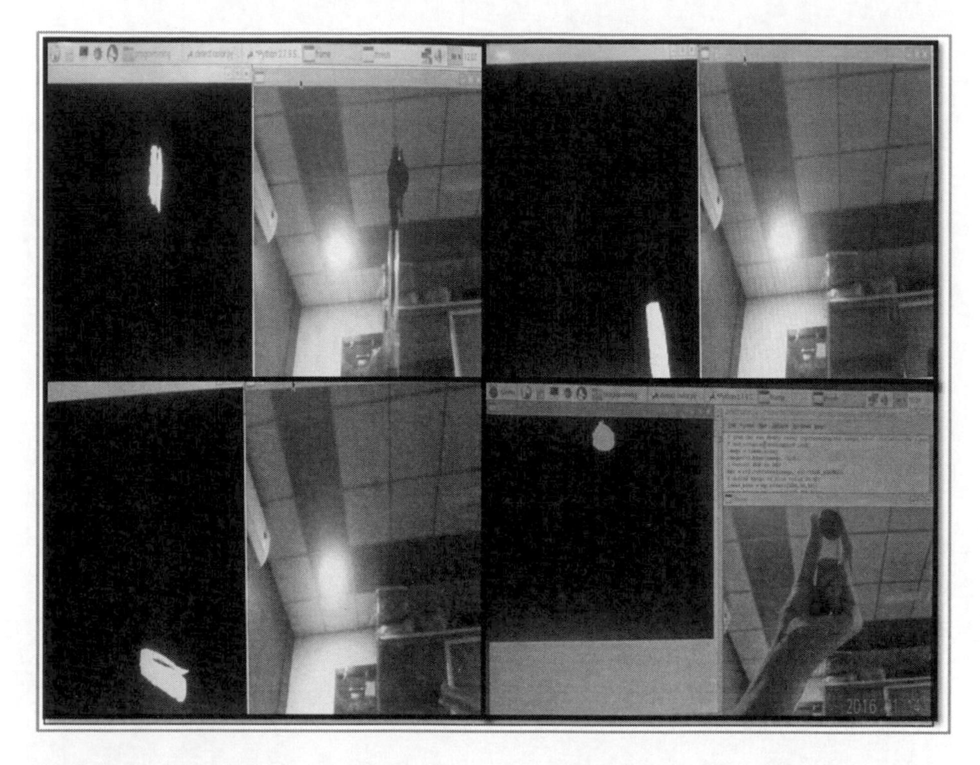

FIGURE 1.7 Color detection in natural lighting.

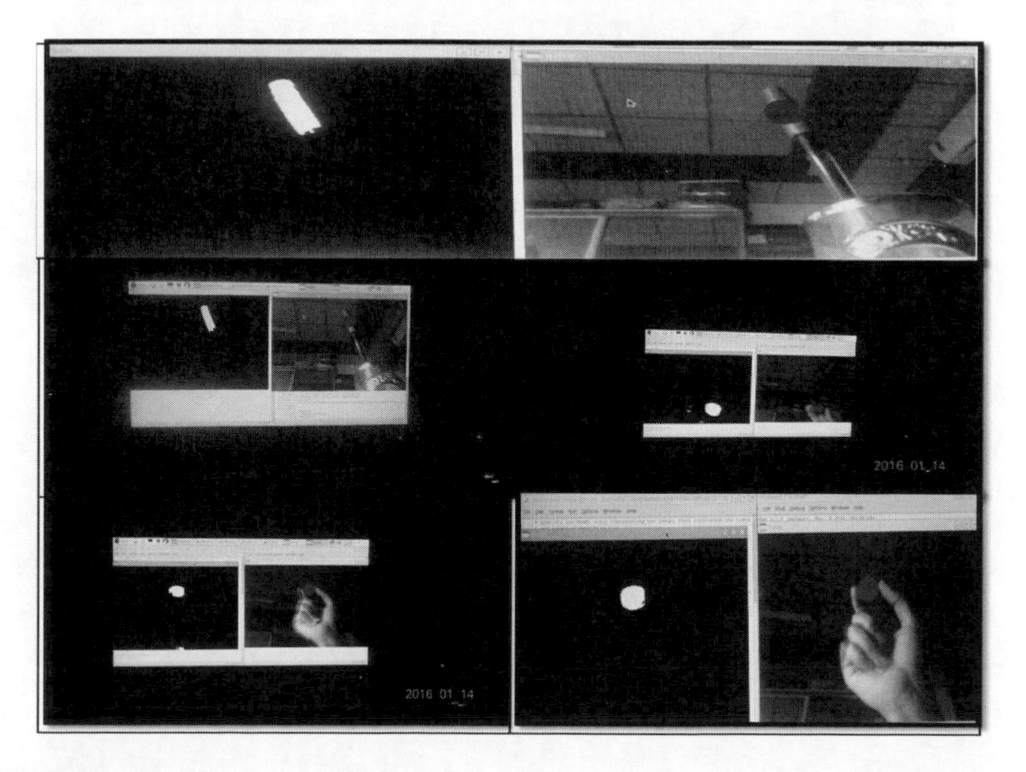

FIGURE 1.8 Color detection in very low light.

FIGURE 1.9 CamShift algorithm in natural lighting.

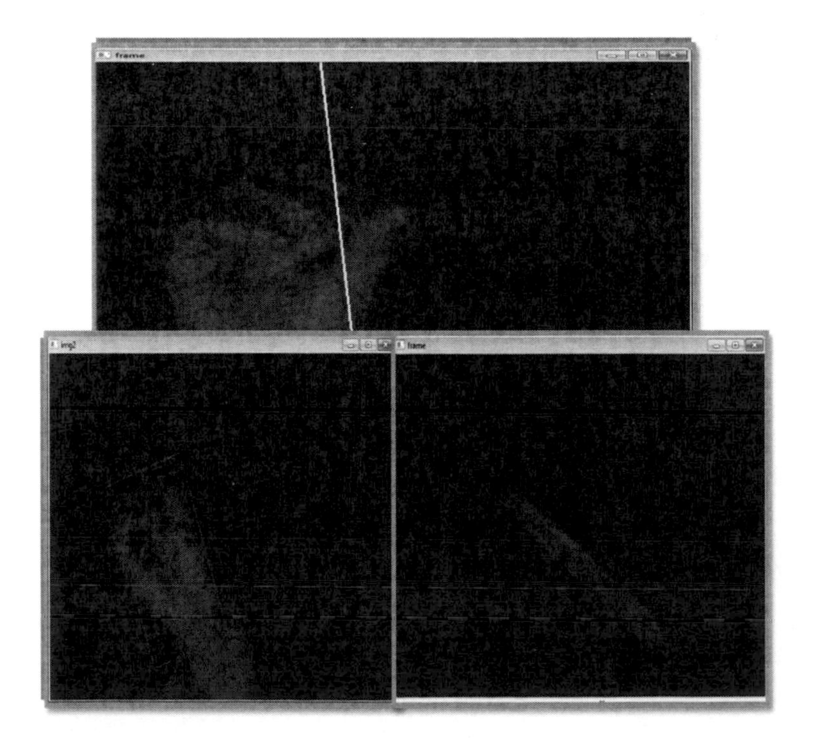

FIGURE 1.10 CamShift algorithm in very low light.

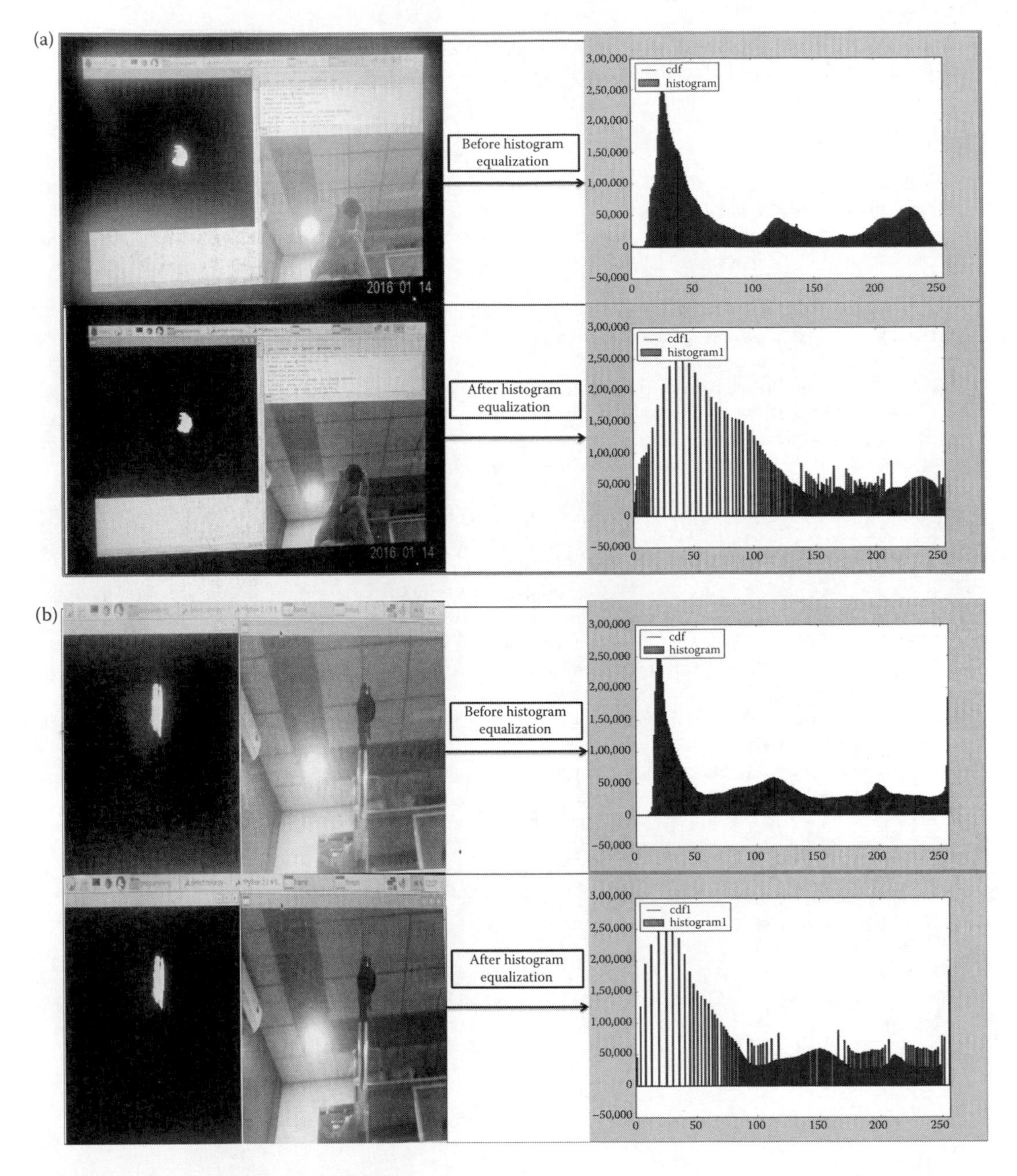

FIGURE 1.11 (a,b) Histogram equalization.

1.7 CONCLUSION AND FUTURE WORKS

In this chapter, a color detection algorithm in very low light conditions is proposed for the enhancement of target tracking. It improved the traditional color detection performance and showed how our technique can solve the flaws of the traditional system. The appropriate color space and the statistical analysis to the various objects of this color space were adopted. The algorithm has the ability to adapt to different lighting intensities. So, the results obtained in the case of daylight were the same in the darkness of night. This is an important and necessary feature of this algorithm and makes it more useful than other algorithms used in the tracking process.

A mobile robot will be designed to track objects in high- and low-light environments to prove the robustness of the proposed algorithms.

REFERENCES

1. Yahya, M. F. and M. R. Arshad, Tracking of multiple light sources using computer vision for underwater docking, *Procedia Computer Science*, 76, 2015, 192–197.
2. Firmanda, D. and D. Pramadihanto, Computer vision based analysis for cursor control using object tracking and color detection, *2014 Seventh International Symposium on Computational Intelligence and Design (ISCID)*, IEEE, 2, 2014, 525–528.
3. Allen, J. G., R. Y. D. Xu, and J. S. Jin, Object tracking using CamShift algorithm and multiple quantized feature spaces, *Proceedings of the Pan-Sydney Area Workshop on Visual Information Processing*, Australian Computer Society, 36, 2004, 3–7.
4. Zhao, C., A. Knight, and I. Reid, Target tracking using mean-shift and affine structure, *19th International Conference on Pattern Recognition*, IEEE, 2008, pp. 1–5.
5. Emami, E. and M. Fathy, Object tracking using improved CAMShift algorithm combined with motion segmentation, *2011 7th Iranian Conference on Machine Vision and Image Processing*, IEEE, 2011, pp. 1–4.
6. Altun, H., R. Sinekli, and U. Tekbas, An efficient color detection in RGB space using hierarchical neural network structure, *2011 International Symposium on Innovations in Intelligent Systems and Applications (INISTA)*, IEEE, 2011, pp. 154–158.
7. Zhang, C., Y. Qiao, E. Fallon, and C. Xu, An improved CamShift algorithm for target tracking in video surveillance, *9th IT&T Conference*, Dublin Institute of Technology, Dublin, Ireland, October 22–23, 2009.
8. Wang, A., J. Li, and Z. Lu, Improved Camshift with adaptive searching window, *International Journal of Soft Computing and Software Engineering (JSCSE)*, 2(3), 2012, 24–36.
9. Li, D., An efficient moving target tracking strategy based on OpenCV and CAMShift theory, *3rd International Conference on Multimedia Technology (ICMT-13)*, 2013, 40–47.
10. Vaitheeswaran, S. M., Leader-follower formation control of ground vehicles using Camshift based guidance, *The International Journal of Multimedia & Its Applications (IJMA)*, 6(6), 2014, 21–33.
11. Kumar, N. and D. Choudhary, Real-time object tracking and color detection using color feature, *International Journal of Computing and Technology (IJCAT)*, 2(02), 2015, 27–30.
12. Garofalaki, Z. I., J. T. Amorginos, and J. N. Ellinas, Object motion tracking based on color detection for android devices, world academy of science, engineering and technology, *International Journal of Computer, Electrical, Automation, Control and Information Engineering*, 9(4), 2015, 893–896.
13. Kapoor, K. and S. Arora, Colour image enhancement based on histogram equalization, *Electrical & Computer Engineering: An International Journal (ECIJ)*, 4(3), 2015, 73–82.
14. Eben, U. and G. Halfacree, *Raspberry Pi User Guide*, Indianapolis Composition Services, Indianapolis, IN, 2014.
15. Jones, D., Picamera Documentation, June 19, 2016.

2 Modeling of the Teaching–Learning Process in E-Learning

Mustapha Bassiri, Mohamed Radid, and Said Belaaouad

CONTENTS

Preamble 16
2.1 Introduction 18
2.2 Emergence, Birth, and Construction of Concept Modeling and Models 18
2.2.1 Typology of Models 18
2.2.2 Construction of the Setting of the Didactic Intervention 19
2.3 The Modeling Notion in the Moroccan Academic Curriculum 20
2.4 Modeling of the Prescribed and Effective Scenarios 21
2.4.1 The Problem and Difficulty of the Systemic Modeling of Effective Courses 21
2.4.2 Hybrid Modeling: Descriptive and Functional 22
2.5 Approach, Method, and Object of Teaching–Training Modeling 22
2.5.1 Reference Setting of the Research 23
2.5.2 The Variables of Research Modeling 23
2.5.3 Model of Modeling Scenaristic of the Teaching for the Training Professionalizing in the Virtual Environment 25
2.5.4 Context of Research 25
2.5.4.1 Population Targets 25
2.5.4.2 Applications 25
2.5.4.3 Organization and Relevance of the Study 26
2.6 Contributions towards Modeling of Pedagogical Scenarios 27
2.6.1 Modeling Levels 28
2.6.2 Theoretical Setting of the Modeling Approach and Perspective of Construction of the Metamodel 29
2.6.2.1 Modeling in the Nonexact "Human Sciences" 29
2.6.2.2 Construction of a Conceptual Setting for Teaching by Modeling and Nonmodeling 29
2.6.3 Present Solutions Proposed: Practical Example of the General Approach of the Metamodel 30
2.6.4 Commentary on the Recommended Approach: Directional Questionnaires and Semidirectional 32
2.6.5 Commentary 33
2.7 Conclusions and Recommendations to Readers 35
Glossary 38
References 39

> Online learning is a catalyst that reflects the image of education links in network the academic institutions, meets government guidelines and requires the teacher to reflect on his new role.
>
> **L'autrice**

PREAMBLE

The globalization of production, the pressure of international competition, changes in the nature and organization of work related to increased demands of professionalizing skills, the emergence of the information society, and technical and technological training are key issues of economic and scientific progress (Glikman, 2002). The traditional models of teaching have been standardized as a normalizing educational and unified approach to learning routes. These training models now seem inadequate to meet the needs and practical experiences of adult learners Knowles (1990). In this context, it makes sense to fit their learning profiles, not to build the instructional sequences that allow them to understand, to make choices, and interact accordingly (awareness and self-management of learning).

During the last decade of education, distance learning has generated much debate in the socio-professional and educational environment. This modeling process is not only expressed in its descriptive and representative component of the complex reality of this virtual environment, but also by the construction of functional scenarios which aims to better understand, characterize and categorize the objects of study, the problems situations equivalent classes, teaching strategies, and personalized experiential learning tailored to adult learners.

The idea is somewhat revolutionary and the introduction of online training is accompanied by a paradigm shift in attitudes, educational concepts, and to andragogical engineering practices. In this framework, a conceptual frame will be shown that illustrates the complexity of the modeling approach to learning in the electronic environment.

An introduction of a chapter is meant to clarify the ideas and the key concept. But at the epistemological level, it is imperative to assume that modeling is not a unique concept, it is polysemic fuel. This notion concerns several model scientific fields, including the mathematical model (applied mathematics), the analysis model in crystallography (chemistry), physics, computer science, meteorology or the sciences life and earth, engineering models, computing (with 3D modeling and data modeling), pedagogy with modeling discipline, and economics. Considering the nomadic nature of Therefore, we encourage readers to understand its complexity and crosscutting principles (Edgar, 1990).

1. Consider "specificity–generality" modeling of a specific object can be likened to a simplistic classification of specific ideas. Similarly too broad modeling makes learning diluted in the nonoperative metaphorical generalities.
2. "Exteriority–interiority" modeling is not unique to a particular scientific field or to study scripted objects or the student's learning environments at home, nor to the profiles of adult learners nor the design of the trainer or tutor in a virtual environment. Now the relevance and effectiveness of the model lies in the reconciliation and harmonization between these endogenous and exogenous factors in the teaching–learning process.
3. Focus on analyzing "quantitative–qualitative." Quantitative occurs when the approach and practice of modeling is linked to the principle of efficiency and academic performance line training projects (based on the objective of terminal integration). The qualitative aspect of modeling learning online would manifest in the establishment of a pedagogical scenario of real professional learning situations and specific to different training paths, and promote the development of social and professional skills of future students.
4. Put the focus on "stability–generality," which starts from the premise that professional skills advanced. Therefore modeling objects will generally reproduce the reality of learning from the most influential people, referring to standards and educational standards. In another, the approach would focus on building a hierarchical representation of complex activities of teaching and learning in the virtual environment. On the other hand, the

modeling approach would focus on the design of a model that is flexible, adaptable, and intelligible in a stable learning development and also has widespread faculties and can be transferred to other areas of social activity.

This preamble serves as a pretext to question the functions of modeling teaching–learning in a virtual training system, and beyond the design of the metamodel that will apply in this teaching. The willingness presupposes the question why and how can the modeling in this teaching be a methodological tool for organization and classification of logic levels of distance learning. What is at stake and what are the consequences?

First, it seems important to introduce the debate on modeling in the field of university education. We believe that this study on modeling is organized first on the problem of designing a metamodel of pedagogical scenario and categorizing classes of meaningful learning situations in the adult learner. Its objective is to describe, structure, characterize, and virtually represent situations and activities of the teaching–learning process. It aims to increase our comprehensive analysis of the concertualization and procedural construction process of objects of studies and learning activities, as well as the mode of interaction between objects in a harmonious learning environment. Also research attempting to formalize the construction of a metamodel advocated two possible axes to structure the didactic preparation of the intervention online:

1. Axis 1 will be conceptualized in terms of the results of theoretical modeling data online learning. This didactic teaching preparation is built from modeling standards including scientific and technical knowledge and institutional requirements from which the teacher can draw.
2. Axis 2 evokes developing a prognosis of adult learners, which emphasizes identification of levels of skills and desires that are estimated *a priori* from the reconstruction of past experience, needs, and interests of the population.

Concretely this didactic preparation step results in a design work and organization of pedagogical action. The purpose of designing the model is to initially define the specific objectives and class situations corresponding to Martinand (1989) learning obstacles, and then it will organize group formation and formalization of forms and modalities of learning, interaction, and interactivity. Thus the methodological approach is formalized so as to establish a constant back and forth between the field and the existing theoretical constructs. It usually requires verification in authentic situations and the characterizations of the "models" used. Another methodological goal is the resilience scriptwriting modeling, that is to say the capacity to produce content relevant training despite the actions of autonomous bodies or of the adult learner freely and independently choosing his learning path.

The intended purpose is at least twofold. On one hand, it speaks directly to teachers, designers, and tutors in these environments seeking to emphasize the need to develop and deliver a variety of training courses. But it is also for learners as motivation through a better understanding of their own learning processes, particularly in the context of collaborative work. This research study highlights the relevance and value of modeling as a process, which describes the simplified representation of a complex reality and measures its effectiveness on academic achievement in various university training courses.

The issue of modeling distance-learning education is to make direct contact with one of the central issues of high dropout rates and educational reflection on the process of pedagogical action "ODMIOSE" (observation, data collection, modeling, interpretation, objective, learning situations, evaluation). Thus, the scientific and axiological analysis (decide the goals) of distance learning is linked to the process of identification and adaptation of training content based on the knowledge and skills of adult learners (personalization of learning pathways).

This chapter addresses the problem of modeling teaching–learning in a virtual context. Our goal is to encourage the reader to the world of scriptwriting modeling in the virtual environment. We present a metamodel modeling scripting objects of study and teaching and learning activities (MSOE-ODA) tailored to the needs of adult learners in a virtual training environment. Modeling models advocated throughout our research allow us to explain and understand the fundamentals that surround the complex reality of virtual education:

1. Design of remote training devices modeled as an adult education and educational organization
2. The analysis and structuring of training content and personalized adult education activity

It is interesting to note that the methods of these models fully realize the functions of categorization and explanatory and heuristic characterization of distance learning content. This construction and reconstruction of units and sequences of learning is never finished because only certain aspects of reality are selected and interpreted (aspects: Reduction of modeling, dynamic and evolving).

2.1 INTRODUCTION

The information and digital communications technologies continue to occupy an increasingly important place in the training of university students: digital textbooks, software, learning games, e-learning, blended learning, massive online open courses (MOOC), reversed classes, and so on.

Modeling of virtual environments for adult learners has grown considerably in recent decades. The construction of a metamodel system of distance learning has become a key issue and important to customizing university training courses. From the point of view of social constructivist and interactive teaching, the new curriculum approach (competence approach; Jonnaert and Vander Borght 2003) of the online education model for adult learners can only be achieved through self-assessments (evaluation diagnosis and positioning test) and building a Personal Action Plan (PAP). Conversely, the teaching—learning model can use the assistance of a custom research object system as a management tool for a set of educational and teaching activities focused on the development of professional skills.

Analysis of distance learning situations brings up the need for modelers to be able to refer to the operating epistemological and methodological study of modeling to construct objects and script-writing to help university students engage in the production of knowledge. This power-act and inter-act is inseparable from the training objectives and the specific nature of the virtual environment.

2.2 EMERGENCE, BIRTH, AND CONSTRUCTION OF CONCEPT MODELING AND MODELS

2.2.1 Typology of Models

Review of the literature on the concept of modeling has enabled us to establish a typology of the different models that inspired our conceptual and methodological framework. In this perspective, it is very important to distinguish a modeling arsenal of models–entities relations, model-learning styles of the adult learner, the adult education model, teaching model, domain model, the model activities and learning tasks, model context, the diegetic model, and oriented model objects. Most have their origin in several scientific fields and are increasingly used in engineering training and information systems (IS). These include the following models: entity–relationship data model describe the reality perceived through the data involved (independently of the operations that will be carried out later on) (Chen, 1976), modern structured analysis (Yourdon, 1989), object modeling technology (OMT) (James Rumbaugh led a team at research labs of General Electric to develop the Object Modeling Technique [OMT] [Blaha and Rumbaugh, 2005]), KADS (Scheiber et al. 1993), and andragogic (Knowles, 1984).

Our proposed metamodel is mainly interested in learning and would place the learner at the heart of the teaching–learning process. One feature of modeling teaching in the virtual context presented here is the integrated treatment of "prototypes" of adult education and learning scenarios based on the engineering model approach to training and skills, the modeling method for intangible object orientation, engineering of the information and communication system, and vocational education (Vergnaud, 1991).

2.2.2 Construction of the Setting of the Didactic Intervention

The work presented here is in line with studies initiated by our doctoral research center "Engineering Training and Teaching of Science and Technology Engineering and Technology Education and Training" and observatory education research and university teaching (ORDIPU). The research is organized around multidisciplinary themes that unite in technical and technological sciences. They enable the development and integration of information technology (IT) and multimedia resources in teaching scientific subjects; the teaching–learning process, teaching of science and technology, experimental aspects, educational aspects, and evaluation; the process of teaching learning and training in the field of adult education engineering devices, and "blended learning."

We are interested in the issue of formalizing a framework (Vergnaud, 1991) methodology and methods of techno-pedagogical instrumentalization of training engineering professionalizing adult learners in the hybrid environment. Furthermore, the study of the determination and characterization of knowledge of adult learner's styles and specificity of the multimedia environment is a powerful lever of personalized learning. Another research study addressed the computational modeling knowledge required in instructional decisions with the aim of promoting the appropriate university course.

We also discussed the explanation of the determinants of the teaching situation necessary for the description of educational activities. They define the elements necessary for pedagogical online training: (1) the educational purpose of the activity, (2) the development of the activity and its interaction elements, and (3) rules of interpretation of student behavior. This is a strong assumption because it addresses the role of adult education activities in the progression of learning for the adult learner. The classic description of these activities goes beyond setting targets likely to reflect on the construction procedures and resource mobilization in complex, novel situations. The analysis of this training content refers to (1) the theories and adult learning principles (Knowles, 1980) on adult learning, (2) the main features of learning in training (blended learning), and (3) the contributions of professional didactics for the professionalization process (Mayen, 1998).

Modeling learning units on distance learning platforms, which highlight the importance of modeling the online training devices, can be divided into several modules. Then we need to identify the various components of each module and connect them to successive aggregations of similar components. The resulting structures are then generic model scenarios as a basis for all modules of modeling. These models of learning scenarios could therefore be the basis for building new collaborative educational scenarios using a similar pedagogical approach (reuse) (Ferraris et al. 2005) to adapt to the new context by adding, modifying, or deleting roles, properties, services, resources, and partitions based on the context and population.

Another study (Bassiri et al., 2017) focuses on the modeling and design of a generic model for understanding the formalization and implementation of observable teaching scenarios (Barré and Choquet, 2005) learning activities, and the understanding of complex correlation relationships between the structure and function of the hybrid education system from the perspective of Pareto's 80/20 rule (face-to-face and remote).

In general the modeling system introduced blended learning models offering new andragogical methods and tool screenwriting innovative technical devices (Brassard and Daele, 2003). It can apply to different levels of descriptions from the strategic objectives of training in small unit of ownerships and evaluation. This mixed andragogical engineering requires the constructive integration of many areas of knowledge and technological sciences modeling of disciplinary knowledge, with a concentration on

- The adult learner's knowledge and status (state of knowledge)
- The process and mechanisms of blended learning
- The national reference framework and institutional foundations for pedagogical action (reflection and action for operationalization)
- Didactic preparation of intervention
- Adaptation procedures and management of customized training courses

We studied the approaches to conceptualizing and teaching scenarios and formalization didaxologic (reflection on didactics and its field of practical application). The research enabled us to structure the knowledge and present the variables and standards of exploitation of the modeling of hybrid teaching–learning situations. In this doctoral research, we presented a framework methodology of knowledge modeling process according to the paradigm–product (Crahay, 2006). And then we analyzed the learning activities in a mixed environment, while providing computer modeling with emphasis on its modular aspects and generic scenarios.

2.3 THE MODELING NOTION IN THE MOROCCAN ACADEMIC CURRICULUM

There are new guidelines for the organization and structuring of research in Morocco at the national level (Ministry of Higher Education and Scientific Research of Morocco) and the regional level (University Hassan II Casablanca). And in partnership with doctoral research teams (Faculty of Sciences Ben M'Sik), a center of research has been created for the development and integration of IT resources and multimedia in higher education, the scriptwriting process of teaching–learning, teaching reflection of science and teaching techniques, experimental aspects of electronic learning books, and educational and evaluative aspects of distance education. These various research structures create doctoral programs in engineering training and teaching of science and technology. It is in this perspective of extension and reflexive continuity on technological innovation of university teaching that is found within the Moroccan higher education system.

Therefore, the education system is more concerned with innovating and renewing its curriculum programs (Macro, meso and micro-vocational training). The course of the modeling approach is at the heart of these debates and the challenges imposed on education systems. The techno-pedagogues recognize that modeling and model concepts are the cornerstone of any education reform based on the objectives and learning content of training programs.

The fundamental challenge is to change the design of the teacher profession and reinvent the paradigm for Generation Y. 4C (Connectivity-Cooperation-Communication and Creativity), this new paradigm would propose a way to normalize pedagogy course training and fight against the problem of online training delay.

In this educational landscape of technical and technological globalization, all institutes of higher education demonstrate the nature of mediated learning, the typology of the recommended research, and socioprofessional impact. Most universities have advocated professionalizing training courses that cover selection of personalized professional development courses and diversification of training programs and a "new school whose main tenets are: fairness, equal opportunity, and promote quality for all the individual and society" according to the Strategic Vision 2015–2030.

This new strategic policy is to strengthen the integration of these technologies at the school in the direction of promoting the quality of academic learning, including

- In the design and preparation of curricula, programs, and materials, and during their implementation
- Use of the software and interactive digital resources across the educational process targeting the self-learning, research, and diversification of learning
- The development and promotion of distance learning as a complement to group lessons

- The promotion of theoretical and practical research in the areas of education and training, in relation to information and communications technology
- The encouragement of excellence and engineering in the areas of research and the use of educational technologies through the establishment of a new dynamic based on research programs and projects in order to identify learners who are distinguished by their excellent results

In this context, the emergence and construction of a curriculum design model and specifically modeling training course in virtual learning environment for adult learners—"human learning" (Burkhardt, 2003)—that deserves attention. If every institution of higher education had the same training strategy, there would be a tendency of differentiation of adult education to fight against the problem of "homomorphism" and "mimicry" of academic learning. Modeling objects of study in this situation would customize modeled teaching in this virtual environment. It allows one to trace, monitor, and management of the individualized training course. Other relevant preferential tasks in this model are to create desire, interest, meaning, and interaction within virtual reality (Fuchs et al. 2006).

2.4 MODELING OF THE PRESCRIBED AND EFFECTIVE SCENARIOS

Systemic modeling aims to balance the orientation of the model prescribed screenplay and optimization of the actual scenario. In effect, the program can deviate substantially from the prescribed scenario. Modeling objects and educational activities all start with the specification of the prescribed scenario (national skills repository) and also monitoring and controlling of the actual scenario (Barot et al. 2013).

The work of our research will focus exclusively on the first stage of specification and educational scenario building. Teaching and learning scenarios include all the learning activities that can be offered to adult learners, including lectures, practical exercises, simulations, case studies, and video conferencing. The metamodel prototype script that we offer describes orchestration of knowledge and professional skills at the level of customization of specific training courses.

Thus the classification and prioritization of objects of study and adult education and pedagogical activities in the sociotechnical environment is part of the quest for creative solutions optimizing a university training course. In these environments, the variability of learning situations, the diversity of needs and expectations of adult learners, as well as the specificity of open distance learning make modeling of the teaching–learning process a powerful methodological design tool, with rationalization and optimization of teaching content that is structured, prioritized, and tailored.

2.4.1 The Problem and Difficulty of the Systemic Modeling of Effective Courses

The problem of modeling teaching–learning is first placed in the overall framework of strengthening the university's offering in the field of engineering education information systems and professionalizing university education. The learning course is collective, individual, and customized, which are inseparable and require harmonization among these training devices. The result is a new design of modeling that is not limited to the functions of classification and categorization of the objects of study, but presides over the identification of the matrix formed by the corresponding logical learning. The degree of membership of each study subject, content, method, and modality of assessment can be defined simply by the iterative learning paths satisfactorily converging to form the fundamental basic or advanced course, whether individualized or shared.

The approach to classification and conceptualization may encounter difficulties in setting up a heterogeneous, comprehensive, and balanced model. This is why it is crucial to use a systems modeling approach (Donnadieu and Karsky, 2002) of estimating with a method of pretreatment and principal components analysis of the teaching situation (ACPSE). This comprehensive modeling can offer great freedom of action and selection of the best training courses. Together, the varied and diverse content and teaching methods to customize and adapt learning situations to each adult learner. However, it is difficult for designers of virtual environments to imagine, design, and classify all actions and sequences of events promoting greater freedom of action for adult learners. This comprehensive functional modeling approach proves somewhat complex to grasp. The solution to this problem requires a means of quantifying the variables of the process and to script the objects of study and adult education activities according to the degree of validity of the various solutions of the process. A validation variable is an objective based "logic learning" used to quantify and compare the quality of academic performance.

2.4.2 Hybrid Modeling: Descriptive and Functional

Hybrid modeling dynamically allows the formalization of training courses and learning within a real learning simulation. The classification of each educational scenario should be in agreement with the model of the learning path chosen. It uses models based on knowledge, communication, and autonomy. For this, we first propose a model of the screenplay that is reconfigurable, flexible, and generic in the teaching–learning process online. This hybrid model allows one to both describe the complex model of the virtual environment, scriptwriting training objectives and the ability to maintain the basic knowledge related to professional training. The learning model proposed is based on the social constructivist approach and interactive virtual learning. This is based on a vector space of a class's inverted position, which is the associated belief of adult learner's ability to handle the situations they describe.

The hybrid modeling that we propose is essentially intradiegetic: it integrates to the world simulated by the virtual environment. For it we propose a method of dynamic, functional, and adaptive generation of situations of training leaning on models of the activity and causality inspired of ergonomic analyses, of the psychology of work, and of professional didactics animated by a humanist and anthropological (capable topic before the effective topic) ideology. The situations of training generated will be articulated under the shape of fiction thanks to the process of inspired diegetization of the structuralist current of the semiology.

2.5 APPROACH, METHOD, AND OBJECT OF TEACHING–TRAINING MODELING

The new modeling approaches for online adult education and learning aims to address the issue of individualization and differentiation of learning paths (Meirieu, 1991). The personalization of a professional training course requires the use of process modeling of learning education as a powerful vector optimization of learning among learners. This means to model the activities of learning targeted in study objects (internal didactic transposition), the content, that is, scripting a training module in many learning activities in order to reconcile the requirements of the computerized context.

The methodology of the recommended model would contribute to the implementation of differentiated instruction in these computerized learning environments. The objects of study in differentiated teaching courses meet concerns related to social constructivist theories (Vygotsky, 1985), which are present in computerized learning environments. The implementation of dynamic custom scenarios in these virtual environments give some reality to differentiated instruction (different learning modes) and differentiated instruction (multiple learning routes), which far too often appears theoretical. Figure presents a synthesis modeling differentiated instruction in computerized learning environments. Modeling of differentiation can manage the complexity of the virtual environment, that is, manage the business areas in which the complexity of sociotechnical systems and the increasing unpredictability of human factors (e.g., motivation and stress management).

Modeled learning objects take into account the heterogeneity of adult learners' profiles, for example, different forms of interactivity (synchronous and asynchronous modes), different forms of communication (e-mail, chat, forum, etc.) that vary the teaching learning media (text, audio, video, PDF, etc.), and multiple construction methods skills (individual work, group synchronous, asynchronous group, etc.).

The proliferation of personalized training courses and management diversity now seem to be realized through systemic computerized learning modeling, which takes into account the different rhythms and student learning strategies. The goal is to place each adult learner in optimal learning path. The emergence of the differential activities of the modeling requires a variety of teaching and learning measures to meet the needs and expectations of learners. From the axiological point of view, the establishment of autonomous and free training is dependent on factors that optimize learning. Multifactor knowledge is the learning profile of the learner, the specificity of the virtual environment, and the modeling approach that will therefore be decisive factors for the optimization of learning environments (Dao Chi, 2000).

After putting our ideas in a wider perspective, namely that of systemic modeling and differentiated instruction, it is now important to further define the key areas of our research, namely that of the teacher's activity designer, which is to refer to design models training units and teaching sequences responsive to the learning profiles of adult learners (converge, diverge, accommodative, assimilation) and axiological requirements (the goals).

In this perspective, the teacher inventor is brought in to categorize the different activities of training to create the favorable conditions of investment, engagement, and mobilization of his or her expertise in a global and harmonious manner (cultural, technological, strategic, methodological, and communicative) National Charter of the Education and the Formation 1999 (the new strategy of the Moroccan educational system). The modeler of the educational and didactic scripts online comes back to put in place the activities governed by actors who use and produce the resources of training (Paquette, 2004). In our case study, it is about numeric resources conceptualized according to the competence reference and the instrument of mediation.

2.5.1 Reference Setting of the Research

The setting that served the analysis of the qualitative and quantitative data of research consists mainly of three main measurements: (1) the modeling of the training in the online formations, (2) the modeling of andragogical and professional theories of training adults (Kolb, 1985), and (3) the academic output of the formation course professionalizing. These measurements are to put directly in relation with andragogical theories and more specifically with the theory of professional didactics.

The questions and the hypotheses that have thus far been elaborated are based on the crossing of these different measurements. The objective of our research is the continuous improvement of the modeling of educational activities in the computerized environment. However, it is important to note that the complexity and the specificity of this training support require the recourse to models of classification to surround the foundations and the terms of references. The setting of a theoretical formalization and operative conceptualization are going to allow us to specify the different classes of objects of study and their functions in term of logic of training that retraces the diversity of courses professionalizing academia.

2.5.2 The Variables of Research Modeling

The recommended variables are supported on the paradigm process product modeling orientation to the virtual object of the act of professionalizing university training (intangible deliverable). The logic of adult education and didactic script was conceptualized from the research process product. The variables included omen, program, context, and product organized around the central variable process (see Figure 2.1).

Analysis of the screenwriting process of teaching–learning among adult learners in the virtual training system took into consideration the following components:

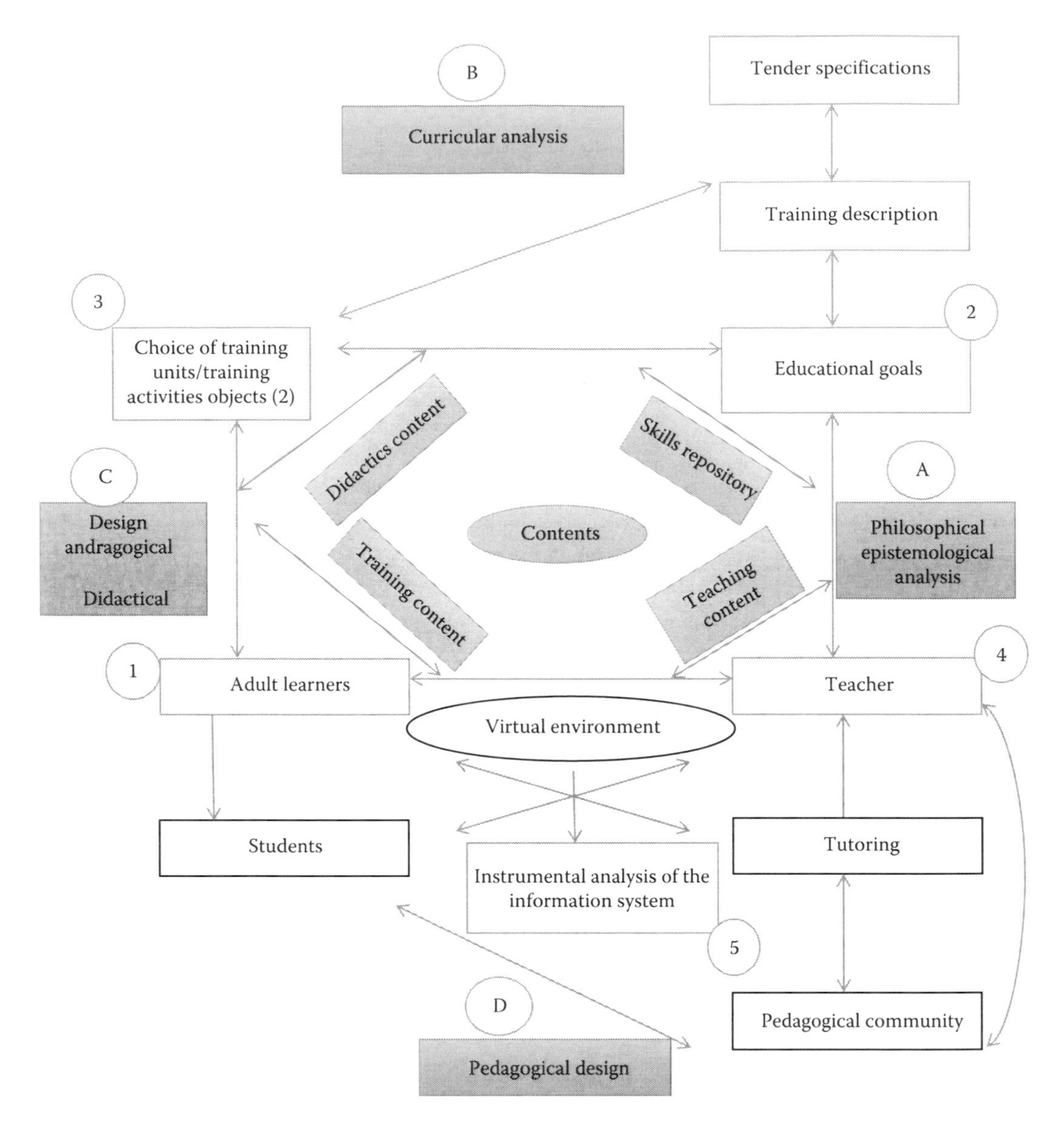

FIGURE 2.1 Construction of a conceptual framework for modeling of teaching learning.

- Area 1 would include the state-of-art-foundations and teaching approaches in computer environments for adult education. This state of the art puts the focus on the referentialization of different knowledge modeling in virtual environments.
- Area 2 would help to describe the learning styles of adult learners that guide the choice of teaching and learning activities and the general approach of custom scripting, that is, the modeling process for creating descriptive scriptwriting goals from contextualized training objectives.
- Area 3 is the present metamodel prototype for knowledge representation modeling for the design and use of virtual environments for training adult learners. It also addresses the problem of modeling the adult learner in the cognitive and social constructivist learning framework in relation to vocational teaching.
- Area 4 is a discussion of our research work, and its strengths and limitations. We also present a number of perspectives, and then we conclude this work.

2.5.3 Model of Modeling Scenaristic of the Teaching for the Training Professionalizing in the Virtual Environment

The aim of this work is to seek to provide a new modeling approach to teaching screenwriting/learning in the virtual environment for professionalizing university training. Online training program content is based on (1) the design patterns of teaching–learning objects (knowledge model, the media model, the pedagogical model of learning activities and resources) and (2) on the educational goals, body of knowledge, and skills.

We wish to offer a more comprehensive modeling of pedagogical scenarios to create the conditions of personalized learning. The descriptive categorization of the different types of learning situations and their sequences are a powerful vector of academic performance with distance education. The difficulty level of adult learners, and their various needs and expectations may differ from the strategic objectives of training and selection of classes. The need for personalization of learning paths requires reliable information on profiles of adult learners (variable omen). The goal is to determine the logic and functions of each class of equivalent scenarios, allowing decisions on modeling training courses, as one of the assumptions of the modeling domain knowledge was to exploit and formalize a dynamic and personalized script based on engineering analysis of needs and tangible training objectives.

The analysis of the profiles of adult learners is crucial. This analysis will be linked to the educational design. We proposed a representation of profiles of students as a process of operationalization of the zone of proximal development. This process is based on a social constructivist and interactive model of the gold-adaptive learning resources mobilized, which relates the objective difficulties of structured educational activities and instrumentalized and logical levels of learners (subjective levels of student resources). The methodology of design is based on structuralist and functionalist theories. It is based on a theory of knowledge and an experiential approach (adult learning styles). This choice facilitates the integration and the proceduralisation of self-directed learning.

2.5.4 Context of Research

2.5.4.1 Population Targets

The target sample was built after the random sampling phase. It consists mainly of university students of the master engineering technology for education and training (ITEF). The objectives of the participants were varied and diverse (Table 2.1). For some the personal action plan was to acquire and develop their technical and technological skills and the development of action research. The result is therefore a pressing need not only to strengthen the system in learning what Dewey calls the "knowledge to maintain the situation," but also to enhance the knowledge and skills of everyone. The consideration of this singularity is the defined vector of construction, interest, and productive engagement. This research aims to participate in the development of the platform of the Faculty of Sciences Ben M'Sik University Hassan 2 Casablanca. In the decision models for collective training courses, individual and personalized will be interdependent on each other, and making the modular modeling approach objects of study would address the problem of complexity and cross-university learning (Morin, 2005).

2.5.4.2 Applications

The special application used for support for the proposals in this research study was the project National Center for Scientific and Technical Research (CNRST), which aims to

- Implement research and technological development programs in the framework of choice and priorities set by the government supervisory authority
- Contribute to the dissemination of scientific and technical information, and publication of research work and to ensure technological monitoring work

TABLE 2.1
Identification and Modeling of Learning Styles in Adult Learners

Student Questioning Master (Engineering and Technology for Education and Training)	Answers
Q1. Do the courses of Master Formation in the conception and the determination of the formation objects take—them in account your needs, your waiting, and your preoccupations of formation professionalizing?	90%: of the respondents consider that the oligoplasts are exclusively prescribed (in accordance with those recommended in the special prescription notebook).
Q2. The language of teaching of the different modules is—it in consistency with the linguistic contextualisation the tasks of trainings recommended	45% of the answers mentioned generally a shift of specific semantic order.
Q3. Does the educational mediation encourage t - it the solicitation of the superior cognitive functions?	76% Of the students reveal difficulties of accompaniment rigorous soliciting the metacognitive process
Q4. Do the information and instructions for the accomplishment of the prescribed learning tasks allow the mobilization of the metacognitive process?	78% of learners think that most of the tasks are defined. These tasks are limited to the mobilization of the processes of understanding and application. And level of factual and conceptual knowledge
Q5. Does the definition and discussion of training objectives through framing (and debrifing) sessions promote reflexive feedback and proactive training strategies?	89% Of respondents explained that the two approaches to adult education are advocated in two training modules out of eight.
Q6. Does the relationship to knowledge delineate a learning environment for dialog, debate and sociocultural conflict?	65% Of the answers consider that the relation to knowledge still determines the relation to power "hierarchical pedagogical relation"
Q7. Does the preliminary analysis of the ITEF modules advocate pedagogical approaches and reflective abstraction?	78% Evoke a discrepancy between the procedures prescribed in the terms of reference and the pedagogical
Q8. The ethical setting and the contract didactics setting up permits them to tie intercourse of Interactive collaboration "Intelligent tutoring" pledge development of the capability of auto and héterorégulation conscientisantes?	82% confirm the lack of a clear and explicit didactic contract where him ya engagement and mutual respect of the different taking parts
Q9. The nature of the course and its sequencing recommended in the multimedia environment allows you to enter the metacognitive paradigm?	83%: Inadequate contribution is limited to the level of the round table discussions in the classroom (absence of community tutoring, reciprocal and bilateral pedagogical relationship).
Q10. Does the design and structure of teaching content in distance learning orient the mobilization of cognitive resources in an optimal way?	95%: The answers collected show negative assertions (informational model unrelated to needs engineering)
Q11. Choices of resources and educational and learning activities are sufficient to create real reflective learning in this computerized learning environment.	98%: Students emphasize the lack of energizing learning activities: learning metacognitive and creative pedagogical practices (case study - simulation and role play).

2.5.4.3 Organization and Relevance of the Study

The choice to study modeling of teaching–learning online allows us to put the focus on structuring and characterizing authentic learning situations. The metamodel conceptualized provides individualization and customization of university training courses adapted to adult learners. The choice and adaptation of learning units is also based on the peculiarity of the computerized learning environment. It is to provide different answers depending on the learning profiles and needs of adult learners (Carré, 2005).

The andragogical modeling engineering advocated is based on the harmonization of all analytical models and methodological approaches to design and exploit content. The disciplinary program content would highlight the valuation of two types of learning: (1) self-based learning modality of self-assessments to address skills acquisition and (2) an updated personal action project where the adult learner can access the training for their prerequisites and his preferred styles. We modeled three scenarios—Level 1 "beginner" basic course or elementary, Level 2 "managed" acquired building courses, and Level 3 "confirmed" personal development courses—that have different adaptive courses. Each lesson plan has a number of educational objectives that are derived from the goals of the training and course prerequisites given at the start of each scenario; these objectives will enable the reuse of scenarios and models regardless of content.

Metamodeling of the teaching process–learning online training is a process of classification and characterization of implanted objects of study in the university training course. It takes as its starting point an *a priori* analysis of the variables of the teaching situation that may influence the professionalization process articulated around three levels that constitute the dynamic axis of our curricular design template: input (student and teacher), the process, and output (the development of skills related to the modeling process).

The axis of functional modeling is completed by a topological axis that defines the various subitems against which the pedagogical scenario modeling process falls. We respect the specificity and adaptability of principles and escalation (from easy to difficult, the single complex, the specific to the general). The subitems of the microsystem consists of the classes of authentic and original learning situations, the meso-system formed by the skills required of the immediate academic environment and the macro-system at which we find the administrative and political managers of the educational system (translated by national educational standards; Law 01-00 of the Moroccan higher education). To these three subsystems that fall within the Moroccan educational system at large, we add the perimodel system that includes all the variables in the immediate environment of the education system that are likely to have an effect thereon (partners, business, etc.).

In another context the formalization of the innovation process through the dynamic axis of our metamodel starts with a very precise characterization of "inputs," that is to say of all the elements that will enter the system to serve as triggers or inhibitors to the innovation process. The second level introduces us to the innovation process itself that we have conceptualized as three distinct phases in the nature and scope of the decisions taken there: the adoption of a metamodel, implementation, and routinization to near adult learners. Variables related to the support of training courses refer to the key role played in the process of change led by the tutor and the educational community. The third level of the model relates to the characterization of "outputs" by analyzing the effects of the implementation of the different subsystems that can act on the innovation process or be affected by it.

2.6 CONTRIBUTIONS TOWARDS MODELING OF PEDAGOGICAL SCENARIOS

The pedagogical scenario-modeling concept is still a wasteland. Paquette (2005) said: "Among the categories of objects and learning scenarios take center stage. The scenarios materialize teaching methods that have proven themselves in various fields of science and literature education.... Reuse quality scenario models, proven in many educational settings can often bring more quality than the simple reuse as content resources of a 'standardized digital press book'." The model of research-teaching situations is the theory of didactic situations (Brousseau, 2011) in which students interact with action strategies, formulating and validating an updated personal learning project.

The first work on the script did not take into account the need to link the scenario models and abstract conceptualization. The work of Laforcade et al. (2005b) on an educational approach where classes of situational problems had begun to clarify the concept of scenario models as a concept to integrate research on the formalization of models of teaching–learning online. Our work has been clearly positioned at this intersection (Laforcade 2005a).

We define a model of educational scenarios as representing a deviation of a scenario category such that it corresponds to the educational approach or learning situation referenced. This model then serves as an example. The lesson plan model may vary in scope:

- Either it is recognized as representing deviation within a community, such as a disciplinary or community level.
- Either it is recognized as such transversely to communities, such as the scenario case study of which the model is recognized by the community of secondary school teachers, the community of teachers in higher education, the engineering education community, and community of work-based trainers.

We define the educational script boss notion from the following definition: "drawn model, paints cut either, of after which work some craftsmen." In this sense, we consider that the boss is the shape of concrete expression of a model of which he contains the operational aspects of artifact, it takes the dimension of the instrument. According to the instrumental genesis, the boss of scripts used in conception can then be adapted to the teacher's particular context and be in the center of the process of technical instrumentation and educational instrumentalization. In this perspective, the boss of script is more than a support; It is a dynamic design tool that evolves according to the learning context, the profiles of the adult learners and the needs of the users. The boss of the script will have an operational dimension, so the inventor must have a good knowledge of this environment.

Meanwhile the study of modeling of pedagogical approaches allows us to go even further in the concept of scenario models. Indeed, for each of instructional approaches, it is possible to find work highlighting the features in the recurrent or habitual activities, and the role of different actors (teachers and students). These features allow identifying invariants to be expressed in an adult education scenario, give it representative-type status, and thereby modeling andrago-learning scenarios. These invariants are mainly at certain groups of activities, which will be built around the screenwriting template. Thus, in a project approach, we will find such learning activities on research methodology or planning action project. The modeling of these approaches would be the guideline that will guide and integrate the relevant activities (debate project, case studies, simulation, video conferencing, application exercises, etc.). We repeatedly found activities on the training path suitable for introductory or familiarity or expertise of such activities.

In this virtual learning environment, modeling the pedagogical scenario could be done at different levels of granularity. Delmas (2009) identifies four levels of granularity (stage, act, episode, series) for which the forms of screenwriting differ. In the context of the script virtual environments, the act level does not seem relevant. Indeed, it is an artificial division, which is not a unit of time or place or from the perspective of the screenplay content or from the perspective of how it is experienced by the user. For the implementation and management of the overall organization of the scripting learning line, the main goal is to provide content adapted to the educated population of adult learners during the academic curriculum.

2.6.1 Modeling Levels

In the literature review we identified three levels of modeling:

- The microscenaristic level is the training runs. The choice and the specification of the trainings consist of ensuring the adequacy and the adaptation of the difficulty of the situations are equivalent to the level of resources of the adult learner.
- The mesoscenaristic level is the training session. The adaptation to the mesoscenaristic level consists in determining some constraints one the decision making ace heart the categories of situations to make meet during has session of teaching organized (familiarization, initiation, backing however perfection), linked and articulate of several sessions author of has theme however title. The mesoscenaristic level pilots the context of the session by objective of formation clarify that year unites of appropriation and the time limits.
- The macroscenaristic level is to produce a particular course of the learner through its use in the virtual environment (classic course, advanced course, and course team).

2.6.2 Theoretical Setting of the Modeling Approach and Perspective of Construction of the Metamodel

The history of contemporary science is action in its approach to analyzing the actors of science and epistemological options. The paradigms or conceptual matrix reflects the "spirit of time," which aims to reinvest the subject and able to mobilize its hidden potential. This type of sociophilosophical history leaves room for modeling knowledge, situations teaching, and the roles of teachers and learners in a learning situation. According to Guy Brousseau: "Modeling mathematical knowledge are built by researchers mathematicians in response to external or internal issues in mathematics. In this paradigm, the new knowledge construction process is not linear, but a labyrinth of reflections, essays, the question of transformations, shifts prior knowledge neighbors, etc. People in a social and historical context of scientific issues construct the mathematical knowledge: it is personalized and contextualized."

2.6.2.1 Modeling in the Nonexact "Human Sciences"

There are at least four ways to write the history of modeling in the nonexact sciences, that is, in the human and anthropological sciences. One approach, often preferred by historians, is to focus on the administrative history of research programs. Regarding modeling, there are few studies on this perspective because the difficulty lies in reducing the appearance and description of the practice of modeling to categorize scientific institutions as complex social activity fields. Recent developments have been made with nonlinear learning modeling techniques (including artificial neural networks). This practice of modeling has invaded science as a whole, beginning to identify its subject of study and to reconfigure this complex reality by simplifying the abstract representations of the mental mapping (synthesis diagram).

A second history of the sciences, more technical and more sectored because it is produced most often by scientists, concentrates on genealogy of supposed relatively autonomous and internal to the scientific productions of the given field. The most recent historic works on modeling have been insofar as scientists producing a certain number of summaries or perspectives aiming to argue, problematize, valorize, root, or merely situate their own scientific contributions. (The educational model of Philippe Meirieu shows modeling the determinants of the axiological pedagogy, scientific, and praxeological.)

A third history of science is philosophical as seeking to make it understandable to some "logic of history" while leaving room for contingency, trying to regain the conditions of possibility as historical, contextual, and social as technical, conceptual, and epistemological science. It determines a range of epistemological productions, including the epistemological obstacle model by Gaston Bachelard (1938), and the theoretical model of scientific knowledge (Lemaine, 1999), especially the epistemology called empirical sciences, which describe the world based on sensitive data from experience. This later can generate knowledge and avoid the repetition or the multiplication of the descriptions, ask always-new interpretative issues in the world, Jean-Claude Passeron Testing and Research on sociological reasoning. The non-Popperian space of natural reasoning (Nathan, 1991).

A fourth history of science focuses on modeling and simulation in science (see Peter Galison for modeling and simulation in nuclear physics). The science modeling was listed by Astolfi and Drouin (1992), who proposed nearly 250 references based on epistemology, cognitive psychology, or science education. Several educational tools were developed: training analysis tools based on the modeling considered can categorize scholarly knowledge, taught knowledge and operation of the student to address this knowledge (Tiberghien, 1994), distinguish pattern registers, and modeling constraints for a modeling teaching project (Martinand, 1994).

2.6.2.2 Construction of a Conceptual Setting for Teaching by Modeling and Nonmodeling

Modeling can in virtual teaching–learning function as a bridge between the scientific theoretical model and the praxeological model real-world modeling process. This is subject to experience (Gilbert, 2004). If it is not interested in status but the functions of the models, the debate can be very constructive. It is a broader vision that determines the educational functions and educational consequences of modeling, and the theoretical importance epistemologically in this teaching process. The instrumentalization of educational tools and techniques to explore the reality may be very different: they can describe, explain, predict or help in decision-making. Legay (1997) distinguishes "hypothesis models," the "mechanism model" (currently the most common), and the "decision and prediction models" (least developed but most requested). In our case of online learning, we will focus our attention on teaching through modeling and nonmodeling, the aim being to help adult learners build their social and professional skills.

Consequently, the approach of teaching engineering–learning computer modeling in the virtual environment would be placed at the heart of the design process and the formalization of training courses, and require the adult learners to identify their personal learning path (promoting the quality of academic learning).

The teaching–learning model is the design of a pedagogical scenario model. The educational and teaching goal is to be able to create an organizational learning in the online environment. The teaching process is to describe, structure, characterize, and virtually represent the situations and activities of the teaching–learning process. It aims to increase our understanding and analysis of the conceptualization process and procedural construction of objects of study and learning activities, as well as the mode of interaction between these objects with a target of autonomy and learning awareness.

The comparison and analysis of the metamodel developed conceptual framework in relation to the modeling results in theoretical data about online learning (theories of industrialization, communications, and autonomy), and allows us to conceptualize by modeling relevant and effective educational actions. Thus, the methodological approach advocated modeling is formalized to establish a constant back and forth between the field and the existing theoretical constructs. They generally require *in situ* verification and characterization of "models" used. One of our goals is also methodological resilience screenwriting modeling, that is to say the ability to produce content-relevant training despite the actions of autonomous bodies or if the adult learner personalizes his own course.

Modeling of teaching–learning online is an abstract representation in a computerized environment, in the sense that the data values of the teaching situation (adult learning, tutor, teacher, teaching material, repository skills, knowledge of the virtual environment) are observed in their symbiotic relationship, dynamic and interactive. The data design model should not only define the data structure, but also what the data should mean (semantics). The modeling is not about the adult education aspect but the didactic, epistemological, and curriculum, as well as the virtual education system (see Figure 2.1). Our challenge is ecologically screenwriting valid content (true to human and material context of the faculty of Ben M'Sik University Hassan 2 Casablanca). We focused our study on the classification and characterization of knowledge and skills training courses. However, with this work, the aim was also to describe knowledge about the virtual environment of teaching support, as the virtual environment has the foundations and theoretical models in which it is possible to both interact and allow the development of skills by interpreting static or functional representations in the computer system.

2.6.3 Present Solutions Proposed: Practical Example of the General Approach of the Metamodel

The approach we propose is to use modeling dynamic scripting and custom objects of study and teaching and learning activities underpinned by dependent metamodels of the online-training context. Metamodeling models of teaching–learning will be directed recommended, first, by the metamodel process product, and second by the andragogical engineering model (Knowles, 1975). This model places the adult learner at the center of the teaching–learning process and makes him a dynamic player and aware of his personal action plan updated. We will focus in this study metamodeling learning path as the translation of preference of an individual for a mode of operation in a virtual environment, the choice of classes of teaching situations, and the equivalent mobilization of adequate learning strategies. The operative demarche methodology allowed us to advocate for a metamodel arsenal:

Model process product: The model offers process product development approaches for objects and scriptwriting activities. It prescribes an explicit methodological approach to achieve the repository is the product (Rodhod, 2010). IT is essential to allow program designers to create or adapt the metaprocess model based on the constraints and characteristics of the learning environment. In addition, the definition of a process model must consist of concepts, rules, and very significant correlations. A process model as a product model must be seen following different approaches and levels of abstraction. However, most of the processes of metamodels allow having monolithic modeling models of teaching–learning payable. They are more often too narrow and too specific to be adaptable to the online environment. The majority of them offer system approaches. To allow a full and harmonious vision of the modeling process of engineering distance education, it would be wise to have the model's uniformalized process multireferenced and adapted to the computerized environment. The objective of this study is to provide a core base for the creation of such a meta-analysis model of the teaching–learning process (see Figure 2.2).

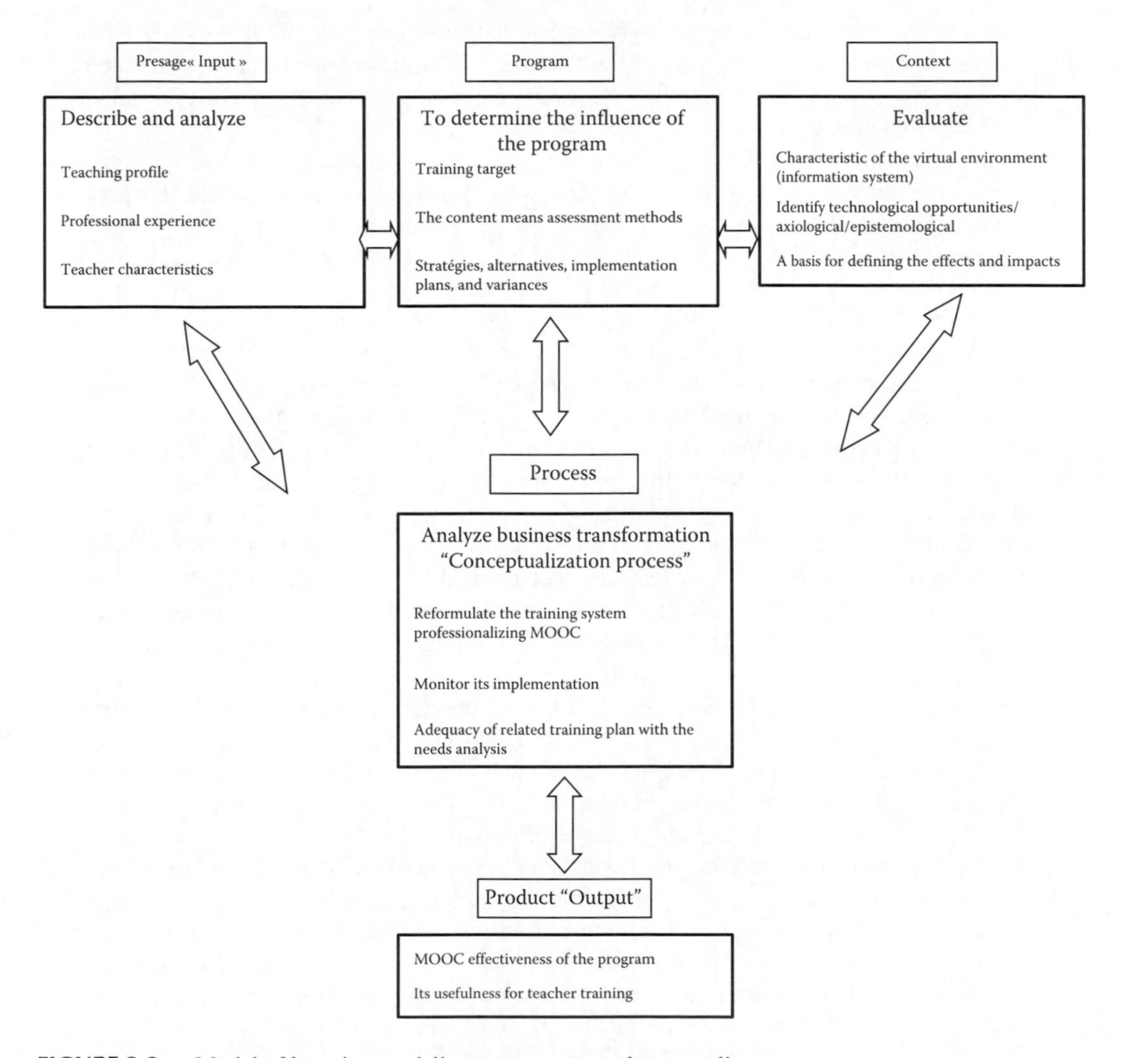

FIGURE 2.2 Model of learning modeling: processes–product paradigm.

2.6.4 Commentary on the Recommended Approach: Directional Questionnaires and Semidirectional

The variable presage: The analysis of the characteristics of teachers is based on the identification of significant variables defined after following random sampling: experiences, profiles, perceived needs, attitude representation, opinion of e-learning education in professional development, values, goals, career management, personal development, emancipation, preferences, interests, experience with personal investigations, style, and preferential professional learning in the e-learning context.

The variable omen: The teacher's characteristics (their initial training, their professional experience and their socioprofessional or personal development goals) are influential in a very significant open, remote vocational training process.

Variable product: Refers to the effects and impact of e-learning devices on developing the professional skills of teachers.

The process variable: Refers to different forms of connection, communication, collaboration, and community interaction (tutor, teacher, teaching, media environment). This variable explicitly explains what is happening in this MOOC device during pedagogical action. It focuses on definition of objects of study and teaching activities. It corresponds to the variable instrumental requirements of e-learning. It is an independent of the variable of teachers, but requires great flexibility and adapting them to a real transformation in their learning and experience.

The variable program: Refers to the training objectives analysis, content, strategies, approaches, teaching methods, tools, and facilitation techniques and modalities. It depends on the decisions of the designer regardless of the platform trained actors.

This research process product approach organizes creates the loop "description–correlation–experimentation." It can accurately describe the online training process, to identify the nature of intervariable correlations and to measure their impact on the development of professionalizing the university career.

Model of the adult learner: The model examines the profiles of the learners. It makes it possible to characterize the learning behaviors in their personal dimension (the regularities) and different (individual differences). The modeling process of learning is influenced online by psychological and socioaffective dimensions specific to adult learners, as well as cognitive dimensions including reports examining learning styles (Peng, 2003). Therefore, studies have come to initiate the issue of learning styles of adult learners as a factor influencing the optimization of academic learning outcomes (Peng, 2003). Referring in particular to Chevrier search results (LeBlanc, 2000; Rouet, 1999), we assume a close relationship between the modeling of learning styles in terms of preferential mode and learning strategies and academic achievement in online training.

The andragogic model: This model informs the learning conditions of adults. The diagnosis model proposed here allows you to select classes of situations and modes of input in accordance with the profile and experience of the learner according to several customized educational goals. According to the social constructivist and interactive learning paradigm online the overall aim is to expand the zone of proximal development (ZPD) of the adult learner. It is interesting to define both what is within the current development area (classes of situations of the adult learner "positioning test and balance of competences") and control situations of proximal development zone (classes od situations that the student could master interacting with the educational community and peers). It is therefore necessary to define both what concerns the current development zone (situations that learners have mastered) and situations of the proximal development zone (situations that learners could master).

2.6.5 Commentary

In this research, we restore the different measurements of the knowledge and cognitive process solicited during the sessions of master formations while using the Anderson and Krathwohl taxonomy model as the priority reference. This model of analysis puts forward the complementary two aspects of the working of the thought. It seems important to highlight the reflection on educational practices by valuing the overall and harmonious development of the cognitive dimension and the process of self-regulation of learning. This survey of research explored by our academic students allows us to deduct that it is also possible to formalize objects of training e-learning, developing these cognitive measurements at a high level in a very meaningful manner. The results open a reflection deepened on the engineering operational anagogical of the metacognitive consciousness in a computerized environment (conceptualization and the formalization of the didactic and educational objects). Students were asked to judge and to describe the preferential situations guaranteeing the quality of the formation metacognitive. In the beginning of the research, the students showed behaviors of fear and anxiety facing this situation of new formation, but that does not generally answer their needs and validate neither their experiences nor their personal action project (Meirieu, 1993).

The examination of this part allows us to point out, in summary, the main characteristics and constants of distance education, such as breaking concepts between teaching and learning acts, the separation between teachers and learners, and especially the role played by the different media.

Subsequently, distance learning engineering in the general Moroccan context has allowed us to highlight a number of variables. The central problem is based on modeling scenarios in the process of teaching-learning online. The formalized meta-models allows the construction and development of study objects and learning activities.

The domain model: Informs knowledge of the areas of concern for professionalizing university training.

The model of activities and learning tasks: For modeling of the nature and difficulty of the tasks of learning, we propose to model the zone of proximal development in relation to the knowledge and skills covered by the choice of teaching situations that involve that knowledge. However, it is impossible to point out the difficulty of judicious assessment of adult learning resources to manage the class of equivalent situations. To solve this problem, a placement test and evaluation of the prerequisites associated with a knowledge modeling (of inhabited level and mastery) allowing learners to self-assess its ability to handle situations that were previously encountered.

Trainer model: First, profile modeling influences phase design learning paths, the scripted content, and the adaptive virtual device context. Indeed, it has a particular impact on the modeling activity of learning objects that reflect the learner's experience and initial and continuing training. It may first act on the educational model. It can act specifically by designing specific learning situations to scriptwriting goals. It can also describe complete scenarios for each course and learning session. The decision-making process will be conditioned to its perception and representation.

Model of the context: The model aims to model the context to determine the teaching situation. This describes both entities that make up the educational work, which is both their relationships, but also the behaviors that govern the evolution of these entities. The model must allow the world to inform the user of various policy options or autonomous entities, the conditions in which they are possible, and the effects they cause.

The pedagogical model (Table 2.2): The pedagogical model informs the learning activities of learners, learning styles, and pedagogical adaptation means. This model contains the elements necessary for a decision in relation to the student profile. In our online training

TABLE 2.2
Grid of Analysis of the Approaches (Steps and Educational Methods) in Relation to the Cognitive Functions

Approaches (Steps and Educational Methods)		Percentages (%)
Approaches 100%	Transmissive	45
	Behaviorist	26
	Constructivist	15
	Socioconstructivist	14
Demarches and strategies 100%	Inductive	22
	Deductive	68
	Dialectic	10
Methods pedagogies	Demonstrative	23
	Analogue	542
	Magistrative	14
	Interrogative	16
	Discovered	

context modeling, personalized pathways screenwriting govern the pedagogical principles: the fundamental choice of direction, adaptation, consistency, customization, generalization, transfer to the next situations to implement or otherwise to avoid, allowing to limit the relevant scenarios of space from the perspective of learning. These rules are separated from the domain model, but they are also closely linked to the learner model. They can be described and set by expert teachers and adapted by trainers during the operation of the device.

The awareness in this virtual environment becomes a powerful lever of autonomy, independence, and self-knowledge permitting learners to know themselves better and to interact within an educational community. The design phase of the didactic and pedagogical approach in distance learning should emphasize the instrumentalisation and implemetation of a learning device. Learning in this virtual environment takes into account the styles of adult learners for the development of professional and individualized path, and adaptating of the training in e-learning (of which the content of teaching, the exercises, the orders of work, modes of assessments, etc.).

Modeling of the theoretical foundations of the virtual environment: We consider here the classification established by Keegan (1986), which brought together the theoretical foundations of distance education into three categories:

The theories of autonomy and independence (administration, coaching, teaching, registration, evaluation) must support the learning process of the learner (e.g., teaching guided conversation theory; enhance the pleasure and motivation).

Theories of industrialization (mass production, automation tools, planning of the production process, formalization of procedures). Standardization of production theories evokes the concepts of specialization of tasks, work organization, structure hierarchical, production control (speed and quality). It is clear that planning and technologies are keys of success.

Theories of interaction and communication. The interaction and communication between the learner and the system; the intermediary (tutor, teacher) as an element reducing the gap between the learner and the system (proactive).

Diegetic model: The diegetic model is the informed knowledge related to the construction of a story. The diegetic model aims to provide creative models of personal and collective history of arranged training courses. It is a diagram showing changes of the progress made, and the sequences of meaningful situations and specific interest. The diegetic model is also independent of the computerized context and must be informed by experts in storytelling as writers or techno-pedagogues. Thus, we believe that the only real context-dependent models of teaching–learning are the model of the adult learner and, of course, the model of the virtual environment. In addition, the context-independent models (adult education model, teaching model, model didactic, and diegetic model) are ways of thinking about the development process, and transmitting and analyzing the strategies and conditions of construction skills.

2.7 CONCLUSIONS AND RECOMMENDATIONS TO READERS

This study highlights the relevance and value of modeling as a vector that describes the most simplified representation of a complex reality of education: distance learning and measuring its effectiveness on academic achievement. Indeed the development of professional training courses skills among university students represents a major challenge in online training. The search for solutions was first brought to the modeling of learning environments. We focus our thinking more on the analysis of screenwriting modeling differentiated learning content adapted to adult learners' profiles (identification and classification of their preferred learning styles).

The modeling, therefore, seems a powerful lens of the input position. It serves not only to rank students according to a normative evaluation process, but to situate learners in a logical and learning function. It thus sets the level at which the adult learner can begin the learning path. Each skill modeled by the student will determine the choice of objectives and resources in the training course to establish. This positioning modeling is designed to be made in a feasibility study. It can also be an excellent benchmark training course adjustment (formative and formative character modeling).

Moreover systemic modeling of the teaching–learning process advocated, focuses more on professionalizing skills that allow adult learners in a personalized and individualized learning path. The dynamic aspect of our metamodeling is completed by the topological model of preferred learning styles (Table 2.3) and levels of classification of cognitive functions (Table 2.4). This axis will

TABLE 2.3
Grid of Analysis and Identification of the Styles of Training of the Studying Master ITEF (Engineering Technological for the Education and the Formation)

Preferential Fashion of Online Formation	Number of Study Respondents
Experience concrete "accommodator"	16
Divergent "reflexive experience"	5
Assimilative "abstract conceptualization"	2
Convergent "active experimentation"	3

TABLE 2.4
The Grid of Analysis Recommended: Anderson and Krathwohd "Training High-Level Cognitive" Taxonomy Revisited of Bloom

	The Cognitive Process Dimension					
The knowledge dimension	*Remember*	*Understand*	*Apply*	*Analyze*	*Evaluate*	*Create*
Factual knowledge	25%–22%	23%	28%	16%	10%	
Conceptual knowledge	38%					
Procedural knowledge	22%					
Meta-cognitive knowledge	15%					

be the starting point of an *a priori* analysis of the variables that may influence the screenwriting process of training objectives. The development of analytical models for a skills training program is implemented by the application of an engineering approach to adult education and vocational teaching focuses on experiential learning theories.

The formalization of a process through modeling the axis of our adult education scripting model and didaxologic begins with a very precise characterization of the variables: Omen "input," background, program, product "output" to implement the reflections on the modeling process, that is, of all the objects of study screenwriting and teaching activities. With this innovation process we have conceptualized three distinct phases in the nature and scope of the decisions making: conceptualization, implementation scenarios, monitoring, and control of the course training. The variables related to readjustment of personal action plans refer to the key role played by the intelligent tutoring process in the educational community. Another aspect of modeling the characterization of "output" through the analysis of the effects of modeling on the academic performance of training courses.

Another aspect of this work is the teaching and learning transposition methods from the field of engineering education and professional skills. Our proposal is therefore in metamodeling process objects of university education studies according to the business model of the student's course, whose main objective is to support teachers in the design and operation of virtual learning scenarios (intelligent tutoring) and are reused in another social field. We consider that the major objectives of our approach are

- Modeling the facilitation of teaching in the virtual environment for adult learning.
- The freedom and autonomy of choice of the updated personal action project.
- The relevance and adaptability of modeled scenarios (andragogical, ZPD) in terms of the adult learner profile.
- The variability of the scripts that ensures the progression of the students to their rhythm within the collective training course. The diversity of the situations of teaching–training allows the adult learners to meet some scripts that are always authentic and unpublished, which permits the learner to maintain interest and the perseverance during maintenance of the incentive and the development of the training.

We have adopted throughout this work a project approach built around the needs analysis phase, design modeled scriptwriting devices, the development of adult education and educational training support activities, and, finally, the procedural classification scenarios adapted to different training profiles.

The work presented in this research deal with the problem of modeling the script of teaching–learning. The specification of the choice of the virtual learning environment as the object of study meets the principles of continuity, complementarity, and extension of classroom teaching. Typological classification and characterization of the objects of study and classroom training situations are equivalent to different custom training courses. In this work of descriptive and exploratory research, we are interested in the conceptualization and formalization of educational scenarios as a student-oriented trade repository by integrating the techniques of learning remotely. For this, we first analyzed the context of virtual environment in its complexity and opportunities in terms of resources and skills (technological, communicational, and interactivity). This remote training system would allow a wide freedom of action and autonomy of choice of training courses in a wide variety and diversity of teaching and learning scenarios. Then we boarded the didactic and pedagogical involvement of social constructivist and interactive learning model on the determination of the ZPD. Modeling the difficulty of "delicious area" learning has led to a prioritization of proposed teaching content in overcoming skills acquisition (elementary basic course of initiation, building, and development). The classification of classes and functions (mobilization of conceptual, procedural, behavioral, and metacognitive resources, etc.) of authentic learning situations is flexible and responsive. Concretely, this translates into a working metadesign and organization of learning modeled units and specific objectives that can serve as medium-term course of training objectives.

To enable effective and appropriate educational activities in a virtual environment, the scriptwriting modeling objects should combine theoretical knowledge, epistemological reflection, conceptualization, and training action. It is to implement from the models the theoretical foundations of teaching and praxeological carried out in educational activities. So we proposed a general approach to modeling the didactic preparation of standards of teaching–learning. These repositories are built on scientific and technical knowledge of the computerized environment, institutional requirements, and learner profiles. In this perspective, we took into account five reference points:

1. Knowledge of the teaching support material. Here activity is divided into spatial and temporal phase (synchronous or asynchronous) and formalization of principles that are built respecting the progressiveness and consistency (intra- and intersequences and spots of learning).
2. The development of a diagnostic "assessment test" of achievements and identification of needs (scoping session, face-to-face). The trainer and the educational community as an assessment of the level of mastery of adult learners, from an overall activity, will model the characteristics of adult learners in the lists into categories more or less homogeneous. Each level of skill and mastery of skills reflects a logical learning (address acquisition) and we will find ourselves in a differentiated teaching university training course.
3. The development of a skills repository professionalizing adult learners in its ecological dimension.
4. Consideration of the application that can be relative to the objectives, activities, and learning the terms of evaluations in training, and the process of mentoring and guided discovery and distributed spots.
5. The evaluation system of learning pathways and requirements of university teaching standards.

The originality of this modeling approach is to allow procedural formalization of teaching and learning scenarios from reference metamodels. These metamodels are inspired by different works from different disciplines of the humanities and science disciplines, "mathematical modeling" of adult learning, and theories of engineering information and communication systems. We have suggested throughout this study a generic modeling approach of educational scenarios and activities that adult learners may encounter in the virtual environment. The association of this modeling as classes and problem situations of families with a model of uncertainty allows a kind of flexibility, adaptation, and progress in the training course (ability of the learner to manage special situations and thus draw the zone of proximal development in connection with the personalized course).

The main contribution of the technical–pedagogical instrumentalization of the virtual environment was also to be based on conceptual modeling based on the model process product. It is possible to change the process of building content and teaching methods recommended in this virtual environment. Our proposals have been implemented and applied in the master training contexts of technology for engineering education and training (ITEF). We had, moreover, scientifically validated a number of variables of our research approach. Although with certain limitations, arising precisely from the distinction between knowledge on the context and the theoretical and institutional models, we believe that the approach to design modeled procedurally the screenplay content from different descriptive models.

The limit of the conceptualized metamodel would be influenced by a set of paradigms, namely those of inflation and obsolescence of knowledge, the dynamics of the needs of adult learners, and the evolution of contexual and institutional requirements. In order to capitalize on these analytical models, it would be necessary to adopt a stance rooted in the paradigm of driven engineering artificial intelligence models, addressing the problem of the link between engineering and modeling theoretical foundations.

This modeling will aim the design and implementation of virtual devices with the flexibility and adaptability of behavior to appear intelligent to adult learners. The implementation of a complex computing device makes the operation script and leads the modeling system to think legitimately that its object of study is guided by an algorithmic reasoning nonlinear vis-à-vis the choice of training courses. This objective, formulated in pragmatic terms, is inseparable from the theoretical goal of operative knowledge modeling: a model that allows the action, communication, and control. This would make it possible to derive and specify the domain models necessary for the virtual environment from the upstream models described. In addition, it could possibly be relevant to use multivariate data modeling methods to generate some of the models in the field from different sources of information, such as plans of facilities, descriptions of procedures, and formal rules.

GLOSSARY

action: An action is a physical or cognitive process that can be implemented by an agent at the end of a decision. An action may be punctual or durative.

adult learner: A subject able to autonomously interact with its environment, that is, both to perceive and perform actions on it.

andragogical engineering: Entire process of adult education. Long known in everyday language trainers and institutions improperly as "adult education," it refers to adult education in its psycho aspects (shared motivations, characteristics of adult learning, and cognitive development) and in its socioeducational aspects (situational, institutional, systemic, sociological), and finally in its educational aspects (methods, techniques, and encyclopedic strategy).

apprenticeship: The apprenticeship describes a stable set of affective dispositions, cognitive and conative, favors the act of learning in all formal and informal situations, experiential or didactic, self-directed or not, intentional or accidental (Philippe Carré).

behavior: A behavior is a way to move the object or system that has one. A behavior can be punctual or durative.

didactic engineering: Born in the 1980s around mathematics education in classroom and vocational teaching on the workplace. Gérard Vergnaud is a central figure, both a Piaget psychologist and mathematician. He would influence Pierre Pastre (professional teaching) but also Yves Chevallard (anthropology of knowledge) and Yves Brousseau (mathematics education).

diegeses: The diegeses is the process of progressive creation of the universe of a story.

diegetic: Diegetic universe is a story that corresponds to the simulated world.

endogenous: An endogenous event is the result of behavior in the field, which fall to user actions, actions of virtual characters, or own behavior to the system itself.

event: An event is a highlight, which occurs at a given time. An event corresponds to an aggregation of changes causally related conditions, signifying the point of view of the observer. An event can be punctual or durative.

exogenous: An exogenous event is an externally triggered event to the system.

history: The story or fabula is the sequence of events taking place in the world, in chronological order.

metamodel: A metamodel is a model that has the particularity to model other models.

model: A model is an object that describes the inherent complexity of a phenomenon, a system, a situation, a domain, and so on for a particular purpose. A template is written in a modeling language. A modeling language focuses on some aspects of what it models.

model of educational scenario: Representing a deviation of a scenario category, it corresponds to an educational approach or a learning situation referenced. This model then serves as an example.

modelization: Modeling involves building and using a model. Which is a simplified representation of complex reality to show the important aspects of the system studied (Quebec Board of the French Language [QOFL]).

object: An object is an entity, abstract, or concrete idea having individual existence, that is to say, can be described or handled without the need to know other objects.

pedagogical scenario: The concept of the pedagogical scenario is expressed in teaching practices through personal initiatives to better explain to learners educational goals and to develop active learning situations. From the educational perspective, the emergence of the concept of scenario is indicative of the displacement of epistemological paradigm of knowledge in a logical transmission to a logical learning (Jonnaert, 2002).

scenarios: "A scenario is defined as a description done *a priori* and a posteriori, the progress of a learning situation for the appropriation of a specific set of knowledge, specifying the roles, activities and the handling of resources knowledge, tools, and services related to the implementation of activities" (Pernin and Lejeune, 2004).

scenarization: The script is a process including both the specification or possible or desirable sequences of simulation and control (implementation and/or monitoring and correction) of the progress of real-time events.

Scenaristicweft: A story line or plot of key points partially ordered not instantiated.

REFERENCES

Astolfi, J.P. and Drouin, A.M. 1992. La modélisation à l'école élémentaire. In J.L. Martinand (dir.) *Enseignement et appren-tissage de la modélisation en sciences*. Paris, INRP.

Bachelard, G. 1938. *The Formation of the Scientific Spirit*. Paris: Vrin, 1986.

Barot, C., Lourdeaux, D., Burkhardt, J.-M., Amokrane, K., and Lenne, D. 2013. V3s: A virtual environment for risk-management training based on human-activity models. *Presence: Teleoperators and Virtual Environments*, 22(1): 1–19.

Barré, V. and Choquet, Ch. 2005. Assistance in re-engineering of a teaching scenario through advocating and observable formalization. In P.Tchounikine, M.Joab, and L.Trouche (eds.), *Conference Proceedings ILE 2005*, NPRI Institute Montpellier II, 141–152.

Bassiri, M. et al. 2017. Modelling of the process teaching-training. In Sampaio de Alencar (ed.), *E-learning in Communication, Management and Information Technology*, London: Taylor & Francis Group, ISBN 978-1-138-02972-9.

Blaha, M. and Rumbaugh, J. 2007. *Object-Oriented Modeling and Design with UML*, Pearson Education.

Brassard, C. and Daele, A. 2003. A reflective tool to design a pedagogical scenario integrating ICT. In C. Desmoulins, P. Marquet, and D. Bouhineau (eds.), *ILE 2003*, Strasbourg: NPRI.

Brousseau, G. 2011. The theory of mathematics learning situations. *Education and Teaching*, 5(1): 101–104.

Burkhardt, J.-M. 2003. Immersion, realism and presence in the design and evaluation of virtual environments. *French Psychology*, 48(2): 35–42.

Carré, P. 2005. *Apprenance: Towards a New Relationship with Knowledge*. Paris: Dunod.

Crahay, M. 2006. A review of the research process-product. In *Can Education Contribute to Student Learning, and If So, How?* Geneva: Book of Educational Sciences.

Dao Chi, K. 2000. Cognitive compatibility in a learning environment hypertext. Unpublished thesis, Educational Sciences, Montreal University.

Delmas, G. 2009. Steering interactive stories and implementation of narrative forms in the context of video play. PhD thesis, Univeristy of La Rochelle.

Donnadieu, G. and Karsky, M. 2002. Systemic: Think and act in complexity. *Liaisons*.

Edgar, M. 1990. On interdisciplinarity, Carrefour Science. In *Proceedings of the Symposium of the National Scientific Research Interdisciplinary Committee, Introduction by François Kourilsky*, Paris: Editions du CNRS.

Ferraris, C., Lejeune, A., Vignollet, L., and David, J. P. 2005. Modeling collaborative learning for classroom scenarios: Towards operationalization within an ENT. In P.Tchounikine, M.Joab, and L.Trouche (eds.), *Proceedings of the Conference ILE 2005*, NPRI Institute Montpellier II.

Fuchs, P., Moreau, G., Arnaldi, B., Berthoz, A., Bourdot, P., Vercher, J. L., Burkhardt, J. M., and Auvray, M. 2006. *The Virtual Reality Treaty: Vol. 1, Man and the Virtual Environment*. Paris: The Press-Mines.

Gilbert, J.K. 2004. Models and modelling: Routes to more authentic science Education. *International Journal of Science and Mathematics Education*, 2: 115–130.

Glikman, V. 2002. *Correspondence Courses in E-Learning: Panaroma Distance Learning*, Paris: PU.

Jonnaert, P. 2002. *Skills and Socioconstructivism*, Brussels: DeBoeck.

Jonnaert, Ph. and Vander Borght C. 2003. Creating conditions for learning. In *Perspectives in Education and Training*. Brussels: De Boeck, pp. 20–24.

Keegan, D. 1986. *Foundations of Distance Education*. London: Croom Helm.

Knowles, M. 1980. *The Modern Practice of Adult Education. From Pedagogy to Andragogy*. New York: The Adult Education Company.

Knowles, M. et al. 1984. *Andragogy in Action: Applying Modern Principles of Adult Education*. San Francisco: Jossey-Bass.

Knowles, M. 1990. *The Adult Learner, to a New Way of Training*. Paris: Editions of Organization.

Kolb, D. A. 1985. *The Learning Style Inventory: Self, Corin Inventory and Interpretation Booklet* (1st ed. 1981). Boston: Mass, McBer and Company.

Laforcade, P., Nodenot, T., and Sallaberry, C. 2005. A modeling language based on UML. In S. George and A. Derycke (eds.), *Concepts and Uses of Training Platforms* [special issue]. Information and Communication Science and Technology for Education and Training, 12.

Laforcade, P., Nodenot, T., Sallaberry, C. et al. 2005. An educational modeling language based on UML. *Science and Information and Communication Technologies for Education and Training (STICEF)*, 89–116.

LeBlanc, R., Chevrier, J., Fortin, G., Théberge, M. 2000. The LSQ-fa: A short French version of the Honey and Mumford-In: The Learning Style, Volume XXVIII Number 1, Spring Summer 2000. ACELf [Online] http: //www.acelfca/revue/XXVIII [18 September 2001].

Legay, J. M. 1997. *The Experience and the Model: A Discourse on Method*, Paris: INRA Editions.

Lemaine, G. 1999. Introduction to the sociology of science and scientific knowledge. *The Journal for the History of CNRS* [Online], 1.

Malcolm Knowles, S. 1975. *Self-Directed Learning: A Guide for Learners and Teachers*. Cambridge: Prentice-Hall.

Martinand, J. L. 1989. Capacity-goals to objectifs-obstacles. In Bednarz, N., and Garnier, C. (eds.), *Construction of Knowledge, Obstacles and Conflicts*. Ottawa: ARC Agency, pp. 217–227.

Martinand, J.-L. 1994. What lessons can be learned from modeling work in the context of curriculum development? In Martinand, J.-L. (ed.), *New Perspectives on the Teaching and Learning of Modeling in Science*, Paris: INRP, pp. 115–125.

Mayen, P. 1998. Developments in the professional race and process of transformation of expertise. In *Acts of the Symposium of the French-Speaking Ergonomics Society (SELF)*. Paris: SELF.

Meirieu, P. 1991. Individualization, differentiation, personalization: From the exploration of a semantic field to the paradoxes of formation. Association of Teachers and Researchers in Educational Sciences (AECSE). [Online]. http://www.meirieu.com/ARTICLES/individualisation.pdf. Accessed February 23, 2006.

Meirieu, P. 1993. *To Learn ... Yes, How Goal?* Paris: ESF, Coll. Pedagogies.

Morin, E. 2005. *Introduction to Complex Thought*. Paris: Seuil.

Paquette, G. 2004. Object-oriented pedagogical engineering and referencing by competencies. *International Journal of Technologies in Higher Education*, 1(3), pp. 45–55.

Paquette, G. 2005. Learning on the Internet: From platforms to knowledge-based object portals. In Pierre, S. (ed.), *Innovations and Trends in Training and Learning Technologies*, Presses of the Polytechnic of Montreal, pp. 1–30.

Peng, I. 2003. Applying Learning Style in Instructional Strategies, *CDTL*, 5(7) [En ligne]. http://www.cdtl.nus.edu.sg/brief/V5n7/default.htm.

Pernin, J.-P. and Lejeune, A. 2004. Models for the reuse of learning scenarios. In *Proceedings of the TICE Méditerranée Symposium*, Nice, France.

Rodhod, R. 2010. Interactive narrative for adaptive educational games. PhD thesis, University of Technology of Compiegne, French.

Rouet, J. R. 1999. Interactivity and cognitive compatibility in hypermedia systems. *Review of the Sciences of Education*, XVII(1): 87–104.

Schreiber, G. et al. 1993. *KADS: A Principled Approach to Knowledge-Based System Development*. London, Academic Press.

Tiberghien, A. 1994. Modeling as a basis for analyzing teaching-learning situations. *Learning and Instruction*, 4(1): 71–87.

Vergnaud, G. 1991. The theory of conceptual fields. *Research in Didactics of Mathematics*, 10: 132–169.

Vygotsky, L. S. 1985. *Thought and Language. Coll. Courses*. Paris: Editions Sociales.

Yourdon, E. 1989. *Modern Structured Analysis*. Englewood Cliffs, NJ: Yourdon Press.

3 Solving the Problems of Linguistically Diverse First-Year University Students Using Digital Learning

Dace Ratniece and Sarma Cakula

CONTENTS

3.1 Introduction 41
3.2 Aim, Hypotheses, and Research Methods 42
3.3 Theoretical Background 42
3.4 Research Methodology 43
 3.4.1 Data Collection Methodology 43
 3.4.2 Description of the Experiment 44
3.5 Data Analyses and Evaluation 44
3.6 Conclusions and Future Work 47
References 47

3.1 INTRODUCTION

Language can be defined as a means of communication that shapes cultural and personal identity and socializes one into a cultural group (Gollnick and Chinn, 2006). It is impossible to separate language and culture. University students from diverse language backgrounds encounter some difficulty every day. Because language and culture are so intertwined, language-minority students are expected to learn and effectively use a tuition language and new cultural dispositions. Often the tuition language and culture are different from what the students have learned at home. In this chapter, the term *linguistically diverse students* will be used to refer to students whose first language (L1) is other than Latvian. The tuition language of state universities in the Republic of Latvia is Latvian.

Data from the 2013–2014 academic year until the 2015–2016 academic year (3 academic years) indicate that first-year linguistically diverse students at Riga Technical University's engineering telecommunications program comprise approximately 40% of all students (Ratniece and Cakula, 2015). Therefore, academic personnel must be aware of diversity in their classrooms and how it may have an impact on students' achievements. (The bachelor study program provides a blend of knowledge from electrical engineering and computer science focusing on communications networks and systems, encoding theory, information/optical processing, and transmission. The aim of the program is to provide an academic education and prepare students for further studies at the master's level.)

Digital learning is any instructional practice that effectively uses technology to strengthen a student's learning experience. It emphasizes high-quality instruction and provides access to challenging content, feedback through formative assessment, opportunities for learning anytime and anywhere, and individualized instruction to ensure all students reach their full potential to succeed in education and a career. Digital learning encompasses many different facets, tools, and applications to support and empower teachers and students, including online courses, blended or hybrid learning, or digital content and resources. Additionally, digital learning can be used for professional learning opportunities for teachers and to provide personalized learning experiences for students.

Machine translation technology as a digital learning tool is constantly being applied by linguistically diverse students in the study process. Here's what a machine translator can do:

- Like a bilingual dictionary, it can match a word in one language with a word in another language. However, the same word may have different meanings. For example, "spirits" can be either souls or alcoholic drinks in English.
- When it has to choose between different possible translations, a machine translator can make statistical "guesses" at the context. For example, in a sentence that talks about both meat and spirit, the machine might guess that the word "spirit" refers to alcohol.

However, the problem with machine translation is that a machine is only a machine. It matches components and follows rules. It does not actually know what it is talking about. A machine cannot assess whether a sentence sounds good or bad. A machine is also incapable of managing nuances, subtexts, symbolism, or wordplay; it cannot control mood or tone. So, machines are not likely to replace human translators in the near future.

Motivation is a powerful force in second-language (L2) learning. Motivation governs a need to communicate, to make friends, to identify with a social group, to become part of a community, and to begin to plan one's future.

3.2 AIM, HYPOTHESES, AND RESEARCH METHODS

The aim of the research was to identify the problems and motivation factors for linguistically diverse first-year university students caused by quality of digital learning.

The hypotheses of the research were

- The use of digital learning in itself does not constitute an enhancement of the quality of teaching and learning, but it is a potential enabler for such enhancement.
- Evaluation of linguistically diverse first-year university students' work using digital learning (many different facets, tools, applications and online course management system) motivates them to study.

As the theoretical framework, the following research methods were applied:

- Evaluation of all homework assignments uploaded in the e-learning environment
- The risk-taking assessment by Schubert's method
- Diagnostics of the person on motivation by T. Elersa methods
- Failure avoidance motivation method
- A survey with the assessment of the course Entrepreneurship (Distance Learning E-Course) developed by the author (D. Ratniece) of this chapter

3.3 THEORETICAL BACKGROUND

Dulay, Burt, and Krashen (1982) in their survey of major findings in second-language research indicate that the most beneficial environment for the learner is one that encourages language learning in natural surroundings for genuine communication. Furthermore, it has been shown that optimal second-language learning takes place in an environment

- That is nonthreatening, in which the learner feels free to take chances and make mistakes
- That is linguistically and nonlinguistically diverse (i.e., no grammatically sequenced syllables, no attempt to homogenize the environment so that learners understand everything)

- In which learners focus on tasks and activities of interest to them, and use language as a tool to get things done (i.e., very little explicit discussion of language)
- In which learners' interests and needs serve as the basis for learning activities
- In which learners' talk is considered to be the task, as in "being on task": small talk, jive, and tall tales are not only tolerated, they are encouraged, and not just at "sharing time" but throughout the day

To become life-long language users, L2 students as well as native Latvian-speaking students need to gain control over language and feel comfortable about using the language. The ensuing principles for second-language instruction can help lecturer create supportive language environments.

Learning by Latvian as a second language (LSL) students should be built on the educational and personal experiences they bring to the educational establishment. In language learning, students should be encouraged to use their previous experiences with oral and written language to develop their second language and to promote their growth to literacy. Students bring to an educational establishment cultural identities, knowledge, and experiences that should be awarded by instructional practices rather than replaced or forgotten as learning takes place (Au and Jordon, 1981; Jordan, 1985; Cummins, 1986; Cummins and Swain, 1986; Edelsky, 1986; Hudelson, 1986; Diaz and Moll, 1987; Enright and McCloskey, 1988). Socializing, learning, questioning, and wondering are some of the many things that one is able to do when one learns a language. However, these things are not quickly learned; it takes many years to develop full-fledged competence (Wong-Fillmore, 1983). Furthermore, rates of development of oral proficiency vary considerably in LSL students. Consequently, lecturers, not just LSL specialists, need to address the learning needs of LSL students and adjust their instruction accordingly to meet the different levels of Latvian proficiency, different learning rates, and styles of their students. Instructional convenience does not mean, however, a watered-down curriculum.

It has long been recognized that if LSL students are to catch up or keep up with their native Latvian-speaking peers, their cognitive and academic growth should continue while the second language is developing. Thematic units (as opposed to exercises in grammatical structures) where language is integrated with academic content appear to be an effective way to simultaneously develop students' language, subject area knowledge, and thinking skills. Thematic units help involve students in real language use—use of language interactively across a variety of situations, modes, and text types. Digital learning and machine translation technology is constantly being applied by linguistically diverse students in the study process and that strengthens the students' learning experience and motivation.

3.4 RESEARCH METHODOLOGY

Methods for diagnoses of the degree of risk preparedness, motivation to success, and motivation to avoiding failures have been tested for all students during the course: at the beginning, in the middle, and at the end.

3.4.1 Data Collection Methodology

Motivation becomes a positive force. Anxiety becomes an inhibitor. Self-confidence is very much related to second-language learning, as is a low anxiety level and a tendency to be risk-takers and do guess work. As the student becomes more secure in the second language, it is entirely likely that the native language recedes to some extent. As vocabulary in the second language increases, words in L1 may well be forgotten. During the second-language learning process, a learner may insert words from each language in the same sentence. Again, this tendency demonstrates a motivation to speak the second language and is a way of permitting precise expressions that carry cultural content and can be stated in a given language.

Web-based e-learning platforms allow educators to construct effective online learning study courses by uploading various categories of study materials. An e-learning platform allows usage of a wide range of online learning tools, such as forums, discussion forums, e-mail messaging, as well as combining face-to-face and online approaches. The purpose of these technologies is to deliver study materials to students, improve students' skills, assess skills and knowledge, and achieve better learning outcomes. Fast and immediate feedback is possible. E-learning platforms produce data logging. Logged data can be used for later analysis. There are two types of data:

- Data produced by students or teachers that represent the content of the learning course
- Data made by the system based on a student's activities, like the time spent in the system, kept sessions, and the number of clicks on items of the content (Ratniece, Cakula, Kapenieks, and Zagorskis, 2015)

At Riga Technical University (RTU), the e-learning platform Moodle has been maintained. In the period from October until December of academic year 2013–2014, academic year 2014–2015, and the academic year 2015–2016 the course Entrepreneurship (Distance Learning E-Course) to first-year students was provided. The course was conducted by RTU professor A. Kapenieks. The author (D. Ratniece), as the assistant to the professor, supplemented the lecture content. Ratniece's study "Use of Social Microblogging to Motivate Young People (NEETs) to Participate in Distance Education" was presented.

3.4.2 Description of the Experiment

The research was carried out during the lectures and the final exam of the course Entrepreneurship (Distance Learning E-Course) with first-year students (respondents) participating on a voluntary basis. The course had two homework assignments:

- The search of business ideas on the Internet
- Your business idea

The author (Ratniece) evaluated all homework assignments uploaded in the e-learning environment. Reviews and comments were added, with the aim to encourage and motivate students to prepare their business plans in time and of good quality. Each comment was prepared according to the results of the content analysis. The feedback comments to students were written in a positive, supportive, and motivational manner, personally addressing each student. The author made notes on students' written-language mistakes. The assessment of the student's homework was done concerning seven criteria:

1. Actuality or viability of idea
2. Technological solution or how to enforce
3. Marketing/promotion of goods or services in the market
4. Competition
5. Financial security, for example, planned revenues, expenses, financial support for the company's start-up and ongoing development (bank loan, other resources, etc.)
6. The ability of a company to realize the idea
7. The potential risks

3.5 DATA ANALYSES AND EVALUATION

Every year the Latvian and linguistically diverse students' success rate increased. Looking at the two groups, it is evident that linguistically diverse students' total proportion increased for

TABLE 3.1
Latvian and Linguistically Diverse Students' Results over Three Academic Years

Academic Year	Language Use Group	Number of Students Who Started the Course Entrepreneurship (Distance Learning E-Course)		Second Homework "Your Business Idea"		Teacher's Comments According to the Incorrect Latvian Language Use		Number of Students Who Successfully Completed Course	
		By Group	Total	By Group	Total	By Group	Total	By Group	Total
2013–2014	Latvians	80	142	62	84	3	21	77	107
	Linguistically diverse	62		22		18		30	
2014–2015	Latvians	78	142	43	80	2	13	53	79
	Linguistically diverse	64		37		11		26	
2015–2016	Latvians	77	132	32	64	0	4	32	56
	Linguistically diverse	55		33		4		24	
Total		416		228		38		242	

those who successfully completed the course Entrepreneurship (Distance Learning E-Course) (Table 3.1).

The quality level of the homework (uploaded in the e-learning environment) for linguistically diverse first-year university students became better each year with the use of a machine translator.

Data relating to the second homework "Your business idea" shows the real-study example of machine translation. Linguistically diverse first-year university students took and used the online translator on Google to translate the homework into Latvian. The results are included in Table 3.2 (see points 5, 6, 7, and 8) as well as in Table 3.3 (see points 2.3 and 2.4).

All respondents (Latvian and linguistically diverse first-year university students) indicate that e-learning and traditional forms of study need to be kept in balance, because e-learning provides a great advantage to learn anywhere, anytime. Successful guidance through the study process, however, is just as important, and can only be ensured when a teacher is present.

TABLE 3.2
Students' Evaluation of the Effectiveness of the Form Practiced in the Course Entrepreneurship (Distance Learning E-Course) (Academic Years 2013–2014 and 2014–2015)

	Low Rating			Average Rating				High Rating			
Form of Study	**1**	**2**	**3**	**4**	**5**	**6**	**7**	**8**	**9**	**10**	**Weighted Mean**
1. Lectures 2013–2014				3	6	16	34	29	11	6	7.30
2. Lectures 2014–2015	1	1	2	1	1	10	25	18	12	8	7.80
3. Discussions 2013–2014			2	3	5	13	18	35	21	8	7.73
4. Discussions 2014–2015		1	1	1	1	3	13	24	18	17	8.40
5. Insertion in ORTUS 2013–2014			1		7	11	19	35	15	17	7.90
6. Insertion in ORTUS 2014–2015			1	3	4	5	15	20	15	16	8.01
7. Teachers' comments 2013–2014			1	2	7	12	15	33	17	18	7.88
8. Teachers' comments 2014–2015	1		1	2	4	4	9	14	12	31	8.54

TABLE 3.3
Students' Evaluation of the Effectiveness of the Form Practiced in the Course Entrepreneurship (Distance Learning Course) (2015–2016)

	Low Rating			Average Rating				High Rating			Weighted Mean
1. Traditional forms of study	**1**	**2**	**3**	**4**	**5**	**6**	**7**	**8**	**9**	**10**	
1.1. Lectures				1		3	18	16	10	8	7.96
1.2. Discussions					1	2	7	16	19	11	8.09
2. Digital Opportunities to Increase Students' Education Quality and Motivation											
2.1. Digital environment interface (cumbersome to easy to learn)				1	1	5	8	15	15	11	8.21
2.2. Digital accessibility (not at all to in full)					1	3	5	8	18	21	8.82
2.3. Assignment preparation and insertion in ORTUS system (unsuccessful to successful)			1			1	5	13	11	25	8.88
2.4. Teachers' comments in ORTUS system (redundant to needed)	1	1		3	3		4	15	11	18	8.14
2.5. Functional content of the course (inadequate to adequate)				1	1	2	6	19	11	16	8.64
2.6. Content structuring (opaque to transparent)		1		1	1	4	5	13	18	13	8.32
2.7. Motivation tests—improve the study process (unimportant to important)		1	1	1	7	6	8	11	10	11	7.57

It should be noted that in the academic year 2013–2014 teacher comments in the online environment (including the correct use of the Latvian language) were written in assessing the students' second homework assignment. In the academic year 2014–2015 teacher comments in the online environment were written in assessing students' first and second homework assignments. In the academic year 2015–2016 teacher comments in the online environment were written in assessing students' first homework assignment. Students' average assessment shows that students appreciated the teacher's job better when the teacher had evaluated both homework assignments.

TABLE 3.4
Independent Samples Test

	t	df	Sig. (2-tailed)	Mean Difference
Activity	2.899	115	0.004	−86.846
First practical evaluation	2.042	115	0.043	−1.519
Second practical evaluation	1.257	115	0.211	−1.033
Final	−0.779	264	0.437	−1.777
Motivation to avoiding failures at the beginning	0.845	139	0.399	0.577
Motivation on success in the beginning	1.644	159	0.102	−0.564
Diagnostics of the degree of risk preparedness in the beginning	−0.183	153	0.855	−0.366
Motivation to avoiding failures at the end	0.922	75	0.359	0.902
Motivation on success at the end	0.137	84	0.891	0.079
Diagnostics of the degree of risk preparedness at the end	−0.314	75	0.754	−1.010

The number of students decreased over the period of one semester; only two-thirds of all students turned in the second assignment. The author (Ratniece) identified that there exists some certain "risk factor" level for students to drop out of the course that correlated with a student's level of preparedness, readiness, and eagerness.

In the 2013–2014 academic year, 107 questionnaires were issued and completed by students, and 107 students finished the course. In 2014–2015 academic year, 79 questionnaires were completed, and 77 students finished the course. In the 2015–2016 academic year, 56 questionnaires were completed, and 53 students finished the course.

The testing mean of activity between the native language-speaking student group and linguistically diverse students are different on 95% probability. Linguistically diverse students are more active (mean of activity for native-speaking students is 159 and for linguistically diverse students is 246). There are no differences on course evaluation and final grade between both groups on probability level 95%. Also all means of psychological test results are equal for both student groups on probability level 95% (see Table 3.4). For example, if H_0 indicates the mean of final course evaluation are equal between native language student group and linguistically diverse student group and Ha indicates the mean of final course evaluation are not equal between native language student group and linguistically diverse student group, and the *T*-test value –0.779 does not increase the critical value on the significant level 0.05, it means that H0 is right. The average evaluation between both student groups is equal on probability level 95%.

3.6 CONCLUSIONS AND FUTURE WORK

Digital learning and machine translation technology emphasizes high-quality instruction and provides access to challenging content, feedback through formative assessment, opportunities for learning anytime and anywhere, and individualized instruction to ensure all students reach their full potential to succeed in education and a career. It confirms that

1. Evaluation of online homework from linguistically diverse students improves students' knowledge, as well as learning the language, if the teacher points to the grammatical errors resulting from the use of machine translation.
2. There will be an increase at higher education establishments concerning the proportion of linguistically diverse students.
3. The quality of homework assignments' (uploaded in the e-learning environment) from linguistically diverse first-year university students concerning the use of a machine translator became better year after year.

Digital learning objects are often considered complete and whole the moment they are uploaded into a digital learning repository. However, these may be versioned over time, and they should be labeled as such. Updates to digital learning objects should generally be done in the following contexts: when the paradigm has shifted in a field, when critical data has changed, when there are important policy changes, and when new learning experiences and methods are possible to enhance the learning. The learning objects that are hosted on sites may be continually updated and revised for quality. It is important to notify an installed base of users' updates to materials if they choose to receive such notifications.

REFERENCES

Au, K., and C. Jordan. 1981. Teaching reading to Hawaiian children: Finding a culturally appropriate solution. In *Culture and the Bilingual Classroom: Studies in Classroom and Ethnography*, edited by K. Au, G. Guthrie, and H. Trueba (pp. 139–152). Rowley, MA: Newbury House Publishers.

Cummins, J. 1986. Empowering minority students: A framework for intervention. *Harvard Educational Review* 56: 18–36.

Cummins, J., and M. Swain. 1986. *Bilingualism in Education: Aspects of Theory, Research and Policy*. London: Longman.

Diaz, R., and L. Moll. 1987. Teaching writing as communication: The use of ethnographic findings in classroom practice. In *Literacy and Schooling*, edited by D. Bloome (pp. 195–221). Norwood, NJ: Ablex.

Dulay, H., M. Burt, and S. Krashen. 1982. *Language Two*. New York: Oxford Press.

Edelsky, C. 1986. *Writing in Bilingual Program: Habia Una Vez*. Norwood, NJ: Ablex.

Enright, D., and M. McCloskey. 1988. *Integrating English: Developing English Language and Literacy in the Multilingual Classroom*. Reading, MA: Addison-Wesley.

Gollnick, D. M., and P. C. Chinn. 2006. *Multicultural Education in a Pluralistic Society* (7th ed.). Upper Saddle River, NJ: Pearson.

Hudelson, S. 1986. ESL children's writing: What we've learned, what we're learning. In *Children and ESL: Integrating Perspectives*, edited by V. Allen and P. Rigg (pp. 23–54). Washington, DC: Teachers of English to Speakers of Other Languages.

Jordan, C. 1985. Translating culture: From ethnographic information to educational program. *Anthropology and Education Quarterly* 16: 105–123.

Ratniece, D. 2014. Use of Social Microblogging to Motivate Young People (NEETs) to Participate in Distance Education through www.eBig3.eu. DATA ANALYTICS 2014 : The Third International Conference on Data Analytics, August 24–28, 2014, Rome, Italy, pp. 7–11. ISBN 978-1-61208-358-2.

Ratniece, D., and D. Cakula. 2015. Digital opportunities for student's motivational enhancement. *Procedia Computer Science* 65: 754–760.

Ratniece, D., S. Cakula, K. Kapenieks, and V. Zagorskis. 2015. Digital opportunities for 1-st year university students' educational support and motivational enhancement. In *The 1st International Conference on Advanced Intelligent Systems and Informatics (AISI2015)*, Beni Suef, Egypt, November 28–30, 2015, edited by T. Gaber, A. E. Hassanien, N. El-Bendary, and N. Dey (pp. 69–78). Cham, Switzerland: Springer International.

Wong-Fillmore, L. 1983. The language learner as an individual: Implications of research on individual difference in the ESL teacher. In *On TESOL '82: Pacific Perspectives on Language Learning and Teaching*, edited by M. Clarke and J. Handscombe (pp. 157–173). Washington, DC: Teachers of English to Speakers of Other Languages.

4 Moving Enterprise Architecture from Professional Certificates to Academic Credentials

Fekry Fouad

CONTENTS

4.1 Introduction 49
4.2 Unique Understanding for Enterprise Architecture 50
4.2.1 What Is an Enterprise Architecture? 51
4.2.2 Enterprise Architecture Frameworks 51
4.2.2.1 The Different Architectural Views within a Framework 51
4.2.3 Integration and Security Architectures 51
4.3 Information Technology and Computer Academic Branches 52
4.3.1 Computer Science 52
4.3.2 Information Systems 53
4.4 Enterprise Architecture in Academic Branches 54
4.5 Application Area and Normative References 54
4.5.1 Learning Goals 54
4.5.2 Competences Obtained during the Studies 56
4.6 Place of This Course in the Educational Program Structure 57
4.6.1 Content of Topics 58
4.7 Course Parts and Types of Classes 59
4.8 Methodical Recommendations for the Course 59
4.8.1 Form of Exam and the Final Mark Structure 59
4.8.2 Possible Questions and Tasks for Exam 59
4.9 Conclusion 60
References 60

4.1 INTRODUCTION

This chapter provides a guideline to move enterprise architecture (EA) from professional certificates to academic credentials for higher education. The main concept of this chapter is that EA is a strategy and business-driven activity that supports management planning and decision making by providing coordinated views of an entire enterprise. These views encompass strategy, business, and technology, which is different from technology-driven, systems-level, or process-centric approaches. Implementing an EA involves core elements, a management program, and a framework-based documentation method, all that should be planned as an academic program given by the universities.

Enterprise architecture is considered as a best practice, because the multidisciplinary, workflow enables planning for a smooth transformation of an organization using a strategic analysis. Also EA is professional work and a practice to assist enterprises in solution designing to achieve the future business objectives. Today's rapid-changing business operating models and technology options are evolving at an ever-increasing rate, and EA enables organizations to move up the curve. However, what does the future really hold for EA as an academic credential. It is clear that there is a major difference between the academic credential and professional practice of EA. The professional practices are certain course of actions to prove the competence or power of expertise, whereas academic credentials for the most part do not. Certification makes someone a more valuable employee. Certification immediately bestows credibility for a certain skill set, and gives the employee an advantage when it comes to a new vacancy or job and career advancement.

This chapter details a way to move enterprise architecture from professional certificates to academic credentials to provide an opportunity for education researchers, practitioners, and professional bodies of enterprise architecture, as well as the information technology (IT) educational staff from around the globe to exchange their experiences, ideas, theories, strategies, and technology-inspired solutions for achieving more engaging, more efficient, more accessible, and more successful IT education globally. This chapter suggests a curriculum for a combination of theoretical knowledge and practical understanding, and the methodology of moving from professional certificates into academic credentials for enterprise architecture.

Credentials refer to specific educational qualifications or academic outcome, for example, a completed (or even partially completed) diploma or degree. *Credentials* can also refer to occupational qualifications, such as professional certificates or work experience [1].

The professional or occupational enterprise architecture credential is possible through different professional entities such as an academic degree or professional designation such as PhD, PEng, or MD, whether this be purely honorary or symbolic, or associated with credentials attesting to specific competence, learning, or skills [2]. The main chapter objectives is to give a clear vision for an EA academic degree such as a college diploma, bachelor's degree, master's degree, PhD, or doctorate, in line with the professional work in the field. The designations of the academic credentials and the related experience required of the academic staff will be in high need to ensure the highest quality education regarding EA to meet the relevant accreditation standards, if any.

The standard minimum qualification for teaching in an EA bachelor degree program will be a master's degree in the discipline of EA or related field. Some colleges may desire the instructional staff to hold the terminal degree in the EA discipline, which often would be a PhD. Professional credentials or certificates are given by professional bodies, such as the Federation of Enterprise Architecture Professional Organizations (FEAPO) [3], which is a worldwide association of professional organizations that have come together to provide a forum to standardize, professionalize, and otherwise advance the discipline of enterprise architecture. The professional credentials that can be given by the FEAPO through its affiliated bodies are professional licenses, memberships in professional associations, apprenticeships, trade certificates, and work experience accredited certificate. This chapter shows the best curriculum for the EA discipline that alternatively is a combination of academic credentials, occupational/professional designations, and experience qualifications combined with demonstrated and documented professional competence in the field of EA.

4.2 UNIQUE UNDERSTANDING FOR ENTERPRISE ARCHITECTURE

Organizations in different disciplines started to look for standard ways to support describing and documenting their processes and systems using EA frameworks such as TOGAF, Zachman, Federal Enterprise Architecture Framework (FEAF), and Treasury Enterprise Architecture Framework (TEAF). An architecture is a framework of principles, guidelines, standards, models, and strategies that direct the design, construction, and deployment of business processes, resources, and information technology throughout the enterprise. Architectures are usually high-level views of the system they describe. An architecture is typically made up of

- A picture of the current state
- A blueprint or vision for the future
- A roadmap on how to get there

4.2.1 What Is an Enterprise Architecture?

An EA is a conceptual blueprint that defines the structure and operation of an organization. The intent of an enterprise architecture is to determine how an organization can most effectively achieve its current and future objectives.

Zachman defines an EA as the "set of descriptive representations (i.e., 'models') that are relevant for describing an Enterprise such that it can be produced to management's requirements (quality) and maintained over the period of its useful life (change)."

4.2.2 Enterprise Architecture Frameworks

Frameworks are commonly used to organize enterprise architectures into different views that are meaningful to system stakeholders. These frameworks, commonly referred to as enterprise architecture frameworks, are often standardized for defense and commercial systems. Frameworks may specify process, method, or format of architecture activities and products. Not all frameworks specify the same set of things, and some are highly specialized.

The practice of enterprise architecture involves developing an architecture framework to describe a series of "current," "intermediate," and "target" reference architectures and applying them to align change within the enterprise. Another set of terms for these architectures are "as-is" and "to-be."

There are many examples of widely used enterprise architecture frameworks across different industries. Some of the most widely know are FEAF, TEAF, EAP, DoDAF (Department of Defense Architecture Framework), TOGAF, Zachman, and Gartner.

4.2.2.1 The Different Architectural Views within a Framework

These frameworks detail all relevant structure within the organization including business, applications, technology, and data. They provide a rigorous taxonomy and ontology that clearly identifies what processes a business performs and detailed information about how those processes are executed. The end product is a set of artifacts that describes in varying degrees of detail exactly what and how a business operates and what resources are required. These artifacts are often graphical.

- A business strategy architecture defines the overall strategic direction of the business, the vision, mission, business plans, and overall business objectives.
- A business process architecture describes the business processes that have to be put in place in order for the business to operate efficiently and support effectively the enterprise business objectives.
- A data/information architecture describes the structure of an organization's logical and physical data assets and data management resources.
- An application architecture provides a blueprint for the individual application systems to be deployed, their interactions, and their relationships to the core business processes of the organization.
- A technology architecture describes the software and hardware infrastructure intended to support the deployment of core, mission-critical applications.

4.2.3 Integration and Security Architectures

Integration architecture can be broadly defined as the discipline of bringing data and information together and sharing it between repositories, applications, business processes, and organizations. Successful integration allows a company's entire organization to be brought into a cohesive, responsive infrastructure for making better decisions, driving the organization toward new business opportunities, and adapting to change in the marketplace.

The integration architecture domain also defines the discovery, interaction, and communication technologies joining business processes across the enterprise, disparate systems, and information sources. It documents the cooperation and interoperability among applications (integration services), the definition of the roles, technologies and standards to protect information assets (security and directory services), and portal requirements to provide access to applications and data within the enterprise (web services).

The purpose of the security architecture is to bring focus to the key areas of concern for the enterprise, highlighting decision criteria and context for each domain.

The security architecture is a unifying framework and reusable services that implement policy, standards, and risk management decisions. The security architecture is a strategic framework that allows the development and operations staff to align efforts. In addition the security architecture can drive platform improvements that are not possible to make at a project level.

Since security is quite often a system characteristic, it can be difficult for enterprise security groups to separate the disparate concerns that exist at different system layers and to understand their role in the system as a whole. The architecture provides a framework for understanding disparate design and process considerations; to organize architecture and actions toward improving enterprise security.

4.3 INFORMATION TECHNOLOGY AND COMPUTER ACADEMIC BRANCHES

The actual functionalities of information technology (IT) versus computer science (CS) and information systems (IS) is still a debatable issue. However, for the sake of the research, we would give their functionalities in brief.

4.3.1 Computer Science

Computer science spans a wide range, from its theoretical and algorithmic foundations to cutting-edge developments in robotics, computer vision, intelligent systems, bioinformatics, and other exciting areas.

We can think of the work of computer scientists as falling into three categories:

- They design and implement software. Computer scientists take on challenging programming jobs. They also supervise other programmers, keeping them aware of new approaches.
- They devise new ways to use computers. Progress in the computer science areas of networking, database, and human–computer interface enabled the development of the World Wide Web. Now, researchers are working to make robots be practical aides that demonstrate intelligence and are using databases to create new knowledge.
- They develop effective ways to solve computing problems. For example, computer scientists develop the best possible ways to store information in databases, send data over networks, and display complex images. Their theoretical background allows them to determine the best performance possible, and their study of algorithms helps them develop new approaches that provide better performance. Computer science spans the range from theory to programming. Although other disciplines can produce graduates better prepared for specific jobs, computer science offers a comprehensive foundation that permits graduates to adapt to new technologies and new ideas.

The shaded portion in Figure 4.1 represents computer science. Computer science covers most of the vertical space between the extreme top and extreme bottom because computer scientists generally do not deal with "just the hardware" that runs software, or about "just the organization" that make use of the information that computing can provide. As a group, computer scientists care about almost everything in between those areas (down as far as the software that enables devices to work;

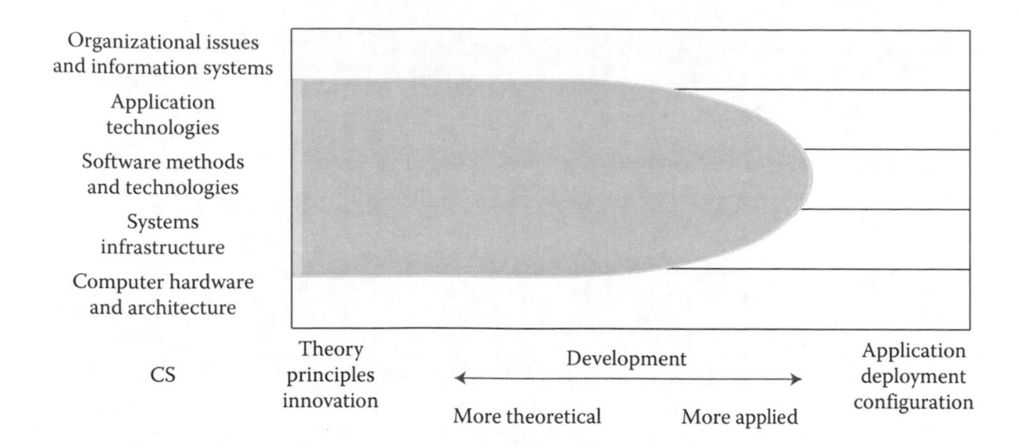

FIGURE 4.1 CS participation for IT field.

up as far as the information systems that help organizations operate). They design and develop all types of software, from systems infrastructure (operating systems, communications programs, etc.) to application technologies (web browsers, databases, search engines, etc.). Computer scientists create these capabilities, but they do not manage the deployment of them. Therefore, the shaded area for computer science narrows and then stops as we move to the right. This is because computer scientists do not help people to select computing products, nor tailor products to organizational needs, nor learn to use such products.

4.3.2 Information Systems

Information systems specialists focus on integrating information technology solutions and business processes to meet the information needs of businesses and other enterprises, enabling them to achieve their objectives in an effective, efficient way. This discipline's perspective on "information technology" emphasizes information, and sees technology as an instrument to enable the generation, processing, and distribution of needed information. Professionals in this discipline are primarily concerned with the information that computer systems can provide to aid an enterprise in defining and achieving its goals, and the processes that an enterprise can implement and improve using information technology. They must understand both technical and organizational factors, and must be able to help an organization determine how information and technology-enabled business processes can provide a competitive advantage.

The information systems specialist plays a key role in determining the requirements for an organization's information systems and is active in their specification, design, and implementation. As a result, such professionals require a sound understanding of organizational principles and practices so that they can serve as an effective bridge between the technical and management communities within an organization, enabling them to work in harmony to ensure that the organization has the information and the systems it needs to support its operations. Information systems professionals are also involved in designing technology-based organizational communication and collaboration systems.

A majority of information systems programs are located in business schools. All information systems degrees combine business and computing coursework. A wide variety of information systems programs exists under various labels that often reflect the nature of the program. For example, programs in computer information systems usually have the strongest technology focus, and programs in management information systems can emphasize organizational and behavioral aspects of information systems. Degree program names are not always consistent.

The shaded portion in Figure 4.2 represents the discipline of information systems. The shaded area extends across most of the top-most level because information systems people are concerned

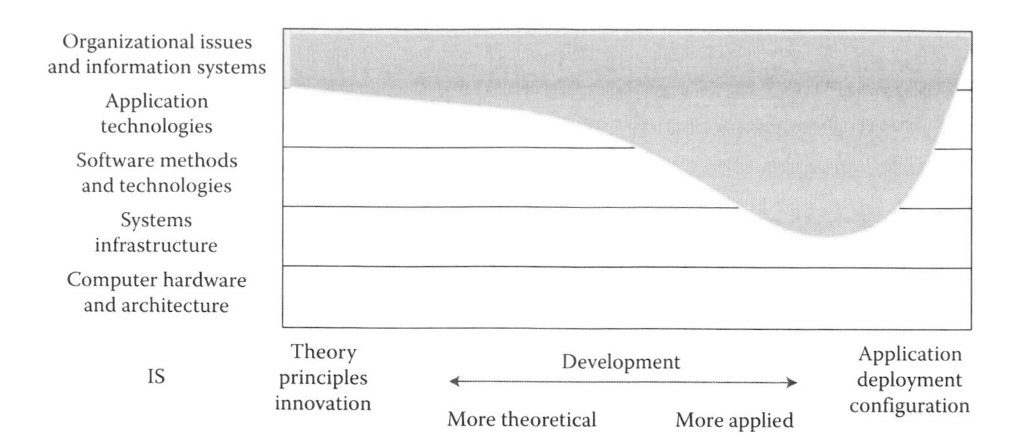

FIGURE 4.2 IS participation for IT field.

with the relationship between information systems and the organizations that they serve, extending from theory and principles to application and development; many information systems professionals are also involved in system deployment and configuration and training users. The area covered by information systems dips downward, all the way through software development and systems infrastructure in the right half of the graph. This is because information systems specialists often tailor application technologies (especially databases) to the needs of the enterprise, and they often develop systems that utilize other software products to suit their organizations' needs for information. (Figure 4.2 does not reflect the attention that information systems programs devote to core business topics.)

4.4 ENTERPRISE ARCHITECTURE IN ACADEMIC BRANCHES

Business information systems that are relevant to the enterprise research area are identified and discovered. These business information systems need to be set within the academic branches and business life cycles as shown in Table 4.1.

The objective of the enterprise architecture effort is to clearly know what the enterprise is, what are its critical components, and to set all the descriptions of these components into their proper relationship with other components. Done completely, but at a reasonably high level, the enterprise's purpose and scope should be quite clear. These products become the foundation from which all databases and business information systems are identified and defined. These products also become the context for all policy specification, implementation, and adherence. Finally, these products become the context for all organizational and functional details.

4.5 APPLICATION AREA AND NORMATIVE REFERENCES

The curriculum for the scientific practical-oriented course Enterprise Architecture Modelling (EAM) declares the minimum knowledge requirements and what will be covered in the course. It determines the content of classes and type of knowledge control (examination).

This curriculum is intended for tutors who are running the EAM as an academic syllabus. The syllabus was developed in correspondence with the best practice of using EA in real business careers.

4.5.1 Learning Goals

The course Enterprise Architecture Modelling is aimed on acquiring the advanced level of knowledge about EA and business modeling. This course provides models and mechanisms of creation

TABLE 4.1
Enterprise Architecture Components

Name	Description	Academic Branch
Business events	A business event is an intersection between a business information system and a business function. A business event is a triggering event. It is invoked by the business function, and the business information systems execute in response. Business events may be set within business event cycles, calendar cycles, or both.	IS
Business organizations	An organization is a unit within an enterprise. It is hierarchical so any quantity of organizational levels can be represented.	IS
Business cycle	A business cycle is a cycle during which business events occur such as financial reports, holidays, business planning, and the like. A business cycle may be simple or complex. If complex, the business cycle actually consists of other business cycles as represented in the business cycle structure.	IT
Business calendar	A business calendar cycle is a set of recurring calendar-based dates that are of interest to the enterprise. For example, quarterly, biweekly, monthly, daily, and the like. Business calendar cycles are linked to business events so that the timing of business event triggering can be known.	IT
Business functions	A business function is a set of hierarchically organization texts that describes the activities performed by a position within an organization. Business functions are entirely human-based and if support is needed from a business information system, then a business event is triggered. Business functions are independent of organizations and may be allocated to more than one business organization.	IS
Business information systems	A business information system is a computer-based business information system that is being managed through the metabase. It is known by its characteristics, its operation cycles (business and calendar), subordinate business information systems, employed databases, views, and associated resource life cycle nodes.	IS
Database domains	A database domain is a hierarchically organized set of noun-intensive descriptions associated with a mission leaf. Analyzed database domains lead to the identification of database object classes, enterprise data elements, and property classes. Property classes, in turn, often become tables in databases.	IS/CS
Database object classes	A database object class is a large collection of data and processes that are tied together for business-based reasons, and when instantiated, proceeds through well-defined states. A database object can exist in two forms: a collection of interrelated database tables or the set of a column-based nested structures within a table. The rows that comprise an object are transformed from one valid state to another via database object table processes and database object information systems. Database objects are related to one or more database domains.	CS
Database object information systems	A database object information system is a collection of processes defined within the domain of the database management system (DBMS) usually as a stored procedure that transforms one or more rows of a database object from one valid state to another. A database object information system accomplishes one or more database object table processes.	IS/CS
Management level	Management level is a named and defined level of bureaucratic management within an organizational setting. Examples are executive, senior, midlevel, and first level.	IT

(*Continued*)

TABLE 4.1 (*Continued*)
Enterprise Architecture Components

Name	Description	Academic Branch
Missions	Missions are hierarchically organized textual descriptions that define the very existence of the enterprise, and that are the ultimate goals and objectives that measure enterprise accomplishment from within different business functions and organizations. An enterprise is incomplete if one of its missions is not defined. Not all enterprises accomplish their missions simultaneously or in an ideal state. Missions are accomplished over time and are subject to revisions.	IT
Organizations performing missions	An organization performing missions, or a mission organization, is the association of an organization with a mission. There can be multiple organizations associated with a mission and an organization can be associated with multiple missions. The description contained within the mission organization may be more refined than the description contained in either the mission or the organization.	IT
Organizations accomplishing functions	An organization accomplishing a function in support of a mission, that is, a mission organization function, is the association of a mission organization with a function. A mission organization can be associated with multiple functions and a function can be associated with multiple mission organizations. One or more mission organization functions may be associated with a business information system. When they are, business events are created.	MIS[a]
Positions	A position is a named and defined collection of work tasks that can be performed by one or more persons. Positions are often assigned to one or more organizations.	MIS
Positions performing missions	A mission organization function position role is the assignment of a position to a particular function within an organization as it accomplishes a mission. Once a position is assigned, its role can be described.	MIS
Resource life cycle analysis node	A resource life cycle node is a life cycle state within the resource. If the resource is an employee, the life cycle node may be employee requisition, employee candidate, employee new hire, assigned employee, reviewed employee, and separated employee.	IT
Resources	A resource is an enduring asset of value to the enterprise. Included, for example, are facilities, assets, staff, money, and even abstract concepts like reputation. If a resource is missing then the enterprise is incomplete.	IT

[a] MIS, management information systems.

of IT architectures that are used in Russian industry companies. Two cases are aimed on practical implementation of the knowledge acquired.

4.5.2 Competences Obtained during the Studies

After passing the EAM exam, students should

- Know
  - Methods and tools of enterprise architecture modeling
  - Methods of EA optimization
  - Benchmarks for (business) IT efficiency

TABLE 4.2
Competences Gained by Students of the Course

Competence	Descriptors—Main Features of Knowledge Acquiring (Indicators of Result Achievement)	Forms and Methods of Learning that Contribute to Formation and Development of Competence
Ability to work with information from different sources	Student uses all available information sources	Seminars, case studies, based on both academic rules and professional practice credentials
Enterprise architecture analysis	Student shows possibilities of selecting types of EA	
Capability to formulate scientific goals in EA research	Student shows the understanding of the practical field of EA	
Application of program components for handling, analysis, and classification of information	Student shows the application of necessary instruments for researchers	
Preparation of scientific (technical) reports, presentations, scientific publications	Student uses reports, presentations during seminars, and prepares publications on the topic	
Develop and integrate the enterprise architecture components	Student analyses the possible ways of EA, provides arguments for and against of application of IA type, shows understanding of architecture development process	

  - Business functions and system of business management
  - Methods of functional business-tasks analysis
- Have an ability
  - To manage enterprise architecture
  - To manage life-cycles of information systems
  - To use methods and instruments for adjusting business models
  - To use methods of innovational and entrepreneurial management

This course is aimed at acquiring the following competences (Table 4.2):

- Understanding the increased role of IT for companies' business
- Newest approaches for EA modeling
- Application of advanced architecture principles, models, and standards
- Methodologies of EA modeling
- Understanding of change management methodology
- Acquiring practical knowledge for EA constructing

4.6 PLACE OF THIS COURSE IN THE EDUCATIONAL PROGRAM STRUCTURE

At universities or higher schools of economics or management or information technology, this course is taught for master's students of business informatics and master's students of big data systems. This course is based on the following subjects in the bachelor program (Table 4.3):

Scientific seminar in enterprise architecture
Bachelor course in system analysis
Bachelor course in the foundations of business process modeling

TABLE 4.3
Duration of Course and Types of Educational Work

		Module			
Type of Work	**All Hours/Points**	**1**	**2**	**3**	**4**
Classes (all)		–	32	–	–
Practical tasks (PT)-1		–	16	–	–
Practical tasks (PT)-2	–	–	–	–	–
Seminars (S)		–	16	–	–
Experimental classes (EC)	–	–	–	–	–
Personal Preparation (all)					
Writing assignments	–	–	1	–	–
Graphical work	–	–	–	–	–
Abstracts	–	–	–	–	–
Home assignments	2	–	2	–	–
Exam	1	–	1	–	–
Overall amount of hours	32	–	32	–	–

4.6.1 Content of Topics

Topic 1—*Strategic management* of enterprise from the viewpoint of enterprise architecture (theory of management: Drucker, Chandler, Zeltsnik, Minzberg, Aacker, Ansoff, Adizes, Traut; interrelationships of key managerial theories and concepts in area of *strategic management* with modern methodologies of enterprise architecture) [4].

Topic 2—*Operational enterprise management* from the viewpoint of enterprise architecture (Drucker, Minzberg, Norton and Kaplan [BSC, ABC]; organizational theory, Lean manufacturing, 6 Sigma, Kaidzen, TOC, ISO9000/ISO20000/ISO27000, TBL, etc.; interrelationship of key managerial and concepts in area of *operational management* with modern methodologies of enterprise architecture) [5].

Topic 3—Information and socio-technical systems from the viewpoint of enterprise architecture (interrelationship of enterprise architecture with standards and practices of system and software engineering ISO15288, ISO15926, RUP, RM-ODP, Agile; with standards and practices of project management PMBoK, IPMA, P2M, MSP, PRINCE2) [6].

Topic 4—Foundations of financial management for enterprise (foundation of budgeting, financial and managerial accounting of IT asserts, liquidity, cost/benefit analysis, shareholders earnings, structure of capital, optimization of IT asserts budgeting, peculiarities of information systems in accounting, construction of financial–information asserts interrelation structures with EA methodologies, *financial architecture* of enterprise) [7].

Topic 5—Review from industry about popular EA methodologies: Zachman, META Group, DoDAF, FEA, LEAD, TOGAF. (EA development, Zachman framework, structure and model of architectural description, META Group, LEAD, DoDAF, FEA, TOGAF; other methodologies: choice of the best in the particular case) [8,9].

Topic 6—EA modeling according to DoDAF. Modeling of capabilities and standards (capability view, standards view).

Topic 7—EA modeling according to TOGAF. Instruments and ways of modeling. (Architecture Development Method (0–7)): preliminary phase, architecture vision, business architecture, information systems architecture, technology architecture, opportunities and solutions, migration planning, implementation governance).

Topic 8—EA modeling according to TOGAF. Instruments and ways of modeling. Architecture change management, requirements management. Enterprise continuum, ACF, and other parts of methodology.
Topic 9—Peculiarities of EA modeling of large-scale holding structures: architectural patterns for appropriate scaling.
Topic 10—Service-oriented architecture (SOA) and its relevance in modern EA. Future concept development, alignment of business processes with cloud solutions.
Topic 11—Practical cases of EA modeling (telecommunication industry, cement, oil and gas industry)
Topic 12—Practical cases of EA modeling (higher education, banking industry, etc.)

4.7 COURSE PARTS AND TYPES OF CLASSES

		Classes (Hours)					
No.	**Topic**	**Practice**	**Seminars**	**All**	**Home Assignments**	**Exam**	**Hours**
1	Business and information technologies	1		1		1	
2	Best IT practices and application areas	1		1			
3	ArchiMate modeling language	2	2	4			
4	Development of enterprise architecture against TOGAF methodology	10	6	16			
5	SOA and cloud solutions in enterprise architecture	1	4	5			
6	Practical cases	1	4	5			
	Sum	16	16	32			

4.8 METHODICAL RECOMMENDATIONS FOR THE COURSE

4.8.1 Form of Exam and the Final Mark Structure

Exam: 50%
Essays, cases, seminar activity: 50%

4.8.2 Possible Questions and Tasks for Exam

Topic 1

1. Enterprise architecture definition.
2. What are the parts of EA? What is their purpose?
3. What is the role of IT in business?
4. What is the role of IT strategy and IT architecture in business changes?

Topic 2

1. What is a balanced scorecard and how it is used for strategic governance?
2. What is the EFQM Excellence Model?
3. What is ISO 9001:2000, ISO 20000, ISO 27000?
4. What is COBIT?
5. What is ITIL?
6. What is CMMI?

Topic 3

1. How many rows and columns does the ArchiMate model have? Why?
2. What elements in Archi are used to model the business architecture?

Topic 4

Which elements are used for information architecture and technological architecture modeling in Archi?

Topic 5

1. Name the structure of architecture described by the Zachman model.
2. What other models do you know?
3. Which architecture methodology you suppose is the best?

Topic 6

1. Name the main phases of TOGAF ADM.
2. What are the main outcomes of the preliminary phase?
3. What are the main outcomes of architecture vision?
4. What are the main outcomes of the business architecture phase?
5. What are the main outcomes of information systems architecture?
6. What are the parts of the information systems architecture?
7. Describe the concept of application architecture and data architecture.
8. Describe the concept of integration architecture.

Topic 7

1. Define technology architecture.
2. What are the main outcomes of the technology architecture phase?
3. What are the main outcomes of the opportunities and solutions phase?
4. What are the main outcomes of the migration planning phase?
5. What are the main outcomes of the implementation governance phase?

Topic 8

1. What are the main outcomes of the architecture change management phase?
2. What are the main outcomes of the requirements management phase?
3. What is the TOGAF Repository used for?
4. What TOGAF extensions do you know and how they are used?

Topic 9

1. Describe the peculiarities of SOA.
2. Describe the connection of the SOA notion with enterprise architecture.
3. What are the advantages of cloud solutions concerning IT architecture development?

4.9 CONCLUSION

In the field of enterprise architecture, the chapter attempts to assist companies, universities, government agencies, and other employers to combine the academic and professional credentials for practicing EA. This research provides an extensive curriculum for the field of EA to join the academic and the professional credentials, and for the academic staff to review, compare, and prepare the proper assessment for the educational track record to determine any gaps in educational progression.

REFERENCES

1. Institute for Credentialing Excellence. Terminology Documents. Retrieved August 02, 2012.
2. Center for Law and Social Policy. Helping Low-Income Adults and Disadvantaged Youth Earn Credentials and Build Careers: Leading Foundations Speak about Policy Priorities. Retrieved August 09, 2011.
3. Ibaraki, S. 2011. Brian Cameron: Professor and Executive Director, Center for Enterprise Architecture, Penn State, Founder FEAPO. stephenibaraki.com. Retrieved March 24, 2015.
4. Schekkerman, J. 2008. *Enterprise Architecture Good Practices Guide*, Chapters 6–8. Victoria, BC, Canada: Trafford Publishing.
5. Hanschke, I. 2010. *Strategic IT Management*, Chapter 4. Berlin: Springer-Verlag.
6. Abrahamsson, P., Salo, O., Ronkainen, J., and Warsta, J. 2002. *Agile Software Development Methods: Review and Analysis (No. VTT 478)*. Oulu, Finland: VTT Technical Research Centre of Finland.
7. Avizienis, A., Laprie, J.-C., and Randell, B. 2004. Basic concepts and taxonomy of dependable and secure computing. *IEEE Transactions on Dependable and Secure Computing*, 1(1), 11–33.
8. Eason, K. 2007. Local sociotechnical system development in the NHS national programme for information technology. *Journal of Information Technology*, 22(3), 257–264.
9. Whelan, J., and Meaden, G. 2012. *Business Architecture: A Practical Guide*. Farnham, UK: Gower Pub Co.

5 Automatic Identification and Data Capture Techniques by Radio-Frequency Identification (RFID) Tags

Reader Authentication and Ethical Aspects

H. Saadi, R. Touhami, and M. C. E. Yagoub

CONTENTS

5.1 Introduction 64
5.2 Radio-Frequency Identification (RFID): Fundamentals 64
5.3 Standardization of RFID 64
5.4 How Does RFID Function? 65
5.5 RFID: Advantages 65
5.6 RFID: Limitations 65
5.7 Ethics and Life Privacy 66
5.8 RFID: Applications 66
5.8.1 Marking of Objects 66
5.8.2 Financial Transactions 67
5.8.3 Marking of Humans, Animals, and Plants 67
5.8.4 Stock Regulation and Inventory 67
5.8.5 History Recording 67
5.8.6 Safety 68
5.9 RFID: Raised Issues 68
5.10 Ethical Aspects of Information and Communication Technology (ICT) Implants in the Human Body 68
5.10.1 European Group on Ethics in Science and New Technologies 68
5.10.2 Opinion No 20 68
5.10.2.1 Classification of ICT Implants 69
5.10.2.2 Applications and Research 69
5.10.2.3 Future Research Directions 69
5.10.2.4 Legal Background 70
5.10.2.5 Ethical Background 70
5.10.2.6 Discussion 71
5.10.2.7 Ethical Aspects of the Evolution of the Information Society 72
5.10.2.8 Other Considerations 73

5.11 Environmental Impact 73
5.12 Security Issues in RFID Systems 73
5.13 Conclusion 74
References 74

5.1 INTRODUCTION

Radio-frequency identification (RFID) technology memorizes and recovers remote data by using radio labels called tags or transponders. It can, for instance, facilitate stock regulation/management in a company; thus, increasing profits and reducing costs while simplifying trade. However, besides its real advantages, RFID technology misuse can negatively affect humans in terms of health and private life protection, for example, by giving access to sensitive information or arbitrarily targeting individuals/groups (Bellaire, 2005). So, the use of RFID technology as automatic identification and data capture technique via wireless radio frequency (RF) signals is raising crucial ethical issues. Investigating this aspect of RFID technology and its impact in our social life is therefore vital to ensure its adequate usage. To do so, we will first have to introduce RFID in order to understand the challenges we can face while using this emerging technology.

This book chapter is structured as follows: After presenting RFID technology, its standardization, and operation, we review its advantages and disadvantages, focusing on collision issues and ethical questions raised by possible misuse of RFID systems because of their rapid growth and wide applications. A conclusion ends the chapter.

5.2 RADIO-FREQUENCY IDENTIFICATION (RFID): FUNDAMENTALS

A typical RFID system is composed of a reader, tags, and data management system, which can be activated by an electromagnetic transfer of energy between the reader and the tags in its reading area. A tag is composed of a chip and an antenna (Finkenzeller, 2003). The reader sends radio waves in a space of a few centimeters to tens of meters, depending on the power supplied. When a tag detects the signal from the reader, it sends it the information it contains. This will later receive the information and convert it to binary data.

The concept of RFID is relatively old since it was used during WWII to distinguish friendly aircrafts from enemy aircrafts. Tags were placed in friendly aircrafts to respond as "friendly" to the interrogation radar. This identify friend or foe (IFF) was the first use of RFID (Stuart, 1947). During the 1960s and 1970s, RFID was considered as a confidential technology for military applications to supervise access to sensible sites. In the late 1970s, this technology was extended to the civil sector. One of earliest commercial applications was the livestock identification in Europe so, in the early 1980s, several European and American companies began to manufacture tags. In 1990, companies started implementing standards for the interfunctionality of RFID equipment with smart cards, leading to a mass marketing of readers/encoders and tags in the fields of logistics and traceability.

5.3 STANDARDIZATION OF RFID

As with any commercial technology, there are several standards aiming to harmonize the RFID industry in order to facilitate market accesses and increasing sale volumes. Two organizations are leading this aspect: the ISO and EPC Global. ISO 18000-x standards for contactless identification specify the characteristics of data physical layer and communication protocols that allow exchange between tags and readers at different frequencies (e.g., 135 kHz, 13.56 MHz, 860–960 MHz 2.45 GHz, and 5.8 GHz) (Finkenzeller, 2003). The Electronic Product Code (EPCglobal) has been defined by EAN–UCC (European Article Numbering–Uniform Code Council) as a standardized solution for RFID systems. It is in charge of creating and monitoring the unique identification number (UID, unique identifier) for each RFID tag worldwide. This code supplies report to identify and recognize the manufacturer, product type, serial number, and other related information for precise tracking of objects along the production/distribution chain (Finkenzeller, 2003).

However, the operating frequency standards of ISO and EPC protocols were incompatible. So, in 2005, the EPC UHF1 standard supported its Generation2 category making it possible to be included as an ISO standard, and in 2006, ISO added that standard into the UHF ISO 18000 class. ISO 18000-6 is split into three classes: ISO 18000-6A; ISO 18000-6B; and the latest standard ISO 18000-6C, also called EPCglobal Class 1 Gen2 (EPC, 2009). The EPCglobal class structure consists of seven levels: Class 0–5 and Class 1 Gen2. The dissimilarity between these classes is the features implemented in the tag (writing/reading options, active/passive tags, etc.).

By assigning to each object a unique number, RFID tags are indeed considered like a means of replacing and improving barcodes under the UPC/EAN standard. However, although they are operational, RFID systems suffer from a lack of normalization. EPCglobal (EPC, 2009) is working toward this aspect to propose an international standard in order to standardize the different technical uses of RFID. The goal is to be able to have a homogeneous international system of distribution of identifiers so that one can have an electronic EPC product code for each object present in the logistic chain of a company worldwide.

5.4 HOW DOES RFID FUNCTION?

Let us target the manufacturing field as example. Each manufactured object has a tag attached to it that contains a unique electronic code called an identification number (ID) and a serial number, which makes it possible to identify it at each stage of the production/distribution process. Once the goods arrive to a retailer, RFID readers question the tags to automatically identify all the delivered products. Each tag provides precise and complete information that is transmitted to the control inventory system, thus facilitating stock regulations and warehouse management (WMS). Such visibility in real time allows companies to devote more time and resources to their customers' needs (Intermec, 2007).

5.5 RFID: ADVANTAGES

RFID is a flexible technology, convenient, easy to use, and perfectly adapted to automated operations. It does not require contact or direct visibility between the reader and the object to be identified. It can operate in harsh environments, allows simultaneous reading of several tags and provides a high level of data integrity. RFID technology makes it possible to store much larger data compared to barcodes, avoid human errors, increase speed and effectiveness, and increase the availability of information and object localization and regulation; thus, increasing profitability and productivity while reducing costs.

Most companies are often unaware of how the functionalities that RFID technology offers can help them. They rather see RFID as an expensive technology to implement and difficult to master, while possibly generating significant downtimes. It is, in fact, a technology of data gathering relatively simple to deploy and easy to integrate to existing data acquisition systems (Intermec, 2007).

5.6 RFID: LIMITATIONS

The reading of radio tags posed on objects located in or close to metallic containers is more challenging. Because of the presence of a ground plan, the performance of the antenna of the tag can be significantly degraded. New tags have been recently developed to address this issue, thus allowing longer reading distances.

Functioning of RFID systems also frequently implicates collisions when various tags are available in the reading area of a reader. In such cases, the communication is scrambled by the simultaneous activity of the tags. We then use one of the two following ways to communicate: (1) the reader–tag communication used to transmit data from a reader to tags (the transmitted data stream is received by all tags simultaneously) or (2) the tag–reader communication where data from tags are sent to the reader. The latter is also called multiple access communication (Finkenzeller, 2003). The issue is that data should be transferred from several tags to the reader without mutual interference. However, when several tags want to communicate with the reader, there might result a collision that must be resolved with specific anticollision algorithms.

Collision issues can be mainly divided into reader–tag and reader–reader collisions. The tag–reader collision occurs when a neighbor reader signal interferes with the responses of tags in the interrogation zone of another reader. The reader–reader collision occurs when a tag receives signals from multiple readers simultaneously. In this case, the tag may be unable to reply to all readers. Different anti-collision methods have been suggested. According to the technique of multiple access to the communication channel, they can be classified into time, frequency, space, or code division techniques (Yu et al., 2008; Lee et al., 2005; Myung et al., 2006; Saadi et al., 2011; Vogt, 2002). Several anticollision algorithms have been presented in recent papers to solve and avoid collision between the tag and reader and to guarantee the correct communication between them (Jia and Feng, 2013; Li and Feng, 2013; Safa et al., 2015; Tonneau et al., 2014; Zhang, 2015). However, because the collision problems are resolved in the arithmetic and MAC protocol (multiple access control channel), their complexity and implementation costs are limiting their use.

5.7 ETHICS AND LIFE PRIVACY

In the 2000s, RFID chips became quickly standardized. In 2010, the implementation of microcells in the human body was initiated (VeriChip chip or "human bar code"), with the correlative risk of the control of individuals (Perret, 2014). This was done before even having a clear legislation and in-depth ethical debates, in particular concerning increasingly miniaturized active or passive devices (in 2006, Hitachi proposed a chip of 0.15×0.15 mm^2, smaller than the diameter of some hairs). Inserted under the skin or put on cloths (wearable computing or cyber-clothing), RFID communicating objects are becoming part of our daily life (GEESNT, 2005).

This is raising ethical issues and risks of new challenge (Council of Europe [COE], 1997; Nsanze, 2005). For instance, in Europe, after a 2005 report about new implants in the human body (EC, 2009) and a round table organized by the GEE (European Group of Ethics of Sciences and New Technologies) (Journal Officiel, 2002), the European Commission asked the BEPA (Office of the Advisers of European Policy) for its opinion. On March 2015, an ethical report titled "Aspects of ITC Implants in the Human Body" was published (GEESNT, 2005). The principal concerns are mainly regarding human dignity, right to privacy, and data security of personal information (Rective, 1990). The question also touches the public health and the security of private life in the sector of communications (COE, 1997), the legislation on medical implantable active devices (UNESCO, 1997), the approval and the right to access information (COE, 1981), the protection of people against automated data processing of personal data (Point, 2003), as well as any possible abusive uses that can result. In May 2009, the European Commission published a recommendation "on the implementation of privacy and data protection principles in applications supported by radio-frequency identification," centered on the systematic deactivation of RFID tags at retailing points. For applications that did not systematically decontaminate the tags, the start of an RFID application is subjected to an evaluation of impacts on private life (EIVP or Privacy Impact Assessment PIA). In July 2014, a European standard was published (INTO 16571) giving the methodology to be followed to carry out an EIVP. The EIVP report must be transmitted to the organization in charge of data protection involving personal information (GEESNT, 2005).

5.8 RFID: APPLICATIONS

5.8.1 Marking of Objects

RFID low-frequency devices (125–135 KHz) are used for the traceability of objects such as books in a library or luggage in an airport as well as for access control done by badges or "free hands" to restricted areas, public transportation systems, buildings, and so forth. They can contain a numerical identity or an electronic certificate, which permit access to an object or its activation. Despite the fact that most RFID badges have a limited reading distance, they allow a read-write of the chip to memorize information (e.g., biometric data). Also their efficiency has been significantly improved. In 2008, the RFID Journal Awards recognized the Omni-ID company that presented an RFID tag readable through water and near metal, with a high level of confidence of 99.9%. RFID tags working at 2.45 GHz allow long-distance reading. These tags are generally active tags used for access control of vehicles or traceability of food (an RFID chip can, for instance, record the temperature changes undergone by a given food during transportation).

5.8.2 Financial Transactions

RFID are also largely present in the financial activities. It involves the systems of contactless payment with for example, credit cards and smart phones. It uses near-field communication properties to make secure payments. An integrated chip and an antenna allow consumers to pay with their contactless card at a point of sale. In Hong Kong and in Holland, tags in the shape of credit cards are widespread like electronic means of payment. They are also used as transport titles on public transportation networks. Some suppliers claimed that such transactions can be almost twice as fast as traditional transactions. No signature nor data capture of a PIN code are necessary for purchases less than US$25, less than CHF 40 in Switzerland, and less than €20 in France.

5.8.3 Marking of Humans, Animals, and Plants

RFID are also involved in the identification of plants, wild animals, or pets. A low-frequency tag (125–135 kHz) is usually installed under the skin in their neck. Originally designed for animal traceability, they can also be implemented in humans without major technical constraints. The company Applied Digital Solutions is proposing subcutaneous radio tags (commercially called VeriChip) as a solution to identify frauds, ensure safe and protected access to sensitive sites, or even store medical data. For instance, Mexico City implemented 160 radio tags under the skin of their policemen to control their access to databases and for better localization in case of emergency. RFID systems can also be an integrated solution for real-time supervision of patients.

5.8.4 Stock Regulation and Inventory

Among existing applications, RFID can identify the nature and quantity of merchandise in a store, with day-to-day updates about stock regulation and inventory. The properties of the radio tags would also make it possible to consider applications for the consumer such as appliances (like refrigerators) capable of not only automatically recognizing and listing all the items they contain, but also to control the optimal date for a given product to be safely consumed.

In some universities, such as Cornell University, RFID cards make it possible for students to have access to the library 24/7. The books are also delivered with radio tags, which eliminate huge administrative workloads. In several Dutch libraries, books are equipped with radio tags (based on the SLI chip of Philips). An academic analysis done by Walmart showed that RFID can reduce sellouts by 30% for products having a rate of rotation between 1 and 15 units/day (Michael and McCathie 2005).

More generally, RFID can simplify the distribution process of an item from A to Z. In fact, a portable reader or a reader fixed in a gate can read and identify the items in a pallet when loaded in a truck from the factory or downloaded from a truck and moved to the customer storage area.

5.8.5 History Recording

An RFID tag usually contains the ID of the device and useful information, such as the origin of the product, the production line, and the production date. The gantries, the embarked units, or the portable devices make it possible to constantly question the RFID tag in order to get the history record of the article (Intermec, 2007). This allows for huge time-savings, increased conformity with regulations, minimization of risks, and increased effectiveness (Intermec, 2007).

Furthermore, information such as the last date of maintenance, the technician who supervised the maintenance, the pieces replaced or to be replaced if deemed necessary, and the actions carried out can be also part of the history of the tagged item.

5.8.6 Safety

RFID badges worn by workers can allow them access to gantries and lock/unlock doors based on their clearance level or within the framework of the tasks assigned to them, thus increasing their safety as well as the safety of industrial or military installation. A wide panel of applications of RFID in various domains, such as financial activities, trade, tracking, asset management, and inventory, are described by Hardgrave et al. (2008) and Michael and McCathie (2005).

5.9 RFID: RAISED ISSUES

By assigning to each object a unique number, RFID tags are indeed considered like a means of replacing and improving barcodes under the UPC/EAN standard. However, although they are operational, RFID systems suffer from a lack of normalization. EPCglobal (EPC, 2009) is working toward proposing an international standard in order to standardize the different technical uses of RFID. The goal is to be able to have a homogeneous international system of distribution of identifiers so that one can have an electronic EPC product code for each object present in the logistic chain of a company worldwide. However, more critical ethical issues are raised by RFID technology when dealing with human rights to privacy and health.

5.10 ETHICAL ASPECTS OF INFORMATION AND COMMUNICATION TECHNOLOGY (ICT) IMPLANTS IN THE HUMAN BODY

5.10.1 European Group on Ethics in Science and New Technologies

The European Group on Ethics in Science and New Technologies (EGE) established by the European Commission is a qualified entity, constituted of 15 specialists hired for their competence. Its mission is to investigate ethical concerns about science and new technologies and to provide recommendations and advice to the European Commission in relation to the elaboration, production, and achievement of community legislation, laws, and policies (Capurro, 2010).

5.10.2 Opinion No 20

The document Opinion No 20 adopted by the EGE in 2005 raised the following issues (Capurro, 2010):

- At first vision and from an ethical point of view, ICT implants are not considered as a source of problem in comparison, for example, with cardiac pacemakers. Even if ICT implants can be employed to address body failures, they can be diverted toward unethical uses, especially when they are reachable via digital networks.
- The concept of allowing ICT devices to be implemented under the skin can enhance human capabilities.
- The strong relationship between physical and psychic functions is fundamental to a person's personal recognition and identification.
- The purpose of this opinion is mainly to sensitize and to raise questions about the ethical difficulties produced by a group of implants in this fast-spreading domain.

5.10.2.1 Classification of ICT Implants

Implantable devices such as ICT implants and tags can be classified based on their application field, mainly as medical or nonmedical, the mode they are powering (passive or active), or as reversible or nonreversible devices.

In this new century of information technologies, one of the critical issues encountered in science and societies is that the development and improvement of human capabilities by new technologies like as nanotechnologies. So, it is primordial to open ethical debates (Lin and Allhoff, 2008) to smoothly cohabit security and ethics in implantable RFID chip commercialization (Kumar, 2007).

5.10.2.2 Applications and Research

The current applications and research directions in this field can be summarized in the following points:

- ICT implants as medical devices
  - Cardiovascular pacers for sick people suffering from blood conduction trouble or heart failure
  - Cochlear and brain implants for sick people suffering from trouble of hearing
  - Implantable medicine pumps for patients suffering from diabetes or multiple sclerosis
  - Implantable devices for neurological stimulation
  - Administration and control of chronic troubles and pains by stimulation of the spinal cord
  - Administration and control of urinary incontinence by stimulation of the sacral nerve
  - In serious depression cases, cessation of epileptic seizures, and mood control by stimulation of the vagus nerve
  - Control of patients suffering from Parkinson disease by deep brain stimulation (DBS)
- Identification and tracking objects by ICT implants
  - Possibility of identifying Alzheimer patients or children by using read-only tags
  - Tracking of medical history of persons by using a set of information stored in read-write tags
  - Wearable ICT (as in Schmidt et al., 2000)
  - Subdermal GPS
- ICT implants as enhancement or commodity devices
  - Prosthetic cortical implant
  - As artificial hippocampus device
  - As cortical implants for blind patients
  - As ocular implants, for example, to put in an artificial retina
  - Audio tooth implant

In a Microsoft patent (patent no. 6,754,472; 2004), the human body is considered as a favorable environment for transmission of data of different nature to various devices around it like cell phones, medical apparatus, giving the possibility of person's localization (Capurro, 2010).

5.10.2.3 Future Research Directions

Future research directions have also been targeted as emerging applications:

- Integrating and miniaturizing of three technologies (Digital Angel™)
- Reading vital signs of people by a simple touch of the skin using biosensors embedded in a wristlet as a watch
- Utilizing cellular packet modules to get data from biosensors and deliver them to pager devices.
- Using radio signal in localization devices to keep contact with a pager device placed in a person for tracking them

5.10.2.4 Legal Background

As mentioned earlier, the usage of RFID devices in our daily life is raising critical issues about using the human body to get financial profits in order to guarantee human dignity, security, and privacy, or to collect data without the express consent of a person/group (Capurro, 2010). In fact, ICT implants allow easy and quick localization of people and their personal information can be remotely exchanged. This can enable unauthorized persons and organisms to modify and read such sensitive information. This is clearly a violation of international standards/regulations regarding data security, collection, and processing.

We, thus, must consider the following points for legal background (Albrecht, 2006; Gasson et al., 2012):

- All kinds of ICT implants inserted in the human body, even the simplest ones, present a legal risk.
- The distinction between medical and nonmedical applications must be considered by the concept of specification of the object.
- ICT implants only oriented to identify patients are considered as illegal by the principle of data minimization (in case where it is possible to replace them with equally secured systems).
- More generally, ICT implants only oriented to facilitate access to public spaces, for instance, are considered as illegal by the principle of proportionality.
- It is insufficient, just by getting authorization from the concerned person, to permit the utilization of any type of ITC implants. This fact is considered as illegal by the principle of integrity and inviolability of the body.
- It is completely forbidden, by the principle of dignity, to deal the human body as an object that may be remotely supervised and monitored.

The possibility of insertion of microcells or ICT implants under human skin for identification, control, and tracking applications is being strongly investigated by legal authorities. For more details, refer to "The Legal Ramifications of Microchipping People in the United States of America: A State Legislative Comparison" (Friggieri et al., 2009).

5.10.2.5 Ethical Background

Modern societies encounter challenges concerning social changes and they have to take into consideration the anthropological nature of individuals. After being detected, spotted, and controlled by video surveillance and biometrics, modifying and linking individuals by smart tags inserted under skin, to be transformed more and more to networked persons, can be no longer seen as science fiction but a reality in a relatively short term. Thereby, we could transmit and receive data permitting the definition and traceability of motions, habits, and contacts of individuals to keep them connected in different ways. This will change the sense of autonomy and independence of an individual, thus influencing his or her dignity.

The context of privacy and related intercultural aspects is well developed by Capurro (2005) in his paper, principally based on differences between Japanese and Western cultures.

The fundamental ethical aspects that are involved are

- Human Dignity
- Privacy
- Nondiscrimination, for example, racial prejudice
- Informed authorization
- Value, rights, and justice (equity)
- The preventive concept

A conflict can exist between personal freedom and utilizing an implantable to improve one's physical aptitudes and capacities, which can be considered as tolerable from the ethical point of view. A further disagreement is about restricting liberties of persons considered as dangerous to the society (by tracking and controlling them) to improve the security of others. Moreover, the obligation to secure and protect the life and health of humans involved in medical research studies may be seen as a restriction to researchers' freedom.

We must also discuss such issues from the point of view of economic competitiveness and development that can have impacts on human dignity. Hence, there are major knowledge insufficiencies in terms of ICT implants in the human body (Capurro, 2010) about

- Human dignity, integrity, and autonomy
  - To which degree can we consider such implants as a menace for human's autonomy, especially when they are embedded into the human brain?
  - To which degree can they irreparably and permanently affect the human body and psyche?
  - Can they affect human memory and how?
  - Can an individual be still seen as human when ICT implants replace vital parts of his/her body, especially the brain?

  Biomedical-based technology is seen as one of the highest emerging fields with significant device miniaturization, while implantable medical technologies can involve health care. As discussed by Schaar and Ziefle (2011), there is a considerable gap to fill to accept these implants from the human point of view.
- Privacy and surveillance
  - To what degree can we consider ICT implants as a menace for private life?
  - To what degree can ICT implants providing specifications grow to the point that they will be considered a menace to society?
- Improvement and consciousness
  - What is the meaning of human being's perfectibility?
  - To what degree can the utilization of such implants be allowed to improve human capabilities and abilities?
- Social sides
  - How can we introduce to people implanted with ICT implants who are linked online to understand that they are controlled?
  - To what degree can ICT implants stay hidden to an external viewer?
  - To what degree can ICT implants be employed to track and follow human beings and in which situations can this be legally permitted?
- Situations for which utilization of ICT implants requires special prudence are (Capurro et al., 2005)
  - ICT implants that are difficult to remove
  - ICT implants that play a dangerous role in influencing, determining, or changing psychic functions and capabilities
  - ICT implants that could be deviated to unethical/unauthorized uses in social surveillance, supervision, and manipulation
  - Military applications

5.10.2.6 Discussion

We have to be aware that ICT implants in the human body do not treat the complete domain of ICT devices, whereas, it is possible to find applications in which such chips may be considered as quasi-implants (Capurro, 2010):

- "We shall not lay hand upon thee" means that we promise to respect the whole body: *habeas corpus*.

- In this modern world, data security satisfies the engagement of guarantee of the habeas data required by unpredicted circumstances.
- The body remains still perfectible; it can be used to perform several functions that were unknown or missing. Also, in sport competitions, smarts prostheses can ensure better social life and better competitive conditions.
- For health objectives, ICT implants must be able to fulfill the following:
  - Significance and importance for patient.
  - The implant is essential and imperative to reach this aim.
  - Absence of other methods that can guarantee to be effective and cost-effective at the same time.

In this domain, the freedom of research must satisfy the fact that researchers should get authorization and advise people ready to volunteer for new trials about the potential risks of such tests.

- Clear authorization and data security (especially privacy and data confidentiality) are necessary for irreversible ICT implants and to inform about the potential damages in a person's body if the implants have to be removed.
- It is absolutely prohibited to manipulate and influence mental functions or modify the identity of persons by the use of ICT devices.
- Filling the condition that ICT devices are set up and inserted according to all principles discussed in this chapter, it is not necessary to make them visible to others. The right to protect private life comprises also the right to own an ICT implant.
- We shall have the right to access ICT implants only to improve them to lead and induce children or adults into better socialization. However, we have to provide them with clear and complete information and get their approval.
- The interests of customers involve an adequate control of ICT implants and all ICT devices in the commercialization step. For example, medical products must be subject to strict and relevant legislations into a legal framework.
- The use of ICT implants in surveillance and supervision are very frequent nowadays. They are employed in locating and tracking persons and to find any kind of information about them. This can be explained by security and safety purposes, but should be strictly legalized. In the case of prisoners, for instance, the legislator can decide about their usage for society's safety but as stated by the EGE, it should not conflict with the item 8 of the Human Rights Convention; in fact, EGE does not support such utilization and prefers that surveillance procedures must be subject to strict legislations and supervised by an independent instance.

Accordingly, the following options must be prohibited:

- Classification of ICT implants as a fundamental factor or a means for cyber-racism.
- The utilization of ICT implants for change or falsification of some personal important information related to a person's identity, memory, self-perception, and perception of others.
- The utilization of ICT implants to abusively control and command individuals.
- The utilization of ICT devices as a constraint toward those who do not utilize such implants.

5.10.2.7 Ethical Aspects of the Evolution of the Information Society

Due to the constant evolution of society in terms of information technology, several ethical aspects should be addressed in regard to ICT implantation in human bodies:

- To favor an information society centered on the persons as affirmed in the Declaration of Principles of the World Summit on the Information Society in 2003.

- To initiate large social and political discussions to address legal issues, for example, what class of applications should be accepted, allowed, and legally approved, especially relating to surveillance and data collection. To this aim, the EGE has suggested a preventative method.
- Institutions, national ethic councils, and governments should develop and establish strict regulations as well as inform the population about the risks of adopting such technology.
- Governments should ensure that the development, spread, and access to ICT implants are determined through democratic ways based on public debates and information to guarantee transparency.

5.10.2.8 Other Considerations

In this domain, it is evident that regulations are required. Actually, current legislations do not cover nonmedical ICT devices, especially regarding private life and information security (Capurro, 2010).

- In the EGE framework, implantable devices for medical applications must be subject to normalization in the same frame as drugs when the medical aim is the same, especially when such implants are not completely covered by the Council Directive 90/385/EEC corresponding to active implantable medical devices.
- Further long-term research must be performed concerning the impact of ICT implants in social and cultural life and health, focusing on all aspects of risk, as well as the characterization, evaluation, administration, and communication about the potential risks it may imply.

Quick development of ICT implants involves increasing fears but also hope. For example, analysis of possible threats for security and privacy of new Federal Information Processing Standard for Personal Identity Verification (FIPS PUB 201) using standardized cryptographic has been investigated by Karger (2006), where technical details for cryptography to guarantee privacy are discussed.

5.11 ENVIRONMENTAL IMPACT

Like all industrial products, RFID chips consume natural resources and produce no recyclable items. It is unfortunate that only very few studies have been conducted to investigate the direct environmental impact of this technology (AFSSET, 2008). However, the RFID industry is making great strides in this direction, in particular to answer the environmental stakes in sensitive areas like production chains, waste management, transportation, and geolocalization. For example, in certain European cities, the residential dustbins are equipped with RFID chips. The trucks' dustbins, equipped with RFID readers, identify the dustbins collected thanks to their RFID chips. This management of waste by RFID allows for better monitoring of its nature and quantity in order to optimize treatment.

5.12 SECURITY ISSUES IN RFID SYSTEMS

RFID technologies could appear dangerous for humans and the society (health and private life protection) (Bellaire, 2005), with: the possibility to access sensitive information of both private entities/companies and governmental structures; use of information contained in tags implemented in passports to selectively target people or groups; abusive use of databases of people having bought or borrowed certain types of sensitive goods (e.g., weapons); potential problems of "numerical/economical sovereignty" related to the infrastructure of the EPCGlobal network; implantation of subcutaneous chips in terms of ethical issues and rights to the physical integrity of a person. Under certain conditions, people refusing these subcutaneous tags could likely be victims of discrimination; identification of people by their signature (e.g., bank cards, mobile phones, public transportation passes); identification and localization of individuals using RFID-tagged objects; beyond a certain threshold, emission of RF signals has proved to be dangerous for health, in particular the multiplication of cancers in the case of experiments on mice or interferences that can disturb the operation of the biomedical apparatuses (Van Der Togt et al., 2008). In a report published on 2009, AFSSET recommended to continue the development of scientific search for biological effects of radiation related to RFID. To protect citizen privacy, some country legislations provide certain protections for citizen private life by forbidding any hidden control or identification. Also, the use of the same apparatuses for the access control (FoeBuD, 2010; Liberation/écrans, 2006) and the control of presence (IPC, 2004).

In 2006, a group of hackers announced at the biannual HOPE convention that they had successfully cracked the safety features of VeriChip's subcutaneous chip (Melanson, 2006). They also claim to have been able to reproduce it. They estimate that the legislation is too flexible with this technology, considering its potential risk to private life.

As an illustration, the French legislation provides certain protections for private life by forbidden/hidden control (any identification must be with visible indication); and the use of the same apparatuses for access and control. According to the German association FoeBuD (2010), the legislation is not restrictive enough for RFID technology and the protection of personal information. Some associations propose tools to be protected from an unauthorized use of radio identification, such as RFID Guardian. Other associations propose the boycott of this technology, which they say is liberticidal. According to them, the pointing of noncontrollable information in an electronic identity card would be prejudicial to the freedom of individuals.

5.13 CONCLUSION

Companies should consider the advantages that RFID technology offers. It is a suitable technology, highly functional, and taken into consideration by the current and emergent standards. Companies of all sizes have profited from using RFID technology and improved their processes, increased productivity, and reduced costs and the errors. Nowadays, RFID has become part of everyday life in many sectors, including access control, asset tracking, and traceability of objects (such as library books) and animals.

However, the characteristics of this technology allow for development that can have many implications for privacy, particularly when the data in a chip can be associated with personal identifying information. This use of RFID is meeting widespread resistance from the public due to its lack of security and privacy, and without considering other impacts on the environment and health. In its current form, RFID technology represents a real danger if we do not secure it and if we do not submit to stricter control. To build trust in RFID applications, we need to ensure that only authentic readers can access the tags.

In RFID systems, the issues of interference and collision between tags and readers commonly occur in the form of tag-to-reader, reader-to-tag, and reader-to-reader collisions. We should then develop enhanced approaches to secure the information stored in RFID tags.

As for ethical aspects of the use of RFID, more advanced research must be performed to master the impact of ICT implants in social and cultural life and health, focusing on all aspects of risks in terms of characterization, evaluation, administration, and communication.

REFERENCES

AFSSET. 2008. French Agency for the Safety of Environment and Labour, p. 98.

AFSSET. 2009. French Agency for the Safety of Environment and Labour.

Albrecht, K. 2006. *The Spychips Threat: Why Christians Should Resist RFID and Electronic Surveillance.* Nashville, TN: Thomas Nelson Inc.

Bellaire, A. 2005. RFID chips: Myths and realities of the miniaturized Big Brother. Futura-Sciences.

Capurro, R. 2005. Privacy: An intercultural perspective. *Ethics and Information Technology*, 7:37–47.

Capurro, R. 2010. Ethical aspects of ICT implants in the human body presentation. *IEEE International Symposium on Technology and Society*. New South Wales: IEEE.

Capurro, R., Hausmanninger, T., Weber, K., Weil, F. 2005. The ethics of search engines. *International Review of Information Ethics*, 3(6): 1–2.

Council of Europe (COE), European Treaties, ETS No. 108. 1981. Convention for the Protection of Individuals with regard to Automatic Processing of Personal Data. Strasbourg, January 28, 1981.

Council of Europe (COE). 1997. Convention of European Council on Human Rights and Biomedicine, April 4, Oviedo. http://conventions.coe.int/treaty/fr/treaties/html/164.htm. (See especially articles 5 to 10)

EC. 2009. Opinion of the French Agency for the Safety of Environment and Labour—AFSSET; epcglobalinc.org., https://ec.europa.eu/research/ege/index.cfm.

Finkenzeller, K. 2003. *RFID Handbook*, 2nd ed. Chichester, England: John Wiley & Sons.

FoeBuD. 2010. German Association FoeBuD to prevent potential abuse radiolabels. Bielefeld, Germany.

Friggieri, A., Michael, K., Michael, M. G. 2009. The legal ramifications of microchipping people in the United States of America: A state legislative comparison. In: *IEEE International Symposium on Technology and Society*. Los Alamitos, CA: IEEE, pp. 1–8.

Gasson, M., Kosta, E., Bowman, D. 2012. *Human ICT Implants: Technical, Legal and Ethical Considerations*. The Hague: T.M.C. Asser Press.

GEESNT (Groupe européen d'éthique des sciences et des nouvelles technologies), 2005. Aspects éthiques des implants TIC dans le corps humain. *Avis du groupe européen d'éthique des sciences et des nouvelles technologies*, PDF, consulted 2016-01-04.

Hardgrave, B., Riemenschneider, C.-K., Armstrong, D.-J. 2008. Making the business case for RFID. In: *First International Conference, LDIC 2007, August 2007, Proceedings*. Bremen, Germany: Springer.

Hitachi, 2006. *World's Smallest and Thinnest 0.15 × 0.15 mm, 7.5 μm Thick RFID IC Chip—Enhanced Productivity Enabled by 1/4 Surface Area, 1/8th Thickness, pdf.* Tokyo: Hitachi, consulted 2016-01-04.

IPC, 2004. https://www.ipc.on.ca/images/resources/up-rfid.pdf.

Intermec Technologies Corporation, 2007. *Aspects pratiques de la technologie RFID dans les applications de la fabrication et de distribution, 2007.* Cedex, France.

Jia, X., Feng, Q. 2013. An improved anti-collision protocol for radio frequency identification tag. *International Journal of Communication Systems*, 28(3): 401–413.

Journal Officiel. 2002. Protection des données dans le secteur des communications électroniques. France, 201 du 31.7.2002, p. 37 à 47. http://eur-lex.europa.eu/legal-content/FR/TXT/?uri=URISERV%3Al24120.

Karger, P.-A. 2006. Privacy and security threat analysis of the federal employee personal identity verification (PIV) program. *Symposium on Usable Privacy and Security (SOUPS) 2006*, July 12–14, Pittsburgh.

Kumar, V. 2007. Implantable RFID chips: Security versus ethics. In: *The Future of Identity in the Information Society: Proceedings of the Third IFIP WG 9.2, 9.6/11.6, 11.7/FIDIS International Summer School on the Future of Identity in the Information Society*, Karlstad University, Sweden, August, S. Fischer-Hübner, P. Duquenoy, A. Zuccato, L. Martucci (eds.), pp. 4–10. New York: Springer.

Lee, S. R., Joo, S. D., Lee, C. W. 2005. An Enhanced Dynamic Framed Slotted Aloha Algorithm for RFID Tag Identification. In: *2nd International Annual Conference on Mobile and Ubiquitous Systems: Networking and Services*, San Diego. San Diego: Springer.

Li, X., Feng, Q. 2013. Grouping based dynamic framed slotted ALOHA for tag anti-collision protocol in the mobile RFID systems. *Applied Mathematics and Information Sciences*, 7(2L), 655–659.

Liberation/écrans. 2006. Interview of Mélanie Rieback.

Lin, P., Allhoff, F. 2008. Untangling the debate: The ethics of human enhancement. *NanoEthics*, 2: 251–264.

Melanson, D. 2006. VeriChip'sHuman-implantable RFID chips clonable, sez hackers. Engadget, July 24.

Michael, K., McCathie, L. 2005. The pros and cons of RFID in supply chain management. In: *Proceedings of the International Conference on Mobile Business (ICMB'05)*, Sydney. Sydney: IEEE Computer Society, pp. 1–7.

Myung, J., Lee, W. and Srivastava, J. 2006. Adaptive binary splitting for efficient RFID tag anti-collision. *IEEE Communication Letters*, 10(3): 144–146.

Nsanze, F. 2005. ICT implants in the human body—A review. The European Group on Ethics in Science and New Technologies to the European Commission, pp. 115–154.

Perret, E. 2014. *Identification par radiofréquence*. Grenoble: ISTE Edition.

Rective 90/385/CEE of Council. June 20, 1990. On the approximation of the laws of Member States relating to active implantable medical devices (JO L 189 du 20.7.1990, p. 17 à 36.)

Saadi, H., Touhami, R., Yagoub, M. C. E. 2011. Simulation of the anti-collision process of RFID system based on multiple access protocols modelling. *IEEE International Symposium on Signal Processing and Information Technology*, Bilbao, Spain, 2011.

Safa, H., El-Hajj, W., Meguerditchian, C. 2015. A distributed multi-channel reader anti-collision algorithm for RFID environments, *Journal of Computer Communication*, 64: 44–56.

Schaar, A.-K., Ziefle, M. 2011. What determines public perceptions of implantable medical technology: Insights into cognitive and affective factors. In: *Information Quality in e-Health*, A. Holzinger, K.-M. Simonic (eds.), pp. 513–531. Berlin: Springer-Verlag.

Schmidt, A., Gellersen, H.-W., Merz, C. 2000. Enabling implicit human computer interaction: A wearable RFID-Tag reader. In: *Fourth International Symposium on Wearable Computers*. Atlanta, GA: IEEE Computer Society, pp. 193–194.

Stuart, L. E. 1947. Identification Friend or Foe-Radar's Sixth Sense. *Tele-Tech Electronic Industries*, January, 60–67.

Tonneau, A.-S., Mitton, N., Vandaele, J. 2014. A survey on (mobile) wireless sensor network experimentation testbeds. *Proceedings of the IEEE International Conference on Distributed Computing in Sensor Systems, DCOSS*, May 26–28, Marina Del Rey, CA, pp. 263–268.

UNESCO. 1997. Universal Declaration on the Human Genome and Human Rights. Adopted November 11. http://portal.unesco.org/shs/fr/ev.php-URL_ID=2228andURL_DO=DO_TOPICandURL_SECTION=201.html.

Van Der Togt, R., Jan Van Lieshout, E., Hensbroek, R., Beinat, E., Binnekade, J. M., Bakker, P. J. M. 2008. Electromagnetic interference from radio frequency identification inducing potentially hazardous incidents in critical care medical equipment. *Journal of the American Medical Association*, 299: 2884–2890.

Vogt, H. 2002. Efficient object identification with passive RFID tags. *Pervasive Computing: 1st International Conference*, Switzerland, vol. 2414/2002, pp. 98. Berlin: Springer.

Xiaowu, Li., Quanyuan, F. 2013, Grouping based dynamic framed slotted ALOHA for tag anti-collision protocol in the mobile RFID systems, *Applied Mathematics and Information Sciences*, 7(2L), 655–659.

Yu, J., Liu, K., Huang, X., Yann, G. 2008. A novel RFID anti-collision algorithm based on SDMA. *Wireless Communications, Networking and Mobile Computing International Conference*. Dalian, China, pp. 1–4.

Zhang, L. 2015. Application of improved anti-collision algorithm of RFID in warehouse management system. *Third IEEE International Conference on Robots, Vision and Signal Processing (RVSP)*. Kaohsiung, Taiwan, pp. 264–267.

6 Artificial Intelligence in E-Learning

Hachem Alaoui Harouni, Elkaber Hachem, and Cherif Ziti

CONTENTS

6.1 Introduction 77
6.2 Chapter Overview 78
6.3 Basic Notions 78
6.3.1 Emotion 78
6.3.2 Classification 79
6.3.3 Multiagent 79
6.3.4 Weka 79
6.3.5 Felder Questionnaire 79
6.4 The ASTEMOI System 80
6.4.1 Architecture 80
6.4.2 Problems Treated by the ASTEMOI System 81
6.5 Determining the Emotion from a Vocal Analysis 81
6.5.1 Acoustic Parameters Characterizing the Emotion 81
6.5.2 Automatic Recognition of Emotions 82
6.5.3 Machines Learning (Support Vector Machine) 83
6.6 ASTEMOI and Motivation of the Learner 84
6.6.1 Self-Determination Theory 84
6.6.2 Improving Motivation through ASTEMOI 85
6.7 The Predictive Model by Exploring Data 85
6.7.1 The Predictive Model 85
6.7.2 Exploring Data and Weka 86
6.7.2.1 Logistic Regression 86
6.7.2.2 Decision Tree Algorithm J48 87
6.7.2.3 IBk 87
6.7.2.4 Naïve Bayes Classifier 88
6.8 Performance and Evaluation 89
6.9 Conclusion 90
References 90

6.1 INTRODUCTION

In a conventional e-learning framework, the learner confronting a computer might be in a circumstance of frustration or obstruction (Després and George, 2001). This requires the usage of a smart computer tutor through the coordination of numerous components (assistance in learning from a human tutor and assistance to navigate and interact with the human machine) for managing learning (Armony and Vuilleumier, 2013).

Numerous inquiries on the investigation and understanding of exercises performed by learners during their association with the learning environment were led to gather data about the learners so that the educator can have a diagram on utilizing the learning framework as a part of request to guarantee individual preparation and improvement.

Many current studies have shown that emotions play an eminent role in the learning process. In order to optimize such processes, we assume that the intelligent tutoring systems (ITS) take both emotions and data mining into account, so as to offer new learning situations.

Take dialogue systems, for example, which determine the emotional state of the user using a dialogic strategy. For example, in call centers, if the user manifests signs of irritation or frustration over the auto responder, a strategy may be to direct it to a human operator (Lee, Narayanan, and Pieraccini, 2002).

However, the applications that are used for emotions recognition are not limited to dialogue systems. They can exist in the health care field as well. Some academic research (Istrad, 2003) has focused on emotion recognition for assistance to elderly or hospitalized people. Regarding security, it allows the machine to diagnose abnormal situations and assist the person in his or her monitoring task. Nowadays, the majority of the existing automatic monitoring systems rely mainly on video mode to detect and analyze abnormal situations (Nghiem, Bremond, Thonnat, and Valentin, 2007).

This work also focuses on collection and interpretation during the training session. The assistance to be provided to the learner is particularly essential to guarantee his motivation and avoid obstruction. The intelligent tutorial system ASTEMOI, based on the notion of agents, is involved.

The ultimate purpose of our research is to suggest characteristics, as independent as possible, for the design of the training environment, enabling tutors to provide a perception of learners' behaviors and to identify their learning styles. Our proposed solutions to this problem are to

- Adapt appropriate learning styles for each learner, without modifying the system to add a new feature based on the Felder questionnaire
- Make the most use of information concerning former students to estimate the strengths and weaknesses of learners using data mining and to assign the proper profile

6.2 CHAPTER OVERVIEW

This chapter states the different stages that are necessary to help learners discover their strengths and weaknesses through analyzing the data using the Weka (Waikato Environment for Knowledge Analysis) tool and determining the emotions based on acoustic cues. In this chapter, we are going to briefly describe the architecture of our system and how to estimate the strengths and weaknesses of a learner from the processing of data, and how to determine the appropriate learning style and improve learners' motivation by using the self-determination theory (SDT).

6.3 BASIC NOTIONS

6.3.1 Emotion

Emotion is a quick process focused on an event and consists of a trigger mechanism based on the relevance that shapes a multiple emotional response (Pasquier and Paulmaz, 2004). The emotion was long regarded as opposed to cognition. Many philosophers, including Plato, Descartes, and Kant, consider this phenomenon as a disturbance of reason that was necessary to correct. For them, rationality and reason should not give way to emotions. In this line of thinking, most theories of education focused on the development of cognitive processes and neglected the emotional dimension (Talhi, 1996), yet emotions color the events of life, give them value, and drive motivation (Pasquier and Paulmaz, 2004). Furthermore, emotions can interfere and make a difference between all the elements. And that is the difference between a "cold" rational and a "hot" cognition emotion (Behaz, Djoudi, and Zidani, 2003).

6.3.2 Classification

Classification is a process of classifying the collected data based on some criteria or similarity. It necessitates the elicitation and selection of a feature that is well described to a particular class. Classification is also referred to as supervised learning, as the patterns are given with known class labels, contrary to unsupervised learning whose labels are unknown. Each pattern in the dataset is represented by a set of features that may be definite or continuous (Kavitha, Karthikeyan, and Chitra, 2010). Classification, therefore, is the process of constructing the model from the training set. The model that is left by the end of the construction is then used to forecast the class label of the testing patterns.

6.3.3 Multiagent

A multiagent system (MAS) is a computerized system composed of multiple interacting intelligent agents within a certain environment. Multiagent systems can be used to solve problems that are difficult or impossible for an individual agent or a monolithic system to solve. Intelligence may include some methodic, functional, and procedural approaches; algorithmic search; or reinforcement learning (Niazi and Amir, 2011).

Agents are endowed with autonomy. This means that they are not directed by commands from the user (or another agent), however, they can only do this by following a set of trends that can take the form of individual goals to satisfy or functions of satisfaction or survival that the agent tries to optimize. So we could say that the motor of an agent is itself. It is active. It has the ability to an affirmative answer or refusal to requests from other agents.

It has a certain freedom of maneuver, which differentiates it from all similar concepts. They are called objects, software modules, or processes. However, autonomy is not only behavioral, it also covers the resources. To act, the agent needs a number of resources: energy, CPU, amount of memory, access to certain sources of information, and so forth.

The agent is thus both an open system (it needs external elements so as to survive) and a closed system (because the exchanges that it has with the outside are closely regulated) (Ferber, 1995).

6.3.4 Weka

Weka is a famous set of machine learning software written in Java, developed at the University of Waikato, New Zealand. The Weka collection contains a compilation of visualization device and algorithm for data analysis. In Weka, the dataset should be formatted to the ARFF format. The Weka Explorer will use this format automatically if it does not recognize a given file as an ARFF file. The preprocess board has facilities that are used to import data from the database and for preprocessing this data using a filtering algorithm. These filters can be used to alter the data (Kamber, Winstone, Gong, Cheng, and Han, 1997).

Weka provides a collection of machine learning algorithms to its users. Among these algorithms, in this chapter we present the most commonly used. These algorithms can be used either for regression or classification. The regression is interested in finding a correlation relationship between inputs (the different independent variables) and outputs (or the dependent variables).

The classification aims at predicting the class of a test set instance. These algorithms are applied in supervised mode, guided by the output (or target class). These algorithms are spread over seven or six families depending on the version of Weka in use (Yazid, 2006).

6.3.5 Felder Questionnaire

The Felder questionnaire has 44 questions. For each question, the learner must choose an answer from a to b. The 44 questions are divided into four groups of 11 questions each. Each group of questions defines a dimension for the cognitive model of the learner that is composed of four dimensions according to Felder (1993):

- First dimension (information processing)—This is the dimension of reflection and information processing by the learner. It varies from reflected to active.
- Second dimension (reasoning)—The second dimension is the reasoning. It varies from deductive to inductive. Deductive learners prefer starting of the principles to deduce the consequences or the applications.
- Third dimension (perception of information)—The third dimension represents the way in which the learner prefers to perceive the information; it is the sensory dimension. It ranges from visual to verbal.
- Fourth dimension (progression to understanding)—This dimension defines the way that the learner prefers to progress in learning a lesson. It varies between global and sequential. A sequential learner prefers to advance in stages; in opposite, a global learner prefers to freely choose their courses to make big jumps in context.

Measuring a dimension: To assign a dimension to a learner, using the Felder questionnaire, just count the number of responses a and the number of responses b to the 11 questions (Felder and Silverman, 1988).

6.4 THE ASTEMOI SYSTEM

6.4.1 Architecture

To guarantee the quality of distance learning we introduce the concept of agents which is presented in an emotionally intelligent tutoring system called ASTEMOI whose architecture is the following (Figure 6.1):

- Tutor agent—Is the agent in charge of managing the courses and cognitive status of the e-learner (Brusilovsky, 2001)?
- Style agent—Using the Felder questionnaire (Derouich, 2011) we can determine the suitable learning style of the e-learner.

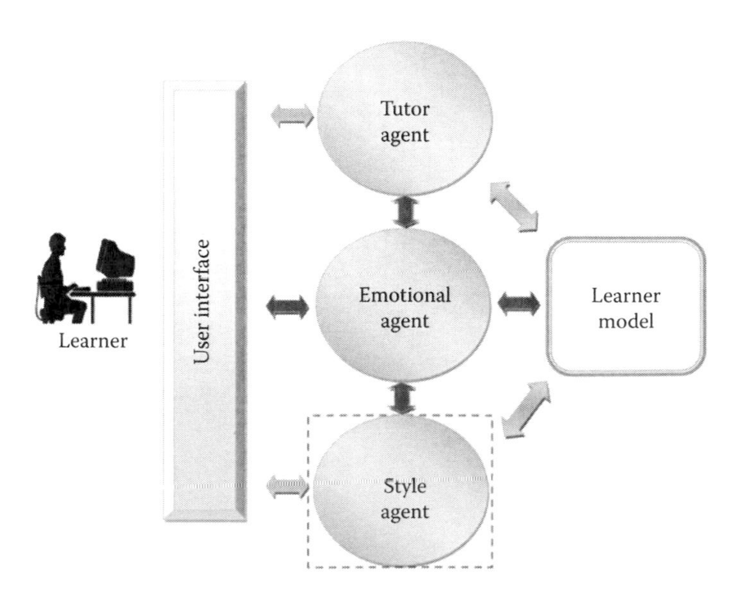

FIGURE 6.1 Global system architecture.

- Emotional agent—The role of this agent is to determine the emotional status of the e-learner using the results of the voice analysis and the feedback.

6.4.2 Problems Treated by the ASTEMOI System

The great majority of digital natives face many problems when it comes to learning using e-learning applications or websites. One of the problems that they face is the fact that these lessons are not tailored to suit neither their cognitive abilities nor their learning style. Another problem that they encounter is blockade and isolation. The online learning resources do not provide their customers with the motivation and interaction they need in order to perform well and they do not take their emotions into account.

Since every learning situation's central objective is to provide the learners with good-quality training, the ASTEMOI system highly considers the learners' needs and also their strengths and weaknesses, because they are the main factors to boosting e-learners' performance and improving the quality of e-learning applications.

6.5 DETERMINING THE EMOTION FROM A VOCAL ANALYSIS

Speaking is a means of communication that contains not only linguistic information, but also provides information on personality traits and the emotional state of the speaker. The information can be used technically so the machine can understand the human speech (Figure 6.2).

6.5.1 Acoustic Parameters Characterizing the Emotion

The results of several studies show that changes in speakers' states modify specifically the acoustic parameters of their words (Picard, 1997).

Acoustic analysis of vocal emotion essentially depends on the following parameters: fundamental frequency (F_0), intensity, and duration of the emotional voice. The values obtained directly reflect the physiological changes of the speaker, who feels particular emotions in particular situations (Skinner, 1935). To synthesize emotional speech, the descriptors of the quality of voice called high level (intensity and frequency) are the most used as they provide a high level of interpretation (Oudeyer, 2003).

- Fundamental frequency F_0 (or pitch)—A particularly relevant index in the expression and perception of emotion. It is related to the pitch of the voice (acute or severe). The signal is modeled as the sum of a periodic signal T and a white noise, such as

$$F_0 = \frac{1}{T} \tag{6.1}$$

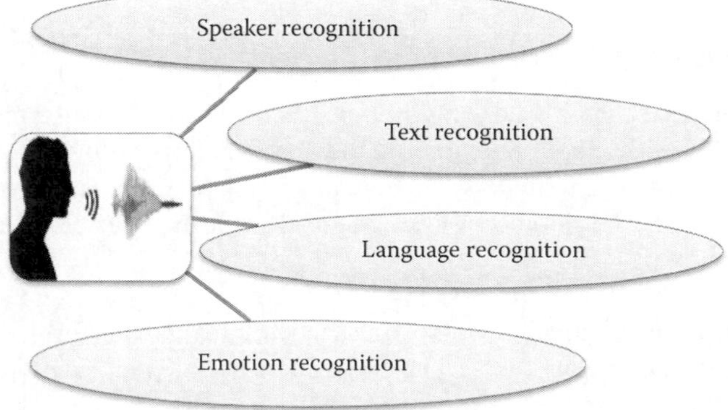

FIGURE 6.2 Information conveyed in speech.

To estimate the fundamental frequency, several methods are available. The method used by the Praat software (Boersma and Weenink, 2005) consists of searching for similarities between the shifted versions of the observed signal, denoted s, and defined as follows:

$$r_s(m) = \begin{cases} \dfrac{\sum_{n=0}^{N-1-m} s(n)s(n+m)}{\sqrt{\sum_{n=0}^{N-1-m} [s(n)]^2}\sqrt{\sum_{n=0}^{N-1-m} [s(n+m)]^2}} & \text{if } m \geq 0 \\ r_s(-m) & \text{if not} \end{cases} \tag{6.2}$$

The period T is estimated by finding the smallest value of m for which $r_s(m)$ is maximum.

The intensity I provides a measure of the loudness of the voice (low or high). It is calculated on a signal portion of length N having the following form:

$$I = 10 \log \left[\sum_{n=1}^{N} w(n)[s(n)]^2 \right] \tag{6.3}$$

where w is the Gaussian analysis window (Amir and Ron, 1996).

- The flow of speech (Q)—One of the parameters calculated with F_0 in the description of the vocal emotion, which is the number of syllables per second.

$$Q = \frac{\text{NSE} \times 1000}{\Delta t} \tag{6.4}$$

where NSE is the number of syllables in the statement, and Δt is the duration of the statement.

The variation of the acoustic parameters of the emotional voice is often described in terms of the degree of deflection of their values that are relative to the values found in the neutral voice. Some acoustic characteristics of emotions, considered primary, are joy, anger, sadness, and fear with different values on the acoustic indices (frequency domain, temporal domain, voice quality) (Chung, 2000). We are going to limit our study to only two states: favorable (joy) and unfavorable (fear, sadness, anger) in order to detect the emotional states that are positive for the e-learning process, using the well-known support vector machine (SVM) method.

6.5.2 Automatic Recognition of Emotions

Speaking is a medium that contains not only linguistic information but also provides information on personality traits and the emotional state of the speaker. Such information can be exploited technically to allow the machine to understand human speech. In recent years, studies on emotional speech begin to address the development of automatic classification system of emotions, based on four main phases (Figure 6.3) (Clavel, 2007):

1. During extraction of acoustic descriptors, the speech signal is transformed into a sequence of acoustic vectors containing the various descriptors used to make a representation of the main acoustic characteristics of the speech signal.
2. During learning, several acoustic vectors corresponding to the same sounds in the same class can be grouped to a representative of this class.

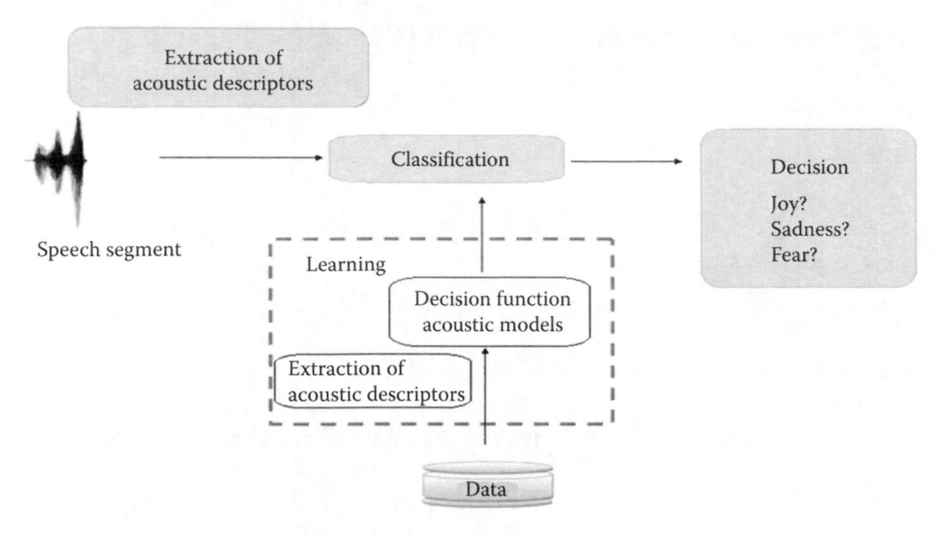

FIGURE 6.3 Principle of an emotion recognition system.

3. During the classification phase, a comparison between the acoustic vectors of the speech signal to be analyzed and representatives of each class or models leads to a probability of belonging to each class for each acoustic vector.
4. The decision phase is the exploitation phase of probabilities calculated to associate a class to a speech segment.

6.5.3 Machines Learning (Support Vector Machine)

The resolution of classification issues may be made by the SVM method inspired by the statistical learning theory; it was introduced by Vapnik in 1995. This method uses a set of learning data represented by a set of pairs of inputs–outputs to learn the model parameters in order to find a linear classifier (hyperplane) in a suitable space able to separate data and maximize the distance initially between two classes, which subsequently had a multiclass generalization (Figure 6.4) (Hsu and Lin, 2002).

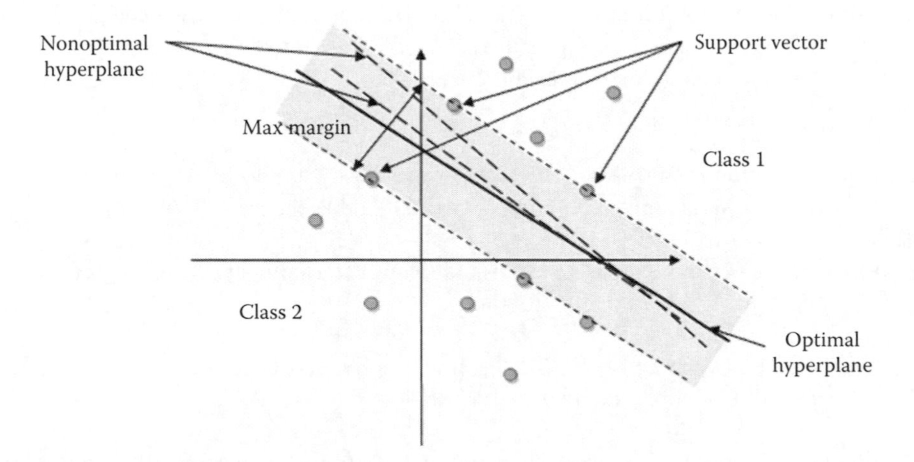

FIGURE 6.4 Hyperplane separating data belonging to two classes.

The hyperplane h satisfies the equation (Antoine and Laurent, 2010):

$$h(x) = \langle w, x \rangle + w_0 = \sum_{i}^{d} w_i \; x_i \tag{6.5}$$

With $X \in \mathbb{R}^d$ and w is a vector orthogonal to $h(x)$. Then we will use the decision function:

$$g(x) = \text{sing}(w, x) + w_0 \quad \text{for (+ or −) the entry } x. \tag{6.6}$$

The plus (+) sign refers to a class and the minus (–) refers to the other class. According to the principle of minimizing the standard empirical risk, a better plan will be one that minimizes the number of samples misclassified by the decision function in the training set

$$S = (x_1, u_1, \ldots, x_m, u_m)$$

w is a vector orthogonal to $h(x)$.

Similarly, to a sample of students, the optimal hyperplane separates the students into two different classes "those who find difficulties" and "those who don't find difficulties" in one module, and we will have a generalization for the different modules in the platform. In our case, this method is going to divide the learners into two states: one is when the learning process is effective and is taking place, and the other one is when it does not.

6.6 ASTEMOI AND MOTIVATION OF THE LEARNER

Motivation is a huge factor when learning. The incentive dimension of the learner profile refers to his motivational abilities directly linked to his emotional and cognitive states (Reuchlin, 1990). Establishing mechanisms supporting the motivational state can promote the learning processes implemented using ASTEMOI.

6.6.1 Self-Determination Theory

There are several theories to better understand motivation, but we take only one, that of self-determination (Deci and Ryan, 1985). According to the theory of self-determination, three psychological needs are the basis of human motivation: the need for autonomy, the need for competence, and the need for social belonging. The satisfaction of these three needs leads to a feeling of well-being for the individual (Vallerand, Pelletier, and Ryan, 1991). There are different types of motivation:

- Intrinsic motivation is considered the highest level of self-determined motivation that can be reached by an individual (Ryan and Deci, 2000a,b), because the activity is done for pleasure.
- If he is intrinsically motivated, he is pushed by an external cause, to get a reward, or to avoid punishment (Ryan and Deci, 2000a,b).

Going from the highest degree of self-determined motivation to the lowest, we find identified regulation, interjected regulation, and external regulation.

- Identified regulation implies that the individual starts to be aware of the interest in an activity and the importance of practicing it freely.

- With introjected regulation the activity is performed to avoid feelings of guilt, anxiety, or to improve his ego (Ryan, Rigby, and Przybylski, 2006).
- External regulation is defined as the individual being motivated by external elements such as material rewards or punishments.

6.6.2 Improving Motivation through ASTEMOI

The aim of our work is to create a learning environment where the data analysis will allow us to understand the behavior of learners and foresee appropriate profiles to improve motivation according to the results found; then the learner will be extrinsically motivated in order to satisfy the three psychological needs presented in SDT:

- Encouraging freedom of decision making in learners to improve the need for autonomy in the sense that they are free to interact as they see it fit
- Provide pedagogical support to guide the learner by the tutor agent who responds to a request for information and congratulates the learner once the activity is completed to encourage the learner's confidence or offer assistance in some failure
- Give the opportunity to share a common learning experience for many students to provide them with ways to interact with each other through competitive activities (Cohn and Katz, 1998)

Note that the degree of motivation given to the learner depends primarily on the style assigned during the training.

6.7 THE PREDICTIVE MODEL BY EXPLORING DATA

6.7.1 The Predictive Model

In the learning environment envisioned, the student will take the time to answer the Felder questionnaire (Felder, 1993), something essential to determine his or her appropriate learning style, as well as give them a personal profile of their strengths and weaknesses in a specific discipline/course, which will be taken into account. The style agent of the ASTEMOI system has a predictive model that tries to achieve this goal. These essential steps must be followed (Figure 6.5):

- Exploitation of alumni data to form the core of the model using logistic regression Beta-Bernoulli method.
- Estimation of the strengths and weaknesses of new students based on the analysis of old data.

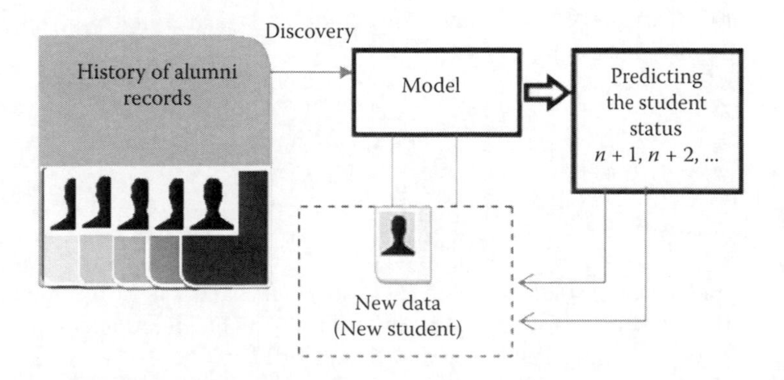

FIGURE 6.5 Predictive model in the ASTEMOI system.

- Student redirection to a suitable profile that responds best to their inspirations and provides additional links to the course proposed by a tutor agent. These courses are presented in several forms (video, audio, text) depending on the learner's style (reflection, reasoning, sensory, or progression).
- Comparison between the final results of new students and those predicted by the model, allowing updating and improving the predictive model.

6.7.2 Exploring Data and Weka

In the field of education, data analysis is becoming more common to predict optimal results. Artificial intelligence provides some software solutions; among them is Weka.

To estimate the appropriate learner profile and the determination of his strengths and weaknesses in distance education, several algorithms dedicated to the profiles classification are developed.

Our work will be limited to a comparison between the following four methods: naïve Bayes classifiers, logistic regression, decision tree J48, and IBk. Knowing which method to choose can enable us to obtain authentic and realistic results to better understand the learner (his/her level, strengths, weaknesses, etc.).

We have as inputs information about 1765 students from Alchariaa Faculty of Fes including the type of baccalaureate, the notes of the baccalaureate subjects, diploma score, and regions (City or province). The classification for each criterion leads to separating the students into groups. A representative profile model will be assigned to each group.

For example, our experience will be based on the following attributes:

- Baccalaureate major: literature, sciences, Islamic studies
- Age: 19–24, 24–30, >30
- Mark: <10, 10–12, >12

However, we just want to know the students who found difficulty in a specific module relying on their yes or no answers; and to assign appropriate profiles to different groups of students.

6.7.2.1 Logistic Regression

We use logistic regression to estimate whether the e-learner faces any difficulty in a certain class. The method is called the logistic regression beta Bernoulli inspired by the mathematical model of the Bernoulli distribution; it adjusts to the experiments in which the results could have two values: 0 or 1. In our work we will identify factors related to a student having problems in a subject student characterized by the following attributes (Bac major, Age, Mark).

The mathematical background is similar to the model in which Y is a binary variable (0 for nonoccurrence of the event, and 1 when the event occurs) with Y random and Xi not random.

Let Y be the variable to predict (explained variable), and $X = (X1, X2, \ldots, Xj)$ be the predictor variables (explanatory variables). So that the expectation of Y takes only two values, we use the logistic function (Boudin, 2012):

$$f(x) = \frac{\exp(x)}{1+\exp(x)} = p \tag{6.7}$$

So $0 < f(x) < 1$ and $E(Y) = 0$ or 1.

We distinguish two cases:

- The first case is that of a single variable. The values of x and y describe each possibility, namely, $X = 0$: no criterion; $X = 1$: existing criterion; $Y = 0$: having problem (no); and $Y = 1$: having problem (yes). So we have (Neji and Jigorel, 2012):

$$Logit[PYi = 1 \mid X = x] = \beta 0 + \beta 1x \tag{6.8}$$

- The second case is the multiple logistic model. The variables are used to establish the link between multiple cases.

$$\text{Logit } P(\text{Problem} = \text{yes} \mid \text{Age, Mark, Type\_Bac}) = \beta 0 + \beta 1 \cdot \text{Age} + \beta 2 \cdot \text{Mark} + \beta 3 \cdot \text{Type\_Bac}$$

So we are able to estimate whether students are going to have problems in a subject.

6.7.2.2 Decision Tree Algorithm J48

The J48 classifier is an extension of C4.5. It creates a binary tree. The decision tree approach is most useful in problem classification. With this technique, a tree is constructed to model the classification process. In a decision tree the internal nodes of the tree denotes a test on an attribute, a branch represents the outcome of the test, a leaf node holds a class label, and the topmost node is the root node (Dunham and Sridhar, 2006).

Algorithm J48:

```
INPUT:
D        //Training data
OUTPUT
T        //Decision tree
DTBUILD (*D)
{
T = φ;
T = Create root node and label with splitting attribute;
T = Add arc to root node for each split predicate and label;
For each arc do
D = Database created by applying splitting predicate to D;
If stopping point reached for this path, then
T' = create leaf node and label with appropriate class;
Else
T' = DTBUILD(D);
T = add T' to arc;
}
```

J48 takes notice of the lost values while building a tree and permits classification through decision trees or the rules produced by them. We can forecast the value for that item by using what is known about the attribute values in the other records. We can also divide the data into series based on the values that are attributed for that one item and originated in the training sample (Spangler, May, and Vargas, 1999).

6.7.2.3 IBk

The IBk algorithm is a k-nearest-neighbor classifier that uses the same distance metric. It can select the appropriate value of *K* based on a cross-validation. It can also do distance weighting. Its distance function is used as a parameter of the search method.

In trial-based learning, each new instance is compared with the existing ones using a distance metric. The closest instance is used to describe the class of the new instance. Therefore, this technique is called the classification method of the nearest neighbor. In some cases, more than one nearest neighbor is used and the majority class of the nearest neighbors (or the average score distances if the class is digital) is assigned to the new instance. This technique is called the method of the k nearest neighbors.

The IBk algorithm represents the implementation of the classifier of k's nearest neighbors using the distance metric. It just uses the nearest neighbor ($k = 1$) by default; the number k can be specified manually or determined automatically using the cross-validation "leave one out." It normalizes the default attributes and can also weight distances (Kavitha, Karthikeyan, and Chitra, 2010).

6.7.2.4 Naïve Bayes Classifier

The naïve Bayes classifier is a simple classifier based on probability. It analyzes a number of probabilities by counting the occurrence, regularity, and mixture of values in a certain data set. The naïve Bayes classifier is also based on the Bayesian theorem and named after Thomas Bayes, who was the first to project the Bayes theorem. Its formula can be written as

$$P(H \mid E) = \frac{[P(E \mid H) * P(H)]}{P(E)} \tag{6.9}$$

The algorithm uses Bayes theorem and assumes all attributes to be independent given the value of the class variable.

6.7.2.4.1 Bayes Theorem

Bayes theorem is used in statistical inference to update or modify the estimation of a probability or any parameter from observations and laws of probability of these observations (Efron, 2013):

$$p\left(\frac{w_i}{x}\right) = \frac{p(w_i) \cdot p(x/w_i)}{p(x)} \tag{6.10}$$

where (w_i/x) is posterior, $p(w_i)$ is prior, $p(x/w_i)$ is likelihood, and $p(x)$ is evidence.

6.7.2.4.2 Naïve Bayes Classifier

The naïve Bayes classifier is a type of simple probabilistic Bayesian classification based on Bayes theorem with strong independence of the hypotheses. A various supervised classification problems implement the naïve Bayes classifier as it performs well and learns rapidly (Dimitoglou, James, Adams, and Carol, 2012).

We express our problem as a probability, in which the variables $x_1, x_2, \ldots, x_n$ are considered independent, so

$$p(x_1, x_2 \ldots x_n) = p(x_1) * p(x_2) * \ldots p(x_n) \tag{6.11}$$

And the naïve Bayes becomes (Layachi, 2007)

$$p\left(\frac{w_i}{x_1, x_2 \ldots x_n}\right) = \frac{p(w_i) * p(x_1/w_i) * \ldots * p(x_n/w_i)}{p(x_1) * p(x_2) * \ldots * p(x_n)} \tag{6.12}$$

We want to figure whether the students have difficulty (w_1 = yes) or not (w_2 = no). We have a new object (student) to classify: x_1, x_2, x_3 which are the attributes type Bac major, Age, and Mark, respectively. We want to know if this object (student) belongs to w_1 or w_2. The student belongs to the class that maximizes this probability. We are not obliged to calculate the evidence because it is a constant and we want to find the maximum of these probabilities.

For example: Student = {Literature, >28, <10}, belongs to which class?

$$p\left(\frac{\text{yes}}{\text{Literature}, >28, <10}\right) \propto p(\text{yes}) * p\left(\frac{\text{Literature}}{\text{yes}}\right) * p\left(\frac{>28}{\text{yes}}\right) * p\left(\frac{<10}{\text{yes}}\right)$$

And of the same for the class "no."

$$p\left(\frac{\text{no}}{\text{Literature}, >28, <10}\right) \propto p(\text{no}) * p\left(\frac{\text{Literature}}{\text{no}}\right) * p\left(\frac{>28}{\text{no}}\right) * p\left(\frac{<10}{\text{no}}\right)$$

So the student belongs to the class that has more probability than the other.

6.8 PERFORMANCE AND EVALUATION

In this study, we are going to compute the trial measures by utilizing the execution components, for example, the order precision and execution time. Furthermore we discover the precision measure and mistake rate to decide the best calculation for the Alchariaa dataset (dataset collected from the database of Alchariaa faculty of fez). The execution components for these arrangement calculations are recorded in Table 6.1 and the exactness measure by class for the classifier calculations is delineated in Table 6.2. From the trial results, it is surmised that for the cross acceptance parameter for naïve Bayes calculation, the precision, F-measure, true positive (TP) rate values, and the receiver operating characteristic (ROC) value gives better results for Alchariaa dataset. The execution elements for the arrangement calculations are demonstrated in Figure 6.6 and the exactness measure for the classifiers is shown in Figure 6.7.

TABLE 6.1
Performance Factors for the Classification Algorithms

Algorithms	TP Rate	Precision	F-Measure	ROC Curve	Kappa Value	Execution Time
J48 decision tree	0.676	0.664	0.667	0.675	0.2533	0.17
IBk	0.646	0.636	0.638	0.643	0.2204	0
Naïve Bayes	0.71	0.702	0.704	0.716	0.3373	0.07
Logistic regression	0.672	0.661	0.658	0.662	0.2634	0.57

TABLE 6.2
Accuracy Measures for Classification Algorithms

Algorithms	J48 Decision Tree	IBk	Naïve Bayes	Logistic Regression (%)
Correctly classified	67.6136%	64.6459%	71.0227%	67.1955
Incorrectly classified	32.3864%	35.3541%	28.9773%	32.8045

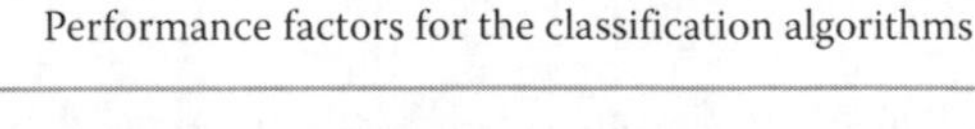

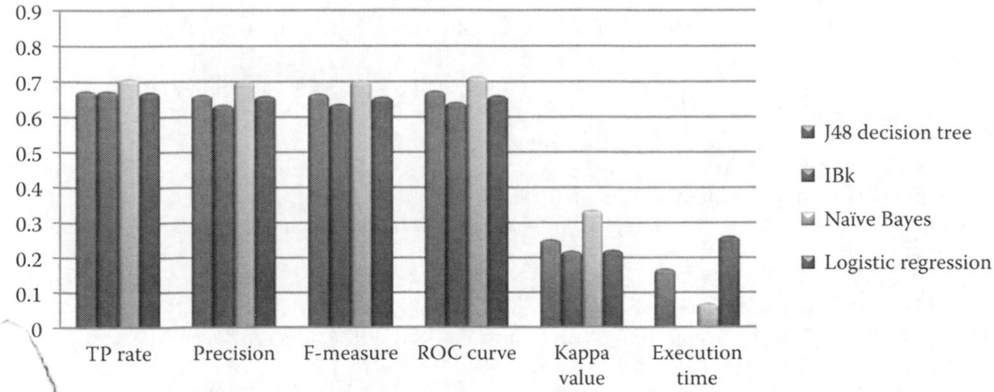

FIGURE 6.6 Performance factors.

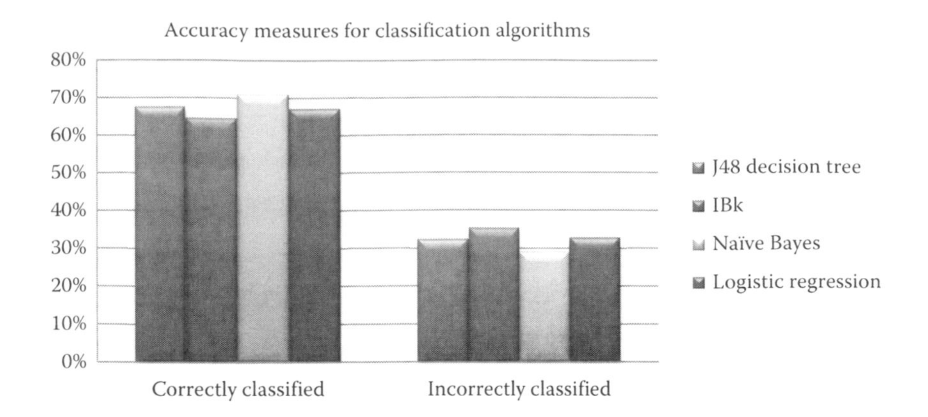

FIGURE 6.7 Accuracy measures.

After the study we conducted on 1765 alumni of Alchariaa Faculty of Fez, we can conclude that the naïve Bayes is the most feasible and reasonable method that has enabled us to reach results close to reality. We are going to make a comparison between the real results (obtained in 3 years or more) and those that have been predicted by the model of our system ASTEMOI. This comparison is going to allow us to update and improve the system to get more reasonable and feasible results in the future.

6.9 CONCLUSION

Thanks to the e-learning multiagent system, we are able to detect learners' emotions, strengths, and weaknesses using automatic systems during distant training. These systems enable us to determine the emotional states of e-learners. This chapter opens the door to future researches that may adopt other systems and methods to provide more feasible results, like treating and analyzing facial expressions using detectors, or even adding other models that would help detect the psychological state of an e-learner through his or her interaction with the online course, which is going to help us find the right motivation for each learner and then improve the result. We can also improve the predictive system ASTEMOI by adding more devices that will help orient learners in their academic careers and even in choosing the topic of their research projects.

REFERENCES

Amir, N. and Ron, S. 1996. Towards an automatic classification of emotions in speech. *Proceedings of ICSLP*, Philadelphia.

Antoine, C. and Laurent, M. 2010. *Apprentissage Artificiel: Concepts et Algorithmes* (2nd ed.) Eyrolles. ISBN: 978-2-212-12471-2.

Armony, H. and Vuilleumier, P. 2013. *The Cambridge Handbook of Human Affective Neuroscience.* Cambridge University Press, New York.

Behaz, A., Djoudi, M., and Zidani, A. 2003. Approche de modélisation et d'adaptation des documents pédagogiques hypermédias en enseignement à distance. *Proceedings of the 6th Seminar CIDE'6*, Caen, France.

Boersma, P. and Weenink, D. 2005. Praat. Doing phonetics by computer [computer program], from http://www.praat.org/. Rapport Technique.

Boudin, F. 2012. Machine Learning avec WEKA module X8II090. Course 1, Department of Computer Science, Nantes University.

Brusilovsky, P. 2001. Adaptive hypermedia. *Journal of User Modeling and User Adapted Interaction*, 11(1–2), 87–110.

Chung, S. 2000. L'expression et la perception de l'émotion extraite de la parole spontanée évidences du coréen et de l'anglais. Institut de Linguistique et Phonétique Générales et Appliquées, Université de la Sorbonne Nouvelle (Paris III).

Clavel, C. 2007. Analyse et reconnaissance des manifestations acoustiques des émotions de type peur en situations anormales. domain other. Telecom Paris Tech. English HAL Id: pastel-00002533.

Cohn, J. F. and Katz, G. S. 1998. Bimodal expression of emotion by face and voice. *Proceedings of the Sixth ACM International Conference on Multimedia: Face/Gesture Recognition and Their Applications*, pp. 41–44. Bristol, UK. ISBN:1-58113-163-1. doi:10.1145/306668.306683.

Deci, E. and Ryan, R. M. 1985. *Intrinsic Motivation and Self-Determination in Human Behavior.* Plenum, New York.

Derouich, A. 2011. *Conception et réalisation d'un hypermédia adaptatif dédié à l'enseignement à distance.* Sidi Mohammed Ben Abdellah University, Fez.

Després, C. and George, S. 2001. Supporting learners' activities in a distance learning environment. *International Journal of Continuing Engineering Education and Lifelong Learning*, 11(3), 261–272.

Dimitoglou, G., James, A. A., and Jim, C. M. 2012. Comparison of the C4.5 and a Naïve Bayes classifier for the prediction of lung cancer survivability *Journal of Computing*, 4(8).

Dunham, M. and Sridhar, S. 2006. *Data Mining: Introductory and Advanced Topics.* New Delhi, India: Pearson Education.

Efron, B. 2013. Bayes' theorem in the 21st century. *Science*, 340(6137), 1177–1178.

Felder, R. M. 1993. Reaching the second tier: Learning and teaching styles in college science education. *Journal of College Science Teaching*, 23(5), 286–290.

Felder, R. M. and Silverman, L. K. 1988. Learning and teaching styles. *Engineering Education*, 78(7), 674–681.

Ferber, J. 1995. *Les systèmes multi-agents, vers une intelligence collective.* InterEditions, Paris.

Hsu, C. W. and Lin, C. H. 2002. A comparison of methods for multi-class support vector machines. *IEEE Transactions on Neural Networks*, 13(2), 415–425.

Istrad, D. 2003. Detection et reconnaissance des sons pour la surveillance médicale. PhD thesis, University of Grenoble.

Kamber, M., Winstone, L., Gong, W., Cheng, S., and Han, J. 1997. Generalization and decision tree induction: Efficient classification in data mining. *Proceedings of 1997 International Workshop on Research Issues on Data Engineering (RIDE'97)*, Birmingham, England, pp. 111–120.

Kavitha, B., Karthikeyan, S., and Chitra, B. 2010. Efficient intrusion detection with reduced dimension using data mining classification methods and their performance comparison. *BAIP* 2010, 96–101.

Layachi, B. 2007. Data mining for scientists Bishop's University, http://aqualonne.free.fr/Teaching/csc/DM.pdf.

Lee, C., Narayanan, S., and Pieraccini, R. 2002. Classifying emotion in human machine spoken dialogs. *Proceedings of ICME*, Lausanne, Vol. 1, pp. 737–740.

Neji, S. and Jigorel, A. 2012. *La régression logistique, exposé statistiques et économétrie.* Rennes University. http://docplayer.fr/3943550-La-regression-logistique-par-sonia-neji-et-anne-helene-jigorel.html.

Nghiem, A. T., Bremond, F., Thonnat, M., and Valentin, V. 2007. An evaluation project for video surveillance systems. AVSS 2007, *IEEE International Conference on Advanced Video and Signal Based Surveillance,* London, UK.

Niazi, M. and Amir, H. 2011. Agent-based computing from multi-agent systems to agent-based. *Scientometrics*, 89(2), 479–499.

Oudeyer, P.-Y. 2003. The production and recognition of emotions in speech: Features and algorithms. *International Journal of Human Computer Interaction, Special Issue on Affective Computing*, 59(1–2), 157–183.

Pasquier, G. and Paulmaz, E. 2004. La gestion des émotions et les implications dans l'apprentissage. *The Notebooks of the Academy of Nantes*, 31, 17–21.

Picard, R. 1997. *Affective Computing.* MIT Press, Cambridge, MA.

Reuchlin, M. 1990. *La Psychologie Différentielle.* PUF, Paris.

Ryan, R. M. and Deci, E. L. 2000a. Intrinsic and extrinsic motivations: Classic definitions and new directions. *Contemporary Educational Psychology*, 25, 54–67.

Ryan, R. M. and Deci, E. L. 2000b. Self-determination theory and the facilitation of intrinsic motivation, social development, and well-being. *American Psychologist*, 55, 68–78.

Ryan, R. M., Rigby, S. C., and Przybylski, A. 2006. The motivational pull of video games: A self-determination theory approach. *Journal of Motivation and Emotion*, 30, 347–363.

Skinner, E. R. 1935. A calibrated recording and analysis of the pitch, force and quality of vocal tones expressing happiness and sadness. *Speech Monograph*, 2, 81–137.

Spangler, W. E., May, J. H., and Vargas, L. G. 1999. Choosing data-mining methods for multiple classification: Representational and performance measurement implications for decision support. *Journal of Management Information Systems*. 16(1), 37–62.

Talhi, S. 1996. Moalim: un système auteur de l'EIAO. *Proceedings of the 18th Symposium DECUS*, France, Paris.

Vallerand, R. J., Pelletier, L. G., and Ryan, R. M. 1991. Motivation and education: The self-determination perspective. *The Educational Psychologist*, 26, 325–346.

Yazid, H. 2006. *Les algorithmes d'apprentissage automatique offerts par l'environnement Weka*. Université du Québec à Montréal Département d'informatique. http://www.info2.uqam.ca/~lounis_h/dic938G-hiv2011/documents_weka/algos_weka.pdf.

7 Wireless Multimedia Sensor Networks

Cross-Layer Approach Protocols

Manal Abdullah and Ablah AlAmri

CONTENTS

7.1 Introduction 93
7.2 Background 94
7.2.1 Wireless Sensor Networks (WSNs) 94
7.2.1.1 WSN Applications 95
7.2.2 Wireless Multimedia Sensor Networks (WMSNs) 95
7.2.2.1 WMSN Architecture 96
7.2.2.2 WMSN Applications 97
7.2.3 Sensor Network Layers and Cross-Layer Architecture 97
7.3 Cross-Layer Protocols for WMSNs 98
7.3.1 Multichannel Routing 98
7.3.2 Multipath Routing 100
7.3.3 Single-Path Routing 101
7.4 Conclusion 107
References 107

7.1 INTRODUCTION

Recently with advanced technology in microelectromechanical systems (MEMS), and communication through wireless and electronic digital, developing sensor nodes with low power, low cost, and the ability to provide different functions have become feasible (Akyildiz and Vuran 2010). A wireless sensor network (WSN) is a network that can be created by deploying a large number of sensor nodes. These nodes sense the environment and send scalar data to the sink (Karl and Willig 2007; Akyildiz and Vuran 2010; Mendes and Rodrigues 2011).

WSN sensors nodes are used in many applications to abstract information about specific areas such as weather information including temperature for that area, level of humidity, pressure, wind speed and its direction, and movement of objects; all these applications required sensing the environment and sending scalar data (Akyildiz and Vuran 2010). More intelligent sensor nodes can be made by integrating a cheap component such as complementary metal-oxide semiconductor (CMOS) cameras and microphones to the sensor nodes to create a wireless multimedia sensor network (WMSN). WMSNs have sensor nodes that are capable of capturing and communicating streams of multimedia data over a wireless channel to the base station (Melodia and Akyildiz 2011; Hamid and Hussain 2014). These new sensor nodes open the door to many new applications that required sensing the environment and sending multimedia data (video, sound, and image), and not only scalar data such as monitoring and surveillance (Mendes and Rodrigues 2011). Sensor nodes have limitations in terms of power provider, processing capability, and storage memory (Karl and Willig 2007; Akyildiz and Vuran 2010; Mendes and Rodrigues 2011; Hamid and Hussain 2014).

WMSNs have many challenges due to (Hamid, Bashir, and Pyun 2012)

- The type of transmission data (multimedia data) which is large in volume, and require quality of service (QoS) in terms of increasing throughput and reduce delay. The packets of multimedia data is very sensitive to delay and losses, because losing packets or arriving after deadline leads to distortion in received multimedia data.
- Used the wireless as a transition medium.
- Capacity limitation of sensor nodes.

Layered architecture such as TCP/IP consists of five layers (application layer, transport layer, network layer, data link layer, and physical layer) (Akyildiz and Vuran 2010). These layers are independent, separated, and encapsulated from each other, and only adjacent layers can communicate directly (Wang et al. 2010). A cross-layer architecture is a new architecture that combines several layers to allow integration and exchange information among them more efficiently than the layered approach (Wang et al. 2010; Farooq, St-Hilaire, and Kunz 2012). Layers in this architecture are dependent and can share variables between nonadjacent layers (Wang et al. 2010). More efficient cross-layer protocols will improve transmission performance and satisfy the stringent quality of service required for multimedia transmission in WMSNs (Farooq, St-Hilaire, and Kunz 2012). Cross-layer architecture has recently gained the attention of many researchers to produce different cross-layer protocols for WMSNs. This chapter focuses mainly on WMSNs; compares the existing cross-layer protocols for WMSNs that join adjacent or nonadjacent layers; and classifies them based on the routing techniques into three categories: multichannel routing, single path routing, and multipath routing.

The rest of the chapter is organized as follows: In Section 7.2 is a brief background about WSNs and WMSNs, Section 7.3 classifies and compares existing cross-layer protocols for WMSNs, then Section 7.4 concludes the chapter.

7.2 BACKGROUND

The advanced technology in low-power circuits—cheap sensor nodes with different functions—open the door to the sensor networks. Thousands of sensor nodes are deployed to cover specific area. These deployed sensor nodes cooperate to create wireless sensor networks (Akyildiz and Vuran 2010).

7.2.1 Wireless Sensor Networks (WSNs)

The sensor nodes have the ability to sense the environment, process the data locally, and communicate giving birth to WSNs by coordinating the effort of a large number of deployed sensor nodes (Akyildiz and Vuran 2010). Sensor nodes are basically a configuration of five components, as shown in Figure 7.1 (Karl and Willig 2007):

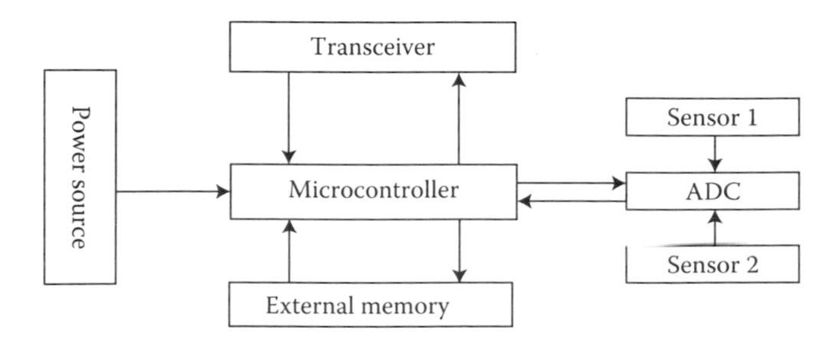

FIGURE 7.1 Main sensor node hardware components. (From Karl, H., and A. Willig, 2007, *Protocols and Architectures for Wireless Sensor Networks*, Chichester, England: John Wiley & Sons.)

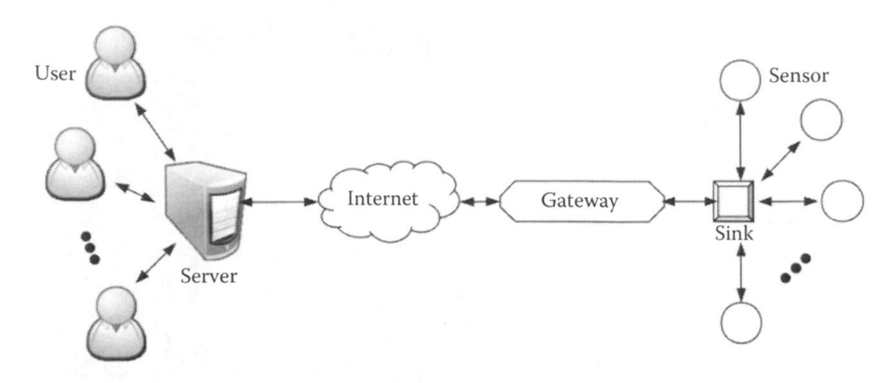

FIGURE 7.2 WSN architecture. (From Mendes, L. D. P., and J. J. P. C. Rodrigues, 2011, A survey on cross-layer solutions for wireless sensor networks, *Journal of Network and Computer Applications* 34(2):523–534.)

1. The *controller* is the core of sensor node, responsible for collecting data from the sensors, processing the data, and deciding where to send it and when.
2. The *memory* stores programs and data.
3. *Sensors* and *actuators* are the interface to the physical environment, and senses and observes the environment parameter. The subunit ADC converts signals from analog to digital.
4. The *transceiver* is responsible for communication through a wireless channel, sending and receiving information.
5. The *power source* provides the energy for the sensor node by some form of battery or perhaps solar cells.

WSNs have sensor nodes that deploy in a physical environment. These nodes are capable of capturing the events and communicating streams of scalar data over the wireless channel to the sink. Sensor nodes are responsible for the fusion and should process the data locally and only transmit the required data. Sensor nodes send the data via a multihop path through the sink. These nodes are usually scattered in a sensor field. Sensor nodes may transmit their packets directly to the sink through a single hope path, or send it to another node in order to forward it to the sink through a multihop path. The sink may use the Internet or a wireless network such as Wi-Fi to communicate with the end user. WSN nodes are limited in resources such as battery, memory, and CPU capability (Akyildiz and Vuran 2010). WSN architecture is shown in Figure 7.2.

7.2.1.1 WSN Applications

WSNs have gained large popularity due to their flexibility in solving problems in different domains of applications and have the possibility to change our lives in many different ways. WSNs have been successfully applied in various application domains such as military applications, area monitoring, in transportation to avoid traffic problem, health care applications, environmental sensing, monitoring building structure, and industrial applications (Akyildiz and Vuran 2010). All these applications require scalar data, but in many different sensitive applications such as monitoring and surveillance to identify the criminals, thieves, or potential terrorists which require sensing the environment and sending video, sound, or even image, not only scalar data. Data contains video, sound, or images called multimedia data, and this type of network is called a wireless multimedia sensor networks (Mendes and Rodrigues 2011).

.2 Wireless Multimedia Sensor Networks (WMSNs)

ISNs are a new branch of wireless sensor networks. The integration of inexpensive components as CMOS cameras and microphones to sensor nodes gave birth to a new branch of wireless r networks, namely, WMSNs. These new nodes are smart devices that have the ability to

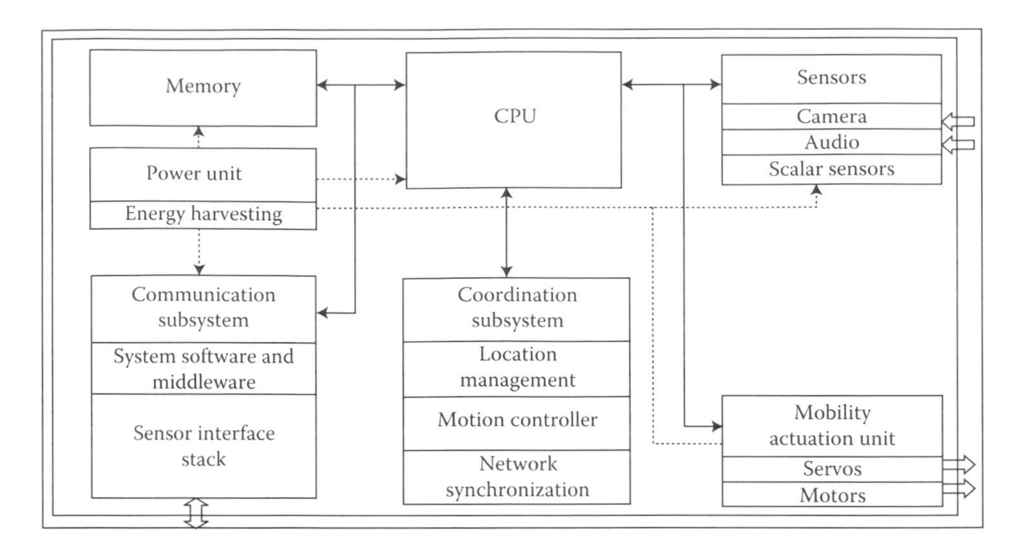

FIGURE 7.3 WMSN sensor nodes components. (From Melodia, T. and I. F. Akyildiz, 2011, Research challenges for wireless multimedia sensor networks, in: *Distributed Video Sensor Networks*, edited by B. Bhanu, C. V. Ravishankar, A. K. Roy-Chowdhury, H. Aghajan, and D. Terzopoulos, pp. 233–246, London: Springer Science & Business Media.)

capture and transmit multimedia data such as video, audio, and even images to the sink (Melodia and Akyildiz 2011). In addition to the components in WSN sensor nodes such as the memory and power units, WMSN sensor nodes, as shown in Figure 7.3, have several new components or more sophisticated components such as (Melodia and Akyildiz 2011)

1. Sensor unit—A unit that contains two subunits: (1) the sensor unit contains a camera, microphone, and scalar sensor; and (2) an ADC converter, which is a unit that converts signals from analog to digital.
2. Processing unit—A small CPU that provides execution of the system software.
3. Communication subsystem—This subsystem contains the communication protocol stack and system software such as middleware, and contains two subunits (a transceiver unit and communication unit).
4. Coordination subsystem—Responsible for coordinating different network operations such as synchronizing the network, controlling the motion, and managing the location.
5. Memory—A storage unit for data and software.
6. Mobility/actuation unit—Optional unit to provide movement ability to the object.

The difference between WSNs and WMSNs is the type of data that transmits through the wireless channel and the design constraints in WMSNs. The design of WMSNs is focused on reducing the end-to-end latency and the speed of delivery of multimedia data packets to the destination, because multimedia data packets are very sensitive to the delay and losses. Losing these packets or arriving after deadline leads to distortion in the received multimedia data (Hamid, Bashir, and Pyun 2012).

7.2.2.1 WMSN Architecture

The architecture of WMSNs consists of multiple components, and each component provides different functions (Melodia and Akyildiz 2011). There are different types of architectures: (1) sing tier flat architecture contains homogeneous nodes where all sensor nodes have the same capabi and provide the same functions; (2) single-tier clustered architecture contains heterogeneous sen nodes, and different sensor nodes with different functionality and capability grouped togethe

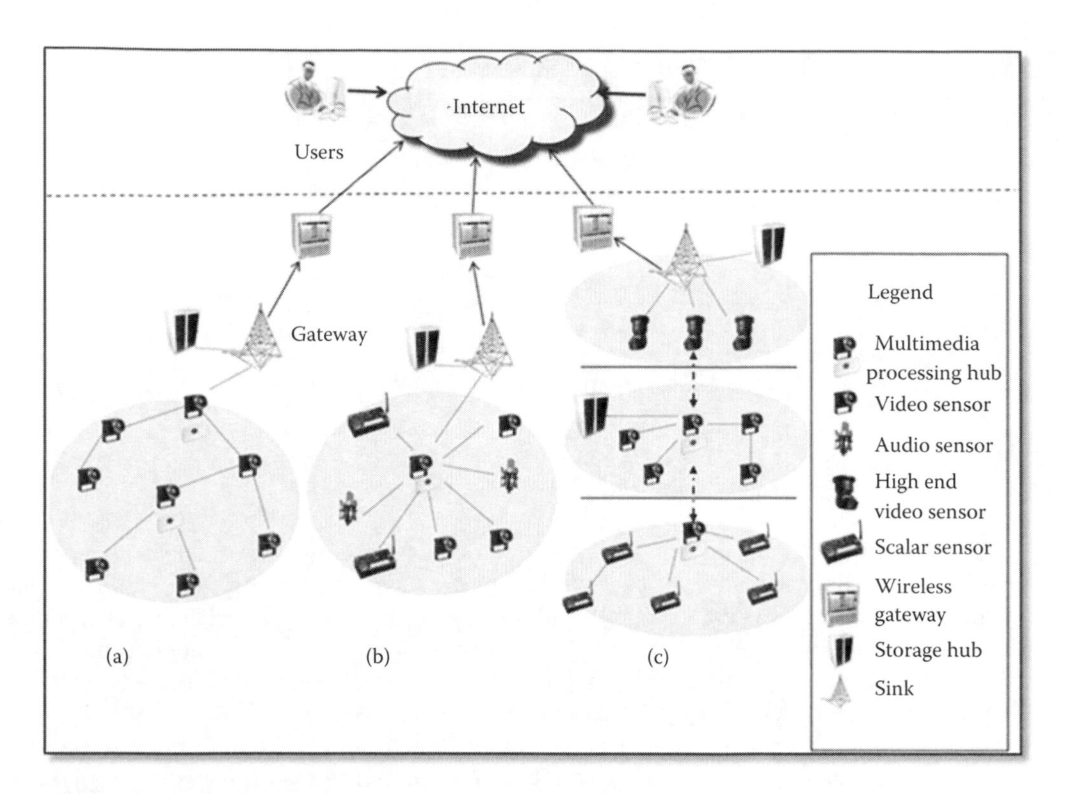

FIGURE 7.4 Architecture of a WMSNs. (a) Single-tier flat, homogeneous sensors, distributed processing, centralized storage. (b) Single-tier clustered, heterogeneous sensors, centralized processing, centralized storage. (c) Multitier, heterogeneous sensors, distributed processing, distributed storage. (From Melodia, T. and I. F. Akyildiz, 2011. Research challenges for wireless multimedia sensor networks, in: *Distributed Video Sensor Networks*, edited by B. Bhanu, C. V. Ravishankar, A. K. Roy-Chowdhury, H. Aghajan, and D. Terzopoulos, pp. 233–246, London: Springer Science & Business Media.)

create a cluster; and (3) multitier architecture contains heterogeneous sensor nodes with distributed processing and data storage to perform complex tasks (Akyildiz, Melodia, and Chowdhury 2007). WMSN architectures are shown in Figure 7.4.

7.2.2.2 WMSN Applications

WMSNs are used in many applications such as multimedia surveillance sensor networks to track the object and take appropriate actions (Al Nuaimi, Sallabi, and Shuaib 2011); to track and locate missing persons; to identify criminals, thieves, or potential terrorists; control systems to monitor traffic to avoid congestion; and also to record activities such as thefts, road accidents, and traffic violations (Melodia and Akyildiz 2011). WMSNs are used in smart homes for energy efficiency to control heating, cooling, and light system based on human activities. WMSNs are used to deliver advanced health care; and in order to remotely monitor patients to infer any emergency situations, patients carry medical sensors to detect their body parameters such as temperature, breathing, and pressure. WMSN sensors are used to monitor the environment and civilian structure such as bridges. Also WMSN use sensors to control industrial processes (Akyildiz, Melodia, and Chowdhury 2008; Melodia and Akyildiz 2011). All these applications require a QoS for multimedia transmission (Al Nuaimi, Sallabi, and Shuaib 2011).

7.2.3 Sensor Network Layers and Cross-Layer Architecture

Layered architecture is a hierarchy; layers are independent, separated, and encapsulated from each other, and direct communication is allowed only between adjacent layers (Wang et al. 2010).

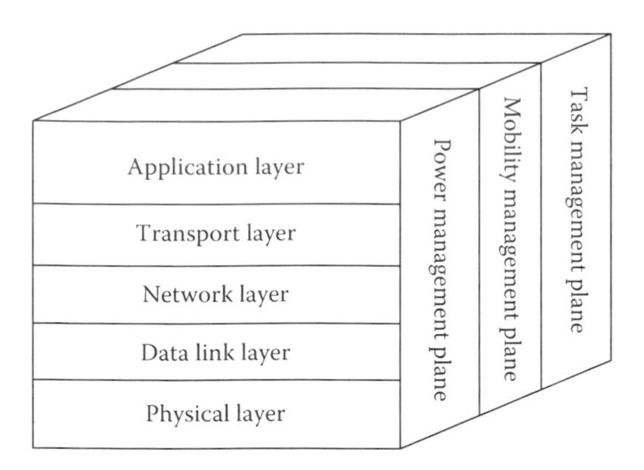

FIGURE 7.5 Protocol stack layers. (From Akyildiz, I. F., and M. C. Vuran, 2010, *Wireless Sensor Networks*, Vol. 4, Chichester, England: John Wiley & Sons.)

The TCP/IP model is an example of this architecture. It is consists of five layers: application, transport, network, data link, and physical. These five layers are separated and only adjacent layers can communicate (Akyildiz and Vuran 2010). In addition, the sensor nodes have different planes that help the sensor nodes to coordinate the task and save energy. The power planes monitor the sensor node power, mobility planes monitor the movement, and the task management plane distributes the task among the sensor nodes. Figure 7.5 shows the protocol stack layers.

The communication protocol plays a major role in the correct functionality for these networks. Limited resources and wireless communication mediums prevent the networks from using the traditional layered architecture in WSNs and WMSNs. For that cross-layer architecture is produced. Cross-layer architecture is a new design that combines several layers—adjacent or nonadjacent—to allow integration and exchange information among them more efficient than in classical layers. More efficient cross-layer protocols will satisfy the stringent quality of service required for multimedia transmission in WMSNs (Farooq, St-Hilaire, and Kunz 2012). Many researchers provide different types of cross-layer architecture. Some researchers combine adjacent layers such as network and media access control (MAC) layer protocols (Hamid, Bashir, and Pyun 2012; Hamid and Bashir 2013), application and transport layers (Paniga et al. 2011), and transport and network layers (Sun et al. 2011). Other researchers combine nonadjacent layers such as application and network (Bae, Lee, and Park 2013). Other researchers combine more than two layers (Çevik and Zaim 2013) where transport, network, and the MAC layers are joint. Researchers also defined performance metrics and used network simulation to evaluate their protocols as described in the next sections.

7.3 CROSS-LAYER PROTOCOLS FOR WMSNs

Many researchers produced different designs and protocols for cross-layer architecture to increase data gathering from WMSNs nodes to the sink, reduce the latency, increase the bandwidth, and reduce the energy consumptions. This is to show how WMSNs can be more efficient depending on the constraints and the requirements of QoS on a specific application. This section summarizes what has been done on cross-layer protocols for WMSNs. We can classify WMSNs protocols based on the routing techniques into three categories as shown in Figure 7.6. These routing categories–multichannel, multipath, and single path—are described in the following.

7.3.1 Multichannel Routing

Multichannel routing divides the bandwidth into different separate channels and dedicate each channel to specific packets types (Çevik and Zaim 2013).

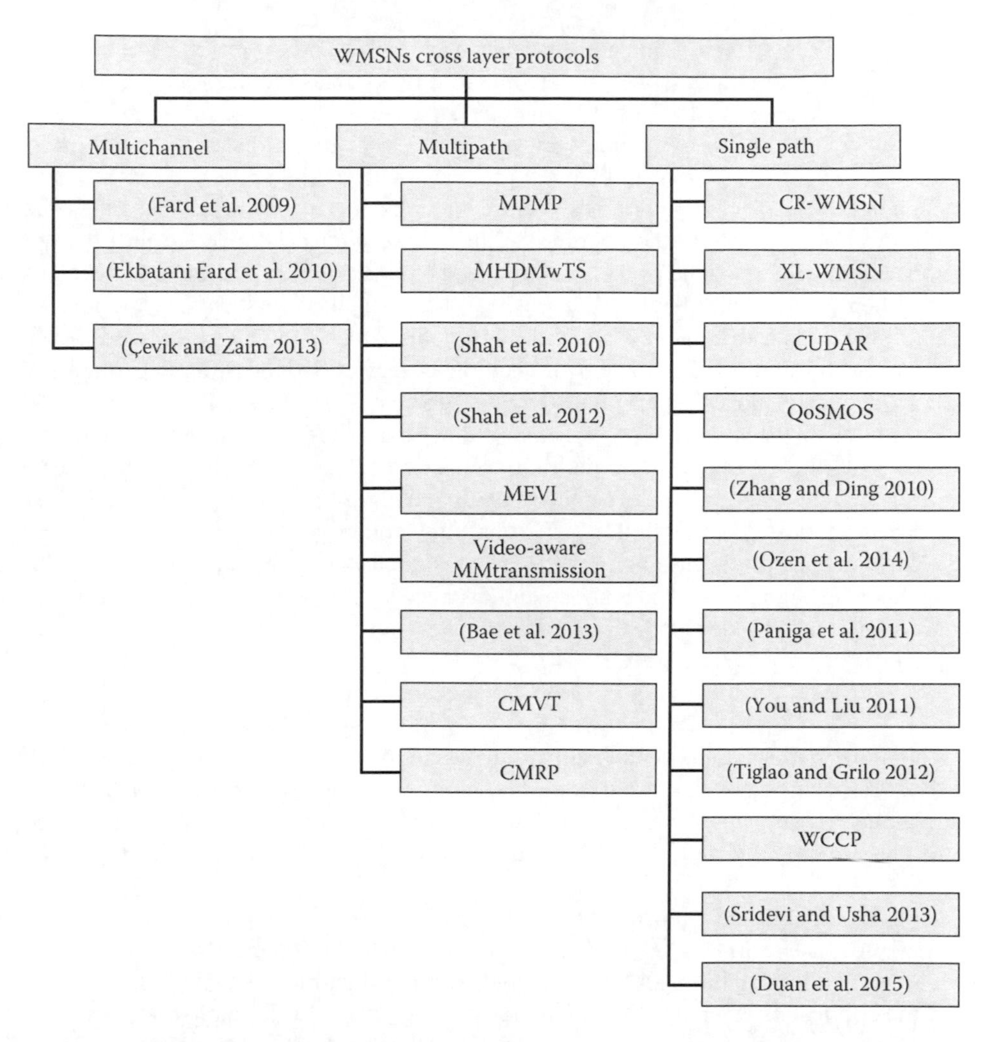

FIGURE 7.6 Cross-layer QoS protocols.

A protocol proposed by Ekbatani Fard, Yaghmaee, and Monsefi (2009, 2010) is a cross-layer multichannel QoS–MAC protocol that supposed a clustered network using any existing clustering techniques, and supposed in each cluster three types of nodes: active nodes defined as nodes need to send data, passive nodes defined as nodes that did not have any data to send, and cluster head. There are three channels where all sensor nodes can transmit or receive through these channels. In an initial stage of network deployment, channel 1 is assigned to the cluster head, the cluster head dynamically assigns channel 2 to sensor nodes, and channel R is shared between all sensor nodes. The MAC protocol proposed is lies in three phases:

1. Request phase where the active nodes send to the head a request message (REQ). This message contains QoS requirements.
2. Scheduling phase where the cluster heads broadcast the scheduling messages to the active nodes. These messages contain the channel and time slot for each node. The scheduling is based on the QoS requirements and the priority. Multimedia is assigned to the highest priority.
3. Transmission phase where an active node starts to send its data on an assigned channel to the cluster head and receive acknowledgement messages (ACK), then the cluster head classifies traffic based on its priority, where video traffic is assigned the highest priority. Finally the cluster head schedules for sending the data to the sink.

A multichannel cross-layer architecture produced by Çevik and Zaim (2013) combined three layers: transport, network, and MAC. The architecture maximizes the network lifetime by fairly distributing the load among all nodes. The incoming packets are classified into three types. The routing packet request reserves the sources; this packet has the highest priority. The real-time packet for an unusual event has less priority than the previous one. The non-real-time packet for delay-tolerant data has the least priority. The architecture contains three schedulers: scheduler 1 for classifying the incoming packets to the appropriate queues, scheduler 3 for the queues that contain real-time packets, and it pulls the packets in a round robin fashion; and scheduler 2 for the request queue first, then for non-real-time packets. In the WMSN application such as surveillance when an unusual event is detected, then a stream of multimedia data should be transmitted to the sink. If this stream is transmitted via a single path, then the energy of the nodes that rely on this path is depleted. To balance the load, the original stream is segmented into flows. The number of flows is defined depending on the number of paths available that satisfy the QoS required. The multichannel structure is used by dividing the bandwidth into separate channels. One channel is dedicated for non-real-time data and control messages; the remaining channels are for real-time data. First, create a route request message for each flow, then send this request via the control channel, using the routing algorithm to select the next hop. Reserve the resource when constructing the path. The path is discovered using a modified ad hoc on-demand distance vector (AODV). The next hop is defined based on load balanced routing algorithm, where the hop count is considered as the QoS parameter.

7.3.2 Multipath Routing

In multipath routing, there are different paths from the source node to the sink (Shu et al. 2010; Sun et al. 2011). Protocols combine transport and network layers (Shah, Liang, and Shen 2010; Rosário et al. 2012; Shah, Liang, and Akan 2012; Rosário et al. 2013); combine the application, network, and MAC layers (Guo, Sun, and Wang 2012; Bae, Lee, and Park 2013); or combine the application and network layers.

An earlier protocol that combined the cross-layer concept with context awareness was proposed by Shu et al. (2010). The aim of this protocol was to maximize the gathering of important information instead of maximizing throughput. An example of important information or information value based on the application is in an ocean monitor, where the sound information is more important than video. But in a fire monitor the video is more important. The protocol splits video streams from audio streams. The protocol is called Multipath Multipriority (MPMP) transmission. At the network layer, Two-Phase Geographic Greedy Forwarding (TPGF) is used to discover the large number of paths from the source node to the base node and the delay for each path. Then at the transport layer a Context Aware Multipath Selection (CAMS) algorithm is used to select the large number of disjointed paths. There is no limit to the number of selected paths as in Sun et al. (2011). The most important stream is assigned the highest priority and to the best routing path that guarantees minimum end-to-end delay.

A Minimum Hop Disjoint Multipath routing algorithm with Time Slice (MHDMwTS) load-balancing congestion control scheme was produced by Sun et al. (2011). They used a minimum hop to reduce the delay and increase the reliability in WMSNs. A Minimum Hop Disjoint Multipath (MHDM) routing algorithm is built of three disjoint paths: primary path, alternate path, and backup path. This algorithm has two phases: path build up and path acknowledgment. In path build up, when the source sensor node is activated, it sends a request package to build a path to the neighbors that have a hop count smaller than the source. The primary path is the first package to reach the sink; the primary path has the least delay. The packages continue to come to the sink from different routes. Every new package is compared with the primary path and discarded if there is a joint node; else the alternate path is built up. At the path acknowledgment phase, the sink sends to the source an ACK. The ACK contains path and time information. To balance the multiple path load, a time slice load balancing congestion control scheme is used. The sink node allocates time, called time slice, for the path. The primary path should take more time than others. After the source node receives the ACK from the sink, it starts to transmit the data. Each path takes a time slice. If it is up, then the sink switches the transmit data to another one.

The MHDMwTS protocol reduces the end-to-end delay, and controls and prevents the congestion. Other protocols (Shah, Liang, and Shen 2010; Shah, Liang, and Akan 2012) consider constraints such as bandwidth, end-to-end delay, and reliability. This architecture consist of multiple components: The traffic classifier module classifies the types of frames, and the application layer encapsulates frame type, frame priority, and group of pictures size (GOP size) to the header and sends the frame to the route classifier module. Then the route classifier module finds three disjoint paths to reach the QoS requirements used in the multipath routing algorithm. The source increases the GOP size when no available bandwidth can provide the bandwidth required. The MAC layer prioritizes scheduling to access the medium.

Two protocols for heterogeneous networks—MEVI and video-aware MMtransmission—have camera nodes (CNs) as the cluster heads and sensor nodes (SNs) as the members (Rosário et al. 2012, 2013). At the network layer, multihop hierarchical routing is used: the intracluster communication between the members and its cluster head (CH) follow the TDMA schedule. And the intercluster communications between CHs and the sink create disjointed path routes to the sink. The routes are classified based on residual energy, hop count, and link quality. At the application layer, the frames are classified as I-frames, P-frames, and B-frames, where I is the high priority frame. High priority frames assigned to best priority paths. The aims of the video-aware MMtransmission protocol are to balance the load and enhance the video quality.

Bae, Lee, and Park (2013) developed a cross-layer QoS architecture. At application layer, a packet-marking algorithm marks each traffic based on its priority. At the network layer, a multipath algorithm classifies the packets into different colors, such as green, red, and yellow, in order to distribute the packets into different paths. The shortest path is assigned to the green packets that require high-quality transmission, assigned to the alternating path are the red packets that consider the level of energy, and yellow packets are assigned to the path considering the level of quality and distance to the sink. Moreover the authors suggest a routing table to keep only information about the nodes in the path to the sink. In order to avoid communication overhead, the architecture stores the important information from each layer at a shared database.

A Cross-Layer and Multipath-Based Video Transmission scheme (CMVT) protocol was produced by Guo, Sun, and Wang (2012). At the application layer MPEG-4 is used (Zhang and Ding 2010), where the video is encoded and the frame marked with the video type. At the network layer, two main components, a route discovery to discover all possible routes from source to the sink, and data transmission where evaluate all the paths and categorize them into three categories (I-frame, P-frame, and B-frame) based on evaluate value, assigned the frame to the path based on frame type and priority, where higher priority frame is assigned to best path. Finally transmit data via selected path transmission.

A Cross-Layer-Based Clustered Multipath Routing protocol (CMRP) was proposed by Almalkawi, Guerrero Zapata, and Al-Karaki (2012) for heterogeneous networks. CMRP combines the network and MAC layers. CMRP defines two thresholds: upper and lower thresholds. The upper threshold selects the cluster heads of the first level and node members, then establishes a link between all cluster heads used in the lower thresholds. The second-level cluster head will select toward the sink, and create a multipath routing. These paths are sorted based on different criteria such as delay, hop counts, bandwidth, and link quality. The best paths are reserved for multimedia data, and the other paths for other data types. The MAC layer uses TDMA with time slot and gives high priority to multimedia.

7.3.3 Single-Path Routing

Single-path routing discovers a single path from the source node to the sink (Hamid and Bashir 2013). Different proposed protocols with different layers combined, for example, between the network and MAC layers (Hamid, Bashir, and Pyun 2012; Hamid and Bashir 2013; Demir, Demiray, and Baydere 2014; Hamid, Hussain, and Pyun 2015), and between the application and MAC layers (Zhang and Ding 2010; Ozen et al. 2014).

Two protocols proposed by Hamid and Bashir (2013) and Hamid, Bashir, and Pyun (2012), namely, XL-WMSN and CR-WMSN, decrease the end-to-end packet latency and increase the throughput of multimedia traffic. But neither enhance the network lifetime. Both protocols consist of multiple components working together to meet the QoS that is required by multimedia applications. They use an admission control scheme that eliminates nodes with less remaining energy during route discovery, and permits only nodes with enough energy into the routing path. To select the appropriate path from source to destination with least delay, Hamid and colleagues (Hamid et al. 2012; Hamid and Bashir 2013) used average packet service time (PSTavg), which is a sum of all possible delays, such as queuing delay, network and MAC layer delay, and transmission delay, to provide information about the node load. Both protocols used channel utilization "Utili," which is a proposed mechanism to regularly check the channel to give an indication of local contention. Both also used a reactive approach to establish the path and used hop count to avoid a long path. There are restrictions on the path length to be close to the shortest path (Hamid, Bashir, and Pyun 2012). Incoming traffic is classified into three classes based on its priority, where video traffic is given the highest priority (Hamid, Bashir, and Pyun 2012). Static duty cycle nodes that have a fixed interval of sleep and awake to listen to the medium is energy consuming and not suitable for WMSN. Hamid et al. (2012) produced dynamic duty cycle assignment (DCA) based on traffic type, where nodes with mainly video traffic expected to have longer duty cycle and consume more energy than nodes with other type of classes. An extended CR-WMSN protocol named CUDAR was proposed by Hamid, Hussain, and Pyun (2015). The CUDAR aims are to guarantee a high throughput, and low delay, jitter, and setup time. At the application layer delay and channel aware routing are used, and at the MAC layer a modified CSMA/CA is used.

A cross-layer QoS architecture (QoSMOS) that developed a cross-layer communication protocol (XLCP) was proposed by Demir, Demiray, and Baydere (2014). This protocol reduces delay and enhances throughput. XLCP contains different elements: a classifier that classifies the packets and mark packets that require QoS, then uses scheduler to broadcast a request to send investigation (RTS-I), and sensor nodes that have enough batteries and are close to the sink node will reply and investigate the route. A cross-layer optimization method has been proposed by Zhang and Ding (2010) to map algorithms that use MPEG-4 standard to define the video frames. It classifies the video frame to three types: the I-frame is the most important type and has highest priority, the P-frame is less important and has less priority, and the B-frame is the least important and has the least priority among them. When needed, the medium IEEE 802.11s is used. IEEE 802.11s uses Enhanced Distributed Channel Access (EDCA) to support QoS. EDCA categorizes the accessing medium into four access categories (ACs) based on the traffic type: background (ACO), best effort traffic (AC1), video (AC2), and audio (AC3). The mapping algorithm maps the packets to access categories based on traffic load using a threshold.

A cross-layer QoS architecture was produced to provide QoS for urgent real-time traffic for emergency situations (Ozen et al. 2014). A two-tiered service differentiation mechanism (TTSDM) is used at the MAC layer to classify the traffic into urgent and nonurgent traffic. Then each one of these is classified to real time traffic (for multimedia traffic), non-real-time traffic, and best effort traffic. There is a predefined threshold. If the urgent real-time queue exceeds the threshold, then the MAC layer interacts with the application layer to lower the data rate, otherwise the MAC layer interacts with the application layer to increase the data rate. The data rate adjustment scheme (DRAS) was used at the application layer to control the data rate.

Paniga et al. (2011) presented a cross-layer architecture that combines the application and transport layers. At the application layer, they used a hybrid DPCM/DCT coding algorithm. It is a predictive compression scheme that provides acceptable compression and low complexity. Using

I-frames with high priority and P-frames in predictive coding, in case of loss frame, only frames with high priority are retransmitted. At the transport layer, a congestion control mechanism is used with two thresholds—stop threshold and restart threshold—at the buffer. This protocol used a static routing at the network layer, and the IEEE 802.15.4 CSMA protocol with clear channel assessment (CCA) at the MAC layer.

A cross-layer solution to maximize the network lifetime by dividing the network lifetime problem into subproblems in different layers, where each layer solved a part of this problem was proposed (You and Liu 2011). At the application layer, pairwise distributed source coding (Pairwise DSC) collects information about neighbor's nodes. To avoid redundancy, Slepian-Wolf distributed source coding (DSC) is used. The transport layer uses source rate adaptation and the network layer solves the routing problem.

An integration between the transport and MAC layers was produced by Tiglao and Grilo (2012), where two mechanisms are provided. Using a negative acknowledgment (NACK)-based repair mechanism, the receivers and intermediate node have the ability to detect loss packets and send back a repair negative acknowledgment (RNACK). If the intermediate nodes have the lost packets in their cache, then they are retransmitted to the destination, or else propagated RNACK to the source.

A WMSN Congestion Control Protocol (WCCP) was proposed by Aghdam et al. (2014) to control congestion and receive high-quality video. WCCP provides interaction between the application, transport, and MAC layers. The application layer defines frame type and packet number. In the transport layer, the WCCP is a two-part protocol with a Source Congestion Avoidance Protocol (SCAP) at the source nodes and a Receiver Congestion Control Protocol (RCCP) in the intermediate nodes. SCAP used a GOP size prediction method to predict the congestion. In case of congestion, the less important frame is dropped and the I-frame is kept to improve the video quality. RCCP detects or predicts the congestion and sends notification if congestion is detected. WCCP controls the congestion and receives high quality video. At the MAC layer, the IEEE 802.15.4 protocol is used to transmit the data.

Sridevi and Usha (2013) proposed a cross-layer framework for heterogeneous WMSNs. This framework provides interaction between the MAC, network, and transport layers, and classifies the traffic into different classes with different priority. The sensor nodes are clustered to different clusters, and each cluster has a cluster head. When the source node needs to send data, it sends the data toward the cluster head using the TDMA schedule. The authors used a dynamic priority for sensitive data and allocated more slots for sensitive data. A congestion detection mechanism was used based on the total number of packets in the queues (Qcw). If the Qcw is less than the threshold, then the network is loaded normally. The authors did not provide a simulation for this framework.

Traditional geographic routing algorithms select the next node based on the short distance between node and destination and ignore the link quality. A cross-layer QoS protocol proposed by Duan, Liu, and Zhang (2015) joined the transport and network layers. The authors produced geographic routing metrics based on link quality and shortest distance to the destination. Link quality was evaluated by the number of Hello packets received by the node from its neighbors. Bit error rate (BER), payload length, and a wireless environment are parameters that affect the packet loss rate. In the traditional algorithm the payload length is fixed, leading to poor video quality because of the dependability between the BER and the packet loss rate. The authors' scheme changed the payload length depending on the transmission quality feedback and hop count. This broke the dependability between the BER and the packet loss rate, and maintained a low packet loss rate when the BER is increased. The authors also encoded the video packet using the short-length LT code technique before transmitting the video packets. In case the packets are lost, the encoded packets can recover the data. This protocol reduces the decoding overhead and the packet loss rate, and enhances the video quality and the network efficiency. A summary of all the discussed cross-layer protocols for WMSNs is shown in Table 7.1.

TABLE 7.1
A Summary of All Discussed WMSNs Cross Layer Protocols

Protocols	Aims	Cluster	Classify Traffic	Energy-Aware	Route Discovery	Joint Layers	Application	Transport	Network	MAC
Multichannel Routing										
Ekbatani Fard, Yaghmaee, and Monsefi (2009)	Energy efficiency, throughput, and data reliability	Yes	Yes	—	—	Network and MAC layers	—	—	Multichannel	Modified 802.11
Ekbatani Fard, Yaghmaee, and Monsefi (2010)										
Çevik and Zaim (2013)	Improved energy efficiency and delay	—	Yes	Yes	—	Transport, network, and MAC layers		Segmentation of the stream into flows	AODV-based routing algorithm	Three schedulers
Multipath Routing										
MPMP	Maximum valuable information	No	Split the video and audio stream	No	—	Transport and network layers	—	CAMS	TPGF	—
MHDMwTS	Reduce delay, prevent congestion	No	No	Yes	—		—	Congestion control scheme	MHDM	—
Shah, Liang, and Shen (2010)	Maximize number of video sources	No	Yes	—	—	Application, network, and MAC layers	Distributed source coding	—	SDMR	802.11e multirate transmission mode
Shah, Liang, and Akan (2012)		No	Yes	—	—			—		
MEVI	Enhance network lifetime, scalability, and reliability	Yes	Yes	Yes	—			—	Multihop hierarchical routing protocol	TDMA
Video-aware MMtransmission	Balance the load and enhance the video quality	Yes	Yes	Yes	—			—		

(Continued)

TABLE 7.1 (*Continued*)
A Summary of All Discussed WMSNs Cross Layer Protocols

Protocols	Aims	Cluster	Classify Traffic	Energy-Aware	Route Discovery	Joint Layers	Application	Transport	Network	MAC
Bae, Lee, and Park (2013)	Improve the transition rate, packet loss, and end-to-end delay	No	Yes	—	Modified routing table	Application and network layers	Packet-marking algorithm	—	Multipath algorithm	—
CMVT	Increase network lifetime, enhance the video quality, and reduce channel conflict	No	Yes	Yes	—	Application and network layers	MPEG-4	—	Route discovery and data transmission function	—
CMRP	Enhance the reliability, throughput, and energy efficient	Yes	Yes	—	—	Network and MAC layers	—	—	Cluster-based multipath routing	TDMA with time slot
Single-Path Routing										
CR-WMSN	Minimize end-to-end delay	No	No	Yes	Reactive approach	Network and MAC layers	—	—	Delay and channel aware routing	Modified 802.11
XL-WMSN			Yes							DCA
CUDAR	High throughput, and low delay, jitter, and setup time		—							Modified CSMA/CA
QoSMOS	Reduce delay, enhance throughput and reliability		Yes	Yes	—				Geographic routing mechanism based on location awareness	CSMA/CA-like
Zhang and Ding (2010)	Forward video sequences	No	Yes	—	—	Application and MAC layers	MPEG-4 standard	—	—	IEEE 802.11s
Ozen et al. (2014)	Lower delay for emergency situation	Yes	Yes	—	—		DRAS	—	—	TTSDM

(*Continued*)

TABLE 7.1 (*Continued*)
A Summary of All Discussed WMSNs Cross Layer Protocols

Protocols	Aims	Cluster	Classify Traffic	Energy-Aware	Route Discovery	Joint Layers	Application	Transport	Network	MAC
Paniga et al. (2011)	Effective multihop streaming video	No	Yes	—		Application and transport layers	DPCM/DCT coding scheme	Congestion control mechanism	Static routing	IEEE 802.15.4 CSMA with CCA
You and Liu (2011)	Maximize the network lifetime	No	No	—	—	Application, transport, and network layers	Pairwise DSC	Source rate adaptation	Overall link rate control	—
Tiglao and Grilo (2012)	Improved performance and energy efficient	No	No	—	—	Transport and MAC layers	Efficient	NACK-based repair mechanism	Efficient	Adaptive retransmission mechanism
WCCP	Control the congestion and receive high-quality video	No	Yes	—	—	Application, transport, and MAC layers	Distributed source coding	SCAP and RCAP	Efficient	IEEE 802.15.4
Sridevi and Usha (2013)	Reduce delay	Yes	Yes	—	—	Transport, network, and MAC layers	—	Congestion detection scheme	Hierarchical routing protocols	TDMA slot assignment
Duan, Liu, and Zhang (2015)	Reduce the decoding overhead and the packet loss rate, and enhance the video quality and the network efficiency	No	No	No	Geographic routing	Transport and network layers	—	Short-length Luby transform (LT)	Geographic routing	—

7.4 CONCLUSION

The advanced technology in low-power circuits, inexpensive CMOS cameras, and microphones gave birth to WMSNs. WMSNs are useful in many applications, especially in surveillance to track objects and take appropriate actions, traffic avoidance and control systems to monitor the traffic to avoid congestion, advanced health care delivery, and smart homes. All these applications are required to transmit video, audio, or even images. The transmitted multimedia data requires constraints on QoS that is defined by the application. Architecture such as cross layer where boundaries between layers are eliminating this, allowing exchange parameters between layers more efficiently to increase the performance. A cross-layer protocol will satisfy the QoS required for multimedia transmission in WMSNs. All proposed cross-layer protocols for WMSNs have different objectives and different approaches to reach these objectives, but still need more effort to satisfy QoS requirements for multimedia with the capability of WMSNs nodes. WMSNs that guarantee no delay and packet delivery are very important to avoid distortion in received multimedia data. All possible delays such as queuing delay and transmission delay must be avoided. The delay also occurs in many layers, for that control the congestion at transport layer, select the shortest path during routing at the network layer, schedule the packets at the MAC layer, and give high priority for packets that are marked as multimedia packets. Sensor nodes with a cache to store data temporarily in case of lost packets, can retrieve the loss packets from the cache. Consider the residual nodes energy during route discovery to avoid path disconnection. Use a reactive approach for routing or a modified routing table, where the table contains only information for nodes on the path.

REFERENCES

Aghdam, S. M., M. Khansari, H. R. Rabiee, and M. Salehi. 2014. WCCP: A congestion control protocol for wireless multimedia communication in sensor networks. *Ad Hoc Networks* 13:516–534.

Akyildiz, I. F., T. Melodia, and K. R. Chowdhury. 2007. A survey on wireless multimedia sensor networks. *Computer Networks* 51(4):921–960.

Akyildiz, I. F., T. Melodia, and K. R. Chowdhury. 2008. Wireless multimedia sensor networks: Applications and testbeds. *Proceedings of the IEEE* 96(10):1588–1605.

Akyildiz, I. F. and M. C. Vuran. 2010. *Wireless Sensor Networks*, Vol. 4. Chichester, England: John Wiley & Sons.

Almalkawi, I. T., M. G. Zapata, and J. N. Al-Karaki. 2012. A cross-layer-based clustered multipath routing with QoS-aware scheduling for wireless multimedia sensor networks. *International Journal of Distributed Sensor Networks* 8(5): article 392515.

Al Nuaimi, M., F. Sallabi, and K. Shuaib. 2011. A survey of wireless multimedia sensor networks challenges and solutions. *2011 International Conference on Innovations in Information Technology (IIT)*. Abu Dhabi, UAE, April 25–27, 2011, pp. 191–196.

Bae, S.-Y., S.-K. Lee, and K.-W. Park. 2013. Cross-layer QoS architecture with multipath routing in wireless multimedia sensor networks. *International Journal of Smart Home* 7(3):219–226.

Çevik, T. and A. H. Zaim. 2013. A multichannel cross-layer architecture for multimedia sensor networks. *International Journal of Distributed Sensor Networks* 2013: article 457045.

Demir, A. K., H. E. Demiray, and S. Baydere. 2014. QoSMOS: Cross-layer QoS architecture for wireless multimedia sensor networks. *Wireless Networks* 20(4):655–670.

Duan, P., L. Liu, and Z. Zhang. 2015. A cross layer video transmission scheme combining geographic routing and short-length Luby transform codes. *International Journal of Distributed Sensor Networks* 11(8): article 140212.

Ekbatani Fard, G. H., M. Yaghmaee, and R. Monsefi. 2009. An adaptive cross-layer multichannel QoS-MAC protocol for cluster based wireless multimedia sensor networks. *International Conference on Ultra Modern Telecommunications & Workshops, 2009*. St. Petersburg, Russia, Oct. 12–14, 2009, pp. 1–6.

Ekbatani Fard, G. H., M. H. Yaghmaee, and R. Monsefi. 2010. A QoS-based multichannel MAC protocol for two-tiered wireless multimedia sensor networks. *International Journal of Communications, Network and System Sciences* 3(7):625–630.

Farooq, M. O., M. St-Hilaire, and T. Kunz. 2012. Cross-layer architecture for QoS provisioning in wireless multimedia sensor networks. *KSII Transactions on Internet and Information Systems (TIIS)* 6(1):176–200.

Guo, J., L. Sun, and R. Wang. 2012. A cross-layer and multipath based video transmission scheme for wireless multimedia sensor networks. *Journal of Networks* 7(9):1334–1340.

Hamid, Z. and F. Bashir. 2013. XL-WMSN: Cross-layer quality of service protocol for wireless multimedia sensor networks. *EURASIP Journal on Wireless Communications and Networking* 2013(1):1–16.

Hamid, Z., F. Bashir, and J. Y. Pyun. 2012. Cross-layer QoS routing protocol for multimedia communications in sensor networks. *2012 Fourth International Conference on Ubiquitous and Future Networks (ICUFN)*. Puket, Thailand, July 4–6, 2012, pp. 498–502.

Hamid, Z. and F. B. Hussain. 2014. QoS in wireless multimedia sensor networks: A layered and cross-layered approach. *Wireless Personal Communications* 75(1):729–757.

Hamid, Z., F. B. Hussain, and J.-Y. Pyun. 2015. Delay and link utilization aware routing protocol for wireless multimedia sensor networks. *Multimedia Tools and Applications* 75(14):8195–8216.

Karl, H. and A. Willig. 2007. *Protocols and Architectures for Wireless Sensor Networks*. Chichester, England: John Wiley & Sons.

Melodia, T. and I. F. Akyildiz. 2011. Research challenges for wireless multimedia sensor networks. In: *Distributed Video Sensor Networks*, edited by B. Bhanu, C. V. Ravishankar, A. K. Roy-Chowdhury, H. Aghajan, and D. Terzopoulos, pp. 233–246. London: Springer Science & Business Media.

Mendes, L. D. P. and J. J. P. C. Rodrigues. 2011. A survey on cross-layer solutions for wireless sensor networks. *Journal of Network and Computer Applications* 34(2):523–534.

Ozen, Y., C. Bayilmis, N. Bandirmali, and I. Erturk. 2014. Two tiered service differentiation and data rate adjustment scheme for WMSNs cross layer MAC. *2014 11th International Conference on Electronics, Computer and Computation (ICECCO)*. Abuja, Nigeria, Sept. 29–Oct. 1, 2014, pp. 1–4.

Paniga, S., L. Borsani, A. Redondi, M. Tagliasacchi, and M. Cesana. 2011. Experimental evaluation of a video streaming system for wireless multimedia sensor networks. *The 10th IFIP Annual Mediterranean Ad Hoc Networking Workshop (Med-Hoc-Net)*. Favignana Island, Sicily, Italy, June 12–15, 2011, pp. 165–170.

Rosário, D., R. Costa, H. Paraense, K. Machado, E. Cerqueira, T. Braun, and Z. Zhao. 2012. A hierarchical multi-hop multimedia routing protocol for wireless multimedia sensor networks. *Network Protocols and Algorithms* 4(4):44–64.

Rosário, D., R. Costa, A. Santos, T. Braun, and E. Cerqueira. 2013. QoE-aware multiple path video transmission for wireless multimedia sensor networks. *Simpósio Brasileiro de Redes de Computadores e Sistemas Distribuídos—SBRC*, pp. 31–44. Brasilia, Brazil.

Shah, G., W. Liang, and O. B. Akan. 2012. Cross-layer framework for QoS support in wireless multimedia sensor networks. *IEEE Transactions on Multimedia*, 14(5):1442–1455.

Shah, G., W. Liang, and X. Shen. 2010. Cross-layer design for QoS support in wireless multimedia sensor networks. In: *Global Telecommunications Conference (GLOBECOM 2010)*. IEEE Globecom 2010 Proceedings, Florida, USA, Dec. 6–10, 2010.

Shu, L., Y. Zhang, Z. Yu, L. T. Yang, M. Hauswirth, and N. Xiong. 2010. Context-aware cross-layer optimized video streaming in wireless multimedia sensor networks. *The Journal of Supercomputing* 54(1):94–121.

Sridevi, S. and M. Usha. 2013. Towards a cross layer framework for improving the QoS of delay sensitive heterogeneous WMSNs. *2013 Fourth International Conference on Computing, Communications and Networking Technologies (ICCCNT)*. Tiruchengode, India, July 4–6, 2013, pp. 1–5.

Sun, G., J. Qi, Z. Zang, and Q. Xu. 2011. A reliable multipath routing algorithm with related congestion control scheme in wireless multimedia sensor networks. *2011 3rd International Conference on Computer Research and Development (ICCRD)*. Shanghai, China, Mar. 11–Mar. 13, 2011, pp. 229–233.

Tiglao, N. M. C. and A. M. Grilo. 2012. Cross-layer caching based optimization for wireless multimedia sensor networks. *2012 IEEE 8th International Conference on Wireless and Mobile Computing, Networking and Communications (WiMob)*. Barcelona, Spain, Oct. 8–10, 2012, 697–704.

Wang, H., W. Wang, S. Wu, and K. Hua. 2010. A survey on the cross-layer design for wireless multimedia sensor networks. In: *Mobile Wireless Middleware, Operating Systems, and Applications* edited by Y. Cai, T. Magedanz, M. Li, J. Xia, and C. Giannelli, pp. 474–486. Berlin: Springer.

You, L. and C. Liu. 2011. Robust cross-layer design of wireless multimedia sensor networks with correlation and uncertainty. *Journal of Networks* 6(7):1009–1016.

Zhang, J. and J. Ding. 2010. Cross-layer optimization for video streaming over wireless multimedia sensor networks. *2010 International Conference on Computer Application and System Modeling (ICCASM)*. Taiyuan, China, Oct. 22–24, 2010, V4-295–V4-298.

8 Survey of E-Learning Standards

Manal Abdullah and Nashwa AbdelAziz Ali

CONTENTS

8.1 Introduction 110
8.2 Categories of E-Learning Standards 111
8.2.1 Types of E-Learning Standards 112
8.2.2 Domains of E-Learning Standards 112
8.2.3 Entities of E-Learning Standards 113
8.3 Standards for E-Learning Styles 113
8.3.1 Kolb Learning Style Indicator 113
8.3.2 Fleming VAK Model 113
8.3.3 Myers Briggs Type Indicator (MBTI) 113
8.3.4 Felder-Silverman Index of Learning Styles 114
8.3.5 Grasha-Riechmann Student Learning Style Scale 114
8.4 E-Learning Personalization Standards 114
8.4.1 Personalized E-Learning Elements 114
8.4.2 Personalized E-Learning Goals 115
8.4.3 Personalized E-Learning Methods 115
8.5 E-Learning Adaptation Standards 116
8.5.1 Adaptation-Oriented Domain Modeling 116
8.5.2 Learner and Group Modeling 116
8.5.3 Adaptation Modeling 116
8.5.4 Standardization of Adaptation Components and Service Levels 117
8.6 E-Learning Management Systems Standards 117
8.6.1 SCORM Standard 118
8.6.2 Metadata and Interoperability Standards 118
8.6.3 T-SCORM Standard 118
8.6.4 IEEE Learning Technology Standards Committee Standard 119
8.6.5 IMS Standard 119
8.6.6 Learning Object Management (LOM) 119
8.6.7 E-Learning Quality Standards 119
8.6.7.1 ISO/IEC 19796-1 120
8.6.7.2 ISO 9126 120
8.7 E-Learning Communities Standards 120
8.7.1 IMS Global Learning Consortium, Inc. (IMS) 120
8.7.2 Aviation Industry Computer-Based Training Committee (AICC) 120
8.7.3 Dublin Core Metadata Initiative (DCMI) 121
8.7.4 Ariadne Foundation 121
8.7.5 Advanced Distributed Learning (ADL) Initiative 121
8.8 Conclusion 122
References 122

8.1 INTRODUCTION

This chapter recognizes definitions of e-learning standards according to many aspects, including personalized e-learning, learning styles standards, and learning adaptation standards. Next the chapter covers e-learning management systems and e-learning qualities standards, and finally the most popular organizations that certify these specifications.

For the purposes of this chapter, e-learning emphasizes the necessity of maximizing Internet and its fields. The process of standardization is experimental, incomplete, and more rapidly evolving. Standards, on the other hand, are much more conclusive, complete, and evolve slowly. E-learning standards are rules, guidelines, or definitions of characteristics, to ensure that materials, products, processes, and services are fitted for their purpose. In the context of e-learning technology, standards are generally developed to be used in systems, design, and implementation for the purposes of ensuring interoperability and reusability. These attributes should apply to both the systems themselves, and the content and metadata they manage. In addition, these standards should capture general acceptance, can serve regulatory purposes, and be used to achieve e-learning efficiency and quality outcomes. One of the main benefits of standardized e-learning is to establish the functionality of interoperability via multiple systems and reuse of learning objects (interoperability of resources). Interoperability and developing efficiency can be considered the two main purposes of standardization of e-learning (Friesen, 2005).

There are requirements that should be considered to achieve these purposes:

- Accessibility—The learning content should be available anywhere in the world, not just on the local network.
- Interoperability—The learning content should work on all conformant platforms, browsers, and learning management systems (LMSs) (not just a handful of products).
- Durability—Components developed in current versions of the reference model should work in later versions without people having to redesign or recode content.
- Reusability—Content can be used not just one time for one course, but whenever it is needed. No need for special codes or links to allow content into a specific course or lesson.
- Adaptability—Adaptability means simply, ways to label content to match learner preferences or skill levels. For a long-term goal, learning progress or preferences are the basis to configure the ability of learning content.
- Affordability—Matching all previous goals will reduce costs for e-learning content and make quality learning widely available at significantly lower costs.

From the literature review, standardization differentiates between common types of standards and specifications as follows:

- Formal standards—Also known as "de jure standards." The International Organization for Standardization (ISO) and International Electrotechnical Commission (IEC) are developed in agreement processes by the official standardization organizations.
- Community specifications—Community specifications are developed by communities or forums. They are open specifications available to public. The Institute of Electrical and Electronics Engineers (IEEE) and the World Wide Web Consortium (W3C) are e-learning communities.
- Industrial specifications—Industrial specifications are developed by closed or open specifications and are available for branches of industrial consortia.

Organizational specifications are internally closed specifications developed by organizational, industrial, or community specifications with global agreement. Microsoft Windows is an example of proprietary organizational, industrial, or community standards.

E-learning standards have different levels of users and providers to address and support their needs, interests, and preferences.

The major advantages of developing e-learning standards are (Ileana, 2007)

- Durability—No need to modify standard when new system version is issued
- Interoperability—Operates across various hardware, operating systems, web browsers, and multiple learning management systems
- Accessibility—Indexing and monitoring
- Reusability—Able to be used by different development tools

E-learning standards can be viewed from two main viewpoints. First, is to use standards as a mean for creating adaptive learning scenarios. These can be imported and run in the LMS to learners with individualized learning experiences. However, the second view supports standards that concern information on the experiences and state of learning of each learner. This vision supports the use of SCORM and xAPI as a provider of information on the learning progress of each learner (Ileana, 2007).

The rest of this chapter is organized as follows: Section 8.2 covers the categories of e-learning standards. Section 8.3 describes the various learning styles. In Section 8.4, the research gives an overview of adaptation e-learning. In the next section the research browses the highly cited quality standards, then lists organizations that issue e-learning standards. Last, the chapter presents e-learning trends.

8.2 CATEGORIES OF E-LEARNING STANDARDS

Categories of e-learning standards may support multiple concepts such as learning content and learning objects, processes, and business units of organization.

The most important purpose of learning standards can be divided into five categories as following (Mohamed, 2015):

1. Metadata—Learning content and catalogs must be labeled in a consistent way to support the indexing, storage, discovery (search), and retrieval of learning objects by multiple tools across multiple repositories. Several initiatives are creating metadata standards: The Learning Object Metadata, or LOM of IEEE Learning Technology Standards, and the Dublin Core Metadata.
2. Content packaging—Content packaging specifications and standards allow courses to be transported from one learning system to another. The initiatives dealing with content packaging include the IMS Global Learning Consortium (IMS) Content Packaging specification, the IMS Simple Sequencing specification, and the Advanced Distributed Learning (ADL) Sharable Content Object Reference Model (SCORM).
3. Learner profile—Learner profile information can include personal data, learning plans, learning history, accessibility requirements, certifications and degrees, assessments of knowledge, and the status of participation in current learning. The most important effort to standardize learner profile information is the IMS Learner Information Package (LIP) specification.
4. Learner registration—Learner registration information allows learning delivery and administration components to know what offerings should be made available to a learner and provides information about learning participants to the delivery environment. There are two initiatives currently dealing with these requirements: the IMS Enterprise Specification and the Schools Interoperability Framework, which supports the exchange of this type of data in a K–9 environment.

5. Content communication—When content is launched, there is a need to communicate learner data and previous activity information to the content. Work going on is the ADL's Sharable Content Object Reference Model (SCORM) project based on the CMI specification of the Aviation Industry Computer-Based Training Committee (AICC).

Categorizing e-learning standards can be identified by three main dimensions of standards: types, domains, and entities. In the following subsections the three dimensions are introduced (Qazdar et al. 2015).

8.2.1 Types of E-Learning Standards

Three types of e-learning standards can be identified. These attributes are the main functions of e-learning standardization:

1. Implementation standards—These types support the functionality of interoperability within all e-learning domains. Metadata, architecture, infrastructure, and interface standards are examples of implementation standards.
2. Conceptual standards—These types of standards support the functionality of quality development by providing reference models.
3. Level standards—These types focus on identifying and addressing the quality level of e-learning systems. So, they are usually used for certifications purposes. Figure 8.1 shows the relationship between types and purposes of e-learning standards.

8.2.2 Domains of E-Learning Standards

E-learning standards may cover one or more of the following e-learning domains:

1. Meaning—This domain focuses on the general concepts of understanding and deals with disciplines semiotics, pragmatics, semantics, and so on.
2. Quality—This domain covers all aspects of the quality management including developing, assurance, and deals with results, processes, and potentials.
3. Didactics—This domain focuses on issues concerning pedagogical questions: methods, learners, and learning environments.
4. Learning technology—This domain focuses on all technological solutions that developed for learning objectives and purposes and deals with data exchange, interfaces, and accessibility questions.
5. Learning content—This domain covers all aspects for e-learning objects and deals with resources, aggregation, and packaging.

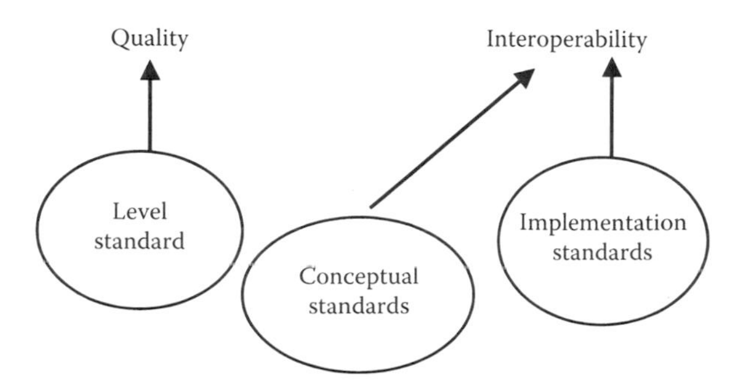

FIGURE 8.1 Relationship between the purpose of e-learning standards and its types.

6. Context—The main purpose of this domain is to include all the disciplines and information regarding e-learning and deals with rights and laws.

8.2.3 Entities of E-Learning Standards

E-learning standards may correspond to more than one entity in combination. The six entities can be addressed as the following:

1. Learning environment—Involves the entire organizational environment including structure, infrastructure, services, and all processes.
2. Roles—Concerns the learner, teacher, and tutor that represent the defined attributes within the e-learning solution.
3. Methods—Concerns the methods used within an e-learning environment.
4. Learning systems—Deals with all technological and conceptual issues including the architecture within the system.
5. Learning resources—Covers all the components of the learning system.
6. Practice—Deals with all information experiences and knowledge that are recognized within the usage of an e-learning offer.

8.3 STANDARDS FOR E-LEARNING STYLES

A learning style refers to characteristics, strengths, and preferences that represent the learner's method of receiving information through the learning process. The following sections described the most cited learning styles (Ileana, 2007).

8.3.1 Kolb Learning Style Indicator

The Kolb Learning Style Indicator is an indicator based on the experiential learning theory, which considers the experience of a learner as an important factor in learning. Therefore, it covers two kinds of experiences, namely, grasping and transforming experiences. Grasping consists of two subcategories, namely, concrete and abstract conceptualization. Similarly, the transforming experience has two subcategories termed reflective observation and active experimentation.

8.3.2 Fleming VAK Model

The VAK learning styles model suggests that most people can be divided into one of three preferred styles of learning, namely, visual, auditory, and kinesthetic.

- The visual learning style involves the use of seen or observed things, including pictures, diagrams, demonstrations, displays, handouts, films, and flip charts.
- The auditory learning style involves the transfer of information through listening to the spoken word of self or others, and sounds and noises.
- Kinesthetic learning involves physical experience: touching, feeling, holding, doing, and practical hands-on experiences.

8.3.3 Myers Briggs Type Indicator (MBTI)

The learning style assessment by the Myers Briggs Type Indicator (MBTI) is resolved using different aspects. The dimensions of this learning style are as follows:

- Sensing judging and perceiving—Attention toward the external world/things or internal world/things.

- Thinking and feeling—Perceiving the world directly or perceiving through impressions/imaging possibilities.
- Sensing and intuition—Learners taking decisions through logic or through mere human values.
- Extroversion and introversion—Learners looking at the world as a structured, planned environment or as a spontaneous environment.

8.3.4 Felder-Silverman Index of Learning Styles

The Felder-Silverman Index of Learning Styles is often used in technology-enhanced learning and designed for traditional learning. Additionally, it defines the learning style of a learner in more detail and distinguishing between preferences on four dimensions as (El-Bakry and Saleh, 2013)

- Active/Reflective—Active learners learn by doing something with information. Reflective learners learn by thinking about information.
- Sensing/Intuitive—Sensing learners favor taking in info that is concrete and practical. Intuitive learners favor taking in info that is original, abstract, and oriented toward theory.
- Visual/Verbal—Visual learners prefer visual presentations of material (pictures, diagrams, flowcharts, and graphs). Verbal learners prefer explanations with words, includes both written and spoken.
- Sequential/Global—Sequential learners prefer to organize information in a linear, orderly fashion. Global learners prefer to organize information more holistically and in a seemingly random manner without seeing connections.

Since most learners fall in the category of either active or reflective from the first dimension, this model is more suitable to evaluate learners in an e-learning environment.

8.3.5 Grasha-Riechmann Student Learning Style Scale

The Grasha-Riechmann Student Learning Style Scale (GRSLSS) focuses on students' interactions among their peers. GRSLSS has a teaching style survey that instructors can complete to see how their instruction matches with learners so they can adapt and modify to meet more learners' needs. The learning styles scale consists of six primary learning styles: avoidant, collaborative, competitive, dependent, independent, and participant. The survey itself consists of 60 items, with 10 questions each. They are averaged together to measure dominance in one or more of the six measured learning styles.

8.4 E-LEARNING PERSONALIZATION STANDARDS

Personalization is wider than just individualization or differentiation in what a learner does with what is learned, and when and how it is learned. Personalized learning is the tailoring of learning environment to meet the needs and preference of individual learners (Patel et al. 2013).

8.4.1 Personalized E-Learning Elements

A variety of elements are involved in personalized e-learning to customize the learning process. Key elements that meet the learner's goals in the personalized e-learning process include the pace of learning; pedagogy; curriculum and instructional approach; and activities that draw upon the student's preferences, skills, and knowledge.

In essence, the personalized e-learning environment enables learners to choose what they want to learn when they want, and even the learning method. This almost leads to improved learning outcomes.

8.4.2 Personalized E-Learning Goals

The main goal to achieve continuing support for every learner at each individual level and monitoring his behavior throughout the learning process is material that is

- Modified to each learner
- Generative rather than precomposed
- Scalable to learning process levels without additional cost

The personalization mechanism is based mainly on feedback. It can be either explicit or implicit, and can be a manually or automatically processed technique. It should be an integral part of the learning system.

8.4.3 Personalized E-Learning Methods

Personalized e-learning focuses on enhancing learning and thinking interaction at the learner's behavioral and physiological level. The knowledge-driven model for personalization e-learning solutions is a part from a sophisticated, stylish multimedia delivery of the learning process.

Another central aspect of enhancing e-learning is an adoption of what is the preferred pace and expertise of the learner. The future direction of e-learning is shifting from a content-oriented approach to a knowledge synthesis approach.

Two main approaches are considered in personalized e-learning methods (Da Silva et al. 2015). First, the knowledge model and concept map approach provides successful personalization of the e-learning process by designing a platform to interact a continuous dialogue between the learner and knowledge resources. Concept maps are used as the graphical representations of knowledge to draw both the learning concepts and its relationships with a human-oriented approach.

In the second approach, tacit knowledge is highly subjective in nature, as it is developed by the learner and based on his cognition and related to context of a specific situation. The tacit (T)-model is used to capture the sequential set of steps that a subject matter expert (SME) would handle to achieve a task or to make a decision. As an important conceptual structure of the T-model, it is identified by using a formal concept analysis (FCA) that is used to restructure and formalize the T-model.

Explicit knowledge is the knowledge that is objective in nature, easily expressed, and shared. Explicit knowledge is modeled in the subject domain master map (M-map), which can be hyperlinked to each knowledge node in the T-map. The M-map is formulated on the concept of learning dependency, which is defined as a dynamic cognitive and pedagogical centered approach for the mapping of a course structure.

Personalization in the e-learning system can be achieved through two levels of personalization. Level 1 allows the personalization of learning contents and structure of the course according to a given personalization strategy and level 2 defines the personalization strategy. The teacher has to choose and apply the personalization strategy that matches the learner's characteristics and specifics of the courses in two steps. First, the teacher selects a subset of personalization parameters for the given courses and then combines the selected personalization parameters and decides how the learning material can be composed with respect to each possible value of the personalization parameters. There are 16 personalization parameters of e-learning scenarios, including information-seeking task, learner's level of knowledge, learning goals, and language preference. Table 8.1 shows personalization parameters and its values according to the Ontology for Selection of Personalization Strategy (OSPS) standard.

TABLE 8.1
Personalization Parameters and Its Values

Personalization Parameter	Values
Information seeking tasks	Tracking activities
Knowledge level	Learner background values are beginner, intermediate, and advanced
Learning goals	Knowledge, comprehension, application
Media preference	Text/image, sound, video, and simulation
Language preference	Represent learning objects in the learner's preferred language
Kolb learning cycle	Grasping, transforming experiences
Felder-Silverman learning	Active/reflective, visual/verbal, sensitive/intuitive, sequential/global
Grasha learning style	Avoidant, collaborative, competitive, dependent, independent, and participant
Participation	Too much, acceptable, not enough
Progress on task	Large, small
Feedback	Significant, medium, low
Motivation level	Components of motivation instructions (attention, relevance, confidence, and satisfaction); set of values: low, moderate, and high
Navigation level	Varying between breadth first and depth first
Cognitive traits	Low, high working memory capacity; low, high inductive reasoning ability; low, high information processing speed; and low, high associative learning skill
Pedagogical approach	Objectivist, competency

8.5 E-LEARNING ADAPTATION STANDARDS

The crucial criterion to providing a responsive learning environment that engages, motivates, and inspires learners, and through this leads to higher learner satisfaction, is adaptation. In the context of this chapter, a learning environment is considered adaptive if it is capable of monitoring the activities of its users; interpreting these on the basis of domain-specific models; inferring user requirements and preferences out of the interpreted activities, appropriately representing these in associated models; and, finally, acting upon the available knowledge on its users and the subject matter at hand, to dynamically facilitate the learning process (Bianco et al. 2004; El Bachari et al. 2011; Fernández et al. 2011). There are numerous adaptive e-learning standards that have adopted various specifications such as IMS1, ADL, SCORM2, and AICC.

8.5.1 Adaptation-Oriented Domain Modeling

Current standards and concepts for educational metadata focus on content-centered approaches and models of instructional design. Standards focus on the search, exchange, and reuse of learning material, often called content items, learning objects, or training components. Examples of Adaptive Learning Environment (ALE) that extend existing standards include OPAL, OLO, and KOD.

8.5.2 Learner and Group Modeling

Learner modeling in existing standards addresses all related specifications to the learner's model, or profile, over time. An example of this type of specification is the IMS Learner Information Package, which incorporates the results of top-level educational activities and other static information about the user (e.g., demographic).

8.5.3 Adaptation Modeling

Two complementary issues of modeling examine the behavior of any adaptive system: the specification of adaptation logic and the specification of adaptation actions (ADL, 2012). The prior is responsible for relating information available in one or more models and assesses whether adaptation is required. The latter refers to specifying the actions needed for a given adaptation to be achieved. The two approaches include simple rule-based engines and case-based reasons. An adaptation logic, adaptation actions are well-researched, especially as far as adaptive hypermedia learning systems are concerned. Furthermore, recent research used an XML language to define and declare adaptation actions. Of the existing standards, the only one that supports the explicit representation of dynamic behavior on behalf of the system is the IMS learning design (LD) specification.

8.5.4 Standardization of Adaptation Components and Service Levels

This type of standard is concerned with utilizing adaptation-oriented component services to/from the "outside world" Personalized Learning System (PLS) and Knowledge Tree. Both of these adaptation techniques are constructed to existing source content and functionality (such as learner management, collaborative tools, and testing services). PLS integrates with the learning management system ADL SCORM based and work within existing courseware. PLS is completely static and achieves adaptation via adaptation services. The Knowledge Tree framework, on the other hand, is designed to facilitate interoperation and reuse at the level of distributed, reusable learning activities. Furthermore, it works on run-time communication and interoperation standards, and standardizes methods to support aspects of the adaptation learning process that can exchange information throughout LMS. Knowledge Tree allows for different kinds of portals; some can be as static as existing content management system (CMS), but the others can be adaptive.

8.6 E-Learning Management Systems Standards

An e-learning management systems is software that records, tracks, and monitors all activities of a learner. In other words, these systems have a user-friendly structure and foundation for users of e-learning, and handle the learning and training process automatically. A powerful and comprehensive learning management system (LMS) provides high performance management of the learning process. It helps learners assess their training and plan their next steps for learning (El Bachari et al. 2011).

All LMSs have their own specifications and properties. Some of these systems aid users throughout the learning process with management tools. Some others enable learners to choose learning objects based on their needs or choose from a list of courses. In addition, these systems supply learners with educational activities in visual form.

LMSs also provide the following functions:

- Structure—Centralization and interoperability by enabling easy and efficient navigation via interfaces.
- Assessment—Creating and administrating assessments and storing assessment data; also includes all functions related to assessment results.
- Tracking—Tracking learning functions into one system.
- Security—Preventing unauthorized access to courses, learner account, and other administrative facilities.
- Registration—Assigning courses and activities to learners or instructors.
- Delivery—On-demand delivery of learning content and experiences to learners.
- Interaction—Learners interact throughout all administrative tools as well as between communicative content and the LMS (i.e., SCORM content).
- Reporting—Extracting and presenting all information about learners and courses.
- Record keeping—Storing and maintaining data about learners including their portfolio.
- Reuse—Searching and composing content for delivery in different learning tracks.
- Personalization—Matching learner preferences with corresponding content.

- Integration—Exchanging learner and content data via multiple systems (i.e., content management systems).
- Administration—Centralized management of all involved functions.

The following sections present the most common standards developed to achieve the preceding functionalities.

8.6.1 SCORM Standard

SCORM is a technical standard that was created and developed by ADL. This standard supports the following keys as high-level requirements: availability, adaptability, economic durability, interoperability, and reusability. In other words, SCORM is a collection of related documents. The three main documents of SCORM handle the following areas and activities (Sivakami and Anna Poorani, 2015):

1. Content aggregation
2. Runtime environment
3. Arrange and conduct

In fact, SCORM is a high-level set of fundamental characteristics and e-learning content standards, technologies and related services. SCORM 2004 introduced a complex idea called sequencing, which is a set of rules that constrain a learner to fix his paths and bookmark learning objects. The standard uses XML to encode a file that describes the components and resources.

8.6.2 Metadata and Interoperability Standards

For economic and efficient benefits, learning organization must be associated and supported by metadata and interoperability standards. Metadata is identification information or information about the main characteristics (El Bachari et al. 2011). In other words, metadata is structured information that describes, explains, locates, and makes information easier to retrieve, use, and manage. Information about the subject, author, title, publisher, edition, and other necessary information in libraries are examples of metadata. MIME is also a part of the standard metadata that is related to information posted on the Internet and informs the receiver about information and the software required to process it.

Interoperability means that standards support different systems or different components or layers of systems. Metadata standards play a role in interoperability standards.

Organizations such as the IEEE Learning Technology Standards Committee (LTSC), IMS, ADL, AICC, and some other European groups develop these standards. In particularly, AICC is working on an independent and unique industry, ADL works on generalizing, and IEEE LTSC looks to create the formal standards.

8.6.3 T-SCORM Standard

For increasing SCORM standard benefits, T-SCORM is an extension of SCORM that improves searching and navigation making learning objects available via iDTV (digital television) platform. One of the benefits of the extension is to enable a system made for T-learning to search information based on learning objects. The T-SCORM extension is an advanced form of metadata information from the SCORM standard based on the learning object metadata (LOM) standard. The addition of new elements should give more emphasis on the metadata information regarding iDTV. These new elements in the LOM structure identify specific information for iDTV (interactivity level, copyright, description on content in digital format, etc.).

8.6.4 IEEE Learning Technology Standards Committee Standard

The IEEE LTSC LOM standard recommends guidelines for educational and training systems, especially software components, tools, and technology solutions that enable development and maintenance. However, it does not support details about implementation of specific technologies. The standard represents a high-level model of e-learning system architecture throughout standardize five categories: general items; learn more about; content; data and metadata; and management systems and applications.

8.6.5 IMS Standard

IMS is a project concerning national infrastructure of higher education in the United States. The project is managed by the union called the EduCAUSE or Educom in forms of hundreds of universities and educational institutions. The project aims to establish standards for dealing with problems associated with the increasing use of new technologies in teaching and learning.

The IMS standard provides the following sections in this regard (Sivakami and Anna Poorani, 2015):

- IMS Learning Resource Meta-Data Specification—Describes learning resources in order to search
- IMS Enterprise Specification—Used for sharing data about learners, courses, and so on
- IMS Content Packaging Specification—Creates and shares content objects and reusable learning content
- IMS Question & Test Specification—Shares test items and other assessment tools
- IMS Learner Information Package Specification—Organizes information so that the learner can learn the system for a specific user and can provide appropriate needed responses
- IMS Reusable Competency Definition Specification—Used for the descriptions, references, and related transactions with key characteristics
- IMS Simple Sequencing Specification—Illustrates how learning is arranged and provided in a specific sequence for a learner
- IMS Accessibility Specification—Provides guidance for other sectors and aims to ensure the availability and ease of use of the standard specification
- IMS Learning Design Specification—Used to introduce the interaction scenarios for the creator of subject and educational courses
- IMS Digital Repositories Specification—Integrates the online learning system with other data sources

8.6.6 Learning Object Management (LOM)

The LOM standard is a formal standardization licensed by the ISO/IEC SC36 with one option being to "fast-track" the standard through a high-level JTC1 committee. As a further result, LOM is subjected as standard issued by the IEEE LTSC and the SC36 subcommittee to work together in the future to develop a "next generation" of this metadata standard. This version is much closer in orientation to the "minimalist" Dublin Core approach to metadata than to the technically demanding "structuralist" approach represented by the LOM.

8.6.7 E-Learning Quality Standards

Quality is a key of learning success in general. The following observations are considered among the most important e-learning quality issues (Ana, 2015; Patel et al. 2013):

- Learner orientation
- Developing quality in the learning process
- Quality must be a key role in education policy
- Quality services should be considered
- Quality standards should be implemented

Following two sections cover the most common e-learning quality standards.

8.6.7.1 ISO/IEC 19796-1

ISO/IEC 19796-1 was published in 2005, and aimed to develop and improve quality systems in the educational processes, activities, and services. The standard is used as a reference to support adaptation-specific requirements of the organization. Since 2007, this standard became a reference model and adaptable to the needs of organizations. In 2012, the official release of an international quality standard for e-learning programs, Open ECBCheck (e-learning in Capacity Building), supported measuring the success of e-learning programs and allows development. ECBCheck supports a set of e-learning quality criteria that helps in the design, development, management, delivery, and evaluation of a program, as well as the quality of learning materials, methodology, media, technology, and e-tutoring.

8.6.7.2 ISO 9126

ISO 9126 is a quality standard proposed as a guideline to evaluate the e-learning systems for teachers and educational organizations. The aim is to support decision making regarding the quality of existing systems and also to develop educational systems by increasing the usability by adding quality attributes such as consistency, simplicity, legibility, and user satisfaction as a global characteristic of the model. In 2010 the ISO 9126 model was used for selecting a standard quality for evaluating course management system to design, develop, and deliver e-learning content and measure e-learning outcomes. Furthermore, the ISO 9126 model is customized to identify acceptance criteria and evaluate business-to-business (B2B) applications by adding additional characteristics to existing quality models. Besides that, ISO 9126 is used for evaluating mobile learning by adding the following characteristics: metaphor, interactivity, and learning content.

8.7 E-LEARNING COMMUNITIES STANDARDS

This section presents an overview of the major organizations that contribute to the development of e-learning standards (e.g., IMS, IEEE LTSC, and the ISO/IEC).

8.7.1 IMS Global Learning Consortium, Inc. (IMS)

IMS is a consortium that develops open specifications for facilitating e-learning. IMS is the only organization that develops standards for K–12. This representation includes governmental representation from education ministries. Contributing members are able to vote on the IMS technical board for the acceptance, rejection, or revision of specification drafts, then for standards.

8.7.2 Aviation Industry Computer-Based Training Committee (AICC)

AICC standard was primarily formed on the need of standardizing of computer training for use in the airline industry, but now, it is used for reusability and interoperability in online learning and applications such as health care, financial services, higher education, and telecommunication. AICC adopted all computer-based training and include supplying, controlling, delivering, and monitoring the results of management systems training and Internet courses. AICC issues three types of documents:

- AICC guidelines and recommendations
- AICC reports and technical articles
- AICC working documents

8.7.3 Dublin Core Metadata Initiative (DCMI)

The Dublin Core Metadata Initiative (DCMI) is adopting an interoperability metadata standard, especially a metadata vocabulary. Dublin Core defines metadata that is developed by, for example, title, creator, subject, description, and publisher. This standard represents XML and RDF languages. There are some available documents issued by Dublin Core, including Dublin Core Template, MyMetaMaker, Reggie-The Metadata Editor, and DC-dot (Blagojević et al. 2015).

8.7.4 Ariadne Foundation

The Ariadne Foundation is a nonprofit association concerned with the area of metadata. The foundation groups metadata into six categories: general information, semantics, pedagogical, technical, indexation, and annotations.

8.7.5 Advanced Distributed Learning (ADL) Initiative

ADL Initiative is established to provide the U.S Department of Defense and White House Office for Science and Technology with development plans for standardization of learning research. The main objective of ADL is to support high-quality education and training tailored to learner's preferences, that is highly cost-effective and accessible. Another main objective is to enrich the SCORM standard in order to be compatible with other systems in interoperability and to encapsulate to its content.

There are several international organizations working on standardizing e-learning technologies. Each develops different learning concepts technology standards, which consist of a set of

TABLE 8.2
Examples of E-Learning Trends Standards

Trend	Standard Name
Architectures	IMS Guidelines for Developing Accessible Learning Applications, IEEE Learning Technology Systems Architecture, Open Service Interface Definitions
Digital repositories	CWA 15454 Simple Query Interface, IMS Digital Repositories Interoperability
Content aggregation	ADL Content Aggregation Model (CAM), IMS Content Packaging (CP), IMS Simple Sequencing (SS), IMS Common Cartridge
Metadata	IMS Learning Resource Metadata Information Model, IEEE LOM, Dublin Core Metadata Element Set (ISO 15836), Metadata for Learning Resources (ISO 19788), Dublin Core interoperability
Accessibility	IMS Learner Information Package, ISO/IEC 24751
Competency definitions	IMS Reusable Definition of Competency, IEEE Data Model for Reusable Competency Definitions
Quality	ISO/IEC 19796
Assessment	IMS Question and Test Interoperability (QTI)
Vocabularies	ISO/IEC 2382, AICC glossaries, IMS Vocabulary Definition Exchange
Run time	ADL SCORM Run-Time Environment, AICC/CMI Guidelines for Interoperability, IMS Shareable State Persistence

definitions, specifications, guidelines, and recommendations. Table 8.2 summarizes examples of e-learning standards trends based on citation and compatibility with LMSs (Mahnane et al. 2013).

8.8 CONCLUSION

This chapter provides a summary of standards used in the e-learning industry, and discussed the most cited organizations that are authorized to approve specifications as standards. Also it presents quality standards and their frameworks, and gives an overview on personalization and adaptation learning. The chapter also covers trends of e-learning standards.

As a future work, it will be wise to explore mobile learning standards, as well as the spread of open-source learning systems. It is important to incorporate the approaches, techniques, tools, and trends in this branch of learning.

REFERENCES

Advanced Distributed Learning (ADL) Co-Laboratories. 2012. *Choosing a Learning Management System Advanced*, Version 4.10, September 25, 2015.

Ana M. M. The use of quality management systems for e-learning, *The Sixth International Conference on E-Learning (eLearning-2015)*, September 24–25, 2015, Belgrade, Serbia.

Bianco A. M., De Marsico M., Temprini M. Standards for E-Learning, QUIS—Quality, Interoperability and Standards for e-learning, 2004-3538/001-001 ELE- ELEB14.

Blagojević M., Micić Ž., Milošević D. Development of standards in e-learning, *Sixth International Conference on E-Learning (eLearning-2015)*, September 24–25, 2015, Belgrade, Serbia.

Da Silva F. M., Neto F. M. M., Burlamaqui A. M. F., Demoly K. R. A., Pinto J. P. F. Providing an extension of the SCORM standard to support the educational contents project for t-learning, *Creative Education*, 6, 2015, 1201–1223.

El Bachari E., Abdelwahed E. H., El Adnani M. Design of an adaptive e-learning model based on learner's personality, *Ubiquitous Computing and Communication Journal*, 5(3), 2010, 1–8.

El Bachari E., Abelwahed E. H., El Adnani M. E-learning personalization based on dynamic learners' preference, *International Journal of Computer Science & Information Tech (IJCSIT)*, 3(3), 2011, 200–216.

El-Bakry H. M., Saleh A. A. Adaptive e-learning based on learner's styles, *Buletin Teknik Elektro dan Informatika (Bulletin of Electrical Engineering and Informatics)*, 2(4), 2013, 240–251.

Fernández J. L., Carrillo J. M., Nicolás J., Carrión, M. I. Trends in e-learning standards, *Proceedings Published by International Journal of Computer Applications (IJCA)*, 353(1), 2011, 49–54.

Friesen N. CanCore Initiative, Interoperability and learning objects: An overview of e-learning standardization, *Interdisciplinary Journal of Knowledge and Learning Objects*, 1, 2005, 23–31.

Ileana A. U. Ţ. Ă. E-learning standards, *Informatica Economică*, 1(41), 2007, 88–91.

Mahnane L., Laskri M. T., Trigano P. A model of adaptive e-learning hypermedia system based on thinking and learning styles, *International Journal of Multimedia and Ubiquitous Engineering*, 8(3), 2013, 339–350.

Mohamed A. K. Adaptive e-learning environment systems and technologies, *First International Conference of the Faculty of Education*, "Education … Future Prospectives," Albaha University, April 13–15, 2015.

Patel C. I., Gadhavi M., Patel A. A survey paper on e-learning based learning management Systems (LMS), *International Journal of Scientific & Engineering Research*, 4(6), 2013, 171–177.

Qazdar A., Cherkaoui C., Er-Raha B., Mammass D. AeLF: Mixing adaptive learning system with learning management system, *International Journal of Computer Applications*, 119(15), 2015.

Roy A., Basu K. A comparative study of statistical learning and adaptive learning, *International Journal of Advanced Computer Research*, 5(21), 2015.

Sivakami R., Anna Poorani G. SCORM/AICC compliance in learning management system and e-learning: A survey, *International Journal of Engineering and Computer Science*, 4(6), 2015, 12894–12897.

9 A Classification Perspective of Big Data Mining

Manal Abdullah and Nojod M. Alotaibi

CONTENTS

9.1 Introduction 123
9.2 Big Data 124
    9.2.1 Big Data Technologies 126
9.3 Data Mining 127
9.4 Classification 128
9.5 Big Data Mining 128
    9.5.1 Issues and Challenges of Big Data Mining 129
9.6 Big Data Classification 129
    9.6.1 Improving Traditional Classification Algorithms 129
    9.6.2 Classification Algorithms Based on MapReduce 130
    9.6.3 Big Data Mining Tools 131
9.7 Conclusion 131
References 133

9.1 INTRODUCTION

With the fast development of Internet communication and collaboration, the Internet of Things, and cloud computing, large amounts of data have become increasingly available at significant volumes (petabytes or more). Such data comes from a wider variety of sources and formats including social networking interactions, web pages, click streams, online transaction, e-mails, videos, audios, images, posts, search queries, health records, science data, sensors, and smart phones and their applications (Zikopoulos et al. 2012). According to the 2014 International Data Corporation (IDC) "Digital Universe Study" (Tuner et al. 2014), 130 exabytes (EB) of the world's data were created and stored in 2005. The amount grew to 4.4 zettabytes (ZB). It is doubling in size every 2 years and is projected to grow to 44 ZB in 2020 (Tuner et al. 2014). In 2012, IBM estimated that 2.5 quintillion bytes of data were created daily (IBM 2012).

The rapid growth in the amount of data led to constitute the big data phenomenon. Since 2004, interest in the search term "big data" has increased exponentially worldwide, according to Google Trends (Google 2015) (see Figure 9.1).

There are three characteristics used to define big data (also called the 3 V's of big data): *volume* as data keeps growing, *variety* as the type of data is diverse, and *velocity* as it is continuously arriving very fast into systems (Zikopoulos et al. 2012). Recently, two other V's have been added to complete the definition of big data: *veracity*, which is an indicator of data integrity, and *value*, which characterizes the business value of the data.

Due to these characteristics, the existing traditional techniques and technologies do not have the ability to handle storage and processing of this data. Therefore, new technologies have been developed to manage this big data phenomenon. IDC defines big data technologies as "a new generation of technologies and architectures designed to extract value economically from very large volumes

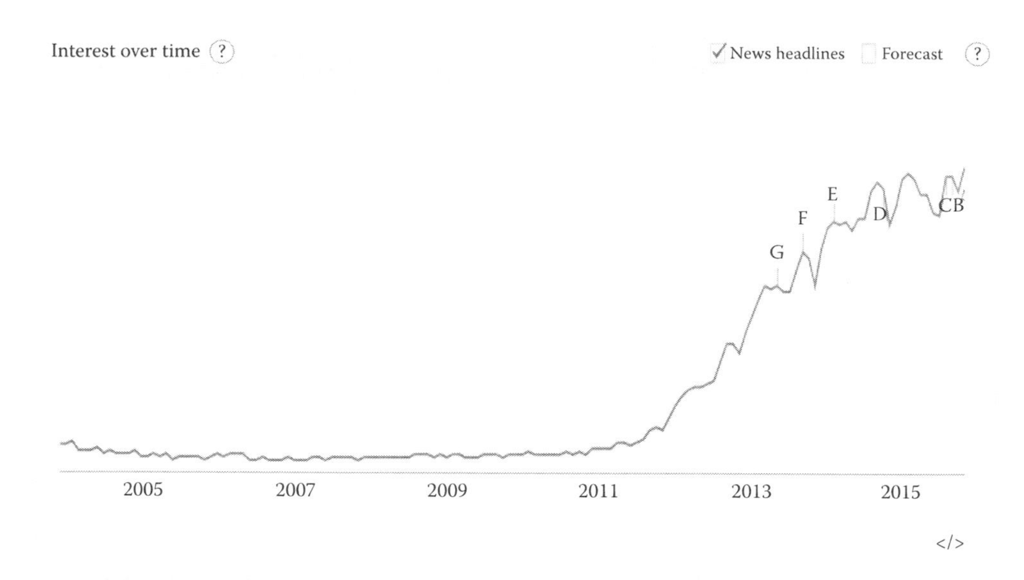

FIGURE 9.1 Worldwide interest: big data. (From Google, 2015, Google Trends, http://www.google.com/trends/explore#q=big%20data.)

of a wide variety of data by enabling high velocity capture, discovery and analysis" (Gantz and Reinsel 2012, 9).

Mining and discovering meaningful knowledge from big data for decision making, prediction, and for other purposes is extremely challenging due to its characteristics. Knowledge discovery (KD) is the process of discovering useful knowledge from a collection of data. Major KD application areas include marketing, manufacturing, fraud detection, telecommunication, education, medical, Internet agents, and many other areas (Fayyad, Gregory, and Smyth 1996; Singh, Gosawi, and Dubey 2014). The KD process may consist of the following steps: data selection, data preprocessing, data transformation, data mining, interpretation, and evaluation. Data mining is the core step of the KD process where algorithms are applied to extract useful patterns from data. Tasks in data mining can be classified into clustering, classification, summarization, regression, association rule, sequence analysis, and dependency modeling.

Supervised classification is one of the most common tasks of data mining, which is concerned with prediction. The aim of the classification is to build a classifier based on the training data with known class labels to predict the class labels of new data (Kotsiantis 2007). There are various methods for data mining classification tasks such as decision trees, support vector machine (SVM), genetic algorithms, and neural networks.

This chapter is organized as follows. In Section 9.2, we briefly review big data definitions and its related technologies. In Section 9.3, an overview of KD and data mining is provided. Section 9.4 presents the concept of supervised classification. Big data mining, and related issues and challenges are described in Section 9.5. Section 9.6 explores some of current works of big data classification. Finally, some conclusions are given in Section 9.7.

9.2 BIG DATA

In recent years, big data has become a hot research topic in many areas where storage and processing of massive amounts of data are required. In March 2012, U.S. President Barack Obama and his administration announced the "Big Data Research and Development Initiative" with over $200 million in research funding (Weiss and Zgorski 2012). The goals of this initiative were to develop and improve technologies needed to collect, store, manage, and analyze big data, to use these technologies to accelerate the pace of knowledge discovery in science and engineering fields, improve national security, and transform teaching and learning, and to expand the workforce required to develop and use big data technologies (Weiss and Zgorski 2012).

According to McKinsey (Manyika et al. 2011) the term *big data* is used to refer to datasets whose size is beyond the capability of existing database software tools to capture, store, manage, and analyze within a tolerable amount of time. However, there is no single definition of big data. O'Reilly defines big data as "data that exceeds the processing capacity of conventional database systems. The data is too big, moves too fast, or doesn't fit the structures of existing database architectures. To gain value from this data, there must be an alternative way to process it" (Dumbill 2012, 9).

As seen from the preceding definitions, the volume of data is not the only characteristic of big data. In fact, big data has three major characteristics (known as the 3 V's), shown in Figure 9.2, which were first defined by Doug Laney in 2001:

- Data *volume* (i.e., the size of data) is the primary attribute of big data. The size of data could reach terabytes (TB; 10^{12} B), petabytes (PB; 10^{15} B), exabytes (EB; 10^{18} B), zettabytes (ZB; 10^{21} B) and more. For example, Facebook reached more than 8 billion video views per day in September 2015 (Facebook 2015). That is double the number of daily video views reported in April.
- *Variety* refers to the fact that big data can come from different data sources in various formats and structures. These data sources are divided into three types: structured, semistructured, and unstructured data (Sagiroglu and Sinanc 2013). Structured data is described as data that follows a fixed schema. An example of this type is a relational database system. Semistructured data is a type of structured data, but it does not have a rigid structure (Pankowski 2002). Its structure may change rapidly or unpredictably (Pankowski 2002). Examples include weblogs and social media feeds. Unstructured data refers to data that cannot be stored into relational tables for analysis and querying. This data represents 80% of the world's data. Files or documents such as videos, images, audio, PDFs, and spreadsheets are examples.
- The *velocity* of data refers to the increasing rate at which data flows into an organization (Dumbill 2012).

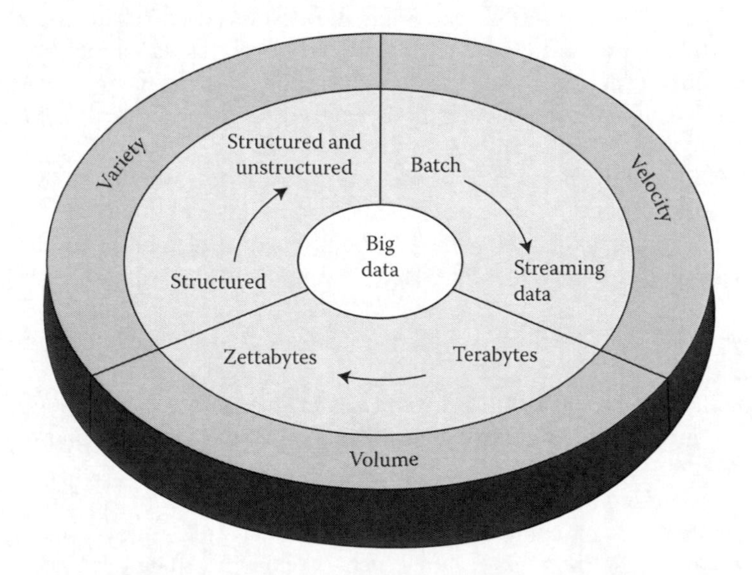

FIGURE 9.2 Three V's of big data. (From Zikopoulos, P., C. Eaton, D. deRoos, T. Deutsch, and G. Lapis, 2012, *Understanding Big Data: Analytics for Enterprise Class Hadoop and Streaming Data*, New York: McGraw-Hill.)

In 2012, Gartner (Beyer and Laney 2012) summarized the big data definition as "high volume, high velocity, and/or high variety information assets that require new forms of processing to enable enhanced decision making, insight discovery and process optimization." More recently, two additional V's have been added to define big data: veracity and value. *Veracity* (uncertainty of data) refers to the accuracy, integrity, and quality of the data being collected, whereas *value* refers to the worth of the data being extracted (Hassanien et al. 2015).

All previous characteristics of big data are considered challenging and this is the reason why we cannot use traditional database management systems (DBMS) for processing and analyzing big data. As a result, new technologies have been developed to meet the challenges. The following sections discuss some of these technologies.

9.2.1 Big Data Technologies

Big data is a new term used to identify the datasets that, due to their large size and complexity, cannot be managed with traditional database systems. In recent years, there have been many technologies developed to process this huge volume of data.

Apache Hadoop (2016) is an open-source software framework that enables the distributed processing of large datasets across clusters of commodity hardware using simple programming models. There are two main components of Hadoop: Hadoop distributed file system (HDFS) and MapReduce. HDFS is a distributed, scalable file system written in java for the Hadoop framework. MapReduce is a programming paradigm that allows users to define two functions—map and reduce—to process large numbers of data in parallel. In MapReduce, the input is first converted into a large set of key–value pairs. Then, the map function is applied in parallel to every pair in the input dataset, which produce a set of intermediate key–value pairs for each call. After that, the reduce function is called to merge together all intermediate values associated with the same key. Companies like Facebook, Yahoo, Amazon, Baidu, AOL, and IBM use Hadoop on a daily basis. Hadoop has many advantages including (Mirajkar, Bhujbal, and Deshmukh 2013) cost effectiveness, fault tolerant, flexibility, and scalability. Hadoop has many other related software projects that uses the MapReduce and HDFS framework such as Apache Pig, Apache Hive, Apache Mahout, and Apache HBase (Apache Hadoop 2016).

Apache Pig (Zikopoulos et al. 2012) was originally developed at Yahoo in 2006 for processing big data. In 2007, it was moved into the Apache Software Foundation. It allows people using Hadoop to focus more on analyzing large datasets and spend less time having to write MapReduce programs. Pig consists of two components: the language itself, which is called PigLatin, and the runtime environment where PigLatin programs are executed.

Apache Hive (2014) was developed at Facebook in 2009. It is data warehouse software for querying and managing large datasets residing in distributed storage. It is built on top of Apache Hadoop. Hive defines a simple SQL-like query language, called Hive query language (HQL), which enables users familiar with SQL to query the data. Hive is optimized for scalability, extensibility, and fault-tolerance.

Apache HBase (2016) is a distributed columnar database that supports structured data storage for very large tables.

Jaql was created by workers at IBM Research Labs in 2008 and released to open source (IBM, n.d.). It is a query language for JavaScript Object Notation (JSON), but it supports more than just JSON such as XML, CSV, and flat files.

Storm (2015) was created at Backtype, a company acquired by Twitter in 2011. It is a free and open-source distributed real-time computation system that does for real-time processing what Hadoop does for batch processing. Storm offers features such as scalability, fault-tolerance, and distributed computation, which make it suitable for processing huge amounts of data on different machines.

NoSQL (not only SQL) is a term used to designate database management systems that differ from classic RDBMS in some way ("NoSQL Databases Explained" 2015). These data stores may not require fixed table schemas, usually avoid join operations, do not attempt to provide ACID (atomicity, consistency, isolation, durability) properties, and typically scale horizontally. There are several types of NoSQL databases:

- Key–value stores—In key–value stores, each single item in the database is stored as an attribute name (or key), together with its value. Examples of key–value stores are Amazon's Dynamo and Oracle's BerkeleyDB.
- Document-oriented database—It is a database designed for storing, retrieving, and managing document-oriented or semistructured data. Examples of these databases are CouchDB and MongoDB.
- Column stores—It stores columns of data together, instead of rows. Examples include Cassandra and Apache HBase.
- Graph database—It contains nodes, edges, and properties to represent and store data. Examples of graph databases are Neo4j and HyperGraphDB.

9.3 DATA MINING

Knowledge discovery (KD) is the process of extracting useful knowledge from huge volumes of data. It can be defined as the nontrivial process of identifying valid, novel, potentially useful, and ultimately understandable patterns in data (Fayyad et al. 1996). The main goal of KD is extracting high-level knowledge from low-level data. The KD process consists of the following steps as shown in Figure 9.3 (Fayyad et al. 1996):

1. Understanding the application domain—It includes learning prior knowledge and the user's goals.
2. Creating a target dataset—In this step, a subset of variables and data are selected that will be used to perform discovery task.
3. Data cleaning and preprocessing—It includes the basic operations of removing noise and dealing with missing values.
4. Data reduction and projection—It involves of finding useful attributes to represent data.
5. Choosing the data mining task—There are several data mining tasks including clustering, classification, regression, and summarization.
6. Choosing the data mining algorithms—In this step, appropriate methods are selected to be used for searching for patterns in the data.
7. Data mining—Searching for patterns in a particular representational form (such as classification rules or trees, regression, and clustering) using the selected data mining methods.
8. Interpretation—Interpreting mined patterns and possibly returns to any of the previous steps for further iteration if the pattern evaluated is not useful.
9. Using discovered knowledge—The final step consists of incorporating the discovered knowledge into another system, or documenting and reporting it to interested parties.

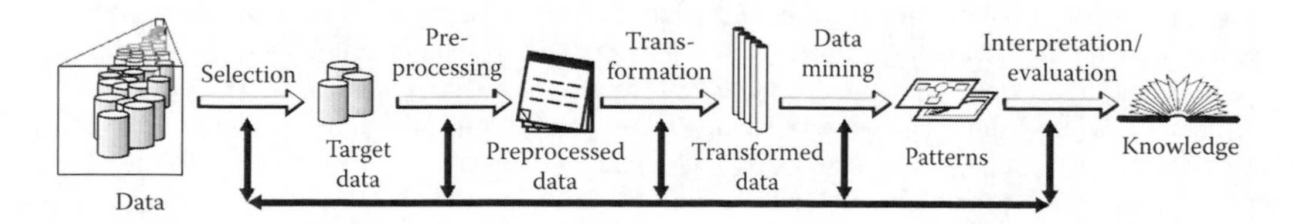

FIGURE 9.3 Typical knowledge discovery process. (From Fayyad, U., P.-S. Gregory, and P. Smyth, 1996. The KDD process for extracting useful knowledge from volumes of data, *Communications of the ACM* 39(11): 27–34.)

Data mining is the core step in the whole KD process. It consists of applying data analysis and discovery algorithms and produce enumeration of patterns (models) over data. It is widely used in the fields of science, engineering, economics, social media, medicine, marketing, and business. Currently, many data mining tools are available for free on the Web such as Weka (http://www.cs.waikato.ac.nz/ml/weka/), RapidMiner (https://rapidminer.com/), Orange (http://orange.biolab.si/), and KNIME (https://www.knime.org/). The major tasks of data mining can be classified as the following (Fayyad et al. 1996):

- Clustering—Maps a data item into one of several clusters, where clusters are natural groupings of data items based on similarity or probability density models.
- Classification—Classifies a data item into one of several predefined categorical classes.
- Regression—Maps a data item to real-valued prediction variable. It is used in different prediction and modeling applications.
- Summarization—Provides compact description for a subset of data. Examples include mean and standard deviation of fields.
- Association rule—Describes association relationships among different attributes.
- Sequence analysis—Models sequential patterns, like time-series data. The goal is to model the process of generating the sequence or to extract and report deviation and trends over time.
- Dependency modeling—Describes significant dependencies between variables.

9.4 CLASSIFICATION

Classification is one of the most common types of data mining, which finds patterns in information and categorizes them into different classes. It is supervised learning, which generates a classifier (model) based on a set of instances with known labels called the training set. Then, the classifier is used for classifying new or previously unseen data (Kotsiantis 2007). The converse of this is unsupervised learning, which involves classifying data into categories based on some similarity of input parameters in the data. Examples of supervised classification are spam detection, credit card fraud detection, and medical diagnosis. There are three different types of supervised classification: binary, multiclass, and multilabel classification. In binary classification, each instance of data may belong to one of two possible class labels. In multiclass classification, more than two class labels are involved and each instance is assigned to only one class label. In the case of multilabel classification, there are more than two class labels and each instance may belong to more than one class label at same time. There are many types of classification algorithms for extracting knowledge from data, which can be categorized into logic-based techniques (C4.5, CART, and RIPPER), perceptron-based techniques (artificial neural networks), statistical learning techniques (naive Bayes classifiers and Bayesian networks), and instance-based techniques (k-nearest neighbor) (Kotsiantis 2007).

9.5 BIG DATA MINING

Useful knowledge can be extracted from big data with the help of data mining. Due to the enormity, high dimensionality, heterogeneous, and distributed nature of data, traditional techniques of data mining may be unsuitable for extracting knowledge from this data. Mining big data is an emerging research area, hence a plethora of possible future research directions arise. The objectives of big data mining techniques go beyond fetching the requested information or even uncovering some hidden relationships and patterns (Prakash and Hanumanthappa 2014). Compared with the results derived from mining the traditional datasets, unveiling the massive volume of interconnected heterogeneous big data has the potential to maximize our knowledge and insights in the target domain. Begoli and Horey (2012) proposed three principles for effective knowledge discovery from big data: first, the architecture should support many analysis methods such as data mining, statistical analysis, machine learning, and visualization. Second, different storage mechanisms should be used because all data cannot fit in a single storage. Also, the data should be stored and processed at all stages of the pipeline. Third, the results should be accessible and easy to understand.

9.5.1 Issues and Challenges of Big Data Mining

There are a number of issues and challenges related to big data mining (Hong et al. 2013):

- Heterogeneity or variety—Existing data mining techniques have been used to discover unknown patterns and relationships of interest from structured, homogeneous, and small datasets. Variety is one of the fundamental characteristics of big data, and comes from the phenomenon that there exists unlimited different sources that generates and contributes to big data. The data from different data sources may form interconnected, interrelated, and delicately and inconsistently represented data. Mining useful information from such data is a great challenge. Heterogeneity in big data also means that it is an obligation to accept and deal with structured, semistructured, and even entirely unstructured data concurrently.
- Scalability or volume—The extraordinary volume requires high scalability of its data management and mining tools. Cloud computing with parallelism can deal with the volume challenge of big data.
- Speed or velocity—The ability of fast accessing and mining big data is highly essential. Processing/mining must be completed within a definite period of time, otherwise, the results becomes less valuable or even worthless.
- Accuracy and trust—With big data, the data sources are of many different origins, not all well-known, and not all confirmable. As a result, the accuracy and trust of the source data quickly become a serious concern, which also affects the mining results.
- Privacy crisis—Data privacy has always been a challenge. The concern has become extremely serious with big data mining that often requires personal information in order to produce relevant/accurate results such as location-based and personalized services. Additionally, there is the enormous volume of big data such as social media that contains tremendous amount of highly interconnected personal information. When all bits of information about a person are dug out and put together, any privacy about that individual instantly disappears.
- Interactiveness—Means the capability of a data mining system to allow fast and adequate user interaction such as feedback, interference, or guidance from users. It relates to all the characteristics of big data and can help overcome the challenges coming along with each of them.

9.6 BIG DATA CLASSIFICATION

Because of big data characteristics, traditional data mining algorithms may not be suitable to mine such huge data. As a consequence, there is an urgent need for developing algorithms and techniques capable of mining big data while dealing with their inherent properties. Several studies attempted to improve the traditional classification algorithms to make them work with big data, to parallelize classification algorithms based on MapReduce, or to develop new software tools to mining big data. The approaches of big data classification are summarized in Figure 9.4.

9.6.1 Improving Traditional Classification Algorithms

Niu et al. (2013) improved the traditional KNN algorithm and proposed a new algorithm called neighbor filter classification (NFC) to realize fast classification operation in big data. Liu (2014) proposed a new improved model for the original random forest algorithm in the big data environment. The proposed model has higher classification accuracy.

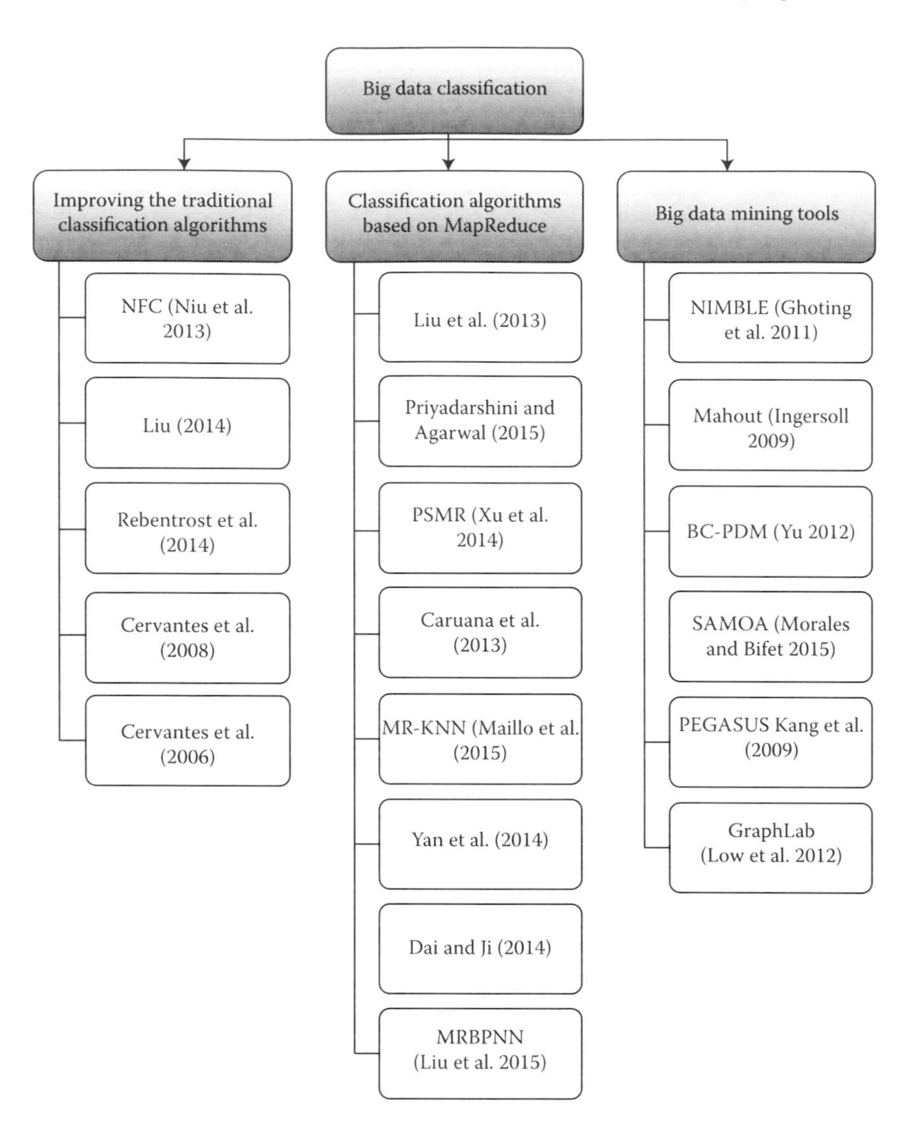

FIGURE 9.4 Big data classification approaches.

Support vector machines (SVMs) is one of the most popular techniques for data classification and regression. Their computation and storage requirements increase rapidly with the size of the dataset, making it unsuitable for big data. Many researchers have tried to find possible methods to apply SVM classification to large datasets. Rebentrost et al. (2014) presented a quantum-based support vector machine algorithm for big data classification. The proposed algorithm can achieve exponential speedup over classical algorithm. Cervantes et al. (2008) presented SVM for classification of large data using minimum enclosing ball clustering. The proposed approach has good classification accuracy compared with classical SVM. Cervantes et al. (2006) proposed an SVM classification algorithm based on fuzzy clustering. The proposed approach is scalable to large datasets with high classification accuracy and fast convergence speed.

9.6.2 Classification Algorithms Based on MapReduce

Liu et al. (2013) designed a big data analyzing system to classify millions of movie reviews using a naive Bayes classifier (NBC). They implemented the NBC on top of the Hadoop framework, with some additional modules. The results show that the accuracy of NBC is improved and approaches 82% when the dataset size increases. Priyadarshini and Agarwal (2015) proposed a MapReduce-based parallel SVM algorithm for big data classification. The software used was lib-SVM. In the proposed algorithm, the training data is divided into subsets and each subset is trained with SVM. Then, the support vectors of two SVMs are combined to be input for the next SVM. This process is repeated until only one set of support vectors is left. Xu et al. (2014) proposed a parallel SVM based on the MapReduce (PSMR) algorithm for e-mail classification. The parallel SVM is based on the cascade SVM model. Caruana et al. (2013) developed a new algorithm for parallelized SVM based on the MapReduce framework for scalable spam filter training. The parallel SVM is built on the sequential minimal optimization (SMO) algorithm. Ontology semantics are used to minimize the accuracy degradation when distributing the training data among a number of SVM classifiers. Maillo et al. (2015) proposed a MapReduce-based k-nearest neighbor approach (MR-KNN) for big data classification. Yan et al. (2014) proposed a parallel KNN-join algorithm using MapReduce for big data multilabel classification. Dai and Ji (2014) suggested a parallel C4.5 decision tree classification algorithm based on MapReduce. Liu et al. (2015) proposed a MapReduce based parallel back-propagation neural network (MRBPNN). In this work, three parallel neural networks are presented to deal with data-intensive scenarios in terms of the volume of classification data, the size of the training data, and the number of neurons in NN. They concluded that the computation overhead of NN can be significantly reduced using a number of computers in parallel.

These big data classification algorithms with their advantages and limitations are summarized in Table 9.1.

9.6.3 Big Data Mining Tools

In big data mining, there are many open-source tools. Some of these tools are summarized in the following.

NIMBLE (Ghoting et al. 2011) is a portable infrastructure that enables rapid development of parallel machine learning and data mining algorithms. It runs on top of the Hadoop framework.

Apache Mahout (Ingersoll 2009) is an open-source project by Apache Software Foundation (ASF). Mahout is written in Java and provides scalable data mining algorithms. It contains implementations for clustering, categorization, collaborative filtering (CF), and evolutionary programming on top of Apache Hadoop.

Big Cloud–Parallel Data Mining (BC-PDM) (Yu et al. 2012) is a cloud-based data mining platform that provides access to large telecom data and business solutions for telecom operators. BC-PDM is based on the MapReduce implementation of cloud computing. It supports parallel ETL processes (extract, transform, and load), statistical analysis, data mining, text mining, and social network analysis.

Apache Scalable Advanced Massive Online Analysis (SAMOA) (Morales and Bifet 2015) is a platform for mining big data streams. It includes distributed algorithms for common machine-learning tasks.

Peta-Scale Graph Mining System (PEGASUS) (Kang et al. 2009) is a graph mining system for very large graphs built on top of the Hadoop framework.

GraphLab (Low et al. 2012) is a high-level graph-parallel system built without using MapReduce. It is an open-source project written in C++.

9.7 CONCLUSION

Big data has become a hot research topic that attracts extensive attention from academia, industry, and governments around the world. In this chapter, we briefly introduce the concept of big data, including its definitions, characteristics, and technologies. This chapter also provides an overview of big data mining and discusses the related issues and challenges. To support big data mining, we

TABLE 9.1
Summarization of Big Data Classification Algorithms

Reference	Algorithm	Limitations	Modification	Advantages
Niu et al. (2013)	KNN	Time cost of modeling is unacceptable. Sensitive to parameter K.	Neighbor filter classification (NFC)	It reduces the computational cost to $O(n)$. It is able to replace or adjust key input parameters automatically. It updates other parameters regularly.
Liu (2014)	Random forest	Accuracy of a random forest will gradually reduce over time.	Improved random forest	It has higher classification accuracy than the traditional random forest.
Rebentrost et al. (2014)	SVM	High computational complexity (long training time) and extensive memory requirements of the required quadratic programming in large-scale tasks.	Quantum least-squares SVM	Achieves exponential speedup over classical algorithm $O(\log N M)$ in both training and classification stages.
Cervantes et al. (2008)	SVM	High computational complexity (long training time) and extensive memory requirements of the required quadratic programming in large-scale tasks.	SVM using minimum enclosing ball (MEB) clustering	It provides good classification accuracy compared with classic SVM, while the training time is significantly shorter.
Cervantes et al. (2006)	SVM	High computational complexity (long training time) and extensive memory requirements of the required quadratic programming in large-scale tasks.	SVM based on fuzzy clustering	It achieves good performance for large datasets and fast convergence speed.
Liu et al. (2013)	NBC	It does not scale up well when the dataset is large.	Implementing NBC on top of Hadoop MapReduce framework	The accuracy of NBC is improved and approaches 82% when the dataset size increase.
Priyadarshini and Agarwal (2015)	SVM	The computation and storage requirement increases tremendously for large dataset.	MapReduce-based parallel SVM algorithm.	It works efficiently on large datasets as compared to the sequential SVM. The computation time taken by the SVM with multinode cluster is less as compared to the single node cluster for large dataset.
Xu et al. (2014)	SVM	The computation and storage requirement increases tremendously for large dataset.	Parallel SVM based on MapReduce (PSMR)	The training time is reduced significantly.
Caruana et al. (2013)	SVM	The computation and storage requirement increases tremendously for large dataset.	Ontology-enhanced parallel SVM based on MapReduce	It reduces the training time significantly.
Maillo et al. (2015)	KNN	The complexity of KNN is $O(n \times D)$, where n is the number of instances and D the number of features. Memory consumption problems.	MapReduce-based k-nearest neighbor approach (MR-KNN)	The reduction of computational time achieved compared to the utilization of the sequential version.

(Continued)

TABLE 9.1 (*Continued*)
Summarization of Big Data Classification Algorithms

Reference	Algorithm	Limitations	Modification	Advantages
Yan et al. (2014)	KNN-join	It needs to spend a lot of time to handle large volume data.	Parallel MapReduce based KNN-join	It achieves higher performance than the serial one.
Dai and Ji (2014)	C4.5	The process of building decision trees can be very time consuming when the dataset is extremely big.	Parallel C4.5 decision tree classification algorithm based on MapReduce	It exhibits both time efficiency and scalability.
Liu et al. (2015)	Back-propagation neural network (PBNN)	The computation process of ANN is slow especially when dealing with large datasets.	MapReduce-based parallel back-propagation neural network (MRBPNN)	The computation overhead of neural network can be significantly reduced.

briefly describe the overview of supervised classification algorithms over big data. In future work, more classification algorithms will be described, as well as their limitations, and efforts made to make them suitable for big data. We will also evaluate the performance of these algorithms using real-world data.

REFERENCES

Apache Hadoop. 2016. Last modified February 13. https://hadoop.apache.org/.

Apache HBase. 2016. Last modified June 13. http://hbase.apache.org/.

Apache Hive. 2014. http://hive.apache.org/.

Apache Storm. 2015. http://storm.apache.org.

Begoli, E. and J. Horey. 2012. Design principles for effective knowledge discovery from big data. *2012 Joint Working IEEE/IFIP Conference on Software Architecture (WICSA) and European Conference on Software Architecture (ECSA)*, Helsinki, Finland, August 20–24.

Beyer, M. A. and D. Laney. 2012. *The Importance of "Big Data": A Definition*. Gartner.

Caruana, G., M. Li, and Y. Liu. 2013. An ontology enhanced parallel SVM for scalable spam filter training. *Neurocomputing* 108: 45–57.

Cervantes, J., X. Li, and W. Yu. 2006. Support vector machine classification based on fuzzy clustering for large data sets. The 5th Mexican International Conference on Artificial Intelligence (MICAI), Apizaco, Mexico, November 13–17.

Cervantes, J., X. Li, W. Yu, and K. Li. 2008. Support vector machine classification for large data sets via minimum enclosing ball clustering. *Neurocomputing* 71(4–6): 611–619.

Dai, W. and W. Ji. 2014. A MapReduce implementation of C4.5 decision tree algorithm. *International Journal of Database Theory and Application* 7(1): 49–60.

Dumbill, E. 2012. *Planning for Big Data*. Sebastopol, CA: O'Reilly Media.

El, M., S. Safwat, and O. Hegazy. 2015. Big data classification using fuzzy K-nearest neighbor. *International Journal of Computer Applications (IJCA)* 132(10): 8–13.

Facebook. 2015. Facebook reports third quarter 2015 results. Last modified November 4. *Techmeme*, November 4. http://www.techmeme.com/151104/p24#a151104p24.

Fayyad, U., P.-S. Gregory, and P. Smyth. 1996. The KDD process for extracting useful knowledge from volumes of data. *Communications of the ACM* 39(11): 27–34.

Gantz, J. and D. Reinsel. 2012. The digital universe in 2020: Big data, bigger digital shadows, and biggest growth in the far east. EMC Corporation.

Ghoting, A., P. Kambadur, E. Pednault, and R. Kannan. 2011. NIMBLE: A toolkit for the implementation of parallel data mining and machine learning algorithms on MapReduce. *The 17th ACM SIGKDD International Conference on Knowledge Discovery and Data Mining*, San Diego, CA, August 21–24.

Google. 2015. Google Trends. http://www.google.com/trends/explore#q=big%20data.
Hassanien, A.-E., A. T. Azar, V. Snasael, J. Kacprzyk, and J. H. Abawajy. 2015. *Big Data in Complex Systems: Challenges and Opportunities*. Cham, Switzerland: Springer International Publishing.
Hong, B., X. Meng, L. Chen, W. Winiwarter, and W. Song. 2013. *Database Systems for Advanced Applications*. Berlin: Springer.
IBM. 2012. What is big data?: Bringing big data to the enterprise. https://www-01.ibm.com/software/data/bigdata/what-is-big-data.html.
IBM. n.d. What is Jaql? https://www-01.ibm.com/software/data/infosphere/hadoop/jaql/.
Ingersoll, G. 2009. Introducing Apache Mahout. http://www.ibm.com/developerworks/library/j-mahout/.
Kang, U., C. E. Tsourakakis, and C. Faloutsos. 2009. PEGASUS: A Peta-scale graph mining system implementation and observations. The 9th IEEE International Conference on Data Mining. Helsinki, Finland, August 20–24.
Kotsiantis, S. B. 2007. Supervised machine learning: A review of classification techniques. *Informatica* 31: 249–268.
Laney, D. 2001. 3D data management: Controlling data volume, velocity and variety. META Group.
Liu, B., E. Blasch, Y. Chen, D. Shen, and G. Chen. 2013. Scalable sentiment classification for big data analysis using Naïve Bayes classifier. *2013 IEEE International Conference on Big Data*, Santa Clara, CA, October 6–9.
Liu, Y. 2014. Random forest algorithm in big data environment. *Computer Modelling & New Technologies* 18(12A): 147–151.
Liu, Y., J. Yang, Y. Huang, L. Xu, S. Li, and M. Qi. 2015. MapReduce based parallel neural networks in enabling large scale machine learning. *Computational Intelligence and Neuroscience* 2015: 1–13.
Low, Y., D. Bickson, J. Gonzalez, C. Guestrin, A. Kyrola, and J. M. Hellerstein. 2012. Distributed graphLab: A framework for machine learning and data mining in the cloud. *Journal of VLDB Endowment* 5(8): 716–727.
Maillo, J., I. Triguero, and F. Herrera. 2015. A MapReduce-based k-nearest neighbor approach for big data classification. *2015 IEEE Trustcom/BigDataSE/ISPA Conference*. Helsinki, Finland, August 20–22.
Manyika, J., M. Chui, B. Brown, J. Bughin, R. Dobbs, C. Roxburgh, and H. Byers. 2011. *Big Data: The Next Frontier for Innovation, Competition and Productivity*. McKinsey Global Institute.
Mirajkar, N., S. Bhujbal, and A. Deshmukh. 2013. Perform wordcount Map-Reduce job in single node Apache Hadoop cluster and compress data using Lempel-Ziv-Oberhumer (LZO) algorithm. *International Journal of Computing Science Issues (IJCSI)* 10(1): 719–728.
Morales, G. and A. Bifet. 2015. SAMOA: Scalable advanced massive online analysis. *Journal of Machine Learning Research* 16: 149–153.
Niu, K., F. Zhao, and S. Zhang. 2013. A fast classification algorithm for big data based on KNN. *Journal of Applied Science* 13(12): 2208–2212.
"NoSQL Databases Explained." 2015. https://www.mongodb.com/nosql-explained.
Pankowski, T. 2002. Querying semistructured data using a rule-oriented XML query language. *The 15th European Conference on Artificial Intelligence (ECAI)*. Lyon, France, July 21–26.
Prakash, B. R. and M. Hanumanthappa. 2014. Issues and challenges in the era of big data mining. *International Journal of Emerging Trends and Technology in Computer Science (IJETICS)* 3(4): 321–325.
Priyadarshini, A. and S. Agarwal. 2015. A map reduce based support vector machine for big data classification. *International Journal of Database Theory and Application* 8(5): 77–98.
Rebentrost, P., M. Mohseni, and S. Lloyd. 2014. Quantum support vector machine for big data classification. *Physical Review Letters* 113(13): 1–5.
Sagiroglu, S. and S. Duygu. 2013. Big data: A review. *2013 International Conference on Collaboration Technologies and Systems (CTS)*. San Diego, CA, May 20–24.
Singh, P., G. Gosawi, and S. Dubey. 2014. Application of data mining. *Binary Journal of Data Mining and Networking* 4(2): 41–44.
Tuner, V., D. Reinsel, J. F. Gantz, and S. Minton. 2014. *The Digital Universe of Opportunities: Rich Data and the Increasing Value of the Internet of Things*. EMC Corporation.
Weiss, R. and L.-J. Zgorski. 2012. Obama administration unveil "big data" initiative: Announces $200 million in new R&D investments. https://www.whitehouse.gov/sites/default/files/microsites/ostp/big_data_press_release.pdf.
Xu, K., C. Wen, Q. Yuan, X. He, and J. Tie. 2014. A MapReduce based parallel SVM for email classification. *Journal of Networks* 9(6): 1640–1647.

Yan, X., Z. Wang, D. Zeng, C. Hu, and H. Yao. 2014. Design and analysis of parallel MapReduce based KNN-join algorithm for big data classification. *TELKOMNIKA Indonesian Journal of Electrical Engineering* 12(11): 7927–7934.

Yu, L., J. Zheng, W. C. Shen, B. Wu, B. Wang, L. Qian, and B. R. Zhang. 2012. BC-PDM: Data mining, social network analysis and text mining system based on cloud computing. *Proceedings of the 18th ACM SIGKDD International Conference on Knowledge Discovery and Data Mining*, pp. 1496–1499. Beijing, China, August 12–16.

Zikopoulos, P., C. Eaton, D. deRoos, T. Deutsch, and G. Lapis. 2012. *Understanding Big Data: Analytics for Enterprise Class Hadoop and Streaming Data*. New York: McGraw-Hill.

10 Streaming Data Classification in Clustered Wireless Sensor Networks

Manal Abdullah and Yassmeen Alghamdi

CONTENTS

10.1 Introduction ... 137
10.2 Data Mining ... 139
  10.2.1 Common Classes of Data Mining ... 139
  10.2.2 Data Mining in Wireless Sensor Networks (WSNs) ... 139
    10.2.2.1 Challenges of Data Mining in WSNs ... 140
    10.2.2.2 Taxonomy of Data Mining Techniques for WSNs ... 140
    10.2.2.3 Application Areas of WSN Data Mining ... 140
    10.2.2.4 Implementation of WSN Data Mining ... 141
10.3 Data Clustering ... 141
  10.3.1 Hierarchical Clustering Structure ... 141
  10.3.2 Designing Clusters in WSNs ... 142
  10.3.3 Clustering Parameters ... 142
10.4 Data Streams ... 143
  10.4.1 Traditional Data Mining and Data Stream Mining ... 143
  10.4.2 Data Stream Characteristics ... 143
  10.4.3 Algorithms of Data Streams ... 144
10.5 Clustering Protocols ... 144
  10.5.1 Clustering Routing Protocols in WSNs ... 144
  10.5.2 Proactive and Reactive Clustering ... 144
  10.5.3 Clustering Algorithms Schemes in WSNs ... 144
  10.5.4 Clustering Protocols for Data Streams ... 147
  10.5.5 Clustered WSNs for Data Streams ... 148
  10.5.6 Taxonomy of Clustering Protocols ... 149
10.6 Conclusion ... 149
References ... 149

10.1 INTRODUCTION

In recent years, the widespread use of wireless sensor networks (WSNs) have been seen in various applications. As known, a WSN is a special kind of ad hoc network that has the ability to sense and process information. They can be used in many fields such as environmental, industrial, military, and agriculture. Specifically, WSNs contain tiny, independent, built-in devices called sensor nodes. These sensor nodes contain four basic components: sensing unit, processing unit, transducer, and energy source (de Aquino et al. 2007b). Sensor nodes are mainly used in data processing and continuously report parameters such as temperature and humidity. Reports transmitted by the sensors are collected by observers called base stations (BS). WSNs depend on their sensors, which consume a lot of battery energy. Unfortunately, the nature of WSNs makes it very difficult to recharge the sensor node batteries. Therefore, energy efficiency is an important objective design in such networks (Younis, Krunz, and Ramasubramanian 2006). WSNs have several resource constraints, such as low computational power, limited energy source, and reduced bandwidth (de Aquino et al. 2007b). Therefore, WSN algorithms should be accurately designed. In more detail, WSNs have three main components: sensing unit, processing unit, and communication unit. The sensing unit senses the surrounding environment and gets the data. The processing unit processes the data and eliminates the redundancy. The communication unit acts as a transceiver, that is, it receives and transmits the data (Thangavelu and Pathak 2014).

WSNs also can be used to observe physical and environmental phenomenon such as heat, vibration, humidity, light, and pressure (Alia 2014). A WSN consists of one or more powerful BSs that serve as the final destination of the sensed data (Alia 2014). Passing sensory data to the BS requires energy. A WSN suffers from four main constraints: energy, memory, computational capabilities, and communications bandwidth (Sabit 2012). One of such constraint is the power consumption. Sensor nodes should be energy efficient. Energy efficiency affects the network lifetime of the entire sensor network (Ali and Rajpoot 2014).

Data are transmitted from the source to the destination with multiple routes in WSNs. The data follows multiple intermediate nodes. In this case, increased traffic means more energy is to be consumed. Hence, intermediate node failure may occur. Therefore, a reliable system topology that provides multiple paths from the source to the destination if required (Khan, 2014). In some applications for sensor networks, data that WSNs process usually arrive in an online fashion. They are unlimited and there is no control on the arrival order of the elements being processed. Such data are called data streams (de Aquino et al. 2007a, b). As a general rule, there are some differences between sensor streams and traditional streams. The sensor streams are only samples of the entire population, imprecise, noisy, and of a moderate size. Whereas in traditional streams, the entire population is used, and the data is exact, error-free, and huge (de Aquino et al. 2007a). The process of extracting knowledge structures from continuous data streams is called data stream mining. A data stream is an ordered sequence of instances that can only be read a few times using limited computing and storage capabilities (Sabit and Al-Anbuky 2014). WSNs generate massive data streams with the spatial and sensor measurements information, and, as known, the energy of sensors is limited. Therefore, reducing sensors' energy expenditure and accordingly extending the network lifetime is the major challenge in such networks (Huang and Zhang 2011c). WSNs can benefit a great deal from stream mining algorithms in terms of energy saving. However, to achieve better energy conservation, the data stream mining has to be performed in a distributed manner, due to their resource constraints (Sabit and Al-Anbuky 2014). Furthermore, transmitting all sensor data to a central location over limited bandwidth exhausts a large amount of energy. This also requires performing in-network distributed data processing (Huang and Zhang 2011c). Algorithms must converge the limited datasets as fast as possible, to ensure that the processor can take on the next set of streams (Sabit and Al-Anbuky 2014).

The widespread deployment of WSNs and the need for data aggregation require efficient organization of the network topology to balance the load and extend the network lifetime. Clustering has proven to be an effective approach in solving the problem of energy consumption, data aggregation, scalability (Boyinbode et al. 2010; Abdullah et al. 2015), and organizing the network into a connected hierarchy. Generally, there are two categories of networks in WSNs: flat networks, and hierarchical or clustered ones (Liu 2012). At any rate, the clustering phenomenon plays an important role in the organization of networks and also affects the network performance. Owing to a variety of advantages, clustering is becoming an active branch of routing technology in WSNs. Clustering algorithms are designed to achieve load distribution among cluster heads, energy saving, high connectivity, and fault tolerance. In WSNs, clustering provides resource utilization and minimizes energy consumption by reducing the number of nodes that take part in long-distance transmission (Huang and Zhang 2011c). In addition, the clustering phenomenon allows for data aggregation. It removes the redundant data and combines the useful ones. It also limits the data transmission.

Clustering systems give an impression of small and stable networks. Generally, they improve a network's lifetime by reducing the network traffic (Thangavelu and Pathak 2014). Obviously, clustering protocols are in widespread use more than any other type of protocols.

In many cases, WSNs are multihop networks formed by a large number of resource-constrained sensor nodes. Each sensor generates a stream of data that are obtained from the sensing devices on the node (Huang and Zhang 2011a, b). Available energy and wireless bandwidth are some limitations found that have required alternative computation and communication approaches. The energy source of sensors is batteries, which is undesirable or even impossible to be recharged or replaced. Therefore, maximizing energy efficiency and prolonging the network lifetime are major challenges in WSNs (Huang and Zhang 2011a, b).

Due to the aforementioned restrictions, sending a large amount of data can take a lot of time when sensor nodes try to reach the wireless medium in a multihop data communication to get to the base station. For this reason, it is required to develop algorithms for the generated data streams to reduce the network traffic that affects the data quality. Trying to solve such limitations in sensor networks, it is the aim for researchers to propose distributed WSN data stream clustering algorithms to minimize sensor node energy consumption, extend the network lifetime, and reduce data traffic to decrease the delay found in such networks. This chapter briefly provides some important concepts in WSNs, data streams, data stream mining, and clustering algorithms. We pay special attention to the aforementioned concepts in WSNs. The rest of this chapter is organized as follows: Section 10.2 represents an overview of data mining. Section 10.3 presents a survey on data clustering. Section 10.4 presents a discussion on data streams. Section 10.5 presents a classification for clustering protocols. Finally, Section 10.6 presents the conclusion of this chapter and some recommendations.

10.2 DATA MINING

Data mining is the computational process of discovering patterns in large datasets involving methods at the intersection of artificial intelligence, machine learning, statistics, and database systems (Chakrabarti et al. 2006). Many databases in various areas have been generated from the development of information technology. Research in databases and information technology have given the importance to store and manipulate with such data for further decision making. Data mining is a process of extracting information and patterns from huge data (Ramageri 2010).

10.2.1 Common Classes of Data Mining

There are six common classes of tasks for data mining (Fayyad, Piatetsky-Shapiro, and Smyth 1996):

1. Anomaly detection, unusual data records identification, or data errors that require further investigation.
2. Association rule learning searches for relationships between variables.
3. Clustering is the task of discovering groups of data that are similar to each other.
4. Classification is the task of generalizing the known structure to apply to new data.
5. Regression allows one to find a function that models the data with the least error.
6. Summarization providing a compact representation of the dataset, including visualization and report generation.

10.2.2 Data Mining in Wireless Sensor Networks (WSNs)

Today many organizations have a lot of large databases that grow without limit at a rate of several million records per day. Mining these continuous data streams brings new challenges (Domingos and Hulten 2000). Managing and processing data in WSNs has become a research topic in several fields of data mining. The main purpose of deploying the WSNs is to make the real-time decision, which has been proven to be challenging due to many resource constraints. This challenge helps the research community to find data mining techniques dealing with extracting knowledge from large, continuously arriving data from WSNs. Traditional data mining techniques are not suitable for WSNs due to the nature of sensor data (Mahmood et al. 2013).

10.2.2.1 Challenges of Data Mining in WSNs

Conventional data mining techniques for handling sensor data in WSNs are challenging for the following reasons (Mahmood et al. 2013):

1. Resource constraint—The sensor nodes are resource constraints in terms of power, memory, communication bandwidth, and computational power.
2. Fast and huge data arrival—The nature of WSNs data is its high speed. In many fields, data arrives faster than they could be mined. The challenge for data mining techniques is how to cope with the continuous, rapid, and changing data streams.
3. Online mining—In WSNs, data is geographically distributed, inputs arrive continuously and so far newer data may change results based on older ones. Most data mining techniques that analyze data offline do not meet the requirement of handling distributed stream data.
4. Modeling changes of mining results over time—Data-generating phenomenon is changing over time, so the extracted model should be updated continuously.
5. Data transformation—Sensor nodes are limited in terms of bandwidth. So, transforming original data over the network is not easy.
6. Dynamic network topology—Sensor networks are deployed in harsh, uncertain, heterogenic, and dynamic environments. This can increase the complexity of designing an appropriate technique.

10.2.2.2 Taxonomy of Data Mining Techniques for WSNs

There are three main classification levels in data mining techniques for WSNs (Mahmood et al. 2013). The highest level is based on general data mining classes used, such as frequent pattern mining, sequential pattern mining, clustering, and classification. For the clustering, it had adapted the k-means, hierarchical, and data correlation based clustering. The second level of classification is based on the ability to process data on centralized or distributed manner. Since WSNs nodes have limited resources, the approach meant for distributed processing requires one-pass algorithms to complete a part of data mining locally, and then gathering the results. The distributed approaches are used to increase the WSNs lifetime, and can extract a large number of data. The third level (Mahmood et al. 2013) is selected based on how to face a specific problem. In WSNs, it has been focused on two aspects of issues: performance and application issues. Mainly, sensor nodes are constrained in some resources, so algorithms that are aware of such constraints are needed to maximize WSN performance. On the other hand, a WSN application requires data accuracy, fault tolerance, event prediction, scalability, and robustness.

10.2.2.3 Application Areas of WSN Data Mining

The following are examples of real-world applications in WSN data mining (Mahmood et al. 2013):

1. In the environmental monitoring, sensors are deployed in an unattended region to monitor the natural environment.
2. For the health monitoring, patients are equipped with small sensors on multiple different positions of their body to monitor their health or behavior.
3. Sensors in object tracking are embedded in moving targets to track them in real time.

4. WSNs are usually deployed in harsh environments. Sensor nodes are resource constrained especially in terms of power. Data mining techniques help to identify the faulty or dead nodes.
5. In data analysis, data mining techniques help to discover data patterns in a sensor network for a certain application.
6. In real-time monitoring, data mining techniques help to identify certain patterns and predict future events.

10.2.2.4 Implementation of WSN Data Mining

Three main techniques are used for data mining implementation in WSNs (Mahmood et al. 2013):

1. Evaluation method—Analytical modeling, simulation, and real deployment (unfeasible) are the most commonly used techniques to analyze the performance of data mining for WSNs.
2. Data source—The dataset used to experimentally validate the proposed technique. Two types of datasets are used: synthetic and real.
3. Optimization objective—WSNs are constrained in different resources. Techniques should consider those constraints, but mostly they cannot efficiently cover all the performance metrics.

10.3 DATA CLUSTERING

Owing to the advances and growth in wireless communication technology, WSNs are becoming increasingly attractive for a lot of application areas. Thus, WSNs connect the physical world, the computing world, and the human society. The clustering phenomenon plays an important role in affecting the network performance and organizing the networks as well. There are several key limitations in WSNs that clustering schemes must consider (Dechene et al. 2006). A good clustering algorithm should be able to adapt to a variety of application requirements. Data clustering can be identified as grouping similar objects. In order to support data aggregation through efficient network organization, nodes can be partitioned into a number of small groups called clusters. Each cluster has a coordinator called a cluster head (Younis, Krunz, and Ramasubramanian 2006).

10.3.1 Hierarchical Clustering Structure

Grouping sensor nodes into clusters has been widely used to satisfy the scalability objective and achieve high-energy efficiency to prolong network lifetime in large-scale WSNs. The hierarchical routing and data gathering protocols imply cluster-based organization of the sensor nodes so that data fusion and aggregation are possible, which leads to a great energy savings. In the hierarchical network structure each cluster has a leader called the cluster head (CH) that performs special tasks (i.e., fusion and aggregation) and several common sensor nodes (SNs) as members (Mamalis et al. 2009). The cluster formation leads to a two-level hierarchy where the cluster heads form the higher level and the sensor nodes form the lower level. The sensor nodes periodically transmit their data to the corresponding cluster heads. The cluster heads aggregate the data (thus decreasing the total number of relayed packets) and transmit them to the base station (BS). This can be directly or through the intermediate communication with other cluster heads. Cluster heads spend a lot of energy more than other sensor nodes due to sending the aggregated data to higher distances. A common solution to balance the energy consumption among all the nodes in the network is to periodically reelect new cluster heads in each cluster (Mamalis et al. 2009). The BS is the point of data processing for the data received from the sensor nodes and where the data is accessed by the end user.

10.3.2 Designing Clusters in WSNs

There are several key attributes that designers must carefully consider when designing clustered wireless sensor networks (Dechene et al. 2006):

1. Cost of clustering—Resources other than the network organization should be considered, such as communication and processing.
2. Selection of CHs—Depending on certain applications, some requirements may play an important role in its operations.
3. Real-time operation—For some applications such as habitat monitoring, simply receiving data is enough for analysis. For other applications like military tracking, the real-time data acquisition is much more vital.
4. Synchronization—Slotted transmission schemes allow nodes to regularly schedule sleep intervals to minimize energy consumption.
5. Data aggregation—In crowded networks, there are many nodes sensing similar data. Data aggregation allows distinguishing between sensed data and useful data.
6. Repair mechanisms—WSNs are often vulnerable to node mobility, node death, and interference that can result in link failure.
7. Quality of service (QoS)—Many QoS requirements in WSNs are application dependent. It is important to consider these metrics when choosing a clustering scheme.

10.3.3 Clustering Parameters

Clustering has many parameters, listed as follows (Abbasi and Younis 2007; Mamalis et al. 2009):

1. Number of clusters (cluster count)—Cluster count is a critical parameter with regard to efficiency of the total routing protocol.
2. Intracluster communication—The communication between a sensor and its CH is assumed to be one-hop communication. However, multihop communication is required when the communication range is limited.
3. Nodes and CH mobility—When CHs or nodes are assumed to be mobile, the cluster membership for each node should dynamically change. On the other hand, stationary CH tends to yield stable clusters and facilitate intracluster and intercluster network management.
4. Node types and roles—In heterogeneous networks, CHs are able to have more computation and communication resources. On the other hand, in homogeneous networks, all nodes have the same capabilities and some are designated as CHs.
5. Cluster formation methodology—Clustering mostly is performed in a distributed manner without coordination. In a few earlier approaches, a centralized approach uses one or more coordinator nodes to partition the whole network off-line.
6. Cluster-head selection—CHs can be preassigned in heterogeneous environments. In most cases, for homogeneous environments, CHs are selected from the deployed set of nodes.
7. Stability—A clustering scheme is said to be adaptive when the cluster count changes and the node's membership evolves overtime. Otherwise, it is considered fixed.
8. Multiple levels—The concept of a multilevel cluster hierarchy provides better energy distribution and total energy consumption.

Many clustering protocols, shown in Table 10.1, are considered to be distributed. They could randomly select their CHs but have a limited node mobility such as the LEACH and TL-LEACH. The LEACHC randomly selects CHs and has a limited node mobility, but is considered to be a centralized clustering protocol. LCA is a distributed clustering protocol with a possible node mobility and an ID-based selection for CHs. Some distributed protocols are based on the highest energy for

TABLE 10.1
Clustering Algorithms with Comparable Parameters

Protocol	CH Selection	Node Mobility	Clustering Methodology	Multiple Levels
LCA	ID based	Possible	Distributed	No
LEACH	Random	Limited	Distributed	No
TL-LEACH	Random	Limited	Distributed	Yes
GROUP	Proximity	No	Hybrid	No
EECS	Energy	No	Distributed	No
WCA	Weight based	Yes	Distributed	No
ACE	Connectivity	Possible	Distributed	No
LEACHC	Random	Limited	Centralized	No

CH selection with no node mobility such as the EECS. On the other hand, some depend on the connectivity in CH selection with possible node mobility such as the ACE. The distributed WCA uses a weight-based CH selection with node mobility. GROUP is a hybrid clustering protocol with no node mobility. Table 10.1 compares some clustering algorithms.

10.4 DATA STREAMS

Sensor networks are the key to gathering information needed by smart environments. In many emerging applications, huge data streams are monitored in a network environment. Each sensor generates a data stream where new data entries keep arriving in a continuous manner (Soroush, Wu, and Pei 2008). Data that WSNs process usually arrives in an online fashion, is unlimited, and there is no control in the arrival order of the elements to be processed (de Aquino et al. 2007a, b). However, there is a difference between a sensor stream and traditional stream. Data streams arrive continuously, thus clustering algorithms have to perform in a single scan. An important characteristic of data streams is mining them in a distributed fashion. Individual processors may have limited processing and memory. Examples of such cases include sensor networks, in which it may be desirable to perform network processing of data stream with limited processing and memory (Aggarwal 2007).

10.4.1 Traditional Data Mining and Data Stream Mining

Traditional data mining is centralized, computationally expensive, and focuses on disk-resident transactional data. It collects data at the central site. On the other hand, a WSN's data flows continuously in systems with varying update rates. It is impossible to store the WSN's entire data or to scan through it multiple times (Mahmood et al. 2013).

10.4.2 Data Stream Characteristics

A data stream has different characteristics of data collection to the traditional database model. Such characteristics are when the data stream arrives, it is not easy to be controlled by the order. A data stream is continuous and dynamic. Additionally, a data stream can be read and processed based on the arrival order (Su, Liu, and Song 2011). Based on that, the processing of the data stream, first, requires each data element should be examined once at most. Second, each data element should be processed as soon as possible. Third, memory usage for mining data streams should be limited even though new data elements are continuously generated. Finally, results generated by online algorithms should be immediately available upon user request (Su, Liu, and Song 2011).

10.4.3 Algorithms of Data Streams

Due to the one-pass constraints on the data set, it is difficult to adapt arbitrary clustering algorithms to data streams. In the context of data streams, it may be better to determine clusters in specific user-defined horizons rather than on the entire data set. The microclustering technique determines clusters over the entire dataset (Aggarwal 2007). To be more general, there are many algorithms found to handle the data streams through various environments. Such algorithms include data stream clustering, data stream classification, frequent pattern mining, change detection in data streams, stream cube analysis of multidimensional streams, load-shedding in data streams, sliding window computations in data streams, synopsis construction in data streams, join processing in data streams, indexing data streams, dimensionality reduction and forecasting in data streams, distributed mining of data streams, and stream mining in sensor networks.

10.5 CLUSTERING PROTOCOLS

Clustering protocols could be generally classified into three main types in our research scope: first, clustering routing protocols in WSNs (without data streams); second, clustering protocols for data streams; and third, clustered WSNs for data streams.

10.5.1 Clustering Routing Protocols in WSNs

Based on network structure, routing protocols in WSNs can be divided into two categories: flat and hierarchical routing. In a flat network topology, all nodes perform the same tasks and have the same functionalities. Data transmission is performed hop by hop using flooding form. The typical flat routings in WSNs include Flooding and Gossiping, SPIN, Directed Diffusion (DD), Tumor, Greedy Perimeter Stateless Routing (GPSR), Trajectory-Based Forwarding (TBF), Energy-Aware Routing (EAR), Gradient-Based Routing (GBR), and Sequential Assignment Routing (SAR). Flat routing protocols are effective in small-scale networks (Liu 2012). On the other hand, in hierarchical topology, nodes perform different tasks and are organized into lots of clusters according to specific requirements or metrics. Generally, CHs have the highest energy in the clusters to perform data processing and information transmission, whereas nodes with low energy act as member nodes (MNs) and perform the task of information sensing (Liu 2012).

Clustering routings protocols in WSNs include LEACH, HEED, Distributed Weight-Based Energy-Efficient Hierarchical Clustering protocol (DWEHC), Position-Based Aggregator Node Election protocol (PANEL), Two-Level Hierarchy LEACH (TL-LEACH), Unequal Clustering Size (UCS), Energy Efficient Clustering Scheme (EECS), EEUC, ACE, BCDCP, PEGASIS, Threshold Sensitive Energy Efficient Sensor Network protocol (TEEN), APTEEN, Two-Tier Data Dissemination (TTDD), Concentric Clustering Scheme (CCS), and HGMR.

10.5.2 Proactive and Reactive Clustering

Proactive clustering algorithms are based on the assumption that the sensors always have data to send; for that reason, they should all be considered during the cluster formation. On the other hand, reactive algorithms take advantage of user queries for the sensed data or of specific triggering events that occur in the WSN. Namely, nodes may react immediately to sudden hard changes in the value of a sensed attribute (Mamalis et al. 2009). Figure 10.1 shows a flowchart for some examples of proactive and reactive clustering protocols.

10.5.3 Clustering Algorithms Schemes in WSNs

Clustering algorithms could be considered under specific schemes, such as hierarchical, grid, heuristic, weighted, Particle Swarm Optimization (PSO)-based. Each scheme will be described in brief. Figure 10.2 shows the clustering schemes in WSNs.

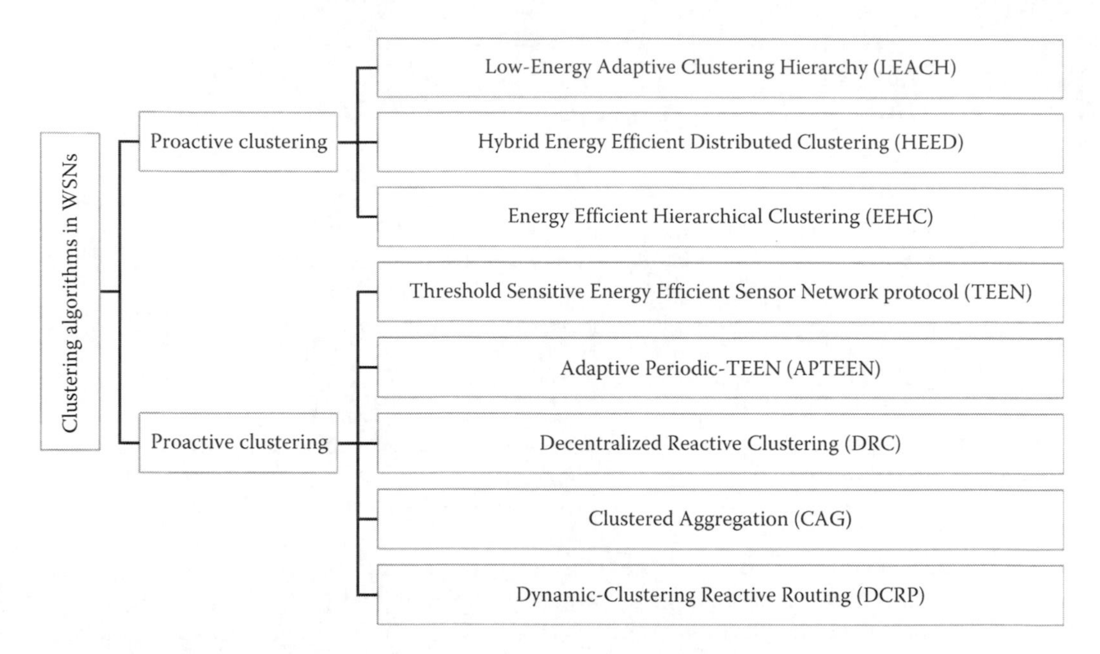

FIGURE 10.1 Examples of proactive and reactive clustering protocols.

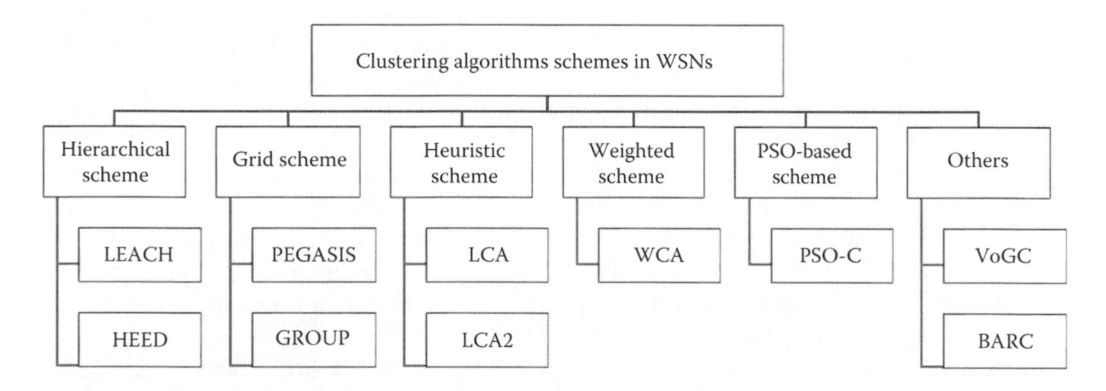

FIGURE 10.2 Clustering algorithms schemes in WSNs.

1. Hierarchical schemes—In LEACH, there are many rounds and each round has two main phases, a setup phase and steady-state phase (Kumar, Jain, and Tiwari 2011). To reduce intercluster and intracluster collisions in LEACH, it uses a TDMA or CDMA MAC. The energy consumption of the information gathered by the sensors node will depend on the number of CHs and radio range of different algorithms (Kumar, Jain, and Tiwari 2011). Table 10.2 gives some examples of LEACH descendants. Another protocol known as Hybrid Energy-Efficient Distributed Clustering (HEED) is a multihop clustering algorithm for WSNs. It focuses on efficient clustering by proper selection of CHs based on the physical distance between nodes. HEED has several objectives, where it distributes energy consumption to prolong network lifetime. It also minimizes energy during the CH selection phase. Moreover, it can minimize the control overhead of the network. In HEED, CH selection is determined based on the residual energy (Dechene et al. 2006; Mitra and Nandy 2012).
2. Grid schemes—Power-efficient gathering in sensor information systems (PEGASIS) is a data-gathering algorithm that establishes the concept that energy savings can result from nodes not directly forming clusters. If nodes form a chain from source to sink, only one node in any

TABLE 10.2
Descendants of LEACH Protocol

Descendant Abbreviation	Descendant Name
LEACH	Low energy adaptive clustering hierarchy
LEACH-C	Centralized—Low energy adaptive clustering hierarchy
LEACH-B	Balanced—Low energy adaptive clustering hierarchy
LEACH-ET	Energy threshold—Low energy adaptive clustering hierarchy
TL-LEACH	Three layer—Low energy adaptive clustering hierarchy
Armor-LEACH	Advance LEACH routing protocol for microsensor networks
O-LEACH	Optical—Low energy adaptive clustering hierarchy
MR-LEACH	Multihop hop routing—Low energy adaptive clustering hierarchy
LEACH-D	Low energy adaptive clustering hierarchy-D

given transmission timeframe will be transmitting to the base station. Data fusion appears at every node in the sensor network allowing for all relevant information to permeate across the network. Moreover, the average transmission range required by a node to relay information can be much less than in LEACH (Dechene et al. 2006; Mitra and Nandy 2012). The group algorithm is another grid-based clustering algorithm. In this algorithm one of the sinks dynamically and randomly builds the cluster grid. Each new CH will then select more CHs along the grid until all CHs have been selected (Dechene et al. 2006; Mitra and Nandy 2012).

3. Heuristic algorithms—These algorithms have one or two goals when solving a problem. First, finding an algorithm with reasonable run time, and, second, finding the optimal solution. There are many types of heuristic algorithms that exist in choosing CHs. The linked cluster algorithm (LCA) was one of the very first clustering algorithms developed. In LCA, each node is assigned to a unique ID and can become a CH if a node has the highest ID number or assuming none of its neighbors are cluster heads, then it becomes a CH (Dechene et al. 2006; Mitra and Nandy 2012). LCA2 was proposed to eliminate the election of an unnecessary number of CHs, as in LCA. LCA2 introduced the concept of a node being covered and noncovered. A node is covered when one of its neighbors is CH. CH election is done by starting with the node having the lowest ID among noncovered neighbors (Dechene et al. 2006; Mitra and Nandy 2012).
4. Weighted schemes—A weighted clustering algorithm (WCA) is a nonperiodic procedure to the CH election, invoked when reconstruction of the networks topology is unavoidable. WCA tries to find a long-lasting architecture during the first CH election. When a sensor loses the connection with its cluster head, the election procedure is invoked to find a new clustering topology. WCA is based on a combination of metrics such as the ideal node degree, transmission power, mobility, and the remaining energy of nodes (Dechene et al. 2006; Mitra and Nandy 2012).
5. PSO-based scheme—In Centralized PSO (PSO-C), nodes that have energy above the average energy resource are elected as CHs. Simulation results show that PSO outperforms LEACH and LEACH-C in terms of network lifetime and throughput (Kumar, Jain, and Tiwari 2011).
6. Other schemes—Voting-on-grid clustering (VOGC) is a combination of a voting method and clustering algorithm developing new clustering schemes for secure localization of sensor networks. VOGC is used instead of traditional clustering algorithms to reduce the computational cost. It is found that the scheme can provide good localization accuracy and identify a high degree of malicious beacon signals (Kumar, Jain, and Tiwari 2011). A mathematical battery model for implementation in WSNs was used in the battery aware reliable clustering (BARC) algorithm. It improves the performance over other clustering

algorithms due to using Z-MAC and rotating the CHs according to battery recovery schemes. Moreover, the BARC consists of two stages per round for selection of CH: initialization or setup, and steady state (Kumar, Jain, and Tiwari 2011).

10.5.4 Clustering Protocols for Data Streams

In 2006, Feng Cao (Jia, Tan, and Yong 2008) proposed the DenStream algorithm for clustering dynamic data stream. It is an effective and efficient method that can discover clusters of arbitrary shape in data streams, but it is insensitive to noise (Cao et al. 2006). The algorithm extends the microcluster concept, and introduces the outlier and potential microclusters to distinguish between real data and outliers. Heng Zhu Wei (Jia, Tan, and Yong 2008) proposed a density and space clustering algorithm called CluStream. It is a data-stream clustering algorithm based on k-means that is inefficient at finding clusters of arbitrary shapes and cannot handle outliers. Further, they require knowledge of k and a user-specified time window (Chen and Tu 2007). The DenStream and CluStream algorithms are not able to reveal clusters of arbitrary shapes effectively and cannot distinguish clusters that have different levels of density (Jia, Tan, and Yong 2008).

The k-means algorithm is used in the offline phase of some algorithms such as CluStream. It is a divide-and-conquer scheme that partitions data streams into segments and discover clusters in data streams. The k-means has a number of limitations. First, it aims at identifying spherical clusters but is incapable of revealing clusters of arbitrary shapes. Second, it is unable to detect noise and outliers. Third, the algorithm requires multiple scans of the data, making it not directly applicable to large-volume data streams (Jia, Tan, and Yong 2008; Chen and Tu 2007). STREAM and CluStream are two well-known extensions of k-means on data streams (Amini, Saboohi, and Wah 2013). Figure 10.3 shows some algorithms based on k-means and fuzzy c-means (FCM).

Many recent data stream clustering algorithms are based on CluStream's two-phase framework. Wang et al. proposed an improved offline component using an incomplete partitioning strategy (Chen and Tu 2007). Extensions of this component include clustering multiple data streams, parallel data streams, distributed data streams, and applications of data stream mining (Chen and Tu 2007).

LOCALSEARCH, STREAM, DenStream, and CluStream are clustering algorithms evolving data streams. They have ignored the problems of the grid border. The data stream is coming with a large number in chronological order, making the original grid no longer adaptable to the new data mapping, so a large number of data is likely to fall on the grid border. But if it is simply discarded, the cost will be greatly increased and the efficiency will be affected (Jia, Tan, and Yong 2008).

D-Stream is a density grid-based algorithm in which the data points are mapped to the corresponding grids and the grids are clustered based on the density (Amini, Saboohi, and Wah 2013). MR-Stream is an algorithm that can cluster data streams at multiple resolutions. The algorithm partitions the data space into cells and a tree-like data structure keeps the space partitioning. The MR-Stream increases the performance of clustering by determining the exact time to generate the

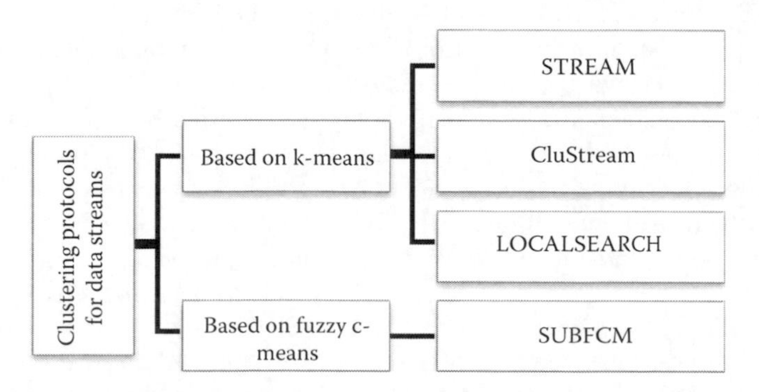

FIGURE 10.3 Clustering algorithms based on k-means and FCM.

clusters (Amini, Saboohi, and Wah 2013). FlockStream is a density-based clustering algorithm based on a bioinspired model. It uses the flocking model where agents are microclusters and they work independently. FlockStream merges online and offline phases. It can get the clustering results without performing offline clustering (Amini, Saboohi, and Wah 2013). DenStream, MR-Stream, D-Stream, and FlockStream are based on density-based clustering. They can effectively detect arbitrary shape clusters and handle noise, but their quality decreases when they are used for clusters with varying densities (Amini, Saboohi, and Wah 2013).

The LOCALSEARCH algorithm uses dividing and conquering to partition data streams into segments, and discovers clustering of data streams in finite space by using the k-means algorithm (Jia, Tan, and Yong 2008). Later on, the STREAM algorithm was proposed by O'Callaghan, which is based on LOCALSEARCH (Jia, Tan, and Yong 2008). It puts equal weight on outdated and recent data, and cannot capture the evolving characteristics of data stream (Jia, Tan, and Yong 2008).

Data stream clustering analysis causes challenges for traditional clustering algorithms. The data can only be examined in one pass. Viewing data stream as a long vector of data is not enough in many applications (Chen and Tu 2007).

Incremental DBSCAN is a method for data warehouse applications. It can only handle a relatively stable environment, and it cannot deal with limited memory and fast-changing streams. The HPStream introduces the concept of projected cluster to data streams (Cao et al. 2006).

A framework to dynamically cluster multiple-evolving data streams called Clustering on Demand (COD) was proposed (Yin and Gaber 2008). It produces a summary hierarchy of data statistics in the online phase, whereas the clustering is performed in the offline phase (Yin and Gaber 2008). It summarizes the data streams using the discrete Fourier transform (DFT). Online Divisive-Agglomerative Clustering (ODAC) was proposed to incrementally construct tree-like hierarchy.

Many density-based clustering algorithms are not suitable for data stream environments. They need two passes of the data and this condition is impossible for data streams. GMDBSCAN and ISDBSCAN use two-pass data. Other algorithms have high execution time, which makes them not applicable for data streams. DSCLU is density-based clustering for data streams in multidensity environments (Amini, Saboohi, and Wah 2013).

DD-Stream is framework for density-based clustering stream data. The algorithm adopts a density decaying technique to capture the evolving data stream and extracts the boundary point of the grid by using the DCQ-means algorithm. DD-Stream has better scalability in processing large-scale and high-dimensional stream data (Jia, Tan, and Yong 2008). D-Stream is a density-based clustering real-time stream data algorithm. It uses an online component that maps each input data record into a grid. It also has an offline component that computes the grid density (Chen and Tu 2007).

10.5.5 Clustered WSNs for Data Streams

A distributed WSN data stream clustering algorithm called subtractive fuzzy cluster means (SUBFCM) was proposed to minimize sensor nodes' energy consumption and extend the network lifetime. Simulations show that the energy efficient algorithm SUBFCM can achieve WSN data stream clustering with significantly less energy than that required by known fuzzy c-means and k-means algorithms (Sabit, Al-Anbuky, and Gholam-Hosseini 2009). The proposed SUBFCM is a result of blending the subtractive clustering and FCM algorithm. Fuzzy c-means is the most widely used algorithm in the field of data mining (Sabit, Al-Anbuky, and Gholam-Hosseini 2009).

In wireless multimedia sensor networks (WMSNs), multimedia clustering protocols use the QoS parameters (Diaz et al. 2014). QoS has several metrics such as delay, bandwidth, reliability, jitter (Abazeed et al. 2013), and packet loss (Diaz et al. 2014). Many multimedia applications are time critical; they need to be reported within a limited time. Multimedia sensors have the ability to capture video, image, audio, and scalar sensor data. A clustering algorithm for WMSNs has been proposed based on field of view (FoV) areas. This algorithm aims to find the intersection polygon

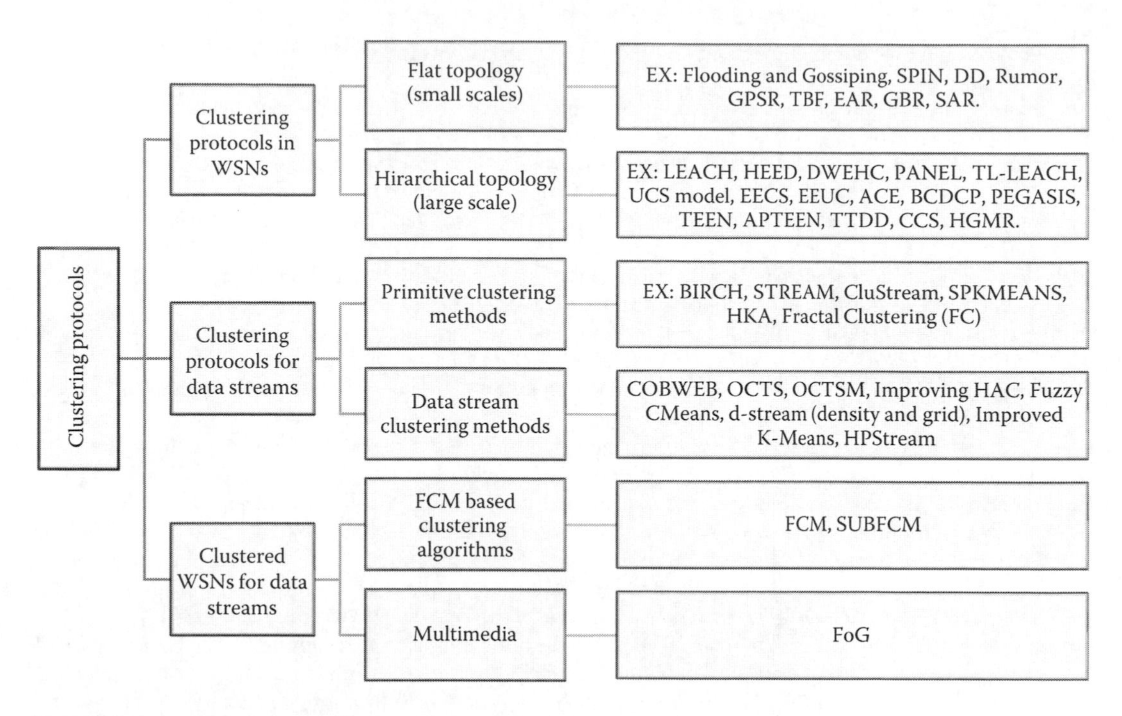

FIGURE 10.4 Classification of clustering algorithms.

and compute the overlapped areas to establish clusters and determine cluster membership. For dense networks, overlapping FoVs causes consuming power of the system (Kumar, Jain, and Tiwari 2011).

10.5.6 Taxonomy of Clustering Protocols

As previously mentioned, clustering protocols are classified in this research in three main types. To summarize, Figure 10.4 shows a structure that describes in brief the classification of clustering algorithms.

10.6 CONCLUSION

The past few years have witnessed increased interest in the potential use of WSNs in a wide range of applications, and they have become a hot research area in the field of data mining. WSNs are special networks that consist of sensor nodes in large numbers and spatial distribution. They have several sensing capabilities and cooperate to accomplish common tasks. This research focuses on the most important concepts of WSNs, data stream mining, and data streams and clustering algorithms. It also provides an overview comparison between clustering routing protocols in WSNs. It gives a classification of some clustering protocols. Generally, they are classified into three main types: clustering routing protocols in WSNs (without data streams), clustering protocols for data streams, and clustered WSNs for data streams. Clustering has proven to be one of the most effective techniques used to solve the problem of energy consumption in WSNs. The optimum cluster protocol depends mainly on the application being used. It is recommended to obtain acknowledgment from the field of clustering protocols in WSNs and algorithms within this research area. In addition, fuzzy clustering in WSNs using streaming data is recommended.

REFERENCES

Abazeed, M., N. Faisal, S. Zubair, and A. Ali. 2013. Routing protocols for wireless multimedia sensor network: A survey. *Journal of Sensors* 2013:1–13.

Abbasi, A. A. and M. Younis. 2007. A survey on clustering algorithms for wireless sensor networks. *Computer Communications* 30(14):2826–2841.

Abdullah, M., H. N. Eldin, T. Al-Moshadak, R. Alshaik, and I. Al-Anesi. 2015. Density grid-based clustering for wireless sensors networks. *Procedia Computer Science* 65:35–47.

Aggarwal, C. C. 2007. *Data Streams: Models and Algorithms*. New York: Springer Science & Business Media.

Ali, M. A., and A. K. Rajpoot. 2014. Development of energy efficient routing protocol using Hop PEGASIS in wireless sensor networks. *International Journal of Computer Science & Engineering Technology (IJCSET)* 5(2):124–131.

Alia, O. M. 2014. A decentralized fuzzy C-means-based energy-efficient routing protocol for wireless sensor networks. *The Scientific World Journal* 2014: article 647281.

Amini, A., H. Saboohi, and T. Y. Wah. 2013. A multi density-based clustering algorithm for data stream with noise. *2013 IEEE 13th International Conference on Data Mining Workshops (ICDMW)*, pp. 1105–1112.

Boyinbode, O., H. Le, A. Mbogho, M. Takizawa, and R. Poliah. 2010. A survey on clustering algorithms for wireless sensor networks. *2010 13th International Conference on Network-Based Information Systems*, pp. 358–364.

Cao, F., M. Ester, W. Qian, and A. Zhou. 2006. Density-based clustering over an evolving data stream with noise. *Proceedings of the 2006 SIAM International Conference on Data Mining*, pp. 328–339.

Chakrabarti, S., M. Ester, U. Fayyad, J. Gehrke, J. Han, S. Morishita, G. Piatetsky-Shapiro, and W. Wang. 2006. Data mining curriculum: A proposal (Version 1.0). *Intensive Working Group of ACM SIGKDD Curriculum Committee*.

Chen, Y. and L. Tu. 2007. Density-based clustering for real-time stream data. *Proceedings of the 13th ACM SIGKDD International Conference on Knowledge Discovery and Data Mining,* pp. 133–142.

de Aquino, A. L. L., C. M. S. Figueiredo, E. F. Nakamura, L. S. Buriol, A. Loureiro, A. O. Fernandes, and C. J. N. Coelho. 2007a. Data stream algorithms for processing of wireless sensor network application data. *21st International Conference on Advanced Information Networking and Applications*, Niagara Falls, Ontario, May 21–23, pp. 869–876.

de Aquino, A. L. L., C. M. S. Figueiredo, E. F. Nakamura, L. S. Buriol, A. Loureiro, A. O. Fernandes, and C. J. N. Coelho. 2007b. A sampling data stream algorithm for wireless sensor networks. *IEEE International Conference on Communications,* Glasgow, June 24–28, pp. 3207–3212.

Dechene, D. J., A. El Jardali, M. Luccini, and A. Sauer. 2006. A survey of clustering algorithms for wireless sensor networks. University of Western Ontario, Canada.

Diaz, J. R., J. Lloret, J. M. Jimenez, and J. J. P. C. Rodrigues. 2014. A QoS-based wireless multimedia sensor cluster protocol. *International Journal of Distributed Sensor Networks* 2014: article 480372.

Domingos, P. and G. Hulten. 2000. Mining high-speed data streams. *Proceedings of the Sixth ACM SIGKDD International Conference on Knowledge Discovery and Data Mining*, Boston, August 20–23, pp. 71–80.

Fayyad, U., G. Piatetsky-Shapiro, and P. Smyth. 1996. From data mining to knowledge discovery in databases. *AI Magazine* 17(3):37.

Huang, J. and J. Zhang. 2011a. Distributed dual cluster algorithm based on FCM for sensor streams. *Advances in Information Sciences and Service Sciences* 3(5):201–209.

Huang, J. and J. Zhang. 2011b. Fuzzy c-means clustering algorithm with spatial constraints for distributed WSN data stream. *International Journal of Advancements in Computing Technology* 3(2):165–175.

Huang, J. and J. Zhang. 2011c. Rough set based generalized fuzzy c-means clustering algorithm with spatial constraints for distributed WSN data stream. *Journal of Convergence Information Technology* 6(11):259–267.

Jia, C., C. Y. Tan, and A. Yong. 2008. A grid and density-based clustering algorithm for processing data stream. *Second International Conference on Genetic and Evolutionary Computing, 2008*, pp. 517–521.

Khan, J. M., and A. Seth. 2014. Performance comparison of FCM and k-mean clustering technique for wireless sensor network in terms of communication overhead. *Global Journal of Advanced Engineering Technologies and Science* 1(2):26–29.

Kumar, V., S. Jain, and S. Tiwari. 2011. Energy efficient clustering algorithms in wireless sensor networks: A survey. *International Journal of Computer Science* 8(5):259–268.

Liu, X. 2012. A survey on clustering routing protocols in wireless sensor networks. *Sensors* 12(8):11113–11153.

Mahmood, A., K. Shi, S. Khatoon, and M. Xiao. 2013. Data mining techniques for wireless sensor networks: A survey. *International Journal of Distributed Sensor Networks* 9(7): article 71.

Mamalis, B., D. Gavalas, C. Konstantopoulos, and G. Pantziou. 2009. Clustering in wireless sensor networks. In: *RFID and Sensor Networks: Architectures, Protocols, Security and Integrations*, Y. Zhang, L. T. Yang, J. Chen, eds., pp. 324–353. Boca Raton, FL: Taylor & Francis.

Mitra, R. and D. Nandy. 2012. A survey on clustering techniques for wireless sensor network. *International Journal of Research in Computer Science* 2(4):51.

Ramageri, B. M. 2010. Data mining techniques and applications. *Indian Journal of Computer Science and Engineering* 1(4):301–305.

Sabit, H. 2012. Distributed incremental data stream mining for wireless sensor network. Auckland University of Technology.

Sabit, H. and A. Al-Anbuky. 2014. Multivariate spatial condition mapping using subtractive fuzzy cluster means. *Sensors* 14 (10):18960–18981.

Sabit, H., A. Al-Anbuky, and H. Gholam-Hosseini. 2009. Distributed WSN data stream mining based on fuzzy clustering. *Symposia and Workshops on Ubiquitous, Autonomic and Trusted Computing*, pp. 395–400.

Soroush, E., K. Wu, and J. Pei. 2008. Fast and quality-guaranteed data streaming in resource-constrained sensor networks. *Proceedings of the 9th ACM International Symposium on Mobile ad hoc Networking and Computing*, pp. 391–400.

Su, L., H.-y. Liu, and Z.-H. Song. 2011. A new classification algorithm for data stream. *International Journal of Modern Education and Computer Science (IJMECS)* 3(4):32.

Thangavelu, A. and A. Pathak. 2014. Clustering techniques to analyze communication overhead in wireless sensor network. *International Journal of Computational Engineering Research (IJCER)* 4(5): 2250–3005.

Yin, J., and M. M. Gaber. 2008. Clustering distributed time series in sensor networks. *Eighth IEEE International Conference on Data Mining*, pp. 678–687.

Younis, O., M. Krunz, and S. Ramasubramanian. 2006. Node clustering in wireless sensor networks: Recent developments and deployment challenges. *IEEE Network* 20(3):20–25.

11 Video Codecs Assessment over IPTV Using OPNET

Eman S. Sabry, Rabie A. Ramadan,
M. H. Abd El-Azeem, and Hussien ElGouz

CONTENTS

11.1 Introduction 153
11.2 Literature Review 154
11.3 IPTV Network Configuration over OPNET 155
  11.3.1 IPTV Hardware 155
11.4 IPTV Networks Protocols 156
11.5 Designed and Implemented IPTV Network Model 156
  11.5.1 IPTV Network Performance Characterization 157
11.6 Simulation Results 158
  11.6.1 Video Coding Traffic Generation (Background Traffic Generation) 158
  11.6.2 Experiment 1: Evaluating Network IPTV with Uncompressed Video Transmission at Different Data Rates 159
  11.6.3 Experiment 2: Evaluating Network Performance in Case of Coded Video Transmission 160
11.7 Conclusion 163
References 164

11.1 INTRODUCTION

Nowadays, there are five primary networking platforms to distribute TV content: wireless off-air, satellite, DSL, fiber, and cable TV networks. The industry is going through a profound conversion, moving from conventional TV to a new era of Internet protocol (IP) technological innovation. Indeed switching to the new trend leads to more consumer choices, Internet applications variety, and household numbers growth, as it is realized that Internet accessibility and download streaming data rates are advancing to multiple of megabits per second (Mbps). Moreover, the difference in received service quality is especially recognized when watching on high-definition (HD) LCD TV sets. In fact, the usage of IP allows repeat of video content material (IP-based television service) from the web known as WebTV. However, WebTV does not ensure high quality of service (QoS). As a result, the telecom organizations released the so-called IPTV (Held, 2007; International Telecommunication Union, 2008; *Wikipedia*, 2015) to overcome the inadequacies of WebTV. IPTV has very stringent QoS and quality of experience (QoE) technology, since video content is delivered over dedicated private and secured fixed geographical area networks.

Lossy compression can be used for audio/video AV in which it approximates the media data rate. H.264/AVC technology is considered the most popular and dominant lossy compression method used by commercial IPTV providers. Nowadays, service providers encounter users demanding highly progressive video services and other entertainment services as three-dimensional (3D) movies, games, and high-video quality. Typically, MV contains two views of a video taken from different perspectives, whereby each view consists of a sequence of video frames (pictures). These two different views could be displayed to give viewers the perception of depth; this is commonly referred to as 3D video. 3D video delivery over transport network poses significant challenges because of the data increase of two views of a video scene compared to conventional SV video. Therefore, it is required to have efficient video compression (coding) techniques and sufficient transport mechanisms to accommodate the large video data volume from the two views on limited bandwidth transmission links. MV is an amendment to the H.264 (MPEG-4 AVC) video compression standard.

Hence, efficient coding techniques for MV video have been extensively researched in recent years, but it is still an open area for research. Therefore, the industry keeps looking for the global benchmark for video compression, which is the International Telecommunication Union (ITU) and its partners, since ITU-T H.264 has underpinned expansion and rapid progress. HEVC or ITU-T H.265 was jointly developed to double the video data compression ratio as compared to its predecessor ITU-T H.264/MPEG-4 Part 10, Advanced Video Coding (AVC) at the same level of video quality or better. HEVC opens the future door for video transmission only using half of the bandwidth (bit rate) compared to its predecessor, which currently accounts for over 80% of all web video. H.265/HEVC, the latest advance video coding, has emerged as the video coding standard and 3D video coding serving multimedia communications.

The goal of this chapter is to give deep insight into IPTV technology with existing protocols, describing how to work with various protocol layers. It describes the literature explaining recent relevant IPTV research. The chapter describes the concepts of IPTV network implementation as a part of the Optimized Network Engineering Tools (OPNET) simulator. It reveals the expected network performance based IPTV. It also includes an explanation of different traffic types modeling compressed and uncompressed TV channels that will be imported over implemented IPTV network. A section contains a set of experiments and scenarios each with a complete simulation focused on performance evaluation of the impact for each configured protocol on their operational environment. In addition, the chapter includes network performance evaluation subjected to uncompressed video delivery at different data rates, highlighting the impact of compression not only on bandwidth saving but also on received video quality and QoE on the user side. The next contribution is to evaluate network performance concerning different application issues related to the delivery of coded SV and MV videos in the currently used H.264 codecs format. Moreover, it includes an IPTV network performance evaluation subjected to the coded video channels delivery in the recently introduced High Efficiency Video Coding (HEVC) standard format at the same resolution and over the same network. The provided performance evaluation includes assessment for both standard codecs, showing that different limitations arise from the behalf of H.264 deployment involved with the delivery increase of HD or MV channels. This will make it easy on researchers as well as the industry to examine their networks before deployment. The last part of this chapter provides complete review of collected QoS parameters from each examined compression standard and hence could presage the quality of received channel (service) and QoE received by the user side at each case.

11.2 LITERATURE REVIEW

Recently, several works have been introduced based on performance studies of video streaming over IPTV networks. For example, Moughit, Abdelmajid, and Sahel (2013) studied the effect of Cisco Group Management Protocol (CGMP) on the IPTV performance in terms of throughput and delay. However, the authors ignored the details of the IPTV architecture including routing and queuing mechanisms; in addition, the article works on a very simplified version of IPTV to utilize only two scenarios with CGMP enabled and disabled in terms of delay and throughput only. Singh and Amit (2013) describe results from an IPTV VoD transportation assessment over WIMAX that packet delay variation (PDV), packet end-to-end (ETE) delay, and delay decreases with little load increase by the increase of WIMAX mobile subscriber mobility. Similarly, Hamodi and Thool (2013) introduce a study for IPTV performance analysis over WIMAX broadband access technology. Indeed, previous efforts were made communicating IPTV assessment only from the perspective of subscriber mobility. System evaluation judgment lacks examination subjected to a variety of practical application effects to assimilate more channels and users in both objective and subjective merits.

Maraj, Shehu, and Mitrushi (2011) assess QoS parameters and QoS requirements for delivering IPTV services transmitting different QoS sensitive services. They attempt to design controlling mechanisms for solving different problems that might occur in network in case of delays, losses, and so forth using a fuzzy logic controller (FLC). Indeed, this assessment and enhancement is based upon the assumption that it lacks the practical network deployment.

Kokoška, Handriková, and Valiska (2014) describe a couple of available network simulators and differentiate among them in order to decide on the proper one for better realization of the whole IPTV QoS process. Unfortunately, the survey did not consider simulators details and network settings. In addition, the survey did not go deep into the implementation of IPTV on such simulators. However, our chapter explains different OPNET capabilities that might help in implementing IPTV such as the available queuing system library, measuring delay and jitter, estimating the throughput, and capability of modifying network elements. In addition, it implements the IPTV features for the purpose of assessment. Our contributions in this chapter are IPTV network performance evaluation subjected to uncompressed multicast video at different data rate, and subjected to coded videos either in H.264 or H.265 formats as to find the best codec format to help in the revolution of the introduced technology.

11.3 IPTV NETWORK CONFIGURATION OVER OPNET

Indeed, it is almost impossible to study the performance of IPTV before deployment. This chapter could be considered as a step forward toward implementing the IPTV network as part of the OPNET simulator. This will make it easy on the researchers as well as the industry to examine their networks before deployment. This section details different components of IPTV that will be implemented using OPNET.

11.3.1 IPTV HARDWARE

Any IPTV system is made up of four major (elements) domains. All are generic and common to any vendor's (or combination of vendors') infrastructure as shown in Figure 11.1. These elements illustrate the high-level functional requirements of an end-to-end (ETE) IPTV system. The four network domains involve the following:

- The IPTV data center or video head end is responsible for capturing or acquiring video from different sources involving receiving, decoding, and decrypting multimedia contents. Then, the data is formatted and encapsulated in IP MPEG stream that is either multicast or unicast.
- The core network (edge network) is located at the network edge connected to the access network to transport the encoded group of channel forwarded from the video head end.

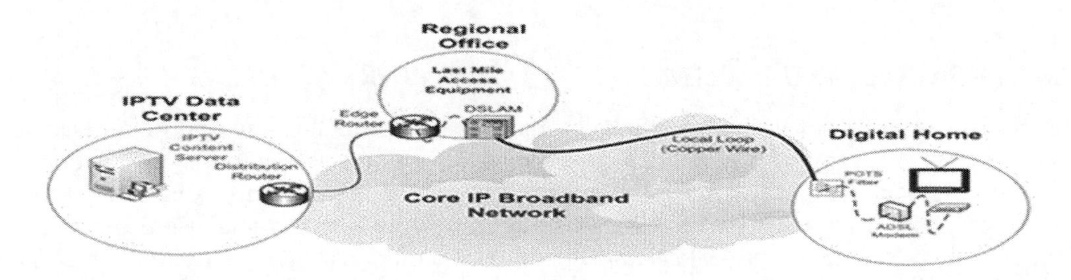

FIGURE 11.1 IPTV main components.

- The access network (IPTVCDs) is responsible for decoding and processing the incoming IP-based video stream, allowing users to access IPTV services. Typically, different technologies used to serve individual subscribers, such as DSL, connect to the Digital Subscriber Line Access Multiplexer DSLAM (DSL Modem).
- The home network is responsible for IPTV service distribution throughout the home. Many different types of home networks exist including TV sets and set-top boxes (STBs). However, IPTV requires a very robust high bandwidth home network that can only be accomplished today using wire line technology.

11.4 IPTV NETWORKS PROTOCOLS

The following are the protocols for IPTV networks:

- Routing protocols—OPNET offers a variety of information routing protocols that could be easily configured whether the network is managed by the same autonomous system (AS) (intra-AS routing protocols) or between ASs (inter-AS routing protocols). This is assigned according to network partitioning. In the project editor all nodes are assumed positioned at the same location; therefore the Routing Information Protocol (RIP) and/or Open Shortest Path First (OSPF) could be assigned.
- IP multicasting—IPTV traffic delivery could be unicast or multicast; however, IP multicast is widely used as it significantly reduces the amount of bandwidth required to transmit high-quality IPTV content across the network, as it is a one-to-many-based network. The OPNET modeler supports IP multicast including the Internet Group Management Protocol (IGMP) and Protocol Independent Multicast-Sparse Mode (PIM-SM). IGMP is used by hosts and adjacent routers to establish multicast group memberships. A TV set transmits IGMP-join/leave messages to notify the upstream equipment by LEAVEing one group and JOINing another channel. PIM-SIM is a multicasting routing protocol that explicitly builds shared trees rooted at a rendezvous point (RP) per group and optionally creates shortest-path trees per source.
- IP compression—Compression decreases the packet size by compressing certain portions of the datagram. The choice of datagram compression type depends upon network size. It must be noted that this differs as compared to the well-known video compression.
- QoS (quality of service)—As IPTV subscribers expect a specific viewing quality level, IPTV service providers use differentiated services (DiffServ) protocol for specifying and controlling network resources and bandwidth for packet transportation. Using OPNET enables examination of IPTV offered QoS support, by traffic marking specify traffic with different classes that are served according to queuing priority. Priority queuing (PQ) and custom queuing (CQ) schemes are deployed to queue traffic of different classes.

11.5 DESIGNED AND IMPLEMENTED IPTV NETWORK MODEL

OPNET is a network simulation that supports modeling of communication networks and distributed systems. It provides a detailed way to model network behavior through calculations of continuous interactions between modeling devices in their operational environment. Both behavior and performance of modeled systems can be analyzed through performing discrete event simulations (DESs). A DES is a typical network simulation method that could be used in large-scale simulation studies providing more accurate and realistic way. However, a DES requires huge computing power and the process could be time consuming in large-scale simulation studies.

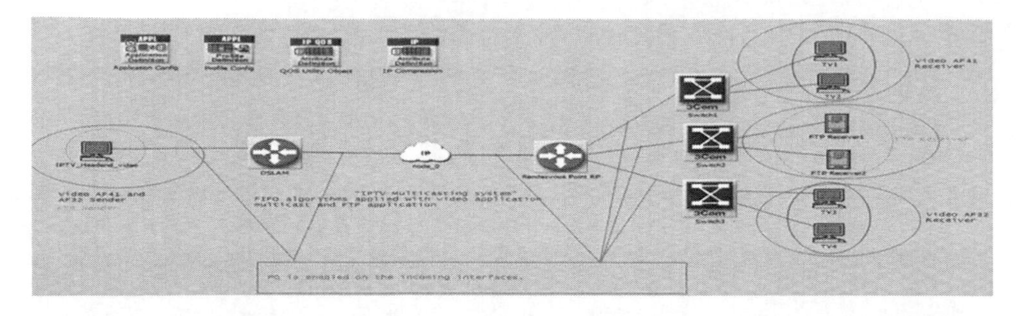

FIGURE 11.2 OPNET IPTV model.

The OPNET simulator is capable of simulating in both explicit DES and hybrid simulation modes, and supports other simulation features like cosimulation, parallel simulation, high-level architecture, and system-in-the-loop interactive simulations. It introduces a huge library of models that simulate most existing hardware devices and provides today most cutting edge communication protocols.

The OPNET environment incorporates tools for all phases of analysis, including model design, simulation, data collection, and data analysis. This chapter explains different OPNET capabilities that might help in implementing IPTV such as the available queuing system library, measuring delay and jitter, estimating the throughput, and modifying network elements, trying to provide a deep practical insight for almost all common IPTV features. The implemented network modeling IPTV is shown in Figure 11.2 using OPNET including the following nodes: IPTV_Headend_Video node, DSLAM node, and a rendezvous point (RP) multicasting node. The remaining architectures nodes in Figure 11.2 are switches, routers, and TV sets.

It must be noted that an RP is a core router at which multicast domain packets from the upstream source and join messages from the downstream routers are "rendezvous." It is important to know that other networks' routers do not need to know the source address of every multicast group after RP configuration. Therefore, for the correct operation, the rest of the networks routers supporting multicast must know the RP address and this is performed through IP addressing of any active RP interface. The RP router could be defined as the distribution/aggregation point for multicast traffic to all other network nodes and routers through the predefined IP address for any active port (router interface) connected to any neighbor node to the RP router. In addition, the QoS Utility Object node is used to assign quality of service schemes and queuing algorithms, and the IP compression node to perform data gram compression.

11.5.1 IPTV Network Performance Characterization

This section assesses different performance metrics key issues related to video transmission over IPTV. Quality of experience (QoE) as defined by ITU consists of subjective and objective quality measures. Generally, there is a correlation between the subjective and objective merits. Regarding Internet services, QoS measurement parameters usually are as follows:

- Packet delay variation (PDV)—Variance in end-to-end delay that video packets experience. OPNET permits collection of a global PDV statistic within a created network that is recording data from all the network's nodes.
- Packet end-to-end (ETE) delay (sec)—Average time counted to send a video application packet to a destination node application layer. It could be computed using Equation 11.1.

$$D_{E2E} = Q(d_{proc} + d_{queue} + d_{trans} + d_{prop}) \tag{11.1}$$

where Q indicates network node elements between IPTV Headend and TV set, d_{proc} is the network node processing delay, d_{queue} is the network node queuing delay, d_{trans} is the packet transmission time between two network elements on a communication link, and d_{prop} is the propagation delay within the network link.

According to Equation 11.1, counted ETE delay is seriously affected by network nodes and links, influencing both QoS and user's QoE. Hence, according to several fulfillments in this tenor ETE delay range has been identified. If the delay is longer than 1 s, it produces bad QoS toward unacceptable service from the end user's QoE. For one way communication, if the delay is shorter than 200 ms, then better QoS is produced and hence acceptable service from the end user's QoE side.

- Packet jitter—Measuring the difference between ETE delays of two consecutive packets, OPNET counts the absolute value of the recorded difference. To satisfying the user's QoE, the jitter delay for one way must be shorter than 60 ms on average and shorter than 10 ms in the ideal case. In other words, the recommended minimum transport layer jitter parameter is to be shorter than 50 ms.

11.6 SIMULATION RESULTS

In this section different experiments are conducted for the purpose of measuring IPTV performance using OPNET over IPTV modeled network in Figure 11.2. The following subsection explains how the background traffic that will be used in later experiments is generated.

11.6.1 Video Coding Traffic Generation (Background Traffic Generation)

As previously mentioned, the OPNET modeler provides the ability to analyze realistic simulated networks to compare the impact of different technology designs on ETE behavior. However, OPNET has several limitations in importation and exportation capabilities. For IPTV performance evaluation, the compressed videos are converted into text in OPNET simulator.

OPNET has the capability to provide conversation pair traffic as a way to traffic modeling over the network that could be imported from outside sources. It is injected at different network layers as IP traffic flows according to the layer where data is deployed. This section suggests the way for coded video ingestion over network using video traces as a text traffic file in .trl OPNET format to be imported as IP traffic flow. Using the huge coded video library traces in "Video Traces" (2015) for different coding standards, each collected trace file includes the number of coded frames within each group of pictures (GOP) and their corresponding bytes per second involved in these frames. These traces are converted into the .trl text file format that is constructed of several traffic lines; each traffic line defines five different header fields. These header fields per each line define the source node, destination node, time window (sec), number of transmitted frames within defined time window, and the corresponding bytes per second included within these frames.

In order to evaluate the network performance subjected to coded SV and MV in H.264 codec format, traces of the same movie sequence are collected and converted into the previously mentioned .trl text file format to be imported through the network as IP traffic flow modeling two coded TV channels. Moreover, to examine and to evaluate network performance subjected to coded video in the new codec HEVC format, H.265 coded video trace is also collected and converted into .trl similarly as its corresponding H.264 format, in order to model the other coded TV channel in HEVC. Thus, three coded TV channels will be imported separately over the IPTV network as an IP traffic flow in the .trl OPNET format. The three collected traces are for the same movie sequence, however SV and MV coded videos are of resolution 1920 × 1088 and of 48 frames per second (FPS), whereas HEVC coded video is of resolution 1920 × 1088 and 30 FPS.

11.6.2 Experiment 1: Evaluating Network IPTV with Uncompressed Video Transmission at Different Data Rates

Through Figure 11.2 network, two video applications are defined through the Application Config node named Video Conferencing (AF41) and Video Conferencing (AF32). The two mentioned video applications are distinguished with a Differentiated Services Code Point (DSCP). Video Conferencing (AF41) has the following characteristics: (1) video packets have lower drop precedence than of that in the Video Conferencing (AF32) application and marked to take higher priority through deployed QoS queuing priority schemes, and (2) Video Conferencing (AF41) models a TV channel with frame size information 352 × 240 that is transmitted with a frame interarrival time of 30 FPS and type of service (ToS) of AF41. On the other hand, Video Conferencing (AF32) models another TV channel with frame size information 128 × 240 that is transmitted with a frame interarrival time of 15 FPS and ToS of AF32. Two corresponding user profiles are defined through the Profile Config node named Video_AF41 and Video_AF32. According to the attributed profile Config node all defined users profiles are configured in simultaneous operation mode as to be delivered together. Moreover, the assigned start time for each defined user profile will start to be collected after the simulation start time by approximately 1 min 75 s as Figure 11.3 defines.

Video_AF41 and Video_AF32 are attributed to the IPTV_Headend_Video node only to represent the IPTV source node. The two defined applications are multicast over IPTV network with two corresponding multicasting group addresses, which are 224.233.24.231 and 224.233.24.232, in which the 224.233.24.231 multicast address is associated with Video_AF41 service and the 224.233.24.232 multicast address is associated to Video_AF32 service. Four receiving nodes are configured in the IPTV network as shown in Figure 11.2 as follows: (1) two TV sets (TV1 and TV2) sharing the same multicasting address group for Video Conferencing (AF41) delivery, and (2) two TV sets (TV3 and TV4) sharing the same multicasting address group for Video Conferencing (AF32) delivery. In addition, the PIM-SM protocol is enabled in network routers (DSLAM and RP) for saving IPTV bandwidth.

IPTV network performance is examined with the delivery of multicast video channels with high frame rates and low frame rates. Then, in other separate DES phase shown in Figure 11.2, network applications are settled to have the Video Conferencing (AF41) frame interarrival time reduced to 15 FPS, and the Video Conferencing (AF32) frame interarrival time reduced to 10 FPS.

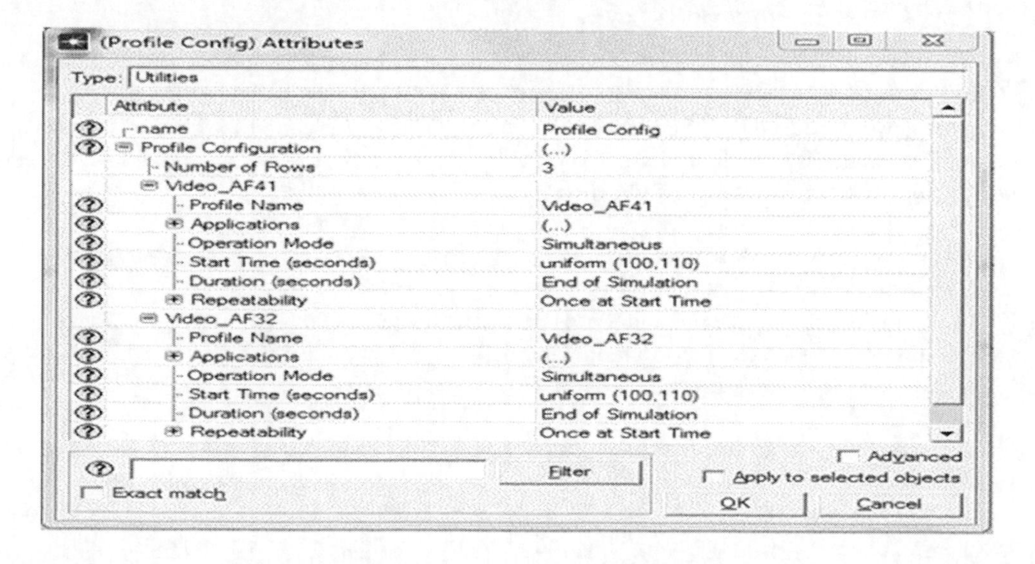

FIGURE 11.3 Profile config assigned attributes.

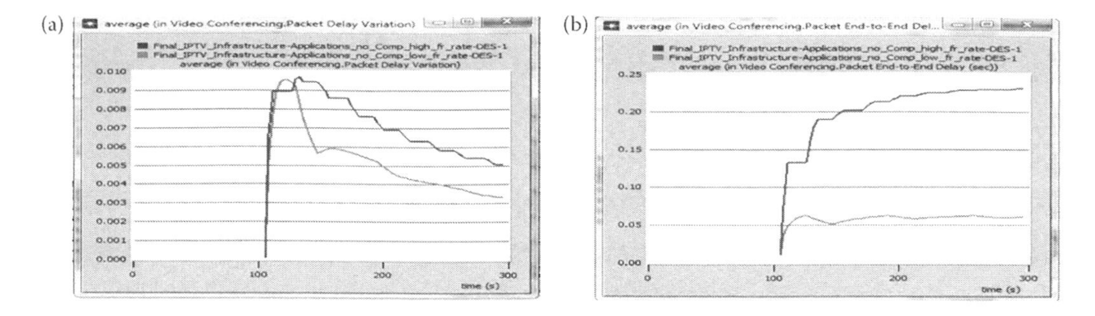

FIGURE 11.4 (a) Global PDV, (b) global ETE associated with the delivery of video channels at high and low frame rates.

Figure 11.4a shows the global PDV with time in both high and low data rates, and Figure 11.4b shows the global ETE with time in both high and low data rates. As given in Figure 11.4, it seems that the PDV for the high data rate is higher than in the low rate by almost 25%. However, at minute 2, the PDV of the low rate shows a peak that quickly disappears later. This is due to the network handling the generated packets in terms of buffering and processing at a low rate. In addition, Figure 11.4b illustrates that the corresponding video packets carrying the high frame rate channel experience higher ETE than of that with low frame rate. For instance, at minute 5 (300 seconds), the video packet of high data rate achieves an increase almost about 180 ms in ETE delay over the corresponding low data rate.

So, to review the results from the preceding figures, it is obvious from ETE delay that users with high data rates might experience bad quality in data transfer. However, the results show that it is almost three times more delay than the one with low data rate. At the same time, due to the high PDV values in high data rates, packets may arrive out of order as well. This reflects the importance of video compression; not only its contribution in nodes buffering reduction since limited memory and speed at any receiving node but also its impact on quality of received TV video channels.

11.6.3 Experiment 2: Evaluating Network Performance in Case of Coded Video Transmission

The main objective behind this experiment is to evaluate IPTV network performance subjected to the delivery of SV and MV video sequence coded in the H.264 format, and to the delivery of coded video in the new introduced H.265 codec format, and modeling three coded TV channels in two different codecs standards over the IPTV network. This experiment uses the same previously created network setup with the following modifications:

- Failing all receiving network nodes in Figure 11.2 except TV1, as it is only a receiving node over the network.
- Admission of previously prepared SV H.264 traffic file (SV H.264 .trl text file) subrogating uncompressed video channels to model the delivery of only one SV TV channel in H.264 format over the same network.

It should be noted that any imported traffic file will be matched with network source and destination nodes by the OPNET importing machine. OPNET traffic center ensures this importation indicating the existence of only one IP traffic flow from the IPTV source node to the only receiving set (TV1) receiving the SV H.264 coded channel. Then, in two separate DES phases the other coded TV channels are imported as the following:

- First, admission of previously prepared MV H.264 traffic file (MV H.264 .trl text file) subrogating the SV H.264 video channel to model the delivery of only one MV TV channel in H.264 format over the same network.
- Second, admission of the previously prepared H.265 traffic file (H.265 .trl text file) subrogating MV H.264 video channels to model the delivery of only one coded TV channel in H.265 format over the same network.

Moreover, the background traffic delay is set to 150 s defining when results will start to be collected. Hence, all coded traffic channel admission will be started over network nodes after simulation start time by approximately 2 min.

Figure 11.5a illustrates average received traffic (bps) in case of the delivery of coded SV and MV channels in H.264 format. Figure 11.5a shows, in case of the delivery either SV or MV coded channels, a peak at 4 min 35 s that is achieved loading the network with maximum traffic received at about 550 and 750 Kbps, respectively, and then goes down again. This entirely reflects a number of variations for transmitted bytes per second included within transmitted frames in these imported traffic files. Moreover, the figure shows that received traffic (bps) for coded MV channels have the same external modality as their corresponding coded SV channel; this could be induced as both channels are for the same movie sequence. However, they differ in data rate, as MV channel loads network with maximum data rate (bps) of about 1.5 times over SV H.264 inspected in terms of traffic received. Figure 11.5b illustrates the corresponding PDV with time in case of the delivery of coded SV and MV channels in H.264 format, showing that the PDV increase is roughly linear with time in the case of the SV coded channel. On the other hand with MV delivery, PDV is increased with time unlike of that in case of SV delivery. This is due to the network capability to handle the generated packets in terms of buffering and processing, reflecting the impact of high data rate involved in MV channel on network performance. Hence, MV packet deliveries are more vulnerable in the arrival of different ETE delays and are out of order as compared with their corresponding SV TV channels.

Figure 11.6a illustrates the average traffic received (bps) for the delivery of coded video channel in HEVC format over the same network. It involves the collected received traffic that has the same external modality as their corresponding coded SV shown in Figure 11.5a. This could be induced as they are used for the same movie sequence; however, they differ in data rate as the H.265 coded channel achieves a reduction in data rate (bps) over that for coded SV and MV channels inspected in terms of traffic received. Figure 11.6b illustrates PDV with time in case of the delivery of the HEVC TV channel. It shows that with the delivery of the HEVC coded channel, PDV is roughly increasing with a low slope as compared to that for H.264 SV shown in Figure 11.5b.

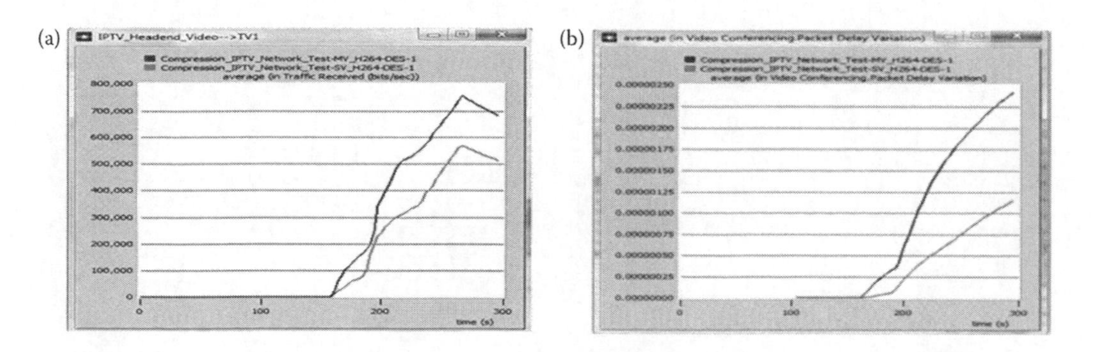

FIGURE 11.5 (a) Traffic received (bps), (b) PDV with the delivery of only one coded SV and MV channels in H.264.

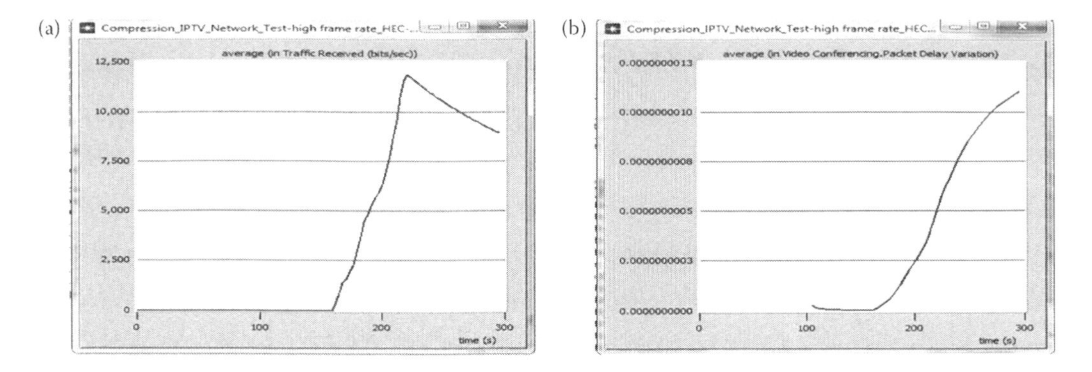

FIGURE 11.6 (a) Traffic received (bps), (b) PDV with the delivery of coded channel in H.265.

As there is correlation between subjective and objective merits, it could be easy to estimate the quality of received video according to the resulting performance networking terms. The most important parameters warranting the estimation of received video quality is ETE delay and packet jitter. Thus, Figure 11.7a shows packet jitter resulting from the delivery of coded SV and MV TV channels. The figure shows that over a majority of time, MV packets form higher picks in packet jitter than SV in which the coded MV channel achieves maximum packet jitter of 38 ms, whereas the coded SV channel achieves 21 ms. The resultant comparison between each case involves the delivery of MV in H.264 format that achieves an increase of packet jitter by approximately 55 ms over of that for the coded SV channel over the same network. Figure 11.7b illustrates that the maximum achieved packet jitter for H.265 delivery is about 22 ms. Showing great reduction in packet jitter is the corresponding H.264 coded channels.

Figure 11.8 shows packet ETE delays for the delivery of SV and MV coded TV channels and coded video channel in HEVC format. The figure shows that all cases have near external modality, as they are for the same movie sequence. However, the coded MV channel achieves maximum ETE delay of 34 ms and the coded SV channel achieves maximum ETE delay of 25 ms. The HEVC achieves maximum ETE delay in about 20 ms. With the comparison of network performance subjected to the delivery of SV and MV coded channels; it is found that the MV coded channel achieves an increase in ETE delay by approximately 75 ms with their corresponding SV coded channel over the same network. So, in conclusion raised highlights video compression impact on the quality of received service with respect to that within the results of the experiment 1 simulation. However, the H.264 video standard has recorded several limitations on the increase of the number of transmitted HD or MV video channels to satisfy user and service provider recommended objectives.

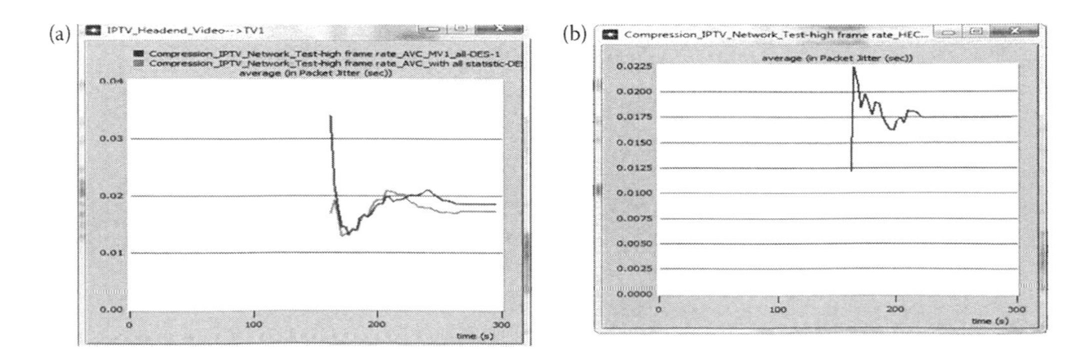

FIGURE 11.7 (a) Average packets jitter for SV and MV H.264 video channels, (b) average packets jitter for coded H.265 channel.

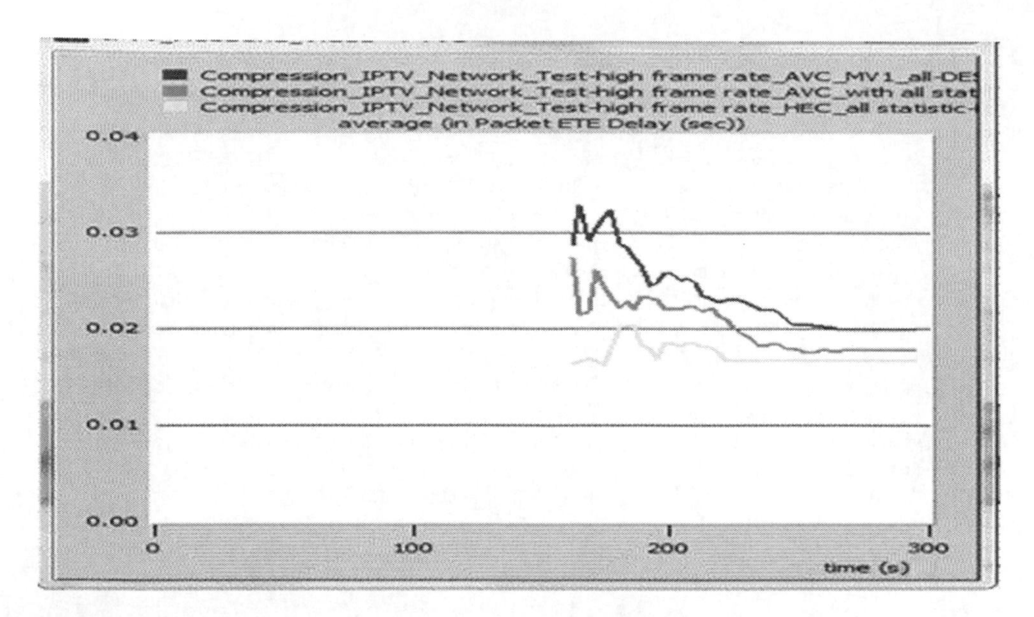

FIGURE 11.8 Average ETE delay included with the delivery of H.265 video, SV H.264 video, and MV H.264 video.

Moreover, evaluating network performance of HEVC in terms of ETE delay and packet jitter, it is found that users receiving TV channels in HEVC format will expect very high video quality as compared to other two cases, comprising the great enhancement achieved by HEVC on network performance.

11.7 CONCLUSION

This chapter introduces a detailed study for evaluating IPTV network performance subjected to multicast uncompressed video with assigned standard protocols behavior at different data rates, highlighting the impact of video compression on QoS. This chapter also provides evaluation for IPTV network performance subjected to the delivery of coded SV and MV in H.264 format and coded video in the new developed HEVC. From the results explained in this chapter, the following remarks could be concluded:

- Coded channel in H.265 format achieves reduction in maximum traffic received (bps) as compared to SV and MV in H.264 format of about 46 times and 63 times, respectively. Moreover, coded channel in H.265 format achieved reduction in maximum PDV as compared to SV in H.264 format of about 92%.
- Coded SV channel in H.264 format achieves an increase in total ETE delay and packet jitter for all packets of about 300 and 260 ms, respectively, over their corresponding HEVC format. Meanwhile coded MV video channel in H.264 achieves an increase in total ETE delay and packet jitter for all packets of about 410 and 300 ms, respectively, over the corresponding HEVC. Moreover, HEVC achieves reduction in maximum ETE delay over MV H.264 of about 16.5 ms, while achieving reduction of about 14 ms over SV H.264 at the same time instance.

Thus, evaluation shows the great enhancement achieved by the new codec on the IPTV system as to satisfy user's recommendations in terms of both QoE and QoS.

REFERENCES

Held, G. 2007. *Understanding IPTV*. Boca Raton, FL: Auerbach.

Hamodi, M. and Thool, C. 2013. Performance evaluation of IPTV over WiMAX networks under different terrain environments. *International Journal of Engineering Inventions*, 2: 21–25.

International Telecommunication Union (ITU). 2008. ITU-T: IPTV Focus Group Proceedings. https://www.itu.int/dms_pub/itu-t/opb/proc/T-PROC-IPTVFG-2008-PDF-E.pdf.

Kokoška, R., Handriková, J., and Valiska, J. 2014. Software network simulators for IPTV quality of services. *Acta Electrotechnica et Informatica*, 14(1): 18–22.

Maraj, A., Shehu, A., and Mitrushi, M. 2011. Studying of different parameters that affect QoS in IPTV systems, *Proceedings of the 9th WSEAS International Conference on Telecommunications and Informatics*. Catania, Italy, pp. 107–112.

Moughit, M., Abdelmajid, B., and Sahel, A. 2013. A multicast IPTV bandwidth saving bandwidth. *International Journal of Computer Applications*, 64(14): 22–26.

Sethi, S. and Hnatyshin, Y. 2013. *The Practical OPNET User Guide for Computer Network Simulation*. Boca Raton, FL: CRC Press.

Singh, G. and Amit, G. 2013. Simulation and analysis: The effect of mobility on IPTV (VOD) over WI-MAX using OPNET. *International Journal of Physical Science*, 8(26): 1401–1407.

Video traces for network performance evaluation. 2015. http://trace.eas.asu.edu/tracemain.html.

Wikipedia. 2015. IPTV. https://en.wikipedia.org/wiki/IPTV.

12 Brain Storm Optimization (BSO) for Coverage and Connectivity Problems in Wireless Sensor Networks

Rabie A. Ramadan and A. Y. Khedr

CONTENTS

12.1 Introduction 165
12.2 Coverage Problem in Wireless Sensor Networks (WSNs) 166
12.3 Problem Statement 167
12.4 Brain Storm Optimization (BSO) 168
12.5 Genetic Algorithms (GAs) 169
12.6 Solution Approach 170
12.7 Simulation Experiments 171
12.8 Conclusion 174
References 174

12.1 INTRODUCTION

A wireless sensor network (WSN) is one of the emergent networks that is used in many critical applications such as battlefield monitoring, health care monitoring, oil and gas tanks monitoring, firefighting, and bridge safety monitoring. A WSN is formed through different steps including the deployment, self-organizing, operation, and maintenance. Sensors can be deployed by many ways. The deployment could be manual where the positions of the sensors are known before the deployment. This also requires the deployment environment to be accessible. On the other hand, the deployment procedure could be random in inaccessible areas and when large numbers of sensors are to be deployed. Many variations of the random deployment method are proposed in the literature including the sequential deployment and using artificial intelligence techniques for efficient deployment. The importance of this phase of forming a WSN is that it is considered as the base of other WSN operations. The coverage and connectivity problems are the results of the deployment process. Therefore, there is huge interest in the deployment problem including research in the deployment with full coverage, partial coverage, point coverage, and border coverage. At the same time, some of the research interest focuses on the deployment of a connected network along with the coverage.

The second phase of the WSN is the self-organization. Self-organization means that the network after deployment will be left unattended and it is the responsibility of the smart sensors to adapt to the environment and its requirements. Therefore, sensors cooperate to reorganize themselves either to be connected or to cover the required areas; even some of the networks might be required to change their topologies to track single or multiple objects in the monitored field. This requires a smart operating system and software to be uploaded to the sensor board, which it is another challenge for the sensors.

The third phase of forming a WSN is the operation of the network. In this phase sensors are required to sense some of the monitored environment features and send them to a centralized node named the sink node. Certainly, this includes routing through multihop nodes. Then, it involves the routing protocols in its operation. The challenge in this phase is how sensors can send their sensed data with less consumed energy. In fact, communication is considered the most consuming energy in the WSNs. Therefore, a large body of the literature targets efficient routing in WSNs. In a more complex environment, multiple sink nodes might be available in the monitored field.

The last phase of forming a WSN is the maintenance where sensors' energy could be depleted or suffer from any type of failure. In some of the networks, recharging or even replacing a sensor is not possible. Therefore, WSNs have to have a self-healing mechanism to maintain itself and handle any bad situation. This is another important field of research in WSNs.

As stated earlier, many artificial intelligence techniques are used in WSNs at different phases. Some of these algorithms include swarm intelligence, genetic algorithms (GAs), fuzzy logic, and neural networks. These techniques have proven efficient in terms of handling many problems of WSNs. Throughout this chapter, the Brain Storm Optimization (BSO) algorithm will be taking into consideration for network lifetime. BSO is an algorithm that appeared in 2011 and was stated by Gao et al. (2015). The work by Rabie (2017) extended the original version of the BSO to include fuzzy functions in the clustering phase of the algorithm and named the algorithm fuzzy brain storm optimization (FBSO). However, the BSO used in this chapter is the modified version of the one proposed by Gao et al. (2015).

The chapter is organized as follows: Section 12.2 surveys the coverage problem in WSNs, Section 12.3 states the problem statement, Section 12.4 explains the solution approach, the simulation experiments are presented in Section 12.5, and finally the chapter concludes in Section 12.6.

12.2 COVERAGE PROBLEM IN WIRELESS SENSOR NETWORKS (WSNs)

The coverage problem in WSNs takes many shapes including target coverage, area coverage, and barrier coverage. The idea behind the point coverage is to cover specific points or moving points in the monitoring field. The problem in its general form is the NP-complete problem (Zhao and Gurusamy, 2008). Zhao and Gurusamy (2008) tried to solve the problem using a heuristic approach as the maximum cover tree problem and showed that it is an NP-complete problem. One more point coverage technique is proposed by Gu et al. (2011) to monitor a moving object in the monitored field. They proposed heuristic approximation algorithms. In area coverage (Singh and Sharma, 2014), the main purpose is to cover the whole area of the monitored field with a minimum number of sensors and prolonging the lifetime of the network as well. To do so, different scheduling algorithms are proposed including learning automata (Dietrich and Dressler, 2009; Cheng and Gong, 2011). The last class of coverage is the barrier coverage, which could be defined as maximizing the detection of penetration of a border. This type of coverage is used mostly in applications like country-border monitoring.

Solutions to these problems also could be classified into centralized algorithms such as in Cardei and Du (2005), Cardei and Wu (2006), and Slijepcevic and Potkonjak (2001), and distributed algorithms such as in Yardibi and Karasan (2010). In Cardei and Du (2005), the authors proposed an algorithm to monitor an object in the monitored field. The main idea behind their algorithm is to divide the sensor nodes into sets where each set can track/cover the target in the whole monitored area. This problem is well known under the title of maximum set cover problem. Cardei and Wu (2006) extended this work and proved that the maximum set cover problem is NP-complete problem. Slijepcevic and Potkonjak (2001) took another approach where they divided the monitored area into fields that can be covered by the same set of nodes. The linear programming techniques were not far from targeting the coverage problem solution. For instance, Pyun and Cho (2009) proposed integer linear programming (ILP) for multiple target tracking with network lifetime extension. In Yardibi and Karasan (2010), a distributed algorithm was proposed for partial target coverage.

In other words, they assume that the full coverage of the monitored field or the target is not required; only the coverage is required with certain percentage. Sensors' residual energy is utilized to check on the performance of the proposed algorithm.

There are many other variations to the coverage problem including coverage and connectivity (Mahmud and Fethi, 2015), k-coverage, connected k-coverage (Ramar and Shanmugasundaram, 2015), and homogenous and heterogeneous coverage problems (Fatemeh and Ahmad, 2013). The target of all of these coverage problems is to prolong the lifetime of WSNs. Some of the used methods are based on scheduling the sensors between sleep and wake modes. Others try to benefit from the characteristics of the sensors such as initial energy and mobility.

The closest study to the work in this chapter is the coverage problem stated in Gao et al. (2015). The problem is to maximize the coverage of a certain area given a limited number of heterogeneous sensors. The heterogeneity of the sensors are in terms of their sensing ranges. The authors proposed a special type of GA with three different normalization methods: Random, MinDist, and MaxDist. Random normalization is the same as the original version of the GA where chromosomes are generated, crossed over, and mutated. In MinDist normalization, the chromosomes are rearranged to have the minimum distance between their corresponding genes then crossed over; on the other hand, in MaxDist, the chromosomes are rearranged to have the maximum distance between their corresponding genes then crossed over. Genes in these cases are the positions of the sensor nodes in the monitored field. The main considered objective function is the coverage. The problem with this approach is that the authors did not consider the energy of the sensors in their deployment model. In fact, the sensors energy could be an important issue since some of the sensor nodes might die in a short period of time.

The coverage problem has many variations including the point, border, partial, and full coverage to the monitored field. Different methodologies as mentioned are used to solve this problem. In this chapter a new proposed algorithm is used inspired from swarm intelligence, called the BSO algorithm. The algorithm is used with some of the normalization methods not only to solve the coverage problem but also the connectivity problem. BSO in this chapter is treated as a multiobjective solution to the coverage and connectivity problems of WSNs. A set of case studies are simulated and compared to the GA. The results show that BSO is outperforming GAs in terms of coverage and connectivity with small overhead.

12.3 PROBLEM STATEMENT

The coverage problem discussed in this chapter belongs to the point coverage class of problems where sensors coverage model is considered as binary model. In the binary coverage model as given in Equation 12.1, a point $p_j(x_j, y_j)$ will be fully covered if it falls within the sensing range (r_i) of a sensor $s_i(x_i, y_i)$.

$$F(x,y) = \begin{Bmatrix} 1 & d(s_i(x_i,y_i), p(x_j,y_j)) < r_i \\ 0 & \text{otherwise} \end{Bmatrix} \tag{12.1}$$

where $d(s_i(x_i, y_i),p(x_i, y_i))$ is the Euclidian distance between the two points $p_j(x_j, y_j)$ and $s_i(x_i, y_i)$, and i is the sensor identification.

This model considers the sensor node sensing range as a circle centered at the sensor itself. When the point $p_j(x_j, y_j)$ is covered, the model value is equal to 1 and 0 otherwise.

Then, the coverage problem could be defined as follows:

> Given a set of heterogeneous sensors (n), where they differ in their sensing range (r_i), communication ranges (c_i), and initial energy (E_i). These sensors are supposed to be deployed in a monitored field (A) with length L and width W. The objective of the deployment is

to maximize the coverage of the monitored field and prolonging the sensor's lifetime by reducing the consumed energy in the network.

12.4 BRAIN STORM OPTIMIZATION (BSO)

BSO is a new algorithm that it is based on the way human beings think of complex problems. It adopts the brainstorming mechanism that human beings use. People come together thinking of a solution or two to a problem. They develop their ideas, share them, evaluate them, select the best ideas, then go for iterations. Figure 12.1 shows the flowchart of brainstorming activities (Lehrer, 2012). The chart shows that the brainstorming process goes for iterations to reach an efficient solution to the problem at hand through collaboration between people that are closely related or not to the problem. This process proved to be efficient in solving real-life complex problems.

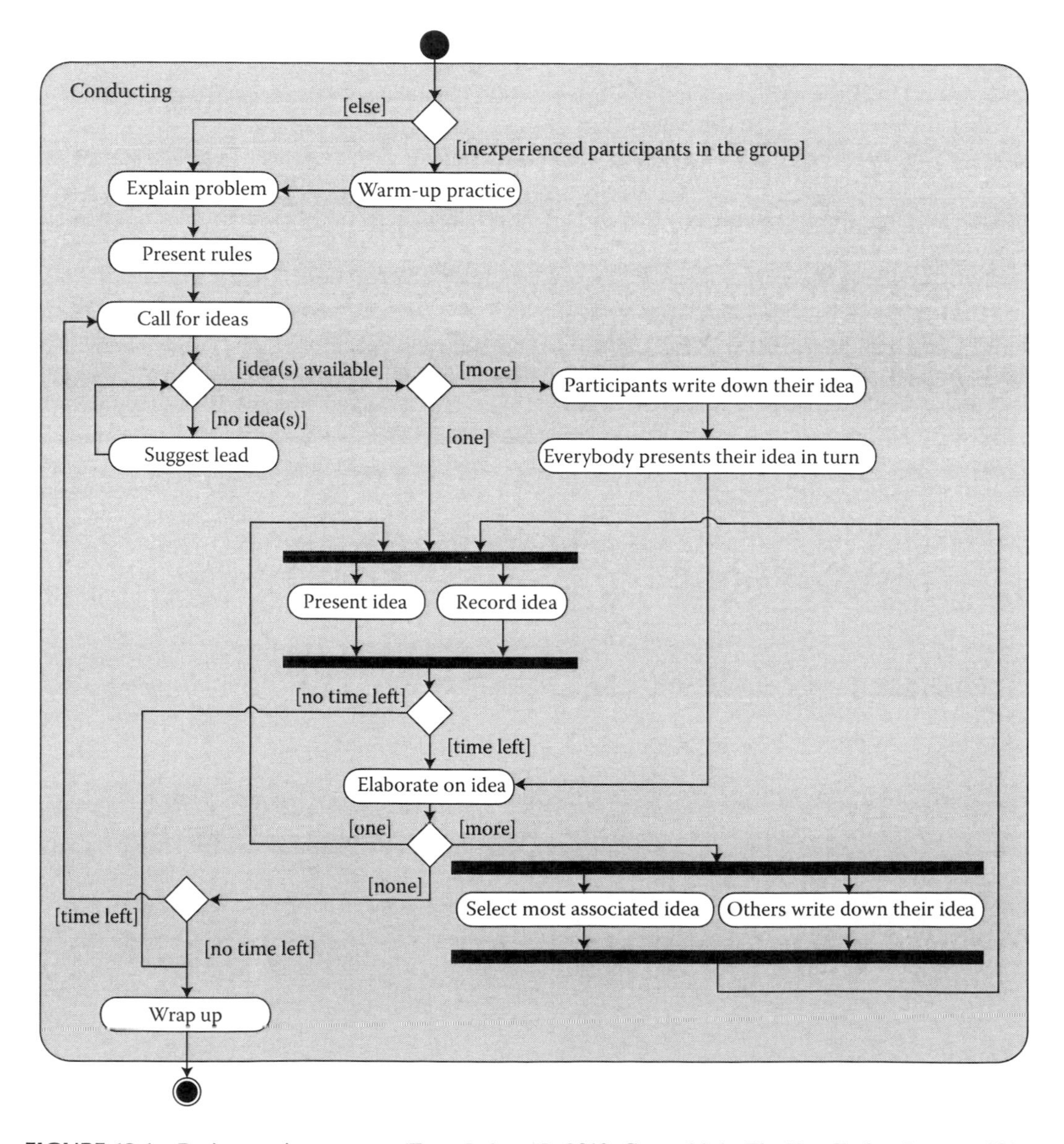

FIGURE 12.1 Brainstorming process. (From Lehrer, J., 2012, Groupthink, *The New Yorker*, January 30.)

TABLE 12.1
Brainstorming Process Steps

Step 1—Get people from different backgrounds as much as possible.
Step 2—People should generate many ideas according to the following rules:
  Rule 1—Suspend judgment
  Rule 2—Anything goes
  Rule 3—Cross-fertilize (piggyback)
  Rule 4—Go for quantity
Step 3—Some people will be selected to be the owners of the problem to pick the good ideas from the gathered people.
Step 4—The selected ideas will be used as a base for generating more ideas according to the same rules stated in step 2.
Step 5—Repeat step 3 to pick the good generated ideas.
Step 6—Randomly pick an object and use the functions and appearance of the object as clues to generate more ideas according to the rules.
Step 7—Let the owners pick several better ideas.
Step 8—By this step, we hope that good ideas have been reached to be considered as a solution to the problem at hand.

In 2011, Shi adapted this method of thinking to solve optimization problems in different fields of science. The author considered human beings as the smartest animal and adapting their way of thinking should lead to efficient solutions. According to Shi (2011), two iterations of the brainstorming process are divided into eight steps summarized in Table 12.1. As can be seen, step 6 works as a divergence to the people from getting trapped in the same ideas.

This process is translated to formal algorithmic steps to be suitable for optimization problems including the generation process, clustering process, and mutation and selector operators. The original BSO uses k-means as the clustering technique. Gaussian random values are added to generate new individuals according to Equation 12.2:

$$X_{\text{new}} = X_{\text{old}} + \varepsilon * n(\mu,\sigma) \tag{12.2}$$

where X_{new} is the newly generated individual and X_{old} is the selected individual to generate new one. $n(\mu,\sigma)$ is the Gaussian function with mean μ and variance σ. ε is the contribution weight of Gaussian random value and it is computed by Equation 12.3:

$$\varepsilon = \text{logsig}\left(\frac{0.5 * m_{\text{iteration}} - c_{\text{iteration}}}{k}\right) * \text{rand}() \tag{12.3}$$

where logsig() is the logarithmic sigmoid function, $m_{\text{iteration}}$ is the maximum number of iterations, $c_{\text{iteration}}$ is the current iteration number, k is a constant for changing the logarithmic sigmoid function slope, and rand() is a random generator function that generates a number between 0 and 1.

12.5 GENETIC ALGORITHMS (GAs)

Genetic algorithms (GAs) is one of the search heuristics that mimics the process of natural selection. GAs consist of four steps: initialization, selection process, genetic operators such as crossover and mutation, and termination process. The initialization phase involves generating the population; this population could be randomly generated or based on some initial processes. The initialization phase depends mainly on the type of the problem to be solved. The second phase is the selection process where new populations will be generated from the old ones based on a fitness or evaluation function. This fitness function measures the quality of each individual chromosome and based on that it is chosen whether to be selected to be part of the next iteration. However, the representation of each population/solution could differ according to the type of problem to be solved. The third phase is related to genetic operators, which are the crossover and mutation. In order to generate a new solution, a pair of the parent solutions are combined generating a new solution hoping it will be better than its parents. Certainly, the new child will have some of the characteristics of its parents. A simple mutation operator is to randomly change some of the genes of the newly generated child to expand the search space. Different methods are utilized for the crossover and mutation operators based on the type of problem at hand. The selection and the genetic operator phase will repeat until one or more of the following conditions are satisfied, which is the termination phase:

- A solution is found that satisfies minimum criteria
- Fixed number of generations reached
- Allocated budget (computation time/money) reached
- The highest ranking solution's fitness is reaching or has reached a plateau such that successive iterations no longer produce better results
- Manual inspection
- Combinations of the above

12.6 SOLUTION APPROACH

This section proposes to apply a modified version of the BSO algorithm for the previously stated coverage problem. The algorithm follows the real brainstorming strategy footsteps and the adapted version of the algorithm can be described as follows:

Step 1—Define the number of iterations (I_{max}), number of ideas (D) to be initially generated, and number of clusters (K_{max}).
Step 2—Randomly generate D ideas.
Step 3—Evaluate the D generated ideas based on the coverage and consumed energy.
Step 4—Apply the Pareto dominance on the generated idea.
Step 5—Apply k-means clustering to cluster the generated ideas into the number of clusters (K_{max}).
Step 6—Within each cluster, apply the crossover operation between the newly generated ideas and the old ones; select the best idea to replace the old one, if any.
Step 7—Rank the selected X solution from all clusters based on their coverage and lifetime and select the best nondominated ones (C_ideas).
Step 8—If the current number of iterations $\geq I_{max}$, go to step 10.
Step 9—If the number of current ideas is less than D, randomly generate other (D- C_ideas) ideas and go to step 3.
Step 10—Report the current idea and terminate.

The crossover methodology used in this chapter are inspired from Chen et al. (2015) where three crossover methods are applied. The first crossover method is the random method where some of the sensor positions in each idea are exchanged based on probability (Probl). The second methodology is the MaxDist where sensor locations in each idea are adjusted to have the farthest distance between the two corresponding sensors in each idea; then the random crossover operator is used. The third crossover method is based on the MinDist normalization where sensor positions are rearranged to have the minimal distance between sensors in corresponding ideas; then the random crossover operator is applied.

K-means clustering is a famous clustering algorithm that tries to group objects based on their features into k classes where k is a positive number. The grouping is done by minimizing the distances between the data and the corresponding cluster centroid.

12.7 SIMULATION EXPERIMENTS

This section shows the performance of the BSO in solving the WSN coverage problem compare to the GA stated in Gao et al. (2015). However, the BSO tries to optimize the WSN not only based on the field coverage but also based on the network lifetime, whereas the GA in Gao et al. (2015) tries only to maximize the coverage. The network lifetime might be affected by sensor connectivity, the used routing algorithm, the number of messages to be sent through the network, and position of the sink node.

Monte Carlo sampling is used for coverage evaluation, as given in Gao et al. (2015) considering X is the coverage area; therefore the coverage could be estimated as follows:

$$\bigcup_{i=1}^{k}\bigcup_{j=n_i}^{n_{i+1}^{-1}} C_{r_i}(x_j,y_j)\cap A \tag{12.4}$$

$$S(X)=\iint_A I_x(x,y)dx\,dy=\lim_{l\to\infty}\frac{S(A)}{L}\sum_{l=1}^{L} I_x(x_1,y_1) \tag{12.5}$$

$I_{x(.)}$ is defined as

$$I_x(x,y)=\begin{cases}1\,(x,y)\in X\\0\,(x,y)\notin X\end{cases} \tag{12.6}$$

In the literature, different energy models can be found. We base our work on the first-order model described in Ramar and Shanmugasundaram (2015), where the transmitter or the receiver dissipates E_{elec} energy per bit to run the digital coding circuit, modulation circuit, and filtering of the signal circuit, which are the radio electronic circuits before it is sent to the amplifier, and dissipates E_{amp} in the power amplifier. E_{amp} varies according to the distance (d) between a transmitter and a receiver: $E_{\text{amp}}=\varepsilon_{\text{fs}}$ assuming a free space model when dis $< d_o$ and $\alpha=2$, while $E_{\text{amp}}=\varepsilon_{\text{mp}}$ assuming a multipath model when dis $\geq d_o$ and $\alpha=4$, where $d_o=\sqrt{(2\frac{X\varepsilon_{\text{fs}}}{\varepsilon_{\text{mp}}})}$. Thus, to transmit a l bit-bit packet over distance dis, the radio expands to

$$E_{Tx}(\text{l bit}*\text{dis})=E_{\text{elec}}*\text{l bit}+E_{\text{amp}}*\text{l bit}*d^{\infty}$$

$$E_{Tx}(\text{l bit}*\text{dis})=\begin{cases}E_{\text{elec}}*\text{l bit}+E_{\text{fs}}*\text{l bit}*d^2 & \text{if dis} < d_o\\E_{\text{elec}}*\text{l bit}+E_{\text{amp}}*\text{l bit}*d^4 & \text{if dis} < d_o\end{cases} \tag{12.7}$$

and to receive this message, the radio expands

$$E_{Tx}(\text{l bit}*\text{dis})=E_{\text{elec}}*\text{l bit} \tag{12.8}$$

In these formulas, E_{Tx} is the transmission power, and E_{Rx} is the receiver power. The radios are assumed to have power control and consume the minimal energy needed to reach the receiver. The monitored areas are assumed to be 800 × 800 m and the number of used sensors are assumed 100. The 100 sensors are assumed to be from three different categories in which they differ in their sensing ranges, communication ranges, and initial energy. The first category involves sensors with sensing range from 30 m, communication range of 60 m, and initial energy 0.5j. The second category includes a 50 m sensing range, 100 m communication range, and 1j as initial energy. The sensing range of the last category is 60 m and the communication range is 90 m and the initial energy is

TABLE 12.2
Common Energy Model

$E_{elec} = E_{DA}$	50 nJ/bit
E_{fs}	10 pJ/bit/m^2
E_{mp}	0.0013 pJ/bit/m^4

assumed to be 1.5j. The number of each category is selected based on a probability P(type) where each experiment runs for 10 times with different settings and a different number of sensors per category; the average over the 10 runs are reported in the following case studies. A common energy model is assumed for all of them, as can be seen in Table 12.2.

Case Study 1: Efficiency of BSO in terms of Converge and Connectivity

This case study examines the performance of the proposed BSO and GA in 100 iterations. Figure 12.2 shows the coverage percentage of the BSO with different normalization methods (Random, MinDist, and MaxDist). It seems that BSO with MaxDist outperforms the other two normalization methods in terms of coverage. However, the difference is not that much noticeable in small networks since it is almost 2%. In large WSNs, this difference becomes noticeable. At the same time, Figure 12.3 confirms the same results in which the GA MaxDist gives the highest coverage performance.

Another set of experiments are conducted to examine the comparison between the GAmax and BSOmax. Out of Figure 12.4, it seems that the BSOmax is outperforming the GAmax by almost 15%. However, BSO still needs more enhancement since its coverage percentage is 75% of the monitored area. The results could be enhanced if the number of iterations is increased. Figure 12.5 examines another issue that is not considered in Gao et al. (2015), which is the connectivity. The connectivity measure in this set of experiment is considered by how many nodes can be reached from any other node out of the total number of nodes. The figure shows that the BSO is much better than the GA by almost 20%. The reason behind that is the BSO is considering the connectivity as one of its objectives.

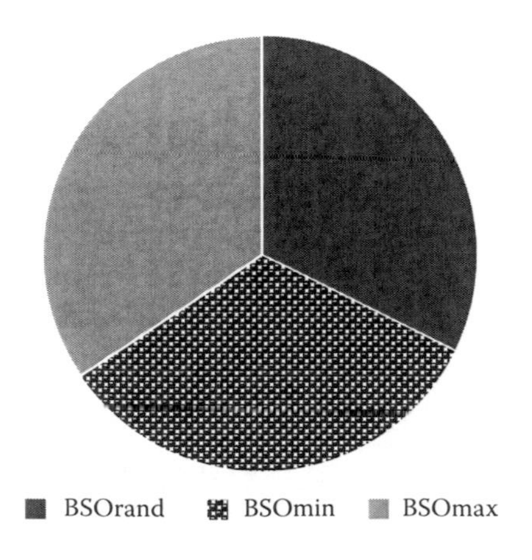

FIGURE 12.2 Coverage performance of BSO algorithms.

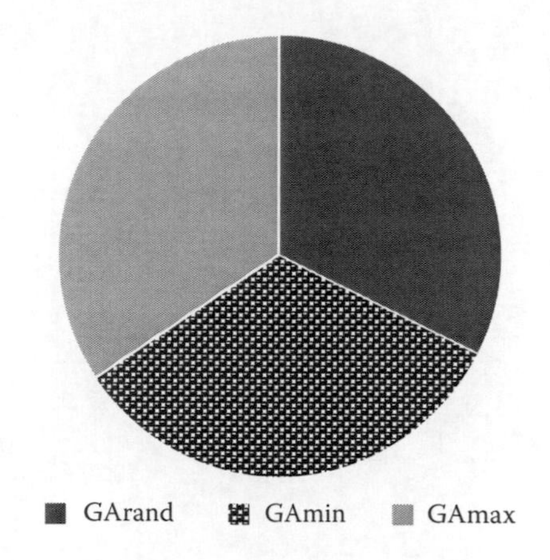

FIGURE 12.3 Coverage performance of GA.

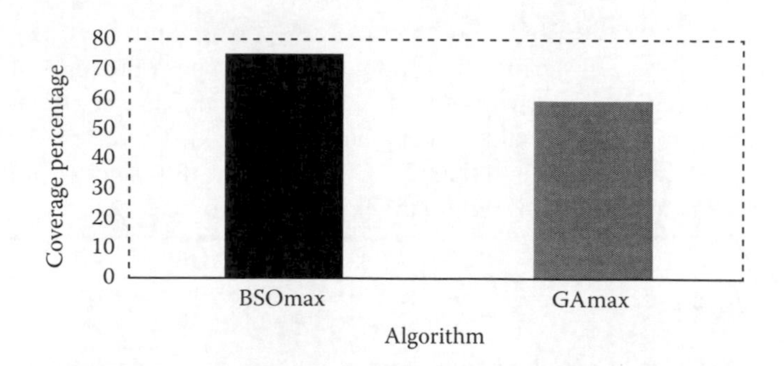

FIGURE 12.4 Coverage performance of the BSO and GA.

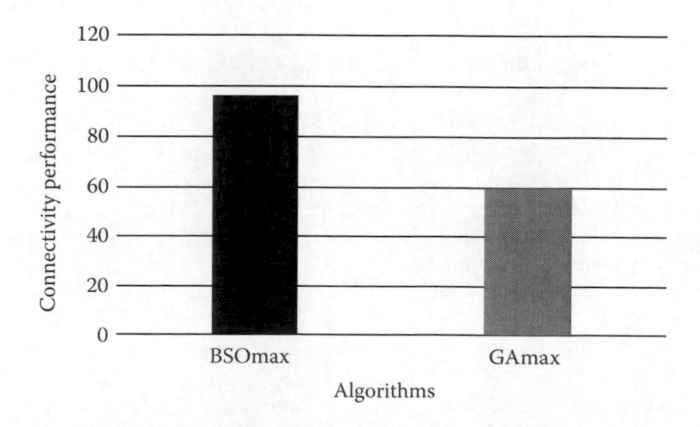

FIGURE 12.5 Connectivity performance of the BSO and GA.

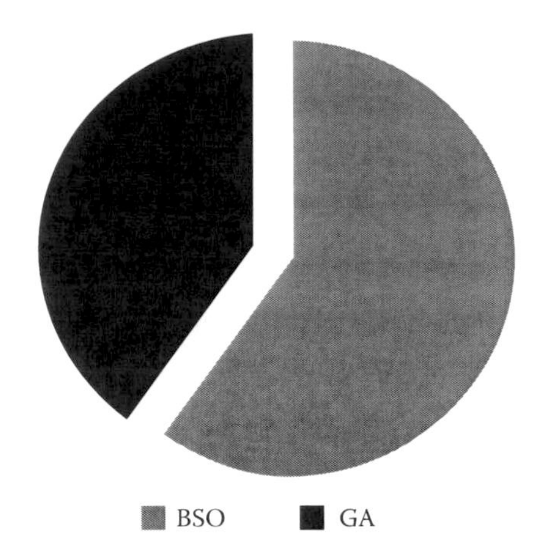

FIGURE 12.6 Running time for the BSO and GA.

Case Study 2: Running Time Evaluation

Looking closely at the proposed BSO algorithm and GA, it is found that BSO needs only two things more than the GA: the clustering and dominating points check process. K-means clustering is used here in the BSO, which costs $O(Kot_{dist})$, where K is the number of clusters, and o is the number of objects, and t_{dist} is the time to compute the centroid. Therefore, the BSO as given in Figure 12.6 takes more running time than the GA. However, this amount of increase in the running time is not that much and it is within the linear complexity.

12.8 CONCLUSION

The research in this chapter targeted solving the coverage problem in wireless sensor networks. The proposed algorithm, known as the BSO, is utilized for the coverage problem. Different variations are considered, especially for normalizing ideas. The BSO solution is compared to the traditional GA and the results show that it outperforms the GA in terms of coverage and connectivity.

REFERENCES

Cardei, M. and Du, D.-Z. 2005. Improving wireless sensor network lifetime through Power Aware Organization. *ACM Wireless Networks*, 11: 333–340.

Cardei, M. and Wu, J. 2006. Energy-efficient coverage problems in wireless ad hoc sensor networks. *Computer Communications*, 29(4): 413–420.

Chen, J., Cheng, S., Chen, Y., Xie, Y., and Shi, Y. 2015. Enhanced brain storm optimization algorithm for wireless sensor networks deployment. In: *Advances in Swarm and Computational Intelligence*. Tan Y., Shi Y., Buarque F., Gelbukh A., Das S., Engelbrecht A. (eds), ICSI 2015. Lecture Notes in Computer Science, 9140. Springer, Cham.

Cheng, M. X. and Gong, X. 2011. Maximum lifetime coverage preserving scheduling algorithms in sensor networks. *Journal of Global Optimization*, 51(3): 447–462.

Dietrich, I. and Dressler, F. 2009. On the lifetime of wireless sensor networks. *ACM Transactions on Sensor Networks*, 5(1): article 5.

Fatemeh, M. and Ahmad, K. 2013. Coverage problem in heterogeneous wireless sensor networks. *European Scientific Journal*, 9(27), 81–96.

Gao, W., Chen, Q., Jiang, M., Li, Y., and Wang, S. 2015. The optimization of genetic algorithm in wireless sensor network coverage. *International Journal of Signal Processing, Image Processing and Pattern Recognition*, 8(1): 255–264.

Gu, Y., Zhao, B. H., and Ji, Y. S. 2011. Theoretical treatment of target coverage in wireless sensor networks. *Journal of Computer Science and Technology*, 26(1): 117–129.

Lehrer, J. 2012. Groupthink. *The New Yorker*, January 30.

Mahmud, M. and Fethi, J. 2015. An iterative solution for the coverage and connectivity problem in wireless sensor network. The Seventh International Symposium on Applications of Ad hoc and Sensor Networks, AASNET'15, Berlin, Germany, pp. 494–498.

Pyun, S. and Cho, D.-H. 2009. Power-saving scheduling for multiple-target coverage in wireless sensor networks. *IEEE Communications Letters*, 13(2): 130–132.

Rabie, A. 2017. Fuzzy brain storming optimisation algorithm. International Journal of Intelligent Engineering Informatics, 5(1): 67–79.

Ramar, R. and Shanmugasundaram, R. 2015. Connected k-coverage topology control for area monitoring in wireless sensor networks. *Wireless Personal Communications*, 84(2): 1051–1067.

Shi, Y. 2011. An optimization algorithm based on brainstorming process. *International Journal of Swarm Intelligence Research*, 4(2): 35–62.

Singh, A. and Sharma, T. 2014. A survey on area coverage in wireless sensor networks. In: *International Conference on Control, Instrumentation, Communication and Computational Technologies (ICCICCT)*, IEEE, Kanyakumari, India, pp. 829–836.

Slijepcevic, S. and Potkonjak, M. 2001. Power efficient organization of wireless sensor networks. *IEEE International Conference on Communications,* Helsinki, Finland, Vol. 2, pp. 472–476.

Yardibi, T. and Karasan, E. 2010. A distributed activity scheduling algorithm for wireless sensor networks with partial coverage. *Wireless Networks*, 16: 213–225.

Zhao, Q. and Gurusamy, M. 2008. Lifetime maximization for connected target coverage in wireless sensor networks. *IEEE/ACM Transactions on Networking*, 16(6): 1378–1391.

13 An Exploratory Evaluation of the Awareness of E-Government Services among Citizens in Saudi Arabia

S. Q. Al-Khalidi Al-Maliki

CONTENTS

13.1 Introduction 177
13.2 Background Literature 178
13.3 Research Problem Statement 179
13.4 Research Objectives and Methodology 180
13.5 Sampling and Data Collection Procedure 180
13.5.1 Interview Questions 182
13.6 Findings and Discussion 183
13.7 Implications 184
13.8 Conclusion 184
References 185

13.1 INTRODUCTION

With rapid developments in information technology (IT), the effects of information systems (IS) on employee performance in organizations are evolving. Organizations are introducing computer technology and developing their own IS for more efficient management. The increasing utilization of IS may encourage employees to use it to help them perform tasks and manage work. This has resulted in the rapid development of the electronic government (e-government) concept. The concept refers to the use of IT/IS by the government to provide citizens and organizations with more convenient access to government services. E-government provides government-related information and services to citizens via Internet and non-Internet applications.

The main purpose of e-government is to build a government that exists everywhere and is ready to serve at any time. Through the use of different information equipment, e-government allows enterprises and the public to receive related services at any time and place. However, the implementation of such new and innovative policies requires consensus among most of the personnel within the concerned organization. Therefore, employee performance can be regarded as an outcome of e-government. Accordingly, employee performance via IS use becomes an important theme in the context of e-government (Luarn and Huang, 2009).

By using information communication technology (ICT) tools and applications, the Internet, and mobile devices to support good governance, government agency and department initiatives strengthen existing relationships and build new partnerships within civil society. These are known as e-government initiatives. As with e-commerce, e-government represents the introduction of a great wave of technological innovation, as well as government reinvention. It represents a tremendous impetus to move forward in the twenty-first century with higher quality, cost-effective government services, and a better relationship between citizens and governments (Fang, 2002; Ndou, 2004). Many government agencies in developed countries have taken steps toward web and ICT use, thereby making all local activities coherent on the Internet, widening local access and skills, opening interactive services for local debates, and increasing the participation of citizens in the promotion and management of territory (Graham and Aurigi, 1997; Ndou, 2004).

In this chapter, the background literature will be presented in the following section, then, in Section 13.3, the research problem statement will be discussed. A research objective and methodology will be presented in Section 13.4, followed by the sampling and data collection procedure and interview questions in Section 13.5. Next, in Section 13.6, the findings and discussion will be presented, followed by the implications and conclusion in Sections 13.7 and 13.8.

13.2 BACKGROUND LITERATURE

This chapter will present a review of the available literature on the topic of e-government adoption. Previous studies by this author (Al-Maliki, 2014) and Abanumy et al. (2005) explored three main topics: website accessibility guidelines, website accessibility tools, and the implication of human factors in the process of implementing successful e-government websites. These studies examined the issues that make a website accessible and explored the importance placed on web usability and accessibility with respect to e-government websites. They also briefly examined accessibility guidelines, evaluation methods, and analysis tools. Alshawi and Alalwany (2009) discussed the ways in which e-government, in both theory and practice, has proven to be important and complex. The chapter is part of research efforts to develop a rigorous and holistic evaluation framework for e-government systems. The main aim of this article was to develop evaluation criteria for an effective, adaptable, and reflective assessment of e-government systems from the perspective of citizens. The criteria can also be used as a means for providing valuable feedback for planning future e-government initiatives. Chatfield and Alhujran (2009) provided insight into the current state of the development of Arab e-government programs. A cross-country comparative analysis of e-government websites and portals involving 16 Arab countries assessed their relative developmental stages in terms of capability of delivering e-government services. The results revealed a wide digital divide among the Arab countries studied, particularly in terms of the development and capacity to deliver advanced e-government services. These results have important implications for developing countries in terms of managing both economic and noneconomic resources effectively for successful e-government development. Luarn and Huang (2009) investigated the implications and consequences of government employee performance via IS. A multiple regression method was used to investigate factors that influence employee performance. The results indicated that three factors affect performance: task–technology fit, computer self-efficacy, and utilization. Utilization was found to have the most positive effect on performance. In addition to verifying prior empirical findings, this study outlined the factors that influence employee performance and IS development in the context of e-government.

In addition, Al-Solbi and Al-Harbi (2008) explored key e-government policies and factors that contributed to the success of e-readiness assessment from the perspective of some of the public and private organizations in Saudi Arabia. With this aim, a questionnaire was developed and distributed, and semistructured interviews were conducted with ICT managers in specific organizations. The findings are very important since they indicate that both Saudi and organizational leadership are important aspects of ICT infrastructure in Saudi society. Al-Jaghoub et al. (2010) conducted a study on awareness and acceptability of using e-government services titled "Evaluation of Awareness and Acceptability of Using E-Government Services in Developing Countries: The Case of Jordan." Their study showed that awareness of e-government in Jordan was not at the required level. They identified factors influencing the usage of e-government services, such as cultural resistance and lack of trust in online services.

In "Challenges of E-Government Services Adoption in Saudi Arabia from an E-Ready Citizen Perspective," Alshehri and Drew (2010) identified technical, ICT infrastructure, organizational, security, social, and cultural barriers, and made recommendations to overcome them.

In another study, Alshehri et al. (2012) investigated some obstacles and challenges in the adoption of e-government services in Saudi Arabia. They identified many important factors that directly affect the adoption process. Also, they made recommendations to help the public sector and government organizations to improve their electronic services.

One study identified the impact of ICT investment in Saudi Arabia and the role of government through a series of 5-year plans (Al-Maliki, 2013). It also outlined and analyzed current IT use and development in Saudi Arabia, as well as the factors affecting economic growth. This research also assessed strategies and policies related to ICT and investment in ICT in Saudi Arabia, and discussed the role of public and private organizations, as well as educational institutions at all levels (Al-Maliki, 2013). Despite progress in the use of ICT in Saudi public and private organizations, there are still challenges to overcome before ICT becomes a viable part of Saudi life. Moreover, another study (Al-Maliki, 2014) determined the factors affecting e-governance adoption in Saudi Arabia. It was found that the main barriers were a lack of skills and human resources, low computer literacy and training capacity, and English language difficulties. The researcher proposed a conceptual architecture for e-portal services. The proposed recommendations addressed the need to understand the adoption of e-government and to help citizens use the available services. The proposed model reiterated that citizens should be helped until they accurately understand the functioning of e-government applications.

Basamh et al. (2014) studied the adoption and implementation of e-government in Saudi Arabia. They explored current practices, obstacles, and challenges affecting the improvement of e-services from the perspective of society. They found that infrastructure costs, computer literacy, privacy issues, accessibility and availability, and trust issues were major obstacles to the implementation and adoption of e-government in Saudi Arabia.

13.3 RESEARCH PROBLEM STATEMENT

E-government is a prominent concept today in both popular and academic discussions on governance reform. Studies on awareness of using e-services in Saudi Arabia show that a number of challenges have hindered the reach and impact of e-government. Several social, economic, and literacy barriers constrict the scope of transformation and restrict the ability of policymakers to effectively use new e-government technologies. There are many potential barriers to the implementation of e-governance:

- ICT skills
- Technology factors
- Cultural differences
- Integration technology
- E-governance application software
- Government support
- Trust and security
- Digital divide

However, the preceding obstacles differ from one country to another according to usage, facilities, and culture. These obstacles and constraints require urgent solutions. In the Saudi context, the main obstacles are related to sociocultural, organizational, and technical factors (Al-Maliki, 2013) and lack of awareness and trust. A review of the literature revealed that there is limited empirical research on e-government services from the citizens' perspective. According to Al-Maliki (2013) cultural differences can result in limitations in IS implementation. For example, different views on logic and reasoning, and limitations in language use can create a barrier to effective communication and understanding.

The Saudi government established a gate to e-government services, called "Saudi," which is an e-government portal. It is a national portal for accessing e-government services in Saudi Arabia from anywhere, available to citizens, expatriates, companies, and visitors. E-services are delivered to users in a highly efficient manner. The e-government program ("Yesser") launched its national portal on March 27, 2006. E-services are accessible through the Saudi portal, either by integration with other government agencies or via links to those agencies or their services on the portal. The Saudi portal has around 2035 e-services of many public and private organizations in Saudi Arabia (Saudi Official portal, 2016). However, Saudis need to be made aware that e-government offers a number of potential benefits. Owing to the lack of awareness, some agencies offer e-services, for example, bookshops, and some have established offices to help the citizens.

13.4 RESEARCH OBJECTIVES AND METHODOLOGY

The objective of this chapter is to evaluate the extent of awareness among Saudis in using e-government services. It also aims to identify the main factors that may inhibit the use of e-services in government organizations.

The research adopted a descriptive and analytical approach, involving the study of the use of e-government services and interviews with citizens. The interviews allowed more in-depth and, in some cases, broader responses from the respondents than would have been the case with other fact-finding approaches. Interviews were carried out with major players in e-government in the cities of Abha and Jazan, located in southern Saudi Arabia, namely, the municipalities, e-government agencies, and some departments in the Ministry of the Interior.

A literature review was also conducted on this topic. Information was collected from articles published by other researchers and from current trends in this area. In addition, data was collected through a website content study. Moreover, observations were carried out at some agencies and service offices. This research approach was considered appropriate for analyzing citizens' awareness of using e-government services in Saudi Arabia.

The interview questions were open questions related to various aspects of using e-government services. The following issues were covered: extent of awareness about the use of e-government services in Saudi, the main factors hindering awareness of the use of e-government services over the Internet, the principal reasons behind Saudis not being very familiar with e-government services, the main factors that could lead to resistance or failure of using e-government services, whether the e-government service providers were satisfied with the extent of citizens' awareness, solutions to the lack of awareness among Saudis, and the appropriateness of e-services websites.

13.5 SAMPLING AND DATA COLLECTION PROCEDURE

E-government has the power to serve and implement good governance, economic growth, and human development though increased efficiency, accessibility, transparency, and accountability of government operations, leading to improved national performance in all aspects.

For this study, 45 citizens were interviewed and 15 service offices visited to observe how citizens seek help and how e-services are executed. The sample was drawn from southern Saudi Arabia, in particular the cities of Abha and Jazan, and can be generalized among Saudis in other cities of Saudi Arabia. The study population was defined as citizens, government employees, and university students. The study sample included citizens who worked at the offices or offered services to use the e-services and citizens who needed the services. Citizens who visited e-government agencies to process e-services were randomly selected. The researcher met the respondents over 3 weeks at different times, at the following locations: passport departments, civil agencies, labor offices, Chamber of Commerce branches, Ministry of Commerce branches, Ministry of the Interior sections, and municipalities. The sample size can be considered adequate for this study.

The Saudi government provides many e-services via its government website portal Saudi; an illustration is shown in Table 13.1.

TABLE 13.1
E-Government Services

Service	E-services	Service	E-services
ICT	(71) e-services	Economy and business	(158) e-services
Islamic affairs	(110) e-services	Labour and employment	(109) e-services
Housing and municipal services	(372) e-services	Health and environment	(171) e-services
Traffic and safety	(372) e-services	Personal documentation	(84) e-services
Traininig, education, and culture	(579) e-services	Travel and tourism	(55) e-services
Social life	(66) e-services	Insurance and pension	(44) e-services
Utilities	(132) e-services	Transportation	(55) e-services

Source: Saudi Official Portal, 2016, Saudi, accessed January 20, 2016, http://www.saudi.gov.sa/.

13.5.1 Interview Questions

To achieve the study objectives, literature on the awareness of using e-governance services among Saudi citizens was reviewed. The following questions on this topic were then addressed:

1. What is the extent of the awareness of using e-government services among Saudi citizens?
2. What are the main factors hindering the awareness of using e-governance services over the Internet among Saudi Arabian citizens?
3. What are the principal reasons behind Saudi citizens being less familiar with the use of e-government services within public organizations?
4. What are the main factors that could lead to resistance or failure of the use of e-government services provided by Saudi public organizations?
5. Are the public organizations implementing e-government services satisfied with the extent of citizens' awareness about the e-services they provide?
6. What are the suitable solutions to address the lack of awareness among Saudis?

According to Alshehri and Drew (2010), most e-government websites are inefficient and provide just basic and general information about organizations, plus often the data is not updated. Therefore, the following specific questions about e-government service websites were posed to interviewees:

1. What is your overall impression of e-service websites?
2. Do e-service websites provide the information you need?
3. Does the content meet your needs?
4. Are you satisfied with the accuracy of e-service websites?
5. Is the help guide of the website clear?
6. Do the related e-service agencies provide up-to-date information?
7. Is the e-service website user friendly?
8. Are the e-service websites easy to use?

In addition, the interviewees were asked to what extent they agreed/disagreed with the statements in Table 13.2 related to the resistance or failure of using e-government services.

TABLE 13.2
Problems Arising When Accessing E-Services

No.	Statement (Factor)	Strongly Disagree	Disagree	Neutral	Agree	Strongly Agree
1	Perceived difficulty or complexity of using e-services					
2	Lack of understanding					
3	Lack of sufficient help					
4	Lack of awareness about using e-services					
5	Lack of organization help desk support					
6	Inappropriateness of the e-service applications					
7	High cost of using e-services					
8	Lack of website support for using e-services					
9	Using e-services is not considered an important process					
10	Using e-services is not required in our daily life					
11	Other factors					

13.6 FINDINGS AND DISCUSSION

E-government supports broad public sector reforms and good governance through the introduction of innovative and sustainable applications of ICT within government administrations, for enhanced interaction with citizens and the private sector. The public sector is increasingly seen as the main element that can bridge the digital divide at national level. Public agencies need to be model users of ICT so that others can follow. The public sector tends to be the biggest provider of local content, and it can nurture and foster further development of the local ICT industry.

Many interviewees stated that the main obstacles to adopting e-government services were a lack of trust in and awareness of using online services. In addition, many interviewees indicated that, in their opinion, e-service departments in all organizations should provide better help and support to improve citizens' awareness. The majority of the interviewees suggested that one of the main barriers to the use of e-services in Saudi Arabia was the lack of awareness about how to execute the e-services. Appropriate help and support files should be included within e-service websites.

In some cases, citizens do not understand how to use e-services owing to a lack of awareness. Furthermore, some citizens complained about the difficulty of using e-services because they were not familiar with e-government services.

It seems that a lack of awareness of using e-services within Saudi public organizations is one of the major barriers to e-governance. Many organizations in Saudi look at e-government services as an important way to improve internal operations and provide quick and more efficient operations and better quality services, but they should educate their clients regarding how to use their e-services.

Saudi public organizations should provide awareness programs and tools to help users learn about how e-government applications work, as well as train their employees so that they have sufficient understanding of how the applications work and, in turn, help citizens.

Concerning training, trainers are an important factor in the implementation and integration of computer technology in education. Without sufficient knowledge, it will be difficult to appropriately use the e-services. Appropriate training is a must for all professional e-services users in order to possess the skills and awareness needed to use these applications.

The major obstacle to successful implementation and usage of e-government applications according to this study is the lack of user awareness and motivation. Weak communication between e-government employees and citizens as clients can explain misunderstandings over the use of e-governance applications. Also, there is a lack of awareness of current services and their usefulness.

Owing to inadequate training support from the vendor, public organizations have had to spend hours learning and training their personnel to use different software applications. Owing to the complexity of these applications, many citizens have to develop their own knowledge by using the Google search engine or by contacting a friend to get help to use these services. Saudi public organizations should address this problem by helping people learn to use e-services.

There is also a lack of e-service expertise within organizations, namely, suitable help and support files or clips showing how to use the services. It seems that some organizations do not think training is important enough to affect their services. Most IT failures are due to inadequate training programs provided either by the system's users or vendors.

The main points of help for citizens to execute e-services are the service office, estate office, friends and colleagues, bookstores, agents working for government and service agencies, and office imaging services.

Other major factors influencing the awareness of using e-services are electronic illiteracy, distrust, fear of making mistakes, absence of an e-mail ID that is required for a user to be able to open an account on the e-service website, lack of awareness of the existence of these services, lack of smart devices, and dependency on colleagues and friends to carry out the e-services.

Lack of updated information on the websites could be a result of the current lack of awareness within e-services management departments. The findings of this study highlighted a number of issues related to e-government applications, appropriate use of e-government applications, availability of up-to-date information, training of users, and provision of support to citizens.

Concerning the e-service websites, the findings indicate that there is a lack of attention given to content quality. The websites do not provide the "Frequently Asked Questions" service, where visitors can get the required information as soon as possible without having to spend a great deal of time searching every part of the website.

Some websites lack contact details ("Contact us"), that is, phone, fax, or e-mail. Some e-service websites are not compatible with specific Internet browsers. In addition, e-service links or icon images are not always in obvious places on websites' main pages, difficulty of website interface, and no possibility in the e-service website to follow-up the transactions or process executed.

13.7 IMPLICATIONS

On examining the main findings of this study, it can be seen that e-government service departments should raise the quality of e-service awareness to help the users, that is, citizens, residents, and visitors, to provide effective and efficient government services and to meet the needs of the beneficiaries of government agencies.

E-government websites should include enough information about their e-services. Also, e-government departments should examine their user-satisfaction forms to evaluate the performance of their e-service applications. Public organizations that provide online services should encourage citizens to use their services by educating them about these services. Also, e-service websites should meet the requirements and desires of users and be easy to understand and use. The findings of this study will help government and private employees and citizens gain a deeper understanding of e-governance use. It will create public awareness about the potential of ICT. Citizen access to government information/services must increase to combat the digital divide. Urgent training in ICT-based systems and services can enhance the knowledge and skills of the concerned parties. Increased awareness of the use of e-services will increase IT literacy and help reduce the internal digital divide. This will also help boost the quality of education, health services, and social security. Further, this will play a major role in increasing the capacity for rational distribution of public funds and strengthening a development-oriented and people-centered service-delivery culture. In addition, e-government agencies should give priority to ensuring that Internet channels are safe and reliable in order to build a reliable ICT infrastructure and to avoid breakdown of services.

In summary, e-governance involves the entirety of society and technology; e-governance is now integrating all the services in a single system. Saudi public organizations should analyze their systems on a regular basis to reassess their readiness for technological progress and ongoing changes in the governance system.

According to Table 13.1, many e-service agencies in Saudi Arabia post relevant information online in an organized and easy-to-access manner for other government agencies, businesses, and citizens. Also, relevant transactions between government agencies and private sector businesses and citizens can now take place online.

13.8 CONCLUSION

In the last few years, there has been considerable infrastructure development in IT, and we need to think about where we will be and where we want to be with regard to e-governance infrastructure to ensure connectivity between the public and government.

The researcher adopted a descriptive survey method involving face-to-face interviews, considered appropriate for collecting sufficient qualitative data on the awareness of using e-services in southern Saudi Arabia. In this current research, many conclusions were drawn, offering the possibility of reshaping the public sector's activities and processes, building relationships between citizens and government, enhancing transparency, increasing government capacity, and providing infrastructure facilities.

The qualitative analysis provided a general overview of current awareness of using e-services by examining the responses of 45 Saudi citizens. The interviews and observations gathered a broad range of ideas regarding how citizens execute e-services. On the whole, the interviewees gave very similar answers about their difficulties and the need to understand the process and overcome specific obstacles in using e-government services. The findings indicate that there is a poor understanding of the use of e-services. There has been tremendous infrastructure development in IT and e-governance in Saudi public organizations to enhance connectivity between the public and government organizations. This is an important finding that reflects the fact that ICT is very heavily invested in Saudi public organizations, but more effort is required with regard to the support and help needed for using e-government services via the Internet.

This study also showed that there is a poor understanding of the process of using e-services and that Saudi citizens' awareness is limited in some areas. There was an evident lack of management support, which many felt hindered the help and support given to citizens, and there was a definite lack of understanding of e-government services, namely, the functioning of public organization websites. In addition, many organizations, unfortunately, showed a lack of willingness to help users. E-government agencies working for many public and private organizations must overcome a number of obstacles to achieve the growth necessary for e-government services to be delivered to all Saudi communities.

Finally, this study placed demands on public organizations that need more attention, and it may lead to in-depth further discussion on the best way to increase awareness of e-government services among Saudis and studies on some aspects of e-government services.

REFERENCES

Abanumy, A., Al-Badi, A., and Mayhew, P. 2005. E-government website accessibility: In-depth evaluation of Saudi Arabia and Oman. *The Electronic Journal of e-Government*, 3(3), 99–106.

Al-Jaghoub, S., Al-Yaseen, H., and Al-Hourani, M. 2010. Evaluation of awareness and acceptability of using e-government services in developing countries: The case of Jordan. *Electronic Journal Information Systems Evaluation*, 13(1), 1–8.

Al-Maliki, S. Q. A. 2013. Information and communication technology (ICT) investment in the Kingdom of Saudi Arabia: Assessing strengths and weaknesses. *Journal of Organizational Knowledge*, 2013, 1–15.

Al-Maliki, S. Q. A. 2014. Analysis and implementation of factors affecting e-governance adoption in the Kingdom of Saudi Arabia. *International Journal of Strategic Information Technology and Applications*, 5(1), 20–29.

Alshawi, S. and Alalwany, H. 2009. E-government evaluation: Citizen's perspective in developing countries. *Information Technology for Development*, 15(3), 193–208.

Alshehri, M. and Drew, S. 2010. Challenges of e-government services adoption in Saudi Arabia from an e-ready citizen perspective. *World Academy of Science, Engineering and Technology*, 66, 1053–1059.

Alshehri, M., Drew, S., and Alfarraj, O. 2012. A comprehensive analysis of e-government services adoption in Saudi Arabia: Obstacles and challenges. *International Journal of Advanced Computer Science and Applications*, 3(2), 1–6.

Al-Solbi, A. N., and Al-Harbi, S. H. 2008. An exploratory study of factors determining e-government success in Saudi Arabia. *Communications of the IBIMA*, 4, 188–192.

Basamh, S. S., Qudaih, H. A., and Suhaimi, M. A. 2014. E-government implementation in the Kingdom of Saudi Arabia: An exploratory study on current practices, obstacles & challenges. *International Journal of Humanities and Social Science*, 4(2), 296–300.

Chatfield, A. and Alhujran, O. 2009. A cross-country comparative analysis of e-government service delivery among Arab countries. *Information Technology for Development*, 15(3), 151–170.

Fang, Z. 2002. E-government in digital era: Concept, practice and development. *International Journal of the Computer*, 10(2), 1–22.

Graham, S. and Aurigi, A. 1997. Virtual cities, social polarisation, and the crisis in urban public space. *Journal of Urban Technology*, 4(1), 19–52.
Luarn, P. and Huang, K. 2009. Factors influencing government employee performance via information systems use: An empirical study. *Electronic Journal of e-Government*, 7(3), 227–240.
Ndou, V. 2004. E-government for developing countries: Opportunities and challenges. *The Electronic Journal of Information Systems in Developing Countries*, 18(1), 1–24.
Saudi Official Portal. 2016. Saudi. Accessed January 20, 2016. http://www.saudi.gov.sa/.

14 An Internet of Things (IoT) Course for a Computer Science Graduate Program

Xing Liu and Orlando Baiocchi

CONTENTS

14.1 Introduction ... 187
14.2 Chapter Outline ... 188
14.3 The Internet of Things (IoT) Course Designed by University of Washington, Tacoma ... 188
14.3.1 Course Design ... 188
14.3.2 Course Content ... 189
14.4 IoT Development Tools ... 190
14.4.1 Development Tool Selection ... 190
14.4.2 Texas Instruments' IoT Kit ... 190
14.4.3 Amazon AWS IoT ... 191
14.4.4 Microsoft Azure IoT Suite ... 191
14.5 Conclusions and Future Work ... 192
Acknowledgments ... 192
References ... 192

14.1 INTRODUCTION

The phrase "Internet of Things" (IoT) has become one of the hottest buzzwords in the technological world nowadays. Not only has IoT become a new research paradigm, but also an important line of products in major technological companies. For example, IEEE has started the IEEE Internet of Things (IoT) Initiative (IEEE, 2015a), which, among other activities, is organizing the leading conference IEEE World Forum on Internet of Things (IEEE, 2015b). In the meantime, IEEE launched a publication titled *IEEE Internet of Things Journal* in 2014 (IEEE, 2014). Microsoft's response to the "IoT movement" is the Azure IoT Suite (George, 2015; Microsoft, 2015a,b), a cloud-based software system with preconfigured solutions that address common IoT scenarios. Amazon's answer for IoT is AWS IoT (Amazon Web Services, 2015a,b,c), a cloud platform that lets connected devices interact with one another and with cloud applications. Google (Google Cloud Platform, 2015), Oracle (2015), IBM (2015), and Salesforce (2015) are also providing tools for IoT data collection and analysis. On the hardware front, Texas Instruments (2015a,b) and Intel (2015) both provide building blocks and enabling technologies for IoT. Cisco (2015) and Mitchell et al. (2013) are actively involved in IoT development as well.

In response to industrial needs for trained graduates, academic institutions have started offering courses dedicated to IoT. For instance, Harvard University's CS 144/244 course is called Secure and Intelligent Internet of Things (Kung, 2015). The Harvard course uses a large number of devices such as depth sensing (Kinect and depth cameras) and Samsung Galaxy and Samsung watch in the course. North Carolina State University (2015) has the course CSC 591 scheduled for spring 2016. The name of the course is Internet of Things: Applications and Implementation. Columbia University (2015) and California Polytechnic State University (Cal Poly, 2015) both have a course titled Internet of Things. Columbia University's IoT course number is E6765. California Polytechnic State University's course is CSC 520. Apart from having IoT courses in standard degree curriculum, universities are offering training either online or via continuing education. For instance, University of California, San Diego and Qualcomm are offering a six-course IoT certificate via Coursera (2015). University of Washington, Seattle (UW Seattle) Professional and Continuing Education is developing a certificate on IoT (University of Washington, 2015), which will be offered in spring 2017.

The Institute of Technology at University of Washington, Tacoma (UW Tacoma) developed its own IoT course (TCSS 573) in autumn 2015. The course is at the graduate level and targets students enrolled in the cyberphysical option of the Master of Science in computer science and systems degree program. It fills the need for content on device connections with the cloud. This chapter describes the details of the course (with some adjustment made to the tentative weekly schedule approved).

Although IoT courses are being taught or are going to be taught at numerous universities, detailed information about the courses, especially course design and development tools used, is rarely available in published literature. This chapter tries to make some contribution in this aspect.

14.2 CHAPTER OUTLINE

The remainder of the chapter is structured as follows. Section 14.3 describes in detail the course designed by UW Tacoma. Section 14.4 explains how IoT development tools are selected for the course. Conclusions and future work are provided in Section 14.5.

14.3 THE INTERNET OF THINGS (IoT) COURSE DESIGNED BY UNIVERSITY OF WASHINGTON, TACOMA

14.3.1 Course Design

The UW Tacoma IoT course (course number: TCSS 573) is designed with certain objectives in mind. The main objective is that, after completing the course, the students will be able to design and develop IoT hardware and software, and utilize cloud tools to collect and analyze data for IoT applications. This objective is formed based on the authors' understanding of the status and needs of the industry.

The course is designed to have adequate breadth and depth in IoT theory. However, great emphasis is placed on practical skills. Students will design and build hardware, write code for embedded systems, and customize and develop IoT cloud services using commercial products. It is envisioned that 40% of the class time will be spent on IoT theory, with the remaining 60% of the time spent learning practical hardware and software development skills. The course has three undergraduate courses as prerequisites: C for system programming, networking and distributed systems, and client/server programming for Internet applications.

The course is worth five credits and will be taught in 10 weeks because of the quarter system adopted by the UW Tacoma. The class will meet 4 hours a week with each meeting being 2 hours. The format of the class will be a mixture of lectures, study of commercial development tools, and hands-on assignments and class discussions. Each week will cover a special theoretical topic in IoT, plus practical examination of one of the key features of IoT device development tools and software development tools. The course will have a project so that students can apply everything they learn in the course in one place and gain practical skills for future employment in industry. A textbook will be recommended (Bahga and Madisetti, 2014). However, a large amount of course material will come from various additional references including those from the Internet. Every student will be required to purchase an IoT device development kit from Texas Instruments (2015a,b) with the cost of around $30. Table 14.1 shows the design of the course.

TABLE 14.1
Design of the UW Tacoma IoT Course TCSS 573

Course number	TCSS 573
Course name	Internet of Things
Credits	5
Duration	10 weeks; 40% theory, 60% hands-on
Format	Lecture, assignments, project, examinations
Prerequisite	Knowledge of C programming, computer networks, distributed systems, client server technologies, Web technologies
Description	Examines physical design and logical design of Internet of Things, functional blocks and architecture, protocols and communication models, enabling technologies, application domains specific to Internet of Things, smart objects, development tools, system management, cloud services, security and data analytics.
Hardware	Texas Instrument CC3200 Development Kit
Software	Amazon AWS IoT, Microsoft Azure IoT

14.3.2 Course Content

In order to give students the overall picture of IoT application development, the course begins with an introduction to the architectures of IoT applications described in Amazon's AWS IoT (Amazon Web Services, 2015a,b) and Microsoft's Azure IoT (George, 2015; Microsoft, 2015a,b). The purpose is to illustrate the physical components of an IoT application, that is, the "things" or the devices, the IoT services, other cloud services, the applications, and how they relate to each other. Therefore, the students will see what they need to understand, and what skills they need to learn in order to be able to develop IoT applications.

Then the course introduces the formal models for IoT architecture, the functional blocks, and the communication models (Bahga and Madisetti, 2014; Carrez, 2015). These models represent an IoT system from different perspectives in a more abstract way and facilitate the students' understanding of the principles and operations of an IoT system.

The Cisco IoT Reference Model (Green, 2014) is introduced in the course next. Cisco conceptualized an IoT application into seven levels and defined "edge" and "edge computing" in the reference model.

The course then studies two of the IoT application domains: smart cities and smart environments (Mitchell et al. 2013; Zanella et al. 2014) so that students can see how IoT technologies can actually be used in the real world.

IoT requires numerous enabling technologies such as wireless sensor networks (Mainetti, 2011; Culler, Estrin, and Srivastava, 2014). The course examines sensors and sensor networks, and wireless technologies such as Wi-Fi, Bluetooth, and ZigBee.

Cloud computing (Botta et al. 2014), security (Roman, Najera, and Lopez, 2011), and big data analytics (Cionanu, 2014) are essential enabling technologies for IoT, too. The course covers these technologies together with the AWS IoT platform and Azure IoT Suite.

IoT communication protocols message queuing telemetry transport (MQTT) and constrained application protocol (CoAP) (Jaffrey, 2015; Stansberry, 2015) are used to address the special characteristics of communications between IoT connected devices. These protocols are discussed in the course. Embedded system concepts are also discussed (Wolf, 2002).

IoT, as a network of connected devices, needs tools that can install, manipulate, and delete devices, as well as tools that model network and device configurations and state data. This is called IoT system management (Schönwälder, 2012). The course dedicates some time to look at this aspect of IoT as well.

The course then moves on to discuss smart objects (Kortuem et al. 2009). IoT can be seen as a loosely coupled, decentralized system of cooperating smart objects. Smart objects are "things"

TABLE 14.2
Weekly Topics

Week	TCSS 573 Topics	Hands On
1	Physical and logical design	Project info
2	IoT architecture and models	IoT hardware
3	Application domains	IoT hardware
4	Wireless sensor networks	IoT hardware
5	Cloud services and security	IoT hardware
6	Data analytics	IoT cloud
7	IoT communication protocols	IoT cloud
8	System management	IoT cloud
9	Smart objects	IoT cloud
10	Ethical and environmental impacts	IoT cloud

that can understand and react to their environment. Smart objects are ready to be connected and exchange information over the web. This part of the course has an in depth look of smart objects.

Finally, the ethical and environmental impact of IoT will be discussed (Weber, 2010) in order to increase students' awareness of the potential complications of adopting the IoT technology.

Table 14.2 is a weekly breakdown of the topics to be taught in the UW Tacoma's TCSS 573.

14.4 IoT DEVELOPMENT TOOLS

14.4.1 Development Tool Selection

The criteria for selecting the tools are twofold: (a) the tools should expose the students to the full spectrum of IoT application development, ranging from the "things" to the "cloud," and (b) the tools should be mainstream products in industry. Based on these criteria, Texas Instruments' IoT kit is selected for device (the "thing" side) development. Amazon's AWS IoT and Microsoft's Azure IoT Suite are selected for development at the cloud side.

14.4.2 Texas Instruments' IoT Kit

Texas Instruments' IoT kit provides an Internet-on-a-chip solution. The main component of the kit is an integrated circuit chip named CC3200 that integrates an ARM Cortext-M4 microcontroller and a Wi-Fi network module, as shown in Figure 14.1. An IoT device can be developed with a single CC3200 chip, which consumes very little power (two AA batteries can power the chip for over a year). The chip includes numerous serial and parallel interfaces and a four-channel analog-to-digital converter. The Wi-Fi network module on CC3200 implements Wi-Fi Internet-on-a-chip, which includes an 802.11 b/g/n radio, a baseband and MAC module, embedded TCP/IP and TLS/SSL stacks, an HTTP server, and multiple Internet protocols. The CC3200 chip can work in station mode or in access point mode. Developers can attach external devices such as sensors to

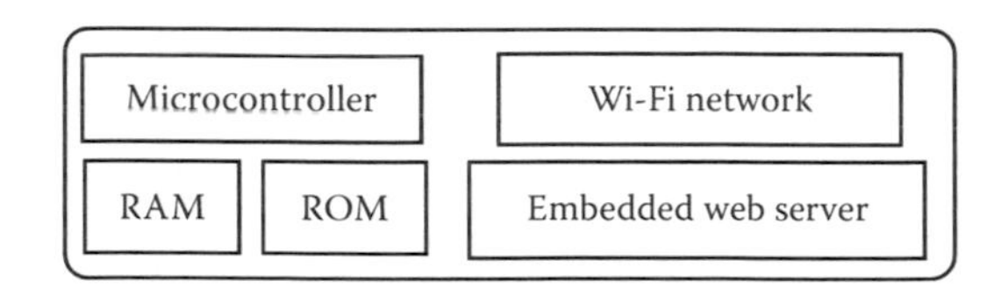

FIGURE 14.1 TI CC3200 Internet-on-a-chip system on chip (SoC) device.

CC3200 and use the CC3200 SDK and the Code Composer Studio IDE to program the chip using the C programming language.

Students will learn to use Texas Instruments' IoT kit to develop the hardware of the "things" or "smart objects" for an IoT system. They will develop the hardware of "things" that are connectable and ready to communicate over the Internet and exchange information over the web. They will attach sensors to the microcontroller on the CC3200, develop embedded C code to implement various functions required by IoT so that the "things" can transmits data wirelessly to the Internet and collaborate with other "things" and cloud services and applications.

14.4.3 Amazon AWS IoT

AWS IoT is the middleman between the "things" connected to the Internet and the AWS cloud (Figure 14.2). Users can collect data from the "things," and store and analyze data in the cloud with the help of AWS IoT. Devices can interact with cloud applications and other devices. Users can also control the "things" via their mobile devices with the help of AWS IoT. With AWS IoT, developers can make use of standard AWS services such as AWS Lambda, a computer service that can receive user code and run the code on behalf of the user; Amazon Kinesis, a service that collects and processes data records in real time; Amazon S3, a simple web service interface for storing and retrieving data; Amazon Machine Learning, which provides data visualization tools, machine learning models, and predictions; as well as the Amazon DynamoDB database service. AWS IoT provides different AWS SDKs for different hardware devices (Amazon Web Services 2015a,b,c).

The learning experience for this tool will enable students to develop AWS IoT cloud services to communicate with the "things" they have developed using the CC3200 hardware, and visualize data and perform data analysis.

14.4.4 Microsoft Azure IoT Suite

Microsoft Azure IoT Suite is a cloud-based platform. It provides preconfigured solutions for common Internet of Things scenarios. Similar to AWS IoT, developers can use the Azure IoT Suite to connect "things," to capture, manage, analyze and present data, and to automate operations (Figure 14.3). The centerpiece of the Azure IoT Suite is the Azure IoT Hub service. This service provides device-to-cloud and cloud-to-device messaging capabilities. It also acts as the gateway to the cloud and to the other key IoT Suite services.

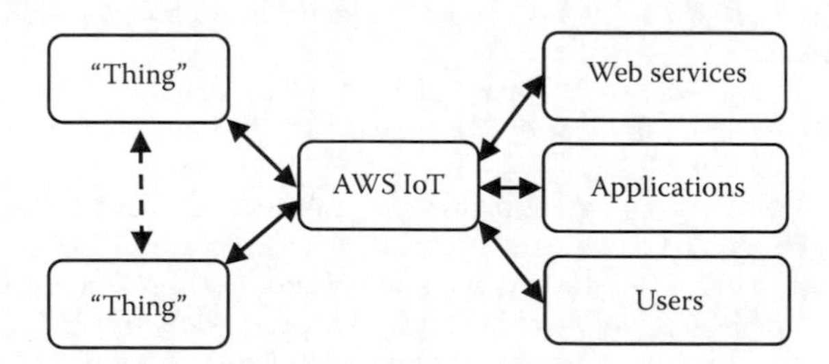

FIGURE 14.2 AWS system structure.

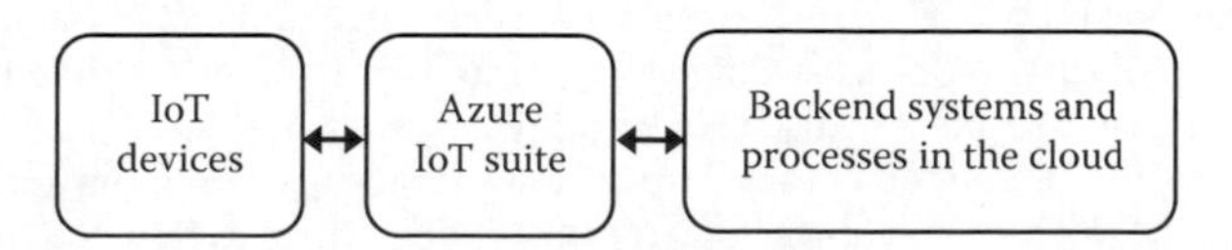

FIGURE 14.3 Microsoft Azure IoT.

Developers can use Azure Stream Analytics to perform data analysis and detect events and device responses to commands. Data can be stored using Azure Storage. DocumentDB can be used to manage device metadata. Data visualization can be achieved with Azure web apps and Microsoft Power BI. The preconfigured solutions and preconfigured services make IoT application development straightforward.

The learning experience for developing applications using the Azure IoT Suite (George, 2015; Microsoft, 2015a,b) will equip the students with skills for developing IoT cloud services to communicate with the "things" they developed using the CC3200, and visualize data and perform data analysis, similar to the AWS IoT case.

14.5 CONCLUSIONS AND FUTURE WORK

This chapter introduced in detail the IoT graduate course developed at University of Washington, Tacoma. The course is worth five credits and lasts for one quarter. It is designed to provide students in-depth training in both IoT theory and IoT application development skills. Commercial hardware and software tools will be used in the course to equip students with practical skills much needed by the local industry.

As a new course, the developed IoT course will be closely monitored for quality control purposes. Feedback will be collected at the end of the quarter from current students and from alumni in the future. New developments in the industry will also be closely monitored and course content will be updated to match the needs of industry. This will include updating course content, as well as hardware and software used in the course.

ACKNOWLEDGMENTS

The authors would like to thank Professor Rajendra Katti for his encouragement in developing the course. Support from all members of the Graduate Curriculum Committee, Institute of Technology, University of Washington, Tacoma is also greatly appreciated.

REFERENCES

Amazon Web Services. 2015a. Getting Started with AWS IoT, https://aws.amazon.com/iot/getting-started/#kits.
Amazon Web Services. 2015b. How the AWS IoT Platform Works, https://aws.amazon.com/iot/how-it-works/.
Amazon Web Services. 2015c. Internet of Things, https://aws.amazon.com/iot/.
Bahga, A. and Madisetti, V. 2014. *Internet of Things (A Hands-on-Approach)*, Arshdeep Bahga and Vijay Madisetti, Johns Creek, GA.
Botta, A., de Donato, W., Persico, V., and Pescape, A. 2014. On the integration of cloud computing and internet of things, *International Conference on Future Internet of Things and Cloud (FiCloud)*, Barcelona, Spain, pp. 23–30, September 27–29.
Cal Poly. 2015. CSC 520: Internet of Things, http://users.csc.calpoly.edu/~foaad/IOTS15.pdf.
Carrez, F. 2015. IoT Architecture: D1.5, www.iot-a.eu/public/public-documents/.
Cionanu, R. 2014. Big data platforms for the internet of things, in: Bessis, N., Dobre, C. (eds.), *Big Data and Internet of Things: A Roadmap for Smart Environments*, pp. 3–34, Springer, Cham, Switzerland.
Cisco. 2015. The Internet of Things, http://newsroom.cisco.com/internetofthings.
Columbia University. 2015. Course EECS E6765: Internet of Things, http://www.ee.columbia.edu/spring-2016-course-list.
Coursera. 2015. Build Your Own Internet of Things, https://www.coursera.org/specializations/internet-of-things.
Culler, D., Estrin, D., and Srivastava, M. 2014. Overview of sensor networks, *IEEE Computer*, 37(8), 41–49.
George, S. 2015. Microsoft Azure IoT Suite—Connecting Your Things to the Cloud, https://azure.microsoft.com/en-us/blog/microsoft-azure-iot-suite-connecting-your-things-to-the-cloud/.
Google Cloud Platform. 2015. Internet of Things, https://cloud.google.com/solutions/iot/.
Green, J. 2014. Building the Internet of Things: CISCO Internet of Things Reference Model, Cisco Connect, https://www.cisco.com/web/PH/ciscoconnect/pdf/bigdata/jim_green_cisco_connect.pdf.

IBM. 2015. Watson Internet of Things: IoT in the Cognitive Era, http://www.ibm.com/internet-of-things/watson-iot.html.

IEEE. 2014. IEEE Internet of Things Journal, http://iot-journal.weebly.com/.

IEEE. 2015a. About the IEEE Internet of Things (IoT) Initiative, http://iot.ieee.org/about.html.

IEEE. 2015b. IEEE 2nd World Forum on Internet of Things (WF-IoT), http://www.ieee-wf-iot.org/.

Intel. 2015. The Internet of Things (IoT) Starts with Intel Inside, http://www.intel.com/content/www/us/en/internet-of-things/overview.html.

Jaffrey, T. 2015. MQTT and CoAP, IoT Protocols, Eclipse, https://eclipse.org/community/eclipse_newsletter/2014/february/article2.php.

Kortuem, G., Kawsar, F., Sundramoorthy, V., and Fitton, D. 2009. Smart objects as building blocks for the Internet of things, *IEEE Internet Computing*, 14(1): 44–51.

Kung, H. T. 2015. CS 144r/244r: Secure and Intelligent Internet of Things, https://www.eecs.harvard.edu/htk/courses/.

Mainetti, L. 2011. Evolution of wireless sensor networks towards the Internet of Things: A survey, *19th International Conference on Software, Telecommunications and Computer Networks (SoftCOM), Adriatic Islands Split*, Croatia, pp. 1–6, September 15–17, 2011.

Microsoft. 2015a. OS Platforms and Hardware Compatibility with Device SDKs, https://azure.microsoft.com/en-us/documentation/articles/iot-hub-tested-configurations/.

Microsoft. 2015b. Tap into the Internet of Things with the Azure IoT Suite, http://www.microsoft.com/en-ca/server-cloud/internet-of-things/azure-iot-suite.aspx.

Mitchell, S., Villa, N., Stewart-Weeks, M., and Lange, A. 2013. The Internet of Everything for Cities Connecting People, Process, Data, and Things to Improve the "Livability" of Cities and Communities, Cisco, http://www.cisco.com/web/strategy/docs/gov/everything-for-cities.pdf.

NC State University. 2015. CSC 591: Internet of Things: Applications and Implementation, http://www.csc.ncsu.edu/courses/.

Oracle. 2015. Oracle Internet of Things, https://www.oracle.com/solutions/internet-of-things/index.html.

Roman, R., Najera, P., and Lopez, J. 2011. Securing the Internet of things, *IEEE Computer*, 44(9): 51–58.

Salesforce. 2015. Salesforce IoT Cloud, 2015, http://www.salesforce.com/ca/iot-cloud/.

Schönwälder, J. 2012. Network configuration management with NETCONF and YANG, *84th IETF Meeting*, Vancouver, July 29, https://www.ietf.org/slides/slides-edu-netconf-yang-00.pdf.

Stansberry, J. 2015. MQTT and CoAP: Underlying protocols for the IoT. *Electronic Design*, October 7, http://electronicdesign.com/iot/mqtt-and-coap-underlying-protocols-iot.

Texas Instruments. 2015a. TI Internet of Things Overview, http://www.ti.com/ww/en/internet_of_things/iot-overview.html.

Texas Instruments. 2015b. TI's SimpleLink™ Wi-Fi® Family: Connect More: Anywhere, Anything, Anyone, http://www.ti.com/ww/en/simplelink_embedded_wi-fi/cc3200.html.

University of Washington. 2015. Certificate in Internet of Things, http://www.pce.uw.edu/certificates/internet-of-things.html.

Weber, R. 2010. Internet of Things—New security and privacy challenges, *Computer Law and Security Review*, 26(1): 23–30.

Wolf, W. 2002. What is embedded computing? *IEEE Computer*, 35(1): 136–137.

Zanella, A., Nicola, B., Castellani, A., Vangelista, L., and Zorzi, M. 2014. Internet of things for smart cities, *IEEE Internet of Things Journal*, 1(1): 22–32.

15 Modeling Guidelines of FreeRTOS in Event-B

Eman H. Alkhammash, Michael J. Butler, and Corina Cristea

CONTENTS

15.1 Introduction 195
15.2 An Overview of Real-Time Operating Systems (RTOSs) 196
15.2.1 FreeRTOS 197
15.3 Event-B 198
15.4 Task Management 199
15.4.1 Process 199
15.4.2 Process Table 199
15.4.3 Process Priority 199
15.4.4 Process States 200
15.4.5 Null Process 201
15.4.6 Timing Behavior 201
15.5 Scheduler States 202
15.6 Scheduling and Context Switching 203
15.6.1 Context Switch 203
15.7 Interrupts and Interrupt Service Routines 205
15.8 Queue Management 206
15.8.1 Waiting Messages 208
15.9 Memory Management 209
15.10 Related Work 211
15.11 Comparison of the Proposed Guidelines with Craig's Models of Operating System Kernels 212
15.12 FreeRTOS Models 213
15.12.1 Task Management 213
15.12.2 Queue Management 214
15.12.3 Memory Management 215
15.13 Conclusions and Future Work 215
References 216

15.1 INTRODUCTION

Formal methods have emerged as an approach to ensuring quality and correctness of highly critical systems. Event-B is a formal method for modeling and reasoning about systems. This chapter presents a set of guidelines for modeling operating system (OS) kernels for embedded real-time systems in Event-B.

Engineering manuals and guidelines for modeling, refinement, and proofs are important in the context of formal methods. They can be used to systematize the modeling and refinement process and aiding the proofs, and therefore reduce the time and cost of the formal development of systems.

Both novices and professionals can gain benefit from adopting such guidelines and modeling patterns on their specifications. The guidelines tell how most effectively to model a system and provide good examples and lessons that aid the formal development of several systems. The guidelines also save time and make a formal method approach more acceptable in industry. Moreover, guidelines can shed light into the requirement of systems and draw attention to some important properties of systems that might go missing in the absence of the guidelines. The guidelines may also give insight on which requirement should be modeled first and how to organize the refinement steps, as this is usually considered a source of difficulty in the process of modeling and refinement.

This chapter provides a set of modeling guidelines of real-time operating systems (RTOSs), including scheduling and context switch, interrupts, queue management, and memory management. Although, the guidelines provided in this chapter are drawing upon our experience with modeling FreeRTOS, it is believed that the presented guidelines can be applied for constructing formal models of different RTOS kernels. This is because RTOS kernels share similar components and features [20].

The presented modeling guidelines are intended to assist users of real-time kernels with a set of modeling steps for the construction of formal models of real-time kernels. Each of these guidelines gives directions on how to model a certain aspect of RTOS kernels. The guidelines focus on the basic functionality of RTOS and represent the primary requirements of RTOS for an Event-B model. Design details that are RTOS specific are left out from the guidelines. Design details of a specific RTOS can be specified through refinement of the abstract model driven by the guidelines. The identified modeling guidelines can be understood as a modeling pattern for the following RTOS features: task management, scheduling and context switch, interrupts, queue management, and memory management.

Table 15.1 describes some of the basic features and concepts of four RTOSs: FreeRTOS [4], UCOS [21], eCos [11], and VxWorks [33].

15.2 AN OVERVIEW OF REAL-TIME OPERATING SYSTEMS (RTOSs)

RTOS is a class of operating systems that is used for applications that have time constraints (real-time applications). These kinds of operating systems are characterized by features such as fault tolerant design and fast task scheduling as they always have specific timing requirements, and they are usually small in size. The structure of an RTOS is shown in Figure 15.1. As can be seen, the RTOS forms the intermediate layer that masks the hardware details to the application level.

The components of most RTOSs are [20]

Scheduler—Special algorithms that are used to schedule objects. Some of the commonly used scheduling algorithms are preemptive priority-based scheduling and round-robin scheduling. In preemptive priority-based scheduling, each task must be assigned a priority. At every clock tick, the scheduler runs the highest priority task that is ready to run. In round-robin scheduling, the tasks with equal priority get an equal share of processing time.

Objects—Entities that are used by the developers to create applications for real-time embedded systems. Some object constructs are tasks, which are objects created by a developer to handle a distinct topic; queues which are used for task–task communication; and semaphores and mutexes, which are used for the synchronization between the tasks and the interrupts.

Services—Operations performed by the kernels, such as task management, intertask communication and synchronization, interrupt handling, and resource management. The task management includes operations such as task creation, task deletion, task suspension, and changing task priority. Intertask communication and synchronization are mechanisms that enable information to be transmitted from one task to another such as message queues, pipes, semaphores, and mutexes. Interrupt handling are software routines used to handle interrupts; an interrupt is a signal to the microprocessor indicating that an event needs immediate attention. Resource management are kernel functions used to manage system resources such as the CPU, memory, and time.

TABLE 15.1
Summary of the Basic Features of FreeRTOS, UCOS, eCos, and VxWorks

Component	FreeRTOS	UCOS	eCos	VxWorks
Process TCB	Yes	Yes	Yes	Yes
Null process	Yes	Yes	Yes	Yes
Process creation and termination	Yes	Yes	Yes	Yes
Process priority	Yes	Yes	Yes	Yes
Clock tick	Yes	Yes	Yes	
Process states	Ready, running, blocked, suspended	Dormant, ready, running, waiting, ISR	Running, sleeping, countsleep, suspended, creating, exited	Ready, suspended, pended (blocked), and delayed
Scheduler state	Start the scheduler, lock the scheduler, and unlock the scheduler	Lock the scheduler and unlock the scheduler	Start the scheduler, lock the scheduler, and unlock the scheduler	Lock the scheduler and unlock the scheduler
Context switch	Yes	Yes	Yes	Yes
Interrupts handling	Yes	Yes	Yes	Yes
Priority manipulation	Changes a process priority and returns the priority of a process	Changes a process priority	Changes a process priority	Changes a process priority and returns the priority of a process
Delay operation	Yes	Yes	Yes	Yes
Scheduling	Priority-based round-robin scheduling	Priority-based scheduling and round-robin scheduling	Priority-based round-robin and bitmap schedulers	Priority-based scheduling and round-robin scheduling
Queue creation and termination	Yes	Yes	Yes	Yes
Process synchronization	Queues, semaphores, mutexes	Semaphores, message mail-box, message queues, tasks, and interrupt service routines	Message box, semaphore, queue	Message queues, pipes, semaphores
Waiting list for tasks waiting to retrieve/post message to a queue	Yes	Yes	Yes	Yes
Memory management	Memory allocation and deallocation	Memory allocation and deallocation	Memory allocation	Memory allocation and deallocation

15.2.1 FreeRTOS

FreeRTOS [4,5] is an open source, mini real-time kernel. It was primarily developed by Richard Barry and written in C with few assembler codes, with the result that it can be modified and extended as required. FreeRTOS code is freely available under GPL license and supports different platforms such as ARM7 and ARM9, MicroB-laze, MSP430, Coldfire V1, V2, and V85078K0R. FreeRTOS has one scheduler that schedules the highest priority thread first and this can be configured for preemptive or cooperative operation.

Data transfer is established by means of queues, semaphores, and mutexes. FreeRTOS is characterized by its simplicity of design, its portability, and scalability.

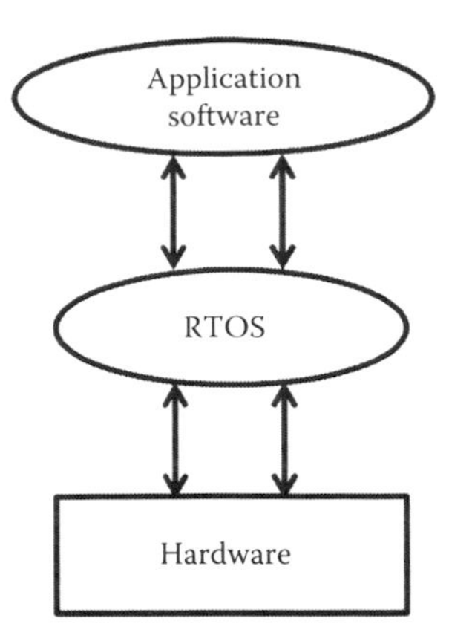

FIGURE 15.1 The RTOS structure.

There are two architectural layers in FreeRTOS design: the "hardware independent" layer and the "portable" layer. The hardware independent layer is responsible for performing the operating system functions and contains four C files, which are Task.c, queue.c, list.c, and co-routines.c. Tasks.c provides functionality for task management, queue.c provides functionality for the task communication of any synchronization mechanism, list.c contains the implementation of list data structure that is used by the scheduler, and co-routines.c contains the coroutine macros and function prototypes. The second layer, on the other hand, contains hardware-specific processing and includes some files such as heap.c, which provides memory allocation functionality. Tasks are implemented using circular doubly linked lists, queues are implemented using circular buffers, and memory management schemes are implemented using arrays and singly linked lists.

15.3 EVENT-B

Event-B [1] is a formal method that uses set theory and first-order logic to provide a formal notation for the creation of models of discrete systems and the undertaking of several refinement steps. In the refinement steps, the models represent different abstraction levels of system design. The consistency between the refined models in Event-B is verified through a set of mathematical proof obligations expressing that it is a correct refinement of its abstraction. The complexity in the design process is managed by abstraction and refinement. Refinement allows for postponing the introduction of some system features to later refinement steps. Rodin [2,3] is a tool used in the development of Event-B models. Rodin improves the quality of models through many features such as the identification of errors, the identification of required invariants, and proofs of consistency.

Event-B notation contains two constructs: a context and a machine. The context is the static part in which we can define the data of the model: sets, constants, axioms that are used to specify assumptions about sets and constants; and theorems that are used to describe properties derivable from the axioms. The dynamic and functional behavior of a model is represented in the machine part, which includes variables to describe the states of the system, invariants to constrain variables, theorems to describe properties that follow from the invariants, and events to trigger the behavior of the machine. The outline of an event used in this chapter is shown in Figure 15.2. The event takes parameters *t*, which satisfies the guard and then executes the body. The guard is predicate on the machine variables and event parameters, and an action updates machine variables automatically.

any *t* where *guard* then *action* end

FIGURE 15.2 The outline of an event.

15.4 TASK MANAGEMENT

15.4.1 Process

A process is an independent thread of execution [4,20]. An RTOS can execute multiple processes concurrently [4,20]. The processes appear to execute concurrently, however, kernels interleave the execution sequently based on the specific scheduling algorithm [4,20]. The term *process* is also called thread or task in some RTOSs.

To define a process, we define a new type in the context *PROCESS* "carrier set" where each process within the kernel is an element of this set.

15.4.2 Process Table

Process table is a data structure consisting of a collection of elements to store the information of a process [19,20]. Each process has its own control block that contains the process information, such as process id and process priority. A process table could consist of more than 10 elements. To deal with the process table, we introduce the following concepts in the Event-B model:

Process set—A set of the possible processes available in the system.
Process attributes functions—(Constant) functions that map processes to the process elements.
Process creation event—An event used to create a process.

Process attributes can be modeled by introducing a constant function for each attribute. For instance, let us use P to identify the set of the created processes, that is ($P \subseteq PROCESS$), $P1$, $P2$, … , PN are sets that define process elements such as name and priority. The constant functions pe_1, pe_2, … , pe_n can be defined as follows:

$$pe_1 \in P \rightarrow P1$$
$$pe_2 \in P \rightarrow P2 \ldots$$
$$pe_n \in P \rightarrow PN$$

The functions that correspond to the process attributes do not need to be introduced in one machine; some of the elements can be postponed until a later refinement level depending on what features the specifier wants to model first and what features the specifier wants to postpone.

When a new process is created, the kernel instantiates the process block of the created process. To model this, we introduce the process creation event as follows:

$$\begin{aligned} PC = \quad & \textbf{any}\ p \\ & \textbf{where}\ \ p \in P \\ & \textbf{then}\ \ pe1(p) := v1 \\ & pe2(p) := v2 \ldots \textbf{end} \end{aligned}$$

The process creation event needs to be extended further in later refinement steps when a new element of a process is introduced.

The definition of process table is similar to the approach to records in Event-B [12]. In fact, the carrier set P can be thought of as a record type. The attributes pe_1, pe_2, … , pe_n are defined using a projection function (function from P to some type $P1$, $P2$, … , PN). **PC** events are used to create new processes. It is possible to extend the record type P by adding more attributes in another refinement step.

15.4.3 Process Priority

In real-time kernels, each process has a priority as defined in the process block. Most RTOSs use a combination of base priority and active priority. The base priority is the original priority specified when the task is constructed. The process priority is the priority that can possibly modified. One possible usage of base and active priority is in priority inheritance protocol. Priority inheritance protocol takes place when a lower priority process blocks some higher priority processes; this problem is called priority inversion. The priority inheritance protocol resolves this problem by raising the priority of the process which caused the problem to that of the highest priority process blocked by it, to release the blocked processes. When the process that caused the problem releases the blocked processes, its priority then returns back to its base priority. For this, it is important to have a base priority that holds the original priority of the process and an active priority that holds the priority that can be changed. To deal with this, we need to define two elements in the process block: one for active priority and one for base priority.

$$ActivePri \in P \rightarrow PRIORITY$$
$$BasePri \in P \rightarrow PRIORITY$$

We need also to introduce the following Event-B events that capture the most common operations related to process priority, which are priority set *PriSet* and priority get *PriGet*:

PriSet = **any** *p np*
where $p \in P$
$np \in PRIORITY$
$ActivePri(p) \; / = np$
then $ActivePri(p) := np$ **end**

PriGet = **any** *p np*!
where $p \in P$
$np! = ActivePri(p)$ **end**

We use the **!** convention to represent result parameters as shown in the *np!* parameter.

15.4.4 Process States

At any time, the process can be in one state [4,19]. There are different states of a process, for instance, in a FreeRTOS a process can be in one of the following states: ready, suspended, blocked, and running.

A state diagram is always useful to show the transition of process states. Figure 15.3 shows possible transitions among the states *RUNNING, READY, SUSPENDED*, and *BLOCKED* during a process life.

To deal with process states, we identify the following concepts in Event-B modeling:

Process states variables—These variables are defined for each possible state. One way of defining process variables states in an Event-B model is to have a set for each state, each set is disjoint from other sets indicating that a process must be in one state at any given

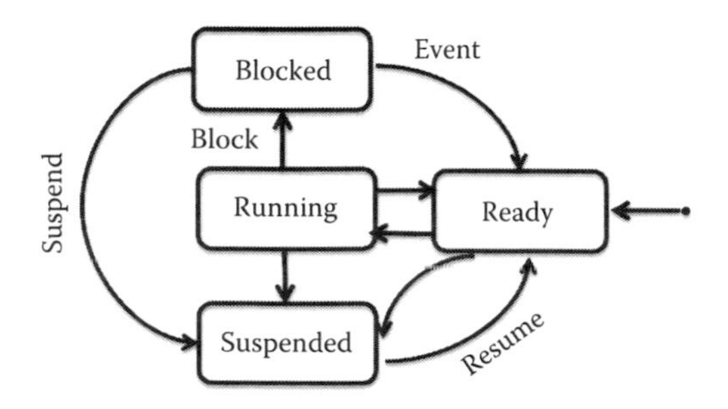

FIGURE 15.3 Process states.

time. Each created process must belong to one of these sets. We use the partition operation to indicate that the collection of the sets is disjoint:

$$partition(P, READY, RUNNING, SUSPENDED, BLOCKED)$$

To check the state of a particular process, we check to which set the process belongs to.

The set *RUNNING* corresponds to the runnable process and demands special treatment. It should be defined as a singleton set if we are dealing with one core processor. Thus, we need to add the following invariant:

$$card(RUNNING) \leq 1$$

Process states event(s)—These events are used to update the state of the process to a new one. Process states events model the transitions between states. According to our diagram, we have eight process states events: *RunToReady, RunToBlock, RunToSuspend, ReadyToRun, ReadyToSuspend, BlockedToReady, BlockedToSuspend*, and *SuspendToReady*

The definition of *RunToReady* would be of the form:

RunToReady= **any** p
where $p \in RUNNING$
then $RUNNING := \Phi$
$READY := READY \cup \{p\}$ **end**

In the *RunToReady* event, the process p is moved from the *RUNNING* set to the *READY* set. In later refinement levels, process state variables "sets" are refined to the appropriate data structure. For instance, ready state is usually implemented as a linked list for each priority. This level of detail can be left while introducing data refinement levels. In data refinement levels, sets are data refined to the appropriate data structure, such as arrays or linked lists, depending on the data structure implementation adopted for the specified RTOS. For instance, the data structure for process states in FreeRTOSs are circular doubly linked list [4].

15.4.5 Null Process

A null or idle process is a special process that is run when there is no process available to run [4,19]. The null process is permanently ready to run and is assigned to the lowest priority.

It is an important process, as the kernel needs always to execute a process or interrupt handler, and there might be a situation where there is no ready process available to run; in this case, the kernel switches control to the null process.

The definition of the null process is $Null \in READY$.

15.4.6 Timing Behavior

The scheduler interrupts at regular frequency to schedule tasks. The RTOS measures time using a tick count variable. A timer increments the tick count with a strict time. Each time the tick is incremented the scheduler checks to see if a task needs to be woken. If this is the case, the scheduler executes the newly woken task.

One important operation used by many RTOS is the *Delay* operation. It is used to suspend a process until a fixed point in the future. Delay operation takes a process and a time as parameters and returns the process delayed until the specified delay time.

To deal with timing behavior, we identify the following Event-B modeling concepts:

Tick variable—A counter that measures the system time ($Tick \in \mathbb{N}$).

Tick length constant—A constant that defines the length of the tick depending on the hardware timer's design.

Sleep time function—Partial function that maps processes to sleep time. The partial function indicates that only some processes in the system have a delay time and not all processes ($Sleep \in P \rightarrow\!\!\!\!\!\!\!+ \; 1\mathbb{N}$).

TimeToWake function—Partial function describes the wake-up time of a delayed task ($TimeToWake \in BLOCKED \rightarrow \mathbb{N}$), where BLOCKED identifies the set of delayed processes. Since all processes that have a wake-up time also have a sleep time, we add the following invariant:

$$dom(TimeToWake) \subseteq dom(Sleep)$$

IncrementTick—An event used to increment a tick counter with a strict accuracy.

IncrementTick=	**when** $Tick \geq 0 \wedge ((TimeToWake /= \varnothing \wedge min(ran(TimeToWake)) > Tick) \vee TimeToWake = \varnothing)$
	then Tick = *Tick* + TickLength **end**

The guard $TimeToWake / = \varnothing \wedge min(ran(TimeToWake)) > Tick)$ ensures that all delayed processes have unexpired delay time. This is because, each time a tick count is incremented it must check if a delayed task needs to be woken.

Delay—An event used to delay a process for a fixed time.

Delay=	**any** p
	where $p \in RUNNING$
	$Sleep(p) > 0$
	then $BLOCKED = BLOCKED \cup \{p\}$
	$TimeToWake(p){:=}Sleep(p){+}Tick$ **end**

15.5 SCHEDULER STATES

The scheduler can exist in one of a few states [4,19]. For instance, the possible scheduler states in FreeRTOSs are not started, running, and suspended.

To deal with scheduler states, we have:

Scheduler states constants—Constants represent scheduler states.

Scheduler status variable—A variable whose value determines the state of the scheduler.

Scheduler events—These events are used to update the value of the scheduler variable to a new one.

Assume that we have a scheduler with three states s_1, s_2, and s_3. We use ss for the scheduler variable; s_1, s_2, and s_3 are constants corresponding to the scheduler states that are defined in the context level; and SE is the scheduler event. To define scheduler states, the partition operation can be used as follows:

$$\textbf{Partition}(ss,\{s_1\},\{s_2\},\{s_3\})$$

A scheduler event would be of the form

SE =	**when** $ss = s_1$
	then $ss = s_2$ **end**

The conditions in which the transitions occurred are considered as guards for the scheduler states events. There might be more than one event that changes the value of the scheduler variable and they would all take the aforementioned form.

Many operations in an RTOS occur when the scheduler is running or suspended, for instance, the *IncrementTick* event given in Section 15.4.6 must occur in a FreeRTOS when the scheduler runs; therefore the *IncrementTick* event is extended by adding the guard $ss = s_1$, where s_1 denotes the scheduler running state.

15.6 SCHEDULING AND CONTEXT SWITCHING

15.6.1 Context Switch

A context switch occurs when the scheduler switches from one process to another. It is the process of replacing the process being executed by another process that is ready to run [19,20]. Context switch takes place when the kernel decides to switch control to other process, so, it exchanges the registers' contents by saving the information of the current executing process to resume its execution later and loading the information of the new process to the processor registers.

There are many situations in which a context switch is performed. The most common context switch performed is tick interrupt. At every tick, the scheduler checks if a new process should run. If such a process is found, the scheduler will save the current process information to resume it later and execute the other one [4].

The kernel schedule processes are based on a specific algorithm. There are different scheduling algorithms; a preemptive priority scheduling mechanism is one of the most common scheduling algorithms. In this algorithm, each process has a priority, and the higher readied priority process runs first. The preemptive priority scheduling algorithm can augment round robin algorithm by giving processes of the same priority an equal share of processor time.

To deal with context switch, we introduce the following modeling concepts:

Context function—A total function that maps each task to its context (attribute in the process table).

$$(ProcessContext \in P \rightarrow CONTEXT)$$

Current context—The physical context that captures the current process that holds the processor.

$$PhysicalContext \subseteq CONTEXT$$
$$card(physicalContext) \leq 1$$

Context switch event(s)—Event(s) replace the current process with another one.

These guidelines deal with the preemptive priority scheduling algorithm. At every tick, a context switch is performed when there is a process with a priority higher than or with equal priority of the current one being executing (equal in this context allows time slicing between processes with same priorities). The guards or the conditions that constrain the context switch *CS* event rely on checking the priority of the readied processes, thus, the context switch event can be of the form

Context Switch= **any** p **cp ph**
where $p \in READY$
$cp \in RUNNING$
$activePri(p) \geq activePri(cp)$
$ph \in physicalContext$
then $RUNNING = \{p\}$
$physicalContext := \{ProcessContext(p)\}$
$ProcessContext(cp) := ph$
end

The actual context switch is done through saving the state of the current context in the process context and sets the current context for the new running process.

Context switch flag—A boolean flag that set to true to indicates the need to perform a context switch. We use *csFlg* for the context switch flag.

There are different cases where a context switch may be performed. For instance, a context switch might be performed when a certain process transmits to the ready state or when a new process is created, because it is possible for these situations to introduce a process with a priority higher than the priority of the current executing one in which case the context switch should perform. Therefore, for every event that may require a context switch, we need to add an additional action that sets the context switch flag to true. And we add an additional guard to ensure that the context switch flag is true in the context switch *CS* event, and finally the context switch flag is assigned back to false in the *CS* event. Thus, we extend the *CS* event to be

$$
\begin{array}{ll}
\textit{Context Switch} = & \textbf{any} \quad p\ cp\ ph \\
& \textbf{where} \quad p \in READY \\
& cp \in RUNNING \\
& activePri(p) \geq activePri(cp) \\
& ph \in physicalContext \\
& csFlg = TRUE \\
& \textbf{then} \quad RUNNING = p \\
& physicalContext := \{ProcessContext(p)\} \\
& ProcessContext(cp) := ph \\
& csFlg = FALSE \quad \textbf{end}
\end{array}
$$

In order to prevent any event to be enabled between calling the context switch event, we add an additional guard *csFlg=FALSE* to any events that might get enabled and affect the execution of the context switch.

Another important issue raised here is what if an event calls a context switch and sets the *csFlg* to true but not all the guards of the context switch *CS* event are satisfied. This situation occurs when the event that called a context switch does not introduce a readied process with higher or equal priority than the executing one. This means that the context switch event is not enabled and consequently the value of *csFlg* is always true, which leads to a deadlock. To prevent this situation, we suggest having two versions of context switch events: the first one performs the context switch if all the guards of the *CS* event are satisfied; the second one, however, triggers only to set *csFlg* back to false when the guards of the context switch are not satisfied to prevent a possible deadlock from occurring. The revised CS events are

$$
\begin{array}{l}
\textit{Context Switch1} = \textbf{any}\ p\ cp \\
\quad \textbf{where} \quad p \in READY \\
\quad cp \in RUNNING \\
\quad activePri(p) \geq activePri(cp) \\
\quad ph \in physicalContext \\
\quad csFlg = TRUE \\
\quad \textbf{then}\ RUNNING = p \\
\quad physicalContext := \{ProcessContext(p)\} \\
\quad ProcessContext(cp) := ph \\
\quad csf := FALSE\ \textbf{end}
\end{array}
$$

$$
\begin{array}{l}
\textit{Context Switch2} = \textbf{any}\ cp \\
\quad \textbf{where} \quad cp \in RUNNING \\
\quad max(activePri[READY]) < activePri(cp) \\
\quad csf = TRUE \\
\quad \textbf{then}\ csf = FALSE\ \textbf{end}
\end{array}
$$

The preceding context switch specification is abstract and only represents the general operations performed in any context switching. Since context switching is a hardware-dependent operation, in order to model the context switch in details, the registers of the target processor needs to be modeled along with the other specific context switch details for the targeted architecture.

15.7 INTERRUPTS AND INTERRUPT SERVICE ROUTINES

Interrupts are hardware mechanisms used to inform the kernel that an event has occurred [4,19]. A process can be interrupted by external interrupts raised by peripherals or software interrupts raised by executing a particular instruction. Interrupts are handled by the interrupt service routines (ISRs), which are stored in interrupt vector table. When an interrupt occurs, the kernel saves the current context of the process being interrupted and jumps to the ISR to handle that interrupt. After processing the ISR, the kernel returns to the process level and resumes the process that was interrupted [4,19].

Interrupts may ready a blocked process for an external device when an event has occurred, so, the kernel might execute different process before completing the preempted one.

To deal with interrupts, we introduce the following modeling concepts:

Interrupt data type—A new data type for interrupts. We will use the *INTERRUPT* "carrier set" as interrupts data type.

Interrupt variable—A set of the raised system interrupts $Interrupts \subseteq INTERRUPT$.

Interrupt handler function—A function that maps each interrupt to its ISR. We will use *interrupt handler* (a total function that maps an interrupt to its corresponding interrupt service routine).

$$ISR \in Interrupts \rightarrow INTERRUPT\ HANDLER$$

Current interrupt—A variable that stores the current executing interrupt.

$$currentISR \subseteq ran(ISR)$$
$$card(currentISR) \leq 1$$

$card(currentISR) \leq 1$ is added in case of single kernel.

Handle interrupt event—An event used to handle interrupts.

Handle Interrupt= **any** *int*
  where $int \in Interrupts$
  then $currentISR = \{ISR(int)\}$
end

Complete interrupt event—An event used to discard completed interrupts, where *INTERRUPT PRIORITY* is a constant set defined in the context level.

The events *Handle Interrupt* and *Complete Interrupt* can be extended as follows:

Handle Interrupt = **any** *int c cint*
where $int \in Interrupts$
$(c \in currentISR \wedge cint = ISR^{-1}(c)$
$\wedge InterruptPriority(int) \geq InterruptPriority(cint))$
$\vee currentISR = \varnothing$
then $currentISR = \{ISR(int)\}$ **end**

Complete Interrupt = **any** *int*
where $int \in Interrupts$
$currentISR = \{ISR(int)\}$
then $currentISR = \Phi$
$Interrupts = Interrupts \setminus \{int\}$
$ISR := \{int\} \leq I\ ISR$
$InterruptPriority := \{int\} \leq I$
InterruptPriority **end**

The processor always gives priority to execute interrupts over tasks, the ISR must complete its execution without being interrupted by tasks. When the ISR is completed, the kernel dispatches the correct task. To deal with this, we identify the following modeling concepts:

Interrupt context—A function that maps each interrupt to its context.

$$InterruptContext \in Interrupts \rightarrow CONTEXT$$

In order to separate a process context from an interrupt context; we add the following invariant:

$$ran(InterruptContext) \cap ran(ProcessContext) := \Phi$$

The *Handle Interrupt* and *Complete Interrupt* are extended as follows:

Handle Interrupt = **any** *int c cint ph*
where $int \in interrupts$
$c \in currentISR \wedge cint = ISR^{-1}(c)$
$\wedge InterruptPriority(int) \geq InterruptPriority(cint))$
$\vee currentISR = \varnothing$
$ph \in physicalContext$
then $currentISR = \{ISR(int)\}$
$physicalContext := \{InterruptContext(int)\}$
$InterruptContext(c) := ph$ **end**

Complete Interrupt = **any** *int*
where $int \in Interrupts$
$currentISR = \{ISR(int)\}$
then
$currentISR = \Phi$
$Interrupts = Interrupts \setminus \{int\}$
$ISR := \{int\} \leftarrow \mathrm{I}\ ISR$
$InterruptPriority :=$
$\{int\} \leftarrow \mathrm{I}\ InterruptPriority$
$InterruptContext :=$
$\{int\} \leftarrow \mathrm{I}\ InterruptContext$
$physicalContext := \Phi$ **end**

An extra guard $currentISR = \Phi$ is needed in *CS* events to prevent any context switch while there is an ISR running.

The timer interrupt given in Section 15.4.6 is an example of an interrupt. In every tick, the tick-ISR represented in the *IncrementTick* event wakes up the blocked process that have expired delay time. If the woken process has a priority higher than the current process, the ISR then will return control to the higher priority process.

15.8 QUEUE MANAGEMENT

Processes often need to communicate. There are several means of communication that most RTOSs offer such as queues and semaphores. Queues are the primary object of process–process communications. Queues can be used to send messages between processes. Semaphores are a special kind of queue that are usually used for mutual exclusion and synchronization (communication between processes and ISRs). Semaphores can be binary or counting. Binary semaphores are queues of length one, whereas counting semaphores are queues of length greater than one.

To define a queue, we define a new type in the context level—*QUEUE* "carrier set"—where each created queue is an element of this set.

$$AllQueue \subseteq QUEUE$$

AllQueue identifies all possible created queues in the system.

We also add the following axiom to indicate that the queue set is finite

$$finite(QUEUE)$$

A queue has a length that defines the maximum number of messages the queue can hold.

$$QLength \in AllQueue \rightarrow \mathbb{N}$$

In order to identify the message (data) that is stored in the queue, we define the following variable:

$$QMessage \in DATA \twoheadrightarrow AllQueue$$

We use a partial function because the queue can be empty.

In order to identify the message (data) that is sent by an object *Process* or *ISR*, we define the following variables:

$$PMessage \in DATA \twoheadrightarrow P$$
$$ISRMessage \in DATA \twoheadrightarrow ISR$$

Again, we use a partial function since some processes or ISR may have no data to send.

The following invariant is added to show that a process message is different from an ISR message:

$$dom(PMessage) \cap dom(ISRMessage) = \Phi$$

Queue messages are usually of fixed size. To define this we add the following constraint invariant:

$$MessageSize \in \text{AllQueue} \rightarrow \mathbb{N}$$

Since there are different types of queues such as general queues, binary semaphores, and counting semaphores, we can define them as follows:

$$partition(AllQueue, queue, binarySemaphore, countingSemaphore)$$

To define the length of each queue kind, we define the following:

$$\forall q \cdot q \in queue \Rightarrow QLength(q) > 1$$
$$\forall q \cdot q \in binarySemaphore \Rightarrow QLength(q) = 1$$
$$\forall q \cdot q \in countingSemaphore \Rightarrow QLength(q) > 1$$

There are three important operations supported by the queues. A queue creates an operation that is used to create a queue. An enqueue or process–send operation is used to send data by a process to another and a dequeue or process–receive operation is used by the other process to receive the data.

To deal with the queue creation operation, the queue send operation, and the queue receive operation, we introduce the following Event-B concepts:

Queue Create=	**any** q
	where $q \in QUEUE \setminus AllQueue$
	then $AllQueue = \text{AllQueue} \cup \{q\}$
	$Qtype := Qtype \cup \{q\}$...**end**

Qtype can be replaced by one of the variables: *queue*, *binarySemaphore*, or *countingSemaphore*. For each queue type, there should be one queue creation event.

The queue send event can be of the following form:

$$
\begin{array}{ll}
\textit{Queue Send}= & \textbf{any}\ \ p\ \ q\ \ m \\
& \textbf{where}\ \ p \in \mathrm{Ob\,j}\ \ \wedge q \in Qtype \\
& \quad card(QMessage^{-1}\{q\}) < QLength(q) \\
& \quad m \in Ob\,jMessage \setminus dom(QMessage) \\
& \textbf{then}\ \ QMessage = QMessage \cup \{m \rightarrow q\} \\
& \quad Ob\,jMessage := \{m\} \leq \mathrm{I}\ Ob\,jMessage\ \textbf{end}
\end{array}
$$

The guard $card(QMessage^{-1}\{q\}) < QLength(q)$ ensures that there is a room in the queue, so the running process can send a message to the queue q. *Ob j* refers to the object that has sent a message to a queue, so it needs to be replaced by *P* or *ISR*.

Ob jMessage denotes the message of the object, so it needs to be replaced by *PMessage* or *ISRMessage*. *Ob j*, *Qtype*, and *Ob jMessage* need to be replaced by appropriately defined variables to show the object-sender and the type of queue the object wants to send the message to.

The queue receive event can be of the following form:

$$
\begin{array}{ll}
\textit{Queue Receive}= & \textbf{any}\ \ p\ \ q\ \ m \\
& \textbf{where}\ \ p \in Ob\,j\ \ \wedge q \in Qtype \wedge m \in QMessage^{-1}[\{q\}] \\
& \textbf{then}\ \ QMessage := \{m\} \leftarrow \mathrm{I}\ QMessage \\
& \quad Ob\,jMessage := Ob\,jMessage \cup \{m \rightarrow \mathrm{o}\} \\
& \textbf{end}
\end{array}
$$

The guard $m \in QMessage^{-1}[\{q\}]$ ensures that the queue is not empty, so that the process can receive a message from a queue.

Similarly, *Ob j*, *Qtype*, and *Ob jMessage* need to be replaced by appropriately defined variables to show the object-receiver and the type of the queue the object wants to receive the message from.

An important point to note is that enqueue and dequeue operations are usually required to be done in first in, first out (FIFO) order. Therefore, one way to deal with this is to define an abstract sequence instead of a set that allows the queue to store messages or refine a set into a sequence to deal with this aspect. It is also possible that a queue message set is refined by a linked list structure to maintain the order between messages.

15.8.1 Waiting Messages

If the queue is full, the sending process will not be successful and has to be blocked and wait before sending its message until the queue has a room to receive the message. Similarly, if the queue is empty, the receiving task will not be successful and has to wait and block until a message arrives.

To deal with this, we introduce the following Event-B concepts:

Waiting to send queue—A function that stores processes that are blocked to send items to a queue:

$$WaitingToSend \in P \twoheadrightarrow AllQueue$$

Waiting to receive queue—A function that stores processes that are blocked to receive items from a queue:

$$WaitingToReceive \in P \twoheadrightarrow AllQueue$$

Waiting messages events—Two events are defined to add blocked processes to the appropriate waiting queues (i.e., waiting to send queue or waiting to receive queue) and two events are defined to remove a process from waiting queues (i.e., waiting to send queue or waiting to receive queue).

The formalization of the operation of blocking a process waiting message events can be expressed as follows:

$$
\begin{array}{ll}
AddToWToS = & \textbf{any}\ q\ c \\
 & \textbf{where}\ q \in Qtype\ \wedge c \in RUNNING \\
 & \quad QLength(q) = card(QMessage^{-1}[\{q\}]) \\
 & \textbf{then}\ WaitingToSend := WaitingToSend \cup \{c \rightarrow q\}\ \ \textbf{end}
\end{array}
$$

The guard $QLength(q) = card(QMessage^{-1}[\{q\}])$ ensures that the queue q is full, so that the process c must be blocked.

$$
\begin{array}{ll}
AddToWToR = & \textbf{any}\ q\ c \\
 & \textbf{where}\ q \in Qtype\ \wedge c \in RUNNING \\
 & \quad card(QMessage^{-1}[\{q\}]) = 0 \\
 & \textbf{then}\ WaitingToReceive := WaitingToReceive \cup \{c \rightarrow q\}\ \ \textbf{end}
\end{array}
$$

The guard $card(QMessage^{-1}[\{q\}]) = 0$ ensures that the queue q is empty and so the process c must be blocked.

$$
\begin{array}{ll}
RemoveFromWToS = & \textbf{any}\ q\ t \\
\textbf{where}\ q \in Qtype & \\
\quad t \in WaitingToSend^{-1}[\{q\}] & \\
\quad card(QMessage^{-1}[\{q\}]) < QLength(q) & \\
\textbf{then}\ WaitingToSend := \{t\} \leftarrow & \\
\quad WaitingToSend\ \ \textbf{end} &
\end{array}
$$

$$
\begin{array}{ll}
RemoveFromWToR = & \textbf{any}\ q\ t \\
\textbf{where}\ q \in Qtype & \\
\quad t \in WaitingToReceive^{-1}[\{q\}] & \\
\quad \wedge card(QMessage^{-1}[\{q\}]) > 0 & \\
\textbf{then}\ WaitingToReceive := \{t\} \leftarrow & \\
\quad WaitingToReceive\ \ \textbf{end} &
\end{array}
$$

The locking queue mechanism and critical sections vary so much from one RTOS to another that it is hard to give universal guidance about how a locking mechanism is used by several RTOSs in any given situation. Therefore, this part has been left out from the presented guidelines.

15.9 MEMORY MANAGEMENT

The RTOS provides memory management techniques to assign memory to objects. The memory usually is divided into fixed-size memory blocks that can be requested by objects. Each object in the system (e.g., process, queue, semaphore) is assigned a private memory space.

To deal with memory management techniques, we introduce the following Event-B concepts:

Block set—A set of all memory blocks:

$$Blocks \subseteq BLOCK$$

Block-size function—A function that defines the size of each block:

$$BlockSize \in Blocks \rightarrow \mathbb{N}$$

BlockAddr function—A function that defines the start address of each block:

$$BlockAddr \in Blocks \rightarrow ADDR$$

where $ADDR$ is a constant set defined as $ADDR = StartAddress..EndAddress$. $StartAddress$ and $EndAddress$ are constants representing the start and the end addresses of the heap structure and are defined in terms of $\mathbb{N}$.

Object—A set of objects in the system:

$$object \in Blocks \rightarrow OBJECT$$

The memory blocks can be further divided into two sets—allocated blocks and free blocks—as follows:

$$partition(Blocks, FreeBlocks, AllocBlocks)$$

To ensure that blocks are consecutive and guarantee that a block cannot be allocated for two distinct objects, we introduce the following invariant:

$$\forall b1, b2 \cdot b1 \in Blocks \wedge b2 \in Blocks \wedge b1 /= b2 \Rightarrow BlockAddr(b1) .. (BlockAddr(b1) + BlockSize(b1) - 1) \cap BlockAddr(b2) .. (BlockAddr(b2) + BlockSize(b2) - 1) = \varnothing$$

In order to model the process of memory allocation and deallocation, we introduce *Alloc1,2* events to allocate memory and assign it to an object and *Free* event to free memory blocks.

```
Alloc1=    any   s b c o
           where   s ∈ ℕ ∧ b ∈ FreeBlocks ∧ s > 0
                   BlockSize(b)=s   ∧ c ∈ BLOCK \ Blocks
                   o ∈ OBJECT \ ran(ob ject)
           then    FreeBlocks := FreeBlocks \ {b}
                   AllocBlocks := AllocBlocks ∪ {b}
                   object := object ∪ {b → o}
                   end
```

Alloc1 is used to allocate enough memory to be assigned to the object. The size of the memory is equal to the size requested by the object. However, if all blocks are not of adequate size to the requested one, the following event can be used to divide the free memory block into two blocks. The first block is of equal size as requested and is used to allocate to the object. The second block is added to the set of free blocks. This allocation process is known as the best-fit algorithm.

```
Alloc2=   any   s b c o k
          where   s ∈ ℕ   ∧ b ∈ FreeBlocks ∧ s > 0
                  BlockSize(b) > s
                  (∀k.k ∈ BlockSize[FreeBlocks]
                  ∧k < BlockSize(b) = ⇒ k < s)
                  c ∈ BLOCK \ Blocks
                  o ∈ OBJECT \ ran(object)
          then    BlockAddr := BlockAddr ← {b → BlockAddr(b) + s, c →
                  BlockAddr(b)} BlockSize ← BlockSize ← {b → BlockSize(b) − s, c → s}
                  AllocBlocks := AllocBlocks ∪ {c}
                  object := object ∪ {c → o}   end
```

The guard $\forall k.k \in BlockSize[\{FreeBlocks\}] \wedge k < BlockSize(b) \Longrightarrow k < s$ guarantees that the free-block b is the best-fit free block.

To free an allocated memory block, the following event is introduced:

Free= **any** b
where $b \in AllocBlocks$
then $FreeBlocks := FreeBlocks \cup \{b\}$
$AllocBlocks := AllocBlocks \setminus \{b\}$
end

The problem of fragmentation occurs when parts of allocated blocks are unused. The fact that the system-allocated memory is more than is requested can be dealt with in an abstract form. However, we have left out this topic from the guidelines as FreeRTOS does not deal with fragmentation problems.

15.10 RELATED WORK

This section examines some of the related work regarding the use of formal methods in operating systems.

Craig's work is one of the fundamental sources in this field [6,7]. He focuses on the use of formal methods in OS development, and the work is introduced in two books. The books contain formal specifications of simple and separation kernels along with the proofs written by hand. The first book is dedicated to specify the common structures in operating system kernels in Z [29] and Object Z [28], with some calculus of communicating systems (CCS) [22] process algebra used to describe the hardware operations. It starts with a simple kernel with few features and progresses to more complex examples with more features. For example, the first specification introduced in the book is called a simple kernel, and involves features such as task creation and destruction, message queues, and semaphore tables. However, it does not contain a clock process or memory management modules, whereas other specifications of a swapping kernel contain more advanced features including a storage management mechanism, clock, and interrupt service routines.

The second book is devoted to the refinement of two kernels: a small kernel and a micro kernel for cryptographic applications. The books contain proofs written by hand and some missing properties resulting due to manual proofs, which have been highlighted by Freitas [14].

Freitas [14,31] has used Craig's work to explore the mechanization of the formal specification of several kernels constructed by Craig using Z/Eves theorem prover. This covers the mechanization of the basic kernel components such as the process table, queue, and round-robin scheduler in Z. The work contains an improvement of Craig's scheduler specification, adapting some parts of Craig's models and enhancing them by adding new properties. New general lemmas and preconditions are also added to aid the mechanization of the kernel scheduler and priority queue. Mistakes have been corrected in constraints and data types for the sake of making the proofs much easier, for instance, the enqueue operation in Craig's model preserves priority ordering, but it does not preserve FIFO ordering within elements with equal priority; this has been corrected by Freitas [14].

Furthermore, Déharbe et al. [8] specify task management, queues, and semaphores in classical B. The work specifies mutexes and adopts some fairness requirements to the scheduling specification. The formal model built was later published [9].

There is also an earlier effort by Neumann et al. [23] to formally specify a provably secure operating system (PSOS) using a language called SPECIfication and Assertion Language (SPECIAL) [13]. This language is based on the modeling approach of hierarchical development methodology (HDM). In this approach, the system is decomposed into a hierarchy of abstract machines; a machine is further decomposed into modules, each module is specified using SPECIAL. Abstract implementation of the operations of each module is performed and then is transformed to efficient executable programs. PSOS was designed at SRI international [32]. The work began in 1973 and the final design was presented in 1980 [23]. PSOS was focused on the kernel design and it was unclear how much of it has been implemented [10]. Yet, there are other works inspired by the RSOS design such as the kernelized secure operating system (KSOS) [25] and the logical coprocessing kernel (LOCK) [27]. The aforementioned examples follow a top-down formal method approach, where the specification is refined stepwise into the final product. On the other hand, there are also some earlier efforts in the area of formal specification and correctness proofs of kernels based on the bottom-up verification approach. The bottom-up approach adopts program verification methods to verify the implementation.

An example of this approach is a work by Walker et al. [32] on the formalization of the UCLA Unix security kernel. The work was developed at the University of California at Los Angeles (UCLA) for the DEC PDP-11/45 computer. The kernel was implemented in Pascal due to its suitability for low-level system implementation and the clear formal semantics [15,26]. Four levels of specification for the security proof of the kernel were conducted. The specifications ranged from Pascal code at the bottom to the top-level security properties. After that, the verification based on the first-order predicate calculus was applied that involves the proof of consistency of different levels of abstraction with each other. Yet, the verification was not completed for all components of the kernel.

Finally, there was an effort by Klein et al. [17,18] on the formal verification of the seL4 kernel starting with the abstract specification in higher-order logic, and finishing with its C implementation. The design approach is based on using the functional programming language Haskell [16] that provides an intermediate level that satisfies bottom-up and top-down approaches by providing a programming language for kernels developer and at the same time providing an artifact the can be automatically translated into the theorem prover. A formal model and C implementation are generated from the seL4 prototype designed in Haskell. The verification in Isabelle/HOL [24] shows that the implementation conforms with the abstract specification.

In this chapter, the modeling guidelines adopt the top-down formal method approach. The guidelines are modeled in Event-B, which is a refinement-based approach for modeling systems. The guidelines outlined in this chapter can be refined later to the appropriate data refinement structure such as linked list.

15.11 COMPARISON OF THE PROPOSED GUIDELINES WITH CRAIG'S MODELS OF OPERATING SYSTEM KERNELS

This section compares our guidelines to Craig's models of operating system kernels. Craig carried out his development in Z and Object Z with some CCS, whereas we use the Event-B formal method to develop our guidelines.

The developed guidelines mostly cover the common concepts of any RTOS, such as process tables, queues, semaphores, and memory. Craig's models, however, are richer and cover different concepts that can be found in some complex kernels, such as virtual storage.

Some of the definitions and data structures used in Craig's models are different from the definitions we adopted in the presented guidelines. Here we are going to compare the definitions of some of the modeling concepts of our guidelines and Craig's models for three aspects: process management, queue management, and memory management.

The definition of process management in Craig's models is similar to the one presented in the guidelines. Craig defines processes as a set of process names and the attributes of the process are mappings from process identifiers to the various attribute types. Process status is defined as a function from process names to process state set, and he also defines a number of operations to change a process from a state to another, including *SetProcessStatusToReady*, *SetProcessStatusToRunning*, and *SetProcessStatusToWaiting*. The context switch specified in Craig's book involves a number of variables for registers and stacks, which we do not specify in our guidelines. The registers and stacks defined in Craig's models are the general register set is *hwregset*, the stack is in *hwstack*, the instruction pointer (program counter) is *hwip*, and the status word is denoted by *hwstatwd*. *STATUSWD* is an enumeration, for example, overflow, division by zero, carry set. He also defines a number of attributes for a process, which map from process identifiers to registers and stacks. They are process stack *pstatcks*, process status words *pstatwds*, process general registers *pregs*, and process instruction pointer *pips*. The context switch then happens, saving all hardware registers used by the running process (e.g., $hwtack'(pid?) = hwstack, pip's(ids?) = hwip$) and restoring all hardware registers for the new process ($hwstack' = pstacks(p?), hwip = pips(p?)$).

Craig modeled the FIFO queue as an injective sequence, whereas we are abstracting the queues as sets. With our investigation on patterns, the abstract set of queues can easily be refined by a sequence pattern or ordered linked list pattern depending on the implementation of the queue.

We define two events to send and receive items from queues: *Queue Send* and *Queue Receive*. Waiting messages are defined in our guidelines as a function from process set to queue set. We define two waiting messages events to store processes that failed to send/receive items to/from queues: *Add ToWToS* and *Add ToWToR*. Craig defines *Enqueue* and *Dequeue* operations using concatenation and extraction operations. Concatenation is used to add an element to a sequence, whereas extraction is used to remove an element from a sequence. Craig also defines waiting messages operations to store processes that failed to send an item to a queue or receive an item from a queue. The *AddWaitingSenders* operation is defined to enqueue the failed process to the sequence of failed processes *waitingsenders*, whereas the *AddWaitingReceivers* operation is defined to enqueue the failed process to the sequence of failed processes *waitingReceivers*.

Memory management in Craig's model has a number of definitions. *ADDRESS* defines the address of memory. It ranges from 1 to maximum constant value. *MEMDESC* describes a region of storage whose address is given by the first component and whose size is given by the second. The free space is called *Holes* and is defined as sequence of *MEMDESC*. The used memory is captured by *usermem*, which is also defined as a sequence of *MEMDESC*. He defines a number of operations, such as *RSAllocateFromHole*, which performs storage allocation from free storage. *FreeMainstoreBlock* is used for freeing of an allocated block. *MergeAdjacentHoles* is used to merge adjacent holes to form larger ones.

Our guidelines define memory blocks as sets. We also define the address of a block as a function from blocks to address set. The size of blocks are defined as a function from blocks to a set of natural number, $\mathbb{N}$. The guidelines cover two types of events: allocation events to allocate memory to the created object and free events to free allocated memory. Unlike Craig's models, the fragmentation issue is not addressed in the presented guidelines.

15.12 FreeRTOS MODELS

The main components of a FreeRTOS are task, queue, and memory. The specification in Event-B starts with an abstract model followed by a number of refinement steps. Each refinement augments more detail than the previous model. The following sections outline an overview of the FreeRTOS specifications. The modeling activity initially was carried out by modeling each set of requirements independently.

15.12.1 Task Management

An abstract specification—Some basic functionality of task management and the kernel: We begin with an abstract model of task management and the kernel focusing on task creation, task deletion, interrupt handling, and context switching. The interrupt handling mechanism handles events that are usually signaled by the interrupts by executing its ISR. The context switch is the mechanism used for swapping the tasks.

First refinement—Scheduler states: The first refinement level specifies the scheduler states. The scheduler can exist in one of the following states: not started, running, and suspended.

Second refinement—Task states: The second refinement specifies the task states. A task can be in one of the following states: ready, blocked/delayed, or suspended.

Third refinement—Hardware clock and timing properties: This refinement level specifies the hardware clock and timing properties associated with a delay task such as the sleep time and the wake-up time.

Fourth refinement—Delay operations: This refinement level distinguishes two kinds of delay operations: "delay" and "delay until." The delay operation places the calling task into the blocked state for a fixed number of tick interrupts without consideration of the time at which the last task left the blocked state, whereas delay until is an alternative operation that delays a task until a specific time has passed since the last execution of that operation. Thus delay until allows a frequent execution of a task so it is suitable for the periodic tasks (arriving at fixed frequency).

Fifth refinement—Clock overflow: This refinement specifies clock overflow. It introduces a collection of overflow-delay tasks to store tasks whose wake-up time has overflowed.

Sixth refinement—Priority: This level introduces priority. FreeRTOS uses a highest priority first scheduler. This means that the higher priority task runs before the lower priority task. The scheduler then uses this priority to schedule the task with the highest priority.

Seventh refinement—Contexts: This level specifies task contexts; the task context represents the state of the CPU registers required when a task is restored. If the scheduler switches from one task to another, the kernel saves the running task context and uploads the context of the next task to run. The context of the previous running task is restored the next time the task runs. Therefore, the kernel resumes the task execution from the same point where it had left off.

15.12.2 Queue Management

An abstract specification—Queue types: The initial model formalizes three types of queues: queues, binary semaphores, and counting semaphores, which are used for communication and synchronization between the tasks, or between the tasks and interrupts.

First refinement—Waiting events: This level of refinement considers additional requirements pertaining to the addition of a task to event lists. The *TaskWaitingToSend* variable is introduced to store the tasks that failed to send items to a queue because the queue was full, and the *TaskWaitingToReceive* variable is introduced to store the tasks that failed to receive items because the queue was empty.

Second refinement—Lock mechanism: This level defines the lock mechanism. Lock mechanism is used to prevent an ISR from updating the event lists *TaskWaitingToSend* or *TaskWaitingToReceive* while a task is being copied to the event lists.

Third refinement—Mutexes: Our third refinement specifies the mutex; mutex is a type of binary semaphore that uses the priority inheritance mechanism to reduce priority inversion. Priority inversion is a problem that occurs when a high-priority task awaiting a mutex has been blocked by a low-priority task that holds that mutex. Priority inheritance works by raising the priority of the lower priority task that owns the mutex to that of the highest priority task that is attempting to obtain the same mutex.

Fourth refinement—Recursive mutex: This level introduces a recursive mutex. A recursive mutex is a type of mutex that enables a token to be "taken" repeatedly by its owner. It only becomes available again when the owner unblocks the mutex the same number of times it has locked it.

TABLE 15.2
Proof Statistics of the FreeRTOS Models

Machines	Total POs	Automatic	Interactive
Task Management (TM)			
M0	32	30	2
M1	10	10	0
M2	72	68	4
M3	24	24	0
M4	39	38	1
M5	28	25	3
M6	41	32	9
M7	34	29	5
Total (TM)	280	256	24
Queue Management (QM)			
M0	53	46	7
M1	28	24	4
M2	109	70	39
M3	70	42	28
M4	19	10	9
Total (QM)	279	192	87
Memory Management-Scheme1 (MM1)			
M0	32	24	8
Memory Management-Scheme2 (MM2)			
M0	32	22	10
Overall	623	494	129

15.12.3 Memory Management

An abstract specification—Memory blocks and addresses

This abstract model defines the memory structure including blocks and addresses. The blocks are classified into a number of allocated blocks and one free block.

The proof statistics of the FreeRTOS models are given in Table 15.2.

15.13 CONCLUSIONS AND FUTURE WORK

In this chapter, we developed modeling guidelines in Event-B for the following RTOS features: task management, scheduling and context switch, interrupts, queue management, and memory management. We chose Event-B to develop FreeRTOS because Event-B is a stepwise formal method that has a platform supported with various plugins. The stepwise methodology allows the complexity to be managed through several refinement steps. The construction of Event-B models for FreeRTOS that can contribute to the Verified Software Repository, and the derivation of a set of modeling guidelines for RTOS based on our experience with FreeRTOS models.

The devised guidelines need to be evaluated through several case studies. As a direction for future research, we are going to choose an RTOS kernel such as UCOS, then evaluate the guidelines by applying them to build UCOS Event-B models. The application of the proposed guidelines to different case studies helps in learning different lessons and provides solid experience about building RTOS kernels using the formal method.

REFERENCES

1. J. R. Abrial. *Modeling in Event-B: System and Software Engineering*. Cambridge University Press, Cambridge, UK, 2010.
2. J. R. Abrial, M. Butler, S. Hallerstede, T. S. Hoang, F. Mehta, and L. Voisin. Rodin: An open toolset for modelling and reasoning in Event-B. *STTT*, 12(6):447–466, 2010.
3. J. R. Abrial, M. Butler, S. Hallerstede, and L. Voisin. An open extensible tool environment for Event-B. In *Formal Methods and Software Engineering*, edited by Z. Liu and J. He, 588–605, Berlin: Springer, 2006.
4. R. Barry. The FreeRTOS project. http://www.freertos.org/, 2010.
5. R. Barry. *Using the FreeRTOS Real Time Kernel: A Practical Guide*. Bristol: Real Time Engineers, 2010.
6. I. Craig. *Formal Models of Operating System Kernels*. London: Springer, 2007.
7. I. Craig. *Formal Refinement for Operating System Kernels*. London: Springer, 2007.
8. D. Déharbe, S. Galvão, and M. M. Anamaria. Formalizing FreeRTOS: First steps. In *Formal Methods: Foundations and Applications*, edited by M. V. M. Oliveira and J. Woodcock, 101–117. Berlin: Springer-Verlag, 2009.
9. D. Déharbe, S. Galvão, and A. M. Moreira. freertosb. http://code.google.com/p/freertosb/source/browse/, 2009.
10. T. der Rieden. Verified linking for modular kernel verification. Dissertation, Saarland University, Saarbruken, Germany, 2009.
11. Micrium. µC/OS: RTOS and stacks. https://www.micrium.com/rtos/ucosii/overview/, 2016.
12. N. Evans and M. Butler. A Proposal for Records in Event-B. In *FM 2006: Formal Methods*, edited by J. Misra, T. Nipkow, and E. Sekerinski, 221–235. Berlin: Springer, 2006.
13. R. J. Feiertag and P. G. Neumann. The foundations of a provably secure operating system (PSOS). In *Proceedings of the National Computer Conference*, pp. 329–334. New York: AFIPS Press, 1979.
14. L. Freitas. Formal methods: Foundations and applications. In *Mechanising Data-Types for Kernel Design in Z*, edited by M. V. M. Oliveira and J. Woodcock, 186–203. Berlin: Springer-Verlag, 2009.
15. C. A. R. Hoare and N. Wirth. An axiomatic definition of the programming language PASCAL. *Acta Informatica*, 2(4):335–355, 1973.
16. P. Hudak, J. Peterson, and J. Fasel. *A Gentle Introduction to Haskell: Version* 98, 2000.
17. G. Klein, P. Derrin, and K. Elphinstone. Experience report: sel4: formally verifying a high-performance microkernel. *SIGPLAN Notices*, 44(9):91–96, 2009.
18. G. Klein, K. Elphinstone, G. Heiser, J. Andronick, D. Cock, P. Derrin, D. Elka-duwe, et al. sel4: Formal verification of an os kernel. In *ACM Symposium on Operating Systems Principles*, 207–220. New York: ACM, 2009.
19. J. J. Labrosse. *Microc/OS-II*, 2nd ed., Lawrence, KS: R & D Books, 1998.
20. Q. Li and C. Yao. *Real-Time Concepts for Embedded Systems*. Gilroy, CA: CMP Books, 2003.
21. Micrium. Micrium embedded software. http://ecos.sourceware.org/, 2016.
22. R. Milner. *Communication and Concurrency*. Upper Saddle River, NJ: Prentice-Hall, 1989.
23. P. G. Neumann, R. S. Boyer, R. J. Feiertag, K. N. Levitt, and L. Robinson. A provably secure operating system: The system, its applications, and proofs. In *Technical Report CSL-116, SRI International*, 1980.
24. T. Nipkow, M. Wenzel, and L. Paulson. *Isabelle/HOL: A Proof Assistant for Higher-Order Logic*. Berlin: Springer-Verlag, 2002.
25. T. Perrine, J. Codd, and B. Hardy. An overview of the kernelized secure operating system (KSOS). In *Proceedings of the Seventh DoD/NBS Computer Security Initiative Conference*, pp. 146–160, 1984.
26. G. P. Radha. *Pascal Programming*. New Delhi, India: New Age International, 1999.
27. S. O. Saydjari, J. Beckman, and J. Leaman. Locking computers securely. In *10th National Computer Security Conference*, Baltimore, pp. 129–141, 1987.
28. G. Smith. *The Object-Z Specification Language*. Norwell, MA: Kluwer Academic Publishers, 2000.
29. J. M. Spivey. *Z Notation: A Reference Manual*, 2nd ed. New York: Prentice Hall, 1992.
30. Sri International. http://www.sri.com, 2008.
31. A. Velykis and L. Freitas. Formal modelling of separation kernel components. In *Proceedings of the 7th International Colloquium Conference on Theoretical Aspects of Computing, ICTAC'10*, pp. 230–244, Berlin, Heidelberg, 2010. Springer-Verlag.
32. B. J. Walker, R. A. Kemmerer, and G. J. Popek. Specification and verification of the UCLA Unix security kernel. *Communications of the ACM*, 23:118–131, 1980.
33. Wind River. VxWorks. http://windriver.com/products/vxworks/, 2016.

16 Generation of Use Case UML Diagram from User Requirement Specifications

Wahiba Ben Abdessalem and Eman H. Alkhammash

CONTENTS

16.1 Introduction 217
16.2 Background 219
16.2.1 Gate 219
16.2.2 Model-Driven Architecture (MDA) 220
16.2.3 UML Use Case Diagram 220
16.3 Related Work 221
16.4 The Proposed Approach 221
16.4.1 Transformation Rules Construction 221
16.4.1.1 Extracting System Concepts 222
16.4.1.2 Extracting the Use Case Concept 222
16.4.1.3 Extracting "Include Dependency" 222
16.4.1.4 Extracting "Extend Dependency" 223
16.4.1.5 Extracting the Actor Concept 223
16.4.2 Text Preprocessing 223
16.4.3 Execution of Transformation Rules 224
16.4.4 Graphical Representation 224
16.5 Evaluation 224
16.6 Conclusion 226
References 226

16.1 INTRODUCTION

The success of software development is based heavily on the ability to develop programs that respond to the needs expressed by its users. The task of transforming user requirements to Unified Modeling Language (UML) models is costly and time-consuming for designers. Finding automatic techniques that support such transformation would be of great benefit. In this chapter, we present an approach that attempts to facilitate the process of mapping user requirements into a UML use case diagram.

Based on the model-driven architecture (MDA) approach (Figure 16.1), a user's requirements are considered as a source model represented in natural language text. The source metamodel represents a description of the concepts of natural language texts (e.g., verbs, nouns). A UML use case diagram is considered as the target model. The target metamodel is an XML file grouping of the concepts related to the use case diagram. Consequently, the approach consists of four steps:

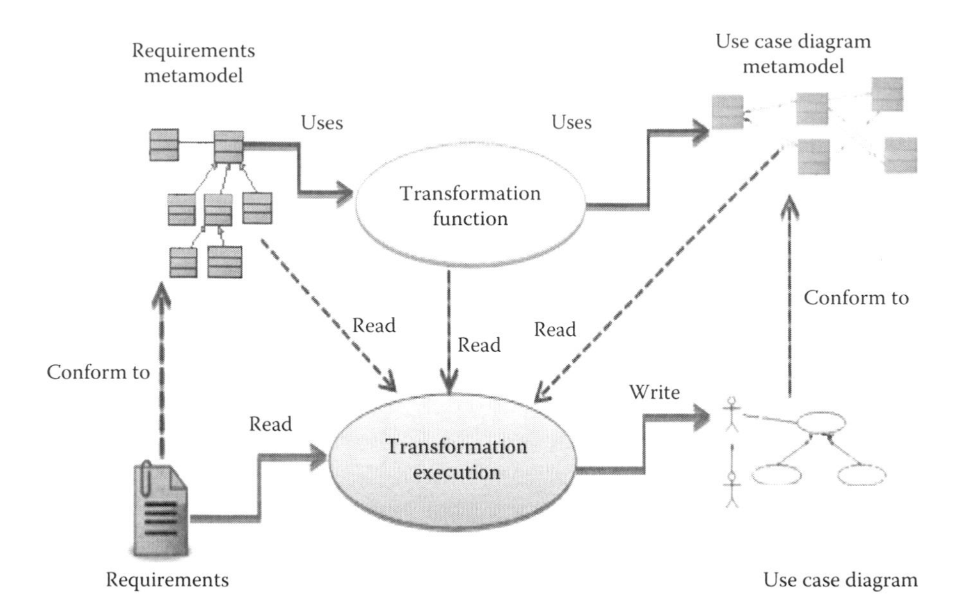

FIGURE 16.1 User Requirement Specifications (URS) to use case diagram transformation framework.

1. Defining a use case diagram metamodel, describing the use case diagram and the required metamodel, and describing the requirements.
2. Defining the transformation function, ensuring the transformation of the required metamodel into the use case diagram metamodel. This function gathers a number of transformation rules. The developed transformation rules are added to Gate components: JAVA Annotation Patterns Engine (JAPE) and Gazetteer lists (Cunningham et al., 2002).
3. Processing the text related to user requirements, using Gate tools (sentence splitter, tokenizer, POS Tagger). The result is processed text with grammatical categories (verbs, nouns, adverbs, etc.).
4. Executing the transformation rules, using as input the results of the previous step, to generate an XML file. Finally, the XML file is transformed into a use case diagram.

The transformation rules developed in the second stage are based on the two metamodels: the use case diagram metamodel, as shown in Figure 16.2, and the user requirement metamodel, as shown in Figure 16.3.

Figure 16.2 describes the concepts of use case diagram where a system is composed of uses cases and actors. The filled diamonds denote the composition relation.

The relationship between the composite (i.e., system) and the components (i.e., use case and actor) is a strong relationship. If the composite is destroyed, all the component parts must be destroyed. An actor is linked to use cases with association. Therefore, each association is created for one use case and one actor. Also, the use cases can have relationships with each other. The relations can be a dependency or a generalization. In the dependency relation, we indicate that a use case may be a source or a target of the dependency. The type attribute refers to the include or extend dependencies. In generalization, we make use of parent and child to describe generalization between use cases.

Figure 16.3 describes the concepts of a natural language text: a text is composed of sentences. Each sentence is composed of terms or concepts, and have a specific type: adverb, verb, or noun. The modal verbs (describing ability, permission, requests, and advice such as can, could) and the possibility verbs (e.g., may, might) are considered as subclasses of the class verb.

The rest of the paper is structured into the following sections: Section 16.2 describes the background. Section 16.3 describes the related work. Section 16.4 gives details about the proposed

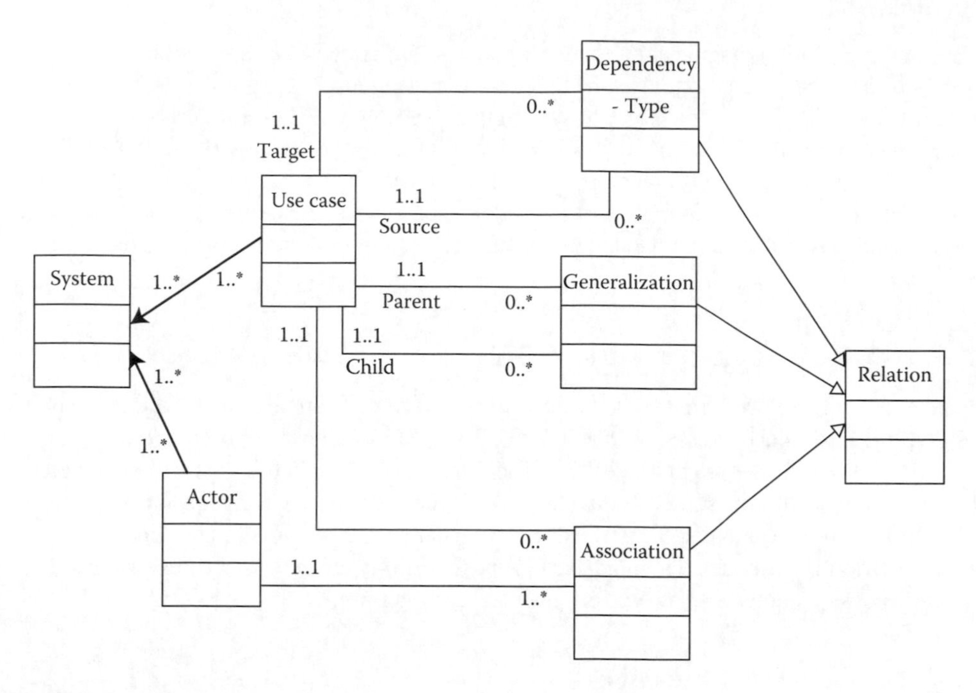

FIGURE 16.2 Use case diagram metamodel.

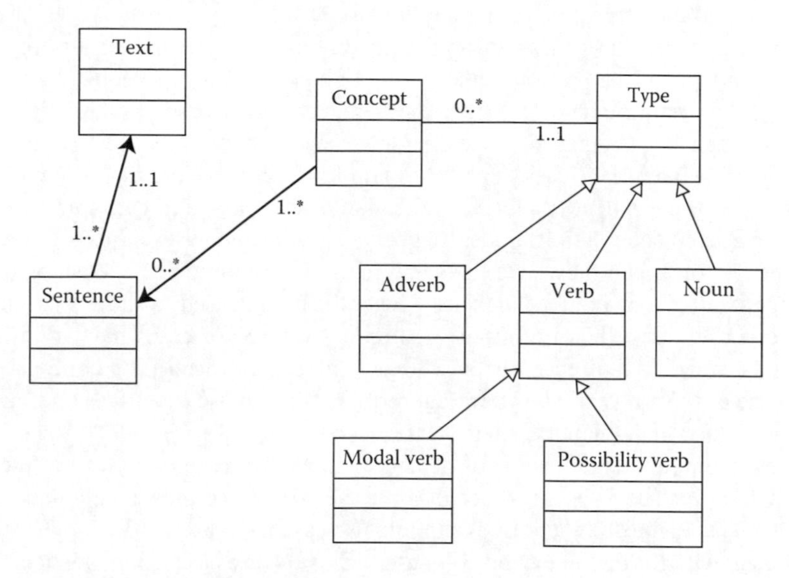

FIGURE 16.3 User requirements specification metamodel.

transformation approach. Section 16.5 is an evaluation of our proposed approach. Finally, the chapter is concluded to discuss future work.

16.2 BACKGROUND

16.2.1 GATE

Gate is text processing platform for language engineering developed at the University of Sheffield, England (Cunningham et al., 2002).

Gate includes several plugins and other components that allow for both annotation and information extraction. Each annotation has start and end offset and a set of features, each of which has a name and value. JAPE is part of the Gate system that executes JAPE grammar phases.

Each phase consists of a set of action rules that consists of two parts. The left-hand side part represents an annotation pattern, whereas the right-hand side part represents the action to be taken when the left-hand side part is matched. Annotations are created by different tools for language processing such as Gazetteers. In Gate, Gazetteers are used to find the occurrence of proper names and keywords in the text.

16.2.2 Model-Driven Architecture (MDA)

MDA provides a framework that uses models for software development. The Object Management Group (OMG) introduced MDA in 2001 (OMG, 2016). Model transformation is one of the prominent features of MDA (Segura et al., 2007). Model transformation is the process of converting one model to another within the same system (Dube and Dixit, 2012). The transformation function uses transformation rules to automatically transform an instance of the source metamodel to an instance of the destination metamodel. The approach proposed in this chapter uses transformation rules to transform user requirements specification to a use case diagram.

16.2.3 UML Use Case Diagram

A use case diagram is a UML diagram used to capture functional requirements and model the dynamic aspects of the system. Use cases play important roles in the early stages of the development of the system. A use case diagram represents an abstract view of the systems and isolates details to better understand the portion of the system of concern (OMG, 2016b). UML use case diagrams consist of four main elements, which are system, actors, use cases, and relations. The actors, depicted by stickman icon, represent system/people that interact with the modeled system. The use cases, depicted by named ellipses, represent main functionalities provided by the system). The relations (association, dependency, and generalization) are used to indicate interactions between components. An association relationship is used between actors and uses cases. Generalization can be used between actors when they have the same roles. The common relationships used between use cases are generalization and dependencies. Generalization is used to show a parent–child relationship between use cases; thus it is used when two or more use cases have similar behaviors. Generalization is shown as a directed arrow with a triangle arrowhead. The child use case is connected at the base of the arrow, as shown in Figure 16.4. The two use cases PIN and Fingerprint are a "child" of the Customer Authentication use case.

There are two common relationships of dependencies in use case diagram: << include >> and << extend >>. The << include >> is used when the base use case is incomplete without the included use case. For instance, in Figure 16.5, the Deposit Fund use case includes the Customer Authentication use case as shown, whereas the << extend >> is used when the use case is independent (optional) on

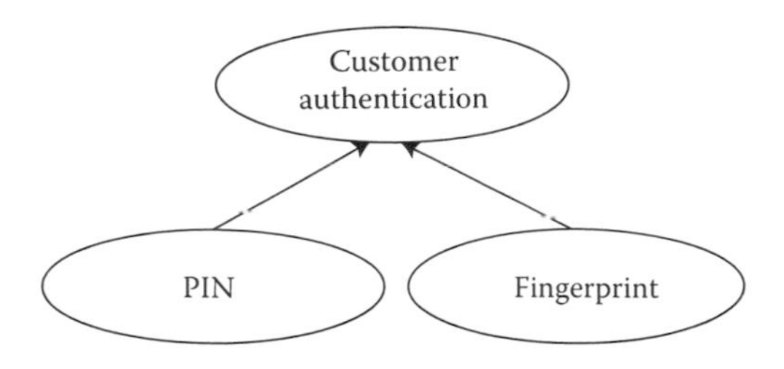

FIGURE 16.4 Generalization use case relationship.

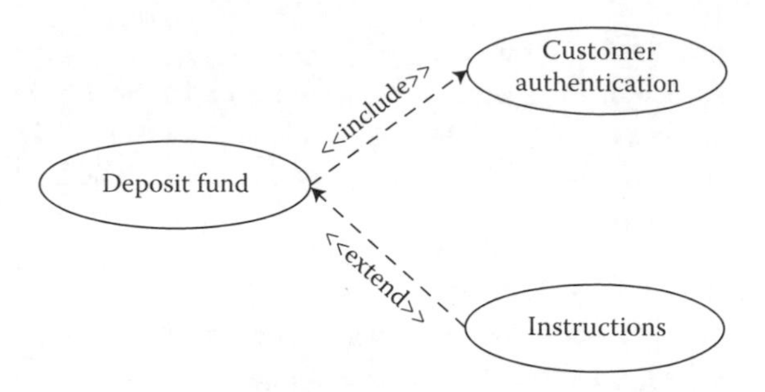

FIGURE 16.5 Use case relationships.

the extended (base) use case. The Instructions use case extends the Customer Authentication use case.

16.3 RELATED WORK

In this section, some works will be described related to automated generation of UML diagrams from user requirements. Examples of such work are NL-OOPS (Mich and Garigliano, 2002), CM-Builder (Harmain and Gaizauskas, 2003), and R-Tool (Afreen et al., 2011; Vinay et al., 2009). NL-OOPS and CM-BUILDER use natural language specifications to construct object-oriented analysis models, but the output is only a preliminary model that necessitates being completed. Segura et al. (2007) proposed graph transformations used to automate the transformations from Web requirement models to Web design models. However, this approach is not working with standard notations, such as UML (Vinay et al., 2009), that describe a tool based on natural language, such as R-TOOL, which aims at supporting the analysis stage of software development. R-TOOL analyzes software requirements texts to generate use cases and class diagrams. The tool demonstrates just an initial experimental work, which has to be improved. In this chapter, a new approach will be presented to map, automatically, user requirements into UML use case diagrams. The proposed approach consists of, mainly, the following steps: First, the requirements and use case metamodels are defined. Second, transformation rules, allowing the transformation of the requirements metamodel into the use case metamodel, are constructed. Third, a text representing requirements is processed to identify its concepts. Then, transformation rules are executed, and finally, a use case diagram is represented graphically.

16.4 THE PROPOSED APPROACH

Transforming a model into another means that a source model is transformed into a target model based on some transformation rules. The basic contribution of this work is in transforming the requirement (source model) into a use case diagram (target model). As known in MDA, the challenge is how to find transformations between two metamodels that can be applied to transform models. Different methods can be used for defining the transformation rules. The key idea is the relationship identification between the two metamodels. Since the two metamodels are defined (Figures 16.2 and 16.3), the rules are created based on the information in the source and target metamodels.

The following sections describe in detail the proposed approach.

16.4.1 Transformation Rules Construction

UML use case diagrams consist of four main elements: system, actors, use cases, and relations (association, dependency, and generalization). Specific rules will be identified to extract each concept from the requirements.

16.4.1.1 Extracting System Concepts

The system indicates the scope of the system. In requirement specifications, the system is mapped to an object type, for example, system or library. We made an extension of Gazetteer lists with new objects related to the context. Then, to extract this concept, we developed a JAPE rule as shown in Figure 16.6.

16.4.1.2 Extracting the Use Case Concept

In requirement specifications, the action denoted as verbs are considered as use cases. The combination of a verb and a noun (e.g., selects a vehicle) constitutes a use case. Similarly, the adverb (e.g., selection) or a combination of an adverb and a noun (e.g., selection of a vehicle) identifies a use case. We extend Gazetteer lists with new lists of verbs and nouns. We developed JAPE rules to identify verbs, nouns, adverbs, and a JAPE rule for the use case extraction. To extract this concept, we developed a JAPE rule as shown in Figure 16.7.

16.4.1.3 Extracting "Include Dependency"

We use "include dependency" whenever one use case needs the behavior of another. In requirement specifications, we have noticed that generally before an inclusion relation, verbs or phrases that express the obligation are used (e.g., should and must). We extend Gazetteer lists with new lists of modal verbs and developed a JAPE rule to extract "include dependency" (Figure 16.8).

```
Phase : System
Input : LookupT oken
 Options : control = applet
 Rule : rule system
 (
 {Lookup.majorT ype == system}
)
 : label
 ->
 : label.System = {rule = "rule system"}
```

FIGURE 16.6 JAPE rule to extract objects.

```
Phase: UseCase
Input: Cod Verbe Adverbe
Options: control = appelt
Rule: rule_usecase
(
 {Verbe} |
 {Verbe}{Cod} |
 {Adverbe} |
 {Adverbe}{cod}
)
:label
-->
 :label.UseCase = {rule = "rule_usecase"}
```

FIGURE 16.7 JAPE rule to extract use case concepts.

```
Phase: Include
Input: Modal UseCase
Options: control = appelt
Rule: rule_include
(
 { Modal } { UseCase }
)
:label
-->
:label.Include = {rule = "rule_include"}
```

FIGURE 16.8 JAPE rule to extract "include dependency."

16.4.1.4 Extracting "Extend Dependency"

An "extend dependency" is a relationship where an extending use case completes the behavior of a basic use case. Verbs and phrases that express possibility are usually used to define extension relationship (e.g., may and might). We extend Gazetteer lists with new lists of those verbs and developed a JAPE rule to extract "extend dependency" (Figure 16.9).

16.4.1.5 Extracting the Actor Concept

An actor specifies a role played by an external entity that interacts with the system (e.g., by exchanging messages). He can be a human user of the designed system, or some other systems or hardware interacting with the system. Actors are generally nouns, for example, customer, supplier, or student. To identify an actor, we identify nouns in the text and we verify in addition if the noun is followed by a use case, since each actor should be associated with one or more use cases. We developed a JAPE rule to extract the concept of Actor (Figure 16.10).

16.4.2 Text Preprocessing

After transformation rules, the text related to user requirements should be preprocessed to detect and tag its components.

The input of this stage is a text related to requirements specification. The text is parsed to have paragraphs, sentences, and tokens. Then, each word (token) is given a grammatical category. This step is performed by applying some GATE API functionalities on user requirements to obtain texts with grammatical categories (e.g., verbs, nouns).

The text preprocessing consists of three steps:

```
Phase: Extend
Input: Optionnel UseCase
Options: control = appelt
Rule: rule1
(
 { Optionnel }{ UseCase }
)
:label
-->
 :label.Extend = {rule = "rule1"}
```

FIGURE 16.9 JAPE rule to extract "extend dependency."

```
Phase: Actor
Input: Lookup Extend Include UseCase
Options: control = appelt
Rule: rule_actor
(
 { Lookup.majorType == actor }
( {  Extend  } |
  { Include }|
  { UseCase }
)+
)
:label
-->
 :label.Actor = {rule = "rule_actor" }
```

FIGURE 16.10 JAPE rule to extract actor concept.

- Sentence splitting—It consists of breaking down the text into a set of sentences. In this step, we use the Gate component *Sentence splitter.*
- Tokenization—This means to split each sentence into tokens, which are words and punctuation. In this step, we use the Gate component *Tokeniser.*
- POS tagging—Each token is given the appropriate word category like a noun, verb, or adjective. In this step, we use the Gate *Pos Tagger* component. However, during its processing, it looks up the Gazetteer lists and JAPE rules to generate the result.

16.4.3 Execution of Transformation Rules

This phase involves the extraction of UML use case concepts. Using the results generated by preprocessing in the user requirements stage, we apply the transformation rules and extended JAPE rules and Gazetteer lists. This phase is considered a semantic annotation, allowing the detection and tagging of the concepts of the use case diagram and the relationships between these concepts. The result obtained from this phase is an XML file that the user or the designer can save, print, or use to visualize the result as a use case diagram in the interface of a given computer-aided software engineering (CASE) tool, such as ArgoUML.

16.4.4 Graphical Representation

In order to transfer the XML file into use case diagrams, we use eXtensible Stylesheet Language Transformation (XSLT), and Simple Api for XML (SAX) to transform the XML file into a Scalable Vector Graphics (SVG) file. Then, we visualize the use case diagram from the SVG file. We developed a tool in Java that performs the aforementioned step that takes user requirements as input and displays use case diagrams that correspond to the input requirements.

16.5 EVALUATION

To validate our proposed approach, we have implemented a tool named Use Case Generator (UCgen). We created our own corpus that is gathered from several documents in several areas such as commercial and educational. Following that, we calculated the following effectiveness metrics: recall, precision, and overgeneration (Chinchor, 1993). The recall is the ratio of the number of correct generated concepts to the total number of relevant concepts. It is calculated as follows:

$$Recall = \frac{N_{correct}}{N_{correct} + N_{missing}}$$

Precision is the ratio of the number of correct generated concepts to the total number of irrelevant and relevant generated concepts. It is usually expressed as follows:

$$Precision = \frac{N_{correct}}{N_{correct} + N_{incorrect}}$$

The overgeneration metric measures the percentage of the actually generated concepts that were spurious. It is calculated as follows:

$$Overgeneration = \frac{N_{overgenerated}}{N_{correct} + N_{missing}}$$

We have compared our tool with CM-Builder, since it is the closest system to ours. Values are given in Table 16.1.

We can note that recall and precision of CM-Builder are much lower than ABCD tool. Moreover, regarding over- generation, UCgen tool is more efficient than CM-Builder, since it makes fewer errors in concepts generation.

We have also compared the functionalities of UCgen tool with other available tools that can perform automated analysis of NL requirement specifications. The results of this comparison are given in Table 16.2. Table 16.2 shows that unlike many tools, the UCgen tool is able to identify information such as associations and dependency from natural language requirements. Also, the UCgen tool does not need user involvement to detect any concept, which makes it a fully automated tool. These results are very encouraging and support very well the approach proposed in this chapter.

TABLE 16.1
Performance Comparison

	CM-Builder (%)	UCgen (%)
Recall	73	88
Precision	66	90.3
Overgeneration	62	29

TABLE 16.2
Concepts Generation of UCgen and Other Tools

Concept \ Tool	CM-Builder	NL-OOPS	R-Tool	UCgen
Generalization	No	No	No	Yes
Dependency	No	No	No	Yes
Associations	No	No	No	Yes
Actor	Yes	Yes	Yes	Yes
Use case	Yes	Yes	Yes	Yes
System	No	No	No	Yes

16.6 CONCLUSION

In this chapter, we presented an approach that transforms user requirements into UML use case diagrams. The approach consists of four steps: the first step is the definition of the requirements and uses case metamodels. The second step is the creation of transformation rules, allowing the transformation of the requirements metamodel into the use case metamodel. The third step is text processing to tag the text components. Then, the last step is execution of the transformation rules on the processed text. The result of the last step is an XML file, used to automatically generate a use case diagram. We developed a tool that supports our approach. In future work, we will evaluate the proposed approach and apply it to several case studies. Moreover, we will try to extend the approach to cover other UML diagrams such as class diagrams (Ben Abdessalem Karaa et al., 2015) and sequence diagrams.

REFERENCES

Afreen, H., Bajwa, I. S., and Bordbar, B. 2011. SBVR2UML: A challenging transformation. In: *Frontiers of Information Technology (FIT)*, Islamabad, December (pp. 33–38).

Ben Abdessalem Karaa, W., Ben Azzouz, Z., Singh, A., Dey, N., S Ashour, A., and Ben Ghazala, H. 2015. Automatic builder of class diagram (ABCD): An application of UML generation from functional requirements. *Software: Practice and Experience*, 46(11), 1443–1458.

Chinchor, N. 1993. The statistical significance of the MUC- 5 results. In: *Proceedings of the 5th Conference on Message Understanding, MUC 1993*, Baltimore, MD, August 25–27 (pp. 79–83).

Cunningham, H., Maynard, D., Bontcheva, K., and Tablan, V. 2002. A framework and graphical development environment for robust NLP tools and applications. In: *ACL*, Philadelphia, July (pp. 168–175).

Dube, M. R. and Dixit, S. K. 2012. Modeling theories and model transformation scenario for complex system development. *International Journal of Computer Applications*, 38(7), 11–18.

Harmain, H. M. and Gaizauskas, R. 2003. CM-builder: A natural language-based CASE tool for object-oriented analysis. *Automated Software Engineering*, 10(2), 157–181.

Mich, L. and Garigliano, R. 2002. NL-OOPS: A requirements analysis tool based on natural language processing. In: *Proceedings of Third International Conference on Data Mining Methods and Databases for Engineering*, Bologna, Italy (pp. 321–330).

OMG. 2016a. www.omg/org. Accessed April 2016.

OMG. 2016b. Documents Associated with Unified Modeling Language (UML), version 2.5. www.omg.org/spec/UML/2.5/. Accessed April 2016.

Segura, S., Benavides, D., Ruiz-Cortés, A., and Escalona, M. J. 2007. From requirements to web system design. An automated approach using graph transformations. *Actas de Talleres de Ingeniería del Software y Bases de Datos*, 1(6), 61.

Vinay, S., Shridhar, A., and Prashanth, D., 2009. An approach towards automation of requirements analysis. In *Proceedings of the International MultiConference of Engineers and Computer Scientists, IMECS 2009*, Hong Kong, Vol. 1 (pp. 1–6).

Section II

Communication Systems

17 Design and Simulation of Adaptive Cognitive Radio Based on Software-Defined Radio (SDR) Using Higher-Order Moments and Cumulants

Ahmed Abdulridha Thabit and Hadi T. Ziboon

CONTENTS

17.1 Introduction 229
17.2 Cognitive Radio Concept 230
17.3 Advantages of Software Defined Radio 231
17.4 Architecture for a Generic Cognitive Radio 232
17.5 Concept of Two Hypotheses 233
17.6 Statistical Features 233
17.7 Proposed Adaptive Cognitive Radio Detection System Based on Statistical Feature 234
17.8 Simulation Results 236
17.9 Conclusion 240
References 240

17.1 INTRODUCTION

The electromagnetic radio spectrum is a licensed resource, carefully managed by governments. Also, the accessible electromagnetic radio spectrum is a constrained natural resource and getting crowded day by day due to expansion in wireless devices and applications. It has been additionally found that the allocated spectrum is underutilized as a result of the static portion of the spectrum [1].

Users' needs of wireless services leads to scarcity of the available spectrum and inefficient channel utilization. The cognitive radio (CR) is the optimum solution for these requirements. The ability to detect a primary user (PU) as well as to avoid any false alarm are important for such a system. The CR has the ability to get the unlicensed user (secondary user [SU]) to use the spectrum for a while.

It becomes very difficult to find a spectrum for either new services or expanding existing services. Currently government policies do not allow access to licensed spectrum by unlicensed users, constraining them instead to use several heavily populated, interference-prone frequency bands. As the result there is huge spectrum scarcity problem in certain bands. In particular, if we were to scan the radio spectrum, including the revenue-rich urban areas, we find that some frequency bands in the spectrum are unoccupied for some of the time, and many frequency bands are only partially occupied, whereas the remaining frequency bands are heavily used.

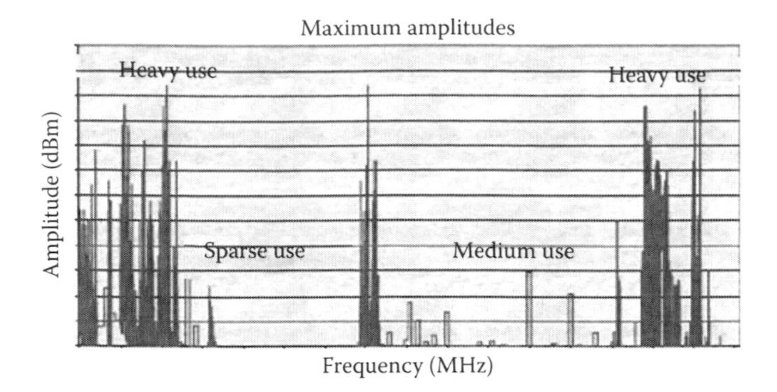

FIGURE 17.1 Frequency spectrum that shows different regions.

Figure 17.1 represents the frequency spectrum and shows the basic goal of the CR to detect the signal and sense whether the spectrum is in use [2,3].

The radio spectrum is a limited resource and is regulated by government agencies such as the Federal Communications Commission (FCC) [3]. Cognitive radio concepts, statistical features, and the proposed system are discussed and explained in the following sections in detail.

17.2 COGNITIVE RADIO CONCEPT

CR, including software-defined radio (SDR) as an enabling technology, was proposed [4–6] to realize adaptable and effective usage of the spectrum. The term cognitive radio is from "cognition." Cognition is a term referring to the mental processes involved in picking up information and comprehension, including thinking, knowing, recollecting, judging, and problem solving. These higher-level functions of the brain include language, imagination, perception, and planning [7].

CR may have significant impacts on both technology and regulation of the use of spectrum leading to a revolution in wireless communication overcoming existing regulatory barriers [8].

CR came from a number of technologies such as the improvement of digital signal processing (DSP); math tools; and source coding of data, voice, and images. CR has turned into a promising strategy to solve the spectrum scarcity problem for supporting evolving wireless services and applications. In CR systems, unlicensed users can use the licensed frequencies when the primary user (PU) is not dynamic. For accomplishing good spectrum sensing performance, a suitable threshold must be selected [9,10].

CR was introduced at 1999 by Mitola [11]. It is supposed to change the operating band, if the currently used band becomes too occupied or the PU puts the band into use. The most important feature of CRs is the capacity to sense the spectrum and whether to take a certain band into use.

CRs can adapt to their environment via varying their transmitter factors to different signaling systems. Depending upon the network and cooperation with other cognitive devices, they can trade information about their location and environment. CR can cooperate with other cognitive radios and offer information between each other. Reconfigurability was used in radio development. A common radio communication system is implemented in hardware. In an SDR, most of the required hardware and required transmitter and receiver algorithms are implemented in software, thus high reconfigurability is achieved [12,13].

A general definition for CR is "a radio of an intelligent wireless communication system that senses and is aware of its surrounding environment and capable to use or share the spectrum in an opportunistic manner without interfering the licensed users" [14].

CR uses different techniques to become aware of the surroundings, has the capacity to learn from the outer environment, and can change the parameters of the transmitted and received data to achieve the goal of effective communication without interference [7]. Figure 17.2 represents the

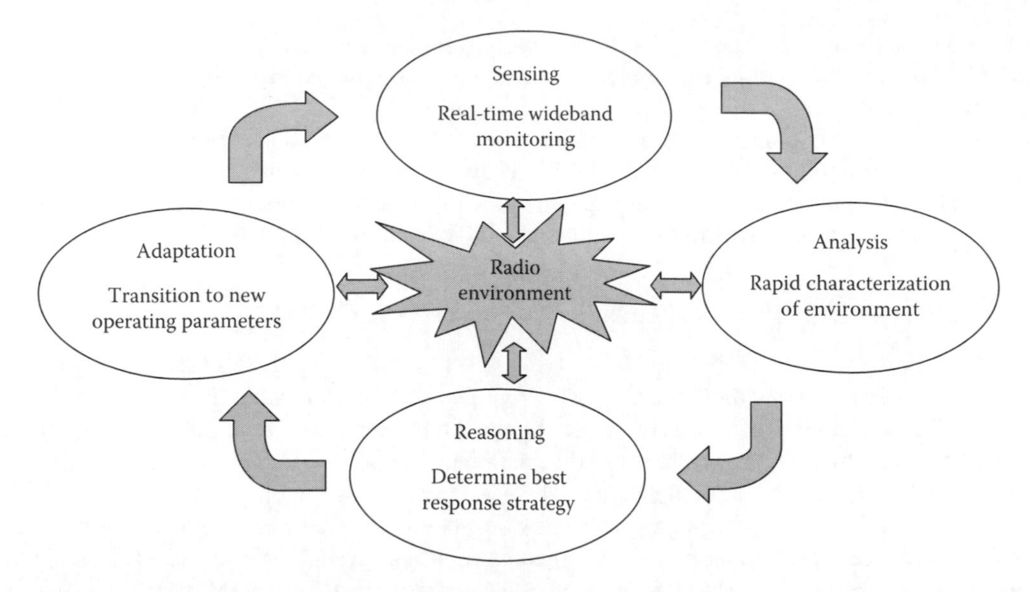

FIGURE 17.2 Detection cycle of CR.

detection cycle of CR that shows the steps of detection process. The sensing stage is the first and the most important stage. The different sensing techniques are covered by Thabit and Ziboon [15].

17.3 ADVANTAGES OF SOFTWARE DEFINED RADIO

SDR is a radio transceiver in which most functions are implemented in software. SDR communication systems place much of the signal manipulation into the digital domain where the computer is then able to process the data. Figure 17.3 depicts the evolution from Hardware (HW) to SDR to SR to adaptive intelligent-software radio (AI-SR). Over time the number of system components performed in software is increasing.

Some of the basic advantages of SDR are the ability to receive and transmit various modulation methods using a common set of hardware; the ability to alter functionality by downloading and

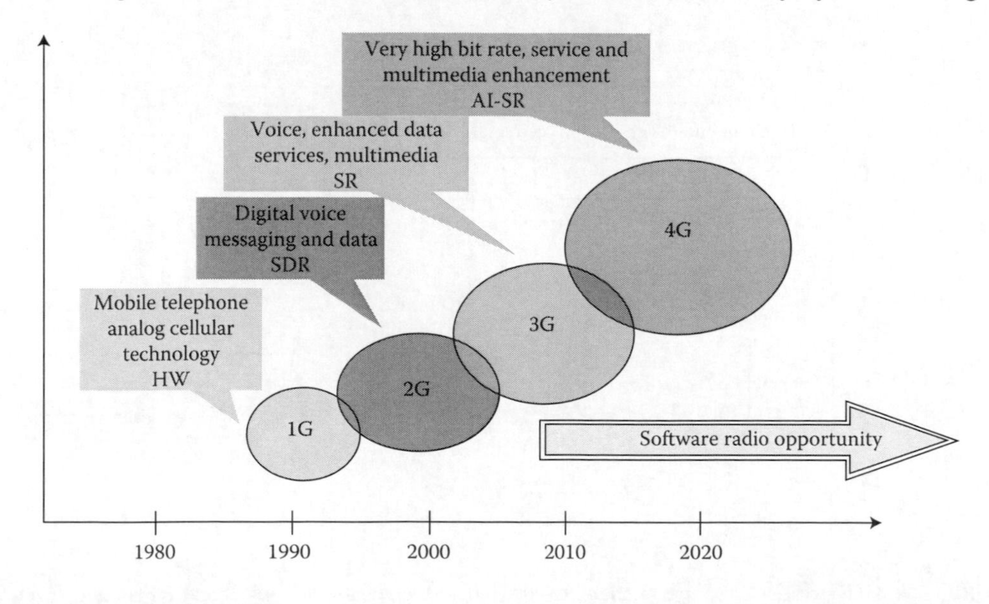

FIGURE 17.3 Evolution of SD.

running new software at will; the possibility of adaptively choosing an operating frequency and a mode best suited for prevailing conditions; and the opportunity to recognize and avoid interference with other communications channels using programmable digital devices to perform the signal processing to transmit and receive baseband information at radio frequency. SDR 3G cellular base stations are now possible and can match the efficiency performance of their HDR. They offer less infrastructure, less maintenance, easier deployment with reduced ownership costs, and support of multiple standards through multimode, multiband radio capabilities [16].

17.4 ARCHITECTURE FOR A GENERIC COGNITIVE RADIO

The architecture for a generic cognitive radio transceiver is shown in Figure 17.4a. As shown from the figure it consists of the radio frequency (RF) front-end and the baseband processing unit. A control bus is used in controlling each component to make the radio adaptive to the RF environment. The RF front-end first amplifies the received signal, then mixes it to a lower band and, finally, the analog signal is converted to a digital signal. The baseband processing unit modulates or demodulates and encodes or decodes the signal depending on whether a signal was transmitted or received. The baseband signal processing unit is like to common transceivers; however the RF front-end is specifically designed to accommodate the need of the cognitive radio. CR transceiver is required to be able to sense over a wide spectrum range and preferably in real time. The RF hardware is needed to be able to tune in to any part of the frequency spectrum. The main components of the cognitive radio RF front-end, shown in Figure 17.4b, are as follows [17]:

- RF filter—Selects the desired operating band by bandpass filtering the received RF signal.
- Low noise amplifier (LNA)—Amplifies the received signal without adding a remarkable amount of noise.
- Mixer—Mixes the received signal with the locally generated RF and then converts it to the baseband or the intermediate frequency (IF).
- Voltage-controlled oscillator (VCO)—Generates a signal at a specific frequency depending on the control voltage. The generated signal is then used to convert the incoming signal frequency to the baseband or intermediate frequency.

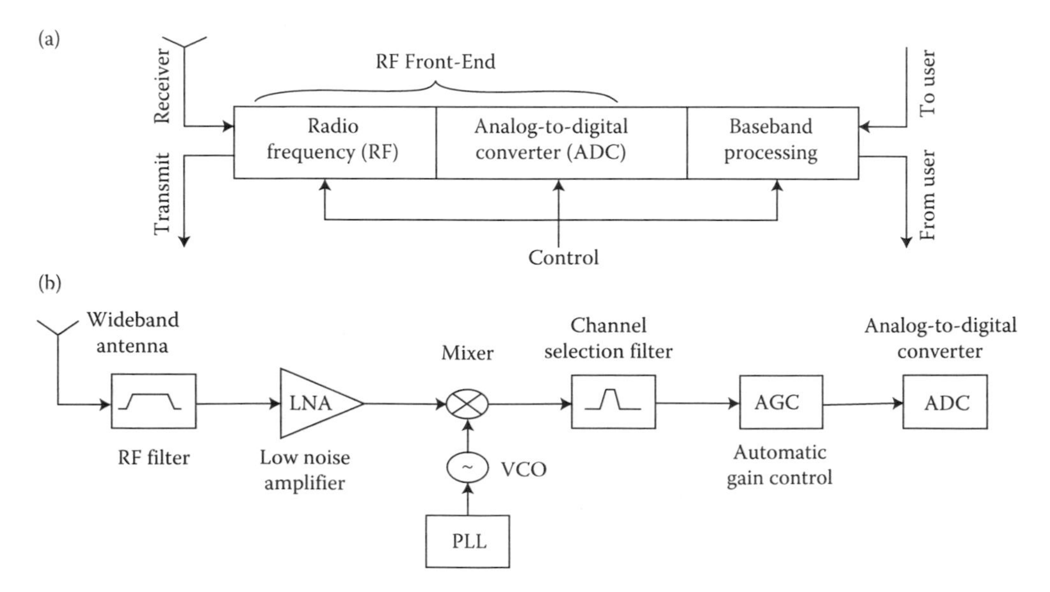

FIGURE 17.4 Physical architecture of the CR: (a) CR transceiver and (b) wideband RF/analog front-end architecture.

- Phase locked loop (PLL)—Ensures the signal of the VCO is locked accurately on the specific reference frequency.
- Channel selection filter—Selects the desired channel and rejects adjacent channels.
- Automatic gain control (AGC)—Keeps the gain or output power level of an amplifier constant over a large range of input signal levels.
- Analog-to-digital converter (ADC)—Converts the analog input signal to a digital signal.

17.5 CONCEPT OF TWO HYPOTHESES

Spectrum sensing can be simply decreased to an identification problem; this can be modeled as a hypothesis test [13]. The sensing equipment has to decide between the following hypotheses:

$$H_0 : y[n] = w[n] \tag{17.1}$$

$$H_1 : y[n] = s[n] + w[n] \tag{17.2}$$

where H_0 hypothesis indicates absence of the primary user, H_1 hypothesis points toward the presence of the primary user, $n = 1 \ldots N$; $s(n)$ is the signal transmitted via the primary users; $y(n)$ is the signal received via the secondary users; $w(n)$ is the additive white Gaussian noise with variances; and N is the length of the message. Hypothesis H_0 indicates absence of the primary user and that the frequency band of interest only has noise, whereas H_1 points toward the presence of the primary user.

Thus for the two-state hypotheses the important cases are as follows: The probability of detection (P_d) is of main concern as it gives the probability of correctly sensing for the presence of primary users in the frequency band. The probability of misdetection (P_m) is just the complement of detection probability. The goal of the sensing schemes is to maximize the detection probability for a low probability of false alarm. But there is always a trade-off between these two probabilities. Receiver operating characteristics (ROC) present very valuable information regarding the behavior of detection probability with changing false alarm probability (P_d versus P_f) or misdetection probability (P_d versus P_m) [8,11,12].

17.6 STATISTICAL FEATURES

Selection of the appropriate features of a signal improves the performance of the detection and reduces its complexity through the implementation. The digital signals have different parameters according to the types of modulations. To distinguish between digital signals, suitable features should be chosen. A feature reflects the digital signal characteristics; it should be sensitive to the received signal and have the discriminating ability to distinguish between digital signals and noise types. Many types of signals features have been suggested in order to identify the schemes of MFSK, MPSK, and MQAM signals [18,19]. The statistical features are represented by higher-order moments and cumulants.

Statistical moments are the expected value of a random variable raised to the power indicated by the order of the moment. A first-order moment $\bar{x}$ is the statistical mean of the random variable x:

$$\bar{x} = E\{x\} \tag{17.3}$$

The received signal is a complex baseband envelope, so the general expression for the $(i + j)$th moment for a complex random variable x is [20]

$$E'_{i,j} = E\{x^i \cdot (x^*)^j\} \tag{17.4}$$

where an asterisk (*) denotes complex conjugate.

The $(i+j)$th moment of the received digital modulation signal is evaluated from its N samples by calculating the numerical mean of these samples after raising each sample to the power according to the moment order and as in Equation 17.5:

$$E'_{x,i,j} = \frac{1}{N}\sum_{k=1}^{N} x_k^i \cdot \left(x_k^*\right)^j \tag{17.5}$$

However, $(i+j)$th central moment is defined as

$$E_{i,j} = E\{(x-\overline{x})^i \cdot ((x-\overline{x})^*)^j\} \tag{17.6}$$

Fortunately, cumulants can also be expressed as functions of equal and lower-order central moments, making it simple to compute numerically. The expressions for the most important cumulants (C) are given as follows [19,21]:

$$C_{1,1} = E_{1,1} \tag{17.7}$$

$$C_{2,0} = E_{2,0} \tag{17.8}$$

$$C_{2,2} = E_{2,2} - (E_{2,0})^2 - 2(E_{1,1})^2. \tag{17.9}$$

$$\begin{aligned} C_{4,4} = {} & E_{4,4} - (E_{4,0})^2 - 18(E_{2,2})^2 - 16(E_{3,1})^2 - 54(E_{2,0})^4 - 144(E_{1,1})^4 \\ & -432(E_{2,0})^2\ (E_{1,1})^2 + 12E_{4,0}\,E_{2,0} + 96E_{3,1}\,E_{2,0}\,E_{1,1} + 144E_{2,2}(E_{1,1})^2 \\ & +72E_{2,2}(E_{2,0})^2 + 96E_{3,1}\,E_{2,0}\,E_{1,1} \end{aligned} \tag{17.10}$$

$$C_{6,0} = E_{6,0} - 15E_{2,0}\,E_{4,0} + 30(E_{2,0})^3 \tag{17.11}$$

17.7 PROPOSED ADAPTIVE COGNITIVE RADIO DETECTION SYSTEM BASED ON STATISTICAL FEATURE

Feature selection is the most important step that should be taken in account at designing a CR. Conventional detection methods are based on instantaneous features such as energy. These methods suffer from noise uncertainty; these systems fail in the detection at low signal-to-noise ratio (SNR) values. Therefore, a new CR system depending on statistical features is proposed. Higher-order moments and cumulants are suggested in this system as a signal feature. Three features were used in this system; they are C_{11}, C_{20}, and C_{60}. These features are selected according to their ability to separate or distinguish between signal and noise, as clearly seen in MATLAB® simulations (Figures 17.5 through 17.7) that show the ability of specified cumulants to distinguish between the noise and the signals. Therefore, these three features are candidates to provide the best response. Equations 17.3, 17.6 through 17.8, and 17.11 are used in the design.

The following cases are plotted to show the performance when single feature was selected. The performance measured by plotting the received signal with noise then checks the ability of selected feature to distinguish between them. Figure 17.5 shows the SNR plotted against the C_{11}. It is seen that there is good separation between the signal and noise. At positive SNR, when C_{11} is less than 1 such as 0.9, the channel is available (free) and there is no signal. For the negative SNR value,

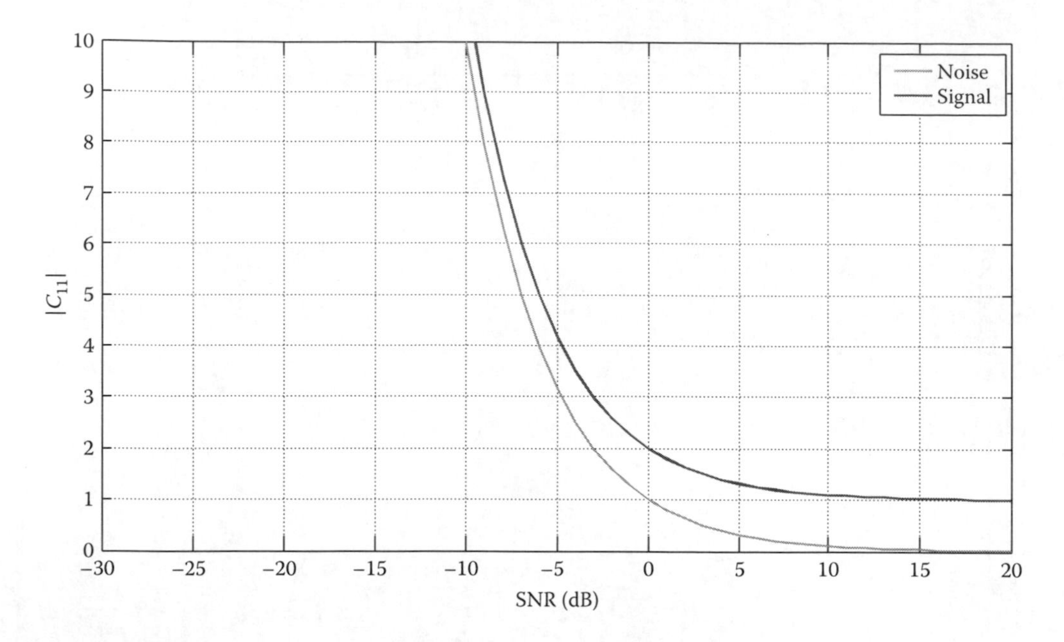

FIGURE 17.5 SNR versus C_{11}.

C_{11} fails in the detection, as clearly seen. In this case, C_{20} is a good choice. Figure 17.6 represents SNR versus C_{20}. As shown in the figure, for negative SNR when $C_{20} < 1.2$ means that the channel is used (not available). But when $C_{20} \geq 1.2$, there are two probabilities. If $C_{20} \geq 25$, then the channel is available (free), else there are two other probabilities, as can be seen in Figure 17.7, which shows if $C_{60} \geq 1.8 \times 10^9$, then the channel is available, else the channel is in use. The system deals with nine types of modulation levels. Each one of them has its frequency and data rate. Due to the periodicity

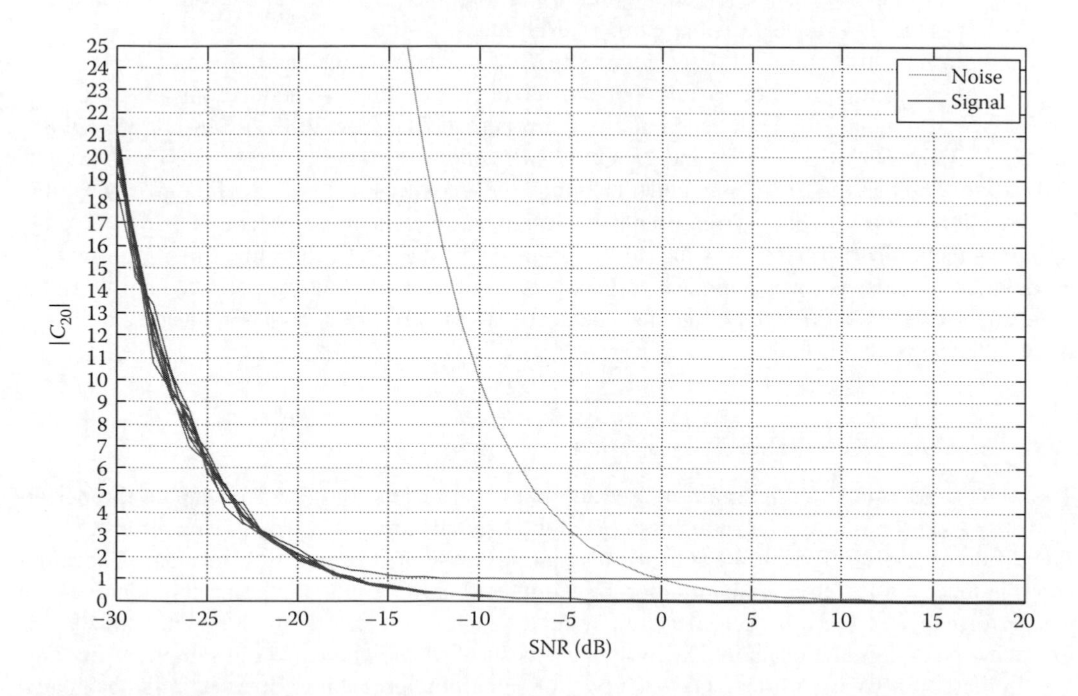

FIGURE 17.6 SNR versus C_{20}.

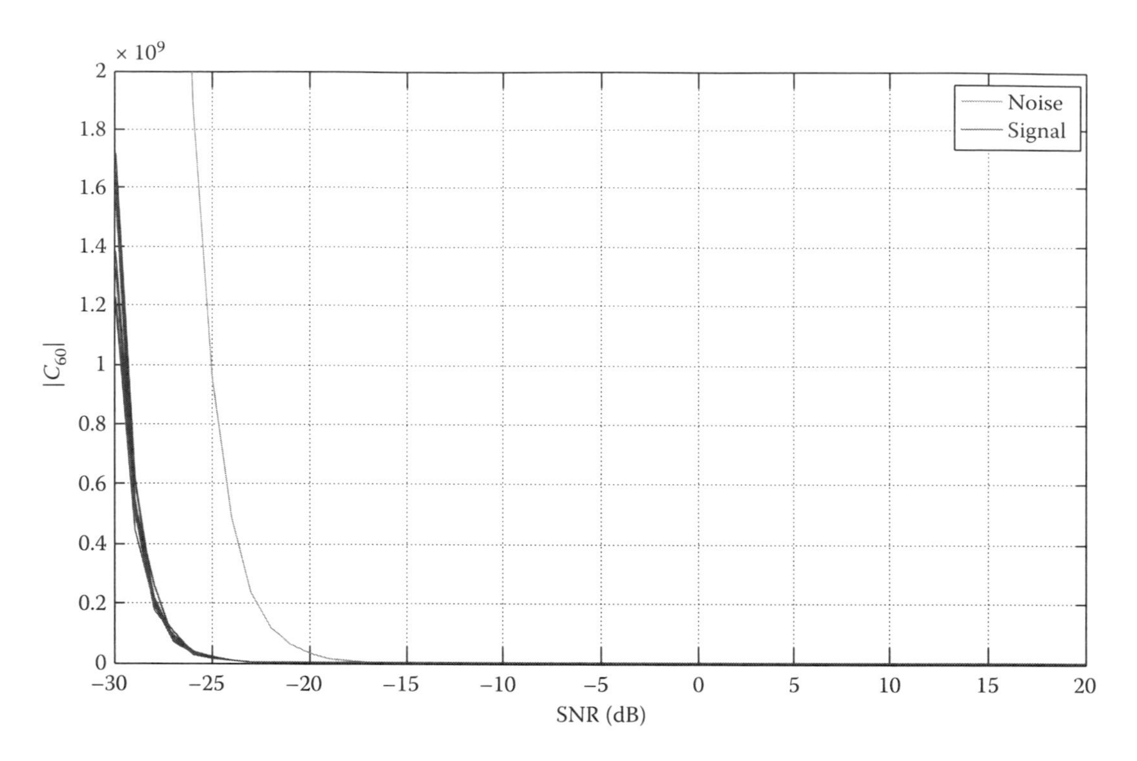

FIGURE 17.7 SNR versus C_{60}.

of the cyclostationary modulated signals, the system will be adaptive to detect each one of these signals and identify between the noise and the signal to submit the CR principle. This is because of using more than one feature (cumulant and moment) at the same time and using more than one threshold instead of a single decision region (hard limiter).

The flowchart in Figure 17.8 describes the baseband stage system. This flowchart explains the detection steps of the proposed system. Nine different types of digital modulations are generated. These types of signals are both coherent and noncoherent. They are 2FSK, 4FSK, BPSK, QPSK, 8PSK, 4QAM, 16QAM, 64QAM, and 256QAM. All generated signals are corrupted with AWGN and fading. The features are plotted under different SNRs to investigate which of them are useful in the detection process. Table 17.1 represents the design parameters of the system.

The detection performance was measured for all types together at the same time. The multiple modulation schemes is a new approach and not used in other works. At low SNR, the noise level is large and this leads to difficulty in the detection process, but the statistical feature has the ability to work in the worse environments.

17.8 SIMULATION RESULTS

The MATLAB simulation program is used to obtain the results. In order to evaluate the performance of the proposed CR detection systems, digital signals are generated at different low SNRs. The different levels of FSK, PSK, and QAM signals are generated at the baseband stage. The simulation is made for different trials in order to guarantee that the results are accurately representing the performance of the proposed detection systems. Each trial consists of 1000 generated signals for each type of digital modulation. Each signal consists of a different number of samples. The trials are simulated for each SNR test. For each type of signal, the probability of detection is evaluated for different SNRs.

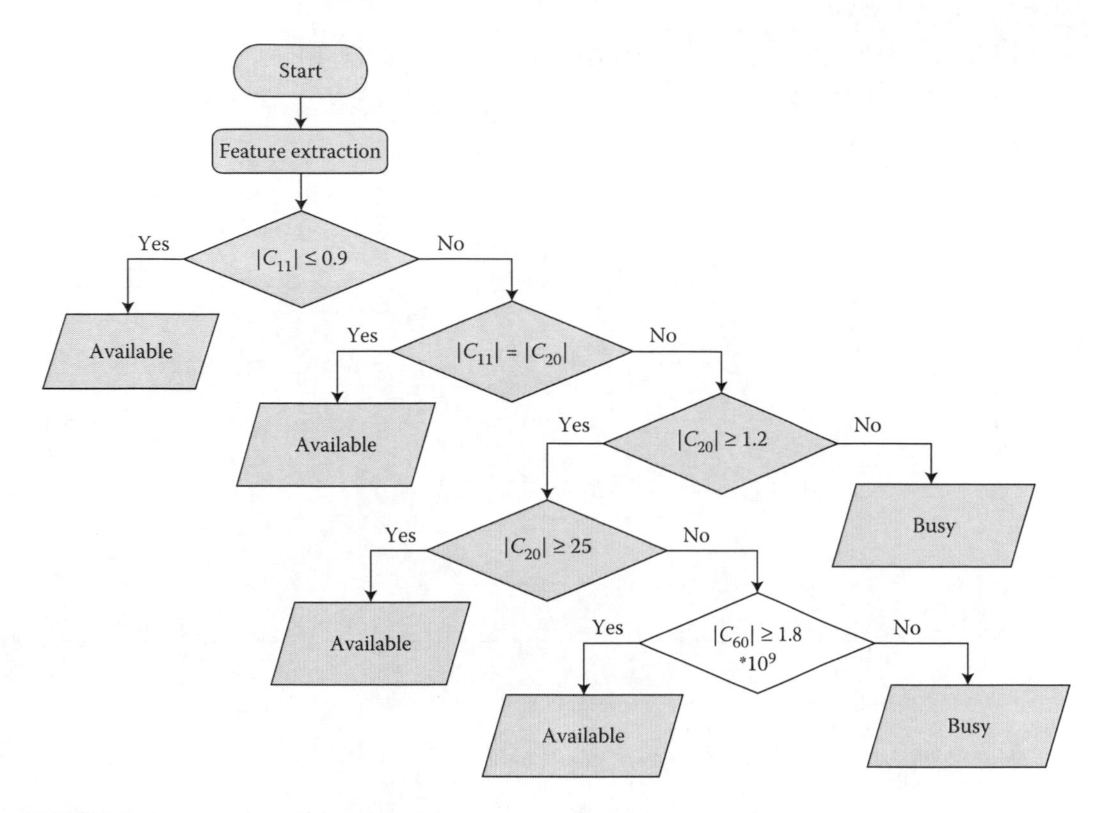

FIGURE 17.8 Flowchart of the proposed system.

High performance is obtained for different modulated signals by a threshold decision system based on statistical features (moment and cumulant) with variable thresholds. The system has a good success rate for these different modulated signals.

Figure 17.9 represents the SNR versus P_d at different P_f values. It is seen that $P_d = 100\%$ for $P_f = 0.1$ at SNR = −25 dB. Another case when P_f is 0.1 and using different SNR values is shown in Figure 17.10, which shows the sensing time versus P_d at different SNR. The case of $P_f = 0.1$ is called the constant false alarm rate (CFAR); this helps to get the required P_d at specified P_f. Figure 17.10 helps to prove that the increase in the sensing time will increase the probability of detection.

Another case of plotting SNR with P_m as shown in Figure 17.11 for message length = 1000.

The threshold value was more than one value to accommodate the soft decision principles. Different channels are used to examine the performance of the CR system. Figures 17.12 and 17.13 represent SNR versus P_d at different noisy channels to check out the performance at different environments at variable Doppler frequency shifts for the baseband stage.

TABLE 17.1
Design Parameters

Parameter	Value
SNR	−40–0 dB
Received signal	2FSK, 4FSK, BPSK, QPSK, 8PSK, 4QAM, 16QAM, 64QAM, and 256QAM
Number of iterations	10, 100
Message length	100–6000 sample
P_f	0.001, 0.005, 0.01, 0.05, 0.1

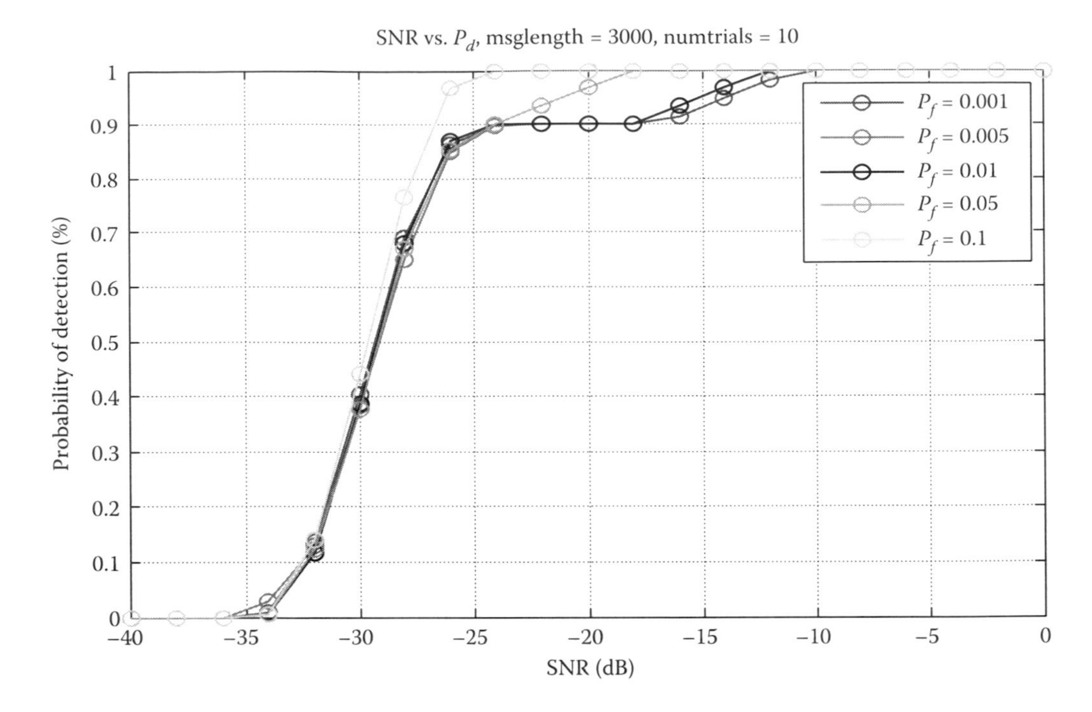

FIGURE 17.9 Probability of detection versus SNR at message length 3000.

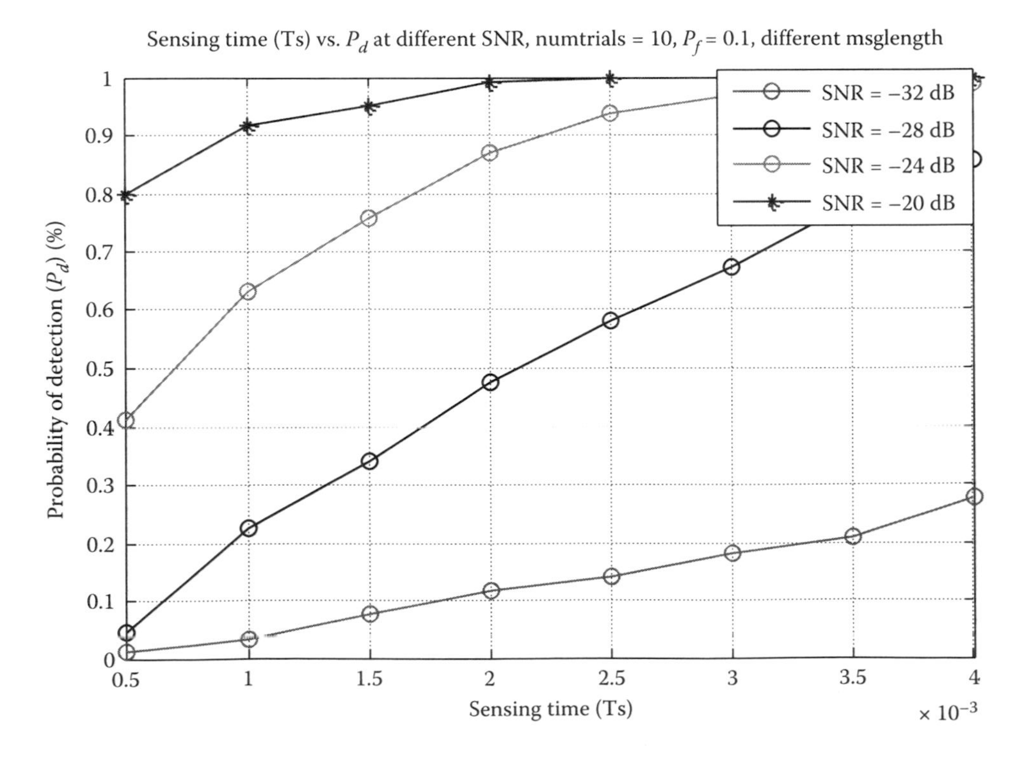

FIGURE 17.10 Sensing time versus probability of detection at different SNR.

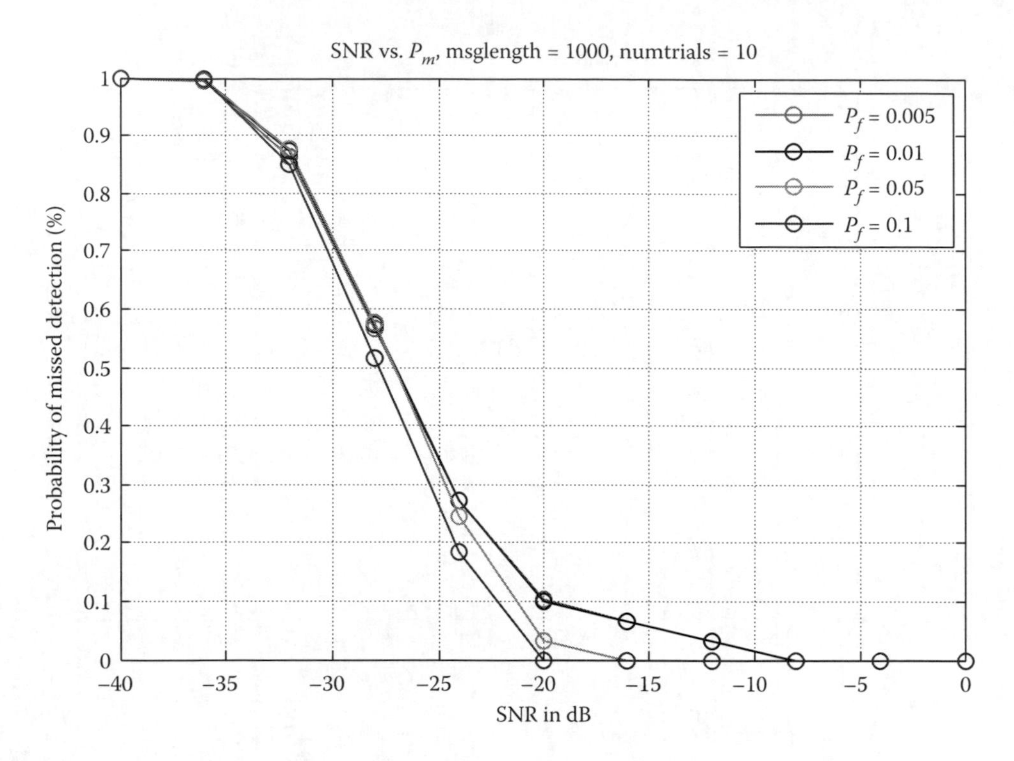

FIGURE 17.11 Probability of missed detection versus SNR at different P_f.

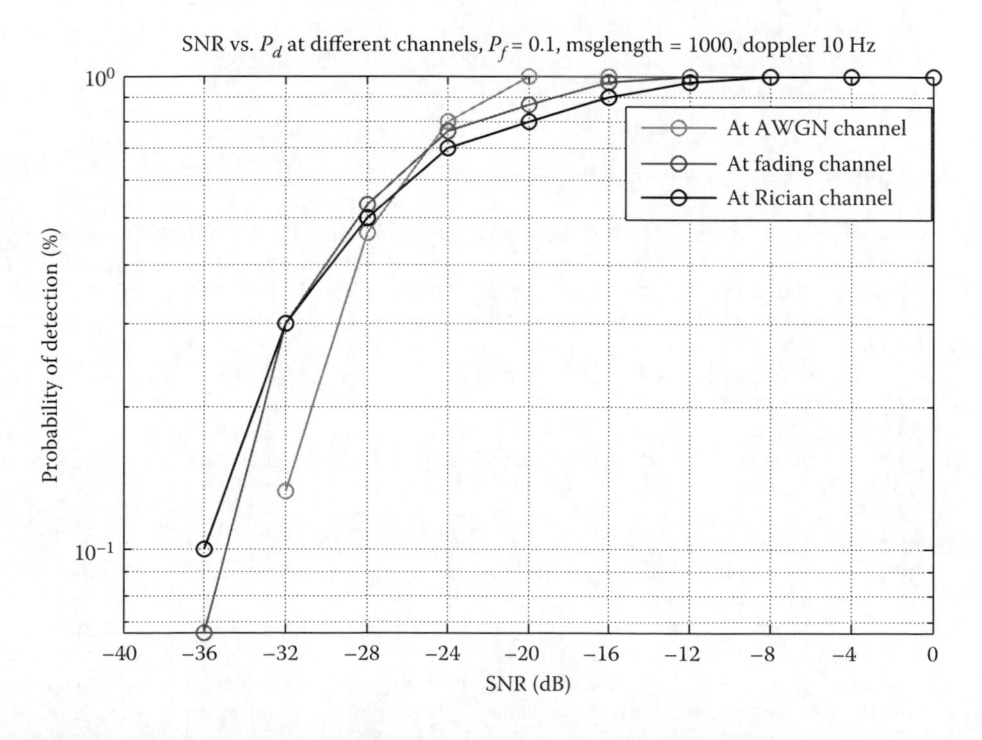

FIGURE 17.12 SNR versus P_d, message length is 1000 for different noisy channels.

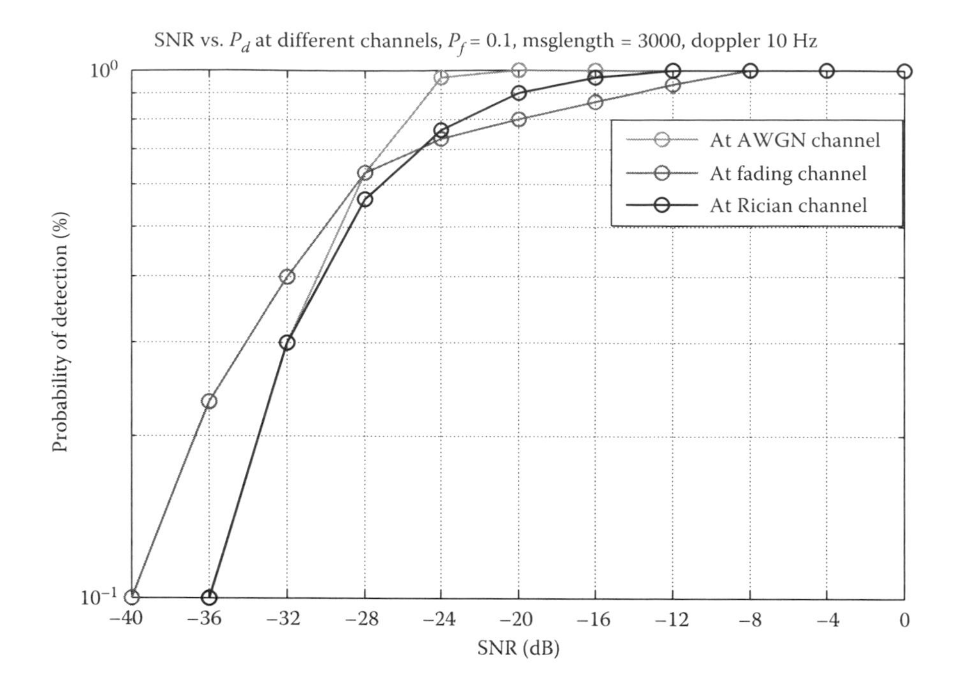

FIGURE 17.13 SNR versus P_d, message length is 3000 for different noisy channels.

17.9 CONCLUSION

It is seen clearly from the simulation results of the system that the proposed system is more powerful than traditional detection systems. This is due to the statistical feature used in this design. Numerical results show that the proposed approach can guarantee a reliable sensing while greatly enhancing the spectrum utilization such as $P_d = 100\%$ for $P_f = 0.1$ at SNR = −25 dB for message length = 3000 samples, another case when P_f is 0.1.

REFERENCES

1. E. R. Lavudiya, K. D. Kulat, J. D. Kene, Implementation and analysis of cognitive radio system using MATLAB, *International Journal of Computer Science and Telecommunications*, 4(7): 2013; 23–28.
2. L. Chu, Project, http://iwct.sjtu.edu.cn/Personal/xwang8/research/CR_zhu/project.html.
3. H. Zahir, Implementation of energy detector for cognitive radio, MSc thesis, University of Baghdad, 2014.
4. V. R. Reddy, Resource allocation for OFDM-based cognitive radio systems, National Institute of Technology, Rourkela, 2011.
5. P. R. Nair, A. P. V. Anoop, K. Kumar, An adaptive threshold based energy detector for spectrum sensing in cognitive radios at low SNR, *IEEE International Conference on Communication Systems*, Singapore, 2010; 574–578.
6. J. O. Neel, Analysis and design of cognitive radio networks and distributed radio resource management algorithms, PhD thesis, Virginia Tech, 2006.
7. S. Haykin, Cognitive radio: Brain-empowered wireless communications, *IEEE Journal on Selected Areas in Communications*, 23(2): 2005; 201–220.
8. A. Puttupu, Improved double threshold energy detection in cognitive radio networks, MSc thesis, National Institute of Technology, Rourkela, 2013.
9. A. Bagwari, G. S. Tomar, S. Verma, Cooperative spectrum sensing based on two-stage detectors with multiple energy detectors and adaptive double threshold in cognitive radio networks, *Canadian Journal of Electrical and Computer Engineering*, 36(4): 2014; 172–180.

10. C. Danijela, T. Artem, R. W. Brodersen, Experimental study of spectrum sensing based on energy detection and network cooperation, *Proceedings of the First International Workshop on Technology and Policy for Accessing Spectrum* (TAPAS '06), Boston, August 5, 2006; article 12.
11. J. Mitola III. Cognitive radio: An integrated agent architecture for software defined radio, PhD dissertation, Teleinformatics, Royal Institute of Technology (KTH), 2000.
12. M. Daniela, M. Plataa, G. Ángel, A. Reátiga, Evaluation of energy detection for spectrum sensing based on the dynamic selection of detection-threshold, *Procedia Engineering*, 35: 2012; 135–143.
13. A. Bansal, R. Mahajan, Building cognitive radio system using Matlab, *International Journal of Electronics and Computer Science Engineering*, 1(3):2009; 1555–1560.
14. M. Chakraborty, R. Bera, P. Pradhan, R. Pradhan, S. Sunar, Spectrum sensing and spectrum shifting implementation in a cognitive radio based IEEE 802.22 wireless regional area network, *International Journal on Computer Science and Engineering*, 2(4): 2010; 1477–1481.
15. A. A. Thabit, H. T. Ziboon, Improving probability of detection using CFAR and adaptive threshold for cognitive radio (CR), *International Journal of Engineering and Advanced Technology (IJEAT)*, 4(5): 2014; 1–4.
16. A. A. Thabit, Design a DPLL synchronizer based on software defined radio, MSc thesis, University of Technology, Baghdad, Iraq, 2008.
17. K. Kokkinen, Implementation of autocorrelation-based feature detector for cognitive radio, MSc thesis, Aalto University School of Science and Technology, 2011.
18. F. B. S. de Carvalho, W. T. A. Lopes, M. S. Alencar, Performance of cognitive spectrum sensing based on energy detector in fading channels, *International Conference on Communication, Management and Information Technology, Procedia Computer Science*, 65: 2015; 140–147.
19. I. A. Hashim, Automatic digital modulation identification for software defined radio based on FPGA, PhD dissertation, University of Technology, Baghdad, Iraq, 2016.
20. J. Eriksson, E. Ollila, V. Koivunen, Statistics for complex random variables revisited, *Proceedings of the 34th IEEE International Conference on Acoustics, Speech, and Signal Processing (ICASSP 2009)*, Taipei, Taiwan, April 19–24, 2009.
21. N. P. Geisinger, Classification of digital modulation schemes using linear and nonlinear classifiers, Thesis, Naval Postgraduate School, 2010.

the existing simulator lacks security issues. In addition, the DTN routing protocols' performance is analyzed. The evaluation of the epidemic routing protocol performance measures is based on three metrics: success delivery probability, message average latency, and overhead ratio to deliver the message to destination. The main objectives can be divided into (1) design and implement black hole attack behavior by extending the existing simulator, and (2) study of the impact of a black hole attack; compare the performance of the epidemic routing protocol when the network is compromised by malicious nodes and then without black hole nodes using different movement patterns and various ranges of black holes. In this chapter, we are interested in studying the performance of some routing protocols in the DTN, where the end-to-end path does not exist and is difficult to predict. However, communications between nodes can occur. In addition, we are also interested in investigating the black hole attack and its influence on the performance of the network.

The rest of the chapter is organized as follows: Section 18.2 represents the related work on DTNs, DTN routing protocols, DTN routing attack, and DTN security. Section 18.3 describes the adopted methodology throughout this research. Section 18.4 includes design and implementation of black hole behavior by the extending ONE simulator, where the proposed work and the extended simulator are explained. Section 18.5 is an evaluation and discussion about the results of behavior of a black hole attack and its influence on DTN performance. Finally, the chapter is concluded in Section 18.6.

18.2 RELATED WORK

In this section, we shed light on the work done by researchers on DTN and security issues. Since DTN is an extension of mobile ad hoc networks (MANETs), we therefore briefly highlight work done on MANETs as well.

Routing protocols in MANETs have been classified into two groups, which are adapted from Dokurer [7] and Kodole and Agarkar [8]. Ad hoc on-demand distance vector (AODV) and dynamic source routing (DSR) are examples of reactive protocols. They maintain routing information only between nodes that want to communicate [9]. Due to lack of preexisting infrastructure, the routing information becomes vulnerable to some attacks, such as a black hole attack. A black hole will try to route false information, disrupt the discovery of the route (path), or advertise itself as having the shortest path using the routing protocol [10]. Therefore, with the need for securing MANETs, many researchers sought to secure routing protocols, such as DSR, AODV, and destination-sequenced distance vector (DSDV), since these protocols were initially developed devoid of security features. Another study proposed two mechanisms to secure the AODV protocol using digital signatures for authentication and hash chains for hop counts [9]. Selvavinayaki et al. have presented a solution to prevent black hole attacks by using security certificates in digital form [11]. Moreover, a survey by Raja Mahmood and Khan presented seven methods to detect a black hole attack in MANETs [12].

Although DTNs and MANETs share common properties, they differ from each other in the way they handle messages. In MANETs, the communication between nodes is possible only if the path exists to the destination; routing protocols, such as DSR and AODV [2,13], assume the existence of the complete path from the source to the destination before sending messages. In addition, nodes have short usage time, which means the message could be easily lost. Moreover, the high mobility of the nodes in MANETs can cause loss of messages. When a message arrives at the node, and if there is no end-to-end contemporaneous path to the destination, the message will be dropped; this results in lost messages. Therefore, some researchers consider that these protocols fail to work properly in a DTN [14–16]. On the other hand, DTNs allow communication between intermittently connected nodes, even if there is no preexisting path to the destination. Using the existing DTN routing protocols, a new approach has been developed in order to overcome the problem of losing messages and to enable routing in the DTNs [3]. The approach is called "store–carry–forward," where the message can be held for a long time by the node (hours and sometimes days) until it encounters another the existing simulator lacks security issues. In addition, the DTN routing protocols' performance is analyzed. The evaluation of the epidemic routing protocol performance measures is based on three metrics: success delivery probability, message average latency, and overhead ratio to deliver the message to destination. The main objectives can be divided into (1) design and implement black hole attack behavior by extending the existing simulator, and (2) study of the impact of a black hole attack; compare the performance of the epidemic routing protocol when the network is compromised by malicious nodes and then without black hole nodes using different movement patterns and various ranges of black holes. In this chapter, we are interested in studying the performance of some routing protocols in the DTN, where the end-to-end path does not exist and is difficult to predict. However, communications between nodes can occur. In addition, we are also interested in investigating the black hole attack and its influence on the performance of the network.

The rest of the chapter is organized as follows: Section 18.2 represents the related work on DTNs, DTN routing protocols, DTN routing attack, and DTN security. Section 18.3 describes the adopted methodology throughout this research. Section 18.4 includes design and implementation of black hole behavior by the extending ONE simulator, where the proposed work and the extended simulator are explained. Section 18.5 is an evaluation and discussion about the results of behavior of a black hole attack and its influence on DTN performance. Finally, the chapter is concluded in Section 18.6.

18.2 RELATED WORK

In this section, we shed light on the work done by researchers on DTN and security issues. Since DTN is an extension of mobile ad hoc networks (MANETs), we therefore briefly highlight work done on MANETs as well.

Routing protocols in MANETs have been classified into two groups, which are adapted from Dokurer [7] and Kodole and Agarkar [8]. Ad hoc on-demand distance vector (AODV) and dynamic source routing (DSR) are examples of reactive protocols. They maintain routing information only between nodes that want to communicate [9]. Due to lack of preexisting infrastructure, the routing information becomes vulnerable to some attacks, such as a black hole attack. A black hole will try to route false information, disrupt the discovery of the route (path), or advertise itself as having the shortest path using the routing protocol [10]. Therefore, with the need for securing MANETs, many researchers sought to secure routing protocols, such as DSR, AODV, and destination-sequenced distance vector (DSDV), since these protocols were initially developed devoid of security features. Another study proposed two mechanisms to secure the AODV protocol using digital signatures for authentication and hash chains for hop counts [9]. Selvavinayaki et al. have presented a solution to prevent black hole attacks by using security certificates in digital form [11]. Moreover, a survey by Raja Mahmood and Khan presented seven methods to detect a black hole attack in MANETs [12].

Although DTNs and MANETs share common properties, they differ from each other in the way they handle messages. In MANETs, the communication between nodes is possible only if the path exists to the destination; routing protocols, such as DSR and AODV [2,13], assume the existence of the complete path from the source to the destination before sending messages. In addition, nodes have short usage time, which means the message could be easily lost. Moreover, the high mobility of the nodes in MANETs can cause loss of messages. When a message arrives at the node, and if there is no end-to-end contemporaneous path to the destination, the message will be dropped; this results in lost messages. Therefore, some researchers consider that these protocols fail to work properly in a DTN [14 16]. On the other hand, DTNs allow communication between intermittently connected nodes, even if there is no preexisting path to the destination. Using the existing DTN routing protocols, a new approach has been developed in order to overcome the problem of losing messages and to enable routing in the DTNs [3]. The approach is called "store–carry–forward," where the message can be held for a long time by the node (hours and sometimes days) until it encounters another relay to forward the message to and it reaches its destination. All the DTN routing protocols follow the store–carry–forward paradigm, which increases the delivery probability of the message. This distinguishes DTN from other traditional networks and makes it more challenging, as its tolerance is quite long.

Jain et al. explain the DTN routing problem, where a message is to be sent end-to-end, on a time-varying directed multigraph [15]. There may be more than one edge between two nodes, because of the availability of different links at different times. Routing decisions can be made based on the information available about the connection and the mobility of the nodes. However, sometimes such information may not be available or known to the nodes.

In recent years, some of the studies started to consider security when designing routing algorithms. However, they did not get adequate attention until now. One of the challenges of DTN is providing efficient routing due to its nature of intermittent connectivity of the nodes, lack of stable connection between source and destination, and low node density [17]. Thus, the DTN becomes vulnerable to a number of attacks, such as black hole attacks [18,19].

The black hole attack is one of the active routing attacks, where a compromised node can use fake information in the routing table, such as message delivery probability to increase its chance of being selected as the next hop node to deliver a message. Once a fake route has been established, it receives messages from other nodes, and it is then able to drop messages or utilize them to launch other attacks [20]. In traditional networks and MANETs nodes utilize evidence, such as temporal leashes [21], to detect such an attack. However, in a DTN such evidence is extremely difficult to collect. A gray hole attack [22] is one of the routing attacks. It has two phases, wherein the first phase, a node advertises itself to have the shortest path to the destination [23]. In the second phase, it drops messages received from a specific source or forwards these to a certain destination. However, this attack is considered more difficult in detecting than a black hole attack, because the drop process occurs with a certain probability.

Burgess et al. explain that a DTN is robust against the presence of malicious nodes, due to the opportunistic nature of DTNs, which can reduce the negative effects of such attacks [24]. The study shows that authentication mechanisms of securing routing are difficult to deploy and function in a DTN. Contrary to research by [24], the work in [25] studies the robustness of DTN against attack without authentication mechanisms. The study shows that replication-based routings can be exposed to attack if the authentication mechanisms are not used. This results in decreasing delivery probability. Another scheme called encounter tickets secures evidence of all contacts in the network [21]. Malicious nodes can provide forged contact history to increase the likelihood of being selected by other nodes.

18.3 METHODOLOGY

This section focuses on the technical tools and the methodology that have been used in this research.

18.3.1 Choice of Simulator

The Opportunistic Network Environment (ONE) simulator includes the traditional mobility models, and map-based model as well, which has the advantage of giving more realistic movement and results. The data provided by the Helsinki map allows simulation of the nodes movement when the nodes communicate with each other, then the contact information can be derived from the result of the simulation. On the other hand, this information can be derived using other approaches based on a random process. This approach gives values that may not be precise and difficult to prove because the behavior of the human is not always random. Moreover, the information can be derived from real-world traces. Although this approach depends on real human interaction, it has problems related to the devices used to track people. Among these problems, the power for the devices is kept low to save battery life; therefore, the information may be not precise as well. In addition, the number of traces is limited [26]. The ONE simulator provides data from a real map (Helsinki map), where nodes can move along roads. Moreover, it provides diverse movement patterns, where trams, pedestrians, and cars can move on the roads. This data provides more realistic movement models. Therefore, the approach based on the simulation of nodes movement will provide more realistic results.

18.3.2 Simulation Scenario

A simulation scenario is constructed by defining node groups and their properties. This may include the movement module, routing models, simulation time, interface type, event generator, and other parameters. The parameters are defined in the configuration file default_setting.txt. Further simulation scenario settings are defined in a separate (optional) configuration file to add more settings or to override (some or all) settings in the default_settings.txt file. The behavior of all modules is implemented using the high-level language (Java). However, the behavior can also be adjusted in the comparison between Batch Mode and GUI Mode. The Batch Mode Time is 431.81 and the GUI Mode Time is 43200.1. The Batch Mode Speed is 147.021/s and the GUI Mode Speed is 72.201/s. Moreover, the simulator has a feature called run indexing, which is defined by the parameters in the configuration file. Thus, it can allow for sensitivity analysis. When the simulator starts, it reads "default settings" and takes the optional file if it exists as an input parameter. The result will be generated in the MessageStatsReport, which contains statistics about the simulation as mentioned before.

18.4 DESIGN AND IMPLEMENTATION

After gaining a deep understanding of the ONE simulator and reviewing its functionality, it is clearly noticeable that the current ONE simulator is capable of simulating a number of scenarios; however, it has some limitations in its design, as it does not provide the behavior required to simulate a black hole attack. Based on our area of interest, which is studying the behavior of the black hole attack, there is a need to extend the simulator functionally to incorporate behavior of the black hole attack so that we can study and analyze the behavior of malicious nodes.

Black hole nodes act as malicious nodes, and in their behavior will drop all the messages received from other nodes. This means that the nodes have to establish a connection first to provide communication with other nodes, and then start sending messages. Thus, as an initial idea for the design, it was suggested to implement the logic of black hole attack behavior when the node starts communication with the other nodes. After the connection is achieved, instead of forwarding a message, it will drop it. The abstract ActiveRouter class is the super class of all active routers, including the epidemic router. This class has a method responsible for checking if the host is ready for transferring based on the information available about the connection. Besides that, it checks if the transferring host is not a black hole so it can start transferring. Otherwise, it will drop the message. However, this approach depends on the connection between nodes, and since the connection can be affected by many factors, it may affect the transfer process and the result from this implementation may not be accurate. Thus, the idea of the design was directed to another approach. The MessageRouter Class has a method called createNewMessage. It is responsible for creating new messages in the router. If the creation process is done successfully, the method will return true. Otherwise, it returns false, for example, if the message size is too big to be carried in the buffer. Therefore, if the host (router) is a black hole, it would not allow the creation of the message. Otherwise, the creation process will be performed normally.

The DTNhost class is responsible for creating new hosts (nodes). Each host belongs to a specific group with a given groupID. Therefore, two methods need to be added to the class. One of the methods is responsible for checking if the created host (router) is acting as a black hole (malicious node), and the other will control whether the message will be dropped. Since a black hole attack should not affect the message creation process and should drop all incoming messages.

18.4.1 Implementation of the Extended Simulator

The ONE simulator lacks the representation of black hole behavior; therefore, a new black hole epidemic routing module has been implemented in order to provide behavior of the malicious nodes. In addition, a new feature in the MessageStatReport module has been added to give statistics about the messages dropped by black holes.

The experiments to be conducted using the simulator include different groups. These groups are defined in the settings file. Each group consists of a number of nodes that vary from one group to another, as defined in the settings file. When conducting experiments, it is usually assumed that all the groups in the network are infected by a black hole attack. Therefore, in order to determine the percentage of the black hole attack by all groups or certain groups, a new property (key) with the percentage value of the black hole (percOfBlackHoles) has been defined in the setting file.

18.4.2 Black Hole Module

The ONE simulator makes extending and configuring of the simulator with different features easy for developers. Therefore, one of the goals when implementing the new code is to provide flexible code to facilitate the use of the extended simulator by other users. For this reason, the IBHRouter interface has been implemented in a separate package. It should be noted that when creating a new model, a new class has to be created [26] and the name of the class has to be defined in the setting file. When the scenario starts, the simulator will automatically load the new configuration from the setting file.

18.4.3 Message Listener

Message Listener is an interface for the classes that want to be informed about message events, for example, the creation and deletion of messages. As in our implementation, the BHEpdiemic.java class contains the logic for handling black hole behaviors, which result in message drops. Therefore, the message listener needs to be informed when a message is dropped.

18.4.4 Black Hole Attack Report

As pointed out earlier, the existing report generates information about the number of messages created, deleted, aborted, dropped, and relayed. Moreover, it gives information about the delivery probability, overhead ratio, and the latency. All these metrics are critical when analyzing behavior of the black hole attack, but do not provide any information about the black hole behavior in terms of the number of public boolean, such as isBlackHole() public, BHEpidemicRouter replicate() public, void messageDroppedByBlackHole (Message aMessage, DTNHost where), and dropped messages. Therefore, this report is extended to include new statistics about the number of messages dropped by black hole attacks. It is notable that the number of dropped messages by black holes is always less than the total of dropped messages. In addition, in the event log panel within the graphical user interface, black hole events are displayed as shown in Figure 18.1.

18.5 EVALUATION AND DISCUSSION

After the design and implementation of the extended simulator, we conducted our experiments. In this section, we aim to focus on two parts. First, testing the performance of DTN routing protocols (Epidemic, Spray and Wait, Direct Delivery, First Contact, MaxProp, and PRoPHET) in the presence of a black hole attack. Second, investigating and evaluating the effects of the black hole attack on DTNs using the epidemic routing protocol.

Event Log

Event	Node		Node	Message
2860.3: Message relay started	c67	< - >	c64	M30
2860.4: Connection DOWN	bus87	< - >	bus85	
2860.4: Message relay aborted	bus87	< - >	bus85	M16
2860.4: Connection UP	policeP139	< - >	bus93	
2860.4: Message relay started	policeP139	< - >	bus93	M17
2860.8: Connection UP	c42	< - >	policeP126	
2860.8: Message relay started	c42	< - >	policeP126	M52
2860.9: Message dropped by Black Hole	**p39**	M3		
2860.9: Message relay started	policeP118	< - >	p39	M24
2862.4: Connection DOWN	c42	< - >	policeP126	
2862.4: Message relay aborted	c42	< - >	policeP126	M52
2863.1: Message dropped by Black Hole	**c64**	m30		
2863.1: Message relay started	c67	< - >	c64	M30
2863.1: Connection UP	p38	< - >	p14	

FIGURE 18.1 Screenshot from GUI event log displaying black hole events.

18.5.1 Experiment Aims and Setup

The main aim of the following experiments is to explore the influence of black hole attacks on DTNs. Therefore, many scenarios and experiments have been conducted to analyze and explore the effect of such attack. It is assumed that all groups communicate with each other, but one (could be more or the entire group) of the groups behaves as a malicious group in order to prevent communication between other groups by dropping messages. For instance, the pedestrian group communicates with the bus group, while the attack source is from the third group (cars). As a result, it is expected that the malicious group will affect the communication between the other groups. Hence, there is no technique to detect this attack.

In order to evaluate the effect of a black hole attack in a DTN, three realistic movement models have been used, namely, MapBasedMovement, ShortestPathMapBasedMovement, and MapRouteMovement models. The movements are in the city of Helsinki to give accurate results closer to reality. The simulation space is an open area with dimensions (4500 m × 3500 m). Six groups have been created with 140 nodes in total. It is highly preferred to use a reasonable density of nodes, as this can affect the overall connectivity of the network, and thus affect message delivery ratio [27]. The groups have both common and specific settings. Groups include two groups of pedestrians with 40 nodes for each group; a bus group with 20 nodes; two groups of trams with 5 nodes in each group; and the last group comprising police patrols with 30 nodes.

The groups move in the map paths, which are roads and tram lines. The roads are specified for the cars, which use MapBasedMovement (road traffic), buses follow MapRouteMovement (with predefined routes and scheduled trips inside Helsinki), and pedestrians move using MapBasedMovement (footpaths). The cars can only drive on the roads at 10–50 Km/s with waiting times of 10–120 s. The police patrol group uses ShortestPathMovement. The ShortestPathMovement is used to represent how patrols move around the city, while they randomly drive around and stop for a few moments. The trams move at 25–36 km/h with pause times of 10–30 s in the predefined routes.

Three metrics have been considered in order to evaluate the performance of the network when it is under a black hole attack and also when there are no attacks. The metrics are (a) success delivery probability; (b) average latency, which is defined as a delay in the message transfer from one node to another in specific time [24,28,29]; and (c) overhead ratio. These metrics are critical in measuring the performance of the DTN routing protocols and the effect of the black hole attack on the network. The overhead ratio is an interesting metric to be evaluated as well. Since we are exploring the effect of the black hole attack, this attack has a significant impact in the number of messages created and delivered.

TABLE 18.1
Message Statistics Report for DTN Routing Protocols

Stat. Metric	Epidemic	SW	MaxProp	DD	FC	PRoPHET
Created	1465	1465	1465	1465	1465	1465
Started	264914	26301	351596	761	34556	293816
Relayed	242173	11012	328294	274	17448	273073
Aborted	22737	15288	23295	487	17106	20739
Dropped	240973	9839	312394	888	914	272001
Dropped by BH	0	0	0	0	0	0
Removed	0	0	14778	0	17448	0
Delivered	419	616	750	274	233	425
Delivery_prob	0.2860	0.4205	0.5119	0.1870	0.1590	0.2901
Overhead_ratio	576.9785	16.8766	436.7253	0.0000	73.8841	641.5247
Latency_avg	5880.1585	4713.4021	5943.1000	6697.8858	6041.3545	0.2901

18.5.2 Black Hole Attack

Our goal from the experiments is to observe changes in the delivery probability and the average latency when the percentage of black hole nodes increases and decreases. We choose malicious nodes randomly to act as black holes with the percentages of 10%, 25%, 50%, and 100%, where 100% indicates that the entire network is infected with the black hole attack.

We are going to test different scenarios where, for example, the attack comes from the pedestrians group, and this group communicates with the other groups in the network. It is expected that delivery probability will decrease as this group will deny forwarding messages to the other groups by dropping all messages. It is worth mentioning that the black hole attack implemented in this project is behaving randomly, where the number of attackers in each group is random. However, the level of the attack can be customized by changing the percentage of the attack.

18.5.3 Protocol Performance

First, the performance of the original DTN routing protocols including Epidemic, Spray and Wait (SW), and Direct Delivery (DD) protocols was studied in the presence of attack. The protocols' performance can be measured by evaluating three metrics: (1) successful delivery rate (2) average latency (3) overhead ratio. Table 18.1 shows a summary of the message statistics report of six routing protocols.

Figure 18.2 presents a comparison of message delivery probability among DTN routing protocols. It is clearly observed that MaxProp has a significantly greater message delivery percentage

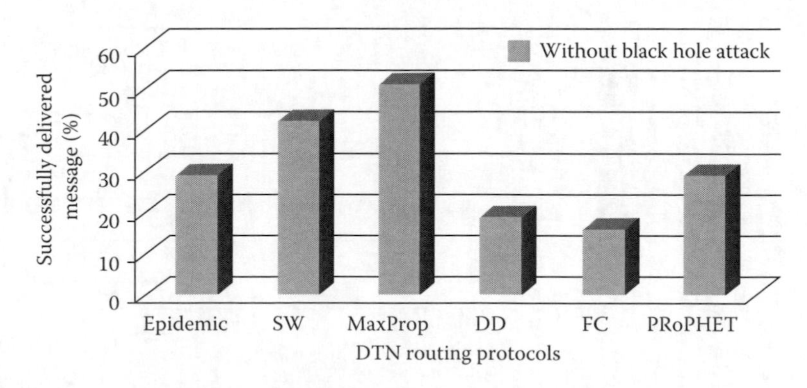

FIGURE 18.2 Delivery probability for DTN protocol.

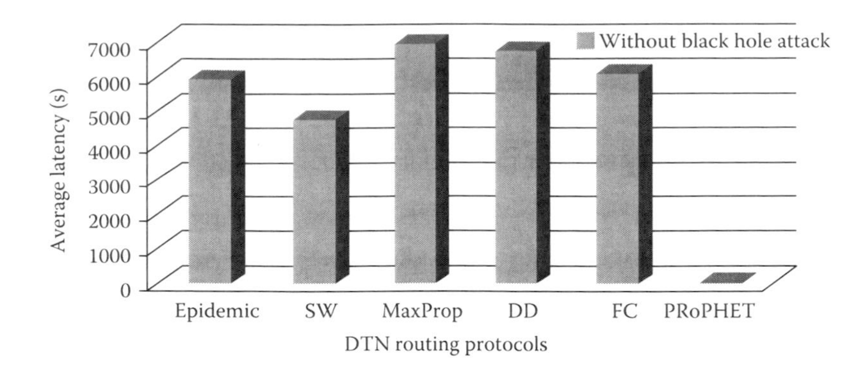

FIGURE 18.3 Average latency for three protocols without an existing black hole attack.

than other protocols at 51%. The reason is that MaxProp uses several mechanisms to increase the delivery rate. It uses a technique of PRioritized message send and delete process. This was closely followed by the Spray and Wait protocol, which restricts the number of message copies in order to increase the delivery percentage. In our experiment, we considered 10 copies in replication, which results in increasing the delivery probability of the protocol to 42%. The PRoPHET and Epidemic protocols are almost the same in delivering messages. However, PRoPHET is slightly superior, as it is able to deliver more messages with a lower communication overhead, because it follows probabilistic routing (Figure 18.4) as it can be clearly seen, the overhead ratio for the Epidemic of 576.9785 decreased dramatically to 0.2901 for PRoPHET. Finally, Direct Delivery and First Contact (FC) are single-copy based protocols, and they had the smallest delivery percentage, 19% and 16%, respectively.

Figure 18.3 shows a comparison of message average delay between DTN routing protocols. It can be observed that the direct delivery protocol has the highest average latency. This protocol is single-copy based, where only one copy of the message exists in the network and is forwarded to the destination only when the source communicates directly with the destination [30]. However, this probability is very low in a DTN. Therefore, the result makes sense. However, it is clear that the Epidemic routing protocol has smaller average latency than Direct Delivery due to unlimited replication of the messages in the network to deliver the message with high probability. On the other hand, the Spray and Wait protocol, with a controlled number of message copies, has the smallest average latency.

Figure 18.4 compares the overhead ratio for a DTN protocol. It can be seen that the Epidemic protocol has the highest overhead (576.97), as this protocol floods an unlimited number of message copies. However, this ratio drops markedly for the Spray and Wait protocol to 16.87, as this protocol floods a controlled number of message copies. Followed by the single-based protocol Direct

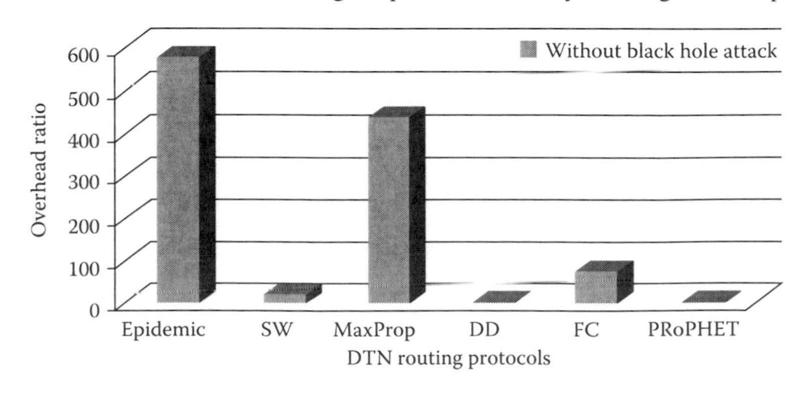

FIGURE 18.4 Overhead ratio for DTN routing protocols.

Delivery with zero overhead, where this protocol uses only one copy to deliver to its destination, and so only one copy is available in the network. The prediction-based protocol (PRoPHET) has the lowest overhead ratio (0.29) after Direct Delivery.

18.5.4 Black Hole Attack Evaluation Criteria

In order to compare the performance of the Epidemic routing protocol in both scenarios, when trust between nodes is limited (under black hole attack) and without attack, it is important to present metrics to evaluate performance. In the following, two important metrics—successful message delivery probability and the average latency—are explained.

18.5.5 Message Delivery Ratio

In order to evaluate the effect of a black hole attack, it is extremely important to measure message delivery probability. It is rare to lose a message in the DTN environment, but the situation will be different under the influence of a black hole attack.

Figure 18.5 shows the effect of the black hole attack on message delivery probability for the epidemic protocol. The results present different percentages of the attacker. According to the graph, it can be observed that there is a significant decrease in delivery probability. If for the entire network is infected by the black hole attack (percentage of the attack = 100%), all the nodes drop all messages and the delivery probability drops dramatically from 0.27 (with 0% attack) to 0.00 (with 100% attack). On the other hand, if 10% to 100% of the nodes in a certain group in the network act as black holes, the average successful delivery rate will decrease. This result shows that the throughput of the network can significantly degrade when a black hole attack exists.

18.5.6 Average Latency

The average latency has a special importance in the DTN. It indicates network speeds and time between message creation and when it is received by its destination. Therefore, average latency can be measured based on groups' buffer size and the number of nodes in these groups. Since DTN suffers from long delay due to its challenging nature in routing messages, it is expected to have a high average latency.

Figure 18.6 illustrates the impact of the black hole attack on the average latency for the Epidemic routing protocol. It can clearly be noticed that the average latency is slightly increased when the percentage of the black hole attack increases. In the case of 25% of node groups acting as black holes, the average latency increases by about 300 seconds, from 5521.0476 seconds for a 10% attack to

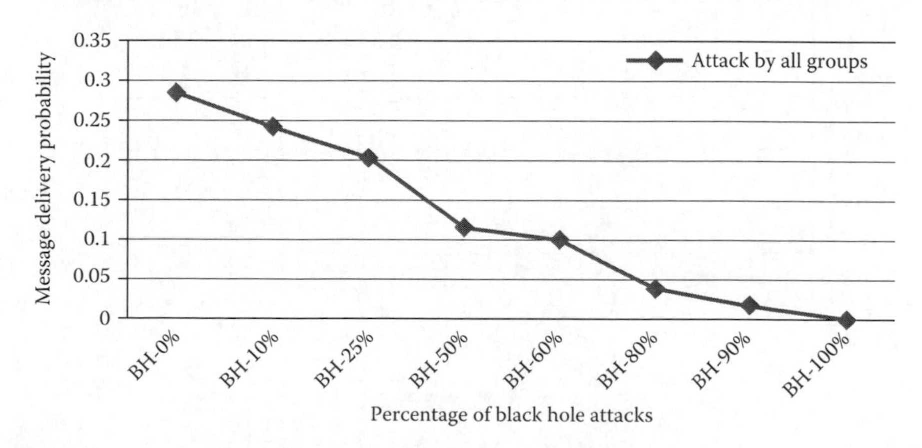

FIGURE 18.5 Delivery probability for epidemic protocol under black hole attack.

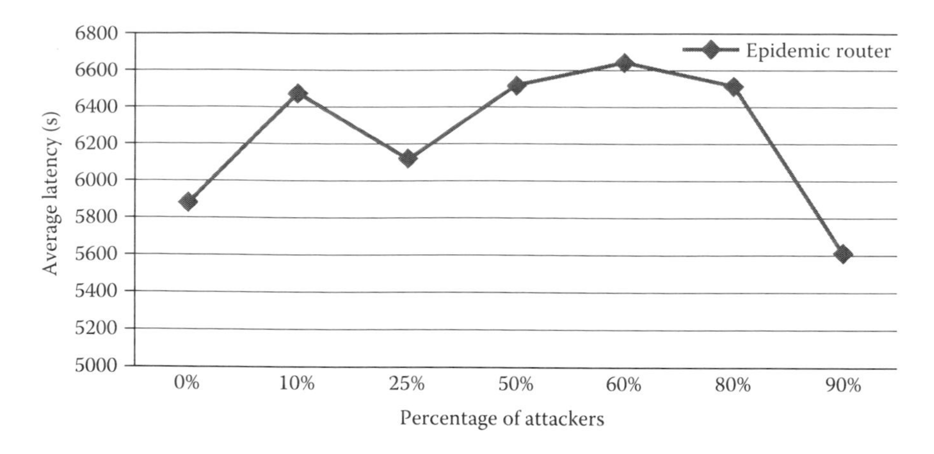

FIGURE 18.6 Average latency of epidemic router under black hole attack.

5884.2181 seconds for a 25% attack. The overall trend in the average latency consistently increases by about 200 seconds for every 10% increase in attack.

It can be argued that the overall trend of the graph for Epidemic routing is fluctuating. There is a peak at 6642 seconds at near 60% attack. However, the delay drops to 5610 seconds at near 90% attack. It has also been noticed that when the percentage of attack is 100%, the result of this attack is that no message is delivered. Therefore, the value for average latency is unavailable, as message average latency is counted based on delivered message.

18.5.7 Impact of Number of Black Holes within All Groups

Figure 18.7 shows the percentage of message delivery when the network is under a black hole attack. When the percentage of black holes is 0% (no attack), the percentage of successful delivery is about 28%. It is observed that there is a slight decrease from 28% to 25% in the percentage of messages delivered when the percentage of black holes is 10%. At 50%, where all node groups are

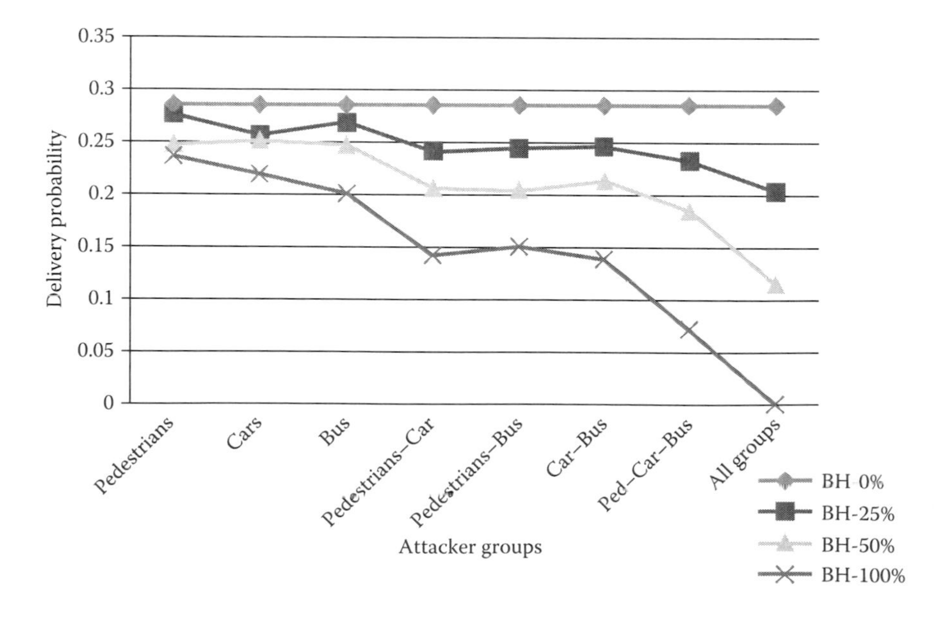

FIGURE 18.7 Delivery probability under black hole attack from certain groups.

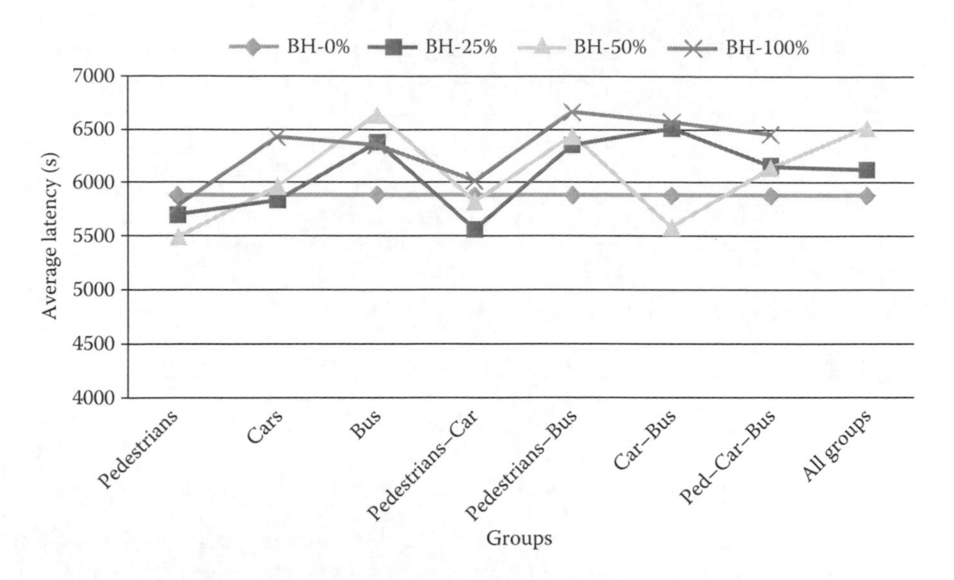

FIGURE 18.8 Average latency under different levels of attack.

compromised by a black hole attack, the delivery percentage drops sharply to 15%. As a result, it can be seen that the Epidemic routing protocol is vulnerable in the face of black hole attack.

Figure 18.8 shows message delay for all groups under different levels of attack. It is observed that when black hole percentage is 100% in all groups, the delay remains the same. However, when the percentage of attack varies within the group, it is clearly observed that the overall trend of the graph of message delay fluctuated, and it has no stable pattern.

18.5.8 Impact of Percentage of Black Holes within Certain Groups

On the other hand, since a black hole attack affects the whole network, this means that when the attack comes from one group within the network all the other groups will be affected by it. Therefore, one of the aims is to analyze and evaluate the effect of such an attack, originating from certain groups in the network, while others are not compromised. As previously pointed out, it is assumed in the first scenario that the source of the attack comes from group1, which represents pedestrians (p) with different percentage levels of attack. The second scenario considers the attack source is the car group (c). The last scenario is where the bus group acts as a malicious group. It was also assumed that the police patrol and tram groups do not attack the network. Figure 18.8 shows the influence of a black hole attack, when it comes from pedestrians, car, and bus groups (each group separately) on the delivery probability. It is clearly observed that the number drops slightly from 0.2771 with 25% of the pedestrian nodes group acting as black holes to 0.2369, where this group is 100% compromised by a black hole attack, which means preventing messages being sent. Similarly, when the attack source comes from the bus group, the delivery probability decreases by slightly more than the pedestrian and car group from 0.2698 (with 25% attack) to 0.2014 (100% attack).

On the other hand, the same figure illustrates the probability of delivery for the Epidemic router under black hole attack from more than one group. It can be clearly seen that whenever the number of black hole groups increase, the decline in delivery probability is far greater. However, the different sources of black hole attack made for a relative decline as a percentage is varied. From this result, it can be concluded that the Epidemic protocol is vulnerable to black hole attack.

Figure 18.9 shows when black hole nodes are selected from different groups (from pedestrians and car groups; or pedestrians and bus groups; or car and bus; or pedestrians, car, and bus groups). When the attack source, for example, is from the car and bus groups, the average latency consistently increased as the percentage of attack increased; from 5946.9803 s with 10% attack

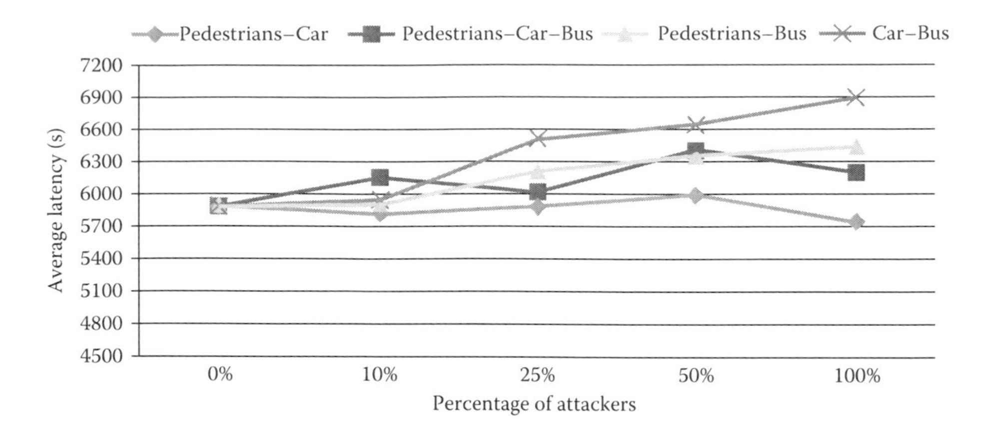

FIGURE 18.9 Average latency for epidemic protocol under black hole attack from certain groups.

to 6889.8568 s with 100% attack. Moreover, there was a slight increase between 20 and 150 s per each level of attack, where the attack source is from pedestrians and bus groups. The delay remains almost the same when the number of attacking nodes grows in pedestrians and car groups, but fluctuates when it comes from three groups.

On the other hand, Figure 18.10 shows the overhead ratio of the Epidemic protocol is not affected by a black hole attack when the source of attack comes from certain groups. It almost remains the same, and may increase or decrease by 1% to 2% with different levels of attack. However, the proportion is dramatically decreased by 8% (from 576.9785 at 0% attack to 74.7176 at 50% attack). However, when the entire network is infected by a black hole attack, the overhead is negligible, as there is no resources consumption to deliver messages to destinations because all messages are dropped.

18.5.9 Total Message Drops and Drops by Black Holes

When a message arrives at a node, the node becomes subject to a number of tests, including the black hole test. If it is a black hole, the message will be dropped. Otherwise, it will be added

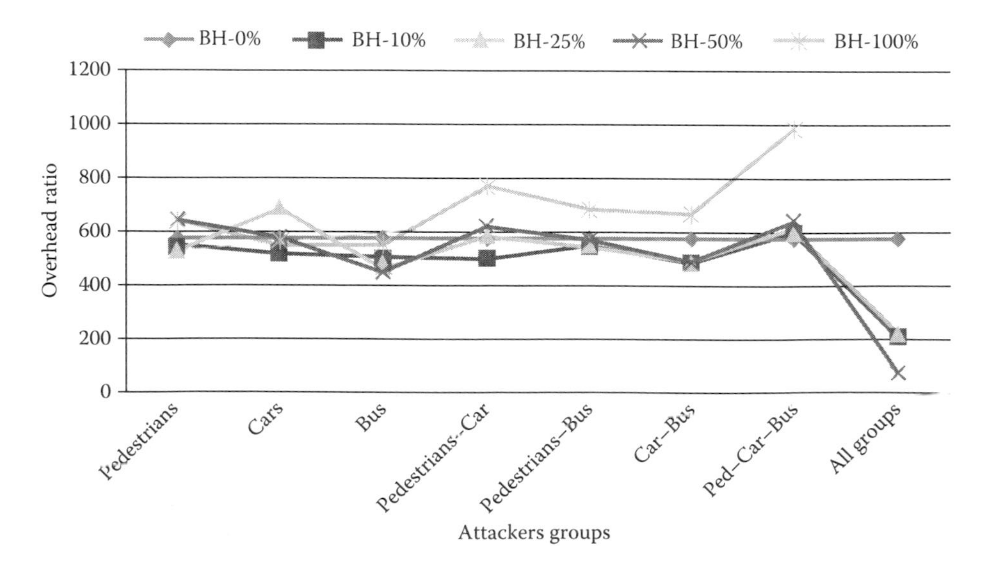

FIGURE 18.10 Overhead ratio under black hole attack from different groups.

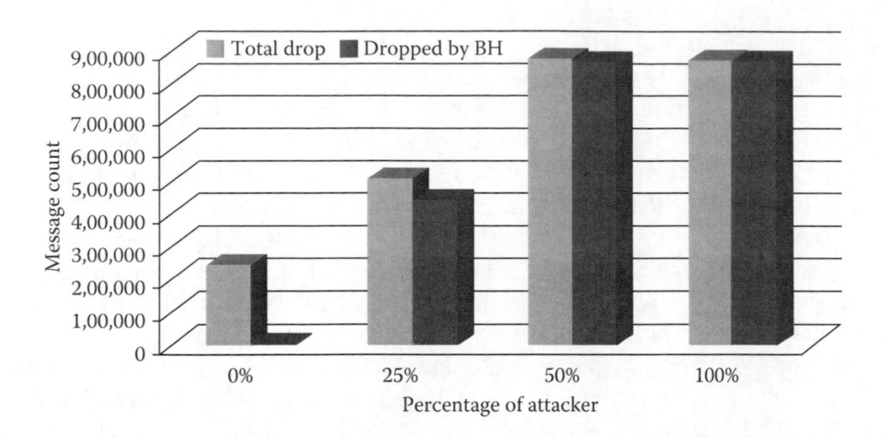

FIGURE 18.11 Total message drop/drop by black hole ratio.

to the buffer. Accordingly, messages dropped by black holes will usually be less than the total dropped. It can be observed from Figure 18.11 that when the percentage of attack is 0, the total drop is 240,973, where the drop by black holes is 0. For a 25% attack, the total drop for other reasons is 508,027, and the drop by a black hole is 440,895. Therefore, it is more likely, when the percentage of attack increases, that the number of dropped messages by black holes rapidly increases from 0 to 867,800 messages (at 100% attack), where the observed total drop is always greater than the drop by black holes.

The simulation results show that the epidemic routing protocol performs well in terms of delivering messages with reasonable average latency and high overhead ratio compared to other DTN routing protocols. However, the performance of the protocol in delivering messages is dramatically degraded under black hole attack. The percentage of messages delivered decreased as the percentage of attack increased. Moreover, the overhead ratio is considerably decreased in the presence of the attack. On the other hand, the latency for the protocol fluctuates under such attack. These results showed that the Epidemic protocol is vulnerable in the face of such attack.

18.6 CONCLUSION

Most DTN routing protocols consider delivery probability as a primary metric in routing. Such a metric can be abused by malicious nodes to launch a black hole attack causing a high proportion of message drop. In this chapter, we have evaluated the performance of DTN routing protocols in the presence of a black hole attack, including the Epidemic protocol. The evaluation is based on three metrics in terms of delivery probability, average latency, and overhead ratio. We have noticed that the MaxProp routing protocol delivers a significantly greater message percentage than other protocols at 51%, followed by Spray and Wait at 42%, while the Epidemic protocol delivers 29%. On the other hand, we have considered the results of the Epidemic routing protocol as a baseline and addressed the black hole attack problem. On this basis, we have developed and implemented a new epidemic routing module using the ONE simulator. A router is supposed to behave as malicious node (black hole) and deliberately drop all the messages. We have investigated the impact of the black hole attack in the DTN environment. A number of experiments and different scenarios have been conducted using the epidemic routing protocol, including an attack from all, as well as specific, groups. We have explored how different mobility patterns of nodes in terms of the speed, movement type, direction of movement, and range of black holes influence the successful delivery ratio, average latency, and overhead ratio. From the results, it has been noted that the epidemic protocol, which floods an uncontrolled number of message copies is susceptible to black hole attack. This attack has a negative effect on the performance of the Epidemic protocol in terms of successful message delivery probability, average latency, and overhead ratio. Moreover, these metrics are also influenced by the percentage of attackers. It has also been observed that this attack dramatically decreased delivery probability when the percentage of attackers increased due to the large amount of message drops. The result showed message delivery percentage declined by about 29% where the entire network is infected by the attack. Moreover, the result showed delivery probability decreased by between 5% and 9% when the source of attack is selected from one group and declined to between 15% and 16% when the attack came from two groups. It also drops by 22% when the attack is originated from the three groups. On the other hand, the message delay for the epidemic protocol under black hole attack remains almost the same when the number of attacking nodes grows in one group. However, it fluctuates when it comes from three groups. The overhead ratio of the epidemic protocol is not affected by a black hole attack when the source of attack comes from certain groups. It remains almost the same, and may increase and decrease by 1% and 2% at different levels of attack. However, the proportion is dramatically decreased by 8% (from 576.9785 at 0% attack to 74.7176 at 50% attack). However, when the entire network is infected by a black hole attack, the overhead is negligible, as there is no resources consumption to deliver messages to destinations because all messages are dropped. The present findings clearly indicate how one of the DTN routing protocols, Epidemic, is vulnerable in the face of a black hole attack.

REFERENCES

1. S. Rashid, Q. Ayub, M. S. M. Zahid, and A. H. Abdullah. 2011. Impact of mobility models on DLA (Drop Largest) optimized DTN epidemic routing protocol. *International Journal of Computer Applications*, 18, 35–39.
2. P. Vinayakray-Jani and S. Sanyal. 2012. Routing protocols for mobile and vehicular ad-hoc networks: A comparative analysis. arXiv: 1206.1918.
3. T. Spyropoulos, K. Psounis, and C. S. Raghavendra. 2008. Efficient routing in intermittently connected mobile networks: The multiple-copy case. *IEEE/ACM Transactions on Networking*, 16, 77–90.
4. A. Omidvar and K. Mohammadi. 2014. Intelligent routing in delay tolerant networks. In: *22nd Iranian Conference on Electrical Engineering (ICEE)*, Tehran, pp. 846–849.
5. M. J. Khabbaz, C. M. Assi, and W. F. Fawaz. 2011. Disruption-tolerant networking: A comprehensive survey on recent developments and persisting challenges. *IEEE Communications Surveys & Tutorials*, 14(2), 607–640.
6. S. Kumagai and H. Higaki. 2014. Intermittent wireless multihop transmission protocol in mobile wireless sensor networks. In: *2014 8th International Conference on Signal Processing and Communication Systems (ICSPCS)*, Gold Coast, Queensland, Australia, pp. 1–8.
7. S. Dokurer. 2006. Simulation of black hole attack in wireless ad-hoc networks. Master's thesis, Atilim University.
8. A. Kodole and P. M. Agarkar. 2015. A survey of routing protocols in mobile ad hoc networks. *Multidisciplinary Journal of Research in Engineering & Technology*, 2(1), 336–341.
9. M. G. Zapata and N. Asokan. 2002. Securing ad hoc routing protocols. WiSe'02, Atlanta, GA, September 28.
10. M. AlShurman, S. Yoo, and S. Park. 2004. Black hole attack in mobile ad hoc networks. In: *Proceedings of the 42nd Annual Southeast Regional Conference*. ACM Press, Huntsville, Alabama, pp. 96–97.
11. K. Selvavinayaki, K. K. Shyam Shankar, and E. Karthikeyan, 2010. Security enhanced DSR protocol to prevent black hole attacks in MANETs. *International Journal of Computer Applications*, 7(11), 15–19.
12. R. A. Raja Mahmood and A. I. Khan. 2007. Survey on detecting black hole attack in AODV based mobile ad hoc networks. In: *Proceedings of 4th International Symposium on High Capacity Optical Networks and Enabling Technologies (HONET 2007)*, Dubai, pp. 1–6.
13. D. Johnson and D. Maltz. 1996. Dynamic source routing in ad hoc wireless networks. In: T. Imelinsky and H. Korth (eds.), *Mobile Computing*, pp. 153–181. Kluwer Academic, Pittsburgh, PA, 1996.
14. S. Burleigh and K. Fall. 2003. Delay tolerant networking: An approach for interplanetary Internet. *IEEE Communications Magazine*, 41(6), 128–136.
15. S. Jain, K. Fall, and R. Patra. 2004. Routing in a delay tolerant network. In: *Proceedings of the 2004 Conference on Applications, Technologies, Architectures, and Protocols for Computer Communications (SIGCOMM '04)*, New York, pp. 145–158.

16. Z. Zhang. 2006. Routing in intermittently connected mobile ad hoc networks and delay tolerant networks: Overview and challenges. *IEEE Communications Surveys & Tutorials*, 8(1), 24–37.
17. E. Bulut and B. K. Szymanski. 2011. On secure multi-copy based routing in compromised delay tolerant networks. In: *20th IEEE International Conference on Computer Communication and Networks ICCN*, Maui, Hawaii, pp. 1–7.
18. A. K. Gupta, I. Bhattacharya, P. S. Banerjee, and J. K. Mandal. 2014. A co-operative approach to thwart selfish and black-hole attacks in DTN for post disaster scenario. In: *2014 Fourth International Conference of Emerging Applications of Information Technology (EAIT)*, Kolkata, pp. 113–118.
19. A. Al Hinai, H. Zhang, and Y. Chen. 2012. Mitigating black-hole attacks in delay tolerant networks. In: *Proceedings of the 13th International Conference on Parallel and Distributed Computing, Applications and Technologies*, Beijing, pp. 329–334.
20. Y. Ren, M. C. Chuah, J. Yang, and Y. Chen. 2010. Detecting blackhole attacks in disruption-tolerant networks through packet exchange recording. In: *11th IEEE International Symposium on a World of Wireless, Mobile and Multimedia Networks (WOWMOM)*, Montreal, pp. 1–6.
21. F. Li, J. Wu, and A. Srinivasan. 2009. Thwarting blackhole attacks in disruption-tolerant networks using encounter tickets. In: *IEEE INFOCOM 2009—The 28th Conference on Computer Communications*, Rio de Janeiro, Brazil, pp. 2428–2436.
22. R. H. Jhaveri, A. D. Patel, J. D. Parmar, and B. I. Shah. 2010. MANET routing protocols and wormhole attack against AODV. *IJCSNS International Journal of Computer Science and Network Security*, 10(4), 12–18.
23. J. Sen, M. G. Chandra, H. S. G., H. Reddy, and P. Balamuralidhar. 2007. *A mechanism for detection of gray hole attack in mobile ad hoc networks*. Embedded Systems Research Group, Bangalore.
24. J. Burgess, G. D. Bissias, M. Corner, and B. N. Levine. 2007. Surviving attacks on disruption-tolerant networks without authentication. In: *Proceedings of MobiHoc '07*, Montreal, pp. 61–70.
25. F. C. Choo, M. C. Chan, and E. Chang. 2010. Robustness of DTN against routing attacks. In: *Proceedings of Second International Conference on Communication Systems and Networks (COMSNETS)*, Singapore, pp. 1–10.
26. A. Keranen. 2008. *Opportunistic network environment simulator*. Helsinki University of Technology.
27. A. Keranen and J. Ott. 2007. *Increasing reality for DTN protocol simulation*. Helsinki University of Technology.
28. J. Kong, X. Hong, Y. Yi, J.-S. Park, J. Liu, and M. Gerla. 2005. A secure ad-hoc routing approach using localized self-healing communities. In: *Proceedings of MobiHoc '05, 6th ACM International Symposium on Mobile Ad Hoc Networking and Computing*, Urban-Champaign, Illinois, pp. 254–265.
29. A. Lindgren, A. Doria, and O. Schelen. 2004. Probabilistic routing in intermittently connected networks. *SIGMOBILE Mobile Computing and Communication Review*, 7(3), 19–20.
30. L. Song and D. F. Kotz. 2007. Evaluating opportunistic routing protocols with large realistic contact traces. In: *Proceedings of ACM 2nd Workshop on Challenged Networks (CHANTS '07)*, Montreal, pp. 35–42.

19 Proposed Adaptive Neural-Fuzzy Inference System (ANFIS) Identifier for M-ary Frequency Shift Keying (FSK) Signals with Low SNR

Hadi A. Hamed, Sattar B. Sadkhan, and Ashwaq Q. Hameed

CONTENTS

19.1 Introduction ... 259
19.2 M-ary Frequency Shift Keying and Mathematical Expressions ... 260
19.3 Feature Extraction ... 261
  19.3.1 Wavelet Transform ... 261
  19.3.2 Moments ... 261
19.4 Adaptive Neural-Fuzzy Inference System (ANFIS) ... 262
19.5 Methodology and Analysis of Simulation ... 263
19.6 Conclusion and Future Work ... 267
References ... 267

19.1 INTRODUCTION

The automatic modulation identification (AMI) technique identifies the modulated signal type at the presence of noise. There are wide applications of AMI. In military applications, it can be utilized for interference recognition, electronic surveillance, and monitoring. In civil applications, it can be employed for software defined radio (SDR), spectrum management, signal conformation, and intelligent modems [1]. Many AMI algorithms have been implemented to identify the modulation type of the digitally modulated signals. These algorithms are classified systematically into five major classifiers: likelihood-based, feature-based, blind modulation, distribution tested-based, and machine learning-assisted classifiers [2]. The existing algorithms for M-ary frequency shift keying (MFSK) identification are mainly classified into decision theoretic (DT) and pattern recognition (PR) [3]. DT requires the parameters of modulation formats of MFSK, which is represented by frequency deviation, carrier frequency, and number of transmitted signal frequencies (M), whereas PR algorithms need less knowledge of the MFSK parameters of modulation format [4]. Thus higher computation complexity is the major drawback of DT, but PR requires less computational complexity and can be classified into two subsystems: the feature extraction and the classifier [5]. Artificial neural network (ANN) is a good candidate for AMI-PR [6], but such signal classifiers are suffering from many shortcomings such as requiring long signal duration, complex algorithms, big size feature vectors for training, and high computer storage capacity. Then performance of AMI is limited by time and cost, and the classifiers suffer from either overfitting or underfitting occasionally. Furthermore there is degradation in performance of ANN classifiers at high signal-to-noise ratio (SNR), when the training of ANN is performed at low SNR [7,8]. The adaptive neural-fuzzy inference system (ANFIS) is proposed as a classifier for multiple frequency shift keying (MFSK) signal identification to overcome some of the shortages of ANN and fuzzy inference system. ANFIS was introduced at 1993 [9]. ANFIS outperforms ANN in many applications like temperature controller [10] and data rate prediction for cognitive radio (CR) [11]. ANFIS was successfully introduced for digitally modulated signal identification [12]. There are many AMI algorithms to classify MFSK signals. Nandi and Azzouz [13] use the standard deviation of the absolute value of the normalized instantaneous frequency as a key feature to discriminate 2FSK, 4FSK, and other digitally modulated signals. The identification is based on two approaches: ANN and DT. They claimed that the performance of ANN outperforms DT in this task. The extracted feature is a function of M and the frequency deviation ratio, then the value of the frequency deviation ratio must be known in advance as well as the value of M. Hatzichristos and Fargues [14] use statistical features, high-order moments (HOMs), and high-order cumulants (HOCs) as a feature extraction; and ANN as a classifier to classify 2FSK, 4FSK, and 8FSK and other digitally modulated signals. The drawback of this approach is the statistical features for 2FSK, 4FSK, and 8FSK are overlapped, especially at low SNR. Yu and Shi [4] use fast Fourier transform (FFT) to classify 2FSK, 4FSK, 8FSK, 16FSK, and 32FSK, but the drawback is to estimate the value of the symbol constellation size (M), as well as the complexity of the FFT algorithm.

The proposed approach overcomes some of the problems of the existing identification systems to identify MFSK signals. In this chapter the features are extracted using fourth-, sixth-, and eighth-order moments for detail coefficients of the Haar-type discrete wavelet transform (DWT) in order to reduce the complexity of the system when it is compared with FFT. (The complexity of DWT algorithm is less than that of the FFT algorithm [15].) This chapter also proposes the identification process does not require the parameters of MFSK modulation format, like frequency deviation and number of transmitted signal frequencies (M). This chapter is organized as follows: Section 19.2 demonstrates the MFSK signals and mathematical expressions. Section 19.3 explains the feature extraction using DWT and high-order moments. Section 19.4 presents a general review of ANFIS. Section 19.5 focuses on the methodology and analysis of simulation, whereas Section 19.6 presents the conclusion and future work.

19.2 M-ARY FREQUENCY SHIFT KEYING AND MATHEMATICAL EXPRESSIONS

Frequency modulation is the process when the baseband signal modulates the carrier frequency of the modulated signal. If the baseband signals are digital signals, then the process is called frequency shift keying (FSK). The digital data is mapped through the variations in the carrier signal frequency, while the amplitude and frequency remain constant. In a binary FSK (BFSK), the two symbols (1 and 0) are mapped through two sinusoidal waves different by a fixed amount from each other. To represent the signal of BFSK:

$$S_{K_{BFSK}}(t) = \sqrt{\frac{2E_b}{T_b}}\cos(2\pi f_k t) \quad 0 \le t \le T_b \tag{19.1}$$

where $k = 1,2$; E_b represents the bit energy, $f_k = (nc + k)/T_b$; and T_b is a single bit period.

In M-ary FSK, the carrier frequency is changed more than two values, and the transmitted signal can be represented by

$$S_{K_{MFSK}}(t) = \sqrt{\frac{2E_S}{T_S}}\cos\left[\left(2\pi f_c + \frac{\pi k}{T_S}\right)t\right] \quad 0 \le t \le T_S \tag{19.2}$$

For $k = 1, 2, \ldots, M$

E_S represents transmitted symbol energy. The carrier frequency $f_c = (nc/2T_S)$, $E_S = E_b\log_2 M$, $T_S = T_b\log_2 M$, where T_S represents the symbol duration and nc is a fixed integer [16]. The main advantage of the FSK system is that it is not susceptible to noise; any voltage spikes introduced by noise affects the amplitude and the frequency is not affected. The physical capabilities of the carrier are the limiting factor of MFSK [17]. In the simulation MATLAB program, we assume $T_b = 1$ ms, and unity bit energy. Each of the MFSK signals have subcarrier frequencies for a certain modulation scheme that must be different from other schemes. The fixed integer nc is set to 2, 1, and 7 in the simulation program for 2FSK, 4FSK, and 8FSK signals, respectively, to prohibit the subcarrier frequencies from being identical for different MFSK signals.

19.3 FEATURE EXTRACTION

In typical AMI systems, the dimension of raw data is reduced before the classifier processing. These new reduced data are represented as a feature extraction. In feature extraction the useful information with discriminative characteristics that differentiate between the modulated signals is extracted from the raw data to be pertinent for classification. The performance of the classifiers depends on discriminatory information of the extracted features, while some of the extracted features are redundant, overlap, unrelated, and noisy. The negative impact on the identification performance will be caused by irrelevant features. For many pattern recognition algorithms, the increase in the number of feature extractions will increase the training time directly or exponentially [18]. In this chapter the extracted features by DWT are reduced by applying the fourth-, sixth-, and eighth-order moments, as well as some useful statements in the MATLAB program are introduced like "if" statements with "and" and "or" logic statements to reduce the number of extracted feature to mitigate the impediment of the classifier.

19.3.1 Wavelet Transform

Wavelet transform (WT) analyzes the signals into different frequency components. Each component can be studied with respect to its scale. WT overcomes traditional Fourier transform when the signal contains sharp spikes and discontinuities. WT outperforms short-time Fourier transform (STFT) for analyzing signals because WT applies multiresolution techniques with different frequencies and resolutions, whereas STFT uses a constant resolution for all frequencies [19], thus WT a is powerful tool for feature extraction with MFSK signals. There are several families of WT, and they include Haar, Daubechies, Coiflets, Biorthogonal, Morlet, Meyer, Mexican Hat, Symlets, and Reverse Bior. DWT is a sampled version of CWT with discrete translation and dilation parameters. DWT can be performed by using a filter bank. High pass (HP) filters and low pass (LP) filters are utilized to decompose the signal. The output gives the approximation coefficients from the LP filter and details coefficients from the HP filter [20]. In this chapter the detail coefficients of Haar DWT, level one are used.

19.3.2 Moments

Statistical moments are the expected value of a random variable raised to the power indicated by the order of the moment [21]:

$$\mu_i = E[(X - E[X])^i] \tag{19.3}$$

$$\mu = E[X] \tag{19.4}$$

$$\mu_i = \frac{1}{N}\sum_{j=0}^{N-1}(X_j - \mu)^i \tag{19.5}$$

where N is the size of data length, μ is the mean, and i is the order of moment.

19.4 ADAPTIVE NEURAL-FUZZY INFERENCE SYSTEM (ANFIS)

A fuzzy inference system (FIS) is a process of using the theory of fuzzy sets and fuzzy rules to map a given input to an output. The performance of FIS depends on the identification of membership functions (MFs) and the fuzzy rules tuned to the requested applications. Usually it is difficult or not possible to transform human knowledge to fuzzy rules and MFs tuning in terms of cost and time. ANN was introduced to overcome these limitations by identifying fuzzy rules and tuning the parameters of MFs automatically [22,23]. ANFIS is an FIS whose parameters are trained by means of any ANN learning algorithms. ANFIS is a multilayer feed forward that uses Sugeno-type FIS [9]. The equivalent ANFIS architecture as an example is shown in Figure 19.1. The system is assumed with two inputs and one output. The individual layers of ANFIS are discussed briefly in the following.

Layer 1—Layer 1 is the fuzzification layer. The fuzzification is performed by the neurons in this layer. The nodes in this layer are adaptive with a function.

$$O_{1,i} = \mu_{A_i}(x) \quad \text{for } i = 1,2 \tag{19.6}$$

$$O_{1,i} = \mu_{B_{i-2}}(y) \quad \text{for } i = 3,4 \tag{19.7}$$

where $O_{1,i}$ is the output of layer 1, μ_{A_i} is the MF of the linguistic variable of fuzzy sets A_i, and $\mu_{B_{i-2}}$ is the MF of the linguistic variable of fuzzy sets B_{i-2}.

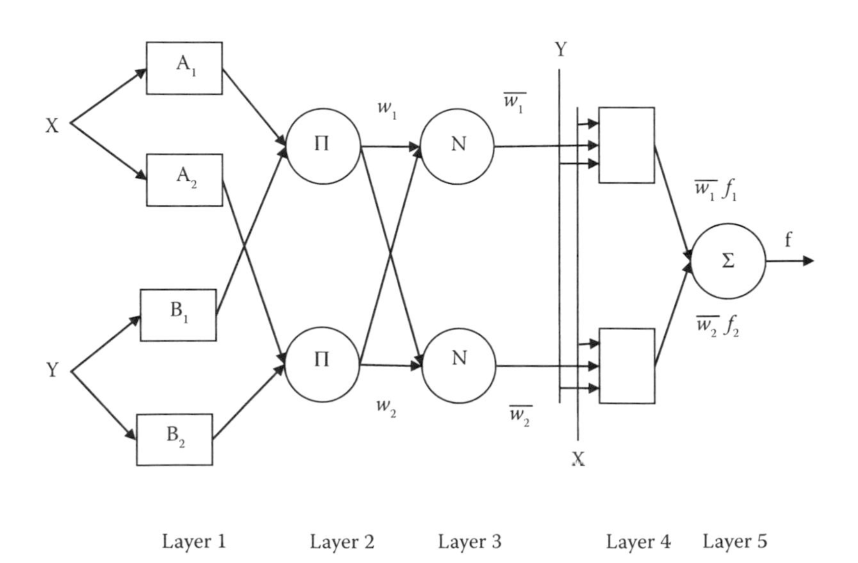

FIGURE 19.1 ANFIS ARCHITECTURE.

Layer 2—This layer receives inputs from layer 1 in the fuzzification form. The output of each node is obtained by multiplying the incoming signals. The nodes in this layer for Figure 19.1 are labeled ∏.

$$O_{2,i} = w_i = \mu_{A_i}(x) * \mu_{B_i}(x) \quad i = 1,2 \tag{19.8}$$

where $O_{2,i}$ is the output of layer 2.

Layer 3—The nodes in this layer are fixed, each node obtains the normalization, that is calculating the ratio of the firing strength of each rule to the sum of all firing strengths of all rules. The output is given by

$$O_{3,i} = \overline{w_i} = \frac{w_i}{w_1 + w_2} \quad \text{for } i = 1,2 \tag{19.9}$$

$O_{3,i}$ is the output of layer 3.

Layer 4—The nodes in this layer are adaptive node and receives the input x and y to compute the output of the rule by evaluating the Sugeno-type linear approximator f_i multiplied by the normalized firing strength:

$$O_{4,i} = \overline{w_i} f_i = \overline{w_i}(k_{0i} + k_{1i}x + k_{2i}y) \quad i = 1,2 \tag{19.10}$$

where $O_{4,i}$ is the output of layer 4; and k_{0i}, k_{1i}, and k_{2i} are the consequent parameter set. The prescribed four layers represent the equivalent Sugeno fuzzy inference system. The premise part is applied to establish the MFs, which are linguistic variables, whereas the consequent part is represented by a crisp equation.

Layer 5—The layer consists of one fixed node that calculates the system output f on the ANFIS architecture of Figure 19.1. The layer is labeled by Σ, which is the sum of all incoming signals:

$$\begin{aligned} O_{5,i} &= \sum_{i=1}^{2} \overline{w_1} f_i = \frac{\sum_i w_i f_i}{\sum_i w_i} \quad \text{for } i = 1,2 \\ &= \overline{w_1}\ (k_{01} + k_{11}x + k_{21}y) + \overline{w_2}\ (k_{02} + k_{12}x + k_{22}y) \end{aligned} \tag{19.11}$$

where $O_{5,i}$ is the output of layer 5. Then the output of ANFIS is determined by the consequent part within the premise parts [24].

19.5 METHODOLOGY AND ANALYSIS OF SIMULATION

The identification of the M-ary FSK system is shown in Figure 19.2. The signals for proposed identification processes are 2FSK, 4FSK, and 8FSK. The signals are digitally generated according to Equations 19.1 and 19.2, using MATLAB Toolbox version 2013b.

The program is designed; 240 signals are generated; additive white Gaussian noise (AWGN) is added at 0 dB, 5 dB, and 10 dB; the sample numbers of each signal is 1000 samples; the identification system consists of feature extraction and classification layers; and the feature extraction plays the key role of identification processes and improves the performance of the classifier, but not all of the extracted feature is important for identification process. The selected features should be as

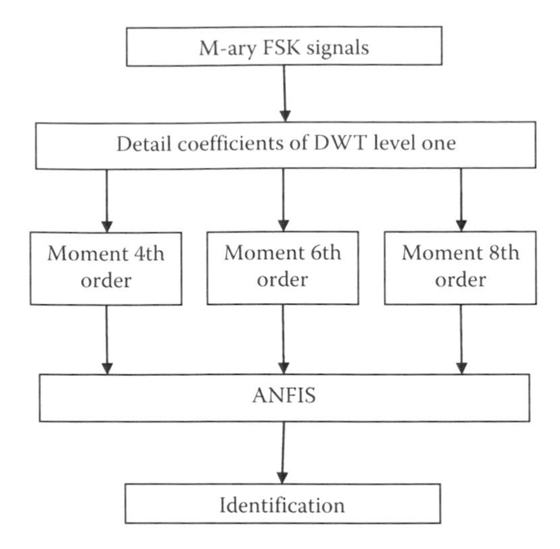

FIGURE 19.2 **IDENTIFICATION PROCESSES.**

few as possible to reduce the complexity of the system. In order to find the appropriate features, computer programs are designed to generate the signals 2FSK, 4FSK, and 8FSK. All the signals are corrupted with AWGN, at 0 dB, 5 dB, and 10 dB; 240 signals of 240000 samples are investigated, the detail coefficient DWT "Haar type" is applied to the samples through the period *Tb*; high numbers of features are extracted; and fourth-, sixth-, and eighth-order moments are applied to the detail coefficient of DWT. Thus the extracted features are transformed into other reduced features that possess the discrimination facilities of the signals. After investigating the results, we find that the extracted feature magnitude has a slight variation at different symbols for the same modulation scheme, moment order, and SNR. Table 19.1 shows this variation as an example for 4FSK, which is four different symbols, namely, S1, S2, S3, and S4 at 5 dB SNR. From the results, the MATLAB program is extended to limit each group of extracted features for all the symbols of a certain modulation scheme, like (0.510 ≤ moment 6th order ≤ 0.533). "if" statement, logical "OR", and logical "AND" are employed to define each group of features. The groups of features having the discriminating properties are used to train the ANFIS classifier.

This approach is used to reduce the complexity of the system, as well as to reduce the training time of ANFIS. The operation of the ANFIS classifier starts with loading the vectors of the training data as mentioned before. The type of the MFSK modulated schemes are distinguished through the output number. The program represents No. 1 for 2FSK, No. 2 for 4FSK, and No. 3 for 8FSK. Then generating the FIS, the input number of membership functions is 3. The generalized Bell MF is utilized as the input membership function. The type of output MF is "linear." The hybrid learning algorithm least square–back propagation is used. Figure 19.3 shows the structure of ANFIS using MATLAB.

TABLE 19.1
High-Order Moments of DWT for 4FSK at Different Symbols and 5 dB SNR

	S1	S2	S3	S4
Moment 4th order	0.294	0.299	0.301	0.323
Moment 6th order	0.510	0.517	0.499	0.533
Moment 8th order	1.181	1.193	1.114	1.185

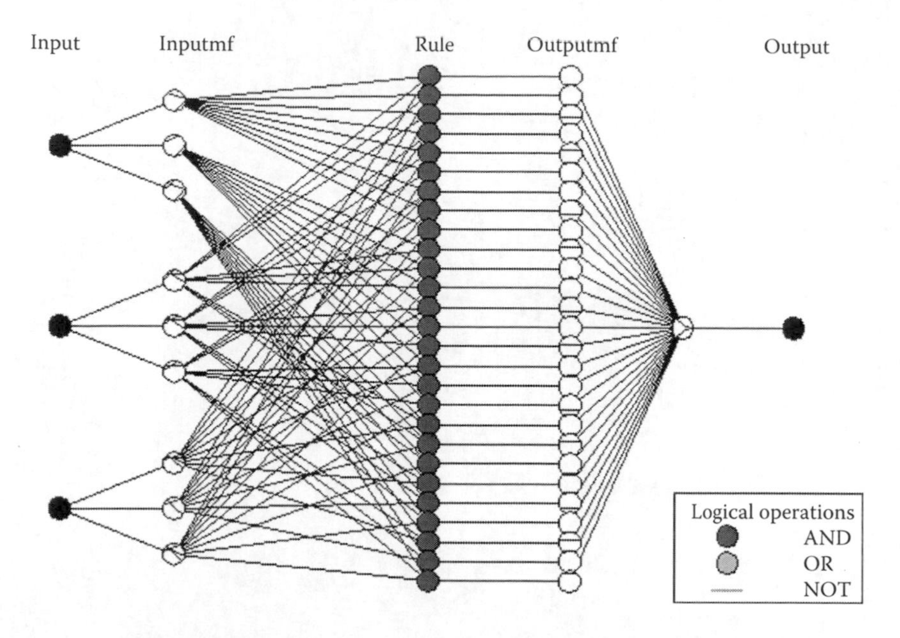

FIGURE 19.3 ANFIS STRUCTURE.

After the proposed number of 25 epochs through the training period is reached, the error minimizing graph shows that the error is 2.7048×10^{-7}, as shown in Figure 19.4.

The MATLAB program could show the three-dimensional surface viewer, which demonstrates the correlation of the inputs to the ANFIS output. The surface graphs for MFSK identification are shown in Figures 19.4 through 19.6, for 2FSK, 4FSK, and 8FSK, respectively. From the graphs it is clearly shown that the output with No. 1 of surface graph of Figure 19.5 represents the correct identification of 2FSK, No. 2 of surface graph of Figure 19.6 represents the correct identification of 4FSK, and No. 3 of surface graph of Figure 19.7 represents the identification of 8FSK signals, while the inputs are the group of extracted features.

The identification approach is tested using different signals of 2FSK, 4FSK, and 8FSK. The system exhibits high correct identification at SNR ≥ 0. The percentages of correct identification is 100% for 2FSK, 4FSK, and 8FSK, at 0 dB, 5 dB, and 10 dB SNR.

When the identification system is trained and tested with low SNR, –3 dB, 0 dB, 5 dB, and 10 dB, the percentage of correct identification is 50% for 2FSK, 100% for 4FSK, and 62.5% for 8FSK. The average identification percentage for the range with low SNR is 70.83%. The performance of identification of MFSK signals is shown in Figure 19.8.

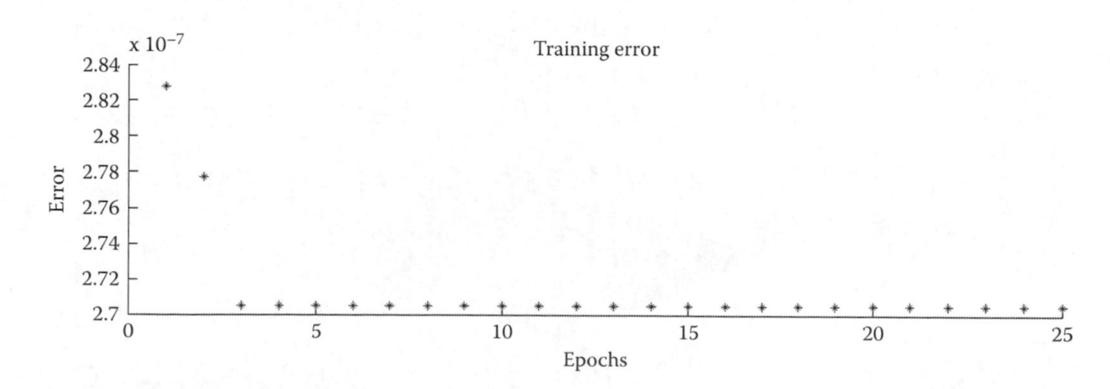

FIGURE 19.4 ERROR MINIMIZING GRAPH.

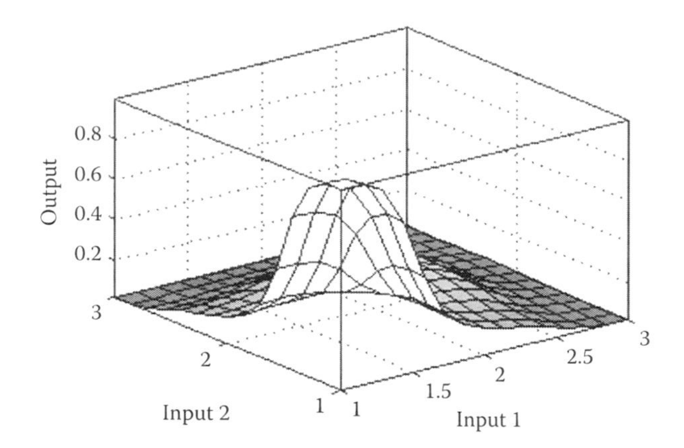

FIGURE 19.5 SURFACE GRAPH OF 2FSK.

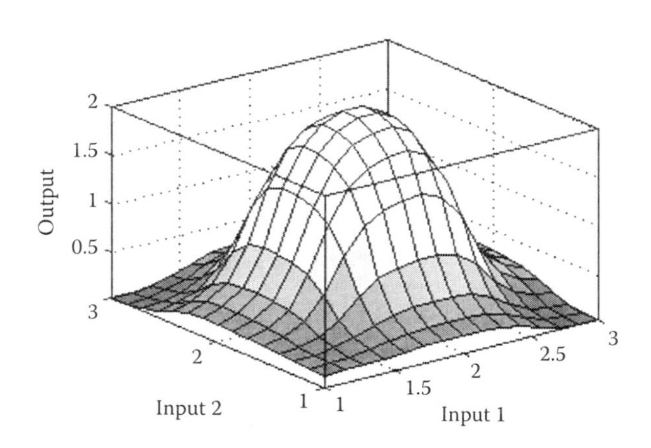

FIGURE 19.6 SURFACE GRAPH OF 4FSK.

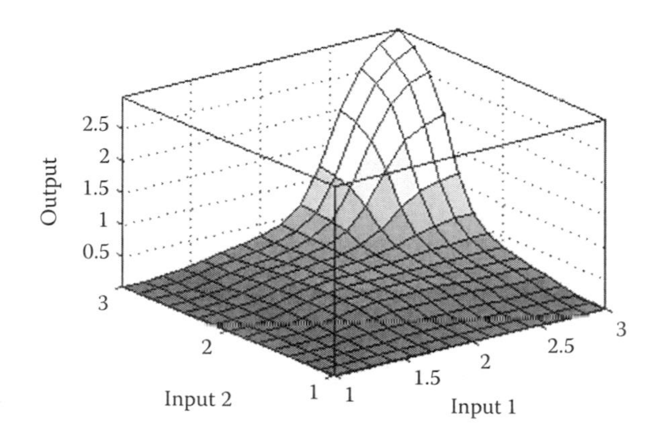

FIGURE 19.7 SURFACE GRAPH OF 8FSK.

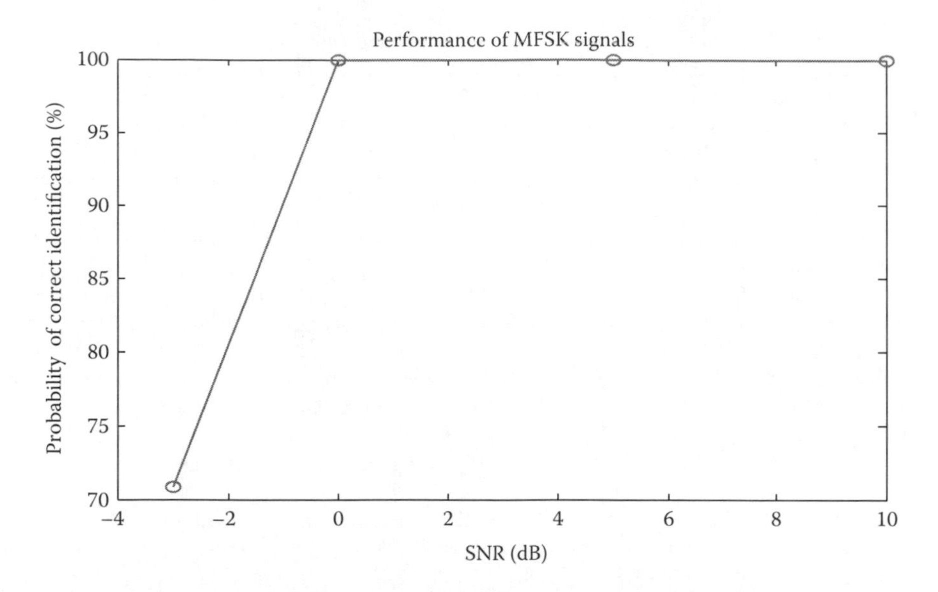

FIGURE 19.8 PERFORMANCE OF THE MFSK SYSTEM.

19.6 CONCLUSION AND FUTURE WORK

The feature extraction identification process is based on detail coefficients of DWT and high-order moments. ANFIS is the classifier. Detail coefficients of DWT, Haar-type of 2FSK, 4FSK, and 8FSK signals are obtained. Fourth-order, sixth-order, and eighth-order moments are applied to the detail coefficient of DWT. The feature extraction parameters are minimized by extending the MATLAB program to include "if" statement, logical "OR", and logical "AND". The identification system is simulated at AWGN with different SNR, the system exhibits high identification performance when SNR ≥ 0, the percentages of correct identification were about 100% when SNR is 0 dB, 5 dB, and 10 dB. There is a reduction in performance at −3 dB, 0 dB, 5 dB, and 10 dB. The main advantage of this method is to reduce the training time and reduce the complexity of the classifier. The method that has been specified is flexible and can be used as a future work for modulation identification of either other MFSK signals or any SNR that is not covered within this chapter. This approach could be modified and extended to identify other types of digital modulation like M-ary phase shift keying (MPSK), and quadratic amplitude modulation (QAM) signals. Other types of DWT could be utilized instead of Haar type to evaluate the performance of the identification.

REFERENCES

1. H. Tao and J. Xiaorong, Modulation classification using ARBF networks, *Proceedings of IEEE International Conference on Signal Processing, ICSP '04*, Beijing, pp. 1809–1812, 2004.
2. Z. Zhu and A. K. Nandi, *Automatic Modulation Classification*, John Wiley & Sons, Chichester, UK, 2015.
3. K. Hassan, I. Dayoub, W. Hamouda, and M. Berbineau, Automatic modulation recognition using wavelet transform and neural networks in wireless systems, *EURASIP Journal on Advances in Signal Processing*, 1–13, 2010.
4. Z. Yu and Y. Q. Shi, M-ary frequency shift keying signals classification based on discrete Fourier transform, *Proceedings of IEEE Conference MILCOM'2003*, Boston, pp. 1167–1172, 2003.
5. A. Ebrahimzadeh and R. Ghazalian, Blind digital modulation classification in software radio using the optimized classifier and feature subset selection, *Engineering Applications of Artificial Intelligence*, 24, 50–59, 2011.
6. S. B. Sadkhan, A. Q. Hameed, and H. A. Hamed, Digitally modulated signals identification based on artificial neural network, *ATTI Della Fondazione Giorgio Ronchi*, 70(1), 59–75, 2015.

7. K. Ahmed, U. Meier, and H. Kwasnicka, Fuzzy logic based signal classification with cognitive radios for standard wireless technology, *Proceedings of IEEE International Conference on Cognitive Radio Oriented Wireless Networks and Communications*, Cannes, pp. 1–5, 2010.
8. E. Avci, D. Hanbay, and A. Varol, An expert discrete wavelet adaptive network based fuzzy inference system for digital modulation recognition, *Expert Systems with Applications*, 33, 582–589, 2007.
9. J.-S. R. Jang, ANFIS: Adaptive network-based fuzzy inference system, *IEEE Transaction on Systems, Man, and Cybernetics*, 23(3), 665–685, 1993.
10. T. P. Mote and S. D. Lokhande, Temperature control system using ANFIS, *International Journal of Soft Computing and Engineering (IJSCE)*, 2(1), 156–161, 2012.
11. S. Hiremath and S. K. Patra, Transmission rate prediction for cognitive radio using adaptive neural fuzzy inference system, *Proceedings of the ICIIS, International Conference on Industrial and Information Systems*, Mangalore, India, pp. 92–97, 2010.
12. S. B. Sadkhan, A. Q. Hameed, and H. A. Hamed, Digitally modulated signals recognition based on adaptive neural-fuzzy inference system (ANFIS), *International Journal of Advancements in Computing Technology (IJACT)*, 7(5), 57–65, 2015.
13. A. K. Nandi and E. E. Azzouz, Algorithms for automatic modulation recognition of communication signals, *IEEE Transaction on Communication*, 46(4), 431–436, 1998.
14. G. Hatzichristos and M. P. Fargues, A hierarchical approach to the classification of digital modulation types in multipath environments, *Proceedings of the Thirty-Fifth Asilomar Conference on Signals, Systems and Computers*, 2, 1494–1498, 2001.
15. A.-J. K. Humady and H. K. Chaiel, Classification of digital modulation using wavelet transform, *Iraqi Journal of Applied Physics (IJAP)*, 1(3), 15–21, 2005.
16. S. Haykin, *Communication Systems*, 4th edition, John Wiley & Sons, New York, 2001.
17. B. A. Forouzan, *Data Communications and Computer Networking*, McGraw-Hill Higher Education, New York, 2006.
18. M. W. Aslam, Pattern recognition for classification of diabetes and modulation data, PhD thesis, University of Liverpool, United Kingdom, 2013.
19. C.-S. Park, J.-H. Choi, S.-P. Nah, W. Jang, and E. Lab, Automatic modulation recognition of digital signals using wavelet features and SVM, *Proceedings of IEEE 10th International Conference on Advanced Communication Technology, ICACT*, Gangwon-Do, South Korea, pp. 387–390, 2008.
20. B. Dastourian, E. Dastourian, S. Dastourian, and O. Mahnaie, Discrete wavelet transform of Haar's wavelet, *International Journal of Scientific and Technology Research*, 3(9), 247–251, 2014.
21. S. Keshav, *Mathematical Foundations of Computer Networking*, Addison-Wesley Professional, Upper Saddle River, NJ, 2012.
22. D. Akay, X. Chen, C. Barnes, and B. Henson, ANFIS modeling for predicting affective responses to tactile textures, *Human Factors and Ergonomics in Manufacturing & Service Industries*, 22(3), 269–281, 2012.
23. M. Panella and A. S. Gallo, An input-output clustering approach to the synthesis of ANFIS networks, *IEEE Transaction on Fuzzy Systems*, 13(1), 69–81, 2005.
24. M. Buragohain and C. Mahanta, A novel approach for ANFIS modeling based on full factorial design, *Applied Soft Computing*, 8, 609–625, 2008.

20 New Image Scrambling Algorithm Depending on Image Dimension Transformation Using Chaotic Flow Sequences

Hadi T. Ziboon, Hikmat N. Abdullah, and Atheer J. Mansor

CONTENTS

20.1 Introduction 269
20.2 The Proposed Scrambling Algorithm 270
20.3 Simulation Results 273
20.4 Conclusions 279
References 279

20.1 INTRODUCTION

Image scrambling is a method for digital image encryption by changing the position of the pixels without changing the gray level of the pixel. The resulting image is an encrypted image that has the same size of the original image and the same histogram, but unreadable because of the changing of the coordinates of each pixel in the original image (Liehuang et al. 2006). In recent years, chaotic sequences have become more helpful in encryption (of images, audio, videos, etc.) because of their characteristics like sensitivity to initial conditions and parameters, random behavior, aperiodicity, and dynamicity. The resulting features suggest that chaos system gives a wide range of numbers (keys) that can be used in the encryption process (Mao et al. 2007; Gao and Chen 2008; Prasad and Sudha 2011).

There are many chaotic maps that are used in image encryption, for example, image encryption based on chaotic maps and memory cellular automata (Bakhshandeh and Eslami 2013) or image encryption based on chaotic maps with using the characteristic of DNA encoding (Wang et al. 2015). Arnold's cat map is a system that is used for image scrambling by swapping or changing the pixel poison or coordinates (Gupta and Singh 2014). The weakness of Arnold's cat map falls in twofold: First, it is applied only to equal dimensional images, that is, N*N images. Second, under descrambling process a number of iterations (m) is needed to get back the original image, and thus, these iterations are increased as the size of image is increased (Peterson 1997, Keshari and Modani 2011, Nance 2011). Table 20.1 shows the number of iterations (m) required to restore an image of size N*N (Tang and Zhang 2011).

In 2011, Tang and Zhang proposed a system that uses Arnold's cat map for an N*M image size by dividing the original image into subimages with equal height and width, then applied Arnold's cat map with different iterations on each subimage, then restructuring the image after scrambling to one image with size N*M (Tang and Zhang 2011). The weakness points in this system are the long

TABLE 20.1
Number of Iterations (m) Required to Restore Image of Size N*N Using Arnold's Cat Map

N	60	100	120	128	256	480	512
M	60	150	60	96	192	240	384

Source: Peterson G., 1997, Arnold's Cat Map, College of the Redwoods.

time needed to apply Arnold's cat map to each subimage and the problems with the overlapping between subimages (Gupta and Singh 2014).

In this chapter, the scrambling process is done by using the chaotic flows instead of using the traditional chaotic scrambling based on chaotic maps. The proposed system allows the user to select one of four chaos flow systems, which are Lorenz (Willsey et al. 2010), Rössler (Gaspard 2005), Chua, or Nien (Nien et al. 2007) to reduce the scrambling/descrambling time and to add more flexibility to the scrambling system. The main contributions of the proposed work are as follows: the use of the chaotic flows increases the immunity against the intruders as compared with the chaotic. Also, the proposed system is applicable to equal and nonequal dimensional images, while the traditional techniques are applied to equal dimensional images only. Many testing techniques have been used to evaluate the system performance comparing with the traditional scheme performance. The used testing techniques are distance scrambling factor (DSF) (Li et al. 2013), average distance change (ADC) (Jolfaei and Mirghadri 2010), 2D correlation coefficient (Merah et al. 2013), and peak signal-to-noise ratio (PSNR) (Jolfaei and Mirghadri 2010).

The rest of the chapter is organized as follows: The proposed scrambling algorithm and the simulation results are presented in Sections 20.2 and 20.3, respectively, while the conclusions drown throughout the work are given in Section 20.4.

20.2 THE PROPOSED SCRAMBLING ALGORITHM

The proposed scheme performs the image scrambling by using Lorenz, Rössler, Chua, or Nien chaotic systems for N*N and N*M image sizes. The algorithm of the proposed scheme is explained throughout the following steps:

Step 1: Start
Step 2: Convert the color RGB of size M*N image into 3 vectors: R, G and B which represent the pixels in Red, Green and Blue respectively (the size of each vector is 1*k, where k = M*N)
Step 3: Create three new vectors: r, g, and b, and set the values of their elements to zero (with same size of R, G, and B vectors)
Step 4: Set an initial values of the used chaotic system
Step 5: Set initial value for a counter i = 1
Step 6: If r(i) = 1 (which means the pixel has been scrambled before) go to step 11
Step 7: Compute x_{n+1}, y_{n+1}, and z_{n+1} using equations 2 or 4 or 5 or 7 depending on the selected chaotic system.
Step 8: $X_{n+1} = (10^{14}*x_{n+1})Mod(k)$, $Y_{n+1} = (10^{14}*y_{n+1})Mod(k)$, and $Z_{n+1} = (10^{14}*z_{n+1})Mod(k)$
Step 9: If $r(X_{n+1}) = 1$ (which means the pixel has been scrambled before) go to step 7
Step 10: Exchange R(i) and $R(X_{n+1})$
Step 11: Make r(i) and $r(X_{n+1}) = 1$

Step 12: i = i + 1, repeat steps 6–10 until i = k
Step 13: Repeat steps 5–11 for G and B vectors
Step 14: Reconstruct the image by converting the vectors (R,G,B) into matrix of size N*M
Step 15: End

Figure 20.1 shows the proposed system steps in flowchart form. The descrambling process is done by following the same steps of the scrambling process but the used image will be the

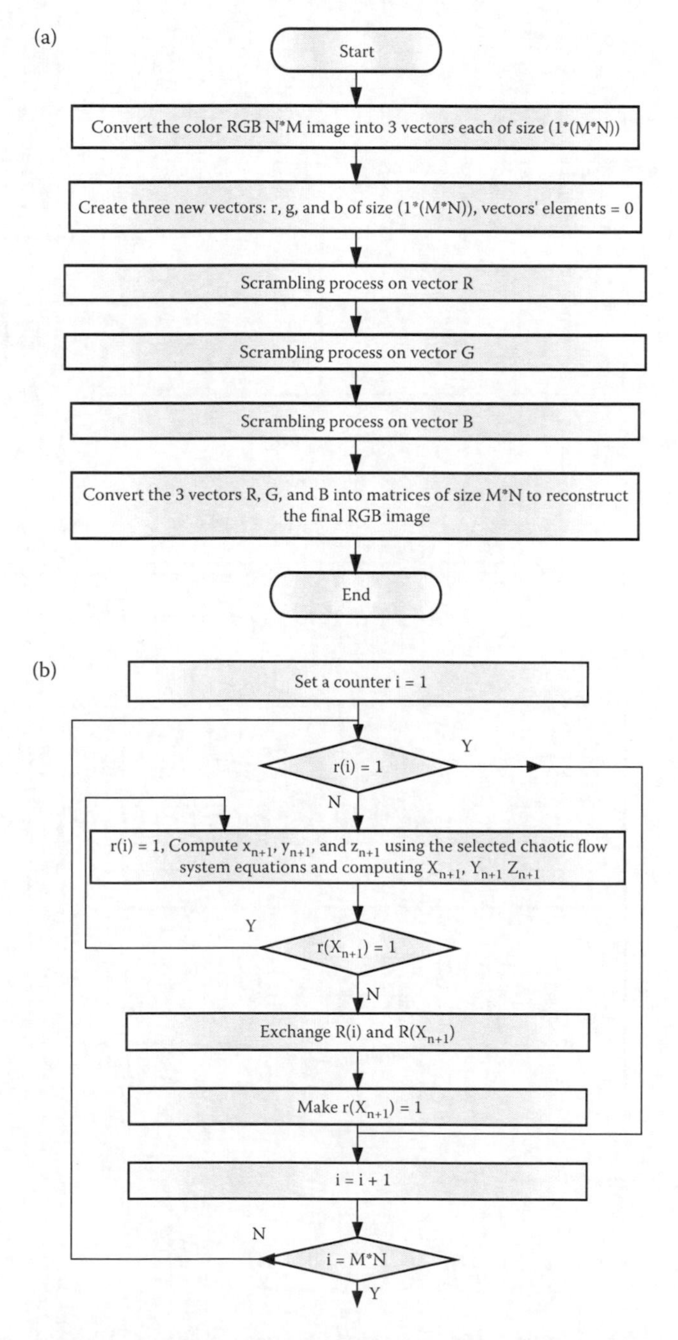

FIGURE 20.1 The proposed system flowchart. (a) The flowchart of the overall proposed system. (b) The scrambling process of vector R.

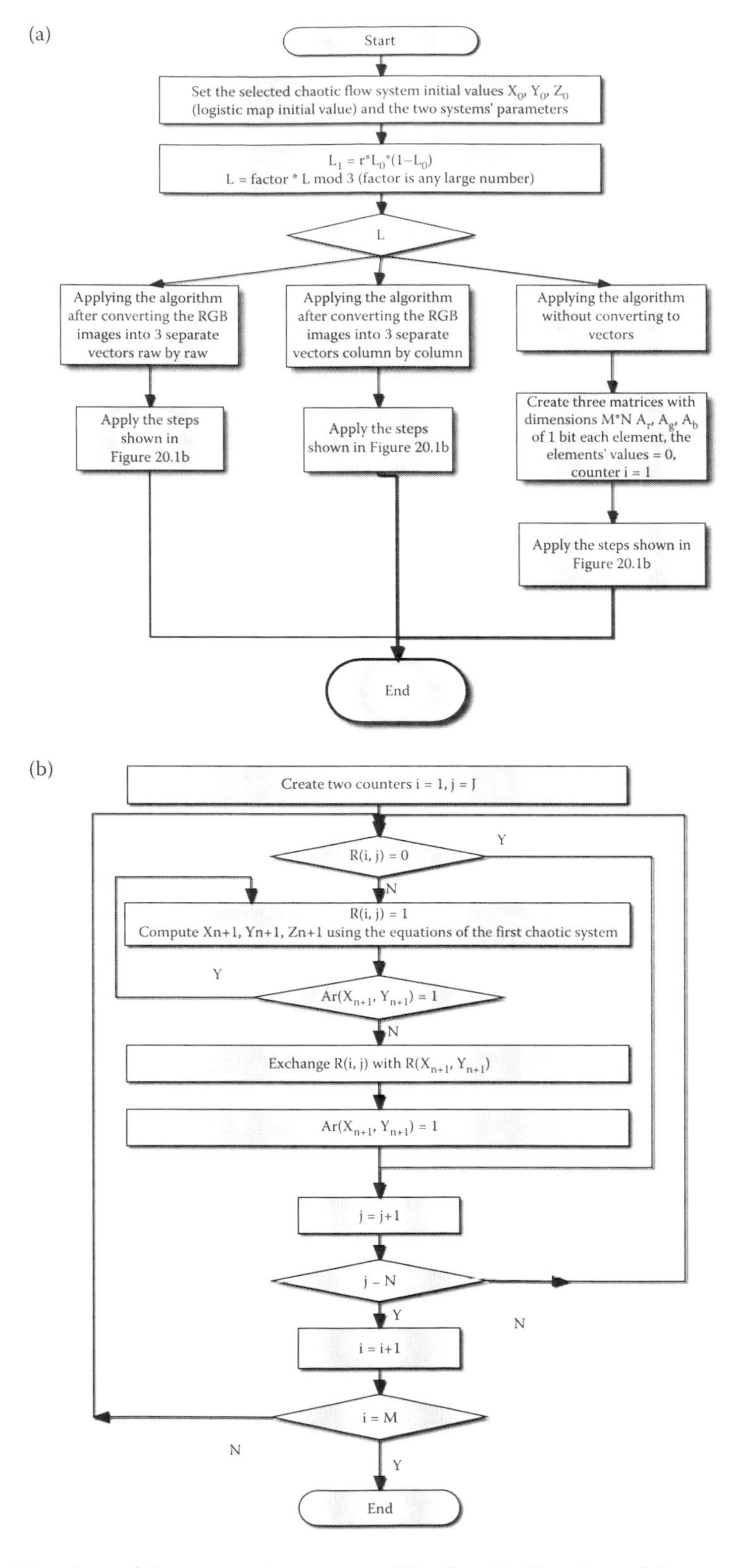

FIGURE 20.2 Flowchart of the proposed system modification. (a) Flowchart of the overall modified proposed system. (b) Detailed steps of right-bottom block of (a).

TABLE 20.2
Parameters of Chaotic Flow Systems Used in Simulations

Chaotic Flow System	Parameters
Lorenz	a = 10, b = 24, c = 8/3
Rössler	a = 0.2, b = 0.2, c = 5.7
Chua	$\delta = 10$, b = 14.78, $\gamma = 0.0385$, $m_1 = -1.27$, $m_0 = -0.68$
Nien	$\delta = 6.3$, $\beta = 0.7$, $\gamma = 7$, b = −0.714, a = −1.143, $I_o = -3$

scrambled image, and the result will be the original image. It is important to use the same initial value and the same parameters of the same chaotic flow system that had been used in the scrambling process. Any tiny difference in one of them and the descrambling result will be incorrect and cannot be descrambled. The scrambling process of the proposed system needs only one iteration to get the final scrambling image result, and the descrambling process of the proposed system needs only one iteration to get the final descrambling image result, which is the original image.

The proposed algorithm is modified to three algorithms to increase scrambling complexity. The modifications are done by converting the RGB images into three separate vectors. The first modification is by taking raw by raw to create the vector, for example, the second value of the vector will be the pixel in position (1,2) of the original image and so on. The second modification is by taking column by column, for example, the second value of the vector will be the pixel in position (2,1) of the original image and so on. The third modification is done without any change (which means without converting the vectors).

Figure 20.2 shows a flowchart of the proposed system modification. The proposed system has been simulated via MATLAB. Because the Arnold cat map is applied only to N*N image, the proposed system is applied to N*N images for comparison purposes. Then the proposed system is applied to a general N*M images. Color image of size 256*256 (RGB) is used for Lorenz, Rössler, Chua, and Nien as well as Arnold's cat map according to the proposed algorithm. The same initial values which are x = 3, y = 3, z = 3 are used for all systems considered. The parameters of chaotic flow systems used are given in Table 20.2.

20.3 SIMULATION RESULTS

Figure 20.3 shows the original image (Figure 20.3a) of size 256*256 with the resultant encrypted images when using Lorenz, Rössler, Chua, and Nien sequences (Figure 20.3b–e, respectively) and the decrypted image (Figure 20.3f) after implementing the proposed system. As is shown in the figure, only one iteration is needed to successfully decrypt the image using any one of the chaotic flow systems considered.

Figure 20.4 shows the corresponding results when implementing the scrambling system using Arnold's cat map using the same original image (Figure 20.4a). A huge number of iterations are required to achieve similar quality scrambling of the proposed system using chaotic flow sequences. The figure shows four values of iterations: 1, 10, 60, and 192 (Figure 20.4b–e, respectively). It can be clearly seen that the decrypted image is recovered successfully after 192 iterations.

Figure 20.5 shows the original image of size 225*400 with the resultant encrypted images after implementing the proposed system. Similarly to the discussion of Figure 20.1, after only one iteration, the encrypted image is decrypted using any type of the chaotic flow sequences

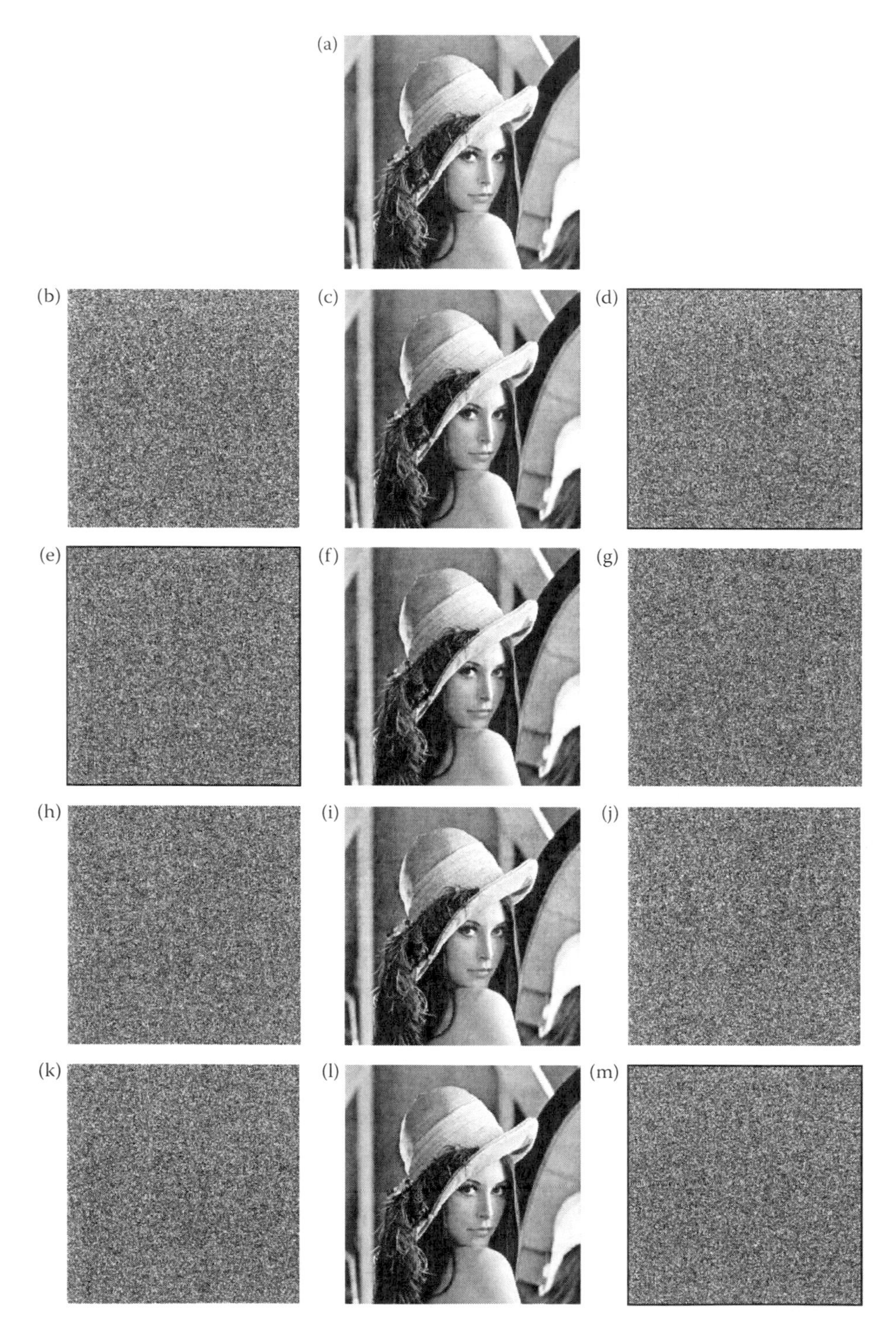

FIGURE 20.3 Results of the proposed system for image of size 256*256 using chaotic flow sequences. (a) Original image. (b) Scrambled image using Nien system $x_o = 0.1$, $y_o - 0.1$, $z_o = 0.1$. (c) Descrambled image. (d) Descrambled image if x_o is 0.100000000001 instead of 0.1. (e) Scrambled image using Chua system $x_o = 0.1$, $y_o = 0.1$, $z_o = 0.1$. (f) Descrambled image. (g) Descrambled image if y_o is 0.100000000001 instead of 0.1. (h) Scrambled image using Rössler system $x_o = 0.2$, $y_o = 0.2$, $z_o = 0.2$. (i) Descrambled image. (j) Descrambled image if z_o is 0.200000000001 instead of 0.2. (k) Scrambled image using Lorenz system $x_o = 0.2$, $y_o = 0.2$, $z_o = 0.2$. (l) Descrambled image. (m) Descrambled image if x_o is 0.200000000001 instead of 0.2.

FIGURE 20.4 Results of the scrambling system for image of size 256*256 using Arnold's cat map. (a) Original image. (b) Image after 1 iteration. (c) Image after 10 iterations. (d) Image after 60 iterations. (e) Image after 192 iterations.

considered. Tables 20.3 and 20.4 show the statistical measurements of scrambling quality and correlation coefficient, respectively, of the proposed system in comparison with the Arnold's cat map for the image size 256*256, whereas Tables 20.5 and 20.6 show the statistical measurements of scrambling quality and correlation coefficients, respectively, when the image size is 225*400.

As is shown in Tables 20.3 and 20.4, for an image of size N*N, the proposed system provides high ADC, high DSF, and a very low correlation coefficient from the first iteration and for the different values for the three matrices (red, green, and blue). Arnold's cat map requires more than 10 iterations to have an ADC, DSF (for the same values of the three matrices [red, green, and blue]), and a correlation coefficient closed to that of the proposed system.

The main disadvantage of ACM is that it requires more than one iteration, which means that the values will be changed, and provides low ADC and DSF and a high correlation coefficient. From other hand, the main advantage of the proposed system is that when using an M*N image, and as it is shown in Tables 20.4 and 20.5, high ADC, high DSF, and a very low correlation coefficient are obtained from the first iteration, yet this is not applicable for ACM, because of the nonequal dimensions of the image.

Figure 20.6 shows the required time for whole scrambling and descrambling processes of the proposed system as compared with that when using the traditional schemes with equal image dimension as a parameter when simulated in MATLAB. A personal computer (HP Elitebook2540p, CPU Intel core i5) is used to carry out simulations. The image dimensions are taken from Table 20.1 to compare the processing time of the traditional algorithm with the proposed one. As shown in Figure 20.6, the proposed system is faster than the traditional scheme by about 13% when the image size is 512*512, whereas for image sizes 256*256 and 128*128 the proposed algorithm is faster by about 49% and 79%, respectively. Figure 20.7 shows the required time for scrambling and descrambling processes for an image of size 225*400 when using different chaotic systems. It can be seen in this figure that the Lorenz system offers the best time performance among other systems, whereas the Nien system acts as the worst one.

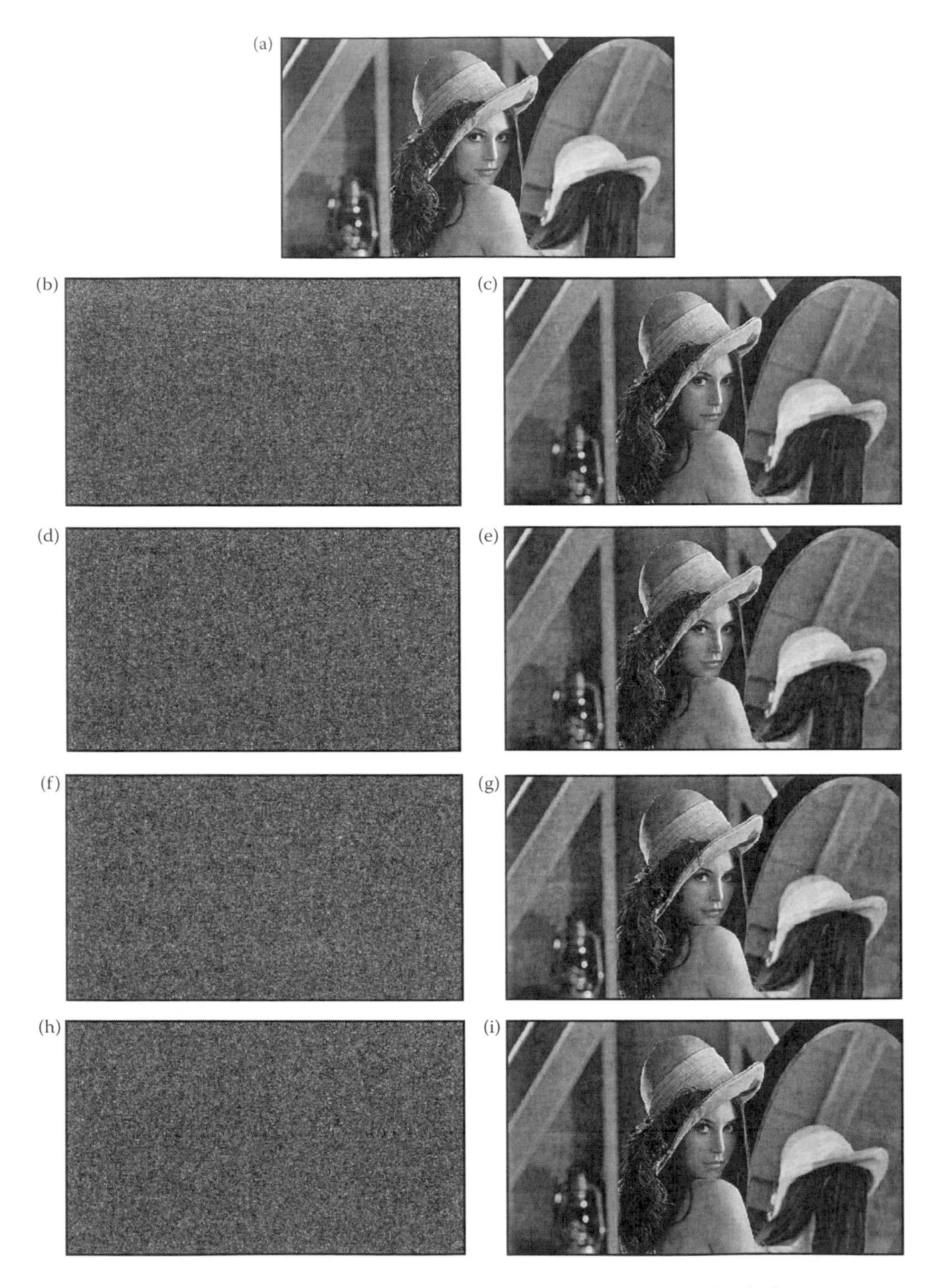

FIGURE 20.5 Results of the proposed system for image of size 225*400 using chaotic flow sequences. (a) Original image. (b) Scrambled image using Rössler system. (c) Descrambled image of Rössler system. (d) Scrambled image using Nien system. (e) Descrambled image of Nien system. f) Scrambled image using Chua system. (g) Descrambled image of Chua system. (h) Scrambled image using Lorenz system. (i) Descrambled image of Lorenz system.

TABLE 20.3
Statistical Measurements of Scrambling Quality of Image 256*256

Chaotic Flow System		ADC				DSF			
		R	G	B	Average	R	G	B	Average
The proposed algorithm	Lorenz	161.9826	163.9418	161.0460	162.3234	0.3695	0.3692	0.3683	0.3690
	Rössler	163.8700	164.9703	161.2583	163.3662	0.3700	0.3721	0.3711	0.3711
	Chua	168.0810	169.0134	160.8444	165.9796	0.3708	0.3693	0.3707	0.3703
	Nien	172.0441	163.6819	167.8763	167.8674	0.3704	0.3703	0.3700	0.3702
The traditional algorithm	Arnold m = 1	129.1744	129.1744	129.1744	129.1744	0.3841	0.3841	0.3841	0.3841
	Arnold m = 10	161.3050	161.3050	161.3050	161.3050	0.3701	0.3701	0.3701	0.3701
	Arnold m = 60	154.4887	154.4887	154.4887	154.4887	0.3694	0.3694	0.3694	0.3694
	Arnold = 192	0	0	0	0	0	0	0	0

TABLE 20.4
Correlation Coefficients of Image 256*256

Chaotic Flow System		Correlation Coefficient				PSNR
		R	G	B	Average	Average
The proposed algorithm	Lorenz	0.0012	0.0068	–0.0101	–0.0007	11.195
	Rössler	0.0002	0.0135	–0.0032	0.0035	11.199
	Chua	0.0069	–0.0087	–0.0066	–0.0028	11.681
	Nien	0.0045	–0.0064	–0.0062	–0.0027	11.720
The traditional algorithm	Arnold m = 1	0.0517	–0.0359	–0.0632	–0.0158	11.6420
	Arnold m = 10	–0.0031	–0.0046	–0.0056	–0.0044	11.6748
	Arnold m = 60	–0.0004	0.0007	0.0023	0.0009	11.6976
	Arnold = 192	0	0	0	0	N/A

TABLE 20.5
Statistical Measurements of Scrambling Quality of Image 225*400

Chaotic Flow System		ADC				DSF			
		R	G	B	Average	R	G	B	Average
The proposed algorithm	Lorenz	222.0945	231.9909	228.0960	227.3938	0.3638	0.3619	0.3651	0.3636
	Rössler	229.8477	233.7200	229.4603	231.0093	0.3628	0.3642	0.3629	0.3633
	Chua	213.1405	225.8605	222.9630	220.6547	0.3631	0.3634	0.3631	0.3632
	Nien	219.6986	223.6542	228.7551	224.0360	0.3637	0.3640	0.3642	0.3640
The traditional algorithm	Arnold m = 1	N/A				N/A			
	Arnold m = 10								
	Arnold m = 60								
	Arnold = 192								

TABLE 20.6
Correlation Coefficients of Image 225*400

Chaotic Flow System		Correlation Coefficient R	G	B	Average	PSNR Average
The proposed algorithm results	Lorenz	−0.0087	0.0042	−0.0012	−0.0019	13.13093085
	Rössler	0.0061	0.0004	0.0058	0.0041	13.13617567
	Chua	−0.0133	0.0024	−0.0046	−0.0052	13.13859637
	Nien	−0.0017	0.0052	0.0002	0.0012	13.12656835
The traditional algorithm		N/A				

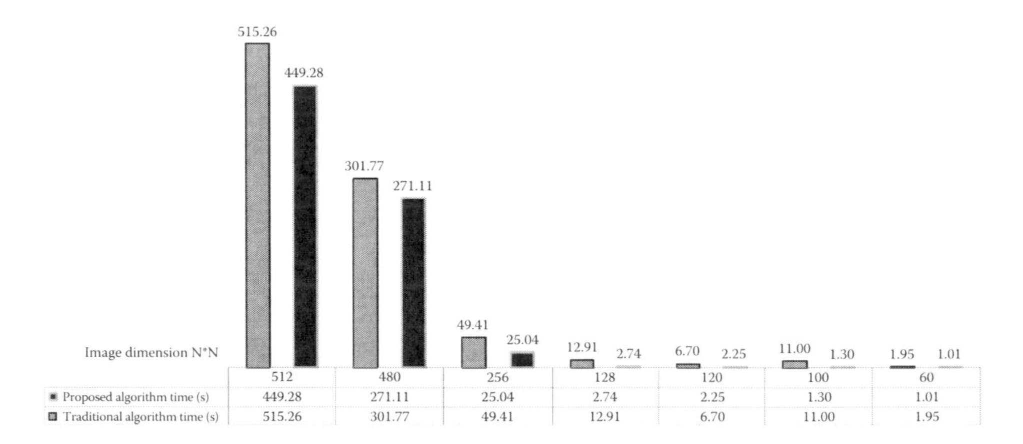

FIGURE 20.6 Required time for scrambling and descrambling processes of the proposed system for equal dimension image.

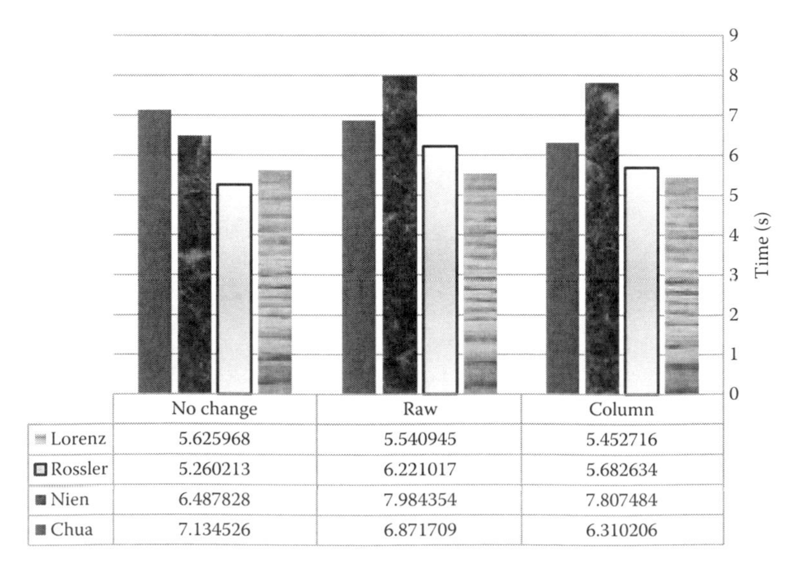

FIGURE 20.7 Required time for scrambling and descrambling processes of the proposed system for non-equal dimension image (225*400).

20.4 CONCLUSIONS

This chapter proposes a new scheme for digital image scrambling using chaotic flow sequence. First, the N*M color RGB image is converted into three 1D vectors where D = N*M. Then the positions of the pixels are changed according to the generated number from the chaotic system that had been used, which define the pixel position for each vector of the image. The image then is reconstructed after changing the locations of all the pixels. The scheme performance is effectively evaluated by applying three types of tests, which are DSF, ADC, and 2D correlation coefficient. The experimental result shows that the proposed system has a higher degree of scrambling from the first iteration compared with the traditional scrambling system such as Arnold's cat map. Another advantage of the proposed system is the implementation on equal and nonequal dimension images. Future improvement of the proposed system can include using hybrid chaotic flows and maps, changing the value of the pixels in the same time when changing their position to obtain a higher degree of security.

REFERENCES

Bakhshandeh A. and Eslami Z. 2013. An authenticated image encryption scheme based on chaotic maps and memory cellular automata. *Optics and Layers in Engineering*, Vol. 51, pp. 665–673.

Gao T. and Chen Z. 2008. A new image encryption algorithm based on hyper-chaos. *Physics Letters A*, Vol. 372, pp. 394–400.

Gaspard P. 2005. *Rössler Systems, Encyclopedia of Nonlinear Science*. A. Scott (editor), New York, pp. 808–811.

Gupta P. and Singh S. 2014. Image encryption based on Arnold cat map and S-box. *International Journal of Advanced Research in Computer Science and Software Engineering*, Vol. 4, No. 8, pp. 807–812.

Jolfaei A. and Mirghadri A. 2010. New approach to measure quality of image encryption. *(IJCNS) International Journal of Computer and Network Security*, Vol. 2, No. 8, pp. 38–43.

Keshari S. and Modani S.G. 2011. Image encryption algorithm based on chaotic map lattice and Arnold cat map for secure transmission. *International Journal of Computer Science and Technology IJCST*, Vol. 2, No. 1, pp. 132–135.

Li M., Liang T., and He Y. 2013. Arnold transform based image scrambling method. In *Proceedings of the 3rd International Conference on Multimedia Technology (ICMT 2013)*, A. A. Farag, J. Yiang, and F. Jiao (editor), Springer, New York, pp. 1309–1316.

Liehuang Z., Wenzhuo L., and Lejian L. 2006. A novel image scrambling algorithm for digital watermarking based on chaotic sequences. *IJCSNS International Journal of Computer Science and Network Security*, Vol. 6, No. 8B, pp. 125–130.

Mao Y., Cao L., and Liu W. 2007. *Design and FPGA Implementation of a Pseudo-Random Bit Sequence Generator Using Spatiotemporal Chaos*. Nanjing University of Science & Technology.

Merah L., Ali-Pacha A., Said V., and Mamat M. 2013. Design and FPGA implementation of Lorenz chaotic system for information security issues. *Applied Mathematical Sciences*, Vol. 7, No. 5, pp. 237–246.

Nance J. 2011. Periods of The Discretized Arnold's Cat Mapping and Its Extension to N-Dimensions. arXiv:1111.2984v1, (Nov) [math.DS].

Nien H. H., Huang C. K., Changchien S. K., Shieh H. W., Chen C. T., and Tuan Y. Y. 2007. Digital color image encoding and decoding using a novel chaotic random generator. *Chaos, Solitons and Fractals*, Vol. 32, pp. 1070–1080.

Peterson G. 1997. Arnold's Cat Map. College of the Redwoods.

Prasad M. and Sudha K. L. 2011. Chaos image encryption using pixel shuffling. *Computer Science & Information Technology (CS & IT) International Scientific and Technical Conference*, Vol. 1, No. 2, pp. 169–179.

Tang Z. and Zhang X. 2011. Secure image encryption without size limitation using Arnold transform and random strategies. *Journal of Multimedia*, Vol. 6, No. 2, pp. 202–206.

Wang Y., Lei P., Yang H., and Caob H. 2015. Security analysis on a color image encryption based on DNA encoding and chaos map. Vol. 46, pp. 433–446.

Willsey M.S., Cuomoy K., and Oppenheimz A. 2010. Selecting the Lorenz parameters for wideband radar waveform generation. *International Journal of Bifurcation and Chaos*. Vol. 21, pp. 2539–2545.

21 Efficient Two-Stage Sensing Method for Improving Energy Consumption in Cognitive Radio Networks

Hikmat N. Abdullah and Hadeel S. Abed

CONTENTS

21.1 Introduction 281
21.2 Energy Detection-Based Spectrum Sensing 283
21.3 The Design System Model 283
21.4 Available Methods to Improve Energy Consumption 283
21.4.1 Time-Saving and Energy-Efficient One-Bit Cooperative Spectrum Sensing (TSEEOB-CSS) Schemes 285
21.4.2 Cooperative Coarse Sensing Scheme for Cognitive Radio Network (CC4C) Method 285
21.5 The Proposed Method 286
21.6 Simulation Results and Discussion 289
21.6.1 Noncooperative Scenario 289
21.6.2 Cooperative Scenario 290
21.7 Conclusions 294
References 294

21.1 INTRODUCTION

The concept of cognitive radio (CR) is that unlicensed users (cognitive radio users) can access the spectrum owned by licensed users (primary users) while they cannot interfere with primary users when they exploiting spectrum. Thus to realize the technique of cognitive radio, a cognitive radio user must have the ability to measure, sense, and learn channel characteristics and availabilities (Kokare and Kamble 2014). One of the main challenges in CR networks is the high energy consumption, especially in battery-powered terminals. In order to identify the unused spectrum portions, the unlicensed users, also called cognitive users (CUs), are enforced to sense it for a specific period, causing energy consumption (Althunibat et al. 2015). Different spectrum sensing techniques are available. In this chapter, energy detection is used, which is a simple method for spectrum sensing and is one of the most commonly employed spectrum-sensing schemes. It does not require any prior knowledge of the structure of the primary users' (PUs) signals (Emara et al. 2015). Two scenarios of sensing spectrum are usually used: noncooperative and cooperative. In the noncooperative scenario, a single user is used to make the sensing, whereas in the cooperative scenario (CSS) multiple cognitive users sense in cooperation to reduce the effect of fading (Emara et al. 2015). In CSS, the local sensing results are reported to a central entity, called the fusion center (FC). The FC is in charge of making a global decision regarding the spectrum occupancy by applying a specific fusion rule (FR) (Althunibat et al. 2015). There are two popular schema used to represent the local result: hard-based scheme and soft-based scheme. In this chapter the hard scheme is used because it consumes less energy when all CUs convey their local decisions as a single bit per CU toward the FC consecutively: bit "1" indicates the PU presence and bit "0" indicates the PU absence (Althunibat 2014). The rule that is employed by the FC is the OR rule, which implies that if at least one CU makes a local decision of busy (or 1), the global decision will be busy (or 1). Otherwise, the global decision will be free (or 0) (Althunibat 2014). Many works have investigated the reduction of energy consumption. Sequential sensing as an approach to reduce the average number of sensors required to reach a decision is studied comprehensively in Kim et al. (2010), Kim and Giannakis (2009, 2010), Zou et al. (2010), and Awin et al. (2015). In Cheng et al. (2012), the CUs are divided into nondisjoint subsets in a way that only one subset senses the spectrum while the other subsets stay in a low power mode. In Maleki and Leus (2013) censoring and sequential censoring methods are presented to improve energy consumption. However, although these methods improve the energy consumption, they have some drawbacks. These methods use two thresholds and this increases the complexity. In censoring methods, the entire spectrum is sensed intensively, which increases the energy consumption. The sequential censoring methods have a weakness in the performance of detection since they stop the sensing process once the accumulated energy crosses the upper threshold without sensing other channels that may have higher accumulated energy, that is, more probable existence of PU signal.

In Ergul and Akan (2014) a coarse–fine sensing method called CC4C (cooperative coarse sensing scheme for cognitive radio network) is presented where coarse sensing is used to find the available channel together with sequential sensing scheme. Although this method reduces the sensing time, it also has a problem that it uses sequential sensing, which does not select the optimal channel available since the sensing is stopped quickly as soon as an available channel is found without checking other channels. Furthermore, for low signal-to-noise (SNR) values (less than 4 dB), this method consumes high energy, because it cannot produce a local decision early and continue to take more samples. In Zhao et al. (2013), a coarse–fine sensing method called time saving energy efficient one-bit based-cooperative spectrum sensing (TSEEOB-CSSS) where the PU presence or absence may be declared in the coarse sensing stage without using the fine sensing stage. This method has two problems: First, it uses a two-threshold comparison process at coarse sensing, which increases the complexity. Second, once the upper threshold is crossed, the sensing process stops without performing fine sensing, which may lead to either selecting the channel of less probability to be vacant or a false alarm state when the SNR is low, which, as a result, increases the energy consumption. Du et al. (2016) also propose a two-stage coarse–fine sensing to improve detection performance. In this method energy detection technique is applied in the coarse stage, while the first-order cyclostationary technique is applied in the fine stage. Although this method gives better detection performance than the previous ones, the cyclostationary sensing technique is complex in implementation and needs a higher sensing time, which implies increasing the energy consumption.

In this chapter we propose a sensing method to overcome the problems mentioned in the previous works. It does sensing in two stages: The first stage is to vastly sense all the channels in the spectrum with a low number of samples and then identify the channel that has the maximum accumulated energy (the channel that is more likely to have a PU signal). Then a fine sensing is done only on this channel with a higher number of samples for verification. The main contributions in this chapter are (1) the improvement of energy consumption based on increasing the detection probability since all channels will be sensed without exclusion and the best one is selected for further verification, which makes the method very efficient at low SNR values; (2) a multiprimary user scenario is assumed, which is not considered in many previous works; and (3) analytical expression for the energy consumption of the proposed method is derived for both sensing and transmission stages assuming the use of ZigBee as a CU.

21.2 ENERGY DETECTION-BASED SPECTRUM SENSING

Energy detection is considered as one of the widely used spectrum sensing techniques due to its simplicity, as it does not need any information about the structure of the PUs signals. In CR networks, the secondary user (SU) checks the spectrum allocated to the PU, and when it detects the absence of PU transmission, it starts data transmission to its receiver. The received samples at the CU receiver are (Emara et al. 2015)

$$Y(n) = h_{ps}\theta X_p(n) + W(n) \tag{21.1}$$

where $X_p(n)$ is the signal of the PU, h_{ps} is the channel gain between the PU and SU, $W(n)$ is the additive white Gaussian noise (AWGN) at the CU receiver, and θ is the PU activity indicator that can take one out of two values as in the following equation:

$$\theta = \begin{cases} 0 & \text{for } H_0 \text{ hypothesis} \\ 1 & \text{for } H_1 \text{ hypothesis} \end{cases} \tag{21.2}$$

When the PU is active it is referred to as hypothesis H_1, whereas the case of inactive PU is referred to as hypothesis H_0. The false alarm and detection probabilities are evaluated by comparing the detector decision metric with a preset threshold λ. The decision metric (E_j) is defined by the energy of the captured samples during the observation window t:

$$E_j = \frac{1}{N}\sum_{n=1}^{N} |Y(n)|^2 \tag{21.3}$$

where N is the number of sensing samples $N = tF_s$, where F_s is the sampling frequency. The probabilities of false alarm and detection are evaluated by

$$P_f = P_r(E_j > \lambda | H_0) \tag{21.4}$$

$$P_d = P_r(E_j > \lambda | H_1) \tag{21.5}$$

21.3 THE DESIGN SYSTEM MODEL

Figure 21.1 explains the system model of the infrastructure-based CRN (centralized CSS), which consists of three stages: sensing stage, transmission stage, and final design stage. It can be seen that from Figure 21.1, the primary user transmitter (PU TX) can communicate with the primary user receiver (PU RX) directly, but when an SU wants to use the license channel, first, it must make a sensing process by using an energy-detection technique through the sensing stage, in which each SU collects a number of samples sensed from the target spectrum. The performance of the sensing process is measured under two channels: an AWGN and an AWGN plus Rayleigh fading. After the sensing process is completed, the SU sends the local sensing results to the FC through the transmission stage, by using a hard-based scheme, assuming no errors are occurred during the transmission. Figure 21.2 shows the procedures to perform the process in Figure 21.1.

21.4 AVAILABLE METHODS TO IMPROVE ENERGY CONSUMPTION

In this section we will present the design of some methods proposed by researchers to improve energy savings in CR, for the purpose of comparison with the proposed method.

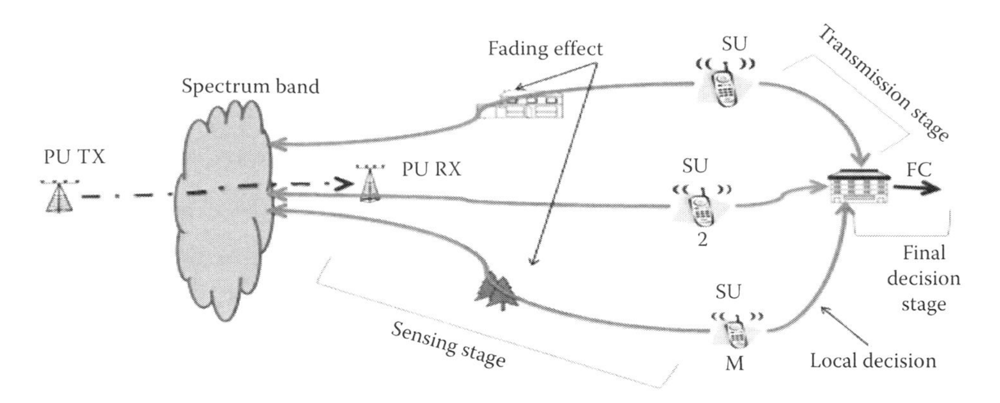

FIGURE 21.1 System design of infrastructure cognitive radio networks.

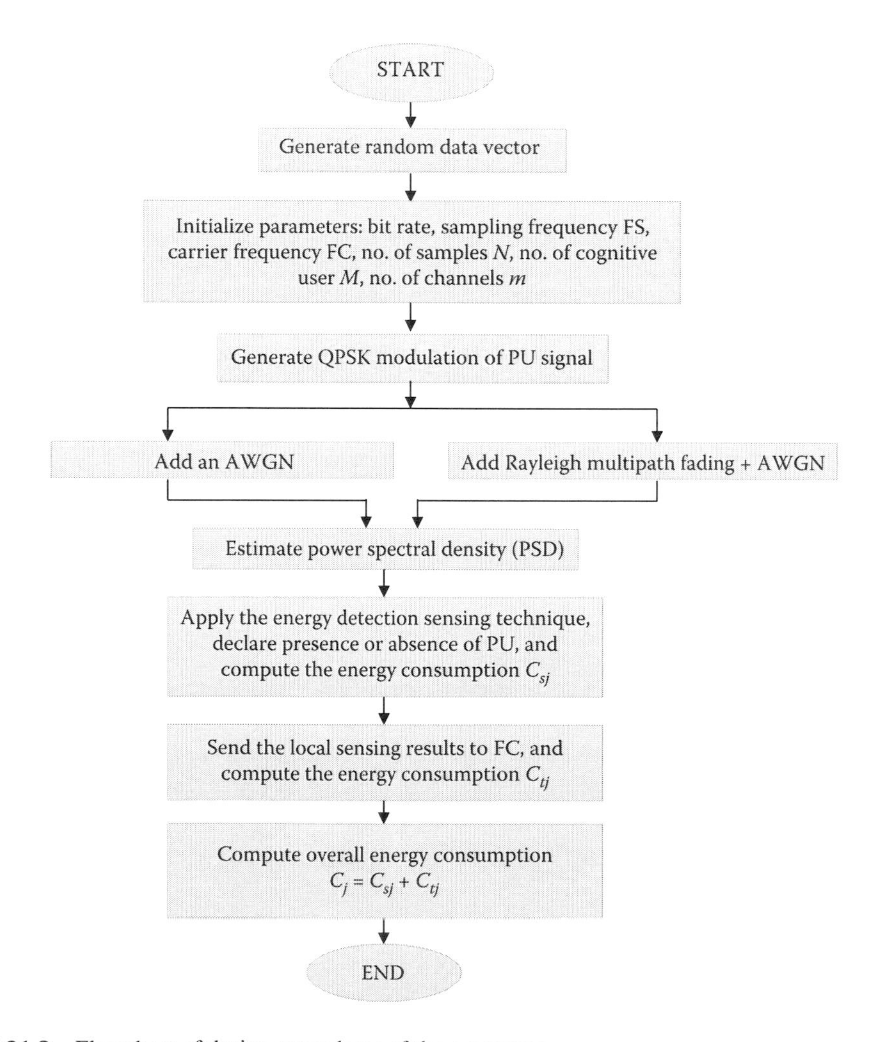

FIGURE 21.2 Flowchart of design procedures of the processes.

21.4.1 Time-Saving and Energy-Efficient One-Bit Cooperative Spectrum Sensing (TSEEOB-CSS) Schemes

In TSEEOB-CSS a two-stage sensing process—coarse and fine stages—is used in a way that applies the coarse sensing stage to the αN samples, and then compares the accumulated energy E_j of each CU with the upper and lower threshold in a way such that if $E_j > \lambda_2$ sends "1" to the FC indicating that the PU is present and stops sensing. However, if $E_j < \lambda_1$ sends "0" to the FC indicating that PU is absence and stops sensing. If $\lambda_1 < E_j < \lambda_2$ no decision is sent to the FC and the fine sensing process is applied to sense $(1 - \alpha)N$, and then a decision is sent to the FC, where $0 < \alpha < 0.5$. Figure 21.3 shows the procedure of design of this method. The amount of energy consumption (C_{sj}) in the sensing is given in Equation 21.6, where C_{ssj} is energy consumption per sample and M is the number of SU (Zhao et al. 2013):

$$C_{sj} = M\alpha N C_{ssj} + ((1-\alpha)N)MC_{ssj} \tag{21.6}$$

21.4.2 Cooperative Coarse Sensing Scheme for Cognitive Radio Network (CC4C) Method

In the CC4C method, a two-stage sensing process—coarse and fine—is used. The procedure of the sensing process is to, first, apply the simple coarse sensing technique based on sequential detection,

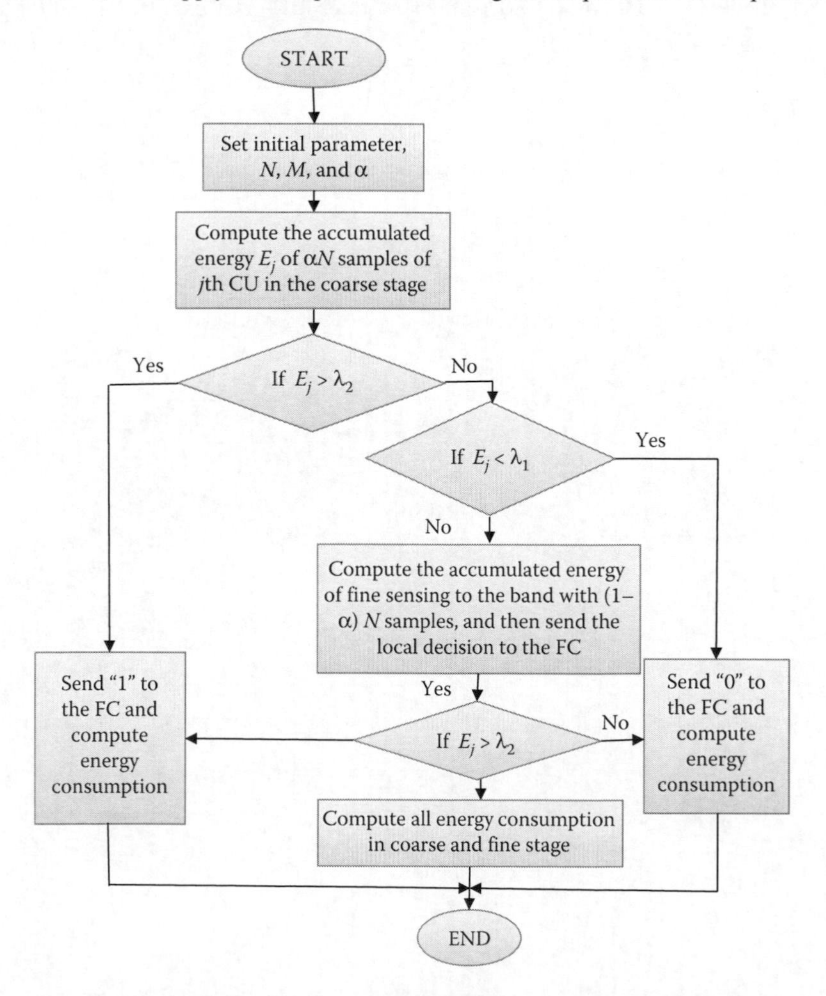

FIGURE 21.3 Flowchart of design of TSEEOB-CSS method.

which yields approximate results for PU occupancy of the spectrum. Based on the results of this coarse sensing, nodes pick the channels that are more likely to be available and then perform fine sensing on these channels. Figure 21.4 shows the procedure of design of the method. The total energy consumption can be computed from Equation 21.7, where C_{sj} is energy consumption in the sensing stage, E_s is energy consumption of one band, and m is the number of bands (Ergul and Akan 2014):

$$C_{sj} = m E_s \tag{21.7}$$

21.5 THE PROPOSED METHOD

The spectrum sensing process consumes energy in both the sensing and transmission stages. The proposed method achieves energy saving in the sensing stage and improves the detection performance. Figure 21.5 shows the flowchart of the proposed method. This method consists of two stages: coarse sensing stage and fine sensing stage. In the coarse stage the CU senses a fixed spectrum length and fixed number of channels with a small number of sensed samples (e.g., 16% of the number of samples is used in the proposed sensing process). The energies obtained from all channels during the coarse sensing are collected (each channel may have a PU signal at the same time) and then a fine sensing is performed to the channel that has maximum energy. Fine sensing senses all samples in the channel selected to confirm. In order to ensure that this signal is a PU signal (not noise), the

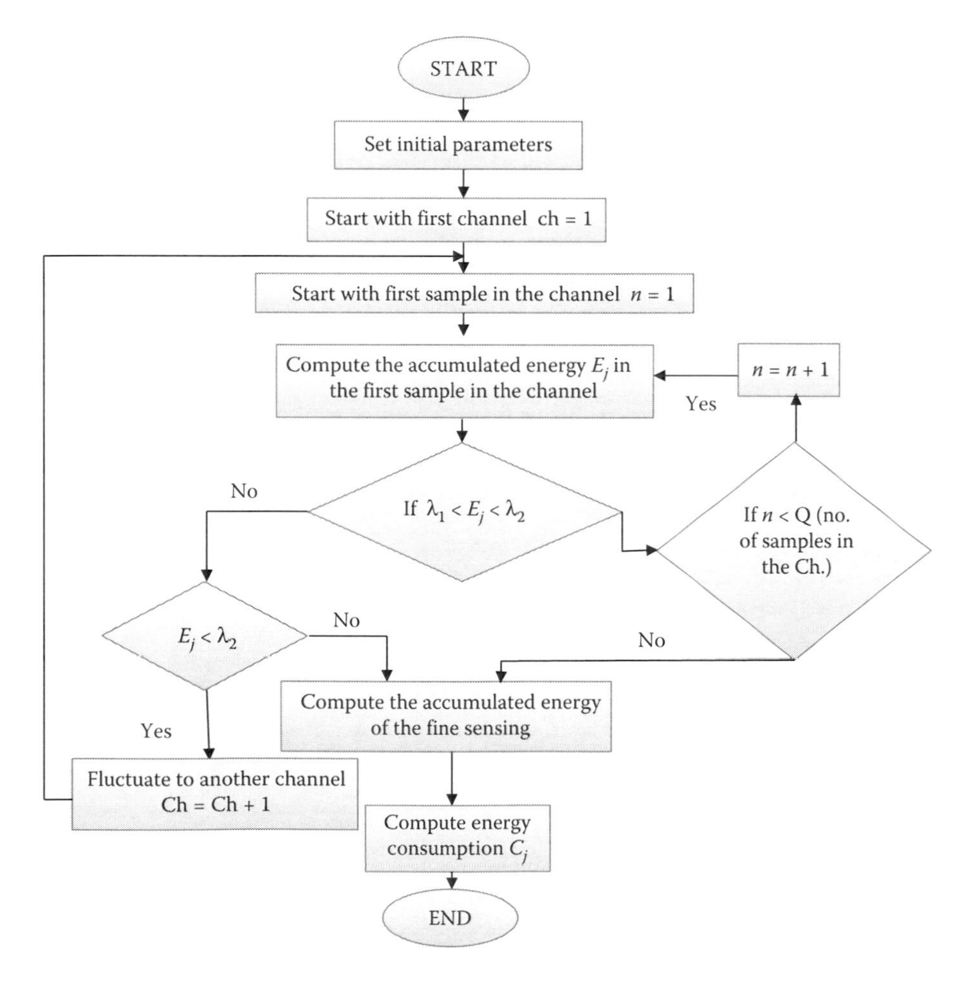

FIGURE 21.4 Flowchart of design of CC4C method.

accumulated energy in the channel is compared with a threshold λ to produce a final decision about PU signal presence or absence. If the requirement is detecting a hole in the spectrum, the same procedure is repeated but fine sensing is employed to the channel that has minimum energy. This method can be applied to find all possible channels to that may have PUs (or holes) by finding the maximum channels (or minimum) and subtracting the accumulated energy contained in these channels from all other channels. If the difference is small, this means that these channels also have (PU or hole).

In general, the amount of energy consumption achieved by the jth CU (C_j) is given by

$$C_j = C_{sj} + C_{tj} \tag{21.8}$$

$$\text{With } C_{sj} = NC_{ssj} \tag{21.9}$$

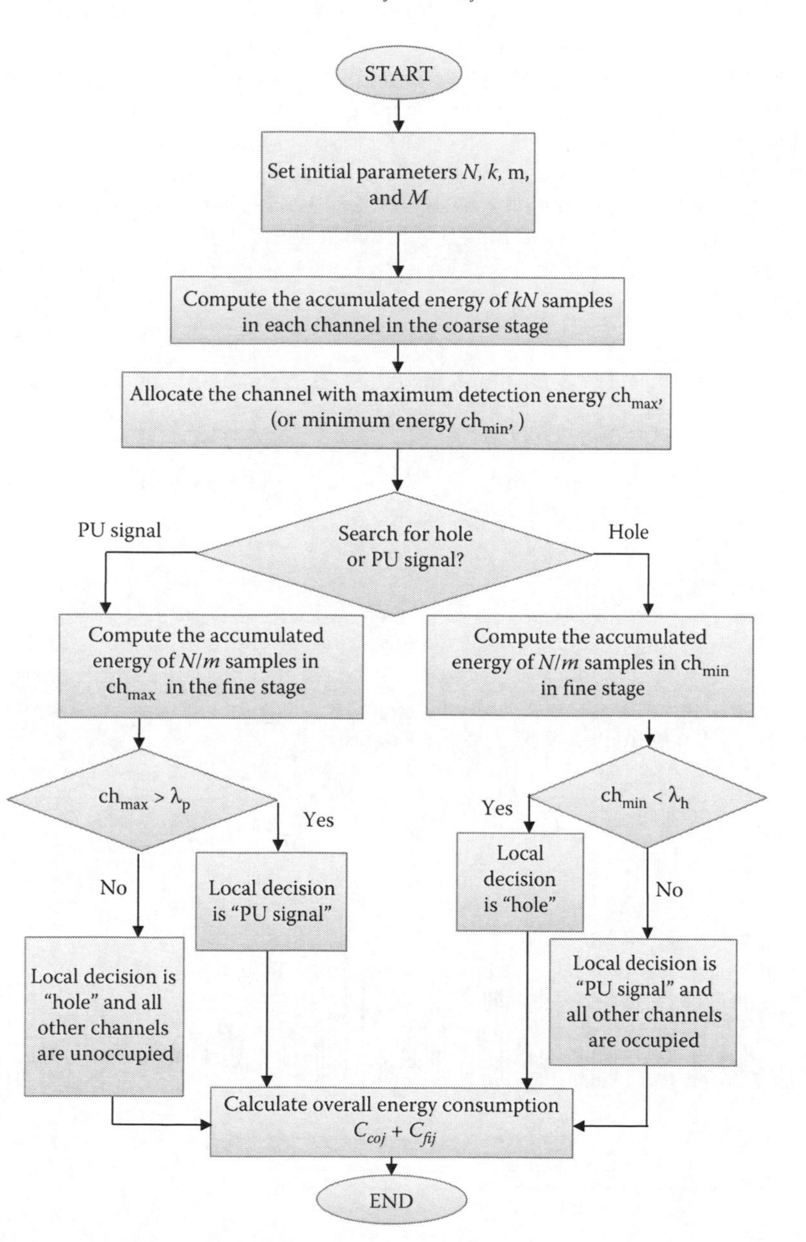

FIGURE 21.5 Flow chart of the proposed method.

where C_{sj} and C_{tj} are the energies consumed by the jth CU in sensing and transmission (per bit) stages, respectively; N is the number of sensed samples; and C_{ssj} is the energy consumption per sample. As stated by Maleki and Leus (2013), C_{ssj} is fixed and depends only on the sampling rate and energy consumption of the sensing module, and computed assuming $P_d = 1$, (i.e., $C_{ssj|Pd=1}$). When P_d is decreased, C_{ssj} is increased since the energy detector will produce a false decision, which causes the sensing process to repeat. According to this discussion C_{ssj} will be

$$C_{ssj} = C_{ssj|Pd=1} + C_{ssj|Pd=1}(1 - P_d) \tag{21.10}$$

considering a sensor used by CU based on IEEE 802.15.4/ZigBee radios. The sensing energy for each decision consists of two parts: (1) the energy consumption involved in listening over the channel and in making the decision, and (2) the energy consumption of the signal processing part for modulation, signal shaping, and so on. The number of samples per detection interval used in our simulation is chosen to be five according to Maleki et al. (2011). This interval corresponds to a detection time of 1 μs. Considering the fact that the typical circuit power consumption of ZigBee is approximately 40 mW, the energy consumed for listening is approximately 40 nJ. Therefore, we conclude that energy consumption per sample is 40 nJ/5 = 8 nJ. So, $C_{ssj} = 8$ nJ will be used. According to the proposed algorithm, Equation 21.9 can be written as

$$C_{sj} = C_{coj} + C_{fij}, \tag{21.11}$$

where C_{coj} and C_{fij} are the energy consumptions in the coarse and fine stages, respectively, with

$$C_{coj} = NkC_{ssj} \tag{21.12}$$

$$C_{fij} = \frac{N}{mC_{ssj}} \tag{21.13}$$

where k is the sensing ratio, which is defined as the number of sensed samples to the total number of samples N; and m is the number of channels in the spectrum. Figure 21.6 explains the division process of the spectrum used in the proposed coarse–fine sensing method. So, the total energy consumption per sensor in both the coarse and fine stages is

$$C_{sj} = C_{coj} + C_{fij} = \left(k + \frac{1}{m}\right)NC_{ssj} \tag{21.14}$$

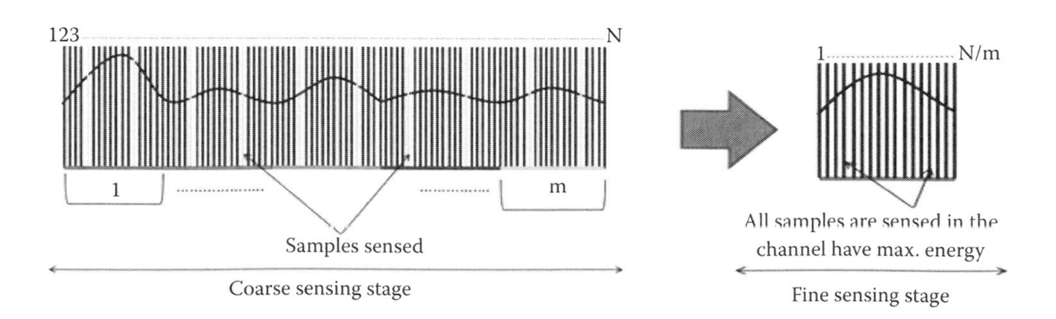

FIGURE 21.6 Division of spectrum using the proposed coarse–fine sensing method.

The amount of energy consumption per bit in the transmission stage is computed as follows: According to Maleki et al. (2011), the consumed energy related to the signal processing part in the transmission mode for a data rate of 250 kb/s, a voltage of 2.1 V, and current of 17.4 mA is approximately 150 nJ/bit. The transmitter dissipates energy to run the radio electronics and the power amplifier. To transmit one bit over a distance (d), the radio spends

$$C_{tbj} = C_{t\text{-}elec\ j} + e_{ampj}d^2 \tag{21.15}$$

where $C_{t\text{-}elec\ j}$ is the transmitter electronics energy and e_{ampj} is the amplification required to satisfy a given receiver sensitivity level. Assuming a data rate of 250 kb/s and a transmission power of 20 mW, $C_{t\text{-}elec\ j} = 80$ nJ. To satisfy a receiver sensitivity of 90 dBm at an SNR of 10 dB, e_{ampj} is 40.4 pJ/m^2. So, the transmitted energy per bit for the jth CU would be

$$C_{tj} = C_{tbj} + 150 \text{ nJ/bit} \tag{21.16}$$

Substituting Equations 21.16 and 21.12 in Equation 21.8 we get

$$C_j = \left(k + \frac{1}{m}\right) NC_{ssj} + C_{tbj} + 150 \,\text{nJ/bit} \tag{21.17}$$

This equation represents the target function with which we can adapt the energy consumed by changing the design parameters k and N. In this chapter, the OR rule is used in FC to make the final decision. This way, the global probability of detection Q_d is (Maleki and Leus 2013)

$$Q_d = 1 - \prod_{j=1}^{M} (1 - P_d) \tag{21.18}$$

where P_d is the local probability of detection and M is the number of cognitive users.

21.6 SIMULATION RESULTS AND DISCUSSION

This section presents the simulation results of energy-detector performance and energy consumption of the proposed scheme under cooperative and noncooperative scenarios. The performance is tested under AWGN and Rayleigh multipath fading channels. All results in this section include the amount of energy consumption of CU in the sensing and transmission stages. To evaluate the performance of the proposed method, it is compared with the traditional sensing method that senses all spectrum samples (i.e., $k = 1$) and with CC4C (Ergul and Akan 2014) and TSEEOB-CSS (Zhao et al. 2013) methods that are based on two stages. The simulation parameters used are shown in Table 21.1. The multipath fading used is "ITU indoor channel model," as shown in Table 21.2 (Pätzold 2000).

21.6.1 NONCOOPERATIVE SCENARIO

Figures 21.7 and 21.8 show the proposed method performance curves of energy consumption of CU versus E_b/N_o in AWGN and Rayleigh multipath fading, respectively, as compared with the traditional method. In Figure 21.7 it can be seen that energy consumption decreases as E_b/N_o increases since fewer sensing samples are required by CU in the sensing process to detect the PU signal when E_b/N_o is high. Also it is noted that the proposed method achieves an improvement in energy consumption when compared with the traditional method. For example, when E_b/N_o equals 0 dB,

TABLE 21.1
Simulation Parameters

Parameter Name	Value
Bit rate	2 Mbps
Number of channels (m)	10
Modulation type	QPSK (PU signal)
Probability of false alarm	10^{-3}
Sensing ratio (k)	16%
Total number of samples in the spectrum (N)	1000
Samples per symbol	100
Distance (d)	10 m

energy consumption decreases by 73% using the proposed method. Figure 21.8 shows the same performance in Figure 21.7 but with more energy consumption due to the fading effect. Figures 21.9 and 21.10 show the performance curves of probability of detection versus E_b/N_o in AWGN and Rayleigh multipath fading channels, respectively. It can be seen that in Figure 21.9 that the detection performance increases as E_b/N_o increases and the proposed method outperforms the traditional method. The improvement in detection probability introduced by the proposed method becomes more significant in the small values in E_b/N_o. For example, when E_b/N_o equals 3 dB, P_d is increased from 0.42 in the traditional method to 0.65 in the proposed method. Figure 21.10 shows the same performance as in Figure 21.9 but with slight degradation in detection performance due to the effect of fading. This improvement in detection performance of the proposed method is referred to with the use of the fine sensing stage that senses the channel with the sensing ratio 100% increasing the accuracy in detection.

21.6.2 Cooperative Scenario

In the cooperative scenario, five CUs and the OR rule in the FC for the final decision are used. Figures 21.11 and 21.12 show the performance curves of energy consumption and Q_d versus E_b/N_o, respectively. The scenarios considered assume that only three CUs out of five CUs suffer from multipath fading. As depicted by Figure 21.11, the average energy consumption per sensor is increased as E_b/N_o is increased. When we compare this figure with the noncooperative scenario given in Figure 21.12, we can see that the reduction in energy consumption is increased in both the traditional and proposed methods. For example, when E_b/N_o equals 0 dB, energy consumption reduces by 72% using the proposed method. The reason behind this improvement is that the CUs will share

TABLE 21.2
Multipath Fading Properties of ITU Indoor Channel Model

Tap	Relative Delay (ns)	Average Power (dB)	Doppler Spectrum
1	0	0	Flat
2	50	−3.0	Flat
3	110	−10.0	Flat
4	170	−18.0	Flat
5	290	−26.0	Flat
6	310	−32.0	Flat

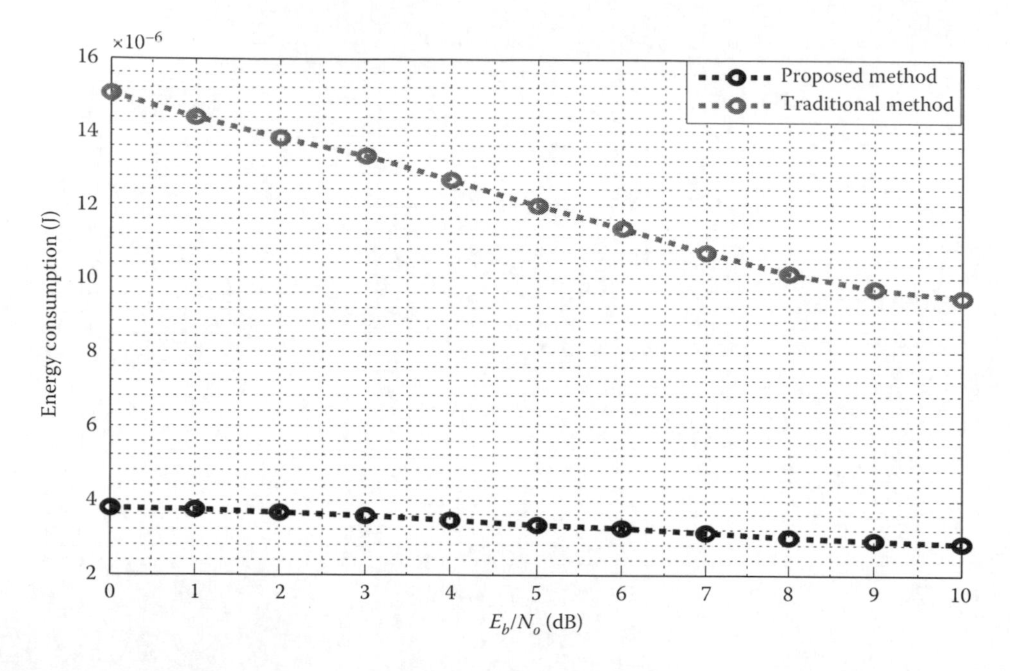

FIGURE 21.7 Energy consumption versus E_b/N_o in AWGN channel.

the statistics about PU existence, which will increase the overall detection probability as shown in Figure 21.12. Also in Figure 21.12 it can be seen that the proposed method has more improvement in detection performance than the traditional method. This is because of the use of two stages in the detection process, which increases detection probability, especially in fine sensing. Figure 21.13 shows the performance curves of energy consumption versus E_b/N_o of the proposed method in comparison with TSSEE-CSS, CC4C, and traditional methods. In this result we used five CUs with the

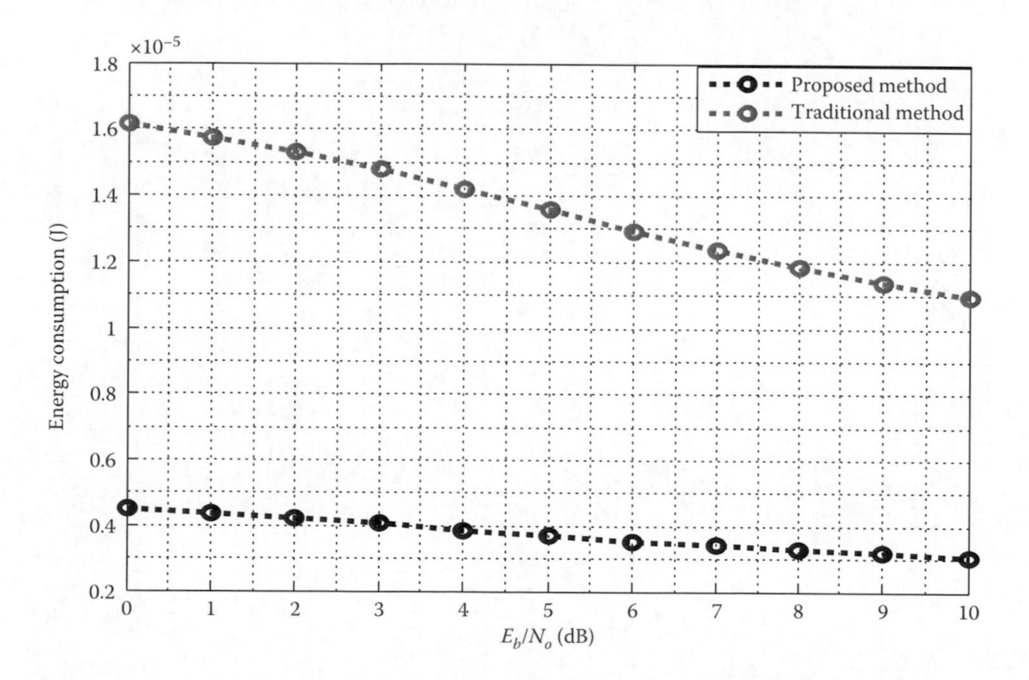

FIGURE 21.8 Energy consumption versus E_b/N_o in Rayleigh multipath fading channel.

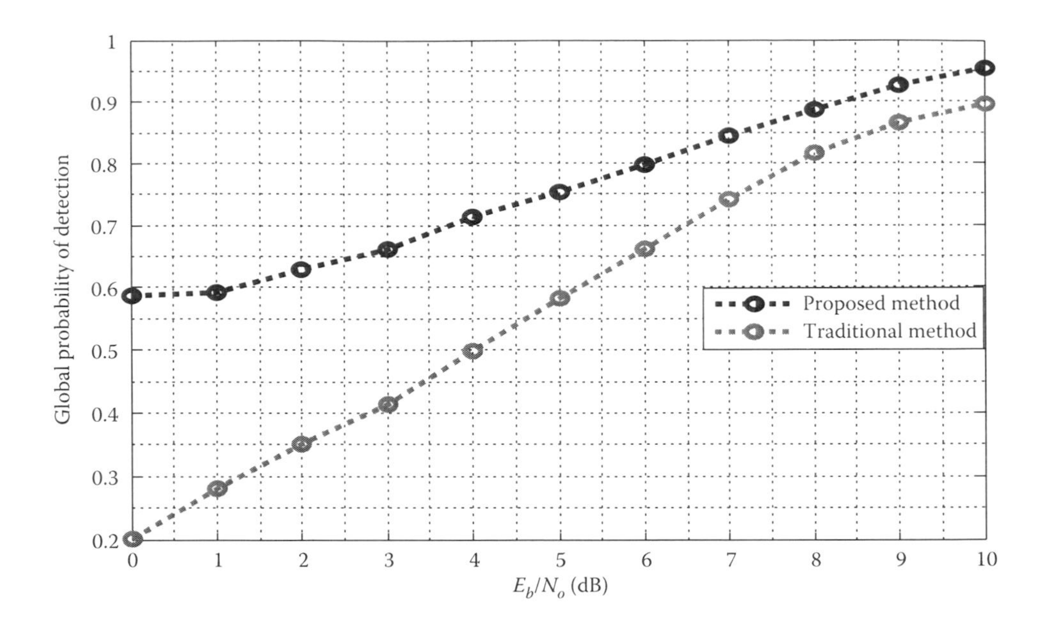

FIGURE 21.9 P_d versus E_b/N_o in AWGN channel.

same simulation parameters in Table 21.1. It can be seen from the figure that the proposed method significantly improves the energy especially at low values of E_b/N_o. For example, when E_b/N_o equals 0 dB, the energy consumption of the proposed method is reduced by 72%, 57%, and 54% as compared with traditional method, TSEEOB-CSS, and CC4C methods, respectively. The reason behind this improvement can be discussed as follows: In the TSEEOB-CSS method, the fine sensing stage is not used upon testing the spectrum channels all the time except when the accumulated energy does not cross the higher threshold or drop below the lower threshold, which reduces the detection performance. In the CC4C method, the sequential sensing scheme is used in the coarse sensing

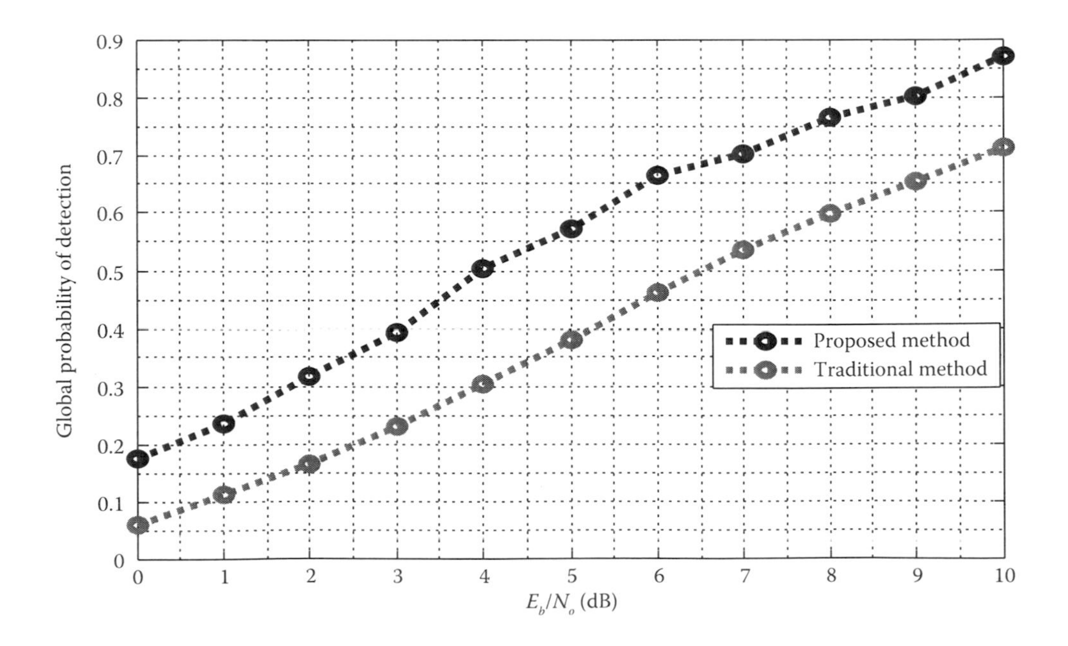

FIGURE 21.10 P_d versus E_b/N_o in Rayleigh multipath fading channel.

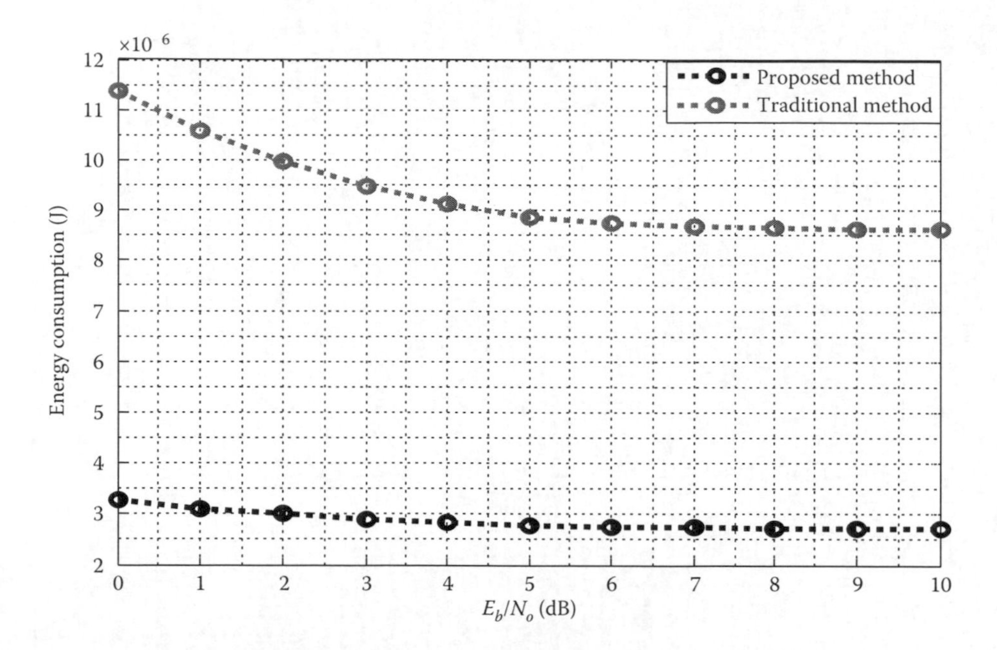

FIGURE 21.11 Energy consumption versus E_b/N_o using five CUs.

stage, which costs more energy at low E_b/N_o values because it continues to apply the sensing process using more and more samples until it can produce a local decision and stop the sensing process. So, the reasons mentioned earlier makes these two methods cost high energy consumption at low value of E_b/N_o. Conversely, the main reason of improvement in energy consumption in the proposed method is the use of fewer samples in the coarse stage, and two-stage sensing is used always for the channel with the highest accumulated energy after passing the threshold check, which gives significant improvement in energy consumption due to improved detection performance.

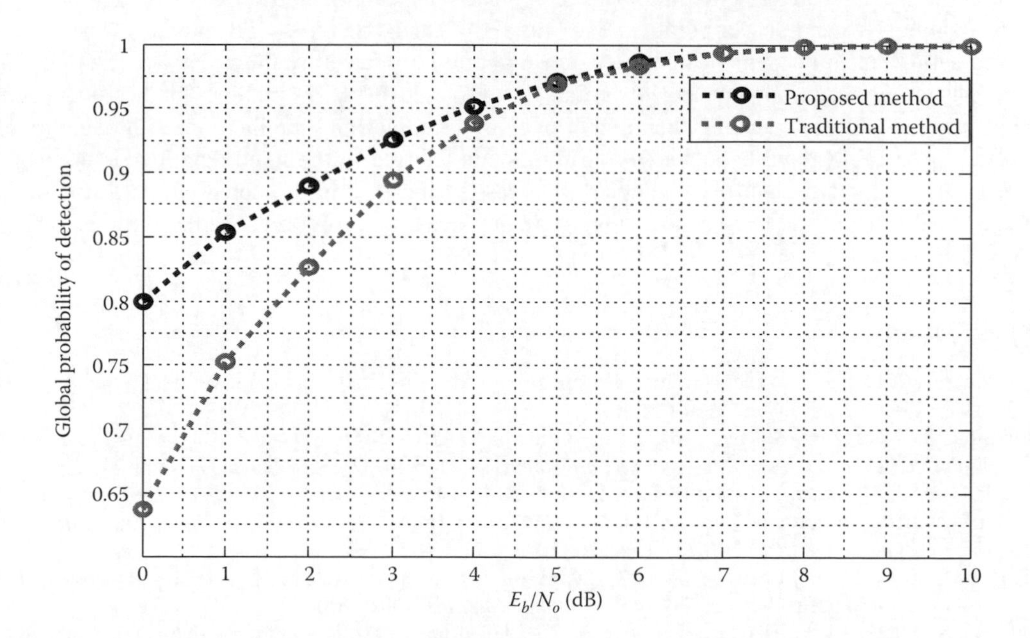

FIGURE 21.12 Q_D versus E_b/N_o using five CUs.

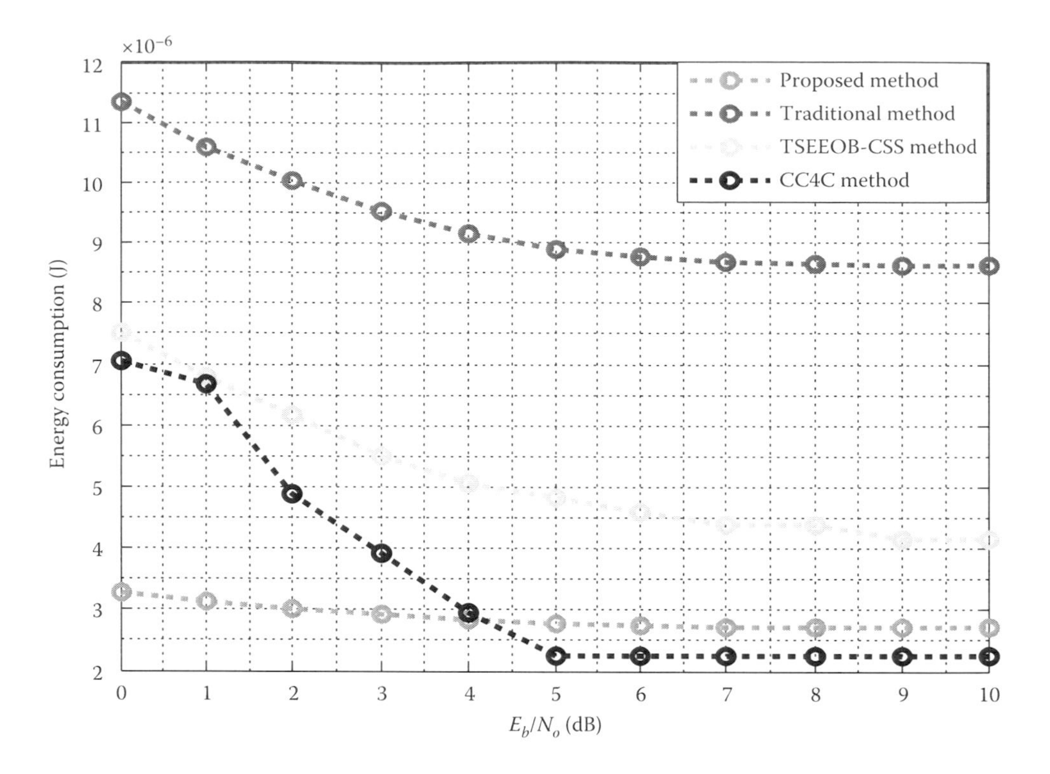

FIGURE 21.13 Average energy consumption per sensor versus E_b/N_o using five CUs.

21.7 CONCLUSIONS

In this chapter we proposed an efficient method to improve energy consumption in CR networks. In simulation, results are presented in forms of comparisons with the traditional method in two scenarios (cooperative and noncooperative) and with CC4C and TSEEOB-CSS methods. The results showed significant improvement in energy consumption introduced by the proposed method. The main reason behind this improvement is that the proposed method combines both reduction in energy consumption by reducing the number of sensed samples and the increase in probability of detection. The improvement in energy consumption is referred to the fixed length coarse sensing stage by using a small number of samples, whereas the increase in the detection performance is referred to the use of the fine-sensing stage. The impact of the proposed scheme becomes clearer at low SNR values.

REFERENCES

Althunibat S., 2014, Towards energy efficient cooperative spectrum sensing in cognitive radio networks, PhD dissertation, University of Trento.

Althunibat S., Narayanan S., Renzo M.D., and Granalli F., 2015, *Software-Defined and Cognitive Radio Technologies for Dynamic Spectrum Management*, DOI: 10.4018/978-1-4666-6571-2.ch004.

Awin F.A., Abdel-Raheem E., and Ahmadi M., 2015, Novel energy efficient strategies for cooperative spectrum sensing in cognitive radio networks, *IEEE International Symposium on Signal Processing and Information Technology (ISSPIT)*, DOI: 10.1109/ISSPIT.2015.7394325.

Cheng P., Deng R. L., and Chen J. M., 2012, Energy-efficient cooperative spectrum sensing in sensor aided cognitive radio networks, *IEEE Wireless Communications*, 19(6): 100–105.

Du H., Fu S., Shi G., Li W., Tian L., and Meng Y., 2016, Performance evaluation for the two-stage cooperative spectrum sensing scheme in cognitive radio, *WSEAS Transactions on Signal Processing*, 12: 68–73.

Emara M., Ali H. S., Khamis S. E. A., and Abd El-Samie F. E., 2015, Spectrum sensing optimization and performance enhancement of cognitive radio networks, *Wireless Personal Communications*, 86(2): 925–941.

Ergul O., and Akan O. B., 2014, Cooperative coarse spectrum sensing for cognitive radio sensor networks, Paper presented at *IEEE WCNC in Track 2 (MAC and Cross-Layer Design)*, New Orleans, Louisiana, March 9–12.

Kim S. J., and Giannakis G. B., 2009, Rate-optimal and reduced-complexity sequential sensing algorithms for cognitive OFDM radios, *EURASIP Journal on Advances in Signal Processing*, Article ID 421540.

Kim S. J., and Giannakis G. B., 2010, Sequential and cooperative sensing for multi-channel cognitive radios, *IEEE Transactions on Signal Processing*, 58(8): 4239–4253.

Kim S. J., Li G., and Giannakis G. B., 2010, Multiband cognitive radio spectrum sensing for real-time traffic, *IEEE Transactions on Signal Processing*, Submitted.

Kokare S., and Kamble R. D., 2014, Spectrum sensing techniques in cognitive radio cycle, *International Journal of Engineering Trends and Technology (IJETT)*, 9(1): 16–20.

Maleki S., and Leus G., 2013, Censored truncated sequential spectrum sensing for cognitive radio networks, *IEEE Journal on Selected Areas in Communications*, 31(3): 364–378.

Maleki S., Pandharipande A., and Leus G., 2011, Energy-efficient distributed spectrum sensing for cognitive sensor networks, *IEEE Sensor Journal*, 11(3): 565–573.

Pätzold M., 2000, *Mobile Fading Channels*, John Wiley & Sons, Chichester, England.

Zhao N., Yu F. R., Sun H., and Nallanathan A., 2013, Energy-efficient cooperative spectrum sensing schemes for cognitive radio networks, *EURASIP Journal on Wireless Communications and Networking*, DOI: 10.1186/1687-1499-2013-120.

Zou Q. J., Zheng S., and Sayed A. H., 2010, Cooperative sensing via sequential detection, *IEEE Transactions on Signal Processing*, 58(12): 6266–6283.

22 Current Status of Information Security Based on Hybrid Crypto and Stego Systems

Rana Saad Mohammed and Sattar B. Sadkhan

CONTENTS

22.1 Introduction 297
22.2 Steganography Techniques 297
22.3 Advantages and Disadvantages of Steganography in Information Security 300
22.4 Chaos in Steganography 300
22.5 Conclusion 305
References 305

22.1 INTRODUCTION

In recent years, people have preferred the use of wireless communication as a primary channel to transmit data around the world. That means data security is important to protect information from theft and modification over the Internet. There are many developed security techniques, including cryptography and steganography, to protect a data [1].

There are three steganography techniques: keyless hiding, hiding using a steganography key, and hiding using a cryptography key. In keyless techniques, it is easy to break and attack them. Steganography key techniques control the process of embedding and extraction of the message. Cryptography key techniques encrypt and decrypt the message [2].

Recently research has moved toward the use of chaotic systems in the steganography. This research uses chaotic maps as the steganography key to select the cover media and embedding locations, or as the cryptography key to encrypt the secret message and embedding it into a cover media by any steganography method. The modern methods are hybrid chaos based on cryptography with steganography.

Both steganography and cryptography are used for concealing information. Steganography does not detect any doubt about the hidden information that helps the attacker. Cryptography is used to change the form of the data. Steganography hides the existence of data itself and the attacker cannot easily find where the message is hidden. Both techniques can be combined to better protect information [3,4].

This chapter describes steganography techniques in Section 22.2, and the advantages and disadvantages of steganography in information security in covered in Section 22.3. Chaos in steganography and the rules for a hybrid structure that combines crypto and stego techniques are given in Section 22.4.

22.2 STEGANOGRAPHY TECHNIQUES

Steganography techniques are based on replacing bits of useless data with bits of important information. There are three methods of steganography: pure steganography, secret key steganography, and public key steganography.

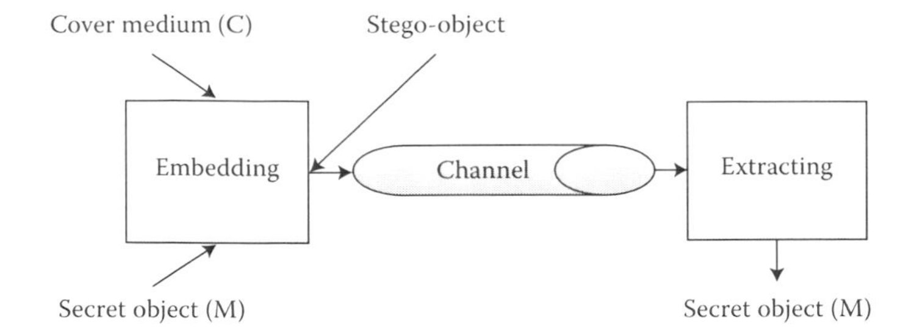

FIGURE 22.1 Pure steganography method.

The pure steganography method does not require any exchange of a secret tool between a sender and receiver like the exchange of a stego-key as shown in Figure 22.1. This method offers less secrecy since the sender and receiver communicate over an open system like the Internet and they suppose that there is no third party [5].

In the secret key steganography method, there is a prior secret session for exchanging of a stego-key. Then the sender can embed a secret message in a cover, which helps a secret stego-key get a stego-object and send it through the Internet. The receiver receives a stego-object and extracts a message using an agreed upon stego-key. The advantage of this method is that even if the message is intercepted, only the authorized parties know the secret stego-key to extract the message.

A general block diagram of steganography is shown in Figure 22.2. It contains three basic components: original message (M), cover medium (C), and stego-key (K) to get a stego-object ($\tilde{c}$) or embedded message (E_m) [6].

$$E_m : C \oplus K \oplus M \rightarrow \tilde{c}$$

To extract the original message from a received stego-object by an authorized receiver, the receiver must have the same key of the sender to get the extracted message (E_x).

A cover medium is the text, image, audio, or video in which the original secret message will be embedded. An original message also has text, images, audio, or video that can be embedded in cover medium to help a stego-key get a stego message [7].

The public key steganography method is similar to public key cryptography. The sender embeds a secret message in a cover using a public key to get a stego message and send it through the Internet. The receiver can extract a secret message using a secret private key. This method is more robust against the attacker since it has multiple levels of security and the attacker needs time to intercept a secret message, as shown in Figure 22.3.

There are two techniques closely related to steganography, but have different algorithms to protect publishing and copyright. These techniques are fingerprinted and watermarking. Fingerprint

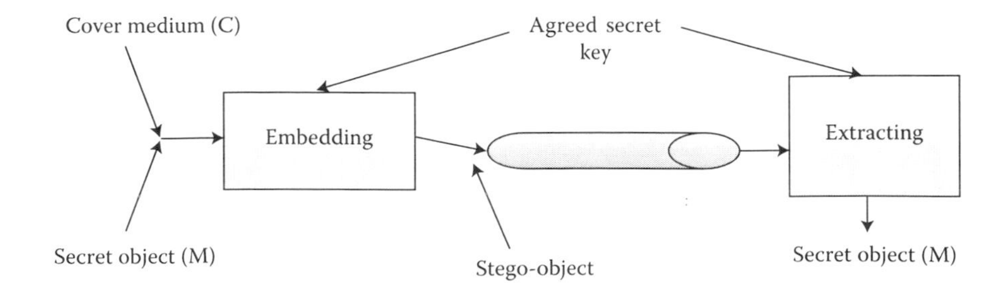

FIGURE 22.2 Secret key steganography.

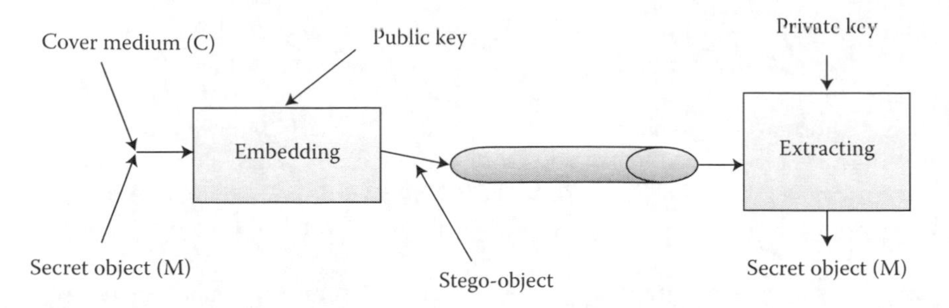

FIGURE 22.3 Public key steganography.

algorithms embed a unique signature of the publisher in his publishing copies using a specific way of embedding for different publishers. The algorithms of watermarking embed a mark or signature in the carrier objects using the same embedded method. Signatures and watermarks and fingerprints are hidden using a specific method, although they are visible [6,8].

Steganography is widely used in many areas including military, banking, market applications, secret communication, copyright protection, feature tagging, and digital watermarking.

There are six attacks against steganography algorithms:

- Stego only attack—Only the stego-object is available for analysis.
- Known cover attack—The original cover object and the stego-object are available for analysis.
- Known message attack—The hidden message is available to compare with the stego-object.
- Chosen stego attack—The stego tool (algorithm) and stego-object are available for analysis.
- Chosen message attack—Takes a chosen message and generates a stego-object for future analysis.
- Known stego attack—The stego tool (algorithm), the cover message, and the stego-objects are available for analysis [9].

Steganography can use all types of digital file formats for hiding information, but works best with those formats that have more redundant bits than necessary to embed secret message bits. The basic types of digital file formats that can be used with steganography are text, image, audio, video, and protocol [8].

- Text file format—A secret message can be hidden in each character of the word in a cover text. This method has decreased in use since the advent of other digital file formats. Also text files do not have much redundant data to hide a message.
- Image file format—A secret message can hide in a cover image using a specific algorithm and stego-key to get a stego image. The receiver can extract the message from a stego image using the same stego-key as the sender. There are several techniques for the image file format:
  - Spatial domain techniques—Least significant bit, pseudorandom permutation, palette-based image, cover regions and parity bits, quantization and dithering, image, downgrading and cover channels
  - Masking and filtering
  - Transform techniques
  - Statistical methods
  - Spread spectrum
  - Distortion techniques
  - Cover generation techniques
  - File and palette embedding

- Audio file format—Examples of digital audio file format are WAVE, MIDI, AVI, and MPEG. This technique hides information in cover sound parts that are unnoticeable by the human ear.
- Video file format—Examples of digital video format are H.264, Mp4, MPEG, and AVI. These formats can be used for hiding any kind of secret message.
- Protocol steganography—The protocols TCP, UDP, ICMP, and IP are used in information hiding. For example, a secret message is embedded in a header of a protocol packet [10].

The performance of steganography algorithms can be measured by comparing the cover medium and the stego. There are statistical measures that help to evaluate the performance, such as Peak signal-to-noise ratio (PSNR), mean square error (MSE), signal-to-noise ratio (SNR), normalized cross-correlation (NCC), and bit error rate (BER).

In addition, steganography algorithms must be robust against statistical attacks and manipulate transformations such as linear and nonlinear filtering, the addition of random noise, sharpening or blurring, scaling and rotations, cropping or decimation, and lossy compression.

The algorithms of steganography also must take into account the capacity of secret information with the choice of a suitable cover medium with the use of various types of formats to make information invisible, which complicates an attack [8,9].

22.3 ADVANTAGES AND DISADVANTAGES OF STEGANOGRAPHY IN INFORMATION SECURITY

- *Advantages*
  - Steganography can hide the existence of the message such that the attacker cannot guess where the message is embedded.
  - It preserves and protects a copyright of publisher copies that contain important information.
  - It keeps confidential data secret, such as credit card, debit card, and bank account numbers.
  - It provides a secrecy service that encourages people to embed their messages in a cover.
  - It is beneficial to publishing and broadcasting industries, which can hide an encrypted copyright mark and serial number in digital films, audio recording, books, and multimedia products [9].
- *Disadvantages*
  - If the algorithm is intercepted, then the attacker can find the hidden messages and read them. Such drawback can be overcome by combining steganography with cryptography (to strengthen the message by encrypting the message before embedding it).
  - A hidden message can be easily destroyed. For example, if the secret message is embedded in the least significant bits of cover, the interceptor can destroy the message by making a slight change.
  - Steganography needs an appropriate size of cover medium and redundant data to hide a message.

22.4 CHAOS IN STEGANOGRAPHY

In the last decade, chaos theory was developed by physicists and mathematicians. It deals with nonlinear functions. It has desirable features such as deterministic, nonlinear, irregular, long-term prediction, and sensitivity to initial conditions.

Studies have used chaos theory with information security techniques like cryptography and steganography to strength the security. These techniques must evolve with the development of communication technologies.

Chaotic maps that are used in cryptography are the logistic map, standard map, piecewise linear chaotic map (PWLCM), Lorenz map, Chen map, Henon map, cat map (Arnold map), Chebyshev map, and beta-transformation map [11].

This chapter classifies research of chaos in steganography into three categories (Figure 22.4).

- *Chaotic maps used as steganography key*
  In 2007, P. Liu et al. [12] performed a discrete cosine transform (DCT) on each 8 × 8 image block, and the secret message was embedded into locations on the image block determined by the chaotic logistic map. Luo et al. [13] embedded the messages in least significant bit (LSB) images randomly using a subsection linear chaotic map according to the perturbing algorithm presented in B. Liu et al. [14]:

$$f(x)=\begin{cases} g(x) & x \notin c_3 \\ g\left(\dfrac{4}{2e}(x-0.5)\right) & x \in c_{31} \\ g\left(\dfrac{x-(0.5+(e/4))}{1-e}+0.75\right) & x \in c_{31} \end{cases} \tag{22.1}$$

Consider a 1D subsection linear chaotic mapping with four subintervals:

$$g(x)=\begin{cases} 4x & 0 \le x < a \\ 2-4x & a \le x < b \\ 4x-2 & b \le x < 1-a \\ -4x & 1-a \le x \le 1 \end{cases} \tag{22.2}$$

Diffuse a third subinterval as $c3 = [b, 1 - a)$: divide $c3$ into two segments according to e: $1 - e$, scale: $c31 = [b, b + e/r)$, and $c32 = [b + e/r, 1 - a]$, where e is the diffused coefficient, r is constant, and x is the initial number. Let $a = 0.25$, $b = 0.5$, and $r = 4$, then obtain a subsection linear chaotic mapping according to the perturbing algorithm as Equation 22.1.

Enayatifar et al. [15] in 2009 used two chaotic signals for specifying the location of the different parts of the message in the picture. In 2010, Yu et al. [16] used logistic map shuffle bits-order of the message whose parameters are selected by the genetic algorithm to improve adaptive LSB steganography.

Alirezanejad and Enayatifar [17] in 2012 found a best position by N-bit mask for hiding one bit of secret data in a cover image (the N value is an image dimension). This mask was made with a hybrid model of cellular automata with a logistic map, and it can be changed for hiding each bit of secret data. Other research has used discrete cross-coupled chaotic maps to embed secret data in LSB image pixels [18].

In 2013, Gabriel et al. [19] used a PWLCM to select pixel positions and used the LSB steganography technique for information hiding. Tayel et al. [2] distributed a hidden-image pixel randomly within the lower byte of the cover image pixels with a logistic map. Then this stego image was processed using a fuzzy logic compressor to invert and compress its color range over a specific threshold color decided by the value of the upper significant bits of the image, that is, 192 to 255.

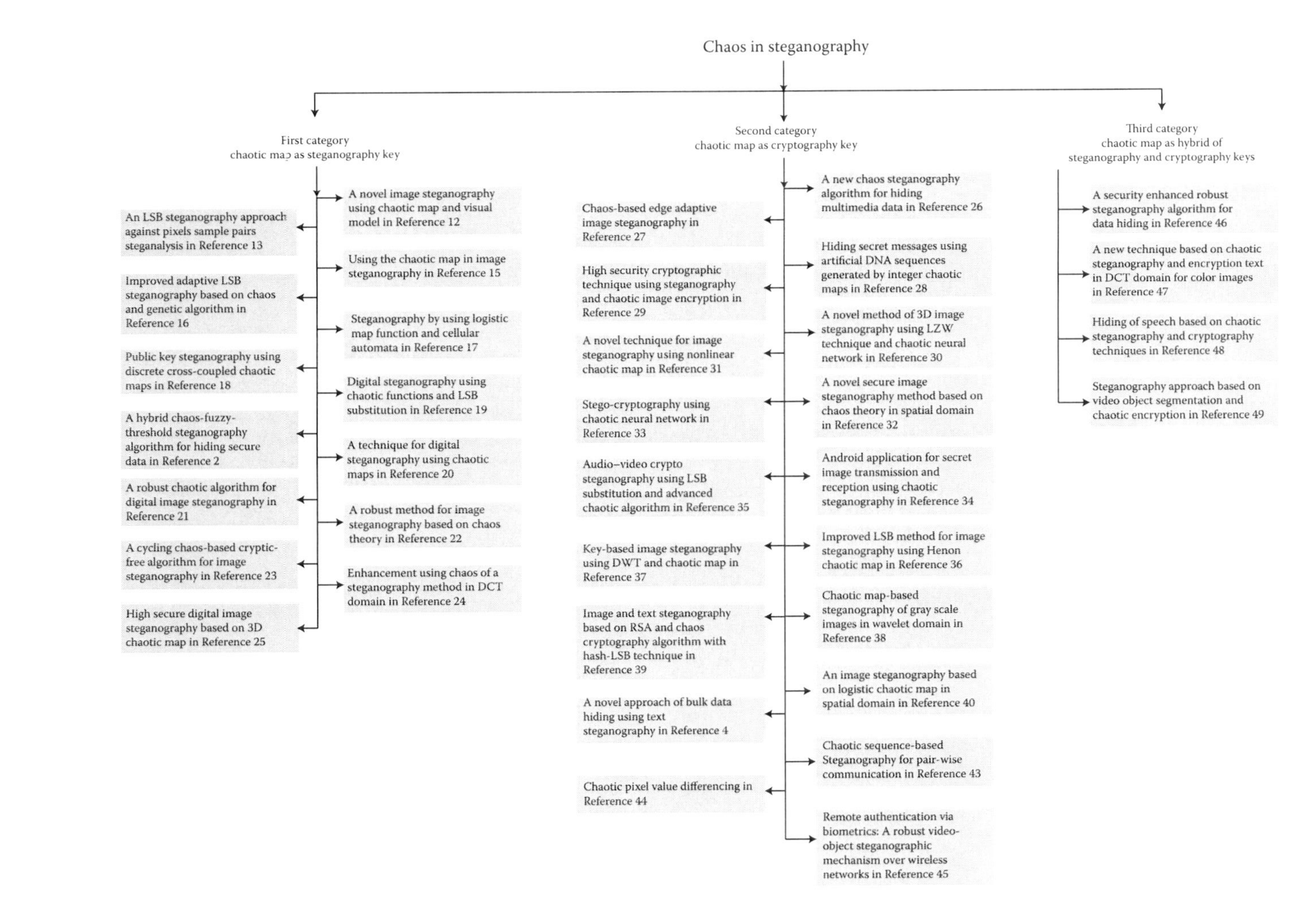

FIGURE 22.4 Chaos in steganography classification.

A study in 2014 proposed a steganography technique in spatial domain for digital images based upon chaotic maps [20]. Another study proposed a steganography algorithm based on a 3D chaotic cat map and lifted discrete wavelet transforms [21]. The irregular outputs of the cat map are used to embed a secret message in a digital cover image.

In 2015, Kumar et al. [22] used a logistic map to chaotically select the image's data and to hide the message's bits. Aziz et al. [23] proposed a cycling chaos-based steganographic algorithm. This algorithm generates the seeds for a pseudorandom number generator (PRNG) by chaos to determine the channel and the pixel positions of the host image in which the sensitive data are embedded. Habib et al. [24] used PWLCM to enhance the LSB–DCT technique. PWLCM was used to generate a series that helps randomly embed a secret image in the DCT coefficients of the cover image. Valandar et al. [25] proposed a 3D chaotic map to develop a spatial steganography method to randomly embed a secret message into a cover image. The equation of the 3D chaotic map is as follows:

$$\begin{aligned} x_{n+1} &\equiv \left[\frac{1}{\alpha_1^2 y_n z_n}\tan^2\left(\arctan\left(\sqrt{x_n}\right)\right)\right]\bmod 1 \\ y_{n+1} &\equiv \left[\frac{1}{\alpha_2^2 x_n z_n}\sin^2\left(\arctan\left(\frac{1}{\sqrt{y_n}}\right)\right)\right]\bmod 1 \\ z_{n+1} &\equiv \left[\frac{1}{\alpha_3^2 x_n y_n}\,|\tan(\arctan(|z_n|))|\right]\bmod 1 \end{aligned} \tag{22.3}$$

- *Chaotic maps used as cryptography key*

  A study in 2012 coordinated the data in the image dimensions using chaos distribution arrangement [26]. The data was embedded with the original image in the pixels least significant bits.

In 2013, Roy et al. [27] used cat mapping to distort the payload before embedding with a combination of matrix encoding and LSB matching. Bashier et al. [28] used an integer logistic chaotic map to generate random sequences of DNA, and then used two steganography algorithms for hiding an encrypted message in artificial DNA sequences. Arun and Joseph [29] embedded a message into the cover image and then encrypted the result using the triple-key chaotic image encryption method by logistic mapping. Vani and Prasad [30] compressed the information using the Lempel-Ziv-Welch (LZW) technique and then encrypted it using the chaotic neural network. The encryption data was embedded inside the 3D image. Thenmozhi and Chandrasekaran [31] used the Henon chaotic map to scramble a secret image and then embedded the result in cover image coefficients of 2D DWT.

In 2014, Bandyopadhyay et al. [32] encrypted the secret bits of the message based on logistic map before embedding into the cover image by a 3-3-2 LSB insertion steganography method. The LSB steganography technique is used for message hiding. Kamila et al. [33] used a 1D logistic map to generate binary sequence to set the biases and weights of neurons in each iteration. Savithri and Sudha [34] scrambled the image using a Henon chaotic map and then embedded the result into redundant positions of the cover image. Praveen and Arun [35] used the chaotic algorithm to encrypt audio and video as separated files and then executed the embedding process using the LSB algorithm.

Another study used the Henon chaotic map to generate random numbers to encrypt the hidden image, which is embedded into a cover picture using the LSB method [36]. Choudhary and Panwar [37] applied a Haar transform on the cover image and extract four subbands from the resulted image. A key was generated by circle chaotic map to XOR with the selected hidden image. Then the LSB technique was used to hide the secret image in two LSB of LH, HL, and HH subbands. Inverse discrete wavelet transform (IDWT) was applied to get the stego-image. Garg and Mathur [38] used an Arnold map to scramble a secret image with key. This method applied fractional Fourier transform and then applied DWT to both the cover and secret images. Then the addition process can be applied using the alpha blending technique between the DWT coefficients of cover and secret images.

Manjunath and Hiremath [39] in 2015 encrypted data using Rivest, Shamir, Adleman (RSA) encryption and then embedded encrypted data in a particular text, image, or file using the hash-LSB technique. The resulting stego is encrypted using the chaos algorithm. Bhargavi et al. [40] used a logistic chaotic map to encrypt the secret message and then embedded it into a cover image using the base embedding technique. Shivania and Yadava [4] used the indexed based chaotic sequence for encryption using a logistic map). Other researchers have used the same technique [41,42].

In 2016, a research team generated key images with a chaotic sequence of numbers [43]. The images are bitmaps (.bmp) having dimensions of 600 × 800 pixels. Next, an XOR operation was performed with plaintext file to get an encryption file. In Pun and Juneja [44] proposed chaotic pixel value difference (C-PVD) approach. It used a 1D logistic map or 2D Arnold's cat map in encryption with traditional steganography to get a secure payload. Ntalianis and Tsapatsoulis [45] proposed a chaotic pseudo random bit sequence (C-PRBG) to generate a key equal in size to the data to be encrypted. C-PRBG can generate a sequence of keys by mixing three different 1D chaotic maps as follows:

$$k(i) = \begin{cases} 1 & F_3(x_1(i), p_3) > F_3(x_2(i), p_3) \\ k(i-1) & F_3(x_1(i), p_3) = F_3(x_2(i), p_3) \\ 0 & F_3(x_1(i), p_3) < F_3(x_2(i), p_3) \end{cases} \tag{22.4}$$

where p_1, p_2, and p_3 are control parameters; $x_1(0)$, $x_2(0)$, and $x_3(0)$ are initial conditions; and $x_1(i)$, $x_2(i)$, and $x_3(i)$ denote the three chaotic orbits. The next values of chaotic orbits x_1, x_2, and x_3 are as follows:

$$\begin{aligned} x_1(i+1) &= F_1(x_1(i), p_1) \\ x_2(i+1) &= F_2(x_2(i), p_2) \\ x_3(i+1) &= F_3(x_3(i), p_3) \end{aligned} \tag{22.5}$$

where $F_1(x_1; p_1)$, $F_2(x_2; p_2)$, and $F_3(x_3; p_3)$ are three different 1D chaotic maps.

- *Chaotic maps that use a hybrid of steganography and cryptography keys*

  In 2012, Singh and Siddiqui [46] converted a cover image into the frequency domain using DCT. On the other side, a secret message is scrambled by Arnold's cat map, and then embed in the middle band coefficient of 2D DCT by using two random sequences with length 22 that are generated by logistic map.

  One study in 2013 used a logistic chaotic map for encrypting a secret text and for embedding it into the DCT cover image [47].

  Another study encrypted a speech message with Lorenz and Chua maps, and used a modified android cat map (MACM) to embed a secret speech message in a digital cover image [48].

  In 2016, a research team [49] encrypted a speech signal by XORing the speech signal with the generated key stream using a chaotic function algorithm and then hid the secret message into the cover selected frame of video using chaos to get a stego frame.

22.5 CONCLUSION

Steganography is an information hiding technique. There has been research to strength the technique. One method is to combine steganography with cryptography to increase its security. Modern researchers oriented their aims toward using chaotic maps in steganography to improve the security aspects resulting in a hybrid security system, and also to increase the cryptanalytic efforts that must be paid to cryptanalysis of the complete hybrid (crypto–stego) system. The performance results of such hybrid systems have been acceptable. The use of different chaotic maps is recommended in the future works.

REFERENCES

1. J. Majumder and S. Mangal. An overview of image steganography using LSB technique. *IJCA Proceedings on National Conference on Advances in Computer Science and Applications (NCACSA 2012)*, Chennai, India, March 10, 2012, pp. 10–13.
2. M. Tayel, H. Shawky, and A. S. El-Din Hafez. A hybrid chaos-fuzzy-threshold steganography algorithm for hiding secure data. *ICACT Transactions on Advanced Communications Technology (TACT)* 2(1), 156–161, 2013.
3. R. Poornima and R. J. Iswarya. Overview of digital image steganography. *International Journal of Computer Science & Engineering Survey (IJCSES)* 4(1), 23–31, 2013.
4. V. K. Shivania and S. B. Yadava. A novel approach of bulk data hiding using text steganography. *Procedia Computer Science* 57, 1401–1410, 2015.
5. J. Ashok, Y. Raju, S. Munishankaraiah, and K. Srinivas. Steganography: An overview. *International Journal of Engineering Science and Technology* 2(10), 5689–5696, 2010.
6. N. P. Kamdar, G. D. Kamdar, and N. D. Khandhar. Performance evaluation of LSB based steganography for optimization of PSNR and MSE. *Journal of Information, Knowledge and Research in Electronics and Communication Engineering* 2(2), 505–509, 2013.
7. N. Hamid, A. Yahya, R. B. Ahmad, and O. M. Al-Qershi. Image steganography techniques: An overview. *International Journal of Computer Science and Security (IJCSS)* 6(3), 168–187, 2012.
8. T. Morkel, J. H. P. Eloff, and M. S. Olivier. *An overview of image steganography*. University of Pretoria, South Africa, 2005.
9. G. Chugh. Image steganography techniques: A review article. *Acta Technica Corviniensis—Bulletin of Engineering* Tome VI–(FASCICULE3), 97–104, 2013.
10. P. C. Mandal. Modern steganographic technique: A survey. *Internationa l Journal of Computer Science & Engineering Technology (IJCSET)* 3(9), 2012.
11. R. S. Mohammed and S. B. Sadkhan. Chaos-based cryptography for voice secure wireless communication. In: *Multidisciplinary Perspectives in Cryptology and Information Security*, S. B. Sadkhan Al Maliky and N. A. Abbas (eds.), IGI Global, Hershey, PA, pp. 97–126, 2014.
12. P. Liu, Z. Zhu, H. Wang, and T. Yan. A novel image steganography using chaotic map and visual model. http://www.atlantis-press.com/php/download_paper.php?id=1452. 2007.
13. X. Luo, F. Liu, and P. Lu. A LSB steganography approach against pixels sample pairs steganalysis. *International Journal of Innovative Computing, Information and Control ICIC International* 3(3), 575–588, 2007.
14. B. Liu, X. Luo, and F. Liu. Perturbing scheme of digital chaos. *Journal of Shanghai Jiaotong University (Science)*, 11(2), 172–176, 2006 (English version).
15. R. Enayatifar, S. Faridnia, and H. Sadeghi. Using the chaotic map in image steganography. *IEEE International Conference on Signal Processing Systems*, Singapore, May 15–17, 2009, pp. 754–757.
16. L. Yu, Y. Zhao, R. Ni, and T. Li. Improved adaptive LSB steganography based on chaos and genetic algorithm. *EURASIP Journal on Advances in Signal Processing* 2010, 6, Article ID 876946, DOI: 10.1155/2010/876946.
17. M. Alirezanejad and R. Enayatifar. Steganography by using logistic map function and cellular automata. *Research Journal of Applied Sciences, Engineering and Technology* 4(23), 4991–4995, 2012.
18. S. Ahadpour, M. Majidpour, and Y. Sadra. Public key steganography using discrete cross-coupled chaotic maps. https://arxiv.org/ftp/arxiv/papers/1211/1211.0086.pdf. 2012.
19. A. C. Gabriel, C. Marian, and R. Ciprian. Digital steganography using chaotic functions and LSB substitution. Conference Paper, January 2013, DOI: 10.13140/RG.2.1.1996.3360.
20. A. Anees, A. M. Siddiqui, J. Ahmed, and I. Hussain. A technique for digital steganography using chaotic maps. *Nonlinear Dynamics, Springer* 75(4), 807–816, 2014.

21. M. Ghebleh and A. Kanso. A robust chaotic algorithm for digital image steganography. *Communications in Nonlinear Science and Numerical Simulation*, 19(6), 1898–1907, 2014.
22. A. Kumar, A. Rajpoot, K. K. Shukla, and S. Karthikeyan. A robust method for image steganography based on chaos theory. *International Journal of Computer Applications* 113(4), 35–41, 2015.
23. M. Aziz, M. H. Tayarani-N, and M. Afsar. A cycling chaos based cryptic-free algorithm for image steganography. *Nonlinear Dynamics, Springer* 80(3), 1271–1290, 2015.
24. M. Habib, B. Bakhache, D. Battikh, and S. El Assad. Enhancement using chaos of a steganography method in DCT domain. In: *2015 Fifth International Conference on Digital Information and Communication Technology and its Applications (DICTAP)*, IEEE, Beirut, April 29–May 1, 2015, pp. 204–209.
25. M. Y. Valandar, P. Ayubi, and M. J. Barani. High secure digital image steganography based on 3D chaotic map. *2015 7th International Conference on Information and Knowledge Technology*. Urmia, May 26–28, pp. 1–6.
26. M. Tayel, H. Shawky, and A. E. S. Hafez. A new chaos steganography algorithm for hiding multimedia data. *14th International Conference on Advanced Communication Technology*, Pyeongchang, South Korea, February 19–22, 2012, pp. 208–212.
27. R. Roy, A. Sarkar, and S. Changder. Chaos based edge adaptive image steganography. *Elsevier Procedia Technology* 10, 138–146, 2013.
28. E. Bashier, G. Ahmed, H.-A. Othman, and R. Shappo. Hiding secret messages using artificial DNA sequences generated by integer chaotic maps. *International Journal of Computer Applications* 70(15), 1–5, 2013.
29. A. S. Arun and G. M. Joseph. High security cryptographic technique using steganography and chaotic image encryption. *IOSR Journal of Computer Engineering (IOSR-JCE)* 12(5), 49–54, 2013.
30. B. G. Vani and E. V. Prasad. A novel method of 3D image steganography using LZW technique and chaotic neural network. *International Journal of Computer Science and Network Security* 13(6), 1–9, 2013.
31. S. Thenmozhi and M. Chandrasekaran. A novel technique for image steganography using nonlinear chaotic map. *7th International Conference on Intelligent Systems and Control*, Coimbatore, India, January 4–5, 2013, pp. 307–311.
32. D. Bandyopadhyay, K. Dasgupta, J. K. Mandal, and P. Dutta. A novel secure image steganography method based on chaos theory in spatial domain. *International Journal of Security, Privacy and Trust Management (IJSPTM)* 3(1), 11–22, 2014.
33. N. K. Kamila, H. Rout, and N. Dash. Stego-cryptography using chaotic neural network. *American Journal of Signal Processing* 4(1), 24–33, 2014. DOI: 10.5923/j.ajsp.20140401.04.
34. G. Savithri and K. L. Sudha. Android application for secret image transmission and reception using chaotic steganography. *International Journal of Innovative Research in Computer and Communication Engineering (IJIRCCE)* 2(7), 5107–5113, 2014.
35. P. Praveen and R. Arun. Audio-video crypto steganography using LSB substitution and advanced chaotic algorithm. *International Journal of Engineering Inventions* 4(2), 1–7, 2014.
36. N. S. Raghava, A. Kumar, A. Deep, and A. Chahal. Improved LSB method for image steganography using Henon chaotic map. *Open Journal of Information Security and Applications* 1(1), 34–42, 2014.
37. S. Choudhary and C. Panwar. Key based image steganography using DWT and chaotic map. *International Journal of Engineering and Management Research (IJEMR)* 4(4), 94–97, 2014.
38. S. Garg and M. Mathur. Chaotic map based steganography of gray scale images in wavelet domain. *2014 International Conference on Signal Processing and Integrated Networks (SPIN)*, Noida, India, February 20–21, 2014, pp. 689–694.
39. N. Manjunath and S. G. Hiremath. Image and text steganography based on RSA and chaos cryptography algorithm with hash-LSB technique. *International Journal of Electrical, Electronics and Computer Systems (IJEECS)* 3(5), 2015.
40. S. Bhargavi, M. J. Shobha, T. N. Swetha, and M. J. Pushpa. An image steganography based on logistic chaotic map in spatial domain. *International Journal of Research in Science & Engineering (IJRISE)* 1(2), 235–240.
41. S. Batham, A. Acharya, V. K. Yadav, and R. Paul. A new video encryption algorithm based on indexed based chaotic sequence. *Confluence 2013: The Next-Generation Information Technology*. IET Digital Library, http://digital-library.theiet.org/content/conferences/10.1049/cp.2013.2307.
42. A. Soni and A. K. Acharya. A novel image encryption approach using an index based chaos and DNA encoding and its performance analysis. *International Journal of Computer Applications (IJCA)* 47(23), 1–6, 2012.

43. S. Harsha, S. Kumar, K. Nazim, A. Sattar, K. Prasanna, and A. D. Shantanu. Chaotic sequence based steganography for pair-wise communication. *Global Journals Inc. (USA)* 16(2), 2016, Version 1.0.
44. N. Pun and M. Juneja. Chaotic pixel value differencing. *International Journal of Image, Graphics and Signal Processing*, 4, 54–60, 2016.
45. K. Ntalianis and N. Tsapatsoulis. Remote authentication via biometrics: A robust video-object steganographic mechanism over wireless networks. *IEEE Transaction on Emerging Topics in Computing* 4(1), 156–174, 2016.
46. S. Singh and T. J. Siddiqui. A security enhanced robust steganography algorithm for data hiding. *International Journal of Computer Science Issues* 9(3 Suppl. 1), 131–139, 2012.
47. M. J. Saeed. A new technique based on chaotic steganography and encryption text in DCT domain for color images. *Proceedings of Journal of Engineering Science and Technology* 8(5), 508–520, 2013.
48. A. S. Hameed. Hiding of speech based on chaotic steganography and cryptography techniques. *International Journal of Engineering Research* 4(4), 165–172, 2015, ISSN: 2319-6890(online), 2347-5013(print).
49. V. Bhagya Lakshmi, T. Ravikumar Naidu, and T. V. S. Gowtham Prasad. Steganography approach based on video object segmentation and chaotic encryption. *IJSETR* 5(18), 3771–3775, 2016.

23 The Status of Research Trends in Text Emotion Detection

Rusul Sattar B. Sadkhan and Sattar B. Sadkhan

CONTENTS

23.1 Introduction 309
23.2 Text Emotion Detection (TED) 309
23.2.1 Importance of Text Emotion Detection 310
23.2.2 Affective Computing 310
23.2.3 Sentimental Analysis 311
23.2.4 Multidisciplinary in Text Emotion Detection 313
23.3 Current Status of Text Emotion Detection 314
23.4 Challenges of Text Emotion DETECTION 316
23.5 Conclusion and Future Works 317
23.5.1 Conclusion 317
23.5.2 Future Works 318
References 319

23.1 INTRODUCTION

The book *Affective Computing* introduced foundational ideas in incorporation of affect for computers, from perspectives of both generation and detection [1]. The book, written by Rosalind W. Picard, is considered as the first explanation of enabling computers with emotions. Emotions such as happiness, sadness, and fear, are important topics in the human Intelligence, and play important roles in human thinking and in decision making. Text emotion (TE) analysis is a rapidly growing field of research [1]. Previous studies have consistently shown emotion regulation to be an important predictor of intercultural adjustment. Emotional intelligence theory suggests that before people can regulate emotions they need to recognize them; thus emotion recognition ability should also predict intercultural adjustment [2]. Emotional intelligence (EQ) is defined as the ability to identify, assess, and control one's own emotions, the emotions of others, and that of groups.

Detecting the emotional state of a person by analyzing a document written by him/her appears a challenging task. But also it is essential because most of the time textual expressions are not only revealed through direct emotion words but also from the interpretation of the meaning of concepts and interaction of concepts that are described in the text document [3].

The main aim of this chapter is to describe the research trends in text emotion detection (TED). Section 23.2 explains text emotion detection and its applications, affective computing, sentiment analysis, and multidisciplinary and its existence in TED problems. Section 23.3 describes the current status of TED through a literature review 2015 and 2016. Section 23.4 explains some of the challenges facing the research field, including technologies used, the length of the texts, and quality assurance. Section 23.5 explains the conclusion and future works.

23.2 TEXT EMOTION DETECTION (TED)

Vaibhav et al. [4] claimed that "text can capture but only a portion of emotions expressed by a human. Even in linguistic studies of emotions, emphasis has been given on intonation and accentuation of speech for emotion analysis." A major part of the human existence and social interactions is emotions. Some people might think that emotions are one of the aspects that make us truly human. The "computing world" seems to struggle with supporting and incorporating the emotional dimension. This section will discuss briefly the following: importance of TED, affective computing, sentiment analysis, and multidisciplinary in TED.

23.2.1 Importance of Text Emotion Detection

TED is considered a recent field of research related to sentiment analysis (SA), which aims to detect positive, neutral, or negative feelings from text. Text emotion analysis (TEA) aims to recognize feeling types, such as disgust, anger, sadness, happiness, fear, and surprise, from written text.

Chew [5] claimed that the "emotion categories that are widely used to describe humans' basic emotions, based on facial expression are: anger, disgust, fear, happiness, sadness and surprise." These are mainly associated with negative sentiment, with "surprise" being the most ambiguous, as it can be associated with either positive or negative feelings. The number of basic human emotions has been recently "reduced," or rather recategorized, to just four: happiness, sadness, fear/surprise, and anger/disgust. In any recognition task, the three most common approaches are knowledge-based, statistic-based, and hybrid. Their use depends on many important factors, such as data availability, domain specificity, and domain expertise.

Texts with rich emotions can be found in personal blogs/journals, websites, social networks, and so on. Many databases are interested in reviews of a product or a type of service, stock market analysis, or political debates. Texts from different articles are very interesting [6].

23.2.2 Affective Computing

Affective computing describes computing that is in some way connected to emotion. It is sometimes also known as emotional artificial intelligence. In their study, Canales and Martínez-Barco [7] conducted a survey about emotion detection from the text. They claimed that their survey described recent works in the field of emotion detection from text. Despite the large amount of work on emotional detection systems, a lot of study had yet to be done. The increase of these systems is due to the large amount of emotional data available on the social web. Detecting emotions from text have attracted the attention of many researchers in computational linguistics because it has a wide range of applications, such as suicide prevention or measuring well-being of a community. Affective computing is the study and development of devices and systems that can detect, recognize, interpret, process, and simulate human emotions and appropriately respond to user's emotions, relative to user emotion from visual textual and auditory sources.

Affective computing is an emerging technology and focused on human studies that investigate computer science, cognitive science, psychology, and behaviors of humans. Such analysis of emotions from humans often involves the use of passive sensors, which capture data about the user's physical state or behavior. Affective computing requires multimodal processing, since there are a wide variety of cues to be captured. However, the used approaches are impeded by their requirements of multimodal data, which may not be readily available. On the contrary, text-based approaches to emotion analysis have become popular [8]. Affective computing is an interdisciplinary field spanning computer science, psychology, and cognitive science. Figure 23.1 shows the interdisciplinary perspective of affective computing.

Of course, as with most areas of scientific research, not all later researchers have agreed with Picard's ideas. This lack of total consensus in itself has helped advancement in the field, as more researchers have tried to expand on and provide proof for their personal theories. The best-known alternative is based on the affective interaction approach. This approach focuses on making emotional experiences available for people to reflect upon that will in some way modify their reactions.

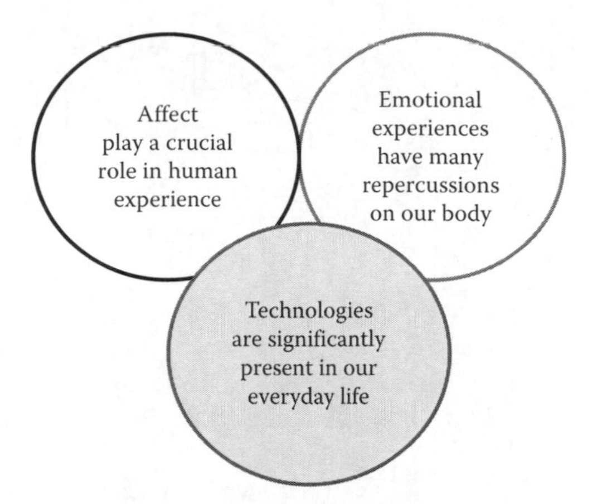

FIGURE 23.1 Interdisciplinary perspective of affective computing.

There are several goals in affective computing. One is to sense and respond respectfully to human emotion. For instance, if a person is communicating with a technology and he/she is frustrated or confused, the technology needs to be able to respond differently to that person. If the humans respond differently to the technology, the technology should to do the same. Another goal of affective computing is to enable people to communicate emotions more clearly. Equally important is to look at the role that emotion plays in intelligence processing. Affective computing is, in part, about

- Understanding how emotions play vital rules in us for regulating our intention
- Helping us make good decisions
- Changing the way we emphasize and prioritize things
- Organizing or figuring out what matters

These roles of emotions are the ones people do not usually think about as emotional, because usually we are not "emotional" when we do that; it is just background regular mechanisms that are important for functioning intelligently.

23.2.3 Sentimental Analysis

Sentiment analysis (SA) or opinion mining (OM) is the computational study of people's opinions, attitudes and emotions toward an entity. The entity can represent individuals, events, or topics. Some researchers note that OM and SA have slightly different notions. Opinion mining extracts and analyzes people's opinion about an entity, whereas sentiment analysis identifies the sentiment expressed in a text, then analyzes it [9]. Therefore, the target of SA is to find opinions, identify the sentiments they express, and then classify their polarity, as shown in Figure 23.2.

Sentiment analysis can be considered a classification process. There are three main classification levels in SA: document, sentence, and aspect. Document-level SA aims to classify an opinion document as expressing a positive or negative opinion or sentiment. It considers the whole document as a basic information unit (talking about one topic). Sentence-level SA aims to classify sentiment expressed in each sentence. The first step is to identify whether the sentence is subjective or objective. If the sentence is subjective, sentence-level SA will determine whether the sentence expresses positive or negative opinions [10].

Some authors propose a novel way to automatically calculate the affective values for emotional words. The affective dimensions are based on the factorial analysis of extensive empirical tests. As a result, researchers not only discovered three major factors—potency, evaluation, and activity—that play a role in the emotive meaning of a word, but also set the basis for the circumflex of affect.

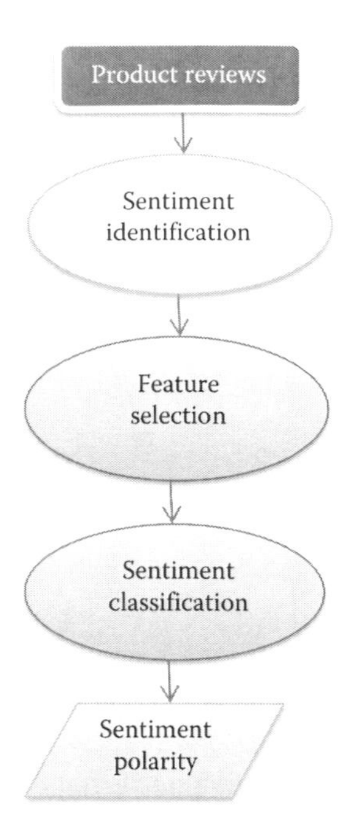

FIGURE 23.2 Sentiment analysis process on product reviews.

Some years later, a novel way for textual affect sensing was proposed, by exploiting commonsense knowledge rather than using keyword spotting techniques that only work when specific keywords occur in the text. As an example, the sentence "I just had a car accident" does not contain any emotional keyword but contains affective information. A person that just had a car accident is certainly not happy, and most probably sad or even frightened. This kind of evaluation of emotional content embedded in text can be extracted by using commonsense knowledge and by reasoning over this knowledge.

The state of the art in sentiment has been studied at three different levels: word, sentence, and document. Methods have been created to estimate positive or negative sentiment of words, sentences, and documents. As an example, some researchers propose machine learning. Next, we present other approaches for assessing sentiment from text. Some of them are based on one or a combination of techniques:

- Keyword spotting, lexical affinity
- Statistical methods
- Fuzzy logic
- Knowledge-base from facial expression
- Machine learning
- Domain-specific classification
- Valence assignment

There are some topics that work under the umbrella of SA that have recently attracted researchers:

- Emotion detection—Sentiment analysis is sometimes considered as an natural language processing (NLP) task for discovering opinions about an entity; and because there is some ambiguity about the difference between opinion, sentiment, and emotion, opinion is defined as a transitional concept that reflects attitude toward an entity. The sentiment reflects feeling or emotion, whereas emotion reflects attitude [11].

- Building resources—Building resources (BRs) aim at creating lexica, dictionaries, and corpora in which opinion expressions are annotated according to their polarity. Building resources is not an SA task, but it could help to improve SA and ED as well. The main challenges that confront work in this category are ambiguity of words, multilinguality, granularity, and the differences in opinion expression among textual genres [12].
- Transfer learning—Transfer learning extracts knowledge from the auxiliary domain to improve the learning process in a target domain. For example, it transfers knowledge from *Wikipedia* documents to tweets or a search in English to Arabic. Transfer learning is considered a new cross-domain learning technique, as it addresses the various aspects of domain differences. It is used to enhance many text mining tasks such as text classification, sentiment analysis, named entity recognition, and part-of-speech tagging [13].

23.2.4 Multidisciplinary in Text Emotion Detection

It is not enough to be within the research field to gather various issues related to that field. The openness in the "field of scientific research" remains necessary and vital. A discipline describes types of knowledge, skills, projects, communities, problems, challenges, studies, approaches, and research areas strongly associated with the academic areas of study or areas of professional practice. Disciplinary advantages are (1) a focus on the limited "mental capacity" of the individual on a specific knowledge domain, (2) removal of science from surface danger and uncertainty, and (3) helps scientists to deepen their research and attention in molecules and atoms of their specializations. However, disciplinary disadvantages are (1) closing within a very narrow scientific particles of the discipline, forgetting that the "thing" they used is a part of "a total," (2) leaving the search for a relationship of that part to other parts of the whole, (3) isolation from the "disciplines that overlap" and intersect with the specialization naturally and essentially, and (4) out of the distribution of work and nature of the higher qualification, most experts seek to become king within his/her specialty [14].

It is well known that a multidisciplinary approach involves drawing appropriately from multiple disciplines to redefine problems outside normal boundaries and reach solutions based on a new understanding of complex situations. Multidisciplinary means knowledge associated with more than one existing academic discipline or profession, or "different disciplinary approaches to the same 'topic' through the different terminologies, methodologies, assumptions, and goals," as shown in Figure 23.3.

Multidisciplinary can be treated in three different approaches: interdisciplinary, transdisciplinary, and cross-disciplinary. Interdisciplinary is a mode of research by teams or individuals that "integrates information, data, techniques, tools, concepts, and/or theories" from two or more disciplines to advance fundamental understanding or to solve problems whose solutions are beyond the scope of a single discipline or area of research. For example, the field of bioinformatics applications combines molecular biology with computer science [15]. Transdisciplinary is an area of research and education that addresses contemporary issues that cannot be solved by one or even a few points of view. For example, there are many good examples related to postquantum

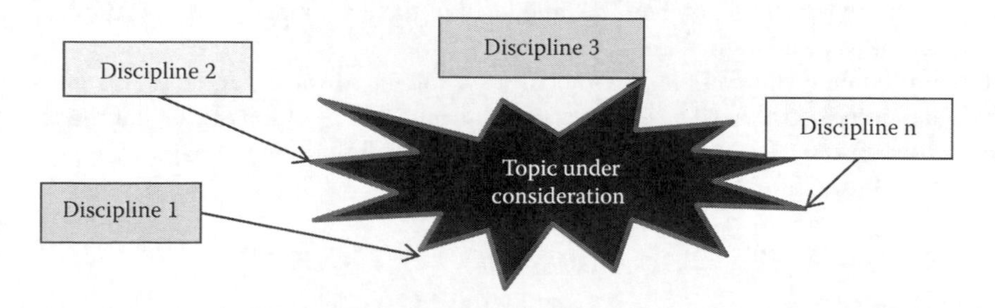

FIGURE 23.3 General representation of multidisciplinary.

cryptography, chaotic-based cryptography, and nano-communication applications [16]. Cross-disciplinary explains aspects of one discipline in terms of another. Examples include studies of the physics of music or the politics of literature [17].

Emotion cannot be understood without relying on a program of multidisciplinary research. Local multidisciplinary cannot be achieved without a programmatic framework that takes three issues into account [18]:

- The relationship of multiple levels of emotions and connected processes
- The mutually informative study of humans and artificial systems
- The dynamic nature of emotions in a dynamic system approach

Although it is true that there are few research programs that actually involve multiple disciplines concurrently, the current multidisciplinary in emotion research can be seen as holistic. Some researchers distinguish between a localist conception of multidisciplinary that implies a "field is multidisciplinary if the individual research efforts of its scientists are, typically multidisciplinary," as opposed to a holist conception of multidisciplinary that implies a field is multidisciplinary if it is "characteristic of the field that multiple disciplines contribute to the execution of its research program. This view raises the following questions [19]:

- How is it possible to find from one discipline information or impact research and theories in another discipline?
- How can a truly multidisciplinary science of emotion begin to integrate findings and theories into a standard theory of emotion that crosses the boundaries of disciplines?

Despite significant progress, affective computing and sentiment analysis are still finding their own voice as new "interdisciplinary fields." Engineers and computer scientists use machine learning techniques for classification of automatic affect from text, voice, video, and physiology. Psychologists use their long tradition of emotion research with their own discourse, models, and methods [18].

The science of emotion today is necessarily multidisciplinary in nature. Of course this is not surprising, as many, if not most, current scientific endeavors are multidisciplinary. In fact, there is formal recognition of this multidisciplinary, such as the foundation of the International Society for Research on Emotions more than 30 years ago that affirms in its mission statement that "the interdisciplinary interests of researchers and the international and cross-cultural scope of much of the recent work on emotions has created a need for a society where researchers from various disciplines can come together to discuss issues of mutual concern." There is also empirical evidence in that researchers in many different domains, such as psychology, biology, anthropology, sociology, ethology, medicine, cognitive science, political sciences, economics, physics, computer science, information theory, digital signal processing, soft computing techniques, communication theory, statistics, and philosophy study, emotional processes.

Figure 23.4 gives an approximate view about the meaning of the multidisciplinary covering the aspect of the emotion detection (generally), and most of the disciplines used in text emotion detection process are illustrated in this figure.

Emotions in humans are complex biological, psychological, social, and cultural processes that must be studied interdisciplinarily. Moreover, it has been clear for a long time that the word "emotion" has no unique and clear meaning. A proliferation of definitions can be found in psychological and philosophical literature.

23.3 CURRENT STATUS OF TEXT EMOTION DETECTION

Parizi and Kazemifard [20] claimed that text can capture only a portion of emotions expressed by a human. Even in linguistic studies, emphasis has been given on intonation and accentuation of

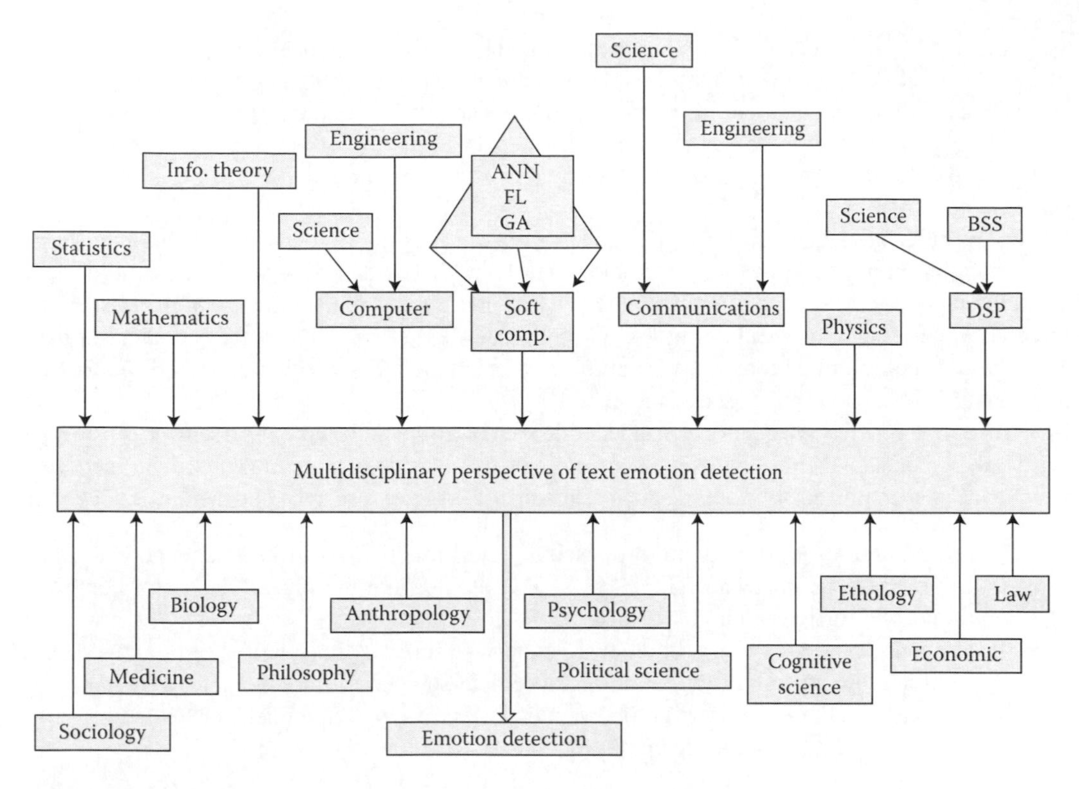

FIGURE 23.4 Multidisciplinary representation of text emotion detection.

speech for emotion analysis. However, these cues are absent from text. There are only two modes, for possible expression of emotion:

- Emotive vocabulary—Express emotions through text by using words from emotive vocabulary, (e.g., love, hate, good). These words refer directly to emotions as symbols.
- Affective items—Sometimes the expression of emotion does not include words from the emotive vocabulary. Interjections such as ugh, uh-oh, and expletives can be useful indicators of emotion, especially in modern social media domains and microblogs.

Chung-Hsien et al. [21] claimed that the conditions for generating emotion can be represented by emotion generation rules (EGRs), which are manually deduced from psychology. The emotional state of each sentence, based on the EGRs, can be represented as sequence of semantic labels (SLs) and attributes (ATTs). SLs can be defined as the "domain-independent" features, whereas ATTs can be defined as "domain-dependent" features.

Marinier and Laird [22] claimed that the automatic emotion classification of users help to understand the preferences of the general public. There are a number of useful applications (within this general public), including opinion summarization and sentiment retrieval. The experimental results indicated quite satisfactory performance regarding the ability to recognize emotion presence in text and also to identify the polarity of the emotions.

Rao et al. [23] claimed that for social emotion classification over short text, a topic-level maximum entropy (TME) model can be used. Topic-level features can be generated by the TME through modeling latent topics, multiple emotion labels, and valence scored by numerous readers jointly. The overfitting problem in the maximum entropy principle is also alleviated by mapping the features to the concept space. An experiment on real-world short documents validates the effectiveness of TME on social emotion classification over sparse words.

Li et al. [24] introduced a multi-label maximum entropy (MME) model for user emotion classification over short text. With the use of MME a rich feature can be generated by modeling multiple emotion labels and valence scored by numerous users jointly. To improve the robustness of their method on varied-scale corpora, the authors further developed a cotraining algorithm for MME and used the limited memory Broyden–Fletcher–Goldfarb–Shanno (L-BFGS) algorithm for the generalized MME model.

Perikos and Hatzilygeroudis [25] presented a sentiment analysis system (SAs) for automatic emotions recognition in text, using an "ensemble of classifiers." The design of such a classifier schema is based on the notion of combining knowledge based and statistical machine learning classification methods. The ensemble schema is based on three classifiers: two of them are statistical (a naïve Bayes and a maximum entropy learner) and the third one is a knowledge-based tool performing deep analysis of the natural language sentences.

Quan and Ren [26] explored an efficient identification of compound emotions in sentences using hidden Markov models (HMMs). In this problem, emotion has a temporal structure and can be encoded as a sequence of spectral vectors spanning an article range. The major research contributions achieved are

- Weighted high-order HMMs are proposed to determine the most likely sequence of sentence emotions in an article. The HMMs take into account the impact degree of context emotions with different lengths of history.
- A compound emotion is represented by a sequence of binary digits, namely, emotion code.
- An architecture (that uses the emotions of simple sentences) is implemented as part of known states in the weighted high-order hidden Markov emotion models for further recognizing more unknown sentence emotions.

The experimental results showed that the proposed weighted high-order HMMs are quite powerful in identifying sentence emotions compared with several state-of-the-art machine learning algorithms, and the standard (n-order) hidden Markov emotion models. And the use of emotion of simple sentences as part of known states was able to significantly improve the performance of the weighted n-order hidden Markov emotion models.

Cernea and Kerren [27] highlighted techniques that have been employed more intensely for emotion measurement in the context of affective interaction. Besides capturing the functional principles behind these approaches and the inherent volatility of human emotions, they presented relevant applications and established a categorization of the roles of emotion detection in interaction.

Yeole et al. [28] presented an approach for emotion estimation from the text entered by users on social networking sites with direct word or indirect emotions like emoji or smiley faces. The work proposed using affective words and sentence context analysis methods for emotion determination.

Gil et al. [29] applied the ontology in the context of EmoCS, a project that collaboratively collects emotion common sense and models it using EmotionsOnto and other ontologies. Currently, emotion input is provided manually by users.

Imam et al. [30] experimented on real-world short text collections validating the effectiveness of the methods applied on social emotion classification over sparse features. They also demonstrated the applications of generated lexicons in identifying entities and behaviors that convey different social emotions.

23.4 CHALLENGES OF TEXT EMOTION DETECTION

It is well known that TED researchers put in their minds an important question: How is it possible to capture the main challenges that the emotion measuring techniques (detection, identification, classification, estimation, and recognition) have to overcome? This question has been addressed by some papers since the start of different research activities related to TED. Interest in TED has increased due to the many challenging research problems and practical applications. In this section we describe some of the challenges faced by researchers [31].

1. The used text length—It is well known that short text differs from the normal text because of its sparse word co-occurrence patterns, which affects the applications of text classification methods. Short texts are prevalent on the web especially in news headlines, instant messages, questions, and tweets [32]. It is well known that most TED research focused on user emotion detection, conveyed by long documents.
2. Context dependence—The "context-dependence" of emotions within text is considered as one of the biggest challenges in determining emotion. A phrase can have element of anger without using the word "anger" or any of its synonyms. For example, the phrase "Shut up!" [33].
3. Labeled emotion database—Within emotion detection, the lack of a labeled emotion database to enable active innovation is well known. Currently, few publicly accessible databases are available [34].
4. Quality of assurance—How is the quality compared by the experts? While ensuring the quality of data labeling using Mechanical Turk could be thought of as equal parts science and art [35], interesting questions remain: When the noise in the data increases with the size of the data, how can we use this fact to draw out insights from our algorithms? How do we calculate the optimum price that is attractive? What are the main metrics that can be used as optimal ones for text emotion detection?
5. Keyword selection—TED usually uses a set of keywords to classify texts in different classes. In sentiment analysis, text is classified into two classes (positive and negative), which are opposite of each other. But coming up with a right set of keyword is not a petty task. This is because sentiment can often be expressed in a delicate manner making it tricky to be identified when a term in a sentence or document is considered in isolation [36].
6. Multiple opinions in a sentence—A single sentence can contain multiple opinions based on subjective and factual portions. It is helpful to isolate such clauses. It is also important to estimate the strength of opinions in these clauses so that researchers can find the overall sentiment in the sentence [37].
7. Negation handling—Handling negation can be tricky in sentiment analysis. For example, "I like this dress" and "I don't like this dress" are different from each other by only one token, but consequently are to be assigned to different and opposite classes. Negation words are called polarity reversers [38].
8. Comparative sentences—A comparative sentence expresses a relation based on similarities or differences of more than one object. Research on classifying a comparative sentence as opinionated or not is limited. Also the order of words in comparative sentences manifests differences in the determination of the opinion orientation [39].
9. Opinion spam—Opinion spam refers to fake or bogus reviews that try to deliberately mislead potential readers or automated systems by giving undeserving positive opinions to some target objects in order to promote the objects and/or by giving malicious negative opinions to some other objects in order to damage their reputations [40].

23.5 CONCLUSION AND FUTURE WORKS

23.5.1 Conclusion

Affective computing is currently one of the most active research topics, having received increasing intensive attention. This strong interest is driven by a wide spectrum of promising applications in many areas, such as affective agents and multiagents or even recommender systems.

- Most existing TED methods focus on either exploiting the social emotions of individual words or the association of social emotions with latent topics learned from normal documents.

- The identification of emotions embedded in user-contributed comments at the social web is both valuable and essential.
- The knowledge-based tool analyzes the sentence's text structure and dependencies, and implements a keyword-based approach, where the emotional state of a sentence is derived from the emotional affinity of the sentence's emotional parts.
- The interest in languages other than English in this field is growing, as there is still a lack of resources and research concerning languages other than English. The most common lexicon source used is WordNet, which exists in languages other than English. Building resources used in SA tasks is still needed for many natural languages.
- Many experiments were tested in the year 2016, and their results were mentioned in Section 23.4. These experiments are very important and produced good results for the progress in this field of research.
- It is very interesting that authors considered different emotion features 7 or 6 or 4. The main important point that must be considered is the calculation accuracy of the values of these features (metrics). It is clear that there are some uncertainties with the different calculations used. For that reason, many researchers aimed toward using the fuzzy logic techniques to improve the calculated values of these features.
- It is important to note that this area requires a broad multidisciplinary background knowledge, such as ethology, psychology, neuroscience, computer science, software engineering, artificial intelligence, and philosophical insight, in the context of creative engineering design.
- Machine learning approaches are a better option for detection emotion tasks, since we obtain a model that is also able to detect emotions in texts that have only an indirect reference to an emotion. It is important use good lexical resource as features in machine learning algorithms to obtain good results.

23.5.2 Future Works

To make the emotion detection system more robust, in the future there are a number of issues we intend to explore further. In regard to future work, there are still many interesting and challenging aspects that need to be investigated in affective computing:

- Research must be developed to make such SA or OM more convenient to treat with the short texts.
- We must concentrate on increasing the labeled database used in TED.
- The research community should establish an annotated corpus and a set of metrics that may be used to evaluate the different existing systems and the future systems.
- Due to the complexity of the data, it is a good proposal to use "rich semantic models based on ontology," since EmotionsOnto is a generic ontology for describing emotions and their detection and expression systems taking contextual and multimodal elements into account.
- Experiments showed that the separable mixture model (SMM) is adopted to estimate the similarity between an input sentence and the emotion association rules (EARs) of each emotional state. Such an approach is promising, and easily ported into other domains, so it will be a good idea to be followed in future works for TED.
- It is very important to train the computer or computational algorithms used within TED about the nature and behaviors of the text authors in order to know the actual behavior of the author, then the emotion determination will be more accurate.
- In many applications, it is important to consider the context of the text and the user preferences. That is why we need to make more research on context-based SA. Using transfer learning (TL) techniques, we can use related data to the domain in question as a training data. Using NLP tools to reinforce the SA process has attracted researchers recently and still needs some enhancements.

- More interest could be given to the use of ensembles of classifiers to benefits from their merits and minimize their drawbacks, and to perform deep analysis of the natural language sentences.
- Experiments must be conducted to automatically measure users' emotional states using brain–computer interface.
- More argument labels within the semantic labeling techniques should be taken into account in the emotion model.
- More adjective groups need to be included in the emotion model to account for more emotions.
- To obtain an accurate emotion for a sentence, the context in which a sentence is used must be taken into account.
- The emotion estimation module must be improved, specifically by combining past emotional states as a parameter for deciding the affective meaning of the user's current message.
- It is very important to take care about orienting the efforts toward using the following techniques:
  - Adaptive neural fuzzy inference system (ANFIS) techniques for enhancing the accuracy of text emotion detection.
  - Blind source separation techniques and independent component analysis and principle component analysis (ICA/PCA). This adaptive signal processing technique will enhance the calculated features from the text, and the emotion will be described in more detail.
  - The proposed direction focuses on deep analysis, since we consider that if we use features based on a deep analysis on the text we could improve the emotional detection systems.

REFERENCES

1. Rosalind W. Picard. *Affective Computing.* MIT Press, Cambridge, MA, 1997.
2. Seung H. Yoo et al. The influence of emotion recognition and emotion regulation on intercultural adjustment. *International Journal of Intercultural Relations* 2006;30:345–363.
3. Shiv Naresh Shivhare and Sri Khetwa Saritha. Emotion detection from text documents. *International Journal of Data Mining & Knowledge Management Process (IJDKP)* 2014;4(6): 51–57.
4. Vaibhav Tripathi et al. Emotion Analysis from Text: A Survey. http://www.cfilt.iitb.ac.in/resources/surveys/emotion-analysis-survey-2016-vaibhav.pdf.
5. Chew-Yean Yam. Emotion Detection and Recognition from Text Using Deep Learning. https://www.microsoft.com/developerblog/real-life-code/2015/11/30/Emotion-Detection-and-Recognition-from-Text-using-Deep-Learning.html.
6. Shadi Shaheen et al. Emotion recognition from text based on automatically generated rules. *2014 IEEE International Conference on Data Mining Workshop*, Shenzen, China, December 14, 2014, pp. 383–392.
7. Lea Canales, Patricio Martínez-Barco. Emotion Detection from Text: A Survey. https://www.researchgate.net/file.PostFileLoader.htm.
8. Rosalind W. Picard. Affective computing: Challenges. *International of Journal of Human-Computer Studies* 2003;59:55–64.
9. Ashish Katrekar. An Introduction to Sentiment Analysis. https://www.globallogic.com/.../Introduction-to-Sentiment-A.
10. Walaa Medhat et al. Sentiment analysis algorithms and applications: A survey. *Ain Shams Engineering Journal* 2014;5:1093–1113.
11. Mikalai Tsytsarau and Themis Palpanas. Survey on mining subjective data on the web. *Data Mining and Knowledge Discovery* 2012;24:478–514.
12. Montoyo Andrés et al. Subjectivity and sentiment analysis: An overview of the current state of the area and envisaged developments. *Decision Support System* 2012;53:675–679.
13. Pang Bo and Lee Lillian. Opinion mining and sentiment analysis. *Foundations and Trends in Information Retrieval* 2008;2:1–135.
14. Sattar B. Sadkhan and Nidaa A. Abbas (editors). *Multidisciplinary Perspectives in Cryptology and Information Security.* IGI-GLOBAL.COM, Hershey, PA, 2014.

15. William H. Newell. A theory of interdisciplinary studies. *Issues in Integrative Studies* 2001;19:1–25.
16. Christian Pohl and Gertrude Hirsch Hadorn. Methodological challenges of transdisciplinary research. *Natures Sciences Sociétés* 2008;16:111–121.
17. Anna C. Evely et al. Defining and evaluating the impact of cross-disciplinary conservation research. *Environmental Conservation* 2010;37(4):442–450.
18. Arvid Kappas. The science of emotion as a multidisciplinary research paradigm. *Behavioral Processes* 2002;60:85–98.
19. Christian V. Scheve and Rolf V. Luede. Emotion and social structures: Towards an interdisciplinary approach. *Journal for the Theory of Social Behavior* 2005;35:3.
20. Ali H. Parizi and Mohammad Kazemifard. Emotional news recommender system. *2015 International Conference of Cognitive Science (ICCS)*, Tehran, Iran, April 27–29, 2015, pp. 37–41.
21. Chung-Hsien Wu et al. Emotion recognition from text using semantic labels and separable mixture models. *Journal ACM Transactions on Asian Language Information Processing (TALIP)* 2016:165–183.
22. Robert P. Marinier and John E. Laird. Toward a comprehensive computational model of emotions and feelings. In *Sixth International Conference on the Cognitive Modeling*, edited by Marsha C. Lovett et al., pp. 172–177. Lawrence Erlbaum, Mahwah, NJ, 2014.
23. Yanghui Rao et al. Social emotion classification of short text via topic-level maximum entropy model Information and management. *Information & Management* 2016; 53(8): 978–986.
24. Jun Li et al. Multi-label maximum entropy model for social emotion classification over short text. *Neurocomputing Journal* 2016;210:247–256.
25. Isidoros Perikos and Ioannis Hatzilygeroudis. Recognizing emotions in text using ensemble of classifiers. *Engineering Applications of Artificial Intelligence* 2016;51:191–201.
26. Changqin Quan and Fuji Ren. Weighted high-order hidden Markov models for compound emotions recognition in text. *Information Science* 2016;329:581–596.
27. Daniel Cernea and Andreas Kerren. A survey of technologies on the rise for emotion-enhanced interaction. *Journal of Visual Languages and Computing* 2015;31(Part A):70–86.
28. Ashwini V. Yeole et al. Opinion mining for emotions determination. *2015 International Conference on Innovations in Information, Embedded and Communication Systems (ICIIECS)*, Coimbatore, India, March 19–20, 2015, pp. 1–5.
29. Rosa Gil et al. Emotions ontology for collaborative modeling and learning of emotional responses. *Computers in Human Behavior* 2015;51(Part B):610–617.
30. Seyyare Imam et al. Performance analysis of different keyword extraction algorithms for emotion recognition from Uyghur text. *2014 9th International Symposium on Chinese Spoken Language Processing (ISCSLP)*, Singapore, September 12–14, 2014, p. 351.
31. Alexandre Denis et al. General Purpose Textual Sentiment Analysis and Emotion Detection Tools. 2013. https://arxiv.org/pdf/1309.2853.pdf.
32. Mohamed H. Haggag. Frame semantics evolutionary model for emotion detection. *Computer and Information Science* 2014;7(1).
33. Jean-Claude Martin et al. Multimodal complex emotions: Gesture expressivity and blended facial expressions. *International Journal of Humanoid Robotics*. http://www.infomus.org/people/niewiadomski/papers/ijhr2006_martinetal.pdf.
34. Yafei Sun et al. Authentic Emotion Detection in Real-Time Video. http://citeseerx.ist.psu.edu/viewdoc/download?doi=10.1.1.4.3064&rep=rep1&type=pdf.
35. Kateri McRae et al. Context-dependent emotion regulation: Suppression and reappraisal at the burning man festival. *Basic and Applied Social Psychology,* 2011; 33: 346–350.
36. Yung-Chun Chang et al. Semantic Frame-Based Approach for Reader-Emotion Detection. http://www.pacis-net.org/file/2015/3083.pdf.
37. Ameeta Agrawal and Aijun An. Unsupervised Emotion Detection from Text using Semantic and Syntactic Relations. http://www.cs.yorku.ca/~aan/research/paper/Emo_WI10.pdf.
38. Nilesh M. Shelke. Approaches of emotion detection from text. *International Journal of Computer Science and Information Technology Research* 2014;2(2):123–128.
39. Kristína Machová and Lukáš Marhefka. Opinion Mining in Conversational Content within Web Discussions and Commentaries. http://link.springer.com/chapter/10.1007/978-3-642-40511-2_11.
40. Sander Koelstra et al. DEAP: A database for emotion analysis using physiological signals. *IEEE Transactions on Affective Computing* 2012;3(1) :18–31.

24 Reaction Automata Direct Graph (RADG) Design on Elliptic Curve Cryptography

Salah A. Albermany and Ali Hasan Alwan

CONTENTS

24.1 Introduction 321
24.2 Related Work on Reaction Automata Direct Graph (RADG) 322
24.3 Multi-Reaction Automata Direct Graph (MRADG) 322
24.3.1 Number of Values in the Design 322
24.3.2 Transition Function 323
24.3.3 Embedding 324
24.3.4 MRADG with Transition Function 325
24.3.5 Example 326
24.4 Authentication 326
24.5 Integrity 327
24.6 Analysis 329
24.7 Conclusion 330
References 330

24.1 INTRODUCTION

The word *cryptography* is a combination of two a Greek words: *kryptos* meaning "hidden" and *graphein* meaning "writing," or the science of encrypting and decrypting text [1]. Cryptography is when the text or message is transferred in an unreadable format (usually called cipher text) done by encryption process. If we divide cryptography into sections according to the encryption key, there are two types: symmetric key cryptography and asymmetric key cryptography. Symmetric cryptography (referred to as secret-key ciphers) uses the same key to encrypt and decrypt [2]. Asymmetric cryptography (referred to as public-key ciphers) requires public and private keys for encryption and decryption, and includes a digital signature to authorize the message's sender. Symmetric cryptography is much faster than asymmetric cryptography, but a public key is safer. Public-key techniques uses an arithmetic operation like modular multiplication and requires exponentiation time when attempting to encrypt. One of the first public key algorithms was RSA. The acronym comes from the first letter of the three scientists Ron Rivest, Adi Shamir, and Leonard Adleman, who published their method in 1977. RSA was one of the first public key cryptography using a public key for encryption and deferent key (private key) for decryption. It is based on two prime numbers and their product before encryption in RSA [3]. The modern public key algorithm is the elliptic curve (EC). There are many fields of EC such as the Galois field of large prime number (GF_{p},) over binary field 2 m, where m is an integer number, and EC over Z, where Z is the integer number. In this chapter, for the discussion about EC over GF_p, the general equation of EC is $y^2 = x^3 + \alpha x + \beta$ mod p $y^2 = x^3 + ax + b$ modp, denoted by $E_p(\alpha, \beta)$ Ep(a,b), where the coefficients α and β are integers in the Galois field, and p is the prime number (where α and β must satisfy the equation $4\alpha^3 + 27\beta^2 \neq 0$). There are two operations in the field F_P. F_P is defined on EC using addition and multiplication (multiplication is repeated addition, e.g., 2P = P + P) [4–6].

24.2 RELATED WORK ON REACTION AUTOMATA DIRECT GRAPH (RADG)

Reaction Automata Direct Graph (RADG) depends on random characteristics; keyless, static internal design; and use in a personal network. And it is useful against statistical attack, based on Automata Direct Graph and reaction states. It applies encryption and decryption without any key and produces several cipher texts from one plaintext. The main design contains R states, Q states, and jump where jump does not have transitions, as illustrated in Figure 24.1.

To explain the encryption process: Suppose the original message to be encrypted using RADG is 1010. Thus we have: T (0, 1) = (3, 14), T (3, 0) = (5, 15), T (5, 1) = (1, 20), and T (1, 0) = (3, 11). The corresponding output is 14, 15, 20, and 11, respectively, where the jump state redirects from state 4 to state 5 (state 5 including in reaction states) [7].

24.3 MULTI-REACTION AUTOMATA DIRECT GRAPH (MRADG)

The main problem in RADG is the static design. If anyone on the network gets the design with cipher text, then he/she can decrypt the cipher text effortlessly and get information from it. The purpose of this chapter is to develop the RADG design to be more secure in wide area networks such as the cognitive radio network (CRN). The RSA transition function is used in RADG and converts the cipher text into a point in specific EC. The new algorithm based on RADG design contains three parts: reaction states, Q states, and jump states. Each state has λ values, except the jump state that just refers to another state in reaction states. The proposed algorithm uses function f instead of the static transition function in RADG [7], as shown in Figure 24.2.

24.3.1 Number of Values in the Design

In general each state, whether it is in Q states or R states, have λ values in single state, expect the jump states that do not have any values. To compute the number of values in the entire design:

$$\Lambda = \lambda * (m + n - k)$$

where m is the number of reaction states, n is the number of Q states, and k is the number of Jump states.

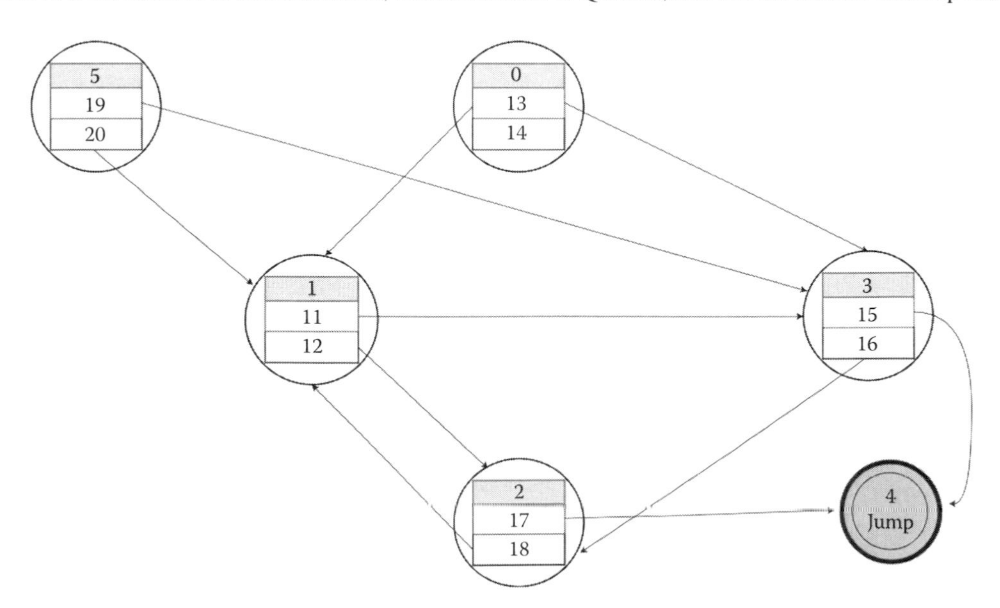

FIGURE 24.1 Illustrates example of RADG implementation.

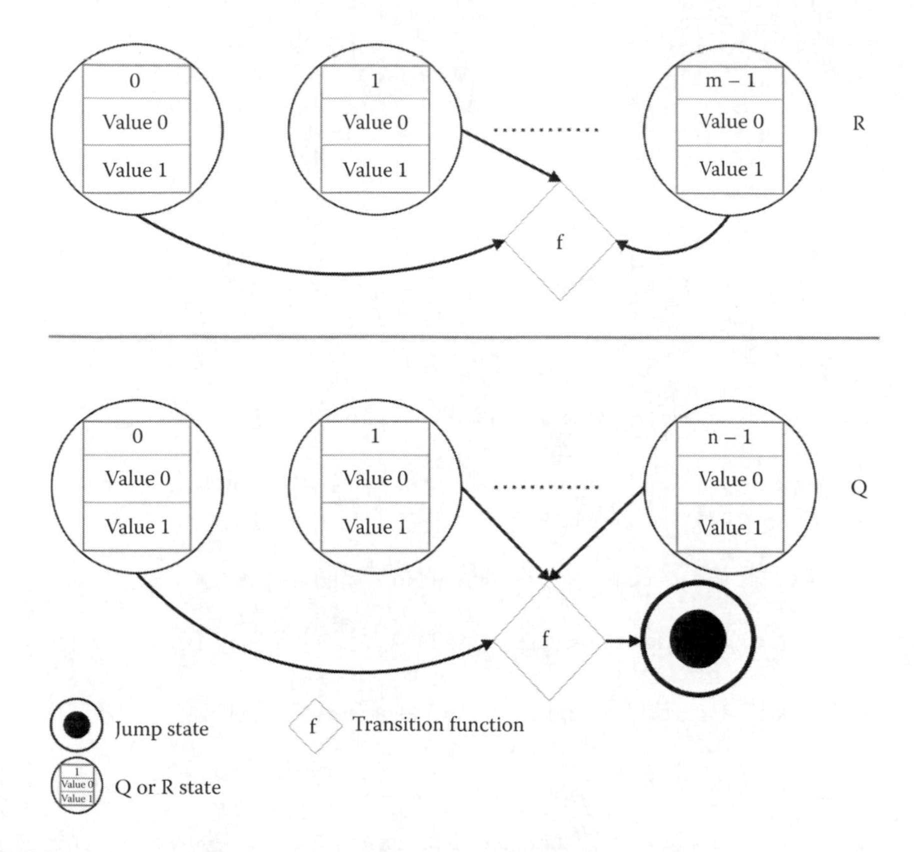

FIGURE 24.2 Illustrates design of proposed algorithm MRADG.

24.3.2 Transition Function

At the beginning the communications, there are a large number of n that consists of two prime numbers, p and q, where $n = p * q$. The number of Q states goes from 0 to $n - 1$. Also the sender who is ciphering must choose a random number e between (1, Φ (n)), where Φ (n) is the Euler's totient function, and the e number represents the public key for the transition function on condition gcd(e, Φ (n)) = 1. There are two types of transition functions: f_Q and f_R. There is no difference in the internal architecture of these functions. The result for them refers to the state within the Q set, but the input of f_Q is from Q states and for f_R is from R states.

To determine the next state, as illustrated in Figure 24.3, where it totally depends on the current state where the cipher progress is located, ensure the distribution of states is not repeated circularly. To avoid this situation one must have a secondary key (the message index represent this key). Suppose the current state number is 1. As known, 1 raised to any power remains 1. If the cipher process starts at 1, then all the next states will become 1 (without secondary key). For example, if the the current state is 11, and the message index is 8, compute the next state:

$$\text{temp} = \text{CurrentStateNo} + \text{key} \bmod n$$

$$\text{NextState} = \text{temp}^{e} \bmod n$$

Then the next state is number 28 in the Q states, and so on to the rest of states. The difference of the two functions f_Q and f_R is the input state either from the R state or Q state; the output is always in the Q state.

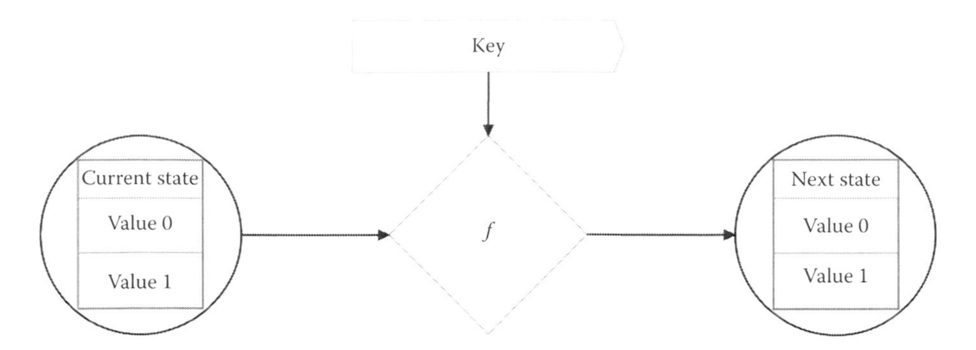

FIGURE 24.3 Illustrates the scheme of transition function inside MRADG.

In the backward process from the last example we have located in the state number 28, and the message index is 8 after calculating the plaintext. Then calculate the previous state:

$$\text{temp} = (\text{CurrentStateNo.})^{d} \bmod n$$

$$\text{PreviousState} = \text{temp} - M_{\text{index}} \bmod n$$

The previous state is 11 (Figure 24.4), and so on for the remaining states.

24.3.3 Embedding

There are several ways to represented data in the EC. Consider a curve $y^2 = x^3 + ax + b \bmod p$, and the message content has numbers (0–35) and alphabet characters. Encode char "A" as m = 10 within a public variable, n = 20.

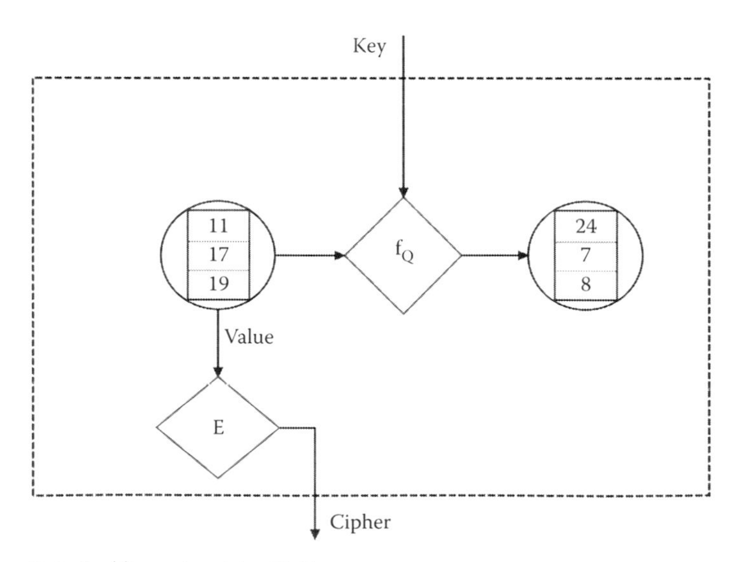

FIGURE 24.4 Illustrates the internal structure of transition function.

1. Compute, ma = m * k + i, where value of i is between {1, … , k − 1}, then try to get the integral value of y.
2. Thus ma encodes as point (x, y).

The decoding is simple: m = floor ((x − 1)/k) [5,6].

24.3.4 MRADG with Transition Function

Using algorithms to describe and understand the encryption and decryption processes, Table 24.1 explains ambiguities, abbreviations, and functions in the algorithms.

- *Key for transition function*

  Step 1 $P \leftarrow Prime_{NO}, q \leftarrow Prime_{NO}$
  Step 2 $n \leftarrow p * q$
  Step 3 $\Phi(n) \leftarrow (p-1) * (q-1)$
  Step 4 $e \leftarrow gcd(random(0, n-1), \Phi(n)) = 1$
  Step 5 $d \leftarrow e^{-1}$

- *Encryption*

  Step 1 $State_{no'} \leftarrow random(0, R_{length})$
  Step 2 $status \leftarrow IN_R$
  Step 3 $while(index < Message_{length})$
  Step 4 $if\ status = IN_R$
  Step 5 $cipher[index] = R[State_{no}].getValue[Message]$
  Step 6 $State_{no'} = next\_state(State_{no'}, index, e, n)$
  Step 7 *end if*
  Step 8 $if\ (State_{no'} = jump_i)$
  Step 9 $State_{no'} = random_{Gi}$
  Step 10 *end if*
  Step 11 $if\ Status = IN_Q$
  Step 12 $cipher[index] = R[State_{no}].getValue[Message]$
  Step 13 $State_{no'} = next\_state(State_{no'}, index, e, n)$.
  Step 14 *end if*
  Step 15 *Embadding(cipher)*
  Step 16 *End while*

TABLE 24.1
RADG Implementation Notations

Notations	Details
random(0, n)	Generate random integer number between 0 and n
IN_R, IN_Q	In R states, in Q states
getValue[Message]	Get the first or second value from the state
$next_{state}()$	Function takes several parameters: previous $state_{no'}$, index as secondary key, e as transition key, and n as number of finite field and return number of next state
$jump_i$	Refers to one of jump states
$random_{Gi}$	In multi-reaction it refers to a random number in one subgroup of reactions
Embadding()	Embedding the cipher value into specific elliptic curve
$perviouse_{state}$	Inverse of $next_{state}$ function

- *Decryption*

Step 1 $Embadding^{-1}()$
Step 2 $[status, State_{No}] \leftarrow perviouse_{state}(State_{no}, Message_{length-1}, d, n)$
Step 3 $while\ (index \geq 0)$
Step 4 $if\ Status = IN_Q$
Step 5 $decipher[index] = Q[State_{no}].getValue[Message]$
Step 6 $State_{no'} = perviouse_{state}(State_{no'}, index - 1, d, n)$
Step 7 $if\ (value\ not\ found\ in\ Q)$
Step 8 $status \leftarrow IN_R$
Step 9 *end if*
Step 10 *end if*
Step 11 $if\ Status = IN_R$
Step 12 $decipher[index] = Q[State_{no}].getValue[Message]$
Step 13 $State_{No} = jump_i(R_i)$
Step 14 $State_{No} = perviouse_{state}(State_{no'}, index - 1, d, n)$
Step 15 $status \leftarrow IN_Q$
Step 16 *end if*
Step 17 *end while*

24.3.5 Example

This section clarifies the "algorithms" via numbers in detail.

Encryption: Before starting in encryption process there are several things that the communication parts must agree to two prime numbers: p and q. In this example p =11 and q = 3.

$$n = p * q = 33$$

$$\Phi(n) = (p - 1) * (q - 1)$$

$$10 * 2 = 20$$

Each part of the communication has its own public key e (random number where $0 < e < \Phi(n)$) and $(e, \Phi(n)) = 1$. Finally the cipher values embedded in this example will use the elliptic curve with the equation $y^2 = x^3 - x + 188 \bmod 751$, and use the design of MRADG with values of the R set and Q set, as shown in Table 24.2 and Table 24.3, respectively. The encryption process is as shown in Table 24.4.

The cipher is {(50, 136), (352, 65), (291, 16), (190, 196), (391, 187), (131, 34), (1, 375), (170, 274)}.

Send the last $state_{no.}$ encryption by e $10^e \bmod n = 24$.

Decryption: To explain the decryption process, start backward. The data the receiver receives was embedding into a specific elliptic curve and cipher of last state number in this example: $State_{no.} = 28, = 17, n - 33, k = 7$.

First, compute the last state number of the decryption process, $State_{no.} = (28^{17} \bmod 33) - 8 = 11$. Table 24.5 explains the next phases of the decryption.

24.4 AUTHENTICATION

Authentication is the method of proving one's identity to somebody else. It is the most important topic in security. The public key techniques are in the following equation:

$$M = D(E(M)) \quad \text{or} \quad M = E(D(M))$$

where M is refers to message, E refers to public key, and D refers to private key [1,8].

TABLE 24.2
Values of R Set

	R Set	
Number of State	**First Value**	**Second Value**
0	38	55
1	58	52
2	45	26
3	1	12
4	10	34
5	—	—
6	57	25
7	31	20
8	63	72
9	—	—
10	78	83
11	66	50
12	37	5
13	93	80
14	35	27
15	36	43
16	56	4
17	68	60
18	7	28
19	11	59
20	77	74
21	62	22
22	30	75
23	32	39
24	15	48
25	16	70
26	67	54
27	47	21
28	73	8
29	61	76
30	14	18
31	51	17
32	2	9

As illustrated in Figure 24.5, first Alice tells Bob she wants to communicate with him. Bob sends a number (R) he chose at random. Alice encrypts the number R with her private key and sends it to Bob. In the final step Bob decrypts the encrypted value with Alice's public key. If she has the same value of R, then she will be able identify the user she is communicating with, and vice versa for the authentication for Bob. The public key also provides a digital signature to ensure nonrepudiation. The receiver verifies the sender's identity, because the message can only be decrypted by the sender's public key, and only encrypted by the private key; then the sender cannot repudiate the message [9].

24.5 INTEGRITY

To get the integrity of the message, the most efficient way is to use one-way hash function h = H(M), where h is smaller than message M, and H is the hash function. The standard on the

TABLE 24.3
Values of Q set

	Q Set	
Number of State	First Value	Second Value
0	40	71
1	33	3
2	19	29
3	46	65
4	46	65
5	23	53
6	84	81
7	85	91
8	90	86
9	89	92
10	44	69
11	88	79
12	87	82
13	49	24

TABLE 24.4
Encryption Process

i	Message	$State_{no.}$	Status	Value	Point on EC
0	0	2	IN R	19	(135, 198)
1	1	29	IN Q	76	(533, 74)
2	1	24	IN Q	48	(337, 178)
3	0	4	IN R	64	(452, 278)
4	1	28	IN Q	8	(57, 332)
5	0	32	IN Q	2	(17, 332)
6	1	16	IN Q	4	(30, 236)
7	1	22	IN Q	75	(529, 254)

TABLE 24.5
Decryption Process

Index	Point on EC	Value	Message	Stat	Status	Search
—	—	—	—	17	—	—
7	(529, 254)	75	1	22	IN R	True
6	(30, 236)	4	1	16	IN Q	True
5	(17, 332)	2	0	32	IN Q	True
4	(57, 332)	8	1	28	IN R	True
3	(452, 278)	64	0	4	IN Q	False
2	(337, 178)	48	1	24	IN Q	True
1	(533, 74)	76	1	29	IN Q	True
0	(283, 54)	19	0	2	IN Q	False

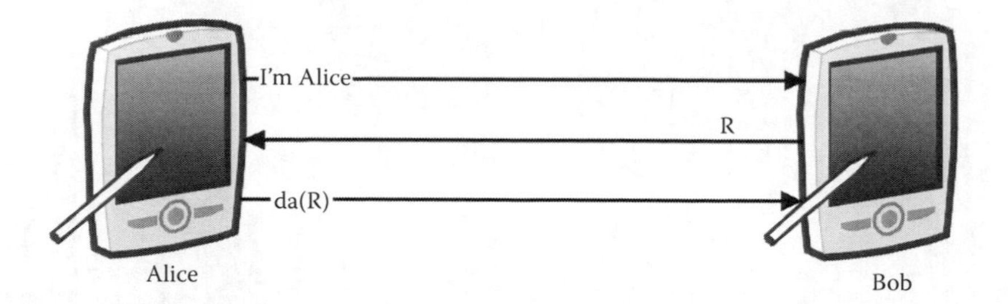

FIGURE 24.5 Illustrates authentication between two users by MRADG.

Internet is the MD5 algorithm. It gives 16 byte as output. It is impossible to find two messages that have the same hash value. The h value can be described as a flag to check whether the message was altered.

24.6 ANALYSIS

The RADG algorithm and MRADG have an important characteristic: they produce random cipher texts in several exactions to same plaintext to find the relation between the variance cipher texts, as illustrated in Figure 24.6, by using the Hamming distance [10]. The algorithm applied 80 times on the same 100-bits plaintext provides different results in each execution. This makes the algorithm stand against a statistical attack compared to general encryption methods. The standard RADG is more efficient with Hamming distance [7] in terms of speed, as shown in Figure 24.7. The algorithm MRADG is twice as fast as RADG. In this figure the black bars refer to the length of message. In this example, the message was sent 20 times and the message length was duplicated in each time. The dark gray bars refer to the time executed by MRADG, and the time required for implementing encryption and decryption is less than in RADG.

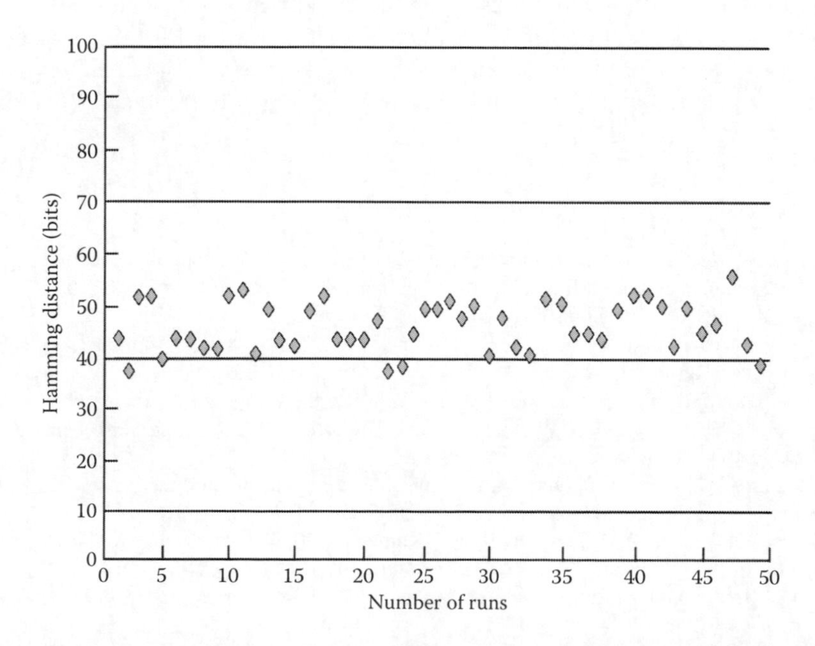

FIGURE 24.6 Illustrates the Hamming weight for different output of MRADG algorithm.

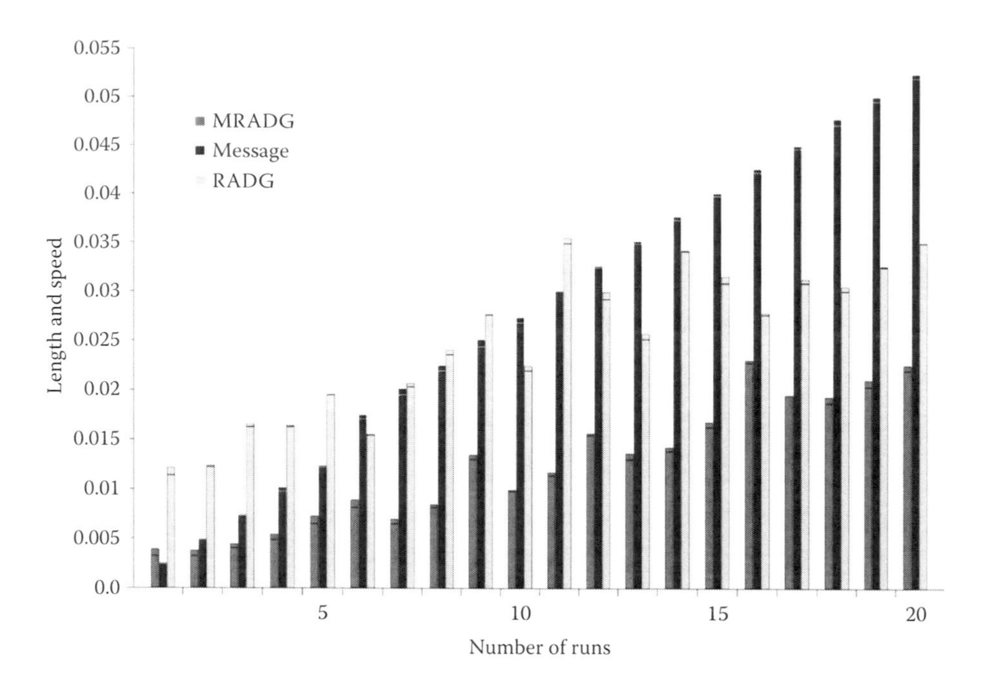

FIGURE 24.7 Illustrates run time comparison of RADG and MRADG.

24.7 CONCLUSION

In summary, the proposed algorithm increasing the range of networks in which RADG is implemented, applying the standard RADG on personal networks, and applying the developed method MRADG on wider networks including intelligent networks by using keys, partitioning the reaction states from single reaction to multireaction states led to a significant increase in the speed of preference of the algorithm while preserving the random property. The whole enhancement works to speed up the algorithm three times higher than it was in RADG. A disadvantage of the algorithm is the length of the output cipher text compared to the input plaintext, because the encryption process takes the plaintext in binary form and each bit represents one point on the EC.

REFERENCES

1. Wireless LAN Medium Access Control (MAC) and Physical Layer (PHY) Specification, IEEE Std. 802.11, 1997; Kaufman, C., Perlman, R., and Speciner, M., *Network Security: Private Communication in a Public World*, 2002, Prentice-Hall, Upper Saddle River, NJ.
2. Menezes, A. J., Van Oorschot, P. C., and Vanstone, S. A., *Handbook of Applied Cryptography*, 1996, CRC Press, Boca Raton, FL.
3. Kaliski, B., The Mathematics of the RSA Public-Key Cryptosystem, 2006, RSA Laboratories.
4. Stallings, W., *Cryptography and Network Security*, 5th ed., 2010, Prentice-Hall, Upper Saddle River, NJ.
5. Cohen, H., Frey, G., Avanzi, R., Doche, C., Lange, T., Nguyen, K., and Vercauteren, F., *Handbook of Elliptic and Hyperelliptic Curve Cryptography*, 2005, CRC Press, Boca Raton, FL.
6. Udin, M. N., Halim, S. A., Jayes, M. I., and Kamarulhaili, H. *Application of Message Embedding Technique in ElGamal Elliptic Curve Cryptosystem*, 2012, IEEE, pp. 1–6.
7. Albermany, S. A., and Safdar, G. A., Keyless security in wireless networks, *Wireless Personal Communications* 79 (3), 2014, 1713–1731.

8. Boyd, C., and Mathuria, A., *Protocols for Authentication and Key Establishment*, 2013, Springer, Berlin.
9. Rivest, R. L., Shamir, A., and Adleman, L., A method for obtaining digital signatures and public-key cryptosystems, *Communications of the ACM*, 21(2), 1978, 120–126.
10. Milenkovic, O., On the generalized Hamming weight enumerators and coset weight distributions of even isodual codes, In *Proceedings of IEEE International Symposium on Information Theory*, Washington, DC, June 29, 2001, p. 62.

25 A Performance Comparison of Adaptive LMS, NLMS, RLS, and AFP Algorithms for Wireless Blind Channel Identification

Sami Hasan and Anas Fadhil

CONTENTS

25.1 Introduction ... 333
25.2 Related Existing Work ... 334
25.3 Mathematical Model of Wireless Channel ... 335
  25.3.1 Time-Invariant Channel ... 335
  25.3.2 Time-Variant Channel ... 336
25.4 Learning Algorithms ... 337
  25.4.1 Least Mean Square (LMS) ... 337
  25.4.2 Normalized Least Mean Square (NLMS) ... 338
  25.4.3 Recursive Least Square (RLS) ... 339
  25.4.4 Affine Projection (AFP) Algorithm ... 340
25.5 Blind Identification Architecture ... 340
25.6 Results and Discussions ... 341
  25.6.1 Time-Invariant Channel Simulation ... 342
  25.6.2 Time-Variant Channel Simulation ... 344
25.7 Conclusion ... 345
References ... 345

25.1 INTRODUCTION

Broadcasts of transmitted signals through wireless channels arrive at the receiver over multiple paths. These paths are created due to echoing, diversion, or spreading in the channel. Multipath propagation produces a received signal by superposition of several delayed wireless digital communications that often require the identification of the channel impulse response and scaled copies of the transmitted signal that can facilitate channel equalization and maximum likelihood sequence detection [1,2].

Adaptive wireless channel identification is typically utilized when simpler techniques [3–6] for received sequence detection cannot be used in telecommunication systems.

The wireless channel distorts the conformity of the transmitted signals making the decoding of the received information difficult. In such cases where the effects of the channel distortion can be modeled as a linear finite impulse response (FIR) filter, the transmitted symbols are known as Intersymbol Interference (ISI). Thus, an adaptive filter can be developed to model the effects of the channel ISI for purposes of decoding the received information in an optimal manner, where the transmitter sends to the receiver a sample sequence $\boldsymbol{x(n)}$ that is known to both the transmitter and receiver. The receiver then attempts to model the received signal, $\boldsymbol{y(n)}$, using an adaptive filter whose input is the known transmitted sequence, $\boldsymbol{x(n)}$, and output signal, $d(n)$. After a suitable period of adaptation, the optimal coefficients of the adaptive filter, $\boldsymbol{h(n)}$, are computed and then utilized in a procedure to decode future signals transmitted across this wireless channel. This blind channel identification is achieved by using only the channel output without using a training sequence.

To simulate this, two types of wireless channels, time invariant and time variant, are mathematically modeled. Then, a generic adaptive wireless channel architecture is designed using adaptive FIR filtering algorithms of least mean square, normalized least mean square, recursive least square, and affine projection.

Blind channel identification methods using the second-order cyclostationary statistics were initiated by Tong et al. [7,8] and have attracted research attention [9–12]. The contributions of this chapter can be shortlisted as the following:

- 5G wireless mobile radio channel identification architecture is described.
- Performance indices comparison of adaptive blind identification algorithms are evaluated.
- Four adaptive algorithms (MLS, NMLS, RLS, and AFFFINE) are analyzed in respect of computational complexity.
- The time-invariant channel and time-variant channel are mathematically modeled and adaptively blind identified using four methods.
- The learning curve of the four adaptive algorithms, iteration = 100, filter order = 10 are plotted using the computer-simulated models.

25.2 RELATED EXISTING WORK

There is no dearth in the literature for system identification using various adapting algorithms. However, the wireless communication channel identification using adaptive techniques are rarely found within the existing works, not to mention the shortage in comparison studies of these adaptive techniques.

Jiang Du, Qicong Peng, and Hongying Zhang [13] investigated a blind algorithm for channel identification and equalization of multiple input, multiple output-orthogonal frequency division multiplexing (MIMO-OFDM) systems by exploiting the second-order cyclostationarity inherent in OFDM with cyclic prefix and the characteristics of the phased antenna. They constructed a practical HIPERLAN/2 standard-based MIMO-OFDM simulator with the sufficient considerations of statistical correlations among the multiple antenna wireless wideband channels, and formulated a new two-stage adaptive blind algorithm using rank reduced subspace channel matrix approximation and constant modulus algorithm (CMA) criteria. The performance of the new method has been justified theoretically and validated through extensive simulations over various common wireless and mobile communication channels.

Rao and Barton [14] developed blind adaptive channel estimation algorithms for system identification on impulse radio channels. These algorithms are based on higher-order statistics, as opposed to conventional second-order algorithms. The objective of this study was to demonstrate that this approach to channel estimation, preferably, for impulse radio systems. Simulation results have verified the advantage of using higher-order algorithms, regardless of whether the channel impulse response corresponds to a minimum phase system.

Chen and Xie [15] constructed signal models of four types of signals in wireless communication: single-user MIMO signal, single user single antenna signal, virtual MIMO signal, and multibeam smart antenna signal. They then proposed a method for blind identification of the single user MIMO signal step by step in noncooperation communication when the unknown source was one of the four types by using the MUSIC algorithm and forward-backward spatial smoothing algorithm to blindly estimate DOAs of the received signal in the first two steps and blind channel estimation in the third step. Simulation results have shown the feasibility of this method.

Alouane [16] proposed a new normalized least mean square (NLMS) algorithm that is, in the steady state, insensitive to the time variations of the input dynamics. The square soot (SR)-NLMS algorithm is based on a normalization of the LMS adaptive filter input by the Euclidean norm of the tap-input. The tap-input power of the SR-NLMS adaptive filter is then equal to one even during sequences with low dynamics. The harmful effect of the low dynamics input sequences on the steady-state performance of the LMS adaptive filter are then reduced. In addition, the square root normalized input is more stationary than the base input. Therefore, the robustness of the LMS adaptive filter with respect to the input nonstationarity is enhanced. A performance analysis of the first- and the second-order statistic behavior of the proposed SR-NLMS adaptive filter is carried out. In particular, an analytical expression of the step size ensuring stability and mean convergence is derived. In addition, the results of an experimental study demonstrating the good performance of the SR-NLMS algorithm are given.

McLaughlin and Cowan [17] considered the performance of an exponentially windowed RLS algorithm carrying out nonstationary channel identification. The two criteria used for assessing the performance of the algorithm are the steady-state mean-squared error (MSE) and the initial rate of convergence. The effect of both filter order and level of no stationarity is illustrated, with the LMS adaptive algorithm used as a reference for comparative purposes.

From the rarely available adaptive channel identification literatures, the research gap of adaptive 5G wireless channel identification architecture necessitates the performance analyses of more than one adaptive algorithm to compute the parameters of wireless time invariant as well as time variant channels.

25.3 MATHEMATICAL MODEL OF WIRELESS CHANNEL

A wireless communication system transmits information through wireless channels. A mathematical model is constructed to reflect the most important characteristics of the transmission channel. This mathematical model is the prelude of the channel's simulation using MATLAB tool.

25.3.1 Time-Invariant Channel

The time-invariant channel may represent the transmission medium for a stationary receiver in one location. The impulse response, ***h(n)***, of this channel can be written as

$$h(\tau) = a_1\delta(\tau-\tau_1)e^{j\theta_1} + a_2\delta(\tau-\tau_2)e^{j\theta_2}\cdots + a_N\delta(t-\tau_N) * e^{j\theta_N} \tag{25.1}$$

which can be expressed in closed form as

$$h(\tau) = \sum_{k}^{L} a_k\delta(\tau-\tau_k)e^{j\theta_k} \tag{25.2}$$

where L is number of paths; and a, θ, and τ is the path loss, phase, and delay respectively. We note that the time parameter is eliminated because the channel is not changing in time, moreover the channel has different delays and attenuations and phase shift. Figure 25.1 relates these parameters (attenuation, path delay, and phase) in one diagram.

The channel output signal is the convolution formula plus the white Gaussian noise, $w(n)$:

$$\boldsymbol{y(n)} = \sum_{k=1}^{L} \boldsymbol{h(n)}x(n-k) + w(n) \tag{25.3}$$

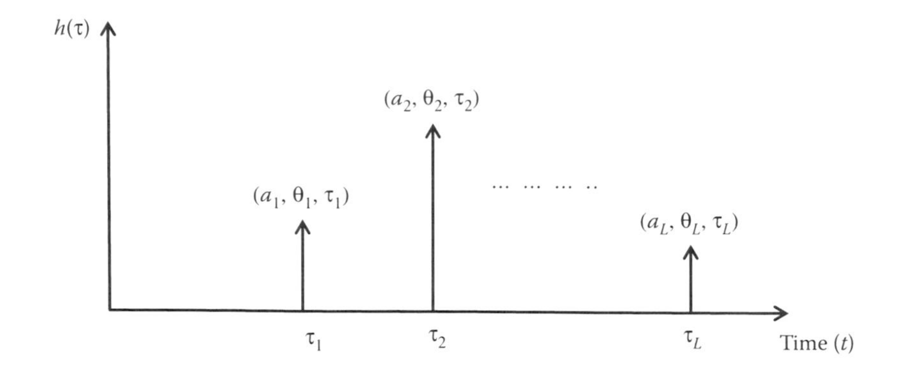

FIGURE 25.1 General impulse response of a time-invariant channel.

25.3.2 Time-Variant Channel

The physical channels that have multipath phenomena such as ionospheric channels, the transmitted signal could show the effect of time-variant linear filter with varying impulse response $h(t,\tau)$ [18]

$$h(t,\tau) = \sum_{n}^{N} a_n(t)e^{-j\theta_n(t)}\delta(\tau - \tau_n(t)) \tag{25.4}$$

where N is the total possible number of multipath components (bins), h is the channel impulse response; a, θ, and τ are the path loss, phase, and delay, respectively. $\delta(\cdot)$ is the unit impulse function, which determines the specific multipath bins that have components at time t and excess delays τ. Figure 25.2

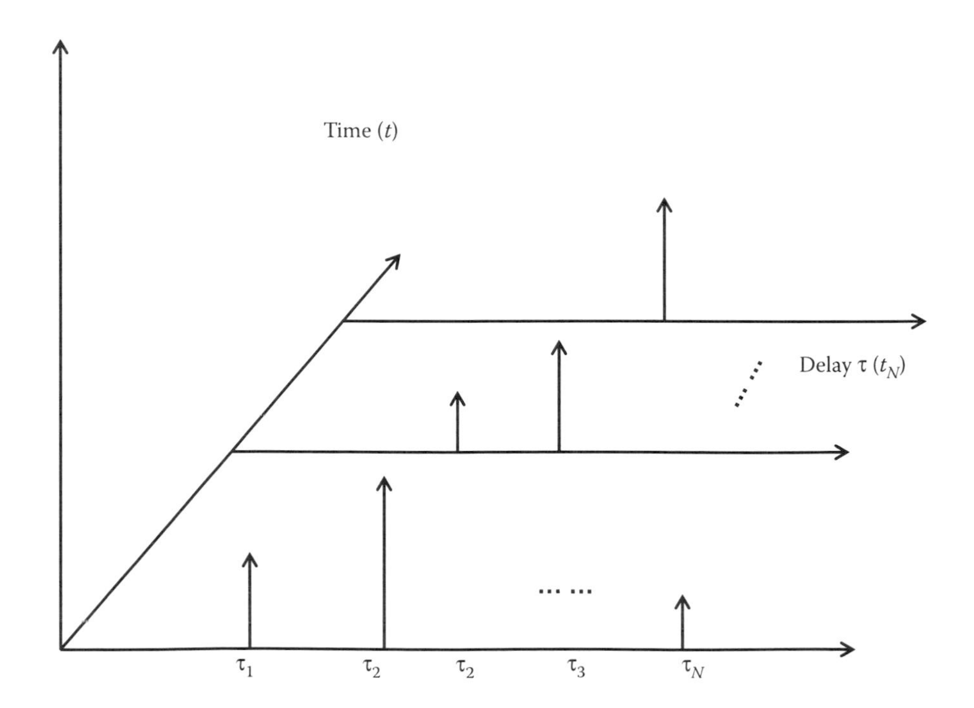

FIGURE 25.2 Time varying impulse response model for a multipath wireless channel. (From G. Proakis and M. Salehi, *Digital Communications*, 5th edition, McGraw-Hill, Boston, 2008.)

illustrates an example of different snapshots of $h(t,\tau)$, where t varies into the page, and the time delay bins are quantized to widths of $\Delta\tau$. Modern wireless communication systems have recently used spatial filtering to increase capacity and coverage.

Depending on the choice of $\Delta\tau$ and the physical channel delay properties, two or more multipath signals arrive within an excess delay bin that vectorially combine to yield the instantaneous amplitude and phase of a single modeled multipath component. These situations fade the multipath amplitude within an excess delay bin over the local area. However, when only a single multipath component arrives within an excess delay bin, the amplitude over the local area for that particular time delay will generally not significantly fade.

The effect of time delay, attenuation, and phase shift on the invariant channel, which has ten paths, are investigated. Assume that the receiver moves toward the transmitter at a velocity of 300 km/s. Moreover, the constant angle of arrival is equal to zero to be the maximum Doppler shift:

$$f_d = f * \frac{v}{c} * \cos(\theta) \tag{25.5}$$

where f_d is the Doppler frequency, V is the velocity of receiver, C is the speed of light, and θ is the angle of arrival. The received signal can be written by the convolution equation:

$$\boldsymbol{y(n)} = \sum_{k=1}^{L} h(k,n)x(n-k) + w(n) \tag{25.6}$$

The input signal, for both models, is

$$x(t) = Re\left\{e^{j2\pi f_c t}\right\} \tag{25.7}$$

$$x(t) = Re\{\cos 2\pi f_c t\} - Im\{\sin 2\pi f_c t\} \tag{25.8}$$

Take into account the whole transmitted and received signals in the real domain as the reason that the modulators are designed to utilize the oscillators that produce a real sinusoidal (to complex exponentials):

$$x(t) = Re\{\cos 2\pi f_c t\} \tag{25.9}$$

25.4 LEARNING ALGORITHMS

The adaptive wireless channel identification architecture is utilized for four adaptive algorithms: least mean square (LMS), normalized least mean square (NLMS), recursive least square (RLS), and affine projection (AFP).

25.4.1 Least Mean Square (LMS)

The LMS algorithm is widely used in different applications due to low computational complexity, and it is part of the stochastic gradient algorithms [20]. This algorithm has two inputs and an output. The inputs are the known signal and the error signal, as the difference between the FIR filter output and the desired signal to be identified. The output signal is the updated filter coefficients.

Definition of the adaptive LMS algorithm:

$$h(n) = [h_0(n), h_1(n), h_2(n), \ldots, h_M(n)]$$
$$x(n) = [x_0(n), x_1(n), x_2(n), \ldots, x_M(n)]$$

where $h(n)$ is the filter coefficients at the nth instant.
$x(n)$ is observed signal vector at the nth instant.

In this LMS algorithm the nth order FIR filter coefficients can be adapted according to the following pseudocode form:

Parameters: N = taps number, μ = step-size

$$0 < \mu < \frac{2}{NS}$$

S is the maximum value of the input power spectral
Initialization: when the tap-weight vector is known, set $h(0) = h(n)$,
Otherwise, set $h(0) = 0$.
Data: Given $x(n)$ is the input $M \times 1$ vector at time n
$d(n)$ is the desired response at time step n.
Computation:
for n = **1:** ***N;*** % N = length (x);
$y(n) = h(n-1)\, x^T(n);$
$e(n) = d(n) - y(n);$
$h(n) = h(n-1) + \mu \times e(n) \times x(n);$
end

where $y(n)$ is the FIR filter output by matrix multiplication.

The choice of the adaptation coefficient, μ, affects the estimation accuracy and the convergence speed of the algorithm. Small values of μ give better final accuracy but slower convergence. Large values do the contrary. Very small or very large values for μ can cause significant errors, and hence, a compromise between these two extremes is desirable. Because quick convergence is achieved by making μ large and good final accuracy is attained by making μ small, many authors have proposed that μ itself be adapted as the filter runs. At the start of the algorithm μ is made large, and after (approximate) convergence is reached, μ is made small. For the sake of simplicity, however, this book will assume that μ is invariant once the algorithm commences. As well as having to select μ, one needs to select an appropriate filter length. A larger filter length $M + 1$ is likely to give better estimation, but it also introduces more delay into the output. Again, a compromise is desirable.

25.4.2 Normalized Least Mean Square (NLMS)

In many applications, the input is huge and the LMS algorithm could not adapt the output because of step size. Thus, NLMS is developed to overcome this problem by normalizing the step size according to input vector energy [21]. The NLMS algorithm can be stated in pseudocode form as follows:

Parameters: N = taps number and μ = adaptation constant

$$0 < \mu < \frac{\mathbf{E}\left[\left|x(n)^2\right|\right]\mathfrak{D}(n)}{\mathbf{E}\left[\left|e(n)^2\right|\right]},$$

Where, $\mathbf{E}\left[\left|e(n)^2\right|\right]$ = error signal power,

$\mathbf{E}\left[\left|x(n)^2\right|\right]$ = input signal power,

$\mathfrak{D}(n)$ = mean-square deviation.

Initialization: when the tap-weight vector is known, set $\boldsymbol{h}(0) = \boldsymbol{h}(\boldsymbol{n})$,

Otherwise, set $\boldsymbol{h}(0) = 0$.

Data: Given: $\boldsymbol{x}(\boldsymbol{n})$ = M×1 tap input vector at time *n*.

$d(n)$ = desired response at time step *n*.

Computation:

for n = 1: N; % N = length (x);

$y(n) = h(n-1)\, x^T(n)$; % Filter output by matrix multiplication

$e(n) = d(n) - y(n)$;

$$h(n) = h(n-1) + \frac{\mu}{\left\|x(n)^2\right\|} \times e(n) \times x(n);$$

end

where $\|\boldsymbol{x}(\boldsymbol{n})\|^2$ is the squared norm.

25.4.3 Recursive Least Square (RLS)

The RLS algorithm is useful when the environment is very dynamic and requires a speedy response [22]. For stationary signals, the RLS filter converges to the same optimal filter coefficients as the Wiener filter. For nonstationary signals, the RLS filter tracks the time diversity of the process [23]. The RLS algorithm can be stated in pseudocode form as follows:

Parameters: N = taps number

μ = forgetting parameter, where, $0 < \mu < 1$,

α = regulation parameter,

$$\text{where } \alpha = \begin{cases} \textit{small positive constant for high SNR} \\ \textit{large positive constant for low SNR} \end{cases}$$

Initialization: set $\boldsymbol{h}(0) = \mathbf{0}$

And, $\boldsymbol{R}(0) = \dfrac{\mathbf{1}}{\boldsymbol{\alpha}}\boldsymbol{I}$

Data: Given, $\boldsymbol{x}(\boldsymbol{n})$ = M×1 tap input vector at sample time *n*.

$d(n)$ = desired response at sample time *n*.

Computation:

for n = 1: N; % N = length (x);

$y(n) = h^T(n-1)\, x(n)$; % Filter output by matrix multiplication

$e(n) = d(n) - y(n)$;

$T(n) = R(n-1)x(n)$;

$$l(n) = \frac{T(n)}{\mu + x^T(n)T(n)}$$

$h^T(n) = h^T(n-1) + l(n)\, e(n)$;

$$R(n) = \frac{1}{\mu}\left(R(n-1) - l(n)x^T(n)R(n-1)\right)$$

end

Since the adaptive filter coefficients are in the range $0 < h < 1$, time varying can be exploited. Note that the coefficients of the FIR filter stay fixed during the observation period for which the error function is defined [24].

25.4.4 Affine Projection (AFP) Algorithm

The affine projection algorithm is a multidimensional generalization of the NLMS adaptive filtering algorithm. Each tap FIR filter coefficient vector update of NLMS is viewed as a one-dimensional affine projection [25]. Thus, AFP uses the projection order, so this algorithm is an extension of NLMS. The AFP algorithm can be stated in pseudocode form as follows:

Parameters: N = number of taps, μ = adaptation constant and
Initialization: set $\boldsymbol{h}(0) = \boldsymbol{h}(n)$, If the tap-weight vector is known
Otherwise, set $\boldsymbol{h}(0) = 0$.
Data: Given $\boldsymbol{x(n)} = M \times 1$ tap-input vector at time step n
$\boldsymbol{d(n)}$ = desired response at time step n.
Computation:
for n = 1: N;
$\boldsymbol{y(n)} = \boldsymbol{x^T(n)\, h(n-1)}$; % Filter output by matrix multiplication
$\boldsymbol{e(n)} = \boldsymbol{d(n)} - \boldsymbol{y(n)}$;
$\boldsymbol{h}(\boldsymbol{n}) = \boldsymbol{h}(\boldsymbol{n}-\mathbf{1}) + \mu \times \boldsymbol{x}^T(\boldsymbol{n})(\boldsymbol{x}(\boldsymbol{n})\boldsymbol{x}^T(\boldsymbol{n}))^{-1}\boldsymbol{e}(\boldsymbol{n})$;
end

The desired update equation of $\boldsymbol{h(n)}$ for the affine projection adaptive filter is uniquely determined by the data matrix $\boldsymbol{x(n)}$ and the error vector, $\boldsymbol{e(n)}$, by acting on the old weight vector, $\boldsymbol{h(n-1)}$, to produce the updated weight vector $\boldsymbol{h(n)}$.

25.5 BLIND IDENTIFICATION ARCHITECTURE

The adaptive learning algorithms of Section 25.4 can be implemented in either digital hardware or software. The time-invariant channel is mathematically modeled in Equations 25.1 through 25.3. This channel model can be identified using adaptive filtering architecture for a wireless time-invariant channel as depicted in Figure 25.3. The input signal samples are injected in the adaptive FIR model as well as in the unknown time-invariant channel. The unknown coefficients of the channel are adjusted to approach the FIR filter coefficients after a certain number of iterations.

The adaptive identification architecture, depicted in Figure 25.4, mathematically emulates the model of the time-variant channel as in Equations 25.4 through 25.6.

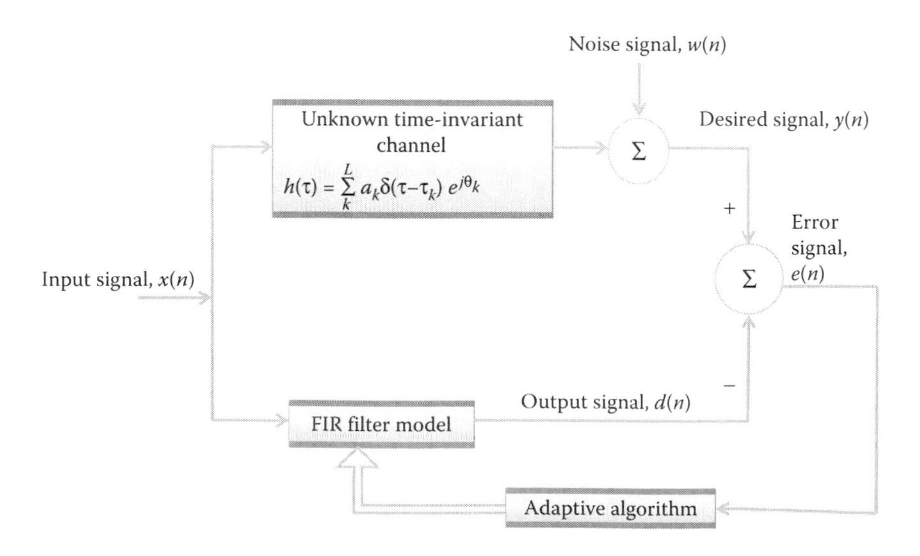

FIGURE 25.3 Adaptive filtering architecture for wireless time-invariant channel.

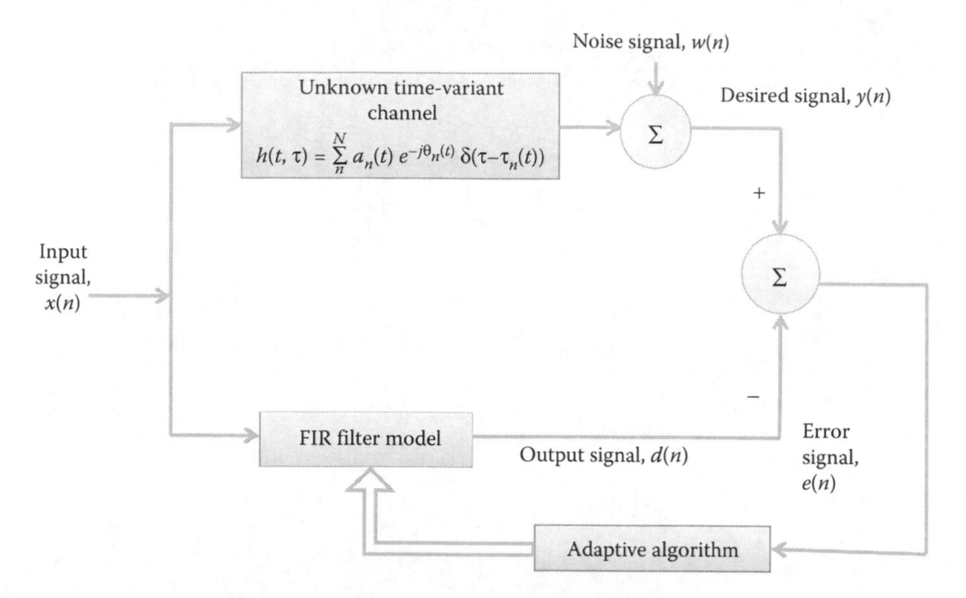

FIGURE 25.4 Wireless time-variant channel identification architecture.

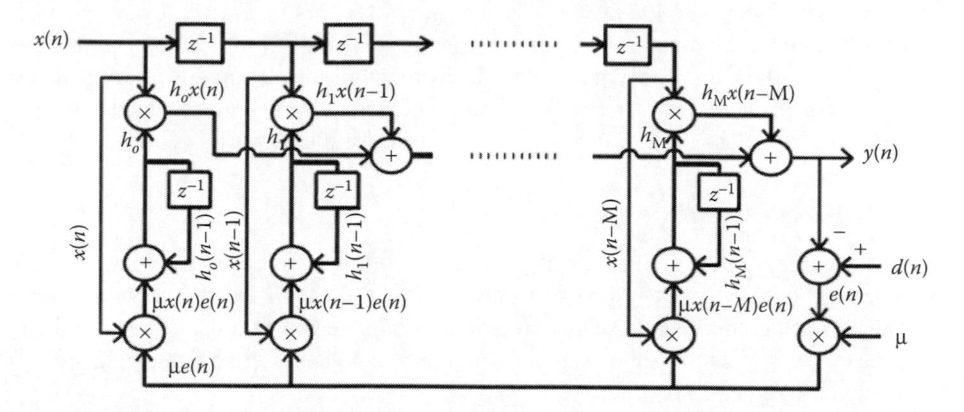

FIGURE 25.5 Hardware implementation of the adaptive LMS filter. (From Z. M. Hussain, A. Z. Sadik, and P. O'Shea, *Digital Signal Processing: An Introduction with MATLAB and Applications*, Chapter 3, 2011, Springer-Verlag, Berlin.)

The detailed direct form-I implementation of the adaptive FIR filter using the LMS algorithm is shown in Figure 25.5.

25.6 RESULTS AND DISCUSSIONS

Computational complexity, mean square error, and convergence time are the performance indices to be analyzed for the four adaptive algorithms (LMS, NLMS, RLS, and AFP). The computational complexity describes the amount of arithmetic operations per iteration and the necessary number of iterations to achieve a desired performance level. The mean square error is the quadratic function of the error, $\boldsymbol{e(n)}$:

$$\boldsymbol{MSE} = \boldsymbol{E}[|\boldsymbol{e(n)}|^2] \tag{25.10}$$

The third performance index is the convergence rate, which represents the number of iterations required for the algorithm to converge to its steady-state mean square error.

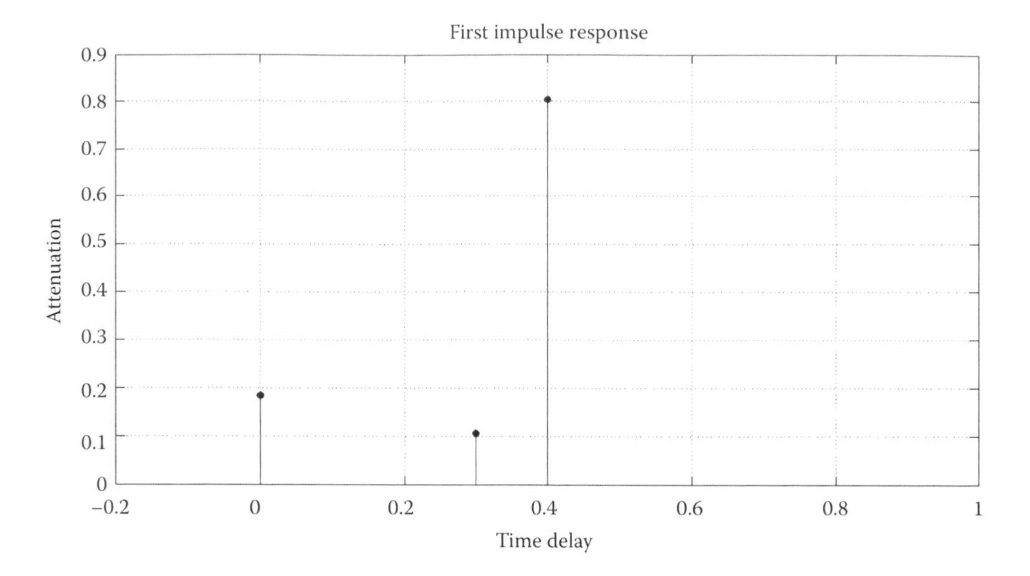

FIGURE 25.6 Impulse response of a time-invariant channel.

The results depend on the adaptation algorithm that has optimum computational complexity, less mean square error, and the fast convergence rate. The simulation result consists of tables and curves representing the performance characteristics of time-invariant and time-variant channels. AFP has a projection order of Equations 25.5 and 25.2 in time-invariant and time-variant channel, respectively.

25.6.1 Time-Invariant Channel Simulation

The time invariant channel is mathematically modeled in Equations 25.1 through 25.3. The input signal samples are injected in the adaptive FIR model as well as in the unknown time-invariant channel. The unknown coefficients of the channel are adjusted to approach the FIR filter coefficients after a certain number of iterations. The impulse response of the channel with 10 different paths is as shown in Figure 25.6.

The performance indices are summarized in Table 25.1. An observation of the RSL algorithm outperforms the other adaptive algorithms in identifying the wireless time-invariant channel.

The computational complexity of the four is further illustrated in the curve of Figure 25.7.

Filter coefficients adapted to the channel coefficients with step size 0.04 for the LMS, and step size 1 for AFP and NLMS, and the RLS forgetting factor of 0.99 with the following mean square error of Figure 25.8.

TABLE 25.1
Performance Comparison of Wireless Time-Invariant Channel Identification Architecture's Adaptive Algorithms, Where N (=10) Is Filter Order Using (100) Iterations

Algorithm	Computational Complexity	Mean Square Error
LMS	$2N + 2$	0.0214
NLMS	$3N + 1$	0.0169
RLS	$4N^2$	0.0116
AFP	$2N^3$	0.0158

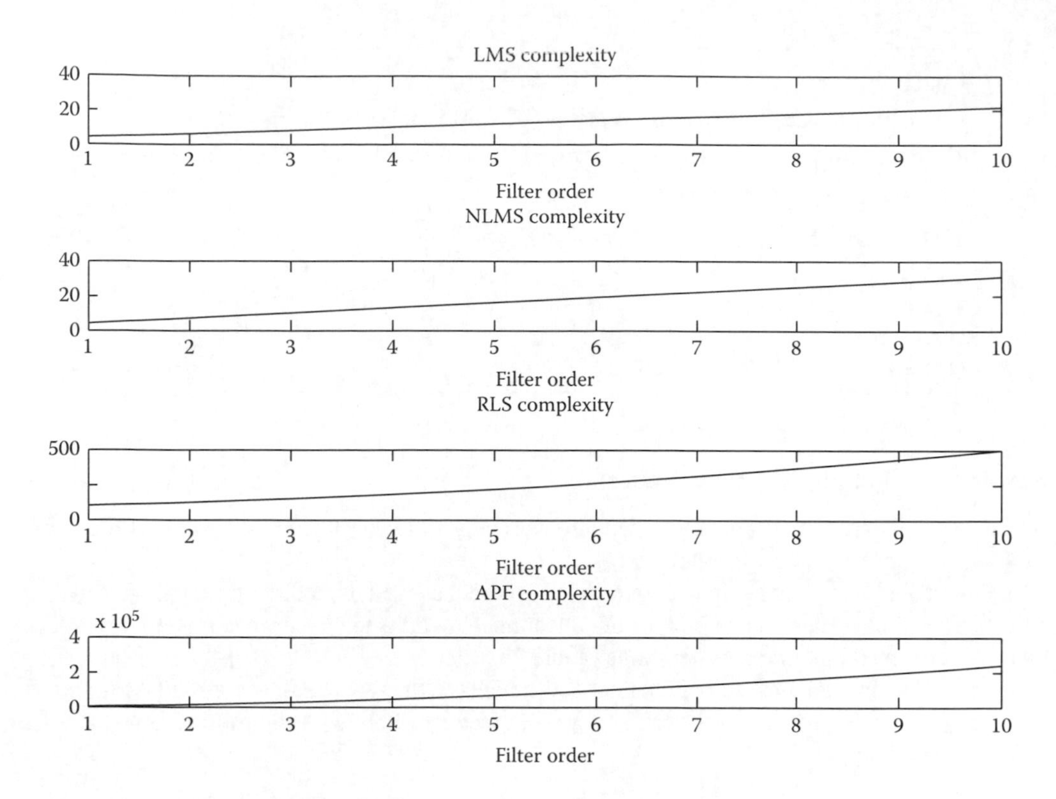

FIGURE 25.7 Computational complexity curve plotted against the adaptive algorithms.

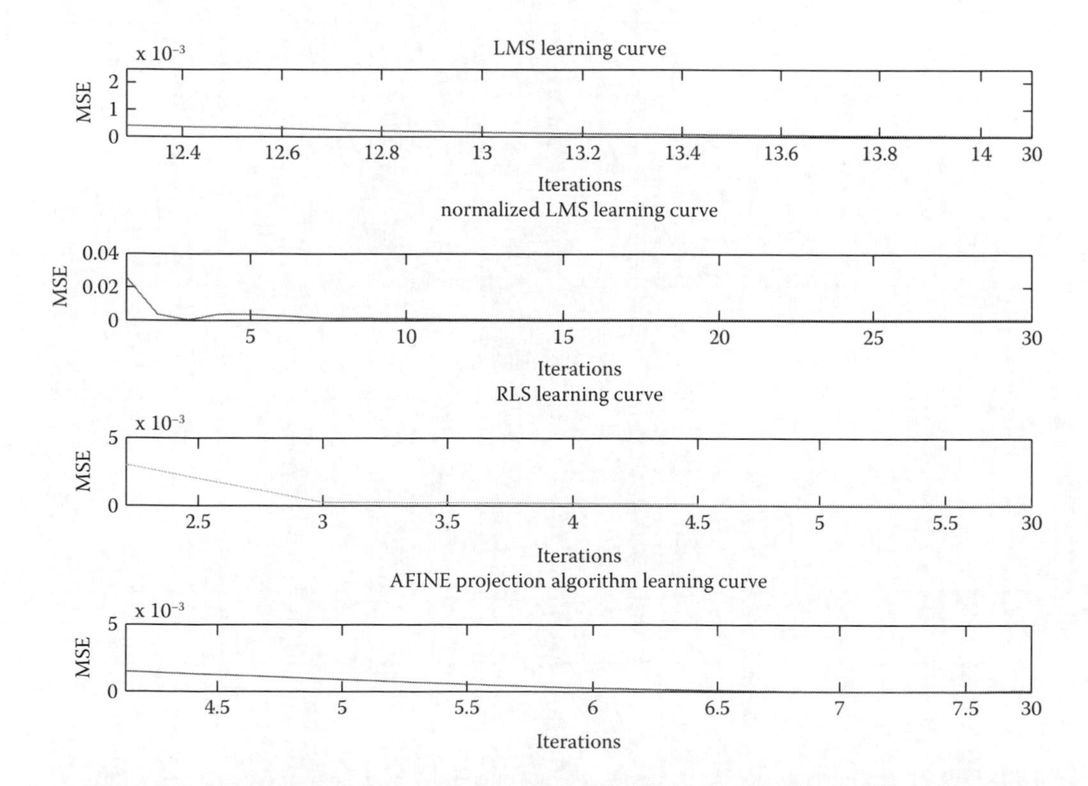

FIGURE 25.8 Learning curve of the four adaptive algorithms, iterations = 100, filter order = 10.

TABLE 25.2
Performance Comparison of Wireless Time-Invariant Channel Identification Architecture's Adaptive Algorithms, Where N (=10) is Filter Order Using (100) Iterations

Adaptive Algorithm	Computational Complexity	Mean Square Error
LMS	$2N + 2$	3.1481
NLMS	$3N + 1$	0.7979
RLS	$4N^2$	0.0454
AFP	$2N^3$	0.0433

25.6.2 Time-Variant Channel Simulation

The adaptive identification architecture mathematically emulates the model of the time-variant channel as in Equations 25.4 through 25.6.

The input signal is sampled in the adaptive FIR model as well as in the unknown time-invariant channel. Then, the unknown coefficients of the channel are adjusted to approach the FIR filter coefficients after a certain number of iterations (Table 25.2).

Filter coefficients adapted to the channel coefficients with step size 0.04 for the LMS, and step size 1 for AFP and NLMS, and the RLS forgetting factor of 0.99 with the following mean square error of Figure 25.9.

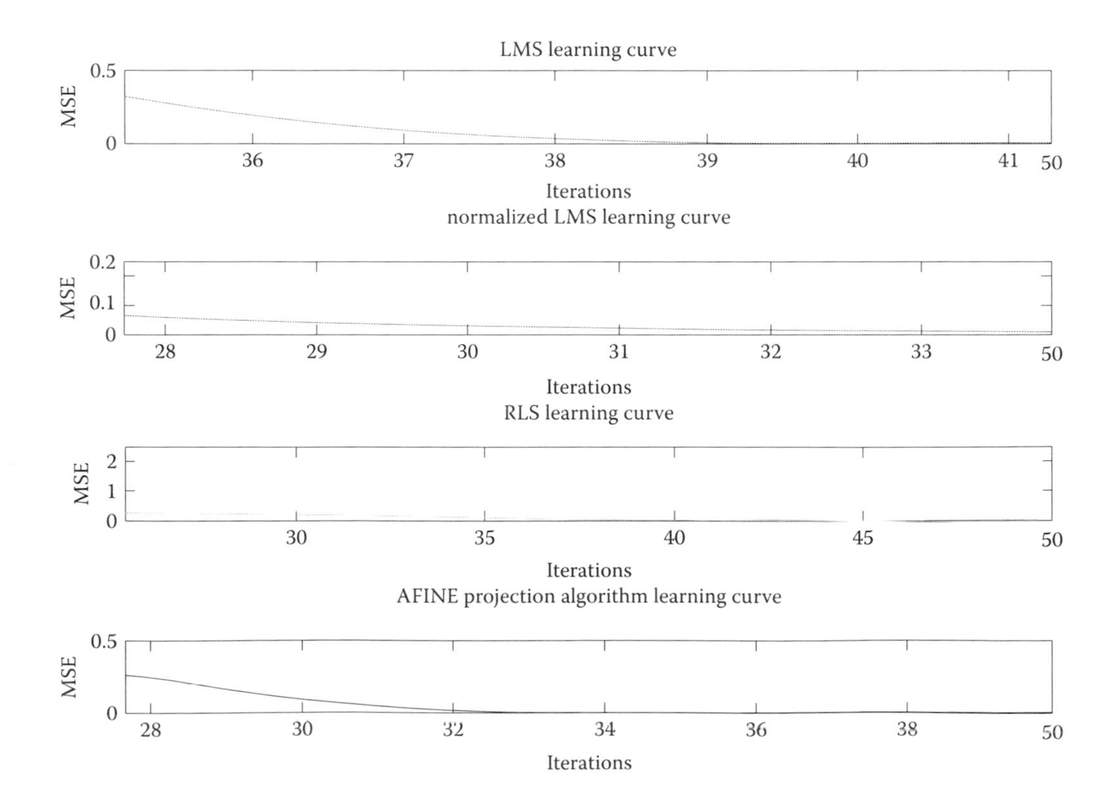

FIGURE 25.9 Learning curve for time varying channel filter order = 10, number of iterations = 100.

25.7 CONCLUSION

This chapter presents a comparative performance study of computational complexity, mean square error, and convergence time for four adaptive blind identification methods of wireless channel. This investigation is efficiently developed through computer simulation of a wireless channel mathematical model for optimum computational complexity, least square error, and fast convergence rate for both time invariant and variant channels.

Improving the current state of adaptive algorithmic aspect of wireless channel, for the next-generation 5G channel model, a novel breakthrough of emerging techniques is needed. A golden ratio-inspired approach that is structured over the emerging union dipole theory [27] may be one of the potential future solutions.

REFERENCES

1. E. Serpedin and G. B. Giannakis, Blind channel identification and equalization with modulation-induced cyclostationarity, *IEEE Transactions on Signal Processing*, 46(7), 1930–1944, 1998.
2. O. Grellier, P. Comon, B. Mourrain, and P. Trebuchet, Analytical blind channel identification, *IEEE Transactions on Signal Processing*, 50(9), 2196–2207, 2002.
3. S. K. Hasan, FPGA implementations for parallel multidimensional filtering algorithms, Agriculture and Engineering Newcastle University, 2013.
4. S. Hasan, Performance-aware architectures for parallel 4D color fMRI filtering algorithm: A complete performance indices package, *IEEE Transactions on Parallel and Distributed Systems*, 27(7), 2116–2129, 2016.
5. S. Hasan, S. Boussakta, and A. Yakovlev, FPGA-based architecture for a generalized parallel 2-D MRI filtering algorithm, *American Journal of Engineering and Applied Sciences*, 4(4), 566–575, 2011.
6. S. Hasan, S. Boussakta, and A. Yakovlev, Parameterized FPGA-based architecture for parallel 1-D filtering algorithms, *2011 7th International Workshop on Systems, Signal Processing and Their Applications (WOSSPA)*, Tipaza, 2011, pp. 171–174.
7. L. Tong, G. Xu, and T. Kailath, Blind identification and equalization based on second-order statistics: A time domain approach, *IEEE Transactions on Information Theory*, 40(2), 340–349, 1994.
8. L. Tong, G. Xu, B. Hassibi, and T. Kailath, Blind channel identification based on second-order statistics: A frequency-domain approach, *IEEE Transactions on Information Theory*, 41(1), 329–334, 1995.
9. Y. C. Pan and S. M. Phoong, An improved subspace-based algorithm for blind channel identification using few received blocks, *IEEE Transactions on Communications*, 61(9), 3710–3720, 2013.
10. H. Murakam, Blind channel identification using interpolation, *International Symposium on Communications and Information Technologies*, Lao, 2008, pp. 652–657.
11. C. E. R. Fernandes, G. Favier, and J. C. M. Mota, Parafac-based blind channel identification using 4th-order cumulants, *2006 International Telecommunications Symposium*, Fortaleza, Brazil, 2006, pp. 771–776.
12. N. D. Gaubitch, M. K. Hasan, and P. A. Naylor, Noise robust adaptive blind channel identification using spectral constraints, *2006 IEEE International Conference on Acoustics Speech and Signal Processing Proceedings*, Toulouse, 2006, pp. V-93–V-96.
13. J. Du, Q. Peng, and H. Zhang, Adaptive blind channel identification and equalization for OFDM-MIMO wireless communication systems, *14th IEEE Proceedings on PIMRC Personal, Indoor and Mobile Radio Communications*, vol. 3, Beijing, China, 2003, pp. 2078–2082.
14. P. V. Rao and R. J. Barton, Evaluation of higher-order techniques for blind adaptive channel estimation on impulse radio channels, *Conference Record of the Thirty-Third Asilomar Conference on Signals, Systems, and Computers*, vol. 2, Pacific Grove, CA, 1999, pp. 1164–1170.
15. Q. Chen and X. Xie, A method for blind identification of single user M MIMO signal in step by step, *The Fourth International Conference on Wireless and Mobile Communications*, Athens, 2008, pp. 144–149.
16. M. T. H. Alouane, A square root normalized LMS algorithm for adaptive identification with non-stationary inputs, *Journal of Communications and Networks*, 9(1), 18–27, 2007.
17. S. McLaughlin and C. F. N. Cowan, A performance study of the RLS algorithm as a channel estimator in a nonstationary environment, *IEE Colloquium on Adaptive Filters*, London, 1989, pp. 1/1–110.

18. B. Su and P. P. Vaidyanathan, A generalized algorithm for blind channel identification with linear redundant precoders, *EURASIP Journal on Advances in Signal Processing*, 2007(1), 1–13, 2007.
19. J. G. Proakis and M. Salehi, *Digital Communications*, 5th edition, McGraw-Hill, Boston, 2008.
20. N. D. Gaubitch, M. K. Hasan, and P. A. Naylor, Generalized optimal step-size for blind multichannel LMS system identification, *IEEE Signal Processing Letters*, 13(10), 624–627, 2006.
21. M. A. Haque and M. K. Hasan, Performance comparison of the blind multi channel frequency domain normalized LMS and variable step-size LMS with noise, *15th European Signal Processing Conference*, Poznan, 2007, pp. 213–217.
22. T. Kimura, H. Sasaki, and H. Ochi, Blind channel identification using RLS method based on second-order statistics, *SPAWC 1999 2nd IEEE Workshop on Signal Processing Advances in Wireless Communications*, Annapolis, Maryland, 1999, pp. 78–81.
23. H. Malani, System identification through RLS adaptive filters, *NCIPET-2012 National Conference on Innovative Paradigms in Engineering & Technology*, Nagpur, Maharashtra, India, January 28, 2012.
24. S. Haykin, *Adaptive Filter Theory*, 5th edition, Pearson, Upper Saddle River, NJ, 2014.
25. S. L. Grant and S. Tavathia, The fast affine projection algorithm, *Proceedings of the International Conference on Acoustics, Speech, and Signal Processing*, Detroit, January 1995, pp. 3023–3026.
26. Z. M. Hussain, A. Z. Sadik, and P. O'Shea, *Digital Signal Processing: An Introduction with MATLAB and Applications*, Chapter 3, 2011, Springer-Verlag, Berlin.
27. A. Al-Mayahi, Union-dipole theory, UDT, *European Journal of Scientific Research*, 118(3), 285–325, 2014.

26 Design of New Algorithms to Analyze RC4 Cipher Based on Its Biases

Ali M. Sagheer and Sura M. Searan

CONTENTS

26.1 Introduction 347
26.2 Literature Review 348
26.3 RC4 Algorithm Description 348
26.3.1 Key Scheduling Algorithm (KSA) 348
26.3.2 Pseudo-Random Generation Algorithm (PRGA) 349
26.4 RC4 Algorithm Weaknesses 349
26.5 The Proposed Single-Byte Bias Algorithm to Analyze RC4 Cipher 350
26.6 The Proposed Double-Byte Bias Algorithm to Analyze RC4 Cipher 352
26.7 The Proposed Single-Byte Bias Attack Algorithm on RC4 354
26.8 Conclusions 358
References 360

26.1 INTRODUCTION

Information security is the process that an organization used to protect and secure its systems [1]. Information can be protected by encrypting it using one of the encryption algorithms. Many factors need to be taken into account including security, the characteristics of the algorithm, time complexity, and space complexity. The main objective of the cryptography is not only used to provide privacy but also to provide solutions to other problems such as integrity, authentication, nonrepudiation, and availability [2]. There are many encryption algorithms widely used in wired networks. In symmetric encryption, when the key size is small, it must be very efficient and encryption time can be quicker. There are various encryption techniques used in wireless devices based on symmetric encryption, such as RC4 algorithm [3]. RC4 stream cipher is an effective and popular algorithm. It is one of the most important encryption algorithms. RC4, or Ron's Code 4, was designed by Ron Rivest in 1987. It is based on the use of random permutation [4] and was a trade confidential to 1994. It is used in commercial software packages such as MS Office, Oracle Secure SQL, and used in network protocols such as IP Sec and Wired Equivalent Privacy (WEP) Protocol [5], and used to protect wireless networks as part of Wi-Fi Protected Access (WPA) protocols and to protect Internet traffic as part of Secure Socket Layer (SSL) protocol and Transport Layer Security (TLS) protocol [6]. RC4 was analyzed by different people and different weaknesses were detected. [7]. Attacks on this algorithm described is by Mantin and Shamir [8] and Fluhrer [9]. The main contribution of this chapter is to design new, efficient, and fast algorithms to analyze the RC4 algorithm based on single-byte bias and double-byte bias. Also, designing an algorithm for a single-byte bias attack that can retrieve the first 32 bytes of the plaintext of RC4 with probability of 100% in a few seconds.

26.2 LITERATURE REVIEW

Several studies of information security analyzed RC4 algorithm based on its weaknesses and suggested different solutions but the ways of bias calculations were slow, inefficient, and used a huge amount of data. Mantin and Shamir showed an essential statistical weakness in the RC4 keystream by analyzing the RC4 algorithm. This weakness makes it significant to discriminant between random strings and short outputs of RC4 by analyzing the second bytes. They also observed that the second output byte of RC4 has a very strong bias that takes the value 0 with twice the expected likelihood (1/128 instead of 1/256 for n = 8). The main result is the detection of a slight distinguisher between the RC4 and random ciphers that needs only two output words under many hundred unrelated and unknown keys to make robust decision [8]. Fluhrer and McGrew were the first researchers to determine a new method to distinguish 8-bit of RC4 from random bits and discovered a new type of bias described as double-byte bias in a consecutive pair of bytes. They discovered long-term biases for RC4; ten conditions stated the positive biases that mean their likelihood are higher than the meant value and two conditions stated negative biases that mean their likelihood is lower than the intended value [9]. AlFardan et al. measured the security of RC4 in TLS and WPA and analyzed RC4 based on its single and double-byte bias and attacking it based on its bias by using plaintext recovery attack. Their results showed that there are biases in the first 256 bytes of the RC4 keystream that can be exploited by passive attacks to retrieve the plaintext by using 2^{44} random keys [10]. Hammood and Yoshigoe determined different biases in the RC4 keystream and analyzed developed algorithms [6] by using C programming and Message Passing Interface environment, and executed experiments using a high execution system with 256 processor units [11]. This work implemented the proposed algorithms on a personal computer and got the same biases shown previously and in less time (seconds only).

26.3 RC4 ALGORITHM DESCRIPTION

Many stream cipher algorithms are based on the use of linear feedback shift registers (LFSRs) particularly in the hardware, but the design of RC4 algorithm evades the use of LFSRs [12]. This algorithm consists of two main components to generate the key: the first is the key scheduling algorithm (KSA) and the second is the pseudo-random generation algorithm (PRGA), implemented sequentially [13]. KSA is more problematic; it was prepared to be simple. At the beginning, few bytes of the output of PRGA are biased or attached to some bytes of the secret key; therefore, analyzing these bytes makes them probable for attacking RC4 [7]. The internal permutation of RC4 is of N bytes; it is a key. The length of the private key is typically between 5 and 32 bytes and is recurrent to form the final keystream. KSA can produce an initial permutation of RC4 by scrambling the corresponding permutation using the key. This permutation (state) in KSA is used as an input to the second step (PRGA) that generates the final keystream [13]. RC4 starts with the permutation and uses a secret key to produce a random permutation with KSA. Based on a secret key, the next stage is PRGA that generates keystream bytes that are XOR-ed with the original bytes to get the ciphertext [14]. The concept of RC4 is to make a permutation of the elements by swapping them to accomplish the higher randomness. The RC4 algorithm has a variable length of keys between 0 and 255 bytes to initialize the 256 bytes in the initial state array (State [0] to State [255]) [15]. The following algorithms show KSA and PRGA steps of the RC4 algorithm.

26.3.1 KEY SCHEDULING ALGORITHM (KSA)

The key scheduling algorithm starts with the permutation of the state table (State [i]). This algorithm consists of two steps: (1) the initialization step where the State[i] is set to the identity permutation that is processed for 256 iterations and (2) the mixing step where a key with N bytes is used to continue swapping values of the state for generating new keystream-dependent permutations [2]. This portion of RC4 equips the state table that is used in the pseudo-random generation algorithm to generate the final key [16].

Algorithm 26.1 KSA of RC4 Algorithm

Input: Key.
Output: State[i].

1. For (i = 0 to 255)
State[i] = i
2. Set j = 0
3. For (i = 0 to 255)
3.1. j = (j + State[i] + Key [i mod key-length]) mod 256
3.2. Swap (State[i], State[j])
4. Output: State[i].

26.3.2 Pseudo-Random Generation Algorithm (PRGA)

The state table from the KSA is swapped with itself by using a known indicator and random indicator. A random index is produced consecutively by using the same values from the prior iteration. Then, the state table is swapped by using these values. The output is generated by taking the modular addition for the values at index pointers [13]. This algorithm continuously mixes the permutation that is stored in the state and selects different values from the state permutation as the output. This algorithm corresponds to the n-bit word as the keystream with set i and j to 0, and loops through four simple operations that use (i) as a counter to be increased, use (j) to be increased pseudo-randomly, swapped the two values of S-box (State[i] and State[j]), and the last is output the value of S-box pointed to by (State[i] + State[j]).

Algorithm 26.2 PRGA of RC4 Algorithm

Input: State[i], Plaintext n.
Output: Key sequence.

1. i = 0, j = 0.
2. While generating output:
    2.1. i = (i + 1) mod 256
    2.2. j = (j + State[i]) mod 256
    2.3. Swap (State[i], State[j])
    2.4. K sequence = State [State[i] + State[j]] mod 256
3. Output: Key sequence.
   Cipher text n = Key sequence n ⊕ Plaintext n [17]

26.4 RC4 ALGORITHM WEAKNESSES

There are several weaknesses found in the RC4 algorithm. Some of these are easy and can be resolved, but other weaknesses are dangers that can be quarried by attackers. RC4 failed in providing a high level of security because of the biases in the bytes of the keystream [11]. Roos [16] found RC4 weaknesses that are a high attachment between the first state table values and generated values of the keystream. The essential cause is the state table that began in series (0, 1, 2, … , 255) and in at least 1 out of each 256 potential keys, the first generated byte of the key is highly attached with a few key bytes. So the keys allow precursor of the first bytes from the output of PRGA. To reduce this problem, it was proposed to ignore the first bytes of the output of PRGA [16]. Mantin and Shamir [8] found the main weakness of RC4 in the second round: the likelihood of zero output bytes. Fluher [9] found a large weakness if anyone knows the private key portion, leading to the potential to fully attack over the RC4 [9]. Paul and Maitra [18] found a private key by using the elementary state table and generated equations on the initial state biases and selected some of the secret key bytes on the basis of assumption and keep private key discovery by using the equation [19]. So the safeness of RC4 is based on private key security and the internal states. Various attacks focus on getting the private key of the internal states [20]. The attack aims to retrieve the main key, the internal state, or the final keystream to access to the original messages [21].

26.5 THE PROPOSED SINGLE-BYTE BIAS ALGORITHM TO ANALYZE RC4 CIPHER

The first researchers that denoted the bias in the keystream of RC4 were Mantin and Shamir after various researchers studied different biases. In this chapter, RC4 and developed algorithms were implemented in C# programming and a new efficient experiment was designed based on proving single bias in the first 32 bytes of the keystream [10,11], which is summarized as Algorithm 26.3. The RC4 output has shown the same biases described in the previous research. The experiment was executed with the generated keystream ranging from 2^{20} to 2^{34} with an independent secret key size of random 16 bytes. The frequents were calculated with the following equation:

$$\text{Frequents} = 2^{34}/\text{State} - \text{length} \tag{26.1}$$

When the likelihood of the frequents of any value is higher than the average, this is taken as positive bias and when the probability is less than the average, this operates as negative bias.

$$\text{Av.} = \text{Frequents}/2^{34} \tag{26.2}$$

Algorithm 26.3 Measuring Distributions of Bytes of Keystream

Input: Key $[k_1, k_2, \ldots, k_{16}]$.
Output: Key position (Kp), key value (Kv), and frequents number in each position (Kf).

1. For (n = 1 to 2^{34}) Do
    1.1. x = 0, y = 0
    1.2. Call Algorithm 26.1: KSA
    1.3. Call Algorithm 26.2: PRGA
    1.4. Deducting new key with a length of 16 bytes from each generated key to be new secret key.
2. For (col = 0 to key Length)
    2.1. For (row = 0 to 2^{34})
        2.1.1. Set key [row] [col] as string
    2.2. For (x = 1 to values. Count)
        2.2.1. If (values [x] = value)
        2.2.2. Increment count by 1
        2.2.3. Key position = col
        2.2.4. Key value = value
        2.2.5. Number of frequents = (count / (2^{34} * 1 6))

Output: Kp, Kv, and Kf for each position of key stream bytes

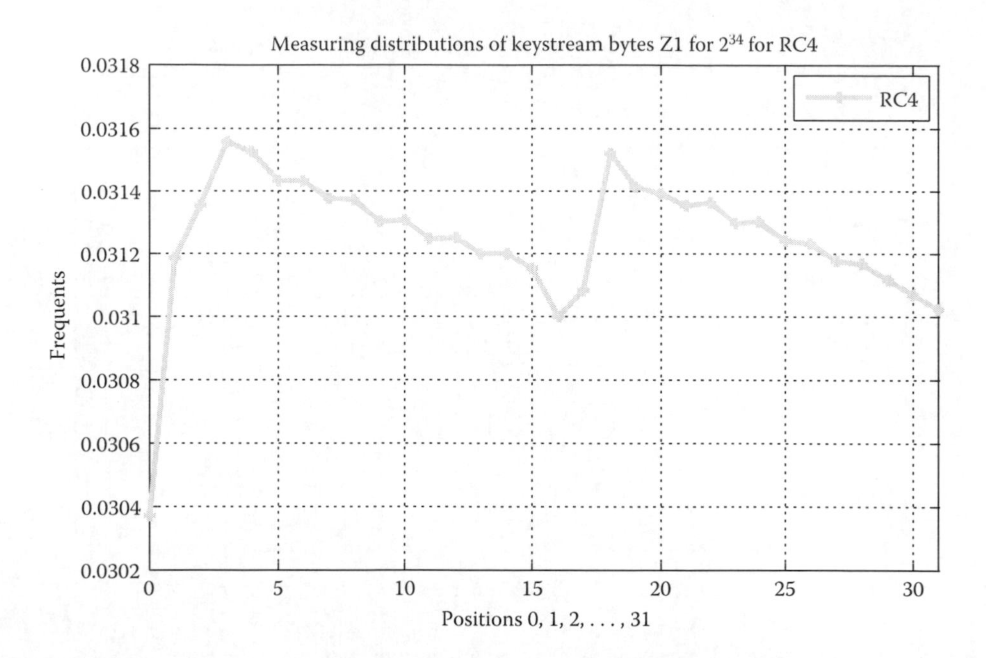

FIGURE 26.1 Distribution of keystream bytes in the 1st position with 2^{34} secret keys.

Different biases in the short-term keystream of RC4 were previously identified. This work successfully regenerated these keystream byte probabilities for the first 32 positions.

Figures 26.1 through 26.4 represent the distribution of keystream bytes in the positions 1, 2, 16, and 32, respectively.

Consider Pr (K_i = n), where i = 1, 2, … , N represents the round number in PRGA phase of RC4, and n = 0, 1, 2, … , N–1 represents the output keystream values as shown in Figure 26.5. The spikes

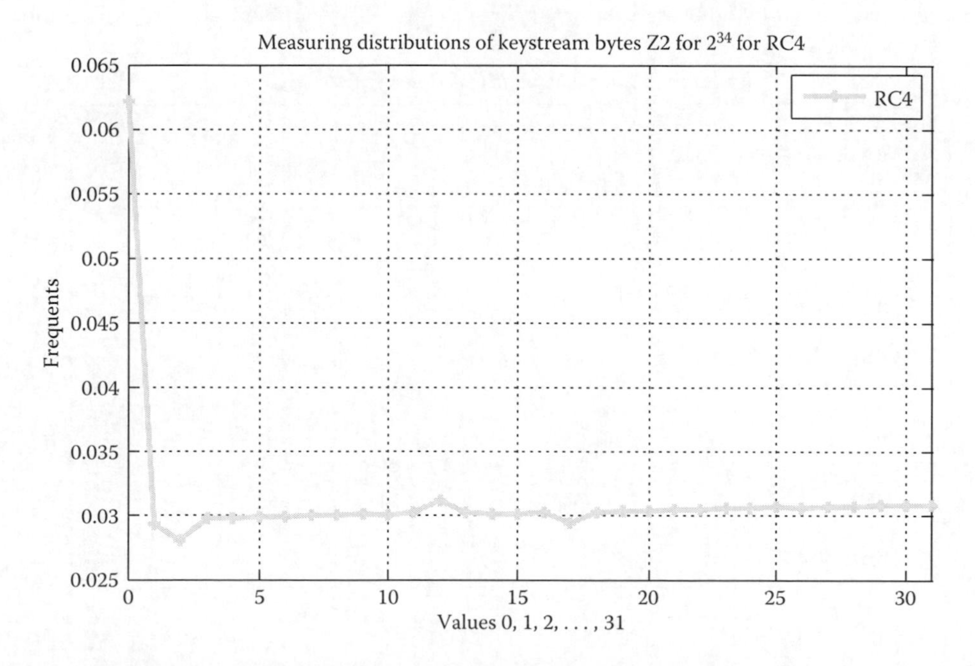

FIGURE 26.2 Distribution of keystream bytes in the 2nd position with 2^{34} secret keys.

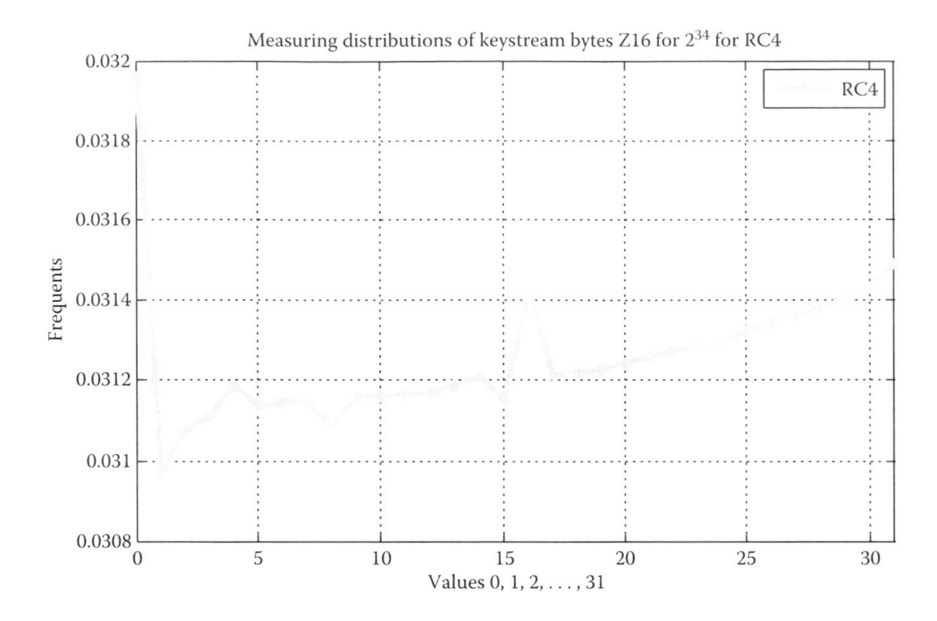

FIGURE 26.3 Distribution of keystream bytes in the 16th position with 2^{34} secret keys.

and apparent vertical walls through the figure represent the short-term bias in the first 32 positions of the RC4 keystream. Particularly, the downward spike and the vertical wall in the front right represent the distributions of keystream bytes for K_1.

26.6 THE PROPOSED DOUBLE-BYTE BIAS ALGORITHM TO ANALYZE RC4 CIPHER

After explaining single-byte biases that are of great benefit to the cryptographic society, the attack simply can be avoided by ignoring the initial bytes. Thus, RC4 with additional configuration can

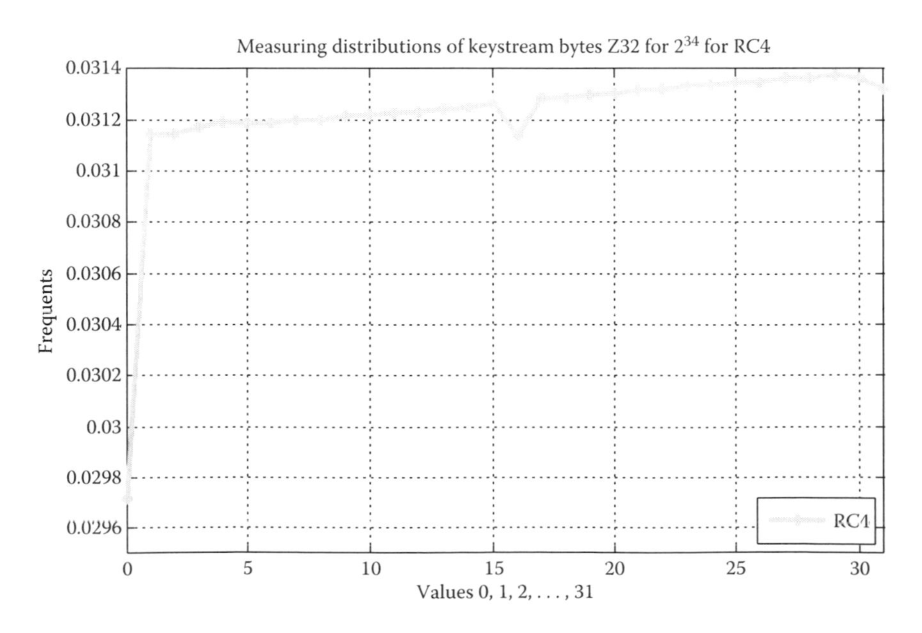

FIGURE 26.4 Distribution of keystream bytes in the 32nd position with 2^{34} secret keys.

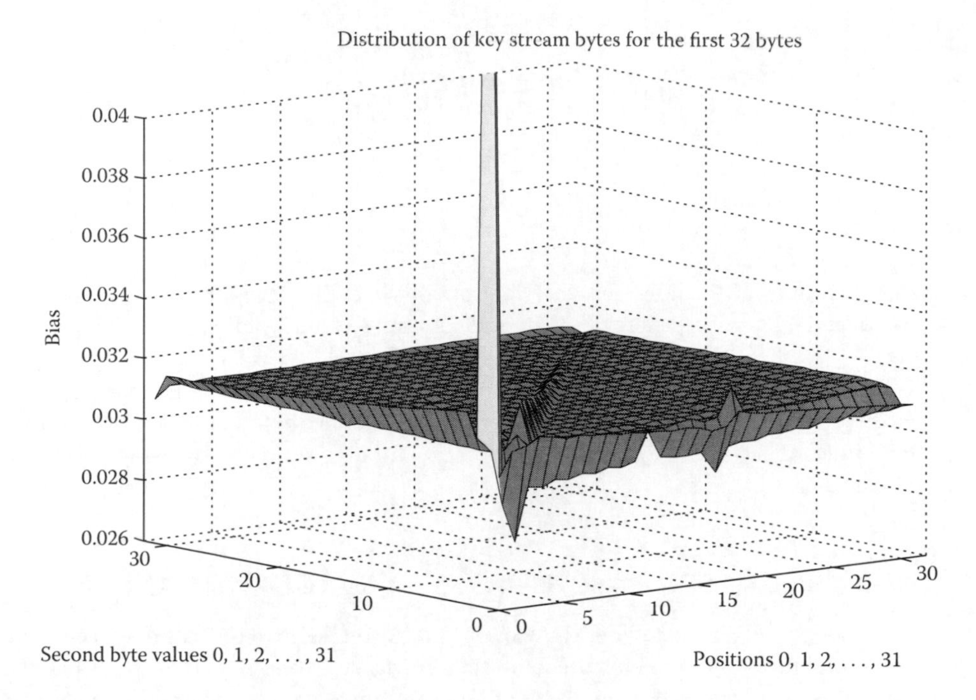

FIGURE 26.5 Measuring distributions of RC4 keystream for the first 32 bytes with 2^{34}.

still be resistant from the single-byte bias attack. However, researchers have studied and investigated biases beyond the initial bytes, and different multibyte biases have been discovered in the keystream of RC4. Fluhrer and McGrew [9] were the first researchers to discover the biases in a consecutive pair of bytes (K_i, K_{i+1}) and detect long-term biases of RC4. They discovered ten positive biases, which means their probability were higher than the desired value, and detected two negative biases, which means their probability were lower than the desired value. Hammood et al. [11] estimated the probability of the cipher for generating each pair of byte values though each 256-byte cycle and got a complete view of the distributions of every pair of byte values at the positions (i, i + 1). They replicated the biases of Fluhrer and McGrew and endorsed the work by AlFardan et al. They found two new positive biases not mentioned by Fluhrer and McGrew [9]. In this chapter, the Fluhrer and McGrew biases and Hammood bias is reproduced with 1024 keys of 16 bytes to generate 2^{32} keystream bytes after discarding the first 1024 bytes. Every key from these 1024 keys produces 2^{32}; therefore, the whole amount of the generated keys is 2^{42}. Algorithm 26.4 is designed to determine the measuring of double-byte bias in a few seconds. The main idea of this algorithm is to measure the appearance of the consecutive pair (Z_i, Z_{i+1}) in each position of RC4 output.

Algorithm 26.4 Measuring Distributions of Keystream Bytes (K_i, K_{i+1})

Input: K [k_1, k_2, …., k_{16}].
Output: Frequents of (K_i, K_{i+1}).

1. i = j = i1 = k = 0
2. For (x = 1 to 2^{10})
    2.1. Call Algorithm 26.1: KSA
    2.2. For (x = 1 to 2^{32})
        2.2.1. i = (i + 1) mod 32
        2.2.2. j = (j + State[i]) mod 32

2.2.3. Swap (State[i], State[j])
2.2.4. Generated Key = State[(State[i] + State[j]) mod 32]
2.2.5. A[k][Generated Key][i1] = A[k][Generated Key][i1] +1
2.2.6. Deducting new key with 16 bytes to be new secret key.
2.2.7. k = Generated Key
2.2.8. i1 = (i1 + 1) mod 32
3. Output: A[k][Generated Key][i1].

Figures 26.6 through 26.9 show the distribution of (Z_r, Z_{r+1}), where $Z_r = 0$ and $Z_{r+1} = 0$, where $Z_r = 30$ and $Z_{r+1} = 31$, where $Z_r = 31$ and $Z_{r+1} = 30$, and where $Z_r = 31$ and $Z_{r+1} = 31$, sequentially.

Figures 26.10 through 26.12 show the results for running Algorithm 26.4 that measures the distributions of keystream bytes (Z_i, Z_{i+1}) to discover possible double-byte biases for RC4 in the first 32 bytes. Figure 26.12 shows the distribution of (Z_r, Z_{r+1}) for all the first 32 bytes, where $Z_r = i$ and $Z_{r+1} = i$ for RC4.

26.7 THE PROPOSED SINGLE-BYTE BIAS ATTACK ALGORITHM ON RC4

Isobe et al. [22] suggested efficient plaintext recovery attacks on the RC4 algorithm that can retrieve all bytes of the plaintext from the ciphertexts in the broadcast setting when the same plaintext is encrypted with different keys. AlFardan et al. [10] and Hammood et al. [11] at the same time used the same concept and determined the plaintext recovery attacks and applied it to single-byte bias attacks on TLS. Their attack successfully recovered the first 256 bytes of keystream with likelihood roughly 1 from 2^{24} ciphertexts encrypted with different random keys. This chapter determines a newly designed fast algorithm for calculating single-byte bias attack on RC4 and retrieving the first

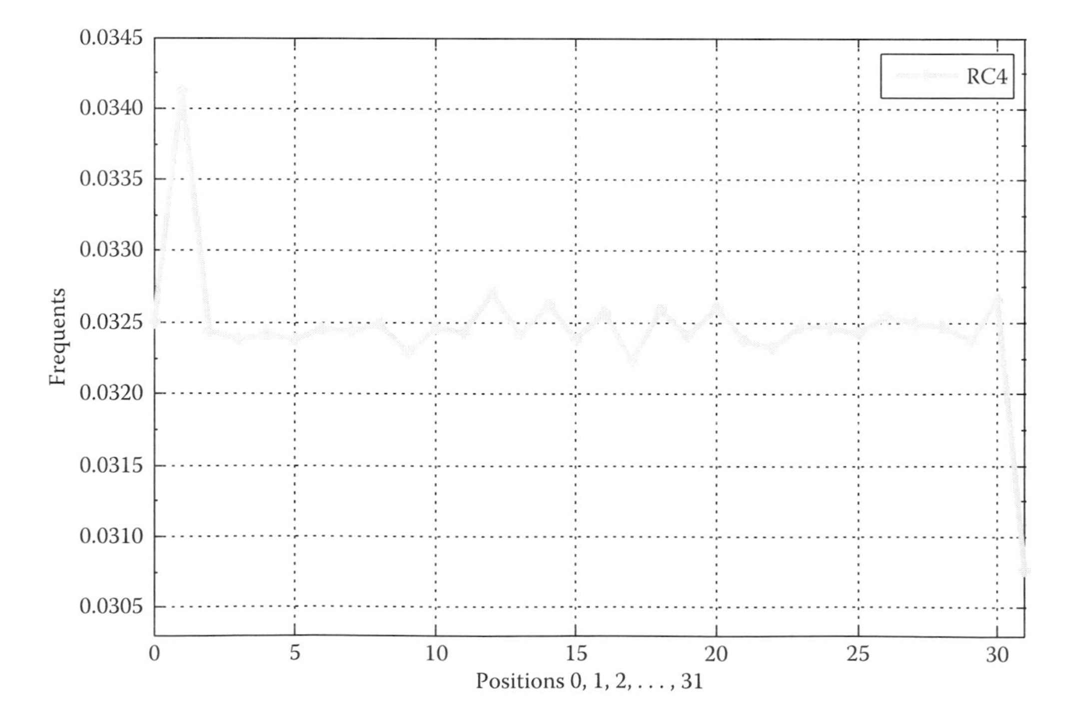

FIGURE 26.6 Double-byte biases (Z_r, Z_{r+1}) where $Z_r = 0$ and $Z_{r+1} = 0$.

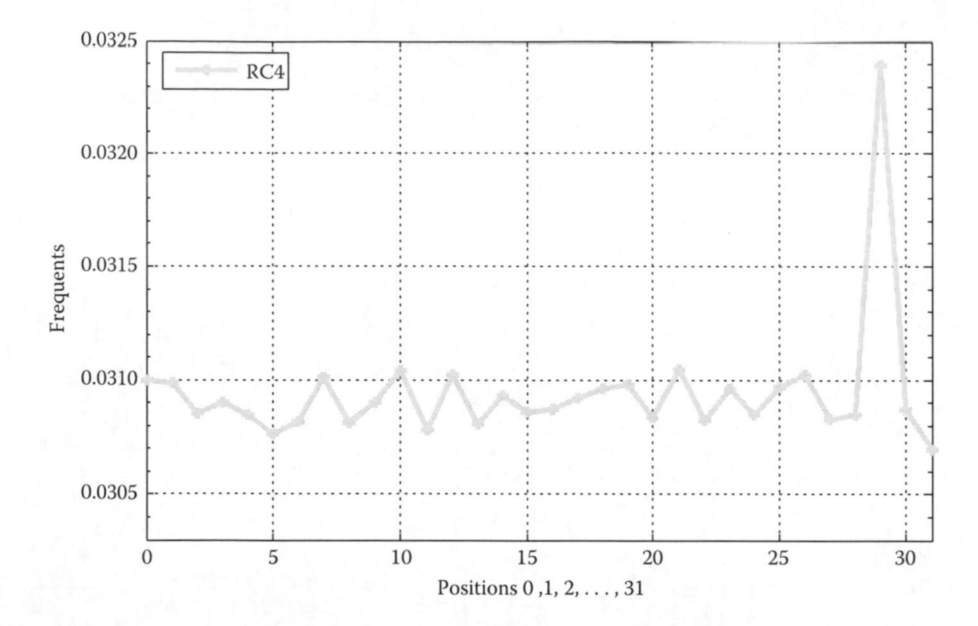

FIGURE 26.7 Double-byte biases (Z_r, Z_{r+1}) for RC4 where $Z_r = 30$ and $Z_{r+1} = 31$.

32 bytes of any plaintext used, illustrated in the Algorithm 26.5. The idea of this algorithm is based on the work of AlFardan et al. [10] and Hammood et al. [11]. The concept of this algorithm aims to quarry the biases in the first 32 bytes of the RC4 keystream by finding the keystream value with the highest bias (K_i) in each position (i). The encryption of the same plaintext (P_i) with various random and independent keys generates many ciphertexts (C_i) that are used to detect the most duplicated byte in each position and to use it as the bias that shifted as the value of the plain text. The most

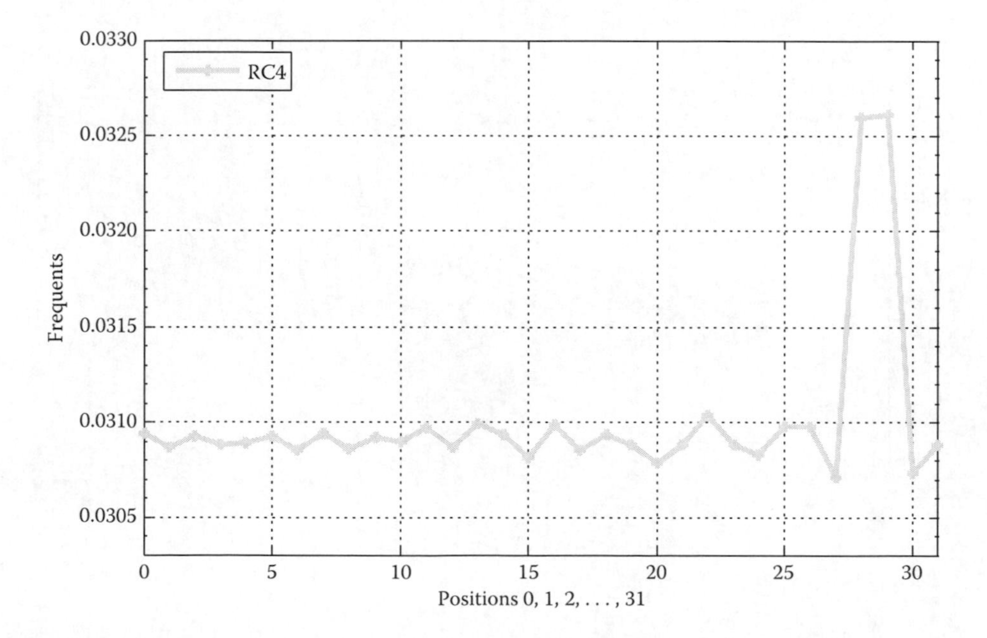

FIGURE 26.8 Double-byte biases (Z_r, Z_{r+1}) for RC4 where $Z_r = 31$ and $Z_{r+1} = 30$.

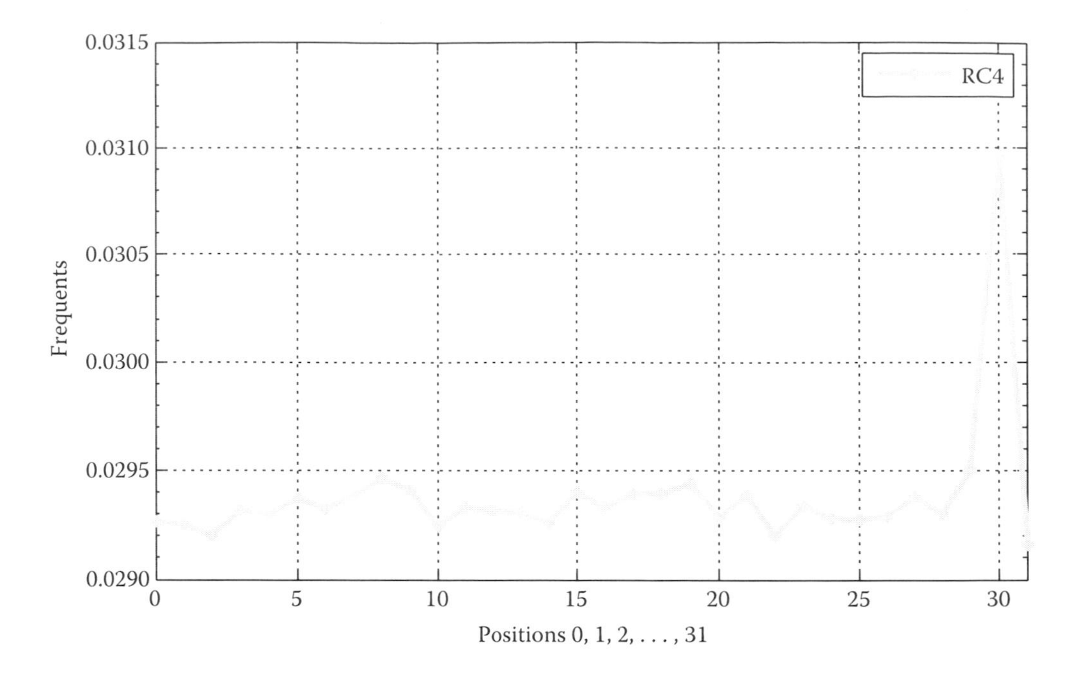

FIGURE 26.9 Double-byte biases (Z_r, Z_{r+1}) for RC4 where $Z_r = 31$ and $Z_{r+1} = 31$.

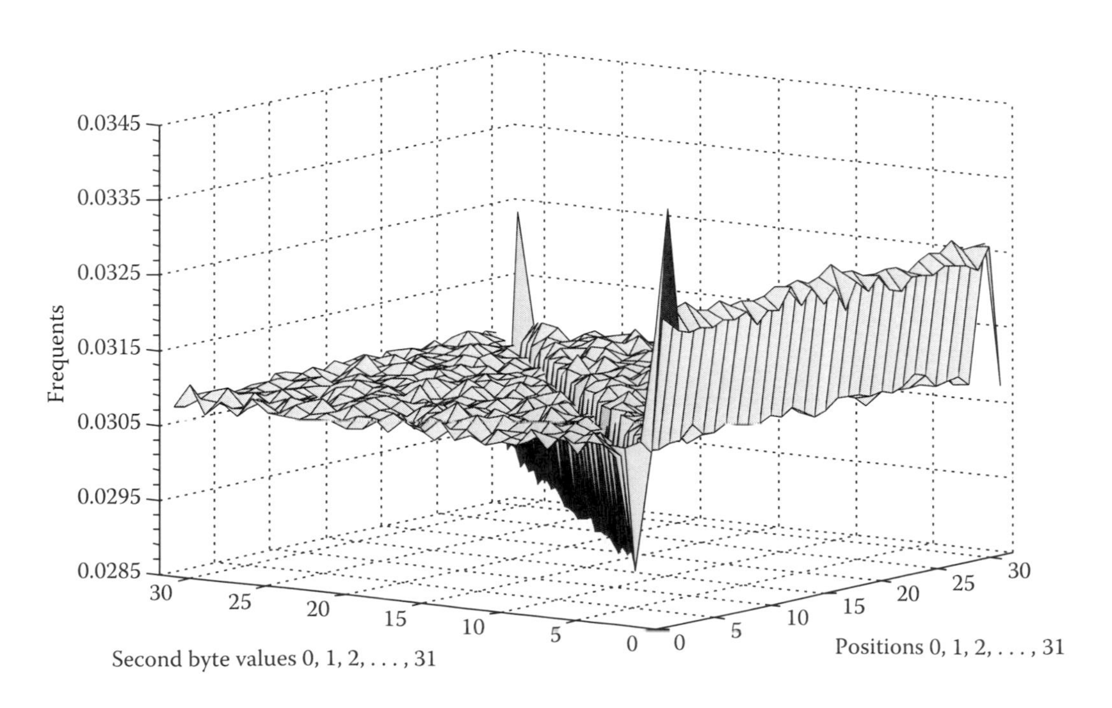

FIGURE 26.10 Double-byte biases (Z_r, Z_{r+1}) for RC4 where $Z_r = 0$.

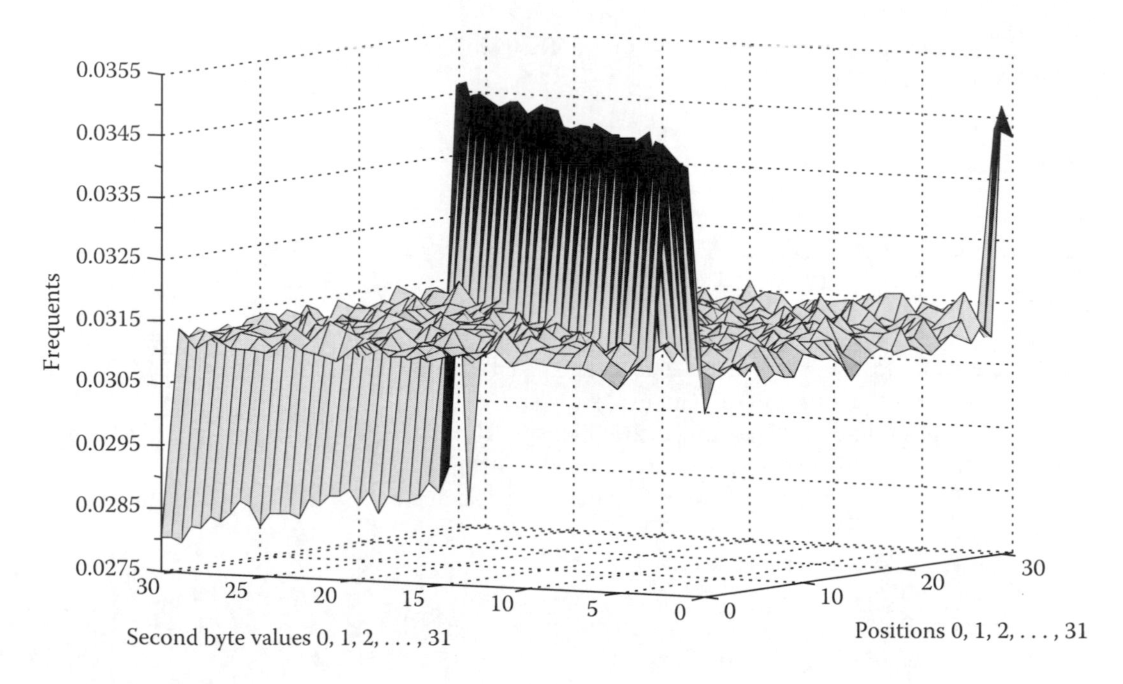

FIGURE 26.11 Double-byte biases (Z_r, Z_{r+1}) for RC4 where $Z_r = 31$.

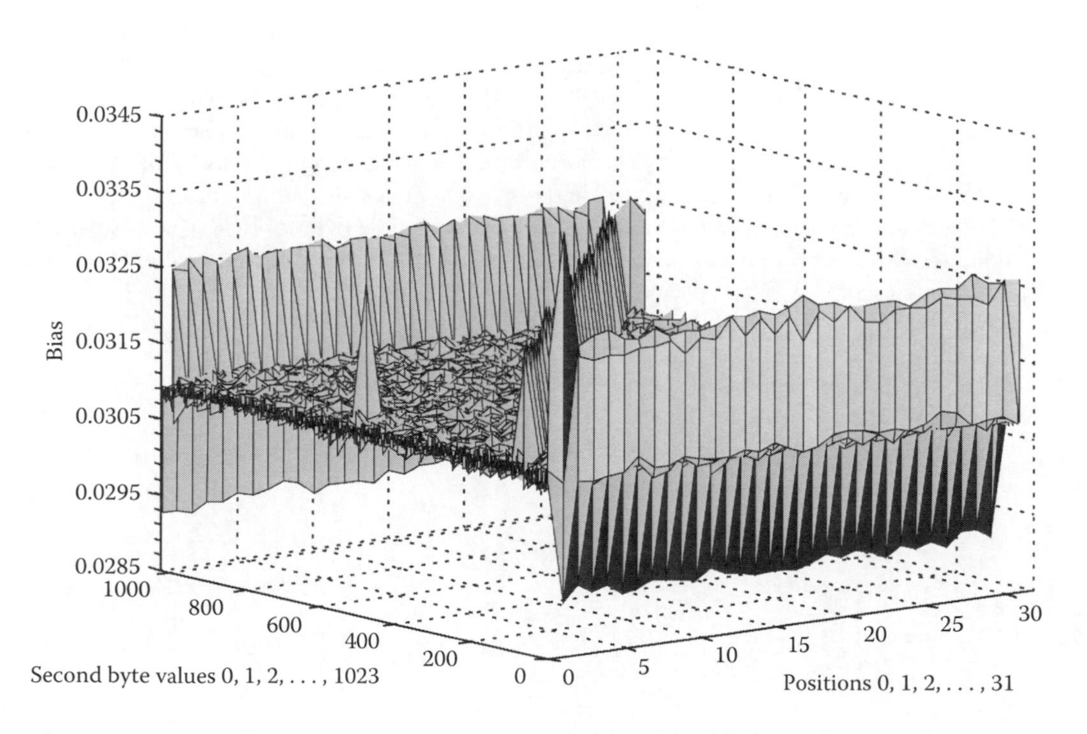

FIGURE 26.12 Double-byte biases (Z_r, Z_{r+1}) for RC4 where $Z_r = i$ and $Z_{r+1} = i$.

duplicate appearing bytes of the keystream is XOR-ed with the most duplicate appearing bytes of ciphertext to retrieve the plaintext.

Algorithm 26.5 Single-Byte Bias Attack

Input: Key [k_1, k_2, …., k_{16}], Plaintext $_i$.
Output: Plaintext*, Frequency of Plaintext*.

1. For (X = 1 to N), where N = 2^{18}, 2^{21}, or 2^{24}.
   1.1. Call Algorithm 26.1: Measuring distributions of RC4 Keystream $_i$. bytes
   1.2. Calculate Max-Frequent [Key sequence $_i$] of each position.
   1.3. Ciphertext $_i$ = Encryption of Plaintext $_I$ with Key sequence $_i$
   1.4. Call Algorithm 26.1: Measuring distributions of RC4 Ciphertext $_i$. bytes
   1.5. Calculate Max-Frequent [Cipher text $_i$] of each position.
2. Plaintext*[X] = Encryption of Max-Frequent [Key-sequence $_i$] with Max-Frequent [Cipher text $_i$]
3. If Plaintext*[X] = Plaintext[X]
   3.1. Counter = Counter+1
4. Frequency of Plaintext * = (Counter * 100 /N)
5. Output: Plaintext *, Frequency of Plaintext *.

The execution time of single-byte bias attack is determined in Figure 26.13. Figures 26.14 through 26.16 show the probability of retrieving the plaintext bytes.

26.8 CONCLUSIONS

RC4 is an important encryption algorithm that can be used for information protection on many communication networks, as it is simple and fast in implementation, but it has weaknesses in its keystream bytes such that these bytes are biased to some different values of the private key. RC4 biases are now quarried for making practical attacks on the TLS protocol. In this chapter, the analysis of the RC4 algorithm is done for the first 32 positions by using newly designed fast algorithms and the same bias as previously shown. Also, a new single-byte bias attack algorithm is designed for attacking RC4 based on its single-byte bias and retrieving all the first 32 bytes of RC4 plain text with the likelihood of 100%. As a future work, the proposed algorithms may be applied on 256 bytes using parallel processors.

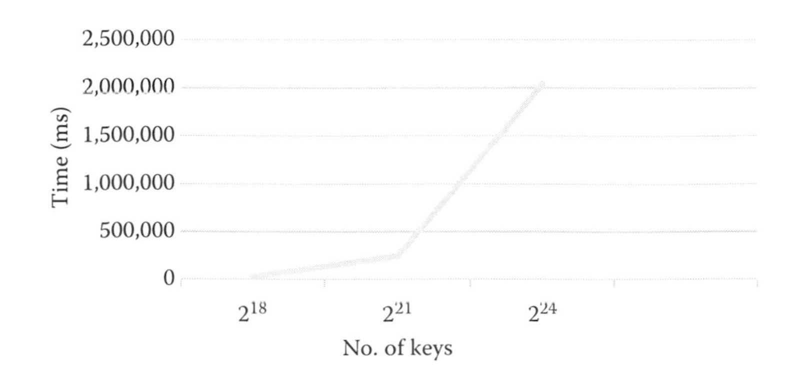

FIGURE 26.13 Execution time of single bias attack.

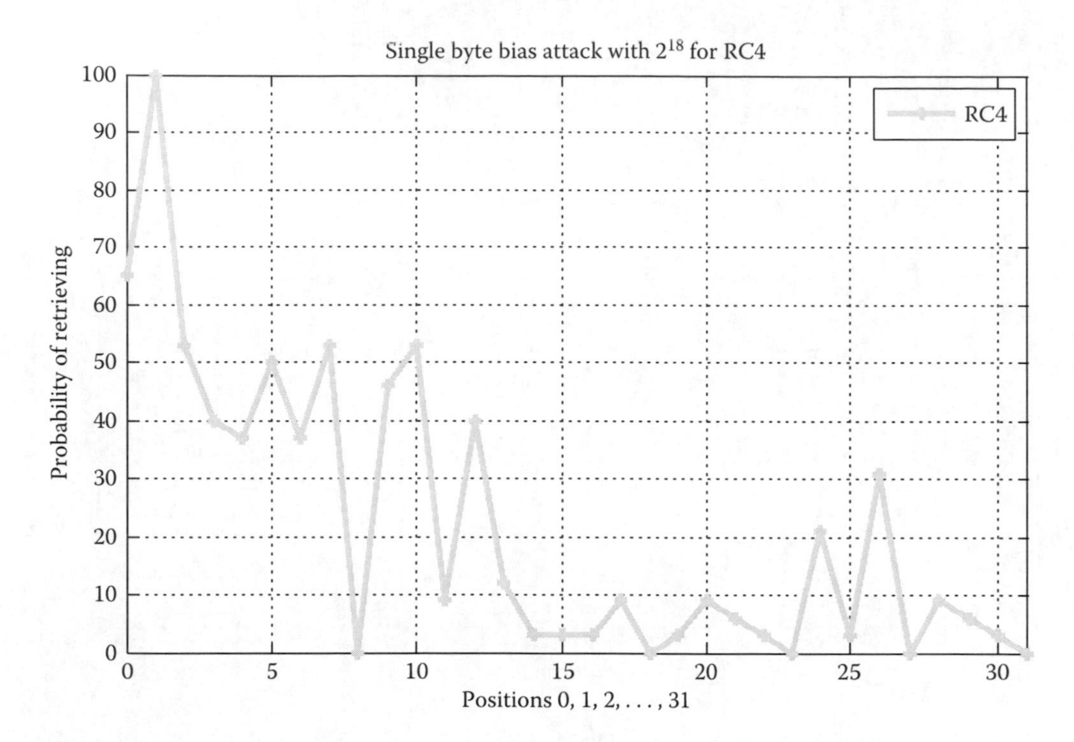

FIGURE 26.14 Recovery rate for the first 32 positions with 2^{18}.

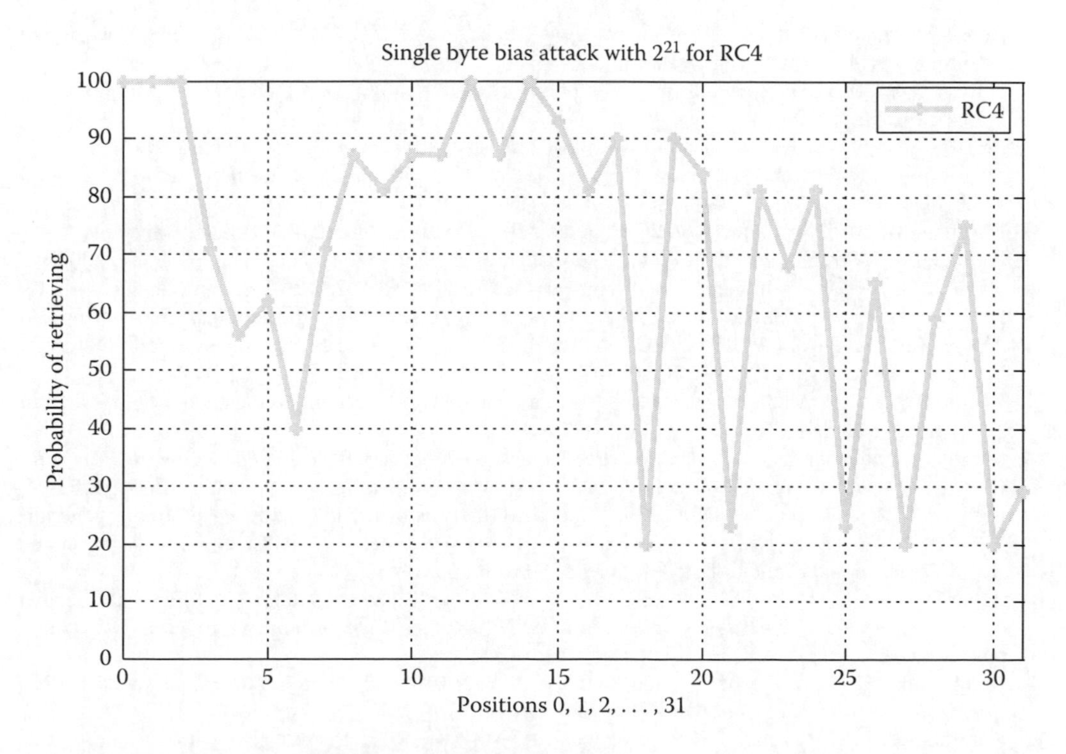

FIGURE 26.15 Recovery rate for the first 32 positions with 2^{21}.

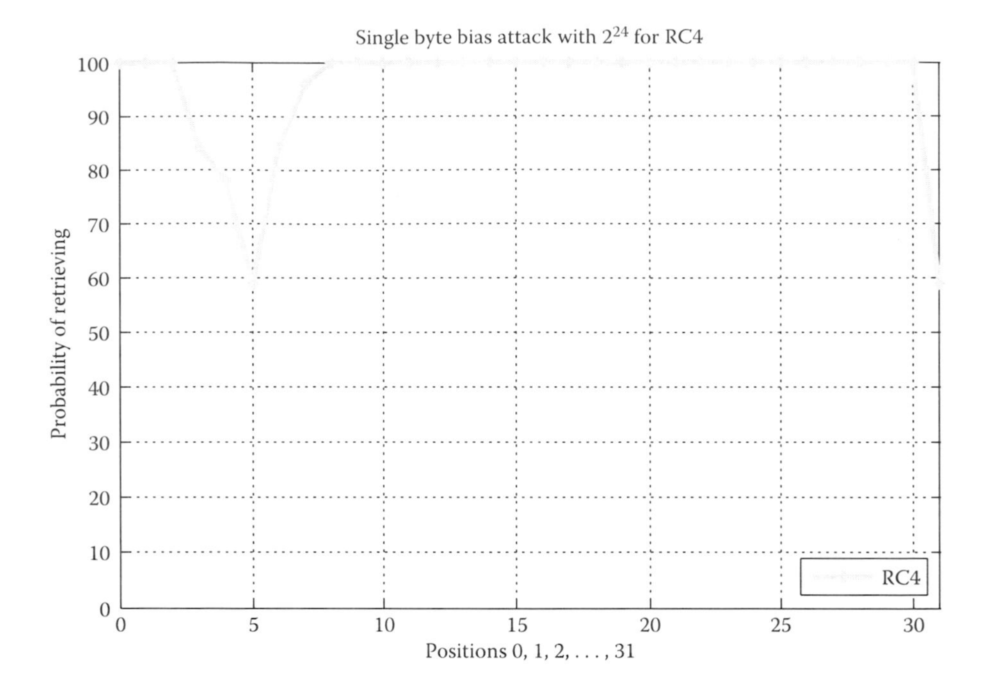

FIGURE 26.16 Recovery rate for the first 32 positions with 2^{24}.

REFERENCES

1. Robshaw, M. and Billet, O. 2008. *New Stream Cipher Designs: The eSTREAM Finalists* (Security and Cryptology, Vol. 4986). Springer, Berlin.
2. Gupta, S. S. 2013. Analysis and implementation of RC4 stream cipher. Doctoral dissertation, Indian Statistical Institute Kolkata.
3. Prasithsangaree, P. and Krishnamurthy, P. 2003. Analysis of energy consumption of RC4 and AES algorithms in wireless LANs. *Proceedings of Global Telecommunications Conference, IEEE*, 1443(3), 1445–1449.
4. Karahan, M. 2015. New attacks on RC4A and VMPC. Doctoral dissertation, Bilkent University.
5. Paul, G. 2007. Structural weakness of the key scheduling of RC4. *IFW*, Jadavpur University, 2007.
6. Hammood, M. M., Yoshigoe, K., and Sagheer, A. M. 2013. RC4-2S: RC4 Stream Cipher with Two State Tables. In: *Information Technology Convergence* (Lecture Notes in Electrical Engineering, Vol. 253), J. Park, L. Barolli, F. Xhafa, and H.Y. Jeong (eds.), Springer Science Business Media, Dordrecht, pp. 13–20, DOI: 10.1007/978-94-007-6996-0_2.
7. Khine, L. L. 2009. A new variant of RC4 stream cipher. World Academy of Science, Engineering and Technology, Mandalay Technological University.
8. Mantin, I. and Shamir, A. 2001. A practical attack on broadcast RC4. In *Fast Software Encryption* (Lecture Notes in Computer Science, Vol. 2355), M. Matsui (ed.), Springer-Verlag, Berlin, pp. 152–164
9. Fluhrer, S. R. and McGrew, D. A. 2001. Statistical analysis of the alleged RC4 keystream generator. In *Fast Software Encryption* (Lecture Notes in Computer Science, Vol. 1978), G. Goos, J. Hartmanis, J. van Leeuwen and Schneier B. (eds.), Springer, Berlin, pp. 19–30.
10. AlFardan, N. J., Bernstein, D. J., Paterson, K. G., Poettering, B., and Schuldt, J. C. 2013. On the security of RC4 in TLS and WPA. In *Part of the 22nd USENIX Security Symposium*, Washington, DC, Vol. 13, pp. 305–320.
11. Hammood, M. M., Yoshigoe, K., and Sagheer, A. M. 2015. Enhancing security and speed of RC4. *International Journal of Computing and Network Technology*, 3(2), 37–48.
12. Hammood, M. M., Yoshigoe, K., and Sagheer, A. M. 2013. RC4 stream cipher with a random initial state. In J. Park, L. Barolli, F. Xhafa, and H. Y. Jeong (eds.), *Information Technology Convergence* (Lecture Notes in Electrical Engineering), Springer, Dordrecht, Netherlands, pp. 407–415.

13. Garman, C., Paterson, K. G., and Van der Merwe, T. 2015. Attacks only get better: Password recovery attacks against RC4 in TLS. In *Part of the 24th USENIX Security Symposium*, Washington, DC, Vol. 15, pp. 113–128.
14. Maitra, S. and Paul, G. 2008. Analysis of RC4 and proposal of additional layers for better security margin. In *International Conference on Cryptology* (Lecture Notes in Computer Science, Vol. 5365), D. R. Chowdhury, V. Rijmen, and A. Das (eds.), Springer, Berlin, pp. 27–39.
15. Orumiehchiha, M. A., Pieprzyk, J., Shakour, E., and Steinfeld, R. 2013. Cryptanalysis of RC4(n, m) stream cipher. In *Proceedings of the 6th International Conference on Security of Information and Networks*, Aksaray, Turkey, pp. 165–172.
16. Roos, A. 1995. A class of weak keys in the RC4 stream cipher. Vironix Software Laboratories, Westville, South African.
17. Sivasankari, N. and Yogananth, A. 2014. Effective and efficient optimization in RC4 stream. *International Journal of Scientific Engineering and Technology*, 3(6), 826–829.
18. Maitra, S. and Paul, G. 2008. New form of permutation bias and secret key leakage in keystream bytes of RC4. In *Fast Software Encryption* (Lecture Notes in Computer Science, Vol. 5086), K. Nyberg (ed.), Springer, Berlin, pp. 253–269.
19. Pardeep, P. and Pateriya, P. K. 2012. PC 1-RC4 and PC 2-RC4 algorithms: Pragmatic enrichment algorithms to enhance RC4 stream cipher algorithm. *International Journal of Computer Science and Network*, 1(3), 36.
20. Ohigashi, T., Isobe, T., Watanabe, Y., and Morii, M. 2013. How to recover any byte of plaintext on RC4. *Lecture Notes in Computer Science*, 8282, 155–173.
21. Sagheer, A. M., Searan, S. M., and Alsharida, R. A. 2016. Modification of RC4 algorithm to increase its security by using mathematical operations. *Journal of Software Engineering and Intelligent Systems*, 1(2), 43–52.
22. Isobe, T., Ohigashi, T., Watanabe, Y., and Morii, M. 2014. Full plaintext recovery attack on broadcast RC4. *Lecture Notes in Computer Science*, 8424, 179–204.

27 Extracting Implicit Feedback from Users' GPS Tracks Dataset

A New Developed Method for Recommender Systems

Tawfiq A. Alasadi and Wadhah R. Baiee

CONTENTS

27.1 Introduction 363
27.2 Geographic Information Systems 364
    27.2.1 Coordinate Systems 365
    27.2.2 ArcMap and ESRI 365
27.3 Proposed System Database 365
    27.3.1 Data Gathering 365
    27.3.2 Preparing Data 366
27.4 System Objectives 367
27.5 Proposed Technique 367
    27.5.1 Splitting Tracks 367
    27.5.2 Creating Rating Matrix for Implicit Feedback 369
27.6 Results 369
27.7 Conclusion 372
27.8 Future Works 373
References 374

27.1 INTRODUCTION

Mobile devices are widely used nowadays, and their popularity is spreading. Communication, applications, functionalities, and challenges of these devices are increasing every year. Users deal with mobile devices to do many activities under many circumstances. Many applications are developed for that reason, and as the datasets become vast and varied, these data must be organized, analyzed, and processed to be meaningful to the user. Traveling is an important area of mobile applications, and an inconceivable number of facilities can serve the users while they travel. It is important to know the capabilities of this field and study the behavior of mobile users [1]. GPS trajectories have presented unique information to understand moving items and places, calling for regular research and improvement of new computing techniques to process, retrieve, and mine trajectory data, and discover its applications [2].

Baltrunas et al. [3] take a new approach for modeling the association among contextual features and user-item ratings. Instead of using the traditional method to collect data, they simulate contextual circumstances to more simply capture data concerning how the context affects user ratings. Zheng et al. [4] model the users' location and activity histories that are taken as feedback to the system. They mine knowledge, such as the position features and activities from the web and GIS databases, to collect additional inputs. Mac Aoidh et al. [5] introduce an approach that observes user's activity and generates a user profile reflecting the user's information desires based on the interactions of the user with the system, physical location, and user movements. The system recognizes the user profile and adjusts to provide appropriate information. Savage et al. [6] implement a more complete universal location-based recommendation algorithm than by gathering a user's preferences and takes into account time geography and similarity measurements automatically.

This chapter introduces a new technique for splitting track lines into numerous sections. Each section characterizes a slice of the track time line that had been taken by the system after 5 years of recording data of mobile GPS use. The proposed method creates an implicit feedback system to facilitate feature extraction from user movement histories instead of depending on social network statuses. This technique generates rating matrices from user tracks depending on the ArcGIS digital map and users' GPS tracks. The rating matrices are spread onto multilevel time periods to be used later as an input to recommender systems.

The proposed system uses the GeoLife trajectory dataset by Zheng et al. [7–9]. This GPS dataset was collected by the Microsoft Research Asia GeoLife project of 182 users in a 5-year period (2007–2012). A trajectory of this dataset is denoted by point sequences; each one has the information of latitude, longitude, and altitude. The system was built by using C#.NET and ArcMap GIS object-oriented programming to generate implicit feedback from GPS tracks.

27.2 GEOGRAPHIC INFORMATION SYSTEMS

Geographic information systems (GIS) are information system that handle, manage, and examine spatial data. GIS are used to create information that is important for decision-making purposes, traditionally in the natural resource controlling area, but increasingly for health, marketing, and other fields. GIS are systems of hardware, software, and algorithms that manipulate, administrate, analyze, model, and display georeferenced data to resolve complex problems concerning planning and resources management [10].

During the 1960s, in the early days of the world's first GIS, the Canada Geographic Information System, which had a strong environmental thrust, there was concern with the analysis of the spatial datasets stored within GIS. Institutions either did not explore the full potential of GIS for analysis and modeling or lacked all these capabilities [11].

In the last years, GIS have developed as a necessary tool for cities and resource management and planning. Their capacity to model, save, retrieve, and map huge areas with extra volumes of spatial data has led to an unexpected spread of applications, but now, GIS are used for land-use organization; services management; planning of landscapes, transport, and infrastructure; analysis of markets; visual effect; facilities organization; tax calculation; and many other applications. GIS functions contain entrance, displaying, and managing of data, information retrieval, and analysis [12].

As mentioned earlier GIS are systems of software and hardware functioned to store, retrieve, map, and analyze geographical data. Experts have opinion that GIS include the working staffs and the data that are used in the system. Spatial data are saved in a coordinate system (longitude, latitude, and azimuth), which positions a certain place on the ground. Data attributes come as tables are related to spatial data. The spatial dataset and its descriptions in the similar coordinate system could be sorted as layers for display and management [13].

GIS functions may be summarized as

Data entrance, data demonstration, data managing, information analysis, image processing functions like satellite and aerial photo processing, geometric and radiometric correction, classification, and compression and decompression of data

Layer-oriented functions such as organizing spatial entities, visual overlaying of entities, and geometric overlaying of entities with algebraic manipulation

Cartographic functions that involve quick and easy map making, effective visual graphics, and communication

Administrative functions defined by graphic and tabular database administration, graphic and tabular database maintenance, registration and upkeep, querying, searching, managing, and updating records [14]

27.2.1 Coordinate Systems

Coordinate systems are location-based reference systems for spatial data on the Earth. Whereas GIS work with geospatial information, the coordinate system acts as the main key in the projects of GIS. Coordinate systems have two main types: geographic and projected. A geographic coordinate system enables every position on the studied area to be stated in three coordinates, using mostly a spherical coordinate system. The projected coordinate system is based on a map projection. For example, the coordinate system of Universal Transverse Mercator (UTM) is built on Transverse Mercator projection.

ArcGIS considers a geographic dataset without any spatial reference data to have an unidentified coordinate system. The existence of the spatial reference information is mandatory for converting a dataset from one coordinate system to another and also significant for spatial analysis [15,16].

27.2.2 ArcMap and ESRI

ArcGIS from the Environmental Systems Research Institute (ESRI) uses a single, scalable architecture. The three versions of ArcGIS (ArcView, ArcEditor, and ArcInfo) share the same applications of ArcCatalog and ArcMap. ArcMap is the application used primarily to examine data, query attributes, conduct spatial analysis, and design maps for output. ArcCatalog is the application used to browse and manage spatial data. The geodatabase data model and ArcObjects provide the foundation for these two desktop applications. A geodatabase is a generic model of geographic information, consisting of topologically integrated feature classes, roughly similar to the ArcInfo coverage model.

The geodatabase data model replaces the georelational data model that has been used for coverages and shapefiles, two older data formats from ESRI. These two data models differ in how geographic and attribute data are stored. The georelational data model stores geographic and attribute data separately in a split system: geographic data ("geo") in graphic files and attribute data ("relational") in a relational database.

The entire system is developed using C#.NET that linked with ArcMap GIS using the ESRI Developer Kit to access the ArcObjects programming libraries and produce the application classes [17].

27.3 PROPOSED SYSTEM DATABASE

27.3.1 Data Gathering

As mentioned earlier, the data that is used in the proposed system is a package or set of user's tracks represented by recorded GPS signals. Each one contains multiple fields like longitude, latitude, azimuth, time, and date. Table 27.1 shows an example of one record of a track's point that is gathered and used by the system.

TABLE 27.1
GPS Point Example

Longitude	Latitude	Azimuth	Time	Date
40.008413	116.319962	13	02:10:04	2008-10-24

The dataset consists of 182 user tracks that are gathered through 5 years from 2007 to 2012. The number of tracks is different from one user to another according to their movements and/or trips.

Users recorded their trips by mobile devices. Each track may represent a movement within a day or less than a day. Each track was saved in PLT files. PLT files have six header lines, which are the track's main information; they are the same in all track files. The rest of the lines are GPS data for each track; each line of data looks like Table 27.1.

27.3.2 Preparing Data

Tracks are read repeatedly by scanning all track files for all users to create a data structure that represents all user histories. We proposed a new data structure using object-oriented techniques with a binary file system to save and retrieve analyzed data repeatedly without needing to build a huge database that costs time when reading and writing data.

Tracks are collected as multiple GPS points that are gained from user movements for 5 years. Each point is represented as illustrated in Table 27.1. Each track is represented in a separated class object and all other data had been saved in a data structure that is shown in Figure 27.1. Tracks are aggregated in one track in a class called TrackCollection. This class defined objects so each one of them represents one aggregated track for one user.

TrackCollection objects have multiple functions. The most important function is dividing the aggregated tracks into periods, which will be defined later in this chapter as well as how can the proposed system tracks this division.

Another important data structure is the MapPlaces class. It defines and stores the digital map data with their relevant attributes and properties with extra metadata. The collected data from the map will be combined with tracks data to be used later in the proposed system to examine implicit feedback to calculate and retrieve a rating matrix for user–places ranking that is used as the base data for our system. MapPlaces has multiple functions that are paired with geographic information system data and applications like ArcMap GIS desktop to retrieve needed data to be processed and combined with our system data and to perform the proposed method and algorithm to recommend information to the users.

The ratings data part contains ratings values that are gained and reinforced from positions of interest (POIs) on digital GIS maps and tracks of user movements, respectively. Ratings are matrices

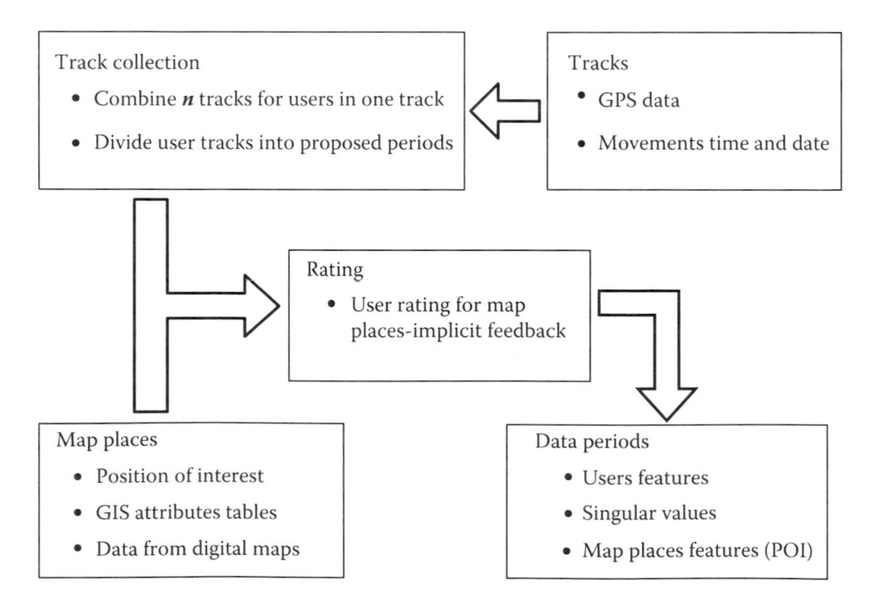

FIGURE 27.1 Data structure of proposed system.

for user–place rates; they are represented by real numbers for each user track segment on the map. Each visit on a place is registered and calculated with other visits to be the rank level for that user on that place.

To save all these data structures we used serialized binary files to avoid time-consuming database engine operations. All objects and their relevant data like maps, places, GPS points, tracks, time, ratings, features, place, matrices, and recommendations have been saved in their designed structure in one mass of binary data. The system can retrieve any object directly to be processed into memory.

27.4 SYSTEM OBJECTIVES

This chapter studies multidisciplinary approaches that are cooperative in one new system for building a location-based recommendation system using pure GIS data and GPS trajectories. So, the objectives can be listed as follows:

- Design and implement an application that recommends desired POIs and routes to the users, according to their preferences.
- Induct user preferences from a user's trajectory histories, relevant to location–time association.
- Show how to build user location-based preferences.
- Create an implicit feedback system to facilitate feature extraction from user movement histories instead of depending on social networks status, creating rating matrices from user tracks.
- Implement and design a cooperative application that is built upon GIS databases and can be an add-on extension to ArcMap GIS software. The results of the recommendations can be shown on desktops and/or mobile devices for each user. The recommendations are displayed on digital maps.

27.5 PROPOSED TECHNIQUE

27.5.1 Splitting Tracks

This method describes the idea of splitting track lines into many segments. Each segment represents a part of the track time line that had been taken by the user along the 5 years of recording data using their mobile GPS. Figure 27.2 illustrates a block diagram of the track splitting method.

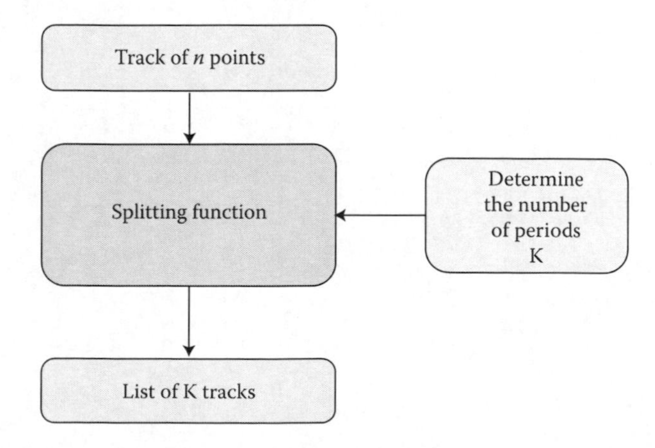

FIGURE 27.2 Block diagram of splitting tracks.

```
Algorithm 1 : Track segmentation :
Inputs : points of track, time periods, case time
Outputs : multiline segments
Process:
Begin
  Time intervals ← case total time/time periods
  crntTime_period ← first time Interval
  For each point in points list do
        Check point's time P_time
        If P_time within crntTime_period then
           Add point to current segment
        Else
           crntTime ← next time interval
           current segment ← new segment
           Add point to current segment
End
```

FIGURE 27.3 Algorithm of splitting tracks.

To detect the features of each user we propose a mechanism to spread the feature space on multiple time periods. Splitting the function is determined by taking all the user's track points and returning a segmented line depending on the time period previously fixed by the user. The resulted segments are not exactly equal because the segmentation process depends on time periods not on spatial points locations. All users' tracks will be segmented in the same strategy so each one of the users has the same number of new tracks according to the number of periods that is determined.

Figure 27.3 shows the proposed algorithm of track segmentation. The input is the points list of the track that will be segmented and the full time of the case we dealt with. First, the algorithm divides the total time on the desired number of periods to get the time interval of each period. Then it repeatedly scans all points of the track to check if the point's time belongs to the tested period. If so, then it will be added to the track segment that represents that period of time.

The process continues unless there is a scanned point that belongs to another period, then the algorithm creates a new track and adds the remained points to it if they match the condition illustrated in the algorithm. The resulting track segments may not have an equal number of points or lines; that means the user may move many times in one period. Figure 27.4 shows a prototype of how track segmentation algorithm works on a long track.

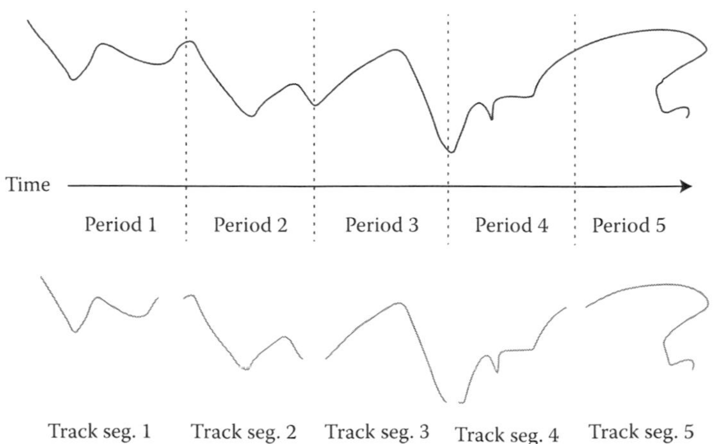

FIGURE 27.4 Sample of track splitting.

In the proposed system we used this algorithm to create tracks within specified periods of time. Twelve periods are supposed in this work so each period contains five months of movement for each user in his history profile. This period of time is calculated according to Equation 27.1:

$$\text{Time period} = \text{Total time of case study/number of periods} \tag{27.1}$$

The resulting tracks will be used later in the system, each one alone to induct the rating matrices that are used in the proposed system to extract features of users against places or POIs that are accessed along their path to be used as historical features for each user ready for the proposed system to predict their right movement. (Note: The proposed system analyzes the data in a main server that is connected directly to the users' mobiles. The mobile application is a client application.)

27.5.2 Creating Rating Matrix for Implicit Feedback

To complete the main job of the proposed system, there should be a rating matrix that is used as an input to the recommender system. The input for this stage of the process is the segmented tracks from the previous stage. For each period of time there are many user tracks, so the inducting process will be applied on these tracks to produce a rating matrix for that period of time.

Positions of interest are the main categories on which the system depends. The link between tracks and POIs is the moving history of users on those positions of interest through points on tracks. In the proposed system we propose a mechanism to link the recurring visits to a digital map of POIs.

Implicit feedback is the key behind our goal. The rating matrix will be the relation between user rates on the places on the map. Rate is a number between two values to represent the likelihood a user will visit the place on a map. The values often may be between the interval of [1..5] and [1..10]. In the proposed system we chose the interval [1..10] to represent likelihood of user–place relation. Figure 27.5 illustrates the process details of implementing a rating matrix from the track history of users and the digital map.

The map is divided into rectangles with rows and columns; 100 rows and 100 columns is proposed here to get 10,000 places to be POIs in the system. The division technique depends on the spatial coordinates on the map, so each rectangle on the map represents 500 m width × 500 m height.

Each track will be examined against the divided map, then each user when he reaches a new point on the track a new access is recorded for that point on that place on the map, the repetition of the accesses of places are accumulated and recoded to the user–place matrix. A user–place matrix will be created and built after scanning all tracks of users for each period alone.

The resulted matrix should be normalized because the user visits have large number on some significant places and little visits on insignificant ones, so a normalization was applied on the values of the matrix to be converted into rating matrix with specified vales of range [1..10]. Each period of time that the system dealt with will have its special rating matrix.

Figure 27.6 illustrates the algorithm for the proposed implicit feedback technique to create and build rating matrices for each period of time. The system depends on the implicit feedback to produce a rating matrix that is collected in other systems by explicit feedback. Here the key is that the proposed system has a unique technique that produces and inducts ratings from scratch without the need for user intervention. This idea will increase systems' abilities to be more independent from user direct choices and decisions.

27.6 RESULTS

The application paths are displayed on the interface map of the application itself as illustrated in Figure 27.7, and the user be able to manage, move, and detect the preferred place on the map. As mentioned earlier the tracks of each user are combined in one track that is represented as the history of one user. The proposed system divides the map into 100 × 100 subregions to be the POIs that are

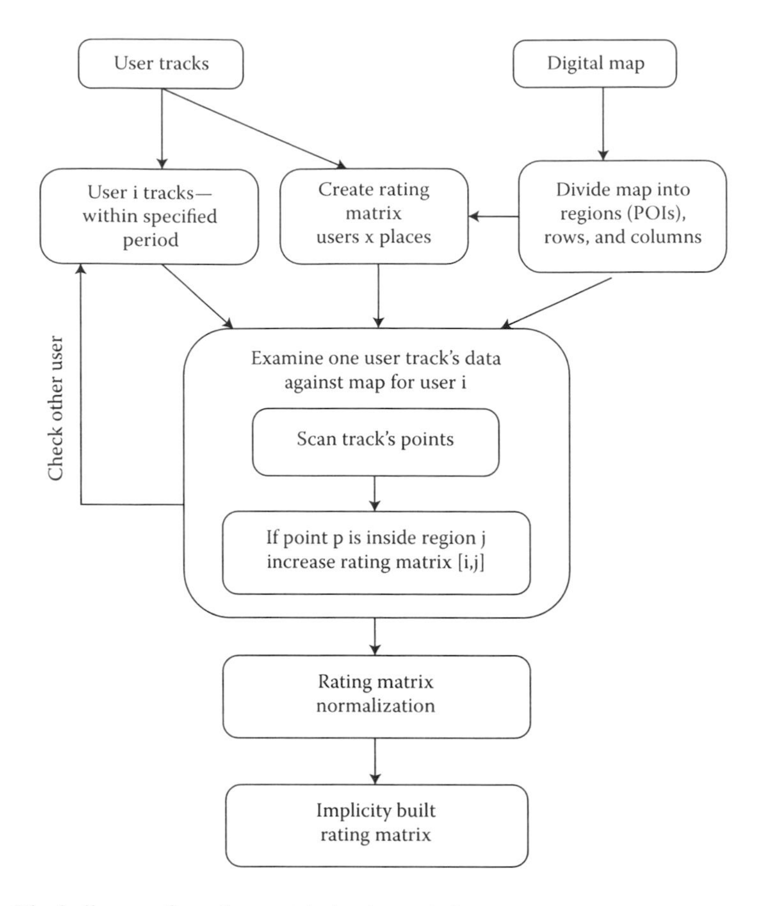

FIGURE 27.5 Block diagram for rating matrix implementation.

used as rating matrices places in the system. The zoomed area is illustrated in the same figure for a region into the map to view the positions that user can visit.

Figure 27.8 shows one track for one user in one day. This track represents user movements in some regions in Beijing. The system scans this track and records each region of the digital map that is walk through into the system database to create rating matrices after data normalization.

Algorithm 2 : Rating Matrices Building
Inputs : User Tracks, Time Periods, Digital Map
Outputs : Rating Matrices (User-Place) for each period
Process :
Begin
Divide the map into 100 * 100 regions
For each period of time do
Create new Matrix m_k = [Users * Places]
Get the users tracks for current period
For each user u_i do
For each user t_j belong to u_i do
Scan all points in tracks t_j
Locate p on map
Calculate the region that p belong to : r
Increase the value of m_k $[u_i,r]$ by 1.0
End

FIGURE 27.6 Algorithm for building rating matrices.

FIGURE 27.7 User path on map.

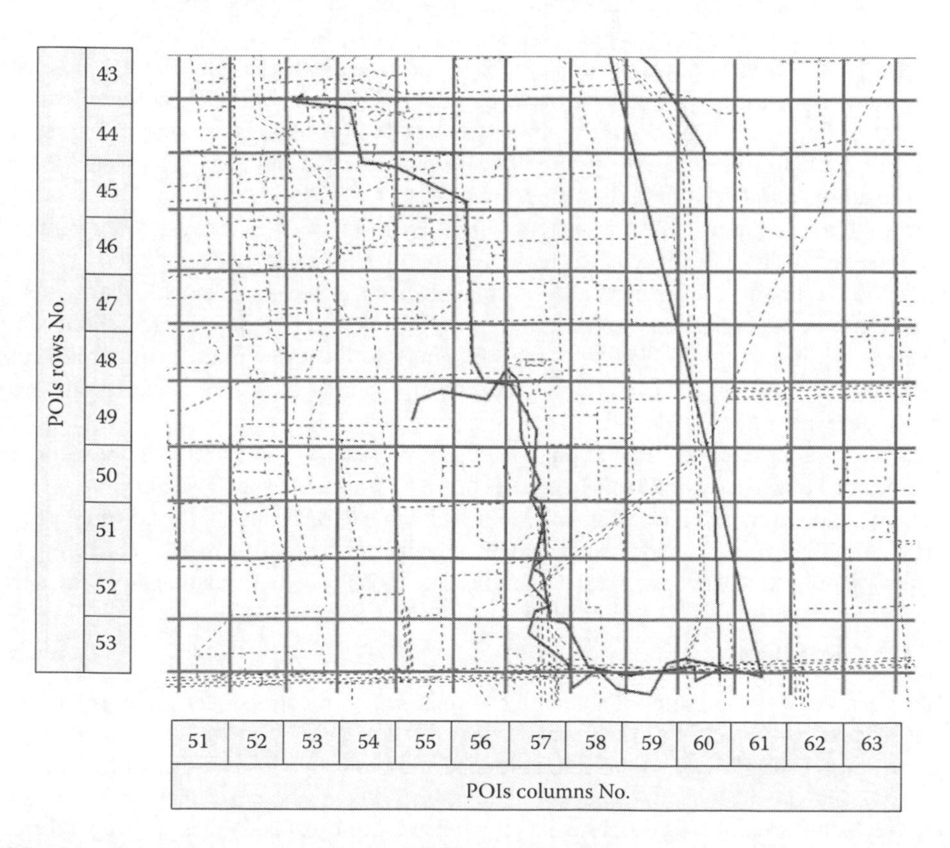

FIGURE 27.8 Track scanning and recording of user movements.

		POIs columns No.												
		51	52	53	54	55	56	57	58	59	60	61	62	63
POIs rows No.	43	0	0	1	0	0	0	0	2	4	0	0	0	0
	44	0	0	2	3	0	0	0	1	2	3	0	0	0
	45	0	0	0	4	4	3	0	0	3	4	0	0	0
	46	0	0	0	0	0	2	0	0	3	0	0	0	0
	47	0	0	0	0	0	4	0	0	3	3	0	0	0
	48	0	0	0	0	0	4	2	0	0	5	0	0	0
	49	0	0	0	0	4	6	9	0	0	4	0	0	0
	50	0	0	0	0	0	0	10	0	0	5	0	0	0
	51	0	0	0	0	0	0	13	0	0	4	0	0	0
	52	0	0	0	0	0	0	14	0	0	4	2	0	0
	53	0	0	0	0	0	0	12	5	2	17	3	0	0

FIGURE 27.9 Sample of filling POIs matrix from track scanning.

To record user movements, each track point will be counted in a matrix. These matrices are two-dimensional matrices; their columns and rows are equal to the map regions, which are 100 × 100. For example if the user visits a region number (51,57) the system will raise the count on the equivalent POIs matrix index (51,57). The visiting time is taken into account, and the more time taken raises the rates. Figure 27.9 shows an example of the system results if the track in Figure 27.8 is entered as an input to the proposed system. The system fills the table and completes the rating matrix for that user to the appropriate time period.

After completing all user tracks the system will unify the POIs matrices into the rating matrix by combining each user's POIs matrix to create a rating matrix for the needed period. The conversion technique begins by converting the POIs matrix from a two-dimensional matrix to a one-dimensional matrix for each user. So, the count of raw fields in the rating matrix becomes 10,000 fields because we have 100 × 100 fields for the POIs matrix. Each raw in the rating matrix is user movement implicit feedback inducted ratings. Figure 27.10 shows the conversion of POIs matrices into rating matrix rows.

After creating rating matrices from implicit feedback by inducting user histories, the rating matrices should be normalized to be within the range of [1..10]. Figures 27.11 and 27.12 illustrate an example of the normalization of rating values. Figure 27.11 shows some of the values of the accumulated rating values matrix and Figure 27.12 shows the normalized values of the rating matrix. The rating matrix is huge; we cannot print it here because, for example, for 40 users it would contain 400,000 entries.

27.7 CONCLUSION

GPS tracks are the most important data source to be used in recommender systems, so the proposed system used user tracks to induct users' behaviors during a 5-year time period. This chapter proposes an implicit feedback system to recommend the best POIs instead of using or depending on social networks applications like check-in databases. It depends on user movement histories by digitizing tracks on a map. The system inducts the behavior of users and creates rating matrices for each period of time according to the system requirements.

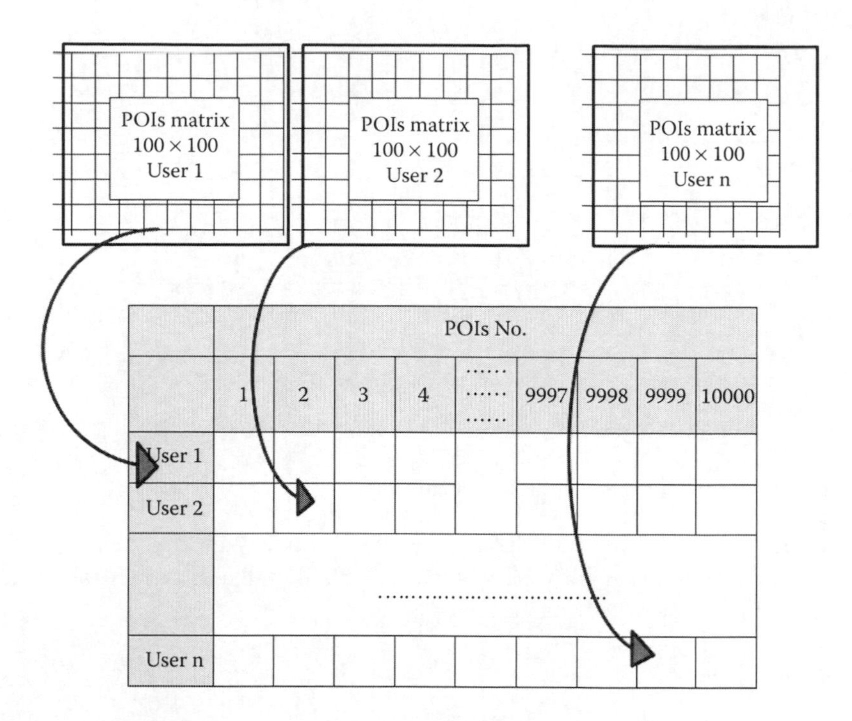

FIGURE 27.10 Creating rating matrix from POIs matrices for N users.

	POIs No.												
	501	502	503	504	505	506	507	508	509	510	511	512	513
User 70													
User 71	0	0	0	172	160	15	200	0	0	0	0	0	0
User 72	0	0	0	0	0	40	310	20	0	0	0	0	0
User 73	56	0	0	0	0	0	0	0	0	122	90	0	0
User 74	0	0	0	0	0	0	0	0	0	0	0	0	10
User 75													

FIGURE 27.11 Example of rating matrix before normalization.

ArcMap GIS and its relevant data with GPS tracks is the core of developing our proposed system and there are two versions of the system: desktop and mobile applications to be used by the users. Recommendations are displayed to the user on the map directly and the POIs are signed directly with short directions and roads to visualize the recommended lists to users.

27.8 FUTURE WORKS

This system still needs more development of its ideas and algorithms. There are many more ideas that could be applied to enhance and improve the proposed system and the algorithm used within, including using social networks to develop a new geosocial recommender system, other techniques like the Relative Motion (REMO) algorithm could be adopted to classify and predict the user's

	POIs No.												
	501	502	503	504	505	506	507	508	509	510	511	512	513
User 20													
User 21	0	0	0	5.3	5.1	0.3	6.5	0	0	0	0	0	0
User 22	0	0	0	0	0	1.2	9.6	0.8	0	0	0	0	0
User 23	1.5	0	0	0	0	0	0	0	0	4.1	2.9	0	0
User 24	0	0	0	0	0	0	0	0	0	0	0	0	0.2
User 25													

FIGURE 27.12 Example of rating matrix after normalization.

positions according to their movement behavior, and creating new vectors of features from similar features of old users. Real-time applications could also be built in the future and we could use GIS servers to reach online databases to enhance our system.

REFERENCES

1. Ricci F, Rokach L, Shapira B. *Recommender Systems Handbook*. New York: Springer, 2011.
2. Lee W, Krumm J. Trajectory preprocessing. In: Zheng Y, Zhou X (eds.), *Computing with Spatial Trajectories*. New York: Springer, 2011, pp. 3–33.
3. Baltrunas L, Ludwig B, Peer B, Ricci F. Context relevance assessment and exploitation in mobile recommender systems. *Personal and Ubiquitous Computing* 2012; 16(5): 507–526.
4. Zheng V, Zheng Y, Yang Q. Collaborative location and activity recommendations with GPS history data. *19th International Conference on World Wide Web*. New York: Association for Computing Machinery, 2010, pp. 1029–1038.
5. Mac Aoidh E, McArdle G, Petit M, Ray C, Bertolotto M, Claramunt C, Wilson D et al. Personalization in adaptive and interactive GIS. *Annals of GIS* 2009; 15(1): 23–33.
6. Savage N, Baranski M, Chavez N, Höllerer T. I'm feeling LoCo: A location based context aware recommendation system. In: Gartner G, Ortag F (eds.), *Advances in Location-Based Services*. Berlin: Springer, 2016, pp. 37–54.
7. Zheng Y, Xie X, Ma W. GeoLife: A collaborative social networking service among user, location and trajectory. *IEEE Data(base) Engineering Bulletin*. 2010; 33(2): 32–39.
8. Zheng Y, Zhang L, Xie X, Ma W. Mining interesting locations and travel sequences from GPS trajectories. *WWW '09 Proceedings of the 18th International Conference on World Wide Web*. New York: ACM, 2009, pp. 791–800.
9. Zheng Y, Li Q, Chen Y, Xie X, Ma W. Understanding mobility based on GPS data. *UbiComp '08 Proceedings of the 10th International Conference on Ubiquitous Computing*. New York: ACM, 2008, pp. 312–321.
10. Clarke K C et al. *Geographic Information System and Environmental Modeling. Fourth International Conference on Integrating Geographic Information Systems (GIS) and Environmental Modeling (GIS/EM4)*, Banff, Canada. Upper Saddle River, NJ: Prentice-Hall, 2009.
11. Harvey F. *A Primer of GIS Fundamental Geographic and Cartographic Concepts*. New York: The Guilford Press, 2008.
12. Verbila D I. *Practical GIS Analysis*. New York: Taylor & Francis, 2003.
13. Decker D. *GIS Data Sources*. New York: John Wiley & Sons, 2001.
14. Information Technology Service. *Introduction to GIS Using ArcGIS*. Durham, England: Durham University, 2006.
15. Sherman G E. *Desktop GIS: Mapping the Planet with Open Source Tools*. Raleigh, NC: The Pragmatic Bookshelf, 2008.
16. Changg K T. *Programming ArcObjects with VBA*. Boca Raton, FL: Taylor & Francis, 2008.
17. Croswell-Schulte IT Consultants. *GIS Design and Implementation Services*, Brochure, 2010.

Section III

Management

28 Management through Opportunities as an Unconventional Solution in the Theory of Strategic Management

Anna Brzozowska and Katarzyna Szymczyk

CONTENTS

28.1 Discussions on the Creation of Opportunities in Management of an Enterprise 378
28.2 Identification of Opportunities in the Context of Strategic Management 384
28.3 Summary 385
References 386

Management through opportunities is one of the ways of seeking increased operation effectiveness of an organization, and skillful use of opportunities is an important element of the modern manager's skill set. The modern manager, regardless of the size of the organization that he/she manages, has to constantly undertake new activities and look for new ways to increase the effectiveness of the organization's functioning. One of the main tasks is to define various plans of activity or functioning, which in the short or long run should give a chance for continuous development, conquering new markets, and individual development. However, given the modern conditions of operation in an environment, in which it is difficult to accurately define and is constantly changing, activity cannot be based on established plans, principles, or patterns. It is important to undertake activities that represent a fast reaction to challenges of the market, competitors, or customers, and respond to emerging expectations from the environment that changes rapidly and instantly. Management, as a field of knowledge, attempts to find methods and present techniques quickly and thoroughly in the most possible ways that will help in moving around the winding roads of modern management. Management in terms of entrepreneurship should focus on external factors of the business environment in order to be able to observe and use the opportunities that are the constant part of that environment and which may determine the enterprise to change and reorganize its internal organization structure and apply new strategies in order to catch the opportunities from the outside.

At the present time, while examining management and analyzing its accomplishments, we can use achievements of various schools of thought (Obłój, 2007), for example, resource, position, and evolution, and various theories, for example, the increasingly popular theory of chaos in management (Brown and Eisenhard, 1996).

This chapter will present the concept of opportunities in management. Numerous publications stress that today's management, on the other hand, have to use possessed potential more often, while on the other hand it has to be able to interpret what is happening and quickly react, therefore opportunity is becoming an important element that can lead to success. In all conditions of its functioning, an organization has to notice or create opportunities and be able to use and appreciate them.

Nowadays, this is becoming increasingly important in the context of what is gradually becoming an imminent characteristic of contemporary times, namely, changeability and uncertainty that accompany almost all activities and decisions that are taken.

Academic literature provides examples of linking many modern concepts and methods of enterprise management (Bubel, 2015) with the use of opportunities, or even treating opportunities as tools or strategies. This context reveals differences in terminology and different perceptions of the relationships between new concepts and methods in enterprises and changes accompanying the concept of opportunity. Both the Polish and foreign academic literature highlights numerous concepts and methods of management, which when examined in the context of opportunities by an enterprise, permits the formulation of the thesis that their use in the process of the implementation of strategic changes may become a factor impacting effective implementation of the concept of opportunities.

At the same time, the use of selected management methods or concepts supporting the implementation of strategic changes is determined by a number of factors (e.g., the current economic situation of an enterprise), and the final choice belongs to the manager of an enterprise (Skowron-Grabowska, 2008).

28.1 DISCUSSIONS ON THE CREATION OF OPPORTUNITIES IN MANAGEMENT OF AN ENTERPRISE

According to the dictionary of the Polish language, *opportunity* is a favorable circumstance that makes something possible, is conducive to something, it is an occasion; thus in the positive context, we can talk about great opportunities, a lot of opportunities, and favorable opportunities, whereas in the negative context we can miss or lose opportunity. The *Merriam-Webster Dictionary* explains the term *opportunity* as "a favorable juncture of circumstances" or "a good chance for advancement or progress."

However, opportunity itself is always something positive, which—if properly used—can bring benefits. It is usually in this context that the problem of closeness (or differences) of meanings of the words *opportunity* and *chance* appears. Although the dictionary of synonymous words uses these two terms in a similar context, we can or even have to talk about subtle differences in meaning (Figure 28.1).

Chance is associated with probability, possibility of success, or achieving success in some area. Thus, we have a chance for success, but we do not talk about opportunity in this context. Moreover, the most important difference in the meaning of these two terms probably results from the fact that chance is a certain probability of success, whereas with opportunity we talk about circumstances

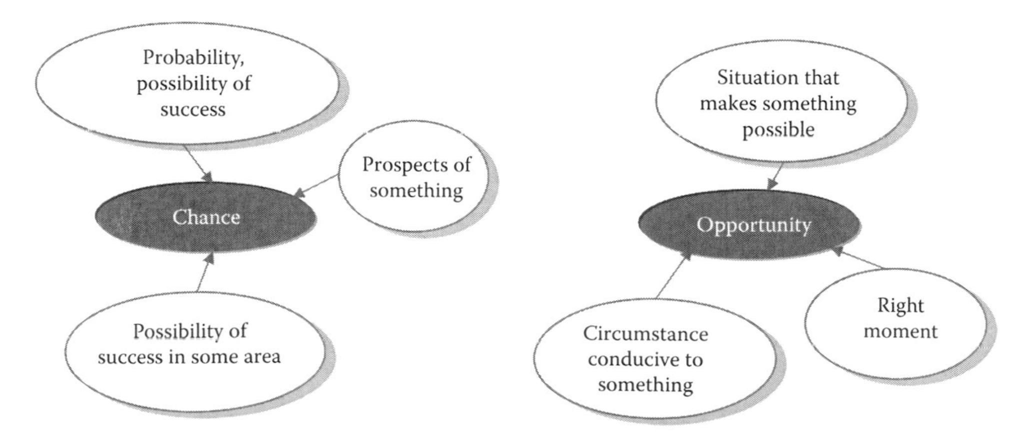

FIGURE 28.1 Theoretical differences in the meanings of *chance* and *opportunity*.

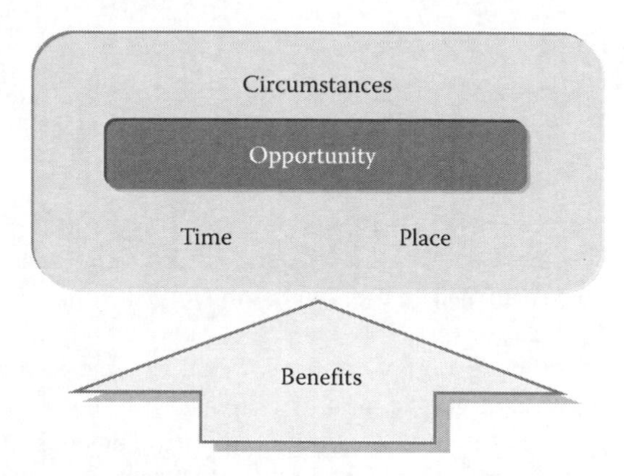

FIGURE 28.2 Identification of opportunities.

that are conducive to success. Opportunity, in order to occur, requires appropriate timing, place, and favorable circumstances. It is this meaning of the word opportunity that should be emphasized in management studies and its use in management of an enterprise. We can assume that opportunity arises for an enterprise as a result of a certain event or series of events that make it possible to achieve additional benefits, both material and intangible ones. An imminent characteristic of opportunity is temporality (momentariness); it appears in certain time and in certain circumstances, and then disappears.

Opportunity is a combination of circumstances, time, and place, which properly used by an enterprise may bring benefits (Figure 28.2) (Sharplin, 2000). We should bear in mind that opportunities appear suddenly; they may come up in the environment or in an enterprise. Whether we assume that opportunity appears or exists, the problem is that somebody has to notice and use it (Alvanz and Borney, 2007). We can assume that opportunity exists irrespective of circumstances. It should be discovered and noticed, as early as possible, before competitors do so; therefore the ability inside the company to observe the environment in order to notice the opportunities is so necessary or even obligatory.

However, we can also use possessed potential (or, more precisely, resources of an organization) to create and use opportunity. Surely, we can talk about internal opportunities generated by an organization itself by skillfully using human resources and so forth and external ones, that is, signals from the environment. An example of opportunity is the collapse of a competitor, as this suddenly gives a chance to significantly increase the number of customers and the scope of impact. Opportunities that give chances for benefits also include the state's activities connected with legal issues, for example, change of tax rates, limits of import from other countries, and changes to insurance regulations (Starostka-Patyk, 2015). Thus, in order to identify opportunities, one has to carefully observe an organization's environment, as it is there that most opportunities can be found and used. It is difficult to talk about any specific typology, divisions, or classifications of opportunities. Krupski (2011) described the opportunities as special situations that appear or happen in the business environment and can be used by the enterprise to develop new strategies to increase its income or to gain better position on the market. These opportunities, according to Krupski (2011), are

- Establishing cooperation with a large and reputable company
- Occasion to merger with another company
- Opportunity to obtain investor
- Opportunity to take over the failing competitor
- Opportunity to possess the clients of the competitor that withdraw from the market
- Opportunity to purchase an element of the infrastructure (square, premises, means of production, building, etc.)

- Opportunity to use European Union funds
- Winning the bid in an unexpected moment
- The emergence of new beneficial system solutions
- Unforeseen and unexpected increase in demand on the foreign market

In Poland in 2004, in a survey on 155 small and midsize enterprises, 54% of entities turned out to use the opportunity as the factor determining the companies' development. Generally, 43 companies used opportunities to buy infrastructure, such as the building, place, or the particular means of production in a moment when unexpectedly the price was relatively low and the investment, at the time, very beneficial. Another 13 companies took the advantage of the opportunity of developing the cooperation with some large and significant players in the market. The survey as well showed that eight companies used the funds from the supporting business programs introduced by the European Union and other seven entities found the new system solutions interesting in terms of occasion to develop their business. The researchers also found out that such occasions like buying a bankrupting competitor was used by five entities as a fortunate opportunity for their position on the market, and in similar way, the competitor's withdrawing from the market was a good occasion for four companies to take over the costumers of the failing company. Also four companies indicated in the survey that they found the moment of getting a new investor as a great opportunity to move forward with their business, two companies admitted that the surprising winning bid was for them a type of occasion to expand functioning on the market and another two enterprises were able to use the occasion of unpredicted demand increase on the foreign market to expand their operations. In 2010 in the publication on the internationalization of Malaysian quantity surveying firms by Aziz et al. (2011), the results of the received questionnaires about the motives to operate abroad indicated that Malaysian companies found the opportunities emerging from the business environment the second important factor to start business or expand business abroad. In China in 2005, the world-known fast food company McDonald's noticed that the rapidly expanding Chinese car market due to an emerging middle class was the great occasion for the American firm to enter the Chinese market with a completely new offer: drive-thru restaurants. The first McDonald's drive-thru restaurant in China appeared in 2005 and a year later McDonald's took advantage of a deal with Chinese giant petrol company Sinopec in which McDonald's could build its restaurants next to more than 30,000 petrol stations. This occasion turned out to be extremely beneficial for the McDonald's company in the Chinese market and it can be considered as valuable proof that the occasion may become a great motive for expanding and developing business activity. The fact that occasion should be taken into account as an important strategy for the company was also noticed by McDonald's for planning expansion over in Italy where McDonald's noticed the lack of quick in-and-out restaurants, especially offering something more or different than traditional Italian pastries or pizzas. Corners near train stations, places where more tourists or businessmen as well as students stop before continuing their travel from one point to another, touristic paths, and main sightseeing spaces were found by McDonald's as the great opportunity to open small food shops and take over costumers tired of the Italian menu. McDonald's used the occasion in a better way and successfully positioned itself in cities like Rome, Florence, and Milan where train platforms or corners just behind cathedrals or art places are filled with McDonald's small bars or food shops (Weyers, 2012). Microsoft even more effectively used occasion for business exploration in 2013 to buy the Finnish company Nokia, which unfortunately was rapidly losing its position and relevance on the market. Both companies had already worked in a strategic partnership, however, the efforts to offer competitive mobile devices to fashionable and very popular smartphones like iPhone or Samsung Galaxy during those years was hard to achieve. Nokia was gradually losing costumers, and Microsoft was losing its computer customers to smartphones. Nokia's mobile phone business cost nearly $8 billion and included not only the business material goods but also a wide range of patents and mapping services. Microsoft, due to such a prominent occasion, could become a competitive mobile and smartphones and software company, which gave the enterprise a great

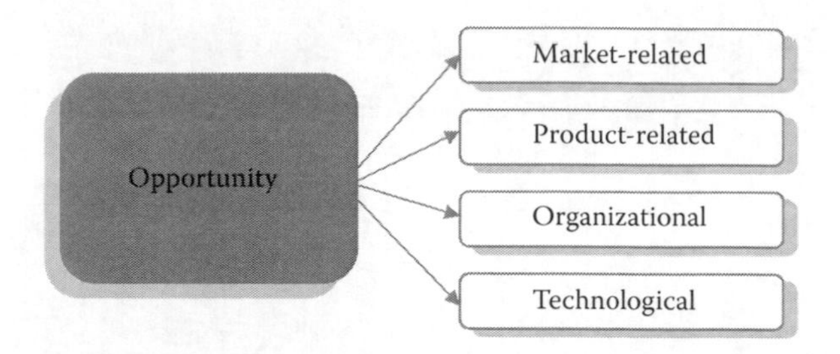

FIGURE 28.3 Ways of creating opportunities. (Based on Obłój K. 2010. *Pasje i dyscyplina strategii. Jak z marzeń i decyzji zbudować sukces firmy*. Poltex, Warszawa, p. 96ff.)

chance to position itself as the competitor for market-leading enterprises like Apple and Google (Goldman, 2013).

Undoubtedly, we can talk about opportunities that come up, for example, with new, unusually effective technologies and possibilities of using large funds obtained from external sources.

Opportunities thus have their sources at different levels of functioning; we can distinguish technological opportunities (improvement, generation of profits, reduced investments), market opportunities (competitor collapses or an investment), legal opportunities (changes to regulations that lead to increased profits), and social opportunities. The use of such opportunities and recognition of their potential, may contribute to development, and constitute an important element of the development strategy of an organization (Figure 28.3).

Opportunities can be created on the market. We also distinguish product, technological, and organizational opportunities. Although opportunity is something temporary that appears unexpectedly, some people think that it can be planned (Obłój, 2010).

Seeking opportunities on the market is a way to develop and to win new customers, which is particularly important nowadays when long-term planning of any activities in quickly changing conditions would be pointless. We noted that it is effective to use two opportunities:

- Generation of opportunities by means of the so-called anchor points, that is, position that a company creates on the market on which it has not started competing yet. It involves launching new products, teasing competitors with price, and depriving competitors of some of their customers (Barney, Ketchen, and Wright, 2011). These points may be generators of opportunities because competitors may respond to them with unsuccessful activities, such as failed investments or even withdrawal from the market.
- Generation of opportunities by means of various activities, such as spreading "rumors" or "organized" activities, for example, on the exchange stock, which may result in a fall in stock rates and purchase of stock at a lower price (Figure 28.4).

The common characteristic of such activity is that opportunities are generated, created, or somebody is the driving force of these situations. We talk here about active activities, where there is action and we have opportunity to use it (Figure 28.5). As was already noted, some people think that opportunities exist independently. They only have to be noticed in the right time and used. It is the human being that gives meaning to the already existing opportunities. Naturally, we can assume that there are also in-between situations, that is, on the one hand, opportunity situations are actively generated, as in the case of the already mentioned expansion of the area of opportunities, while on the other hand, existing opportunities are used in a passive way.

Summing up the discussion, we can distinguish an active approach to opportunities, that is, creation of such events that can be seen as opportunities, and a passive one in which opportunity is what already exists. Krupski (2013) described the active and passive approach to opportunities in the area of external and internal determinants, which are shown in Figure 28.6.

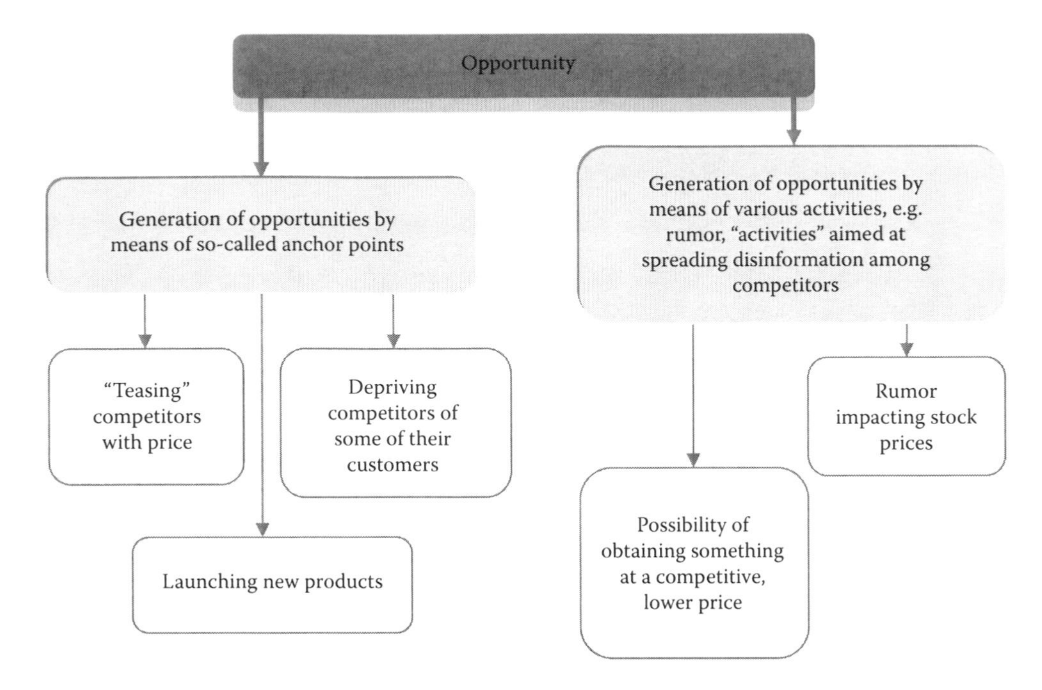

FIGURE 28.4 Ways of generating opportunities.

FIGURE 28.5 Types of opportunities.

Trzcieliński (2011) proposed another very detailed typology of opportunities according to which the opportunities can be divided from the point of regarding the opportunity as objective or rather subjective, or the opportunities might be quite consciously or unconsciously stated. Apart from that, the same author suggested distinguishing the opportunities as those dependable on the company and those that are not dependable on the company, which generally means that such a distinction allows for the division according to the abilities inside the company to create the opportunities. Furthermore, the author indicates the importance of the company's involvement that results in opportunities that have not been noticed, or the opportunities that have been taken or untaken, as well as the essential role of the company's professional action thanks to which the opportunities can be used or not used. The final types of opportunities according to Trzcieliński (2011) are distinguished due to the overall aim of the company, namely, the opportunities that were taken and accomplished with success and the opportunities that ended in failure.

It has been repeatedly indicated that opportunities come up, exist for a certain period of time, and then they disappear. According to those assumptions, one may conclude that the opportunities possess a certain life cycle, namely (Krupski, 2011):

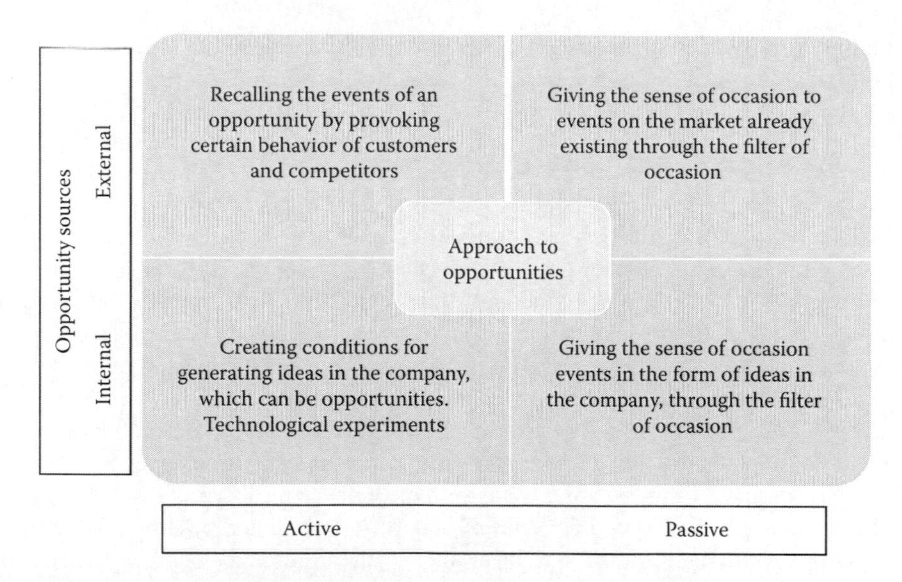

FIGURE 28.6 Approaches to opportunity. (Based on Krupski R. 2013. Rodzaje okazji w teorii i praktyce zarządzania, In: Mroczko F. (ed.), *Prace naukowe Wałbrzyskiej Wyższej Szkoły Zarządzania i Przedsiębiorczości*. Wydawnictwo. Wałbrzyskiej Wyższej Szkoły Zarządzania i Przedsiębiorczości w Wałbrzychu, Wałbrzych, p. 12.)

- Noticing opportunity
- Defining opportunity
- Financing the use of opportunity
- Using opportunity
- Managing the life cycle of opportunity
- Supervision over the use of opportunity

In the case of noticing opportunity, a large role is played by subjective perception of the problem and the ability to use favorable circumstances, which may include knowledge of new technologies, proper interpretation of relationships on the market, or cooperation with partners. It is also important to define in the precise way the far and near environment of an enterprise, which can be a kind of source of opportunities (Figure 28.7).

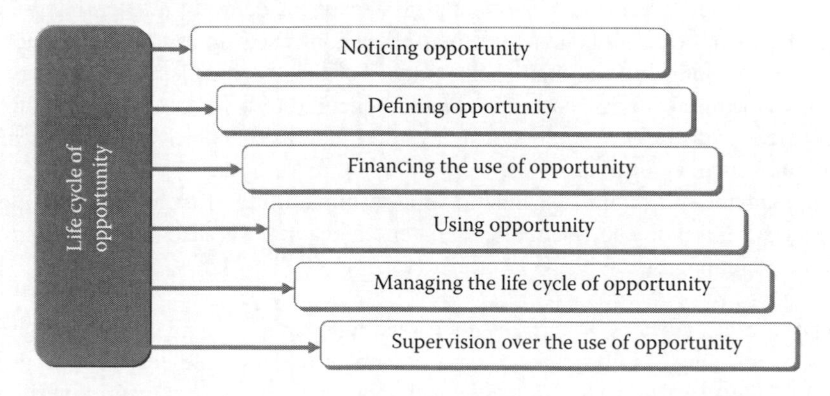

FIGURE 28.7 Life cycle of opportunity.

Noticing opportunity can be understood as skillfully interpreting signals from the environment and describing them in such a way that they will enable recognition of the opportunity. This stage of the life cycle of the opportunity also involves the ability to assess (estimate) the value of the expected opportunity, that is, what can be considered as added values (material and intangible) achieved thanks to using the opportunity. When defining opportunity, it is necessary to think of what activities are required to implement an idea. It is also important that the attractiveness of opportunity is always and relatively promptly assessed, that is, what can be gained if it is used. When financing opportunity, the enterprise has to use already possessed resources either through their transitions or possessing already planned funds earmarked for financing the opportunities. The use of opportunities requires mobilization and, above all, focuses on learning. One cannot constantly repeat old patterns. At different levels of activity, it is advisable to experiment. Management of the life cycle of opportunity engages employees' knowledge and skills; it is necessary to know how to use this potential at every level of activity (Lis and Bajdor, 2015). For the supervision of the implementation of opportunity, it is important to treat opportunity as an independent, autonomous entity which has its place in an organizational system, with identified processes and guaranteed funds for them. Taking into account life cycles of opportunity, it should be stressed that in order to create and use opportunity, an enterprise engages all its capabilities and resources, which is directly connected with some concepts of modern management, such as the resource-based school and rule-based school (Obłój, 2007).

28.2 IDENTIFICATION OF OPPORTUNITIES IN THE CONTEXT OF STRATEGIC MANAGEMENT

All the concepts listed at the end of the preceding section constitute one of the elements of strategic management and are an attempt to combine this management and principles of entrepreneurship to increase the efficiency of an enterprise. We should highlight here the concept (theory) of the so-called simple principles (rules) (Eisenhardt and Sull, 2001), which include

1. Manner of operation, rules for the functioning of a process (hot to rules)
2. Limitations of the choice of emerging opportunities (boundary rules)
3. Priorities hierarchizing opportunities and the related system of resources (its use) (priority rules)
4. Synchronization of people's activities and emerging opportunities (timing rules)
5. Completion of activities connected with implementation of a given project (task) (exit rules)

These principles are implemented as a result of a manager defining a few strategic objectives (processes) and creating rules of conduct for them. We should bear in mind that the more unstable and turbulent the environment, the smaller the number of these principles, as the ability to plan in such conditions is limited. In this case, one should concentrate on a small number of short-term actions. The purpose of this is to use opportunities and increase flexibility of processes (Krupski, 2008) taking place in an enterprise.

Strategic management is a joint, systematic, long-term planning of issues that are important for an enterprise and it refers to future effects of current decisions. The use of principles of strategic management helps managers to

- Predict the consequences of current decisions and changes in the environment and to define the hierarchy of importance of the elements of the environment
- Define an enterprise's position in its environment, identify success factors on the market and specify the ways of gaining competitive advantage
- Specify the objectives and directions of activity

- Coordinate the acquisition and use of resources
- Improve the communication within and between an enterprise and its environment

In the context of opportunities for strategic management, two approaches emerge. The first one ascribes strategies to enterprises (prescriptive approach), and the second one describes why strategies are such and not other (descriptive approach). In these approaches, it is important to specify areas characterizing the position of an enterprise for using opportunities and to define ways of their operation on this basis. The first approach dominates in the mainstream works on formulation and implementation of strategies. Descriptive approaches significantly enrich the content of formulated strategies. They are based on different epistemological assumptions than the prescriptive approaches, therefore many theoreticians think that linking them with opportunities is methodologically incorrect.

Despite differences in the interpretations of strategic management using opportunities, some similarities can be indicated:

- Strategic management, defined as a set of actions forming a process, can be described in terms of two stages (phases): strategic planning and implementation of a strategy.
- The aforementioned phases can be divided into a set of such stages: mission identification, analysis of the environment and industry, analysis of an enterprise, choice of basic strategies, strategy formulation, description of a business plan, description of the organizational structure, and assessment and selection of a strategy, which are followed by the implementation and use of opportunities, and then control of results.
- Process of strategic management, which is focused on remote future without using opportunities and mission of an enterprise; also refers to operational activities.

Once a strategic plan has been defined and assessed, it should remain unchanged throughout the whole period of implementation. This approach would work well if the environment had been stable, but it usually is not. That's why only a small number of corporations implement their strategies without using unexpected opportunities. Strategic changes are often defined as rapid and fundamental, as if there was a tendency to look for a ground-breaking point where opportunities are used. However, when an organization adopts a certain strategy, it starts to develop in line with this strategy, and changes occur gradually, step by step. Some functional strategies are changed; others remain the same, and there is increasing evidence that an enterprise's strategic changes are better described in the convention of continuity. The so-called learning approach states that an organization adapts to changing conditions of operation through small changes or an organization recognizes and responds to small changes by using opportunities (issue management), or recognizes and responds to the so-called weak signals. Therefore, new strategies bring solutions within the already applied organizational structures and concepts of problem solving in strategic management, suggesting approaches to change using the concepts of opportunity.

28.3 SUMMARY

As it had been repeatedly stressed in the preceding discussion, the use of opportunities is associated with something momentary and immediate, in opposition to everything that is planned or predicted. One may understand the management as the process of planning and organizing the work to achieve certain expected results, but one may as well observe the business environment and the competitors in the enterprise surroundings to realize that planning and preparing can be substituted by adequate and successful usage of chances and opportunities emerging from this business environment. This will rather lead to the process of planning the conditions that have to be met when opportunity comes up and can be used. Therefore, the business entity, that wants using and relying its success on opportunity coming from the business's surroundings, should focus on the nature and scope of project, design of the plan, and strategy to prepare the company to react quickly and use the opportunity that appears in the business environment. That should generally be understood or regarded as a new way of managing by formulating the certain type of strategy for using opportunities. In such a situation, the entrepreneurs may frequently refer to all the resources possessed by an enterprise, but they will use them to plan conditions to use opportunities and they will rely on these resources to assess whether the opportunities in the surroundings are worth pursuing and could bring the positive outcome for the enterprise. In order to use opportunity, it seems to be obvious that it is absolutely essential to have access to resources, their own and others, and possess intangible resources, at an appropriate level and appropriately varied, that is, knowledge, skills, attitudes, and culture. In general, one may refer at this point to the classical resource-based school, but it should be as well stated how important it is to use and plan intangible resources, because they determine the use or identification of opportunities, as in this context specific opportunities, assuming the characteristics of projects are taken into account (Krupski, Niemczyk, and Stańczyk-Hugiet, 2009). Implementation of projects has long been an important element of activity of every enterprise and every manager. Thanks to adequate evaluation and verification of resources inside the company to use them as the basis to prepare the strategy for using the opportunities in the business environment, the company can determine whether the new project is payable back in the initiation stage or assess its final financial value on an ongoing basis when the managers monitor the project itself which can give the chance to reevaluate the payback period.

Management through opportunities is still an unconventional solution that appeared in the theory of strategic management and is designed to help in effective management of an enterprise. It puts emphasis on skillful and appropriate use of emerging, favorable circumstances (opportunities), which should be recognized early enough and in which potential benefits for an enterprise should be noticed. This approach will certainly be monitored and developed, which is connected to changes that take place within management itself. Opportunities put pressure on skillful use of emerging favorable circumstances and resignation from long-term earlier planning in favor of "catching opportunities." Although the temporality and transitory nature of opportunities is highlighted in management, one should take into consideration that opportunities have their value, timing, and probability of success that can be forecasted and analyzed in terms of win or loss. Also in order to recognize and use opportunities, it is necessary to have extensive knowledge, and ability to correctly assess situations and interpret signals coming from the environment of an enterprise and from within it, as well as the knowledge and proficiency to evaluate the resources of the company and how they should be used in order to employ the opportunity as the next project of the company. It is necessary to know the principles governing the market.

REFERENCES

Alvanz S. A., Borney J. B. 2007. Discovery and creation: Alternative theories of entrepreneurial action. *Strategic Enterpreneurship Journal*, 1(4), 49.

Aziz A. A.-R., Pengiran N. D., Lim M. Y., Nuruddin A. R. 2011. Internationalization of Malaysian quantity surveying firms: Exploring the best fit models. *Construction Management and Economics*, 29, 49–58.

Barney J., Ketchen D. J. Jr., Wright M. 2011. The future of resource-based theory: Revitalization or decline? *Journal of Management*, 37(5). Published online, March 10.

Brown S., Eisenhard K. M. 1996. The art of continuous change. *Administrative Science Quartely*, 42(1), 1ff.

Bubel D. 2015. A modern system of enterprise management in the concept of an intelligent organisation. Zeszyty Naukowe Uniwersytetu Przyrodniczo-Humanistycznego w Siedlcach. Administracja i Zarządzanie, Seria, 106, 9–21.

Eisenhardt K. M., Sull D. N. 2001. Strategy as simple rules. *Harvard Business Review*, 79(1): 106–116.

Goldman D. 2013. Microsoft to buy Nokia's phone business for $7.2 billion. *CNNMoney (New York)*. Retrieved October 7, 2013, from http://money.cnn.com/2013/09/03/technology/mobile/microsoft-nokia/index.html.

Krupski R. (ed.). 2008. *Elastyczność organizacji* [Flexibility of organization]. Wydawnictwo Uniwersytetu Ekonomicznego we Wrocławiu, Wrocław, pp. 24–25.

Krupski R. 2011. Okazje w zarządzaniu strategicznym przedsiębiorstwa [The opportunities in strategic management of enterprise]. *Organizacja i Kierowanie*, 4(147), 11–14.

Krupski R. 2013. Rodzaje okazji w teorii i praktyce zarządzania [Types of opportunities in theory and practice of management]. In: Mroczko F. (ed.) *Prace naukowe Wałbrzyskiej Wyższej Szkoły Zarządzania i Przedsiębiorczości*. Wydawnictwo. Wałbrzyskiej Wyższej Szkoły Zarządzania i Przedsiębiorczości w Wałbrzychu, Wałbrzych, p. 12.

Krupski R., Niemczyk J., Stańczyk-Hugiet E. 2009. *Koncepcje strategii organizacji* [The concepts of strategies of the organisations]. PWE, Warszawa, 2009, pp. 23–52.

Lis T., Bajdor P. 2015. Wiedza i kultura organizacyjna a optymalizacja wykorzystania zasobów ludzkich w przedsiębiorstwach [Knowledge and organizational culture and optimization of the use of human resources in enterprises]. In: Perechuda K. and Chomiak-Orsa I. (eds.) *Wiedza i informacja w akceleracji biznesu*. Wydawnictwo Wydziału Zarządzania Politechniki Częstochowskiej, Częstochowa, pp. 62–65.

Obłój K. 2007. *Strategia organizacji* [The strategy of organisation]. PWE, Warszawa.

Obłój K. 2010. *Pasje i dyscyplina strategii. Jak z marzeń i decyzji zbudować sukces firmy* [Passions and a discipline of strategy]. Poltex, Warszawa, p. 96ff

Sharplin A. 2000. Strategic management. *Supernet J., Management*. Tezaurus Kierownictwa. Kolonia, p. 47ff.

Skowron-Grabowska B. 2008. Development strategy in production enterprises and logistics centers. In: Havrila M. and Modrák V. (eds.) *Manufacturing and Industrial Engineering*. Technical University of Košice, Preszow, Slovakia, pp. 41–43.

Starostka-Patyk M. 2015. Potrzeba zmian restrukturyzacyjnych w zakresie zarządzania produktami niepełnowartościowymi na przykładzie branży AGD [The need for restructuring changes in the management of defective products on the example of household appliances]. In: *Współczesne oblicza i dylematy restrukturyzacji*. Fundacja Uniwersytetu Ekonomicznego w Krakowie, Kraków, p. 403.

Trzcieliński S. 2011. *Przedsiębiorstwo zwinne* [The agile company]. Wydawnictwo Politechniki Poznańskiej, Poznań.

Weyers B. 2012. Global expansion of U.S. fast food restaurants: A case study of McDonald's in Italy. Honors scholar theses, University of Connecticut.

29 Concept of Supply Chain Management in the Context of Shaping Public Value

Dagmara Bubel

CONTENTS

29.1 Introduction 389
29.2 Identification of Logistic Processes in Organizations Providing Social Services 389
29.3 Premises of Public Logistics in the Supply Chain and Determinants of the Multilevel Character of Networks in the Public Sector 391
29.4 Use of the Supply Chain Concept in the Public Sector: Theoretical Case Study 393
29.5 Summary 399
References 400

29.1 INTRODUCTION

The process approach to management has been present in literature and economic practice since the 1990s. Implementation of this approach resulted in improved effectiveness and efficiency of the functioning of enterprises and whole supply chains. Process orientation sets the framework for the functioning of enterprises, whose main objective is to offer the highest possible value to customers. Recently, process orientation has also become a focus of attention of the public sector, as a result from the pressure of higher effectiveness, permanent financial shortfalls, and growing number of public tasks performed by self-governments. Therefore, the aim of this chapter is to indicate the main characteristics of public supply chains that should be taken into account during implementation of process orientation in value creation for customers as recipients of public services.

29.2 IDENTIFICATION OF LOGISTIC PROCESSES IN ORGANIZATIONS PROVIDING SOCIAL SERVICES

Public organizations perform a range of processes that significantly impact the quality of provided services. Of importance here are not only activities connected with human resources management (Jelonek, 2012), but also the quality of services determined by processes of selecting suppliers, systems for providing organizations with goods, availability of products, systems of control over equipment performance, and safety and used means of transport and communication. In such organizations, the cost aspect plays a less important role in decision making, with more importance given to the interests of the beneficiary, measured by humanitarian considerations (Simsa and Patak, 2016). Thus, the social element is beginning to increasingly dominate in public organizations, shaping the issues of costs of ensuring availability of products or services in a different way than in other organizations. When using methods, tools, and concepts known from economic logistics, one should adapt them to the character and nature of social services. Thus, the use of the concept of logistic management, understood as a process of comprehensively planning, organizing, and controlling activities of logistic processes and logistic actions performed to ensure efficient and effective flow of materials, semifinished products, and final goods in organizations, logistic chains or supply chains can make it significantly easier for public organizations to find answers to several important questions (Blaik, 2012):

- What procedures should be used? (technical effectiveness)
- What is the best combination of expenditures? (cost effectiveness)
- How many services should be produced/provided? (allocation effectiveness)
- How do you form contacts between customers and service providers? (transaction costs)
- How do you change the production/offer in time? (dynamic effectiveness)

Logistic management in organizations providing public services is also aimed at

- Constant care about customer service quality
- Calculation of the unit cost of provision of services
- Planning supply sources
- Controlling stock level, minimization
- Controlling the number and structure of physical and human resources
- Controlling the number and type of transport fleets
- Location of potential warehouses
- Defining the demand for services
- Organizing promotion of services based on market research
- Introducing the principles of Lean management to a public organization
- Defining principles and procedures of waste treatment
- Minimization of material flow time

Table 29.1 presents examples of the main logistic processes and operations that can be identified in organizations providing social services. Their comprehensive planning, organization, and control can be useful and bring measurable economic and organizational benefits.

TABLE 29.1
Examples of Logistic Processes and Actions in Public Services

Control of stocks	• Regulation of the structure and level of stocks • Decisions concerning change of the level of stocks • Coordination of the level and structure of stocks between the links of supply chain • Use of ABC analysis in accordance with the 80/20 (Pareto) rule • Use of two-level control of socks • Establishing the level of buffer stocks
Transport	• Decisions concerning the type of transport • Choice of means of transport • Optimisation of the tasks of transport basc • Choice of the optimal route • Scheduling of transport tasks • Analysis of transport costs and their share in the structure of total costs
Control of the information system	• Evaluation of the current capabilities of an organisational unit in the area of data processing • Monitoring of operational activity • Creation of a database based on various information sources • Creation of a logistic information system that will fulfil the functions of planning, coordination, control, communication and database
Warehousing	• Defining the needs of storage space • Designing the location of stocks • Admission, picking and release of materials to and from the warehouse

In recent years, customer orientation in logistics has been increasingly highlighted, which can be an important element justifying an attempt to implement logistic concepts and tools in public services sector. Four basic determinants of customer orientation in logistics were indicated: marketing perspective, market segmentation, service quality, and logistic strategies subject to customers' interests (Witkowski et al., 2015). The key determinant of customer orientation is logistic support—the basic instrument for winning and retaining customers. Its aim is to ensure continuity, speed, and reliability of supplies and services.

A socially important area of using logistic concepts and tools is the organization of humanitarian campaigns, for example, to help those affected by disasters and cataclysms. Decisions of government and local authorities concerning city logistics affect day-to-day quality of life. Today, an increasing phenomenon is congestion, which is usually caused by (Erd, 2015)

- Disproportion between transport and the size and structure of cities
- Conflicts between collective transport and individual transport
- Disproportion between the level of infrastructure development and traffic

Activities undertaken in modern cities to reduce congestion are addressed by Carsten Deckert (2016). To ensure the appropriate quality of life, sustainable development of civilization, natural environment protection, and minimization of costs of achieving these objectives, decision makers have to use interdisciplinary tools including solutions from various research areas (e.g., economics, statistics, psychology, sociology, jurisprudence, management, marketing, transport engineering, town planning, architecture, political science). Due to this multidimensional character of decisions made in public services, their effectiveness and success often depend on the ability to cooperate in a task team.

An important task in the provision of public services is coordination and control of the flow of people. There are many interesting proposals to use logistic management tools in medical and educational services or the penitentiary system (Gołembska, 2012). Research into the quality of public services shows that the basic factor determining subjective satisfaction with services is their availability. Ensuring availability should thus become the basic task of public organizations. An issue often examined in this context is queues for admission to hospitals, diagnostic tests, planned surgical procedures, and issuance of opinions and permissions.

In improving the quality of public services, it is also worth using conventional logistics tools (histogram, Ishikawa diagram, Pareto diagram, correlation diagram, check sheet, control chart, flowchart) and unconventional logistics tools (affinity, relationship, systematic, matrix, and arrow diagrams; matrix data analysis; process decision program chart). In quality measurement and assessment, one can use quality function deployment (QFD) and the SERVQUAL method. Using the system and process approach, a public organization should operate in accordance with the so-called Deming circle (PDCA), that is, plan, do, check, and act (Simsa and Patak, 2016). This is an activity model that, if followed continuously, systematically, and consistently, may increase an organization's effectiveness, including the quality of provided services.

29.3 PREMISES OF PUBLIC LOGISTICS IN THE SUPPLY CHAIN AND DETERMINANTS OF THE MULTILEVEL CHARACTER OF NETWORKS IN THE PUBLIC SECTOR

Public logistics refers to transformation in time and space connected with performance of public tasks and coordination of points of contact between involved institutions. As many organizations, public and private, participate in performance of public tasks, public logistics should be examined in the context of the supply chain. It means flows-oriented management of public chains and networks of value creation (Mesjasz-Lech, 2013). In this context, the public supply chain is a network of organizations and government and self-government agencies, as well as private institutions and enterprises, aimed at best performance of public tasks bringing added value to the public sector and beneficiaries.

Thus, the public supply chain involves any activity connected with transformation of goods in time and space, regardless of their form of ownership or character of the organization undertaking these actions. Within the supply chain, tools and machines necessary for performance of public tasks are provided, made available, and maintained (Friedländer and Röber, 2016). The public sector strives to provide services at the highest level at acceptable costs.

As the term *public* is mainly used to distinguish the character of tasks performed in supply chain, its characteristics can be regarded as analogical to the classical supply chain. These are mainly customer orientation and development of partnership-based relations with supply chain participants. Customer orientation means society orientation and conscious focus of the offer of services on beneficiaries. Partnership-based relations refer to forms of private sector's involvement in performance of public tasks. Moreover, supply chains are oriented toward optimization of activities of all links of the chain and efficient functioning of whole supply networks.

The public supply chain is a multilevel network (Schönsleben, 2016) comprised of a range of entities connected with and between each other (Figure 29.1). The multilevel character results from involvement of various decision-making levels.

The type of public tasks performed by different administrative levels is determined at the legislative level. The political sphere generates impulses for shaping public value creation chains. This value is the sum of unit benefits provided to beneficiaries by the public sector and expenditures necessary to acquire the service.

The type of public tasks postulated by the sphere of politics is usually determined by political parties' programs and coalition agreements. However, they should be in line with social interest, as political elections are subject to social assessment during parliamentary and local government elections. Political objectives determine development of action programs reflected in laws and ordinances. Implementing rules constitute guidelines for activities of public administration at different levels. Value creation at the legislative and administrative levels refers to shaping the foundation of performing tasks serving the public interest.

In the supply chain, these levels are links preceding creation of real value for society. Real value is created by public sector entities independently or by entities to which delivery of services was contracted out (Bogaschewsky et al., 2015).

The public supply chain is characterized by the network structure of relationships between its participants and decision-making levels. The relationships are between

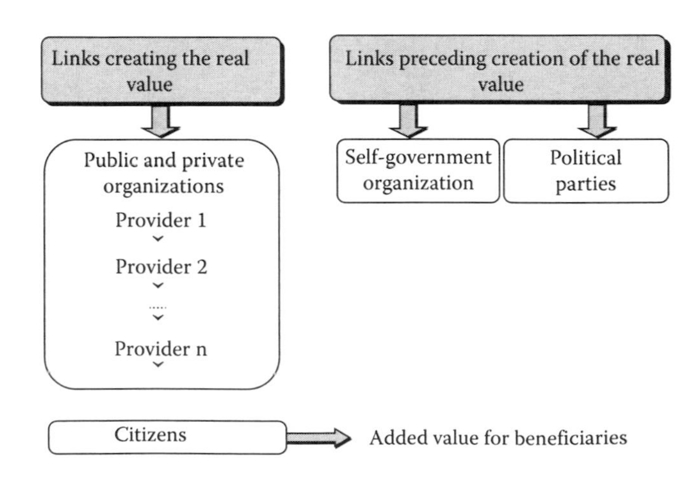

FIGURE 29.1 Public value creation chain as a multilevel network.

- Entities at the legislative level
- Administrative entities
- Entities performing public tasks
- Legislative level and public administration
- Entities providing public services and their beneficiaries

Due to the multilevel character of value creation networks in the public sector, adaptation of the classical concept is neither simple nor clear. Difficulties result, for example, from the fact that value is delivered to beneficiaries when the amount they are ready to pay for the service exceeds the cost of its generation. The public sector is financed from taxes, which disturbs exchange relationships and prevents direct dependencies between a public service and payment for it. Consequently, beneficiaries do not pay directly for a service.

29.4 USE OF THE SUPPLY CHAIN CONCEPT IN THE PUBLIC SECTOR: THEORETICAL CASE STUDY

Classical value creation chains are controlled by customers ready to pay for a product. The starting point for analysis of such chains are all processes necessary for movement and transformation of goods, from customers to suppliers of raw materials. Customers' inclination to make a purchase determines success or failure of different links of the chain. Exchange relations within a classical value chain are also characterized by freedom of purchases made by customers and their direct relations with suppliers. These relations can be simply illustrated using two links: supplier and customer.

In the classical value chain, the recipient is an independent customer who takes advantage of freedom of choice, making choices with the right not to participate (Stadtelmann, Lindner, and Woratschek, 2015). The customer is a person purchasing a product of his or her own free will and making payment for it. With public chains, identification of the customer is not so simple, as the recipient of public services does not always incur costs for using services (Figure 29.2).

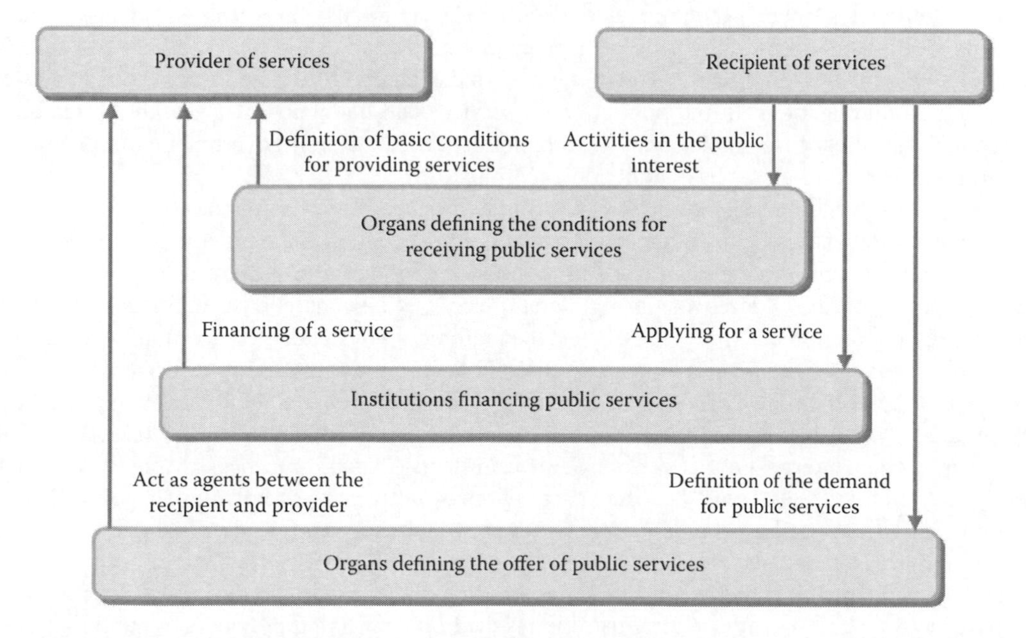

FIGURE 29.2 Exchange relations in the public sector.

Consequently, there are intermediaries, for example, financing institutions, between the recipient of public services and the entity offering them. Therefore, in order to illustrate exchange processes in the public sector, it is necessary to present the network of relationships. Exchange relations involve not only suppliers and recipients, but also institutions responsible for defining the type of services provided to citizens, developing action programs, and performing tasks. The network of relationships in public supply chain includes the supplier and recipient of services, institutions defining the types of public services, organs setting terms of providing public services, and financing institutions.

The recipient is a person that needs and uses public services, whereas the supplier is an institution that generates services and delivers them to recipients. The concept of the recipient and customer of public services is ambiguous. It can be understood as a passive agent that needs assistance from the state, who is unable to solve their problems or meet their needs. It may also be a person that can make rational assessment of a public service, rational choice of a supplier, and efficient purchase of a service. The recipient may also be understood as the cocreator of models and values that should be part of a specific public service (Schauer, 2015). Public service suppliers are organized in hierarchic structures within which problems of a services' recipients are identified and appropriate ways of solving the problems are determined. Organs defining the offer of public services constitute the link connecting suppliers with their recipients; they help recipients in specification of demand for services. These are, in particular, persons able to objectively assess the social situation of a potential recipient of services (e.g., social welfare workers). In relations between the supplier and recipient, there are also organs defining the conditions for receiving public services that at the same time develop action programs. They are the links preceding creation of real value for society, that is, formulating political goals (e.g., crime reduction) at the legislative level. They also include the level of local-government administration, which is responsible for putting into effect decisions taken at the legislative level. Institutions financing the performance of public tasks, in turn, take over costs of services, which should be paid by recipients.

The assumption of costs by the public sector means that citizens "pay" for public services through taxes and charges. However, a taxpayer is not a recipient of specific services financed from individual taxes. For the political sphere and self-government administration, this means that exchange processes occur between a number of entities connected in one network. From the perspective of the classical supply chain, it is the self-government administration providing social and public services that can be regarded as the recipient and customer, not the citizen that is the actual recipient of services (Baars and Lasi, 2016). The citizen cannot be treated as the final recipient also due to the compulsory character of public services, for example, charge for waste disposal, residence, or car registration obligations. In the public sector, services can be monopolized, with the recipient deprived of the right of free choice and the right not to participate. Self-government administration offers then necessary services as a monopolist (Mroß, 2015).

However, from the perspective of self-government organs, the citizen und resident is the customer, who by indicating the need for specific public services initiates the process of their delivery. This requires citizens' readiness to pay (through taxes and charges). Failure to pay may reduce the number of services. Moreover, in democratic elections citizens choose political representatives, expressing their needs and wishes. Thus, they indirectly impact the formulation of objectives. Because citizens, as customers, often cannot directly decide about the use of a service, the Polish legal system gives them indirect ways of controlling the value creation chain, for example, lobbying, protest, civil disobedience, public education, and advocacy (Bednarek and Chruściel, 2015). Moreover, citizen interest groups can participate in decision-making processes. Thus, although public value creation chains lack the direct act of purchase, citizens indirectly impact political decisions and activities of self-government administration, which thoroughly analyzes suggestions from citizens and tries to adapt the offer of public tasks to changing social needs. Therefore, citizens and residents may be treated as recipients within the meaning of the classical value creation chain. From this perspective, self-government administration is the link connecting providers of public services and final recipients.

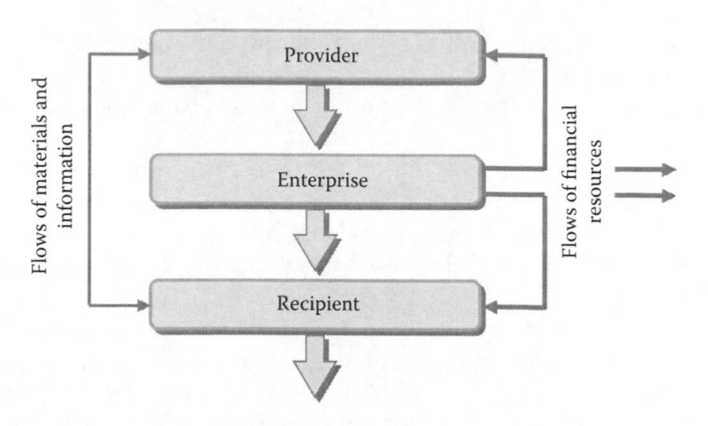

FIGURE 29.3 Flows occurring within a supply chain.

Summing up, although recipients and citizens only indirectly control value creation, in the conditions of democracy, they cocreate values that public services should include. Thus, it seems reasonable to also use the concept of the supply chain in the public sector.

The classical value creation chain is strongly related to logistics. This makes supply chain management concentrate on optimization of flows of goods and services as well as accompanying information. In this context, flows orientation is a link connecting all entities involved in value creation (Figure 29.3).

From the perspective of public supply chains, the flows presented in Figure 29.3 appear insufficient. This results mainly from the multilevel character of the chain and indirect relations between suppliers and customers. Classical supply chains are directly controlled by final recipients who by purchase or resign from its impact on the economic situation of entities involved in value delivery. In public supply chains, this mechanism is disturbed. In shaping the offer of public services, the public sector acts as an entity financing services. Recipients and citizens pay for a public service through charges and taxes, but without a specific, direct mutual consideration. Indirect relations between the provider of services and their recipients lead to separation of services and payment, which affects flow of financial resources. For instance, social support to single mothers in the public supply chain is financed through self-government organs. They define terms for applying for and allotting social benefits. It is similar with benefits in kind, for example, issuance of an ID. The recipient does not make payment for the ID directly in the Polish Security Printing Works, which produces such documents, but at the relevant local authority.

It is different with the flow of public services. Their provider is obliged to prepare a service and make it available to appropriate recipients and residents, for example, addicted people. As in the classical supply chain, a service flows between the supplier and final recipient. When the public sector uses third parties to perform public tasks, exchange processes (flows) are performed in the public supply chain between the links delivering the service. It is different when public bodies are involved in delivery of a service, for example, in the case of IDs, where self-government bodies act as agents between the recipient and supplier (i.e., the Polish Security Printing Works). The recipient does not receive a document directly from the producer but via the relevant office. Thus, indirect relations between the supplier and recipient impact not only the flow of financial resources, but also the flow of services.

In the public supply chain, there is no mechanism of direct purchasing decisions, and consequently no direct flow of information between the supplier and recipient. Public services are provided not on the initiative of the final recipient but political circles. The information exchange between residents and the political sphere is indirect and reflected, for example, in interest groups. For instance, the sphere of politics postulates increase in the number of places in nursery schools. Self-government administration is responsible for implementation of political decisions. Thus, information flow takes place between the political sphere and implementing bodies. These, in turn, inform entities providing services. As the recipient applies for a public service in a public organization financing such a service, the information flow is between the recipient and the institution financing the service.

The variety of entities involved in provision of public services also complicates coordination of flows. Coordination is required both in flows between self-government administration and services providers, and entities involved in the process of the delivery of services. In public supply chains, flows are multilevel in character and take place through agents, which makes their coordination much more difficult.

The supply chain in public sector is a network of organizations, government and self-government agencies, private institutions, and enterprises designed to perform public tasks, bringing added value to public sector and beneficiaries. It involves any activity connected with transformation of goods in time and space, regardless of their form of ownership or character of the organization undertaking these actions (Göpfert, 2016).

The public supply chain is a system, that is, a set of organized elements identified from the environment that cooperate with each other to achieve common objectives (Razak et al., 2016). The systemic approach indicates the process approach to activities undertaken by public organs, that is, transformation of elements of resources into final effects, that is, public services. Effectiveness of the process of their delivery determines achievement of public organizations' objectives. A process is thus the basic and most important manifestation of the operation of organizations involved in providing public services. It can be defined as a chain of actions aimed at creating value for customers (Haynes, 2015).

In the public sector, political goals serving the social interest, expressed in laws and ordinances, are the input. At the output of a process we receive a result in the form of a public service, for example, an educational service (Figure 29.4).

The use of the concept of the supply chain in public sector should lead to increased effectiveness of local self-governments and improved quality of offered services. This can be achieved by closer integration (cooperation) of the involved links or a more efficient flow of materials, information, and finances (Walker, 2016). Thus, management of the public supply chain can be defined as the

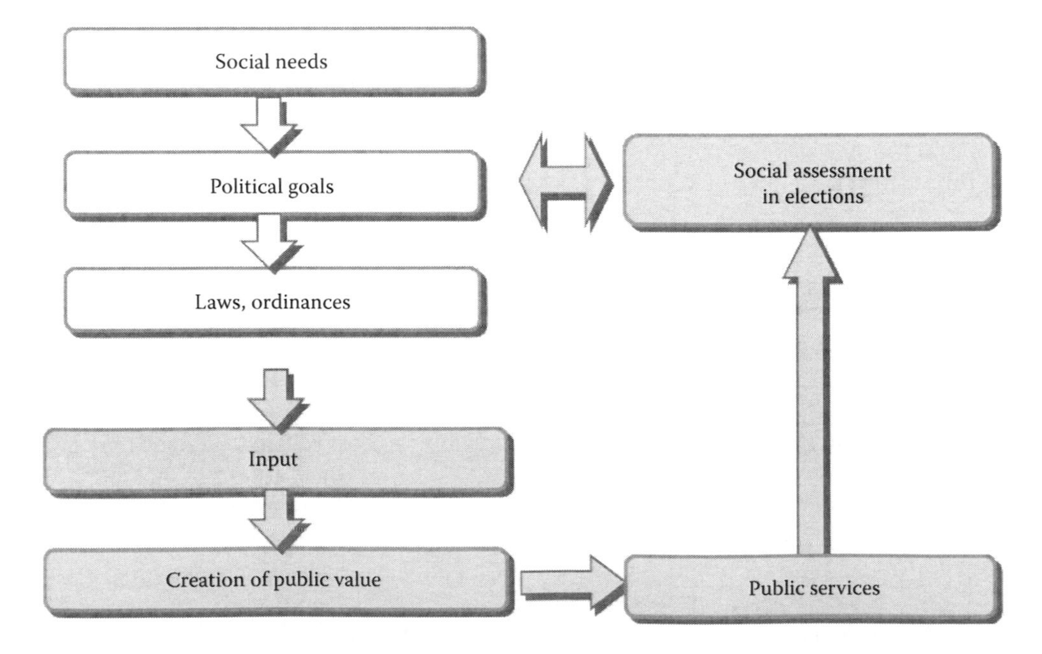

FIGURE 29.4 Environment of the process of providing public services.

integration of government, self-government, and economic entities along the whole supply chain and coordination of all flows to deliver the public services expected by citizens. Thus, the aim of supply chain management is integration of key processes, from the final recipient (citizens) to self-government bodies and economic entities creating real value for stakeholders.

The basic characteristics of the public supply chain, important from the perspective of process orientation, include network character and a multifaceted system of objectives and structures. This is because value shaping in the public sector takes place at political and execution levels. At the first level, objectives, types of public tasks, and instruments for their performance are defined. This level constitutes an impulse for shaping the public value chain and is a link preceding creation of real value for recipients. Real value is delivered by public sector entities or those to which provision of a service was contracted out (e.g., private enterprises) (Deasy et al., 2014).

Moreover, a number of entities related with each other and at decision-making levels are involved in public value creation—political parties, administration organs responsible for implementation of objectives set at the political level, and all entities performing public tasks. Multidimensional relations between involved organizations indicate network structure of the public supply chain with vertical and horizontal relations. The multitude of relationships results in a high complexity of coordination and instability of partnership in terms of long-term cooperation. It is a consequence of the character of performed public tasks and restrictions resulting from public procurement law.

The actual creation of public value occurs at the execution level; its character and stages of the implementation process do not differ from those in economic practice. However, political and legal conditions in the public sector significantly determine the choice of supply chain partners, the degree of freedom of activity, and character of performed processes. The distinction between public value creation levels leads to the conclusion that supply chain links at the different levels of value transformation have different competences in terms of control and delivery of public services. The parliament is the link financing public services, while self-government administration and economic entities are the provider, and citizens the recipient (Bojar, 2005).

It is thus reasonable to identify characteristics of processes taking place in public organizations (Berman et al., 2015):

- Input initiator—Usually the legislator guided by social interest
- Measurable objectives—Creation of value expected and verified by recipients (citizens)
- Sequential character of actions—Changing input measurable elements (e.g., documents, decisions) into output elements (public services)
- Repeatability—Enables development of models that are adequate in every process of delivery of public services

The process orientation in public supply chain management can be examined from wider and narrower perspectives. The former perspective includes processes performed at political and execution levels and beyond, referring thus to all participants in performance of public tasks. The narrower perspective concentrates on economic processes connected with the actual delivery of public services. In this context, process orientation includes supply chain links involved in the process of real value creation for recipients of services and administration organs that support the provision of public services. Such understanding of the process orientation in the public supply chain is analogous to the classical one, known from economic practice. This suggests that effective management of the supply chain in the public sphere requires integration of (business) processes with key participants of the chain.

Process orientation enables efficient planning and control of delivery of public services (Osborne et al., 2015). It has the potential of improvement and restructuring, in particular in the aspect of customer orientation. From the perspective of effects, processes are connected with the use of resources, impact on recipients, and provided services. Thus, positive effects occur only when the

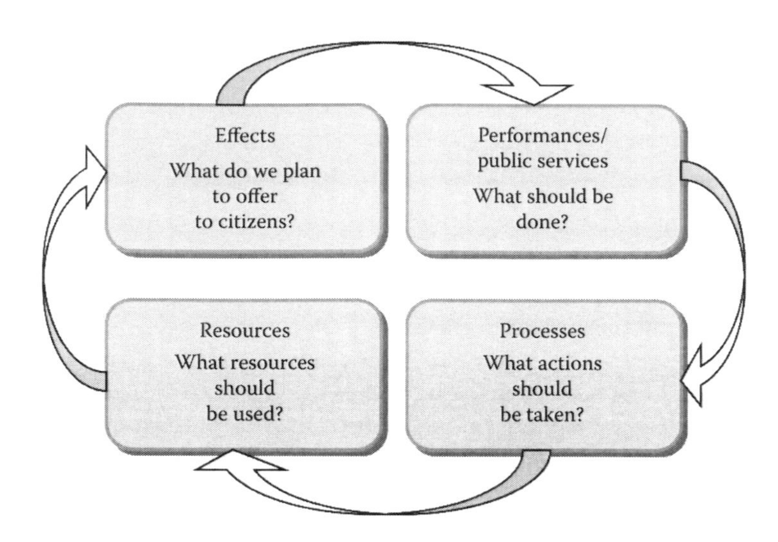

FIGURE 29.5 Effects-oriented control of the public supply chain.

offered public services meet social needs. This, in turn, requires appropriate planning of value creation processes and use of proper resources (Figure 29.5).

Performances delivered by the public sector usually include services, perceived as a real performance offered to a customer (citizens). The process of services delivery is characterized by intangible character and interactions between entities providing services and their recipients. Intangibility means that fulfillment of recipients' expectations is not limited to creation of material conditions for provision of services. Interactions highlight coparticipation of citizens in the process of delivery of performances. Participation of citizens can be active and passive, and the scope of their coparticipation in value creation defines the degree of integration in the supply chain. Customers (citizens) evaluate satisfaction with the process of public services provision based on relations between received benefits and incurred costs, which include price, waiting time, physical effort, and stress. It seems that the use of process orientation in management of public supply chain changed how the customer is perceived. From a passive applicant, they became a person with the right to their needs and expectations.

It seems that the process approach to management of supply chain in the public sector makes it easier to adapt services to the needs of the modern environment and provide citizens with desired public services. However, implementation of process orientation in supply chain management in the public sector is not simple. It requires taking into account a number of conditions of its functioning. The public sector is responsible for shaping policies, administration and delivery of goods, and meeting citizens' needs to the greatest extent. Processes performed in public supply chains, due to their multilevel structure and character of offered performances, are subject to different conditions compared to classical chains.

Some similarities are visible only in the narrower perspective of process orientation in supply chain management in public sector. That is, referring to only those supply chain links that are involved in real value creation for recipients. From this perspective, implementation of process orientation in public supply chain management requires (McCormack and Johnson, 2016)

1. Identification of external (political and legal) conditions determining performance of public tasks
2. Definition of a system of objectives important for performance of public tasks
3. Distinction of processes
    a. Strategic (megaprocesses)—Key to social and economic development

b. Basic—Performed to deliver a service directly connected with the main type of a public organization's activity
c. Supporting—For ensuring efficient performance of basic processes
4. Configuration of points of contact in public supply chain
5. Indicating persons responsible for performance of a process

Processes performed in the public sector differ in character. They may refer to abstract activities, for example, support for policy development or economic regulations, or very specific ones, connected with delivery of public services to citizens. Regardless of their character, a public organization should be able to identify basic processes aimed at bringing expected effects and results.

29.5 SUMMARY

The extent of implementation of process orientation in supply chain management in the public sector is still small, but application of the process idea requires performance of a range of tasks to design (identify and model) processes, and implement, perform, control, and improve them. The starting point for process management in this sense is to identify and understand processes, but also analyze points of contact between the links of the public supply chain involved in the process of value creation for citizens. Identification of processes requires definition of criteria for their distinct, general structure of processes important from the perspective of the implementation of public organs' objectives and social requirements (Romanowska, 2011).

It is particularly important at this stage to set objectives of different subprocesses composing the architecture of processes; their basis is a system of objectives formulated by the political sphere. The aspect of modeling processes is about creating clarity in terms of the number and type of points of contact, which makes it possible to thoroughly analyze and identify disturbances and undertake prevention activities. In the context of defining basic and supporting processes, it is important to define responsibility for this process. In the case of public organizations, it is reasonable to entrust responsibility for the process to the party ordering delivery of public services, that is, self-government organs. They would fulfill the role of the superior organ for planning and controlling the public supply chain without the obligation to actively participate in the process of public value creation.

The emphasis of modeling processes is configuration of the public supply chain, that is, defining the architecture of different processes—structuring subprocesses, specifying their parameters, and indicating expected effects.

From the perspective of public logistics and the supply chain, the most important aim is to guarantee supply continuity. In terms of process modeling, it is necessary to plan instruments for securing uninterrupted implementation of processes along the whole supply chain (Dorobek, 2013). It is necessary, for example, to plan parallel sequential changes for crucial points of processes. However, modeling processes in the public sector encounters organizational and legal limitations at various points, which significantly reduces the freedom of shaping processes.

The key aspect of implementation of process orientation is implementation of supraorganizational process thinking. This requires public supply chain partners to closely cooperate and consciously form internal relations between recipients and suppliers. This phase starts permanent monitoring of the course of processes. In control of processes, it is important to take into account the relationships between the objectives of the different processes and the system of priority objectives. It is not sufficient to rely only on evaluation of the extent of achievement of the objectives of processes; evaluation should be made in the context of general objectives connected with delivery of public services. The same applies to improving solutions and adapting them to changes in the environment.

The public supply chain is a complex network of relationships between organizations designed to deliver public services serving social interest. The basic challenge facing management of networks is to shape value creation processes in such a way that will guarantee uninterrupted achievement of public organizations' objectives. This can be achieved by implementing the process orientation that takes into account the specificity and conditions of public logistics. In implementation of the process orientation in public supply chain management, the basic role seems to be played by political and legal conditions determining the system of objectives and creating the framework for performance of public tasks. It is important to identify and specify basic and supporting processes, and coordinate points of contact in the public value creation chain. It is also important to appoint people responsible for implementation of the process.

In summary, implementation of process orientation in supply chain management in the public sector is still not obvious. However, due to its potential for modernization and improvement of the effects of public organizations' functioning, it will be increasingly widespread. This is mainly due to the constantly growing number of public tasks performed by private entities. Therefore, it seems reasonable to develop a concept of process management that could be applied in public logistics.

The indicated attributes of the public supply chain show that the classical concept of supply chain management can be applied in the public sector and represent an instrument for analysis of points of contact between different decision-making levels and entities involved in the process of performing a public task. Adaptation of classical concepts to the needs of the public sector requires, however, significant extension and analysis of elements characteristic only of the sector analyzed. They include indirect relations between suppliers and recipients, and lack of indirect relationship between performance and payment.

The author is aware that the attributes presented do not exhaust the issues of public supply chain management, but hopes that the chapter will become inspiration for further discussion on this relatively new concept in the field of management.

REFERENCES

Baars H., Lasi H. 2016. Innovative business-intelligence-anwendungen in logistik und produktion. In: P. Gluchowski and P. Chamoni (eds.) *Analytische Informationssysteme*. Springer, Berlin, pp. 283–302.

Bednarek M., Chruściel T. J. 2015. Doskonalenie systemów zarządzania w sektorze usług komunalnych. *Ekonomika i Organizacja Przedsiębiorstwa*, 4, 22–32.

Berman E. M. et al. 2015. *Human Resource Management in Public Service: Paradoxes, Processes, and Problems*. Sage, Thousand Oaks, CA.

Blaik P. 2012. Tendencje Rozwojowe I Integracyjne Logistyki. *Gospodarka Materiałowa I Logistyka*, 6, 4–10.

Bogaschewsky R. et al. (ed.). 2015. *Supply Management Research: Aktuelle Forschungsergebnisse*. Springer-Verlag, Wiesbadan, pp. 3–32.

Bojar E. (ed.). 2005. *Ekonomia w dobie globalizacji*. Wydawnictwo Politechniki Lubelskiej, Lublin, pp. 87–112.

Deasy M. et al. 2014. Asymmetric procurement in the public sector. *Strategic Change*, 23(1–2), 21–29.

Deckert C. 2016. Nachhaltige logistik. In: C. Deckert (ed.) *CSR und Logistik*. Spannungsfelder Green Logistics und City-Logistik. Springer, Berlin, pp. 3–41.

Dorobek S. 2013. *Public Supply Chain Management: Steuerung öffentlicher Wertschöpfungsketten nach privatwirtschaftlichem Vorbild*. Springer-Verlag, Wiesbaden, pp. 71–73.

Erd J. (ed.). 2015. Betrachtung ausgewählter Projekte zur City-Logistik. In: *Stand und Entwicklung von Konzepten zur City-Logistik*. Springer Fachmedien, Wiesbaden, pp. 80–84.

Friedländer B., Röber M. 2016. Rekommunalisierung öffentlicher Dienstleistungen. *VM Verwaltung & Management*, 22(2), 59–67.

Gołembska E. 2012. Współczesne trendy w kształtowaniu wizji logistyki przyszłości. *Gospodarka materiałowa i logistyka*, 6, 11–15.

Göpfert I. (ed.). 2016. *Logistik der Zukunft-Logistics for the future*. Springer-Verlag, Wiesbaden, pp. 39–97.

Haynes P. 2015. *Managing Complexity in the Public Services*. Routledge, Abingdon, England.

Jelonek D. 2012. *Wybrane problemy zarządzania wiedzą i kapitałem intelektualnym w organizacji*. Wydawnictwo Politechniki Częstochowskiej, Częstochowa.

McCormack K. P., Johnson W. C. 2016. *Supply Chain Networks and Business Process Orientation: Advanced Strategies and Best Practices*. CRC Press, Boca Raton, FL.

Mesjasz-Lech A. 2013. Wykorzystanie technologii informacyjnych w zarządzaniu łańcuchem dostaw. Zeszyty Naukowe Uniwersytetu Szczecińskiego. Ekonomiczne Problemy Usług, 105 Europejska Przestrzeń Komunikacji Elektronicznej. T. 2, pp. 543–552.

Moß M. 2015. *Betriebswirtschaft im öffentlichen Sektor: Eine Einführung*. Springer-Verlag, Wiesbaden, pp. 44–68.

Osborne S. P. et al. 2015. The SERVICE framework: A public-service-dominant approach to sustainable public services. *British Journal of Management*, 26(3), 424–438.

Razak A. A. et al. 2016. Public sector supply chain management: A triple helix approach to aligning innovative environmental initiatives. *Foresight and STI Guidance*, 10(1), 43–52.

Romanowska M. 2011. Przełomy w praktyce zarządzania-przesłanki i przyczyny. *Przegląd Organizacji*, 3, 16–20.

Schauer R. 2015. *Öffentliche Betriebswirtschaftslehre-Public Management: Grundzüge betriebswirtschaftlichen Denkens und Handelns in öffentlichen Einrichtungen*. Skriptum. Linde Verlag GmbH, Wien.

Schönsleben P. 2016. Supply chain design: Standortplanung und Nachhaltigkeit. In: P. Schönsleben (ed.) *Integrales Logistikmanagement*. Springer, Berlin, pp. 116–167.

Simsa R., Patak M. 2016. *Leadership in Non-Profit-Organisationen: Die Kunst der Führung ohne Profitdenken*. Linde Verlag GmbH, Wien.

Stadtelmann M., Lindner A., Woratschek H. 2015. Wertschöpfungskonfigurationen. In: H. Woratschek, J. Schröder, T. Eymann, M. Buck (eds.) *Wertschöpfungsorientiertes Benchmarking*. Springer, Berlin, pp. 25–32.

Walker W. T. 2016. *Supply Chain Architecture: A Blueprint for Networking the Flow of Material, Information, and Cash*. CRC Press, Boca Raton, FL.

Witkowski J. et al. 2015. Logistyka w warunkach kryzysu ekonomicznego i w innych sytuacjach kryzysowych. *Prace Naukowe Uniwersytetu Ekonomicznego we Wrocławiu*, 382, 154–165.

30 Current State of Information and Communication Provision of Ukraine and Spread of Information Technologies in Agricultural Sector

Antonina Kalinichenko and Oleksandr Chekhlatyi

CONTENTS

30.1 Formulation of the Problem 403
30.2 Analysis of Recent Research and Publications 403
30.3 Main Material of the Research 404
30.4 Conclusions and Suggestions 410
References 410

30.1 FORMULATION OF THE PROBLEM

Under market conditions, the issue of the development of communication technologies is highly important for the agricultural sector of Ukraine, as possessing information and its use in the production process is directly related to ensuring the food security of the state.

Rural area development is largely determined by applying more advanced forms of management that provide efficient use of economic mechanisms under specific production conditions. For this purpose, diverse informational means of processing and analysis must be considered. Currently, information transmission, and software and hardware processing are equally important resources as material and energy. Information transmission in the system of management of enterprises in the agricultural sector requires effective organization (Khudyakov, 2016).

Therefore, improving the quality of information and communication software can play a crucial role in improving the efficiency of agrarian enterprises of Ukraine. It will enable a more clear focus in the legislative area, forecast rates of production and marketing, the regional variation of prices for the products and the resources in order to define a strategy for economic development, implement and use new technologies, storage, and sales, and to build financial relationships in a tactically proper way.

30.2 ANALYSIS OF RECENT RESEARCH AND PUBLICATIONS

Under current conditions, one of the crucial factors in the effective development of the agricultural sector is the design of an effective system of information and communication security. The systematic study on the formation of the information and communication provision of the agricultural sector as well as the analysis of information needs of an entity has not been covered. The importance of these issues, and the need for their thorough theoretical study and practical specification predetermined the topicality of the subject of our study, its goals, and objectives.

The main purpose of the chapter is to study the current operating conditions of the information and communication provision in Ukraine, underscore the main results of the information and consultancy services in the Poltava region, and their international cooperation and contribution to the spread of new information technologies in the agricultural sector.

30.3 MAIN MATERIAL OF THE RESEARCH

The implementation of information and communication technologies (ICTs) and their widespread use in various areas of human life, society, and state is one of the most important tools that promotes the increase of the level of economic, social, cultural, and technological development (Larin and Rudenko, 2013).

However, it may be stated by now that the level of technological development defines not only the economic potential of the country and the quality of life of its citizens, but also the role and the place of Ukraine in the global community, and the scales and the prospects of its economic and political integration with the whole world.

The ICT area has a significant place in the economy of Ukraine. Its component is the sector of information and telecommunications, in which at the beginning of 2015, there were over 114,300 business entities of different types of ownership, of which 13,300 were enterprises and 101,000 were individual entrepreneurs. There were 306,300 ICT workers, which is nearly 3.5% of the total number of workers involved in the business entities of Ukraine, from them 192,700 people worked at enterprises. The gross domestic product, produced by ICT business sector in 2014, was equal to 3% (47.4 billion of UAH) of the GDP of Ukraine, and during the first 9 months of 2015 3.1% (42.8 billion of UAH) (Resolution VRU, 2016).

The enterprises, which worked by the type of economic activity "information and telecommunications," in 2015 came out on top for annual increase in direct foreign investments. The amount of attracted direct foreign investments as of the end of 2015 was US$2.3087 million, about 5.3% of the total amount of foreign investments. The amount of capital investments, which were done during 2014, was 8.2 billion of UAH (during the first 9 months of 2015 it was 17.6 billion of UAH).

During 2014, ICT goods (works and services) earned 105.7 billion of UAH, and from them by the enterprises that provide services, 74.3 billion of UAH (during the first 9 months of 2015 it was 62.0 billion of UAH).

In 2015 the export of services in the area of telecommunications, computer, and information services was equal to $1.5 billion; the import was almost three times less at $537 million. In comparison with 2014, the export of the specified services decreased by 9.5% ($98.5 million), and the import increased by 4.9% ($39.3 million) (Resolution VRU, 2016).

One of the main indicators of development in ICT are the indicators that characterize the level of penetration of landline telephone communications, mobile communications, cable television, access to broadband Internet, and the share of population that uses the Internet:

- The level of penetration of mobile communications at the beginning of 2015 (the quantity of subscribers per 100 people) was equal to 142.4% (as of October 1, 2015, 142.9%).
- The level of penetration of landline telephone communications at the beginning of 2015 (the quantity of subscribers per 100 people) was 24.4% (as of October 1, 2015, 21.1%).
- An estimated 18.2% of Ukrainian households had cable television at the beginning of 2015 (as of October 1, 2015, 17.1%).
- At the beginning of 2015, 22.2% of Ukrainian households had access to fixed broadband Internet (as of October 1, 2015, 29.8%).
- According to the independent expert organizations, at the beginning of 2015, 57% of the total population of Ukraine regularly used the Internet, and at the end of the third quarter of 2015, more than 58% (Resolution VRU, 2016).

More interaction between the state, business, and citizen is happening with the use of ICT, and for this very reason public sentiment in the country and the creation of prerequisites for the stable growth of the economy depend on the stability of functioning and development of ICT.

The value of ICT has been witnessed in the development of many other countries, in particular, the strategy of socioeconomic development of the European Union for the period until 2020, "Europe 2020." According to it, the economic growth is based on the accomplishment of the initiative "The plan for development of digital technologies in Europe" (Resolution VRU, 2016).

Taking into consideration global trends of the last 10 years, Ukraine embarked on the development of an information society. This was confirmed in a number of conceptual and strategic documents, primarily in the Law of Ukraine "On the Fundamentals of Information Society in Ukraine for 2007–2015" (Larin and Rudenko, 2013). At the same time, a set of unresolved legal, organizational, technical, scientific and methodological, analytical, resource support issues for the information society development remains. A large number of government decisions on these issues has nonsystematic and declarative character, is not financially supported, and is largely "borrowed" from other countries without considering the peculiarities of the current state and trends of Ukraine. The official confirmation of this point of view is the annual report on the state of information and communication provision in Ukraine, which is developed and submitted for the Parliament by the government along with the draft of the state budget for the next year according to the National Informatization Program.

Due to the economic crisis and the war in Ukraine, in the last year there was a decrease in the rate of computerization of enterprises in various sectors of the economy. Financing of information projects from the state budget has decreased almost 2 times and 10 times in the National Informatization Program (Larin and Rudenko 2013).

Unfortunately, as of today in Ukraine there remains the substantial problems on formation and implementation of effective state policy, particularly, in the area of development of an information society. There has been no further action on the implementation of the Law of Ukraine "On the Main Fundamentals of Information Society Development in Ukraine for 2007–2015." However, the level of computer literacy is increasing, albeit slowly. The level of implementation and use of ICT is relatively low in the areas of education, science, culture, and health care. In agro-industrial complex and other sectors of the economy, there are no corresponding sectoral and cross-sectoral programs of ICT implementation (Resolution VRU, 2016).

In the area of e-governance development there is no action plan on implementation of the concept of e-governance development in Ukraine, which was approved by the Ordinance of the Cabinet of Ministers of Ukraine on September 26, 2011. There are no programs of e-governance development in the system of executive bodies and bodies of local self-government, in particular, the programs on provision of electronic administrative services.

It should be acknowledged that there are problems with the creation of favorable conditions for conducting of ICT business. There exists the cumbersome procedures of entrance to the market of telecommunications, in particular, the business entities, in addition to inclusion in the Register of Operators, Providers of Telecommunications, depending on the type of activity, as a rule, are obliged to obtain licenses on the type of activity in the area of telecommunications, on the use of radio-frequency resources of Ukraine, and permission for the use of resources. Relations on provision of economic competition at some markets of telecommunication services remain insufficiently adjusted (Resolution of Verkhovna Rada of Ukraine [VRU], 2016).

There exists the need of restructuring the modern regulatory and legal framework of ICT sector reforms and development of information space, many provisions of which are not meeting the modern requirements and require improvement.

As of today, harmonization of the national legislation with the statutory regulations of the European Union according to the "Association Agreement between Ukraine and European Union," have not been completed, particularly concerning the issues of information society, electronic services, development of telecommunications, and intellectual property protection.

Despite the potential for implementation of modern ICT into all spheres of activity of the country, the notable public request for such implementation, the state of development of an information society and ICT in Ukraine, in comparison with the rest of the world, is insufficient and does not correspond to the strategic objectives of development of Ukraine.

The international ratings quite clearly reveal the state of ICT development in Ukraine:

- WEF Networked Readiness Index 2015—71st place out of 143 countries (in 2014, it was 81 out of 148 countries)
- WEF Global Competitiveness Index 2015—79th place out of 140 countries (in 2014, it was 76 out of 144 countries)
- WEF Technological Readiness Index 2015—86th place out of 140 countries (in 2014, it was 94 out of 148 countries)

According to "The Global Report on Development of Information Technologies, 2015" (The Global Information Technology Report), which, beginning from 2002, is annually published by the World Economic Forum, Ukraine took the 71st position among 143 countries of the world in the rating of the level of ICT development. As the basis of rating assessment is the Networked Readiness Index, which determines the level of ICT development in the countries of the world. It consists of the four subindexes, distributed by the components (indicators), which characterize the role of the government, business, and the society in formation of an environment for ICT development. Ukraine had shown the highest rating position by the Networked Readiness Index in 2009 (62nd place). However, during the next 2 years Ukraine lost 28 points, resulting in the fact that Ukraine moved to the 90th position among 138 countries of the world for 2011. In 2015, with taking into account the expansion of the number of participating countries of the rating, Ukraine is now number 71 and giving ground to the Commonwealth of Independent States and Eastern Europe (World Economic Forum [WEF], 2016).

In correspondence to the UN Global E-Government Development Index, in 2014, Ukraine was ranked the 87th place in the world out of 193 UN member states (in 2012, Ukraine was 68 out of 190 countries). Despite the loss of positions in the rankings, in particular in the index of online services, Ukraine joined the group of countries with a high index of e-government in 2014, which is considered to be a positive trend for the country (Information Society, 2016).

According to the report of the International Telecommunication Union "Measuring the Information Society 2015," which includes the rankings of 167 countries in the index of ICT development, Ukraine was rated 79 (according to the ITU Report for 2014, it was 73 out of 166 countries). One of the reasons for the low rating of Ukraine in that ranking is the uneven access to ICT in some regions, confirmed by the results of the analysis on the development of information and communication infrastructure and ICT in different areas of the country.

According to the data of the World Wide Web Foundation on the Internet development ranking in 2014, Ukraine was given the 46th place out of 86 countries in the web index.

According to the annual report "The State of Broadband 2015," prepared by a joint initiative of the International Telecommunication Union and UNESCO, the level of Internet penetration in Ukraine was ranked 95th out of 191 countries (in 2013, it was 94 out of 191 countries) (Information Society, 2016).

Unfortunately, nowadays in Ukraine, both a strategy and effective mechanisms for the information society development are absent (Larin and Rudenko 2013). However, recently there appeared to be hope on improvement of the ICT situation due to scientific and technical and legal reforming.

Thus, by the Resolution of Verkhovna Rada on March 31, 2016, there were accepted Recommendations of Parliament hearings on the subject: "Reforms in the area of information and communication technologies and the development of information space of Ukraine." The People's Deputies of Ukraine, representatives of the Cabinet of Ministers of Ukraine, specialized government authorities, public organizations, and the heads of industry organizations took part in the hearings. The total number of participants was more than 500 people. Also involved in the hearings were international experts on the issues of information and communication technologies and representatives of the European Union (Resolution of VRU, 2016).

It was recommended to the Verkhovna Rada of Ukraine to provide top-priority consideration and adoption of the legislative initiatives intended to create an integral legal system on the development of information and communication technologies, in particular, the bills on electronic communications and amendments to some legislative acts of Ukraine. It was suggested to create more favorable terms of taxation for the business entities of the ICT area and to encourage the development of small businesses by implementing of economic inducements for ICT businesses.

It was also recommended to the Cabinet of Ministers of Ukraine to create a central executive body to provide formation and implementation of national policy in the areas of ICT and communication, development of an information society, informatization, telecommunications, programming, information security and cybersecurity, adoption of technologies of e-governance, electronic document flow, electronic signature, and so on. It was suggested to delegate to the specified authority responsibility of other executive authorities, concerning ICT and communication, to clearly distinguish responsibility between the executive authorities in the specified areas according to the European Union law.

It was also recommended to the Cabinet of Ministers to develop and submit for approval by the Verkhovna Rada of Ukraine the bills on the government program of development of information and communication technologies, on electronic identification, concerning the electronic administrative services, provision of medical services with the use of information and computer technologies, distance education, free and joint use of public information resources and data, and use of radio-frequency resources, which must be developed on the basis of implementation of the European Union law and international law with the involvement of the leading scientists and specialists (Resolution of VRU, 2016).

In addition, it was recommended to the Cabinet of Ministers to develop and submit for approval by the Verkhovna Rada the number of corresponding legislative acts on development of information and communication technologies and information space of Ukraine to develop an action plan on implementation of a national strategy to develop an information society in Ukraine, engaging the representatives of scientific institutions and public organizations, and the leading experts and businessmen (Resolution of VRU, 2016).

The execution of tasks assigned to the Cabinet of Ministers and the Verkhovna Rada of Ukraine concerning the reforming and streamlining of ICT and development of information space will provide the opportunity to accelerate development of the national information infrastructure, e-governance, and electronic administrative services; to provide information security; and to increase the level of use of ICT capabilities in the areas of education, science, health care, the agro-industrial complex, and other sectors of economy.

We will next highlight important problems in the development and formation of ICT in Ukraine, such as the software piracy. The volume of unlicensed software that is illegally used in Ukraine is about 82%. Though it is 1% less than what it was one year ago, Ukrainians continue to use a significant amount of unlicensed software, which carries threats and increases the probability of cyberattacks. Companies can reduce the risks of cyber attacks by buying software from the legal suppliers and implementing software asset management (SAM) programs. Organizations that are actively implementing SAM have complete understanding of what software is installed in their network, and whether it is legal and licensed; they optimize the use of software according to their needs; they have policies and procedures to regulate the acquisition, deployment, and writing-off of software; and they completely integrate SAM into their business (O.UA, 2016).

Also, international organizations expressed displeasure with three key directions of the Ukrainian government: the high level of Internet piracy, the use of pirated software by government authorities, and the problems with administration of royalties. In the last year Ukraine appeared in the group of 11 countries, on the so-called Priority Watch List. Ukraine ranked first among the countries that violate the intellectual property right most often. In the annual report of the International Intellectual Property Alliance (IIPA) it is recommended to the countries of this group to take the active measures for the fight against "piracy." Other countries in this group were China, Argentina, Chile, and Russia (IIPA, 2016; O.UA, 2016).

According to the IIPA, in Ukraine there continues to be websites on which "pirated" content is posted, including the large torrent trackers, and during the last few years "Internet pirates" from the other countries have intentionally transferred the servers to Ukraine.

The status of "pirate № 1" threatens Ukraine with exclusion from the Generalized System of Preferences (GSP), the American government program that supports economic growth of developing countries, within the framework of which goods from 140 countries are imported duty-free into the United States. Similar sanctions for the high level of piracy were applied against Ukraine in 2001, when Kiev was officially excluded from the GSP for 5 years (IIPA 2016).

One-time U.S. Ambassador to Ukraine Geoffrey Pyatt appealed to Ukraine to root out Internet piracy. Piracy also complicated the work of the leading companies in Ukraine, which could not be confident in the safety of their author's rights. According to Pyatt, the export market of software in Ukraine is equal to $1.5 billion, while the domestic software market is worth $3.5 billion. Mistakes made by Ukrainian authorities in relationships with the companies, regarding accumulating royalties, have alienated Ukraine (O.UA, 2016).

According to an annual report that is devoted to protection of intellectual property rights in the world, the U.S. government has agreed that Ukraine has taken some steps in the fight against piracy, particularly with the creation of a special division in the Ministry of Internal Affairs, and the work of the Ukrainian government in this direction has become more transparent (IIPA, 2016). Also on May 10, 2016, the government of Ukraine submitted to the Verkhovna Rada a draft bill to introduce legal amendments concerning the protection of author's right and the neighboring rights on the Internet. Such an initiative by the Cabinet of Ministers is aimed at improving the position of Ukraine among intellectual property rights violators (O.UA, 2016).

As for the process of informatization in the economy of Ukraine, it is the worst in the agricultural sector. This is explained by the peculiarities of the agro-industrial complex. Agriculture is an ideal environment for the use of modern information technology. However, the lack of funds in the area of agricultural science and production does not assure widespread use of information technology (Tzyferova, 2012). There is a lack of facilities in the majority of modern computer technologies, unpreparedness or lack of qualified experts in information technology, and a lack of appropriate information and software to automate the management of enterprises in the agricultural sector (Tzyferova, 2012). Among these reasons, the latter is the most important.

Despite a very large number of problems in the implementation and the provision of the agricultural sector with the latest information and communication technologies, constant work is being performed to improve the current situation. First, the training of information specialists for the agricultural sector is being done by educational institutions. It is noteworthy to draw attention to the National University of Life and Environmental Sciences of Ukraine (NULES), the largest agricultural university in Ukraine with more than 26,000 students. Besides the Information Technology Department, the university includes the Ukrainian Education and Research Institute (ERI) that supports information and telecommunication in the agricultural and environmental sectors of the economy (Ukrainian ESI, 2016). The Research Institute for Information Technologies in Environmental Management, which is a part of the ERI, combines the scientific activity of research and innovative laboratories. During the first years of its functioning, the experts completed a number of IT application layouts (Ukrainian ESI, 2016).

A contribution of the Institute in the improvement of the access to agricultural information became the creation of the Internet portal "The Agricultural Sector of Ukraine" AgroUA.net (http://agroua.net/). The purpose of its creation was the development of a universal and comprehensive information resource to meet the needs of agricultural producers, commercial organizations, advisory services, researchers, teachers, students, and other users (Ukrainian ESI, 2016). Recently, in the collaboration with leading advisory services and their leaders, the staff of the institute is successfully creating a system of electronic counseling, eXtension.

All the aforementioned issues will make it possible to improve the state of information, consultation, and communication in the agricultural sector of Ukraine. It is worth noting that over the last decade the advisory services made a significant contribution to the development and the establishment of an information system of the agricultural sector of Ukraine.

In Ukraine, the process of creating advisory services became most prevalent after the adoption of the Law of Ukraine "On Agricultural Advisory Activities" of June 17, 2004. According to the latest register of the Ministry of Agrarian Policy of Ukraine, in the beginning of 2016, 71 agricultural advisory service had been founded in Ukraine (in 2008 there were only 26) (Chekhlatyi, 2008).

In the Poltava region, the first advisory service, called the Poltava Regional Agricultural Advisory Service (PRAAS), was established in 2004. This project was a result of the intergovernmental agreement of June 15, 2004, signed between the Federal Ministry of Food, Agriculture and Consumer Protection and the Ministry of Agrarian Policy of Ukraine. The founding of PRAAS was the result of the efforts of teachers at the Poltava State Agrarian Academy (PSAA) (Chekhlatyi, 2008).

Due to the increase in the demand for qualified advisory services, the need to solve the lack of effective mechanisms for agricultural science cooperation, education, and agriculture was faced in the foundation of advisory services that would be an alternative to the existing ones in Poltava region. Applying the existing technical, scientific, and organizational potential of PSAA could provide effective social-focused advisory services to agricultural producers and the rural areas.

This service was established in 2007 on the basis of PSAA, and it was given the right to provide the socially oriented advisory services using state budget. This body was called Poltava Regional Public Organization Official Agricultural Advisory Service (TRPO OAAS) (Kalinichenko, Chekhlatyi, and Gorb, 2011).

In order to disseminate the information on the activities of a new advisory service, an Internet resource was established. TRPO OAAS has gained a considerable experience in the international cooperation and participation in various projects, which makes it possible not only to adopt best practices, but also the necessary funds for a variety of educational events. One of the results of such projects was the creation of a training manual for the distance learning in advisory services (Kalinichenko, Chekhlatyi, and Kostohlod, 2009).

PSAA and TRPO OAAS successfully completed another joint ecological project "Tempus-Tacis 2006" (JER_27168_2006), which was to find the "agro-ecological center of the Poltava region." It had to address such major challenges in environmental areas as

- Improvement of environmental education
- Gaining experience by Ukrainian experts in the area of environmental problems
- Study of the possibility for the introduction of the international standards for environmental protection in Ukraine
- Dissemination of environmental information and results of studies (Kalinichenko and Chekhlatyi, 2009)

In order to disseminate information about the work of the project, the website "International Agro-Ecological Center" was designed. For agricultural producers, both in certain regions and Ukraine as a whole, the information posted on the website is very valuable.

The main task of advisory services in the agricultural sector of Ukraine is to disseminate information among agricultural producers. Their efficient functioning is possible primarily due to the use of modern information and communication technologies by agricultural manufacturers. The use of information technology in advisory services significantly reduces management costs, expands the access of agricultural producers and rural populations to information sources and communications, and facilitates profitable farming.

30.4 CONCLUSIONS AND SUGGESTIONS

Nowadays, due to the absolute cost, Ukraine occupies the last place in Europe in all indicators of information (e.g., density coverage area and capacity of telecommunication and computer networks, Internet users, the proportion of broadband Internet to the total population). In the area of the access to the Internet, Ukraine considerably lags behind developed countries.

Development of the information and communication infrastructure and the broadband access in Ukraine, first must provide the solution of accelerated development of telecommunication infrastructure in the rural areas and on depressed territories of Ukraine.

The experience of developed countries demonstrates that the use of the latest informational technologies and information support systems is a prerequisite for high-tech agricultural production and management. It is of common knowledge that Ukraine is one of the world's largest potential agricultural producers. Improving information and communication for the agricultural sector provides significant opportunity to increase the production of agricultural products and to become one of the largest food manufacturers. Of course, fulfilling these tasks should help agricultural producers and the state. Its funding should be one of the main priorities of the state agricultural and information policy. Considering the fact that advisory services make a significant contribution to the awareness of agricultural producers, expanding their competitiveness and solving the problem of employment of the rural population, there is a need to support them at the state level.

Currently, the most serious challenge to the existence of agricultural advisory services in Ukraine is primarily the practical lack of state funding. The only way to solve this problem is to attract local budget funds, international grants, and financial assistance from foreign investors. The situation may change for the better only if the financial crisis is overcome, and the economic situation in the country is improved. Only after improving the financial security of advisory services in Ukraine, will it be possible to state the prospects of providing them with innovative technologies. In turn, it will improve the efficiency of the work of agricultural advisory services, the quality of information and consultative support, and promote the organization of competitive production.

REFERENCES

Chekhlatyi O. M. 2008. The development and the establishment of agricultural advisory services in Poltava region. *Scientific Bulletin of S. Z. Gzhytsky National University of Veterinary Medicine and Biotechnology in Lviv*, 489–493.

Draft concept of the Design of an Advisory Services Electronic System (e-Extension) in NULES of Ukraine. http://edorada.org/ (accessed March 23, 2016).

IIPA. 2016. IIPA Highlights Challenges in Opening Foreign Markets for U.S. Creative Works. http://www.iipawebsite.com/ pressreleases/2016spec301pressrelease.pdf.

Information Society. 2016. Information Society Kyiv: National Commission for State Regulation of Communication and Informatization. http://www.nkrzi.gov.ua (accessed March 25, 2016).

Kalinichenko A., Chekhlatyi O. 2009. Development and international cooperation of the informational services and consultation to ensure the AIC in Poltava region. *Bulletin of KhAI Series, Economics of Agriculture and Natural Resources*, Kharkiv, pp. 60–68.

Kalinichenko A., Chekhlatyi O., Gorb O. 2011. The formation and the development of the advisory services system at Poltava State Agrarian Academy. In: *Collection of Scientific and Methodical Works Science and Technique* [ed. by Ischenko T. D.]. Kyiv: Agricultural Education, Vol. 23, pp. 52–58.

Kalinichenko A., Chekhlatyi O., Kostohlod K. 2009. Features of information support of agricultural enterprises in Poltava region. *Proceedings of TSAU*, Melitopol, pp. 354–361.

Khudyakov H. O. 2016. The role of information security in the management of agricultural enterprises. *International Internet-Conference Formation and Development of the Economy under Current Economic Conditions*. http://www.wp.viem.edu.ua/ (accessed February 25, 2016).

Larin N. B., Rudenko O. M. 2013. *Information and Communication Provision of Efficient Functioning of Power: Educational and Methodological Materials*. Kiev: NAPA, pp. 5–7.

O.UA. 2016. Users became sad: Ukraine declared the war to the Internet piracy. http://my.obozrevatel.com/politics/41396-yuzeri-zasumuvali-ukraina-ogolosila-vijnu-internet-piratstvu.htm (accessed June 1, 2016).

Resolution of VRU (Verkhovna Rada of Ukraine). 2016. About Recommendations of Parliament hearings on the subject: "Reforms in the area of information and communication technologies and the development of information space of Ukraine." http://zakon3.rada.gov.ua/laws/show/1073-19 (accessed May 30, 2016).

Tzyferova N. H. 2012. Information management mechanism of state regulation of a agriculture complex. *Scientific Notes National University Ostroh Academy*, Series: Culture and Social Communication, 3: 72–78.

Ukrainian ESI. 2016. Ukrainian ESI Information and Telecommunication Support of the Agricultural and the Environmental Sectors of Economy. http://nubip.edu.ua/node/9488 (accessed March 22, 2016).

World Economic Forum (WEF). 2016. Global Information Technology Report 2015. http://reports.weforum.org/global-information-technology-report-2015/economies/#indexId=NRIandeconomy=UKR (accessed March 22, 2016).

31 Selected Issues of Management of Green Logistics in Transport Sector

Marta Kadłubek

CONTENTS

31.1 Introduction 413
31.2 Areas of Green Logistics and Transport 413
31.3 Proposals of Analytical Grounds for Green Logistics Management in the Transport Area 415
31.4 Green Logistics Effective Operation Area for Transportation 416
31.5 Prospects of Rail Transport in the Era of Environmental Green Logistics 421
31.6 Conclusions 422
References 423

31.1 INTRODUCTION

In reference to the World Economic Forum and Accenture (2009) data, logistics determines about 5.5% of greenhouse gas emissions of the world, which is designated to the division between modes of freight transport and logistics buildings. Greenhouse gas emissions of the world with the origin in logistics as its footprint seems to be fairly unpretentious. However, in contrast to a large amount of other sectors that need to reduce greenhouse emissions, the transport sector, especially freight transport, increases its productivity of these gases.

In many industries the term *green logistics* is used to refer to implementation of a proactive environmental protection strategy (Romanowska, 2010) and accomplishment of operational decisions. In the area of transportation the companies are under pressure to develop costly responsible and effective activities that will also be in correlation with environmental awareness and green logistics solutions. The aim of the chapter is analysis of the importance of green logistics and its transport area management with particular emphasis on possibilities of building a proper analytical ground and green logistics effective operation area.

31.2 AREAS OF GREEN LOGISTICS AND TRANSPORT

Nature of logistics is functionally related, intensely integrative and various logistics activities influence the environment. Environmental objectives are closely aligned to economic objectives in the sphere of logistics designated as green logistics.

Mesjasz-Lech (2012) points out that green logistics includes all the activities interrelated to the eco-efficient management of the flows of products and information from the point of origin to the point of consumption with the aim to meet or exceed consumer demand at minimum cost. In other words, green logistics combines the environmental attributes and logistics activities to manage them in the approach that takes into consideration the importance of the environment in all decision-making activities, operations, and processes of logistics (Pishvaee et al., 2012). On the foundations of acknowledgment that a company's environmental influence expands well beyond its borders,

Klassen and Johnson (2004) settled their vision of green supply chain (Brzeziński, Brzozowska, and Korombel, 2014) management as the arrangement and integration of environmental management within supply chain management. Aksoy et al. (2014) identify green logistics with producing and distributing products in a sustainable way, taking into account the environmental and social factors. Green logistics forces all users of logistics system to consider how their actions affect the environment. The main objective of green logistics is to coordinate all activities in such a way to use their supply chains in the most efficient way, while minimizing the cost borne by the environment (Bajdor, 2012).

The extent of green logistics themes is growing as the theory and practice of its functions, processes, and relationships are developing. The most widespread and frequent themes of green logistics according to McKinnon et al. (2015) are

- Reducing freight transport externalities
- City logistics
- Reverse logistics
- Logistics in corporate environmental strategies
- Green supply chain management

Progressive environmental degradation, due to reckless management of the population of our planet, has become a global problem (Dziuba and Ingaldi, 2015). Such a situation has been brought about, among others, by transport. Nowadays, the enterprises of this sector ought to take into account responsibility for the natural environment and therefore search for solutions aimed at the protection of the ecosystem. In the transport development strategies of the countries of the European Union, there is underlined the role of the integrated transport system, among others: "Better exploitation of opportunities provided by the network and the use of relative strengths of each type of transport significantly contributes to the reduction in congestion, emissions of pollutants and number of accidents. Therefore, there is a need for optimization of the network and its smooth functioning as a whole" (European Commission, 2016).

In the era of sustainable development and widespread concern for the environment of enterprises, they should include the environmental aspect in each area of their activity to achieve the appropriate competitive position (Cohen, 2011). Many enterprises have begun to take into account green logistics while noticing the positive impact of these activities on relationships with customers. Green logistics is a relatively new concept, which is equivalent to, among others, the reduction in emissions of carbon dioxide and energy, and their low consumption in transport as well as the reduction in costs incurred by customers (Angheluta and Costea, 2011). The aspect of green logistics is therefore inextricably linked to the area of transport.

According to Hajdul et al. (2015), transport is the activity consisting in the provision of services against payment, the result of which is the service connected with the movement of cargo/people from the point of origin to the point of destination and all support services associated with transport.

According to Michałowska (2009) and Żak (2009), the provision of transport services is connected with appropriate organizational, economic, and technical preparation.

In transport activity, it is very important not only to transport companies, designers, or vehicle manufacturers to pay attention to pro-environmental aspects but also to customers and users themselves since both automotive and transport affect the natural environment most negatively (Sachs and Ban, 2015). Nowadays customers more frequently use the services of enterprises protecting and positively influencing the environment. Green and sustainable transport should have the least negative impact on the environment, and carriers can achieve this by

- Avoiding empty runs
- Using the full loading capacity of a vehicle
- Controlling vehicles

- Controlling energy consumption
- Controlling emissions of pollutants
- Using reusable packages during transportation
- Using efficient, environmentally friendly, and not burdensome for the environment means of transportation

From the economical and managerial perspective, the logistics and transportation area commonly are in contradiction with sustainable practices of logistics and environmental responsibility related to green logistics (Kadłubek, 2015).

Since sustainability is becoming a more imperative business feature (Grudzewski et al., 2010), the consequences are also in building proper analytical grounds looking for the measures and methods to facilitate evaluation of the complete image of environmental effects related to the activities of the transport area and in green logistics.

31.3 PROPOSALS OF ANALYTICAL GROUNDS FOR GREEN LOGISTICS MANAGEMENT IN THE TRANSPORT AREA

Green logistics activities include measuring the environmental impact of logistics areas. As the freight transport on average determines about 80% to 90% of carbon dioxide emissions related to logistics, this is the main reason why it is also the major area of carbon-reduction efforts and other activities focused on identification and improvement of its environmental impact.

Nowadays, the whole transport in the European Union is the generator of 24.7% of the greenhouse gas emissions and 24.6% of carbon dioxide emissions, which is an upward trend. The European Commission's Transport White Paper (European Commission, 2011) requests the transfer of the total increase in carriages to environmentally friendly means of transport. According to the report "Climate for a Transport Change" (European Environment Agency, 2008), such a change (so-called modal shift) is the most effective way to reduce emissions in the transport sector.

Due to climate protection, the pricing policy also ought to include external costs, that is, the impact on the environment and the society. The principle of "polluter pays" should constitute the integral part of the transport policy.

The external costs of transport include the following components: accidents, air pollution, external costs of energy production and infrastructure, impact on climate change (mainly carbon dioxide emissions), congestion, and noise. In transportation of goods, external costs are evaluated in the following way (Cieślakowski, 2012):

- Truck, 10.5 Euro/100 ton-km (all the elements)
- Containership, 7 Euro/100 ton-km (air pollution 6.5 Euro; impact on climate 05 Euro)
- Inland waterway, 3.5 Euro/100 ton-km (air pollution 3 Euro; impact on climate 05 Euro)
- Freight train, 2.5 Euro/100 ton-km (accidents, costs of energy production, impact on climate, noise—25% each)

In March 2011 the European Commission announced the plan to reduce carbon dioxide emissions in the transport sector by 60% by 2050. To achieve the planned reduction in exhaust emissions, the commission set 10 targets, among others, the transfer of 30% of cargo from road transport to, among others, rail transport with distances of more than 300 km by 2030 (European Commission, 2011).

The activities focused on identification and improvement of its environmental impact should take into account environmental issues integrated with logistics in the transport area in order to change the environmental performance, reduce negative effects, and create ecofriendly solutions. First, they have to be properly determined and analyzed, therefore measurements of this area appear to be initial in constructing the methodical basement. A few proposals of formations of the analytical grounds for green logistics in the transport area are introduced next.

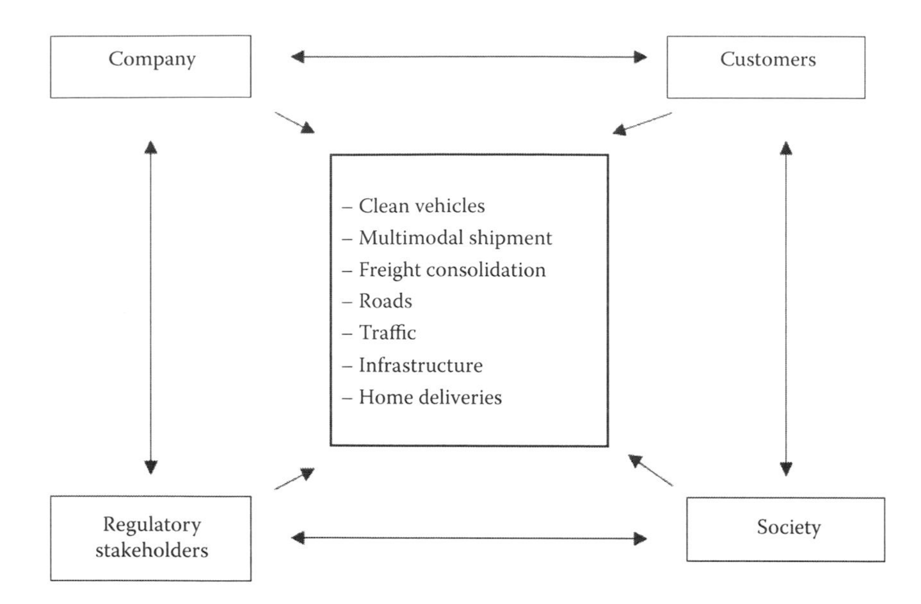

FIGURE 31.1 Four factors affecting green logistics management and indications for the transport area. (Based on Schmied, M., 2010, *Aktuelle Entwicklungen zur Standardisierung der CO2-Berechnung*. Hannover: Institute for Applied Ecology.)

Schmied (2010) makes a distinction of four general factors that affect green logistics and the most important indications for the analysis of the transport area. The factors are company, customers, regulatory stakeholders, and society, which in their environmental consciousness have requirements due to the transportation for home deliveries of the products with clean vehicles, safe for the green nature, developing operations of multimodal shipment and freight consolidation with the aim of emissions reduction, and other ecofriendly solutions for transport infrastructure, roads, or traffic (Figure 31.1). These indications in the transport area should be essential impose for measure assortment in green logistics.

McKinnon et al. (2015) propose key parameters in analysis related to green logistics management in the transport area, which are presented in Table 31.1.

In the context of sustainable development, the transport area involves three dimensions while building analytical ground: economic, social, and environmental. In reference to Russo and Comi (2012) typically used measures are focused on several vital aspects with examples indicated in Table 31.2.

In the category of logistics systems, universal cost trade-offs involving areas of transport, warehousing, and inventory are also sensitive to environmental impact and greenhouse gas. Orienting these cost trade-offs adequately to stimulate a direction to more decentralized outlines of production area and supply chain would force huge extension in freight transport costs. Establishing of the carbon dioxide influence is also possible by accomplishing a green logistics trade-off analysis, along the theoretical and practical background that is functional in the management optimization of logistics systems but related to CO_2 emissions (Figure 31.2).

31.4 GREEN LOGISTICS EFFECTIVE OPERATION AREA FOR TRANSPORTATION

Over the past years green logistics has represented a lot of nature trails, and one of the most distinguishable is reduction of transport costs by choosing cheaper and more eco-effective types of transport, for example, rail transport. Unfortunately in recent years, the share of road transport in the provision of green logistics services has significantly increased (Bubel, 2015). The fact that the

TABLE 31.1
Parameters for Analysis Related to the Green Logistics Management in the Transport Area

Parameter	Depiction
Modal split	Indicates the proportion of freight carried by different transport modes and can be expressed as the ratio of tonne-km carried by more carbon-intensive modes such as road and air to tonne-km carried by greener modes like rail, barge, ship, and pipeline
Average handling factor	Ratio of the weight of goods in an economy to freight tonnes-lifted, as products passed through the supply chain are loaded often several times
Average length of haul	Mean length of each link in the supply chain; coverts the tonnes-lifted statistic into tonne-km
Vehicle utilization	Can be measured by the ratio of vehicle-km to tonne-km, in other words how much vehicle traffic is required to handle a given amount of freight movement; if the vehicles are well loaded on outbound and return journeys this ratio is minimized
Energy efficiency	The ratio of energy consumed to vehicle-km traveled; it is a function mainly of vehicle characteristics, driving behavior, and traffic conditions
Emissions per unit of energy	The amount of CO_2 and noxious gases emitted per unit of energy consumed either directly by the vehicle or indirectly at the primary energy source for electrically powered freight transport operations
Other externalities per vehicle-km and per unit of throughput	Environmental effects such as noise irritation, vibration, accidents, etc.
Monetary valuation of externalities	Physical measures of logistics-related externalities converted into monetary values

Source: Based on McKinnon, A. C., 2010, Green Logistics: The Carbon Agenda, *LogForum*, 6(3), 1–9.

road transport has a number of advantages, such as the ability to deliver small consignments "door to door," which is an important transport logistics technology, is irrevocable. On the other hand it should also be noted that each mode of transport has its own optimum range of action. For short distances (150–200 km), small cargo transportation may be difficult and often even impossible for the railway transport to compete with road transport. But in many cases the choice between the mode

TABLE 31.2
Dimensions and Their Aspects for Building Analytical Ground of Transport Area in the Context of Sustainable Development

Dimension	Aspect
Economic	• Route length • Delivery times • Infrastructural costs • Transport congestion (i.e., additional times spent in journey, journey time, journey speed)
Social	• Decreasing interference between individual segments of mobility • Decreasing number of vehicles engaged in their tasks • Decreasing number of traffic accidents
Environmental	• Decreasing pollution • Decreasing the noise level • Loss of residential space

Source: Based on Russo, F., and Comi, A., 2012, City characteristics and urban goods movements: A way to environmental transportation system in a sustainable city, *Procedia—Social and Behavioral Sciences*, 39, 61–73.

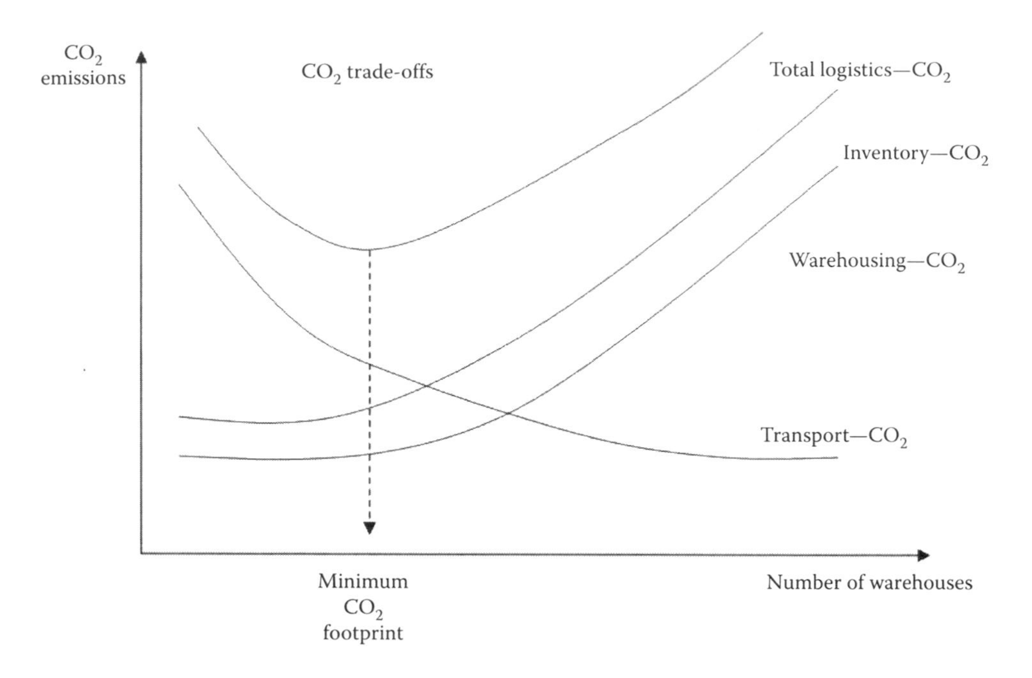

FIGURE 31.2 Logistic system with respect to CO_2 emissions as relation to green logistics. (From McKinnon, A. C., 2010, Green Logistics: The Carbon Agenda, *LogForum*, 6(3), 1–9.)

of transport is less obvious. Then the premises of the companies to realize green logistics management activities may be primarily motivated by environmental considerations while taking into the consideration cost effectiveness of the operations (Man and Nowicka-Skowron, 2010). Somehow in the range of, for example, rail or road transport effective operation is determined by the transport service change point (CP), which is calculated by the formula (Brusyanin et al., 2013)

$$CP = \frac{F_1 - F_2}{V_2 - V_1}, \tag{31.1}$$

where F_1 and F_2 are fixed costs of the first and the second type of transport; and V_1 and V_2 are variable costs of the transport types.

The graph in Figure 31.3 shows that the area within the range of 0 to the point CP may be more or less cost-effective in accordance to the first or second type of transport. The problem of green logistics to reduce the adverse impact of transport on the environment may be transformed into a problem of various types of transport integration, and the implementation of their cooperation with a minimum of vehicles, particularly into the task of the organization of intermodal transport.

In reference to the intention to determine the area of effective operation of the transport types, the proposal of the area called the "GreenLogistics" may be introduced as in the following. To realize it, into Equation 31.1, one needs to insert the green area, followed by the development of a fine system for each type of transport, both on variable and fixed costs or the green start (green benefit). It is proposed to designate it with the same term: GreenLogistics. If the GreenLogistics (fine) has a value greater than 1 and GreenLogistics (benefit) is located within the range of 0 to 1 as

$$\begin{aligned} &\text{GreenLogistics} > 1 \Rightarrow \text{Area} \\ &0 < \text{GreenLogistics} \leq 1 \supset \text{Start} \end{aligned} \tag{31.2}$$

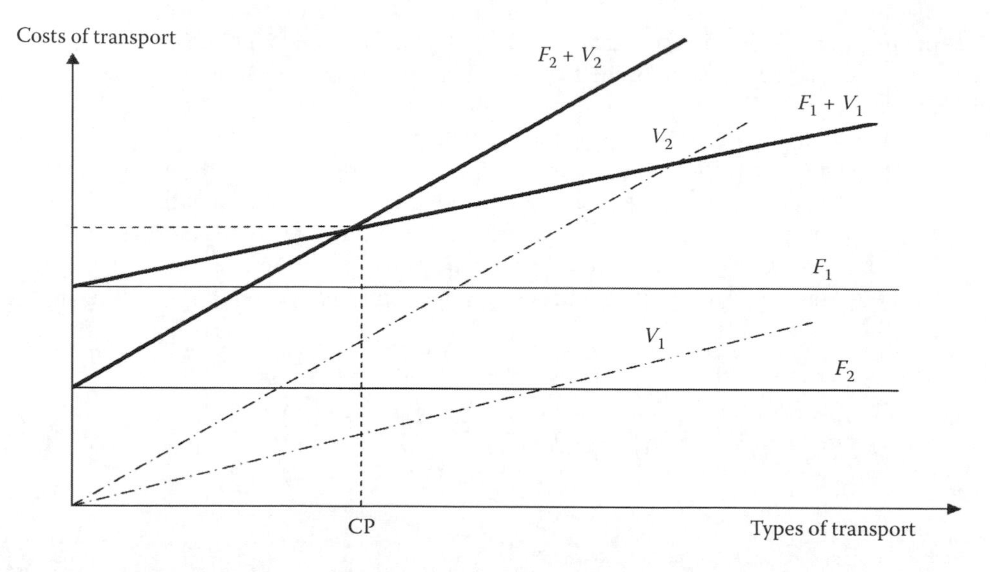

FIGURE 31.3 Correlation between the transport costs when choosing a type of transport. (From Brusyanin, D. A., Say, V. M., and Vikharev, S. V., 2013, Validation of vehicles choice for the route of regular passenger road and rail transport, *Herald of USRT*, 1(17), 50–64.)

then the amount of fines–benefits of each type of transport will give a coefficient (called "green"); the calculation results are presented in Table 31.3.

Then the formula 31.1 takes the form

$$CP = \frac{a_1 F_1 - a_2 F_2}{b_2 V_2 - b_1 V_1} \tag{31.3}$$

where a_1, a_2 is the GreenLogistics coefficient of fixed costs of the first and second transport type

$$a_{1,2} = \sum_{j=1}^{k} \text{GreenLogistics}_j \tag{31.4}$$

and b_1, b_2 is the GreenLogistics coefficient of variable costs of the first and second type of transport:

$$b_{1,2} = \sum_{j=1}^{k} \text{GreenLogistics}_j \tag{31.5}$$

TABLE 31.3
Calculation of "GreenLogistics" Coefficients for Determining Effective Operation Area of Various Types of Transport

GreenLogistics		Fixed Costs, F_i	Variable Costs, V_i
Benefit	GreenLogistics_1	…	…
	$\text{GreenLogistics}_{j-1}$	…	…
Fine	GreenLogistics_j	…	…
	GreenLogistics_k	…	…
Total		a_i	b_i

Getting parameter values in the right-hand sides of Equations 31.4 and 31.5 can be the subject of a separate study. They can be determined by the expert assessments method or mathematical modeling (Kazakov et al., 2013). One can visualize how the transport service CP can be changed while implementing the GreenLogistics by using the example graphs in Figure 31.4. The graphs show that with the implementation of the GreenLogistics effective operation area, the variable costs of the second type of transport is sharply reduced, and the effective operation area of the fixed costs of the second type of transport is increased. This approach to the transport types evaluation in terms of their effectiveness seem to be efficient in the use of the principles of green logistics management as an obligatory direction of development. Accordingly, the proposed path in the beginning of creation

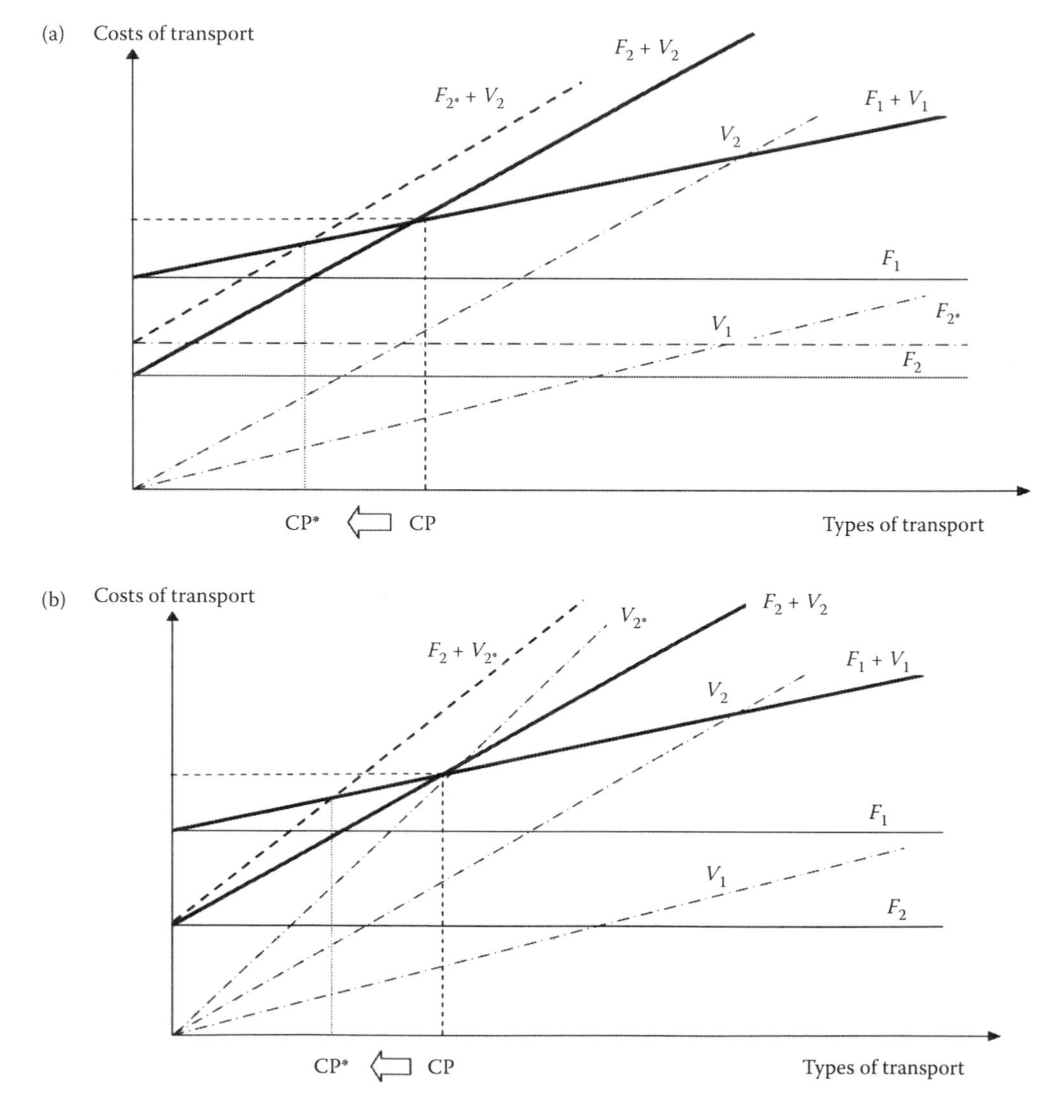

FIGURE 31.4 Examples of shifting the transport service change point (CP) with the implementation of green fines for fixed costs (a) where F_{2*} is the fixed costs of the second type of transport with the implementation of green fine and CP* is the transport service change point with the implementation of green fines for fixed costs and variable costs and (b) where V_{2*} is the variable costs of the second type of transport with the implementation of green fines and CP* is the transport service change point with the implementation of green fines for variable costs of the transport types. (Based on Kazakov, A. L., Lempert, A. A., and Bukharov, D. S., 2013, On segmenting logistical zones for servicing continuously developed consumers, *Automation and Remote Control*, 74(6), 968–977.)

of the GreenLogistics area, based on the promotion of transport types implementing green technologies and fining those types of transport that do not pay enough attention to environmental issues, may allow for estimating the contribution of each type of transport in the solution of environmental problems for the benefit of future generations.

31.5 PROSPECTS OF RAIL TRANSPORT IN THE ERA OF ENVIRONMENTAL GREEN LOGISTICS

It is commonly believed that rail transport does the least harm to the natural environment and, in spite of this, it is necessary to search for solutions increasing the extent of greening of both the carriage of passengers and goods with reference to many issues. The implementation of these assumptions is to positively affect an increase in competitiveness of railway with reference to other means of transport. The changes in the previous strategy of management of enterprises providing railway services are evidently necessary and, as Günter (2008) claims, there are three keys leading to the strategic success: costs, quality, and the way of providing services. In the era of green logistics, carriers need to skillfully combine the economic aspect, the business aspect, and the environmental aspect. Therefore, railway managers, not only in Poland but also in other European countries, must recognize how important it is that the environmental factor is implemented into the strategy of the company and increasingly extensive knowledge in this field. Romanowska (2004) noted that "full and well-conducted strategic analysis is half the battle of the strategy of enterprises," whereas Obłój (2000) underlines that "success of the company strategy depends on careful selection and creating its own market," and railway still has a lot to do with regard to this issue.

To increase environmental performance of rail transport, it is necessary to follow the guidelines given, among others, in the White Paper (2011):

- New technologies in the area of vehicles and traffic management are the key to reduce emissions of pollutants to the ecosystem
- Improvement in energy efficiency via development and introduction of fuels and propulsion systems compliant with the principle of sustainable development
- Full development of the integrated European rail market
- Gradual introduction of the new rolling stock with quiet brakes and automatic couplings
- Infrastructure should be planned in the way maximizing the positive impact on economic growth and minimizing the negative impact on the environment
- The transport sector must consume less energy in a more environmentally friendly way

The pro-environmental policy in the railway sector will force appropriate units to search for funds for the implementation of the adopted assumptions and it will include not only the modernization of the rolling stock but also the infrastructure of railway stations and stops, improving the quality of services and travel comfort in the case of passenger transport, whereas in transportation of goods, full control of technical conditions, particularly wagons carrying materials with different degrees of environmental impact.

For years, there has been observed the interest of customers in rapidity and timeliness of carriages and, for some time, customers have preferred carriages in the area of logistics by environmentally friendly means of transport such as railway.

Nowadays—in the era of environmental green logistics, which came after the industrial age when, as Nogalski (2011) indicates, "the man must have environmental sensitivity, environmental awareness and use environmental mind"—no organization can any longer be indifferent to the pro-environmental aspect in its management strategy. Sbihi and Eglese (2009) noted that "in the modern economy, there dynamically change both environmental and social requirements of enterprises, which require flexible adjustment, simultaneously creating an opportunity for their development." Organizations, due to the application of the strategy of environmental innovativeness, prove that they have achieved a certain level of environmental responsibility, which positively affects their competitive position.

There is underlined the essence of intermodal transport in which, in the case of long distances, the railway, providing full comfort and freedom to travelers, would work, whereas, in the case of transport services, providing greater safety of delivery to transshipment points and polluting the environment to a lesser extent than other means of transport. Environmental and energy benefits of railway are more often recognized by customers and railway must be significantly faster. Igliński (2006) recalls that high speeds of the European trains are 200–300 km/h and in Asia up to 350 km/h. In Poland, passenger trainers can move at the speed of 160 km/h and freight trains rarely exceed 120 km/h, constituting the railway "open-air museum" among the countries of the European Union, and the improvement of this situation will involve major financial investments and time.

For comparison, Kruczkowska (2014) recalls the data on the Chinese train (called the Silk Road of the twenty-first century) consisting of 30 wagons with a load of 1,400 tons, which has the longest route in the world (13,000 kilometers), which is covered in up to 14 days. Interestingly, such a containership would travel the distance from Eastern China to Madrid within the time of 45–60 days by sea. It is worth mentioning that Asian railways dominate others. For example, Japanese railways "nowadays belong to the most modern and comprehensively developed in the world" (Skowrońska, 2014). They are characterized by high level of energy efficiency, and are extremely fast (up to 300 km/h) and punctual (the delay observed over the last years amounts to 6 to10 seconds, which, for example, in Poland is not a delay).

Nowadays, the authorities of Polish railways hope that it will be possible to keep passengers due to investments in railway tracks and the modern rolling stock. The necessity becomes innovative solutions consisting in increasing operational efficiency and reducing costs. This can be achieved by the implementation of modern information systems improving the sale of services and customer service, which are also useful in the management of the rolling stock and terminals. The applied information tools and systems contribute to the optimization of logistic processes both in passenger movement and goods movement. Polish State Railways have also invested in drones used for the protection of trains.

After modernization, electrification of locomotives and, for the last few years, after the introduction of trains with increasingly modern design solutions of the pair of wheel and trail, there has been reduction in the level of noise and vibration. Experts have also been working on the parameters of the ground. A well-known solution is laying rails on sleepers lagged with rubber and those on "rugs"—shock absorbers with the ability to reduce vibrations. Weaker force of vibrations produced by the speeding train set automatically translates into less impact on the ground, buildings, bridges, or tunnels, which also have their own resistance.

There are also works in progress on the reduction of toxic fuels and design of pneumatic brake pads. Pads, during braking while rubbing wheels may cause sparking that could cause a fire of railway subgrade. The change of block brakes for disk brakes allows for reduction of noise by 10 dB but also provides a higher level of safety both to passengers and the vehicle itself.

Summing up, the changing transport market and the necessity to adjust the organization of railway traffic, taking into account modern rolling stock, to new challenges will contribute to increasing competitiveness of this branch of transport on the market of transport services.

In the assumptions of the transport policy of the EU (Mynarzova and Kana, 2014), there is underlined the issue of multiplication of the volume of railway transport with the least energy consumption, lowest costs, and least impact on the environment. The desire of all entrepreneurs, carriers passengers, and local and state authorities to create a sustainable transport system in the area of green logistics provides humanity with opportunities to stop the pace of degradation of the environment in which we live.

31.6 CONCLUSIONS

In view of profitability aspects as a primary requirement to exist despite exceedingly growing competition, green logistics, specially its transport area and its management, has a significant position in the national and international sustainability objectives (Nogalski and Szpitter, 2014) since they essentially are inclined toward environmentally friendly transportation types and modes. Elimination of contamination, transfer of high volume freight moving from roads to rail, and preservation of appropriate environmental management standards are valuable factors in terms of green logistics management in the transport area (Altuntas and Tuna, 2013). Realizing its aims among others through building the proper analytical ground of the area, measures or searching for effective operations areas, moreover follow the line of international conventions concerning sustainable development (World Commission on Environment and Development [WCED], 1987), the Kyoto Protocol (United Nations Framework Convention on Climate Change [UNFCC], 2012), European Union initiatives such as the White Paper on transport (European Commission, 2011), and others.

Carbon dioxide emissions determined by logistics should be reduced due to numerous negative consequences, especially in the sphere of transport. Reduction possibilities are also in the areas of logistics management operations, beginning their analysis with the use of the right parameters of carbon dioxide emissions related to the transport logistics. The proper decarbonization analysis, measures, or methods should diminish emissions, reduce the costs, and produce flow of financial and ecological benefits.

Carbon intensity differs extensively among transport types and modes. The shift of freight from types and modes of transport with comparatively high carbon amounts, such as air and road transport, to those with much poorer carbon dioxide emissions, such as rail and waterborne transport, can considerably decarbonize freight transport operations. This is also the reason to follow the other possible solutions in realization of green logistics objectives. Such a proposal may be determination of the area of effective operation of the transport types, which may assist decision makers and managers in acquiring an in-depth understanding of environmental impacts and associated costs.

REFERENCES

Aksoy, A., Kucukoglu, I., Ene, S., and Ozturk, N. 2014. Integrated emission and fuel consumption calculation model for green supply chain management. *Procedia—Social and Behavioral Sciences*, 109, 1106–1109.

Altuntas, C. and Tuna, O. 2013. Greening logistics centers: The evolution of industrial buying criteria towards green. *The Asian Journal of Shipping and Logistics*, 29(1), 59–80.

Angheluta, A. and Costea, C. 2011. Sustainable go-green logistics solutions for Istanbul metropolis. *Transport Problems*, 6(2), 59–67.

Bajdor, P. 2012. Comparison between sustainable development concept and green logistics—The literature review. *Polish Journal of Management Studies*, 5, 236–244.

Brusyanin, D. A., Say, V. M., and Vikharev, S. V. 2013. Validation of vehicles choice for the route of regular passenger road and rail transport. *Herald of USRT*, 1(17), 50–64.

Brzeziński, S., Brzozowska, A., and Korombel, A. 2014. Tools for integrating enterprises in a supply chain, Parts 1, 2. *Logistyka*, 4, 5.

Bubel, D. 2015. Configuration of flexibility of logistic services. *AD ALTA Journal of Interdisciplinary Research*, 5, 10–16.

Cieślakowski, S. 2012. Proekologiczny transport ładunków koleją [Environmentally friendly transport of goods by rail]. *Logistyka*, 5, 55–58.

Cohen, S. 2011. *Sustainability Management*. New York: Columbia University Press.

Dziuba, S. T. and Ingaldi, M. 2015. Segregation and recycling of packaging waste by individual consumers in Poland. In: *Proceedings of 15th International Multidisciplinary Scientific Geoconference (SGEM 2015)*. Sofia: STEF92 Technology Ltd., pp. 545–552.

European Commission. 2011. *White Paper. Roadmap to a Single European Transport Area—Towards a Competitive and Resource Efficient Transport System*. Brussels: European Commission.

European Commission. May 26, 2016. http://www.europa.eu.

European Environment Agency. 2008. *Climate for a Transport Change*. Luxembourg: Office for Official Publications of the European Communities.

Grudzewski, W. M., Hejduk, I. K., Sankowska, A., and Wańtuchowicz, M. 2010. *Sustainability w biznesie, czyli przedsiębiorstwo przyszłości. Zmiany paradygmatów i koncepcji zarządzania* [Sustainability in business as company of the future. Paradigm and management concepts changes]. Warszawa: Poltext.

Günter, E. 2008. *Ökologieorientiertes Management* [Ecology-oriented management]. Stuttgart: Lucius und Lucius.

Hajdul, M., Stajniak, M., Foltyński, M., Koliński, A., and Andrzejczyk, P. 2015. *Organizacja i monitorowanie procesów transportowych* [Organization and monitoring of transport processes]. Poznań: Instytut Logistyki i Magazynowania.
Igliński, H. 2006. Czy w Polsce będą koleje dużych prędkości? [Will in Poland be high speed rail?]. *Logistyka*, 2, 17–19.
Kadłubek, M. 2015. Transport sector and the assumptions of low-carbon transformation of Poland. In: B. Skowron-Grabowska (ed.), *Logistics and Marketing Determinants of Enterprises Management*. Ostrava: Vysoka Skola Banska—Technicka Univerzita Ostrava, pp. 62–70.
Kazakov, A. L., Lempert, A. A., and Bukharov, D. S. 2013. On segmenting logistical zones for servicing continuously developed consumers. *Automation and Remote Control*, 74(6), 968–977.
Klassen, R. D. and Johnson, F. 2004. The green supply chain. In: S. J. New and R. Westbrook (eds.), *Understanding Supply Chains: Concepts, Critiques and Futures*. Oxford: Oxford University Press, pp. 229–251.
Kruczkowska, M. 2014. Chiński pociąg do Europy [Chinese train to Europe]. *Gazeta Wyborcza*, 298(21), 31.
Man, M. and Nowicka-Skowron, M. 2010. Costs related to the functions of company logistics. *Polish Journal of Management Studies*, 1, 23–33.
McKinnon, A., Browne, M., Piecyk, M., and Whiteing, A. 2015. *Green Logistics: Improving the Environmental Sustainability of Logistics*. London: Kogan Page.
McKinnon, A. C. 2010. Green logistics: The carbon agenda. *LogForum*, 6(3), 1–9.
Mesjasz-Lech, A. 2012. *Efektywność ekonomiczna i sprawność ekologiczna logistyki zwrotnej* [Economic effectiveness and ecological efficiency of return logistics]. Częstochowa: Wydawnictwo Politechniki Częstochowskiej.
Michałowska, M. 2009. *Efektywny transport—konkurencyjna gospodarka* [Effective transport—competitive economy]. Katowice: Wydawnictwo Akademii Ekonomicznej w Katowicach.
Mynarzova, M. and Kana, R. 2014. Theory and practice of industrial policy of the EU in the context of globalization challenges. In: I. Honová et al. (eds.), *Proceedings of the 2nd International Conference on European Integration 2014*, Ostrava, Czech Republic. Ostrava: VŠB—Technical University of Ostrava, pp. 499–507.
Nogalski, B. 2011. Modele biznesu jako narzędzia reorientacji strategicznej przedsiębiorstw [Business models as a strategic reorientation tool for companies]. In: W. Kiezun (ed.), *Krytycznie i Twórczo o Zarządzaniu* [Critically and creatively about management]. Warszawa: Oficyna Wolters Kluwer Business, pp. 445–460.
Nogalski, B. and Szpitter, A. 2014. Koncepcja sustainability jako determinanta rozwoju przedsiębiorstwa [The concept of sustainability as a determinant of enterprise development]. In: I. Hejduk and A. Herman (eds.), *Dla przyszłości* [*For the Future*]. Warszawa: Difin, pp. 197–210.
Obłój, K. 2000. *Strategia sukcesu firmy* [*Startegy of the firm's success*]. Warszawa: PWE.
Pishvaee, M. S., Torabi, S. A., and Razmi, J. 2012. Credibility-based fuzzy mathematical programming model for green logistics design under uncertainty. *Computers and Industrial Engineering*, 62, 624–632.
Romanowska, M. 2004. *Planowanie strategiczne w przedsiębiorstwie* [*Strategic planning in the enterprise*]. Warszawa: PWE.
Romanowska, M. 2010. Przełomy strategiczne w przedsiębiorstwie [Strategic breakthroughs in the enterprise]. *Studia i Prace Kolegium Zarządzania i Finansów SGH*, 98, 7–15.
Russo, F. and Comi, A. 2012. City characteristics and urban goods movements: A way to environmental transportation system in a sustainable city. *Procedia—Social and Behavioral Sciences*, 39, 61–73.
Sachs, J. D. and Ban, K. 2015. *The Age of Sustainable Development*. New York: Columbia University Press.
Sbihi, A. and Eglese, R. W. 2009. Combinatorial optimization and green logistics. *Annals of Operations Research*, 175(1), 159–175.
Schmied, M. 2010. *Aktuelle Entwicklungen zur Standardisierung der CO2-Berechnung*. Hannover: Institute for Applied Ecology.
Skowrońska, A. 2014. Polityka transportowa Japonii w kontekście stanu i perspektyw rozwoju japońskiego transportu [Japan's transport policy in the context of the state and prospects of Japanese transport development]. *Logistyka*, 5, 7–8.
United Nations Framework Convention on Climate Change (UNFCC). 2012. *Kyoto Protocol*. Retrieved from http://www.unfcc.int.
World Commission on Environment and Development (WCED). 1987. *Our Common Future*. Oxford: Oxford University Press.
World Economic Forum and Accenture. 2009. *Supply Chain Decarbonisation: the Role of Logistics and Transport in Reducing Supply Chain Carbon Emissions*. Geneva: World Economic Forum and Accenture.
Żak, J. 2009. Transport. In: D. Kisperska-Moroń, S. Krzyżaniak (eds.), *Logistyka*. Poznań: Instytut Logistyki i Magazynowania, pp. 140–175.

32 HR Analytics as a Support of High- and Mid-Level Managers in Contemporary Enterprises in Eastern Poland

Monika Wawer and Piotr Muryjas

CONTENTS

32.1 Introduction 425
32.2 Business Analytics in Today's Organizations 426
32.3 Human Resources (HR) Analytics Approach: Fad or Necessity? 427
32.4 Research Methodology 430
32.5 Findings and Discussion 430
32.5.1 Frequency of Utilization of Analytics in Different HR Areas Depending on the Level of Management Position 430
32.5.2 Types of Analytics Utilized in Particular HR Areas Depending on the Level of Management 434
32.5.3 Level of Benefits of HR Analytics Use in the Enterprise Management 436
32.6 Conclusions 436
References 437

32.1 INTRODUCTION

"If you can measure something, you can manage it. If you can manage something, you can achieve objectives." This sentence is as true today as ever before. The necessity to measure and analyze different processes and the people who are involved is a key factor that has a critical influence on organizations' outcomes.

The analytics in today's modern organizations play a crucial role in the management processes. Many business functions such as production, supply, sales, marketing, and customer relationships management apply analytical approaches utilizing various metrics, key performance indicators (KPIs), scorecards, and information technology (IT) tools to improve general performance (Kaplan and Norton 2005; Parmenter 2010). These areas are often perceived as traditional places where the decision-making process is supported by the results of analytics. Nevertheless, today the analytical approach should be broadly applied in all areas that influence an organization's outcome. One of them is human resource management (HRM), which seems to fall behind the aforementioned areas.

In many organizations the HR function has traditionally been seen as a cost center rather than a value-creating source (Pemmaraju 2007). Nowadays it is necessary to shift this point of view and treat HR as a valuable strategic partner that helps the company achieve its goals (Philips and Philips 2008). The key to success is to identify and measure the HR "deliverables" that support corporate strategy—and the HR systems that create those deliverables (Becker et al. 2001). That means that contemporary managers have to apply the new IT tools and methods of data analysis, which enable one to find the relationships between people and the organization's outcomes more effectively.

The exceptional importance in this process has the high- and middle-level managerial staff (Song et al. 2014). They are directly responsible for the results of the work of their subordinates. Their main task is to realize the most important functions of human resource management. According to Armstrong (2006), the key HRM areas are human resource planning, recruitment and selection, introduction to the organization, formulating and implementing learning and development strategies, performance appraisal and performance management, talent management and career management, motivation and compensation, reward management, employee benefits, pensions and allowances, and release from the organization.

The role of the modern high- and middle-level managers in each of the aforementioned areas is fundamental. The more often they use HR analytics, the more effective they will acquire, motivate, develop, and retain employees (Kapoor and Sherif 2012). By understanding the past performance, current results, and future possibilities, analytics-driven managers can help companies achieve better business outcomes.

The aim of this chapter is to evaluate the level of utilization of the human resource analytics by high- and middle-level managers in contemporary organizations.

32.2 BUSINESS ANALYTICS IN TODAY'S ORGANIZATIONS

A digital universe is probably the best description of today's world. Data is generated constantly and everywhere by people, processes, and devices. According to a study conducted by International Data Corporation (IDC), the amount of data will increase 50-fold from the beginning of 2010 to the end of 2020 (Gantz and Reinsel 2012). But possessing data does not create its value. Only information, that we can discover using this data, has the real value. IDC estimates that by 2020, as much as 33% (8% more than today) of the digital universe will contain information that might be valuable if analyzed.

Along with the growth in the volume of data, its complexity and variety have risen as well. Decision making in these conditions has become a very difficult task and requires support that will allow an effective and efficient utilization of this resource. The capabilities of managers do not allow them to identify which data is significant and to discover relationships between that data as well as create the multidimensional views of the organization. This is why in the analysis of such huge resources the specialized IT tools have to be used in order to transform data into useful and valuable business information. The active and widespread data utilization in decision making is strong evidence of applying data-driven management. One of its most significant elements is business analytics (BA).

BA is defined as a set of methods that transform raw data into action by generating insights for organizational decision making (Liberatore and Luo 2010). Watson (2009) enhanced this definition about the technical aspects and claims that BA is a broad category of applications, technologies, and processes for gathering, storing accessing, and analyzing data to help business users make better decisions.

Deloitte Corporation confirms the growing interest in analytics and its critical importance in decision-making processes. In a survey, 84% of respondents expressed the opinion that the utilization of analytics has increased the competitiveness of their organizations, 25% of them stated that the growth is very high, and 30% gauged it as high (Davenport 2013). Moreover, the analytical approach and BA in managing the contemporary organization have been indicated by 77% of chef information officers (CIOs) in Deloitte's latest survey as the most important area of technology that will have a significant impact on business in the next 2 years (Kark et al. 2015). There has never been a better time to understand the importance of analytics in the contemporary business.

From a historical point of view, the analytics were primarily utilized to create the quantitative description of the past and to answer the following questions: What happened, how often, and why (Fitz-enz 2009)? This type of analytics is mainly supported by data stored in the enterprise resource planning (ERP) systems that have dedicated modules to carry out the business processes (Parry

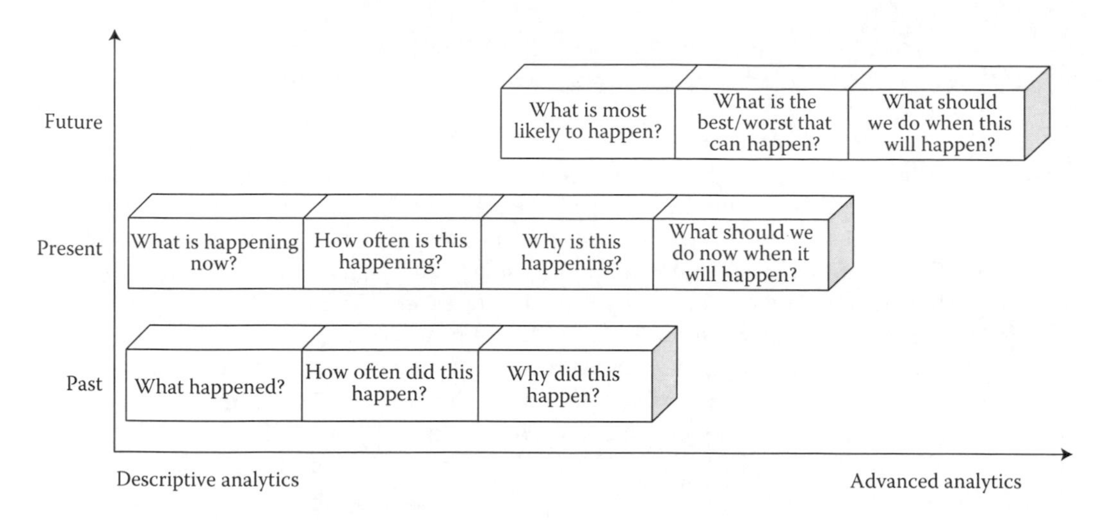

FIGURE 32.1 Dependence between the level of advancement of analytics and the time horizon of their use.

2011). Along with the shrinkage of the decision window and IT development, the analytics evolved to deliver the information about the present and predict the future (Fitz-enz 2010; Fitz-enz and Mattox 2014; Corne et al. 2012). Depending on the maturity of the organization, the managers can analyze business facts using techniques with various levels of advancement. Figure 32.1 presents the questions that can be asked in the frame of different types of analytics depending on the time horizon of their use.

Creation of deep and full insight requires the consideration of a broad range of data and its sources. The ability to generate information from both internal and external data is crucial in today's world (Drucker 1995; March and Hevner 2007). Although much has already been achieved through the use of analytics in the contemporary organizations, there is still much that can be achieved with existing data collected in the organization's IT systems (Angrave et al. 2016).

The utilization of analytics in many business areas (mentioned in the Introduction) gives much evidence that it is in the right direction of activities that lead to performance improvement on the whole organization level as well as on the level of particular departments. The key question is how analytics can help make HR management more effective and how to increase the HR contribution to the success of the firm.

32.3 HUMAN RESOURCES (HR) ANALYTICS APPROACH: FAD OR NECESSITY?

The analysis of the subject's literature (e.g., Smith 2013; SAP 2014; Demetriou et al. 2015; Moon and Prouty 2015) indicates that HR analytics are still in their infancy stage and very few organizations are actively implementing people analytics capabilities to address complex business and talent needs. Angrave et al. (2016) notice that many firms have begun to engage with HR data and analytics, but most of them have not progressed beyond operational reporting. According to Moon from the Aberdeen Group (2015b), 44% of organizations cite that one of the main reasons their HR function struggles with systematically using analytics is a lack of people who understand how to interpret analytics and how to turn them into actionable insights. There are few examples of more advanced utilizations of analytics (predictive and/or prescriptive analytics) in HR management (e.g., Fitz-enz 2010; Fitz-enz and Mattox 2014; Kapoor and Kabra 2014).

Some authors (e.g., Rasmussen and Ulrich 2015) note that the HR analytics in the current form has a risk of being a fad that fades. The main pitfalls mentioned by them, which testify to it, are a lack of analytics about analytics, mean/end inversion or data fetish, academic mindset in business setting, running HR analytics only from HR department, and a journalistic approach to HR analytics. Boudreau (2014) claims that the implementation of HR metrics and analytics systems is one of the arenas reflecting HR's role in shaping strategy and building effective HR skills, where changes are the slowest.

However, the literature of the subject (Levenson 2011; Aral et al. 2012) and many surveys conducted by leading companies such as Aberdeen Group, Deloitte Consulting, PwC, KPMG, and McKinsey & Company deliver the evidence that the awareness of importance of analytics is growing in organizations and the need of analytics in HR is much stronger today than in the past. According to Deloitte, 75% of surveyed companies believe that using people analytics is important (Demetriou 2015). The Aberdeen Group states that nearly 80% of all respondents indicated that having analytics about their workforce was critical to their organization's business strategy (Moon 2015a). KPMG informs that 56% of HR functions report an increase in using data analytics compared to 3 years ago, 31% plan to implement new technology to support this, and for 23% of the respondents the adoption of data analytics would be their main focus in the next 3 years (KPMG 2013). Moreover, 65% of organizations have applied advanced analytics to improve efficiency of the HR functions and 70% expect to begin using or increase their use of advanced analytics to inform HR decisions in the next three years. The proportion of respondents who say their organization's HR function "excels" at providing insightful and predictive analytics increased from 15% in 2012 to 23% in 2014. Over the same period, the percentage who say the function excels at measurably proving the value of HR to the business has increased from 17% to 25% (KPMG 2015).

More evidence of rising interest in HR analytics comes from PwC. Approximately 86% of PwC Saratoga participants reported that creating or maturing their people analytics function is a strategic priority over the next 1 to 3 years. And nearly one-half (46%) of those organizations already have a dedicated people analytics function (PwC 2015). In turn, McKinsey reports that HR analytics will be one of the top ten critical future priorities in the HR area in the next 2 to 3 years. Among these respondents who represent this point of view, 36% of them are already doing this and another 36% have plans with a high priority to do analytics (McKinsey & Company and The Conference Board 2012).

The Chartered Institute for Personnel and Development emphasizes that analytics is a "must have" capability for HR managers that creates value from people and a pathway to broadening the strategic influence of the HR function (CIPD 2013). Moreover, Ulrich and Dulebohn (2015) mention that creating HR analytics focused on the right issues and gaining the skills to comprehend how to use metrics to support decision making are important domains for HR investments. This point of view is also shared by Cohen (2015) who emphasizes the key role of HR analytics in the contemporary business acumen that influences the ability to understand and to apply information to contribute to the organization's strategic plan.

HR analytics is strongly supported by business intelligence (BI) systems, which can be seen as the technological foundation to conduct BA (Chiang et al. 2012; Lim et al. 2013). New information technology tools and methods of data analysis enable HR professionals to find the relationships between people and organization's outcomes more effectively (Muryjas and Wawer 2014). Advances in technology are creating opportunities for managers to start a new kind of dialogue about the link between the people and their performance (Gardner et al. 2011). This dialogue allows one to determine the impact of HR activities on achievement and it will create business value for the enterprise. The relationship between BA and HR management has been mentioned by Laursen and Thorlund (2010). They indicate four scenarios of BA use in the organization:

- BA and HR management separation BA does not deliver data to the strategic level, it is only used to answer some questions on the operational level.
- Passive support of the HR management by BA—The only role of BA is to produce reports to support the strategy performance.
- Dialogue between BA and HR management—The results of BA may modify the management activities.

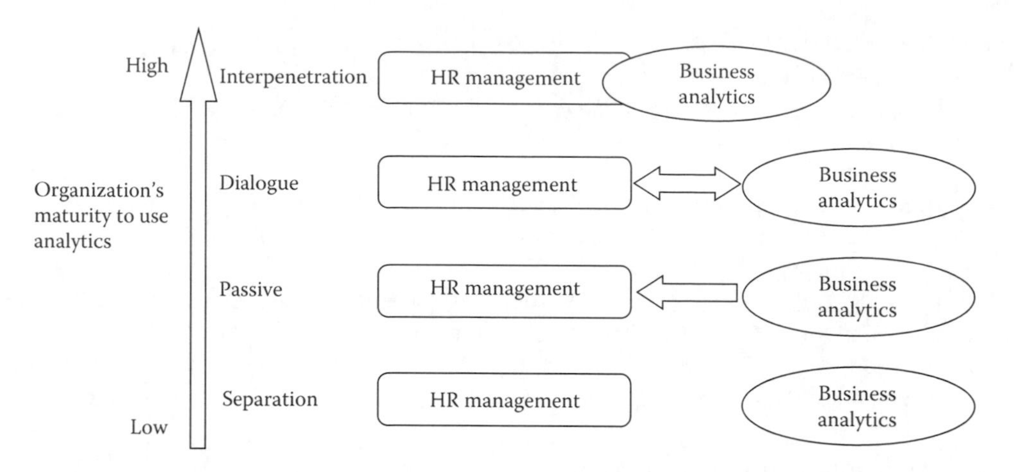

FIGURE 32.2 Dependency between scenarios of BA and HR management and the organization's maturity.

- Interpenetration of BA and HR management—Results of BA are treated as a crucial resource of the organization, which determines the HR management.

The adoption of a scenario depends on the maturity of the organization to use BA in defining and realizing of the HR management strategy. Figure 32.2 shows the dependency between the type of BA scenario and the maturity of the organization.

The importance of BA in HR management has been emphasized by Fitz-enz, who says that "the human capital analysis and predictive measurement can provide this information and are, therefore, critical for business success in this global marketplace" (2010).

There are many arguments for implementing HR analytics in the modern enterprise, but the main problem is that HR professionals do not understand the potential that analytical thinking offers. One of the reasons is that there is little evidence of successes of HR analytics implementation. However, to be a successful catalyst for change, HR should not only be capable of analyzing and interpreting human capital metrics and analytics, but also be capable of recommending and implementing interventions to drive organizational effectiveness (Moon 2015b).

Based on the previous discussion, it is possible to state that the achievement of organization's aims with the utilization of HR analytics requires an engagement of managers at different levels. According to Gary and Wood (2011), every manager has knowledge structures that impact the perception, information processing, problem solving, and decision making, influencing the organizational learning capability and firm performance. It is also worth emphasizing that high-level managers should share the expert knowledge with mid-level managers and propagate the analytical approach among them to increase their analytical competences to utilize the analytics in HR management. However, in many organizations middle managers do not receive the right information or do not take appropriate actions aimed at the increase of the efficiency of HR management.

The aim of our research is to find whether high- and mid-level managers are engaged in the same way in the utilization of analytics in the HR management. The following main hypothesis has been defined:

H: There is a dependence between utilization of HR analytics in the organization and the level of a managerial position.

In order to verify this hypothesis, three detailed hypotheses have been formulated:

H1: There is a dependence between the frequency of using the analytics in different HR areas and the level of a managerial position.

H2: There is a dependence between the types of analytics used in different HR areas and the level of a managerial position.
H3: There is a dependence between the appraisal of the level of benefits of HR analytics utilization in the enterprise management and the level of a managerial position.

To verify these detailed hypotheses, managers have been asked the following questions:
1. How often do you utilize the analytics in different HR areas? (Possible answers: very often, often, rarely, never)
2. Which types of analytics do you use in particular HR areas? (Possible answers: descriptive analytics, predictive analytics, both of them, none of them)
3. In your opinion, what is the level of benefits of utilization of HR analytics in the enterprise management? (Possible answers: high, middle, low, none)

32.4 RESEARCH METHODOLOGY

In November 2015, we sent e-mails with a questionnaire to 237 potential organizations that reside in Eastern Poland. The survey was conducted until January 20, 2016, on the territory of three provinces, namely, Lubelskie, Podlaskie, and Subcarpathian. We received a total of 73 responses to this study. Every organization was represented by one respondent. To ensure the quality of data, we carefully scrutinized and verified all respondent's entries to ensure that the study includes only fully completed questionnaires. We excluded those that did not contain answers to every question and accepted 61 questionnaires for further analysis.

The surveyed managers represented organizations in the financial and insurance services (16.39%), consulting (8.20%), technology/telecommunication (6.56%), entertainment (4.92%), health care (9.84%), education (8.20%), consumer products/retail (21.30%), manufacturing (18.03%), and public administration (6.56%) sectors.

Most of the surveyed participants were mid-level executives (75.41%), and 24.59% represented high-level managers. The detailed description of the respondents is presented in Table 32.1.

The statistical analysis of the survey results was performed using the R 3.1.2 environment. The dependence between categorical variables was examined using Pearson's chi-square test for independence with Yates' correction for discontinuity. The accepted statistical significance level is $p < 0.05$.

32.5 FINDINGS AND DISCUSSION

32.5.1 Frequency of Utilization of Analytics in Different HR Areas Depending on the Level of Management Position

The survey confirms that HR analytics are not yet commonly utilized in contemporary organizations. Among all the surveyed firms, taking into consideration the total results, only 17% use HR analytics very often, and 29% often. Unfortunately, 34% said they utilize HR analytics rarely and 20% never apply them. Such distribution confirms that HR analytics are in the area of interest of managers only in few firms nowadays.

Considering the aim of this survey it is important to analyze the utilization of HR analytics on different levels of management. The key question is whether the mid-level and high-level managers utilize the HR analytics with the same frequency. The results of this survey related to this question are presented in Table 32.2.

The analysis of these results allows us to state that 28% of the high-level managers and 14% of the mid-level managers utilize HR analytics very often. This proportion may indicate that the decision making of high-level management requires the support more often than the mid-level managers. However, other data representing the overall view on the HR area does not confirm the

TABLE 32.1
Respondents' Description

Category	Total (%)	High Level (%)	Middle Level (%)
Gender			
Woman	47.54	46.67	47.83
Male	52.46	53.33	52.17
Age			
20–35 years	34.43	20.00	39.13
36–45 years	39.34	20.00	45.65
46–55 years	26.23	60.00	15.22
Seniority			
Up to 3 years	31.15	20.00	34.78
4–8 years	34.43	13.33	41.30
9–13 years	22.95	26.67	21.74
14–18 years	3.28	6.67	2.17
19 and more years	8.20	33.33	0.00
Organization Size			
10–49 persons	11.48	0.00	15.22
50–249 persons	45.90	46.67	45.65
250–499 persons	16.39	0.00	21.74
More than 500 persons	26.23	53.33	17.39
Core Business			
Production	19.67	20.00	19.57
Services	54.10	46.67	56.52
Trade	14.75	13.33	15.22
Other	11.48	20.00	8.70
Place of Organization			
City up to 10,000 citizens	4.92	0.00	6.52
City 10,000–50,000 citizens	9.84	0.00	13.04
City 50,000–250,000 citizens	13.11	0.00	17.39
City more than 250,000 citizens	72.13	100.00	63.04
Property			
Private with foreign capital	39.34	40.00	39.13
Private with Polish capital	40.98	46.67	39.13
State Treasury	18.03	13.33	19.57
Community	1.64	0.00	2.17

TABLE 32.2
Frequency of HR Analytics Utilization Depending on the Management Level

	Very Often (%)	Often (%)	Rarely (%)	Never (%)
Middle level	14	31	34	21
High level	28	22	31	19

statistically significant dependence between the frequency of utilization of the analytics and the level of management positions.

The analysis, which takes into consideration 10 main HR areas, delivers very interesting results. Figure 32.3 shows how often the high- and mid-level managers apply the analytics in the following areas:

1. Human resource planning (A1)
2. Recruitment and selection (A2)
3. Introduction to the organization (A3)
4. Formulating and implementing learning and development strategies (A4)
5. Performance appraisal and performance management (A5)
6. Talent management and career management (A6)
7. Motivation and compensation (A7)
8. Reward management (A8)
9. Employees absenteeism (A9)
10. Release from the organization (A10)

The chi-square test does not demonstrate the statistically significant differences in answers concerning areas A1 to A6 and A8. It means the structure of the answers to the question about the frequency of utilization of analytics in these HR areas does not depend on the level of managerial position.

However, results of this test show that this type of dependence exists in areas A7 ($p = 0.049^*$), A9 ($p = 0.012^*$), and A10 ($p = 0.007^{**}$).

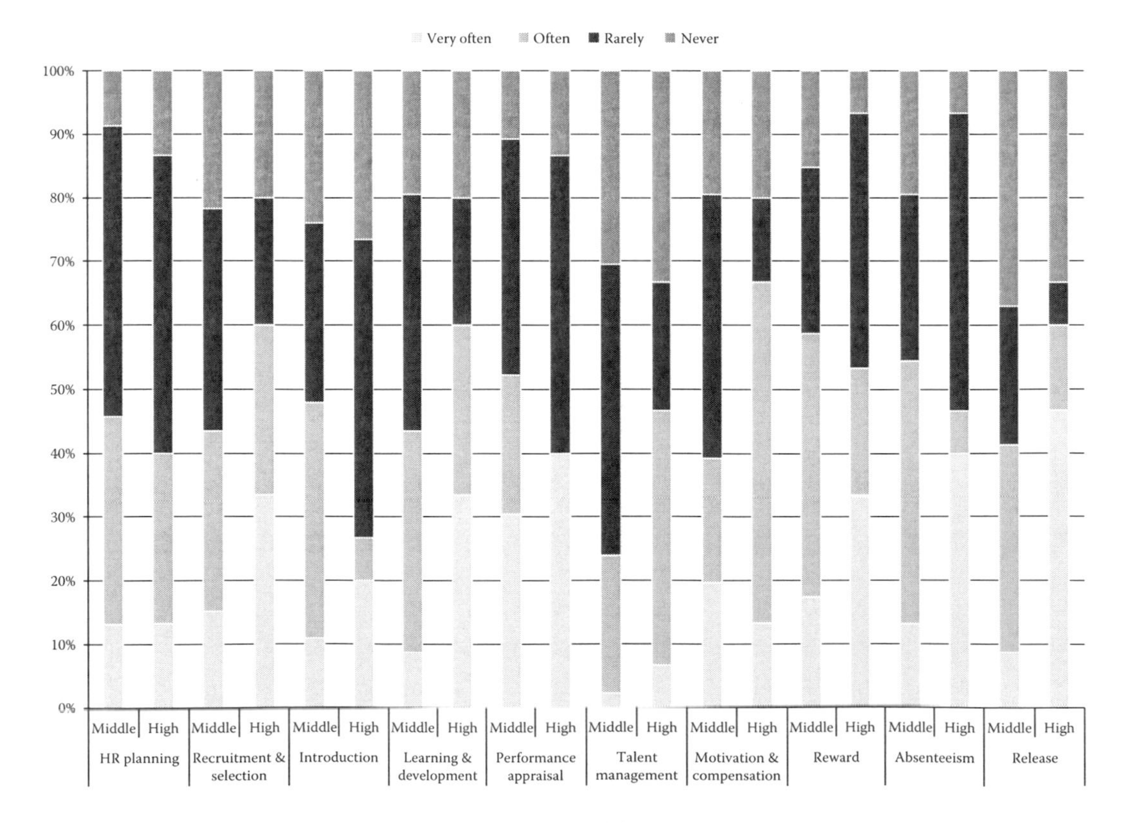

FIGURE 32.3 Frequency of utilization of analytics within each HR area depending on the level of the managerial position.

TABLE 32.3
Frequency of Utilization of Analytics in Motivation and Compensation Area Depending on the Level of the Managerial Position

Motivation and Compensation	Total (%)	High Level (%)	Middle Level (%)	Chi-square *p*
Very often	18.03	13.33	19.57	$p = 0.049$
Often	27.87	53.33	19.57	
Rarely	34.43	13.33	41.30	
Never	19.67	20.00	19.57	

In area A7, 53% of high-level managers answered "often," while 41% of the mid-level executives said "rarely." The detailed distribution of answers is shown in Table 32.3.

For the A9 area, the most answers for high-level managers were "very often" (40%) and "rarely" (47%), whereas on the mid-level it was mainly the answer "often" (41%). The percentage share of particular answers is presented in Table 32.4.

The survey results in area A10, presented in Table 32.5, indicate that the most frequent answers in the group of high-level managers were "very often" (47%) and "never" (33%), while for the mid-level management answers were "often" (33%) and "never" (37%).

The analysis of the statistical dependencies indicates that in three HR areas the visible and significant differences in utilization of analytics by managers of high- and mid-level have been observed.

The first area is motivation and compensation. Here the high-level managers, first, have the main impact on the organization's strategy and they define directions of HR shifts that shape this strategy.

TABLE 32.4
Frequency of Utilization of Analytics in Absenteeism Area Depending on the Level of the Managerial Position

Employee Absenteeism	Total (%)	High Level (%)	Middle Level (%)	Chi-square *p*
Very often	19.67	40.00	13.04	$p = 0.012$
Often	32.79	6.67	41.30	
Rarely	31.15	46.67	26.09	
Never	16.39	6.67	19.57	

TABLE 32.5
Frequency of the Utilization of Analytics in Employee Release Area Depending on the Level of the Managerial Position

Release from the Organization	Total (%)	High Level (%)	Middle Level (%)	Chi-square *p*
Very often	18.03	46.67	8.70	$p = 0.007$
Often	27.87	13.33	32.61	
Rarely	18.03	6.67	21.74	
Never	36.07	33.33	36.96	

The mid-level executives in this area have limited privileges and therefore they are less engaged in these processes.

The second area is the employee absenteeism. The middle managers have direct contact with the employees and they are responsible for the results of their work to the greatest extent. This is probably the reason why they utilize the analytics to evaluate absenteeism, which has a strong influence on the outcomes.

High-level managers are more involved in the strategic activities than in the operational processes. Therefore they are aware of the capabilities and benefits of analytics and utilize them very often in the long-term analysis of the reasons of absenteeism as well as its prevention. The managers, who do not have adequate knowledge and awareness in this area, do not feel the need and do not perform these analyses.

The third HR area, which differentiates the approaches of managers to utilize analytics, is the employee release from the organization. Perhaps the reason behind this fact is the same as the previously mentioned one.

In summary, due to strategic reasons, high-level managers utilize HR analytics very often and middle managers are doing this often due to their engagement in the operational activities with subordinates. The explanation of these differences requires further research.

32.5.2 Types of Analytics Utilized in Particular HR Areas Depending on the Level of Management

As mentioned in Section 32.2, analytics should be applied so as to consider two main time horizons. The first one concerns the past, which can be analyzed using techniques of descriptive analytics. The second one is the future that can be foreseen by using predictive analytics. Both types of approaches are very important in every organization. The effective support of achievement of organization's aims by HR management will be possible only when HR analytics will be utilized from the point of view of both time perspectives.

A conducted survey confirms that HR analytics are not properly used in organizations. Among all the firms, taking into consideration the total results (high- and mid-level managers), both types of analytics are adopted only by 8% of those surveyed, whereas the descriptive analytics is performed by 28% of managers and the predictive by 35%. Unfortunately, 28% of respondents stated that they do not apply the analytics for any purpose mentioned here. This distribution of answers confirms that managerial staff does not fully utilize the capabilities of HR analytics.

From the point of view of the essence of our research, it is important to recognize the type of HR analytics that are used by the managers on different managerial levels. The key issue is why the high- and mid-level managers are using the HR analytics. What is the time horizon of analytics for every manager group? The results of the conducted survey are presented in Table 32.6.

As results from this survey show, HR descriptive and predictive analytics are utilized only by 6% of the high-level managers and 8% of the mid-level managerial staff. Also 47% of high-level and 30% of mid-level executives declare they use the data analytics to predict the future activities in the HR areas. This difference shows that the high-level managers have a better

TABLE 32.6
Type of HR Analytics Depending on the Level of the Managerial Position

	Descriptive (%)	Predictive (%)	Descriptive and Predictive (%)	None (%)
Middle level	32	30	8	30
High level	21	47	6	26

understanding of the true meaning of the use of this type of analytics. However, the disturbing fact is that 26% of high-level and 30% of mid-level respondents never use the HR analytics in their work.

Enhancing the aforementioned reflections and taking into consideration the individual areas of HR, we can observe an interesting relationship between them. Figure 32.4 presents the type of analytics used in particular HR areas depending on the level of managerial position.

The middle managers utilize both types of analytics most often for the analysis of motivation and compensation (15%), performance appraisal (13%), recruitment and selection (13%), and HR planning (10%), whereas the high-level executives use analytics in the following areas: learning and development (15%), motivation and compensation (14%), and absenteeism (13%).

Descriptive analytics are utilized by about the same percentage of respondents in the group of middle- and high-level managers in the following HR areas: performance appraisal (48% versus 40%), reward and absenteeism (32% versus 33%), and HR planning (26% versus 27%). In other HR areas, the middle executives use the descriptive analytics more often than high-level managers: learning and development (39% versus 13%), release (37% versus 20%), talent management (31% versus 7%), recruitment and selection (23% versus 13%), motivation and compensation (22% versus 13%), and introduction (20% versus 0%).

The survey results also show that predictive analytics are performed by high-level managers mainly in the following HR areas: recruitment and selection (67%), motivation and compensation (60%), and talent management (53%).

Summing up the preceding results, we can state that although there is no statistically significant dependence between the types of HR analytics and the level of the managerial position, both groups of managers utilize them to analyze the activities of employees in different time horizons and in different HR areas.

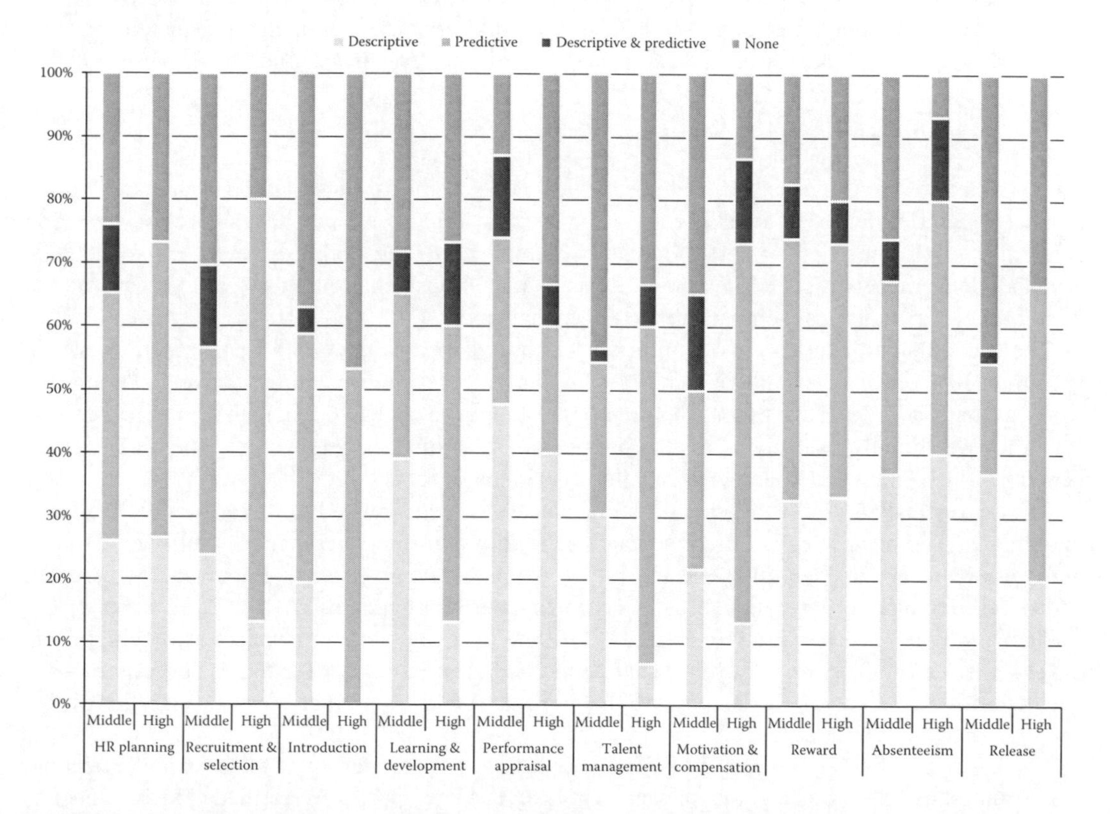

FIGURE 32.4 Types of analytics in HR areas depending on the level of management.

TABLE 32.7
Level of Benefits of HR Analytics Utilization in Enterprise Management Depending on the Level of the Managerial Position

Level of Benefits	Total (%)	High-Level Managers (%)	Mid-level Managers (%)	Chi-square p
High	50.82	73.33	43.48	$p = 0.042$
Average	40.98	13.33	50.00	
Low	8.20	13.33	6.52	

32.5.3 Level of Benefits of HR Analytics Use in the Enterprise Management

The third research area concerns the opinion of the managerial staff about the level of benefits of using HR analytics in enterprise management. The results allow us to conclude that all surveyed managers have a high awareness in this area and see the benefits of HR analytics use in the enterprise management. Half of the respondents (50.82%) reported that using analytics ensures the high level of benefits, followed by 40% who said that it is average. Another 8% stated that the benefits are very low.

Because of the purpose of our research and defined hypothesis, it is important to compare the opinions of managers about the level of benefits taking into account their managerial position. The analysis performed with the use of chi-square tests revealed that there is a statistically significant dependence ($p = 0.042$*) between the answers of both groups of managers, as shown in Table 32.7.

Most high-level managers (73%) see a high level of benefits delivered by data analytics in management, whereas 13% of respondents in this group reported the average and low level, respectively. The situation looks different in case of mid-level managers. Only about 43% of them stated that the benefits of use of the analytics are very high, and one-half of those surveyed determine benefits as average.

32.6 CONCLUSIONS

In the context of presented subject literature review it is possible to state that the frequency of utilization of analytics in the HR area is still very low in many organizations. The results of our research confirm that HR analytics are not used at all by managers in 20% of the surveyed firms, and 30% of respondents rarely utilize HR analytics. It means that almost a half of all surveyed organizations do not apply the analytical approach in the HR management.

The second significant conclusion from our research concerns the type of performed analytics. Less than 10% of managers utilize both types of analytics (descriptive and predictive), and unfortunately as many as 28% of the respondents do not apply any type of analytics in management processes.

Third, we should positively assess the opinions of 50% of managers who see the high level of benefits of utilization of HR analytics in enterprise management.

These conclusions are of a general nature but are directly related to three specific research hypotheses. The conducted research indicates that high- and mid-level managers utilize analytics with various frequencies in three HR areas: motivation and compensation, employee absenteeism, and release from the organization. This fact confirms our first hypothesis.

The survey results describing the types of HR analytics applied by managers of varying levels did not disclose the statistically significant difference. Therefore, the second hypothesis has not been confirmed. However, we can observe that high- and mid-level executives use analytics differently in particular HR areas.

The last hypothesis concerning the dependence between the appraisals of the level of the benefits of HR analytics utilization and the management level has been positively verified. The results confirm the statistically significant dependence between the answers and the managerial level.

Summing up the preceding conclusions, it can be stated that the main hypothesis about the various approaches of managers of different levels to the utilization of analytics in HR areas has partially been confirmed.

While formulating these conclusions it is necessary to be aware of the occurrence of certain limitations. The analyses performed in this chapter did not take into account the specifics of the surveyed organizations that are different in size, the sector, the type of core business, and the form of ownership. Moreover, they are located in Eastern Poland that has an agricultural character and is less economically developed than Western Poland.

These circumstances could affect the survey results and they should be the base for further research that might also focus on the dependences, mentioned earlier, in a larger group of organizations located throughout country.

REFERENCES

Angrave, D., Charlwood, A., Kirkpatrick, I., Lawrence, M., and Stuart, M. 2016. HR and analytics: Why HR is set to fail the big data challenge. *Human Resource Management Journal*, 26(1): 111.

Aral, S., Brynjolfsson, E., and Wu, L. 2012. Three-way complementarities: Performance pay, human resource analytics, and information technology. *Management Science*, 58(5): 913–931.

Armstrong, M. 2006. *A Handbook of Human Resource Management Practice*. London: Kogan Page Limited.

Becker, B., Huselid, M., and Ulrich, D. 2001. *The HR Scorecard: Linking People, Strategy and Performance*. Boston: Harvard Business Press.

Boudreau, J. 2014. Will HR's grasp match its reach? An estimable profession grown complacent and outpaced. *Organizational Dynamics*, 43: 189–197.

Chiang, R. H. L., Goes, P., and Stohr, E. A. 2012. Business intelligence and analytics education, and program development: A unique opportunity for the information systems discipline. *ACM Transactions on Management Information Systems* 3(3): 1–13, Article No. 12.

CIPD. 2013. Talent analytics and big data—The challenge for HR. Research Report, November 2013. London: Chartered Institute of Personnel and Development. https://www.cipd.co.uk/binaries/talent-analytics-and-big-data_2013-challenge-for-hr.pdf (accessed June 10, 2016).

Cohen, D. J. 2015. HR past, present and future: A call for consistent practices and a focus on competencies. *Human Resource Management Review*, 25: 205–215.

Corne, D., Dhaenens, C., and Jourdan, L. 2012. Synergies between operations research and data mining: The emerging use of multi-objective approaches. *European Journal of Operational Research*, 221: 469–479.

Davenport, T. H. 2013. The analytics advantage. We're just getting started. Deloitte Analytics. http://www2.deloitte.com/content/dam/Deloitte/global/Documents/Deloitte-Analytics/dttl-analytics-analytics-advantage-report-061913.pdf (accessed June 10, 2016).

Demetriou, S., Kester, B., Moen, B., and O'Leonard, K. 2015. HR and people analytics: Stuck in neutral. In *Global Human Capital Trends 2015*. Deloitte University Press, pp. 71–77. http://www2.deloitte.com/content/dam/Deloitte/at/Documents/human-capital/hc-trends-2015.pdf (accessed June 10, 2016).

Drucker, P. F. 1995. The information executive's truly need. *Harvard Business Review*, 73(1): 54–62.

Fitz-enz, J. 2009. *The ROI of Human Capital: Measuring the Economic Value of Employee Performance*, 2nd ed. New York: American Management Association.

Fitz-enz, J. 2010. *The New HR Analytics. Predicting the Economic Value of Your Company's Human Capital Investments*. New York: American Management Association.

Fitz-enz, J., and Mattox, J. R. II. 2014. *Predictive Analytics for Human Resources*. Hoboken, NJ: John Wiley & Sons, Inc.

Gantz, J., and Reinsel, D. 2012. The digital universe in 2020: Big data, bigger digital shadows, and biggest growth in the far east. http://www.emc.com/collateral/analyst-reports/idc-the-digital-universe-in-2020.pdf (accessed June 12, 2016).

Gardner, N., McGranahan, D., and Wolf, W. 2011. Question for your HR chief: Are we using our "people data" to create value? *McKinsey Quarterly*, March: 1–5.

Gary, M. S., and Wood, R. E. 2011. Mental models, decision rules, and performance heterogeneity. *Strategic Management Journal*, 32(6): 569–594.

Kaplan, R. S., and Norton, D. P. 2005. The balance scorecard. Measures that drive performance. *Harvard Business Review*, July–August: 1–10.

Kapoor, B., and Kabra, Y. 2014. Current and future trends in human resources analytics adoption. *Journal of Cases on Information Technology*, 16(1): 1–10.

Kapoor, B., and Sherif, J. 2012. Human resources in an enriched environment of business intelligence. *Kybernetes*, 41(10): 1625–1637.

Kark, K., White, M., and Briggs, B. 2015. 2015 *CEO Global Survey: Creating Legacy*. Deloitte University Press. http://www2.deloitte.com/content/dam/Deloitte/at/Documents/technology-media-telecommunications/cio-survey2015.pdf (accessed June 12, 2016).

KPMG. 2013. People are the real numbers. HR analytics has come of age. https://www.kpmg.com/NL/nl/IssuesAndInsights/ArticlesPublications/Documents/PDF/Management-Consulting/People-are-the-real-numbers.pdf (accessed June 12, 2016).

KPMG. 2015. Evidence-based HR: The bridge between your people and delivering business strategy. https://www.kpmg.com/Global/en/IssuesAndInsights/ArticlesPublications/Documents/evidence-based-hr.pdf (accessed June 12, 2016).

Laursen, G., and Thorlund, J. 2010. *Business Analytics for Managers: Taking Business Intelligence beyond Reporting*. Hoboken, NJ: John Wiley & Sons.

Levenson, A. 2011. Using targeted analytics to improve talent decisions. *People and Strategy*, 34(2): 34–43.

Liberatore, M., and Luo, W. 2010. The analytics movement: Implications for operations research. *Interfaces*, 40(4): 313–324.

Lim, E. P., Chen, H., and Chen, G. 2013. Business intelligence and analytics: Research directions. *ACM Transactions on Management Information Systems* 3(4): Article No. 17.

March, S. T., and Hevner, A. R. 2007. Integrated decision support systems: A data warehousing perspective. *Decision Support Systems*, 43(3): 1031–1043.

McKinsey & Company and The Conference Board 2012. The state of the human capital 2012. Research report. http://www.mckinsey.com/business-functions/organization/our-insights/the-state-of-human-capital-2012-report (accessed June 12, 2016).

Moon, M. M. 2015a. *Talent Analytics: Where Are We Now?* Boston: Aberdeen Group. http://v1.aberdeen.com/launch/report/research_report/10459-RR-talent-analytics-insights.asp (accessed June 14, 2016).

Moon, M. M. 2015b. *Five Foundational Metrics for Meaningful Workforce Measurement Insight*. Boston: Aberdeen Group. http://v1.aberdeen.com/launch/report/research_report/11114-RR-hr-measurement-maturity.asp (accessed June 14, 2016).

Moon, M. M., and Prouty, K. 2015. *Productivity: Managing and Measuring a Workforce*. Boston: Aberdeen Group. http://v1.aberdeen.com/launch/report/research_report/10143-RR-Productivity-WFM.asp (accessed June 14, 2016).

Muryjas, P., and Wawer, M. 2014. Business intelligence as a support in human resources strategies realization in contemporary organizations. *Actual Problems of Economics*, 152(2): 183–190.

Parmenter, D. 2010. *Key Performance Indicators. Developing, Implementing and Using Winning KPIs*. Hoboken NJ: John Wiley & Sons.

Parry, E. 2011. An examination of e-HRM as a means to increase the value of the HR function. *The International Journal of Human Resource Management*, 22(5): 1146–1162.

Pemmaraju, S. 2007. Converting HR data to business intelligence. *Employment Relations Today*, 34(3): 13–16.

Phillips, J. J., and Phillips, P.P. 2008. *Proving the Value of HR: How and Why to Measure ROI*. Alexandria, VA: Society for Human Resource Management.

PwC 2015. Trends in people analytics. https://www.pwc.com/us/en/hr-management/publications/assets/pwc-trends-in-the-workforce-2015.pdf (accessed June 14, 2016).

Rasmussen, T., and Ulrich, D. 2015. Learning from practise: How HR analytics avoids being a management fad. *Organizational Dynamics*, 44: 236–242.

SAP 2014. 100 critical human capital questions: How well do you really know your organisation? http://go.sap.com/docs/download/2015/08/2e95bcfd-377c-0010-82c7-eda71af511fa.pdf (accessed June 12, 2016).

Song, L. J., Zhang, X., and Wu, J. B. 2014. A multilevel analysis of middle manager performance: The role of CEO and top manager leadership CEO. *Management and Organization Review*, 10(2): 275–297.

Smith, T. 2013. *HR Analytics: The What, Why and How*. Charlotte: Numerical Insights LLC.

Ulrich, D., and Dulebohn, J.H. 2015. Are we there yet? What's next for HR? *Human Resource Management Review*, 25: 188–204.

Watson, H. J. 2009. Tutorial: Business intelligence—Past, present, and future. *Communications of the Association for Information Systems*, 25(1), 487–510.

33 Effects and Impact of Playing Computer Games on Gamers

W. Chmielarz and O. Szumski

CONTENTS

33.1 Introduction 439
33.2 Assumptions of Research Methodology 440
33.3 Analysis of the Obtained Results and Relevant Discussion 441
33.4 Conclusions 451
References 453

33.1 INTRODUCTION

The main aim of this work is to analyze the use of computer games as one of the alternative forms of entertainment in the selected group of users under the circumstances of a dynamic development of devices and mobile applications running on them. The aim of this chapter is to analyze the situation where computer games are used by people who treat them not only as a form of entertainment but also as a kind of sport. The popularity and specific universal nature of the access to computer games facilitates a fast development of information technologies. A broadly defined concept of mobility also impacts the use of computer games, moving the focus from using PCs to the use of smartphones and tablets.

According to the statistics of the Newzoo service (GRY-OnLine 2014), in Poland in 2013 the number of gamers amounted to 13.4 million, out of which 98% used their PCs to play computer games (together with other platforms). Poland takes the second position in Europe among the examined countries. The market of computer games in Poland is growing every year; in the end of 2014 it was worth about $280 million and it was predicted to grow by 3.8% a year, thus increasing the value of the entire market to $437 million at the end of 2016 (Akcjonariat Obywatelski 2013). Hence, undoubtedly the subject matter is worthy of attention.

Unfortunately, the phenomenon itself is difficult to define and examine taking into account formalized scientific analyses. First, there is no clear definition of computer games (GRY-OnLine 2004; Wiedza i Edukacja 2009; PTBG 2010; Zając 2014; Chmielarz 2015a; IT-Pomoc.pl 2016; KIPA 2016; *Wikipedia* 2016). In its narrow sense, this concept is treated literally as games in the form of software running only on traditional hardware such as desktops, microcomputers, laptops, or palmtops. In its broad, historical approach, the group encompasses also games running on devices such as a console, TV, gaming machines, smartphones and tablets (which are in fact communication and application computers). As the games running on all kinds of devices were being developed in parallel, and, in fact, there are PC equivalents of all kinds of games, we sometimes use the term computer games in its broad meaning. Thus, for the needs of this study, the authors assumed that computer games is defined as a generic term (hypernym) encapsulating the whole class of all kinds of games presented as a homogenous phenomenon. Second, there is no one generally accepted definition of a person playing computer games (e-gamer). Thus, in the narrow sense of the word, an e-gamer is a person who plays computer games every day or a few times a week, individually or as part as a multiplayer game. Sometimes, the scope of the term e-gamer is limited to include only those players who treat massively multiplayer online (MMO) class games as a sport, and they try to play them professionally. However, we observe a common tendency to expand the term to include also any individual who plays any kind of game from time to time, perceiving it as just one more alternative kind of entertainment. This chapter treats the concept of e-gamers in such a way. Third, there is no (specific or clear) classification of computer games: there are a number of typologies based on various criteria, most frequently taking into account the type of activity required from the e-gamer playing games (e.g., logic, strategic, arcade, RPG, etc.), with a number of varying kinds and versions.

The phenomenon of computer games has been examined in numerous studies (Mijal and Szumski 2013; Żywiczyńska 2014a), including large-scale studies (Żywiczyńska 2014b); nevertheless, they were carried out before the recent period of extreme popularity and growth in the number of applications running on smartphones and tablets. The authors hoped to establish certain implications of the new phenomena with regard to the direction of computer games development. Therefore, the authors have undertaken the studies whose main aim is to analyze the use of such applications among users. The findings presented in this chapter constitute a brief report on the second stage of the research conducted among gamers in Poland in 2015.

33.2 ASSUMPTIONS OF RESEARCH METHODOLOGY

Due to limited and fragmentary research concerning the area of Internet computer games and e-gamers, both from the point of view of an individual client and a group of customers, in Polish and foreign literature, the studies have been based on the authors' own approach (Chmielarz 2015b) consisting of the following steps:

- Analysis of a selected group of players on the basis of a quantitative and qualitative survey, divided into the following parts:
  - Characteristics of a computer player and identifying his or her preferences in computer games
  - Identification of potential effects and consequences of playing computer games for e-gamers
- Placing an Internet version of a survey on the servers of the Faculty of Management of the University of Warsaw, conducting functionality test and its verification
- Carrying out the survey among the users, analysis, and discussion of the findings
- Drawing conclusions from the obtained results concerning the current situation and possible directions of the future development of Internet computer games on the basis of the users' opinions

The chapter presents the results of the analysis of the first and second part of the completed survey. It allowed for identifying a particular group of people who play various types of games, using different kinds of hardware and software, with a varying level of skills and expectations concerning the organizational and technical aspects of playing games. Only after selecting the group of best, "professional" players, we may proceed to specify the implications and psychophysical effects of their involvement in individual and multiplayer games. The latter aspect was examined in the second, sequentially conducted, stage of the survey, whose results and conclusions are presented in this chapter, too.

The questionnaire surveys were conducted near the end of December 2015 and in the beginning of February 2016. The selection of the study sample was not accidental: it belonged to the category of convenience sampling. The respondents were mainly students of selected universities in Warsaw (University of Warsaw and Vistula University [Akademia Finansów i Biznesu Vistula]), of full-time and part-time BA, BSc, and MA studies. The survey was also completed by some members of university staff who declared playing computer games. The surveys were circulated electronically, and the response rate did not exceed 70%. Students are particularly open to all kinds of innovation, especially if it concerns their private life or entertainment.

A specific limitation concerning this particular sample was an anticipated high percentage of smartphone, tablet, laptop, and mobile phone users, devices of lower quality but with a longer durability. The survey was completed by 274 people, out of which 254 participants submitted correctly completed questionnaires (which constitutes 92.70% of the sample). Among the respondents, 59.45% were women and 40.16% were men; 0.39% respondents did not answer this question. The average age of the respondent was 20.62 years, and the median value was 19 years. The age is typical of students of the first years of BA and BSc study and the first years of studies of the second cycle—the group asked to complete the questionnaires. The oldest person taking part in the survey (member of the university staff) was 37. Among the survey participants 63.39% were students, 35.83% were working students, and 0.79% were employees. Of the respondents, 70.87% indicated secondary-level education and 20.08% postsecondary education; the survey was primarily conducted among the students of BA studies. Also, 8.66% declared holding a BA degree or a certificate of completion of studies, and only one person indicated having a PhD.

Over 45% of survey participants indicated that they are inhabitants of cities with over 500,000 residents, over 14% came from cities with 100,000 to 500,000 of inhabitants, over 21% from towns with 10,000 to 100,000 residents, almost 5% from towns up to 10,000 residents, and 12.6% declared that they come from rural areas. The simplicity of the survey did not cause many distortions during its completion; few respondents (17) also completed additional sections of the survey.

33.3 ANALYSIS OF THE OBTAINED RESULTS AND RELEVANT DISCUSSION

The respondents provided responses to 41 substantive questions. The first group of questions concerned the characteristics of e-gamers and the scope of the use of computer games. On this basis the authors formulated more difficult questions concerning the effects and consequences of participating in computer games.

Research shows the generic characteristics of gamers. The research indicates the following findings: Of the respondents, 40% confirmed frequent game playing, that is, every day (20%) and a few times a week (over 19%). This is 10 percentage points lower than rare use of e-games, with results higher than 49%. The preliminary interview with respondents gave the impression that more respondents are interested in frequent computer gaming.

High indication of rare frequency of game playing (22% respondents confirms use of such games several times a month) showed that those respondents treat computer gaming as one among other manners of entertainment.

Another element of the research was related to technical aspects of gaming including platforms that e-gamers use for playing games. During the last 12 months, a shift toward mobile solutions especially smartphones had been observed. Last year over 35% of e-gamers (80.75% including other platforms) used mobile platforms (mostly Android). The second rank was assigned to the PC platform (65.24% including other devices), and the third position was taken by the console (e.g., PS, Xbox) with a result of 27.38% (63% including other devices). Smart TV games are not as popular and received the lowest score in the ranking with a result of 2.09% (4.81% including other devices).

Authors noticed discrepancies amounting to 33 percentage points with perception of particular e-gaming platforms. Research showed that a great number of e-gamers use both smartphones and PCs as their favorite gaming platforms. The dispersion of the results is almost 76 percentage points.

The reason for that is guided by datum that the majority of e-gamers 59.87% (including other types of access to games, 97.33%) prefer to use games installed on a device (either PC or smartphone). The second position is assigned to platforms such as Steam, and Origin amounting to 26.32% (including other types of access to games, 42.78%). Those two main sources of games together give over 86% of sources where e-gamers were playing games in the last year. The remaining origins of games, for example, Facebook (6.25%) and low score for browsers (e.g., Quake Live) of 7.57% are considered to be of marginal importance in this relation.

Very interesting results were brought by questions related to the age of starting the e-gaming experience by respondents. Research shows that most of the e-gamers started their experience during their childhood. For example, 17.11% of e-gamers declare starting e-game playing at 5 years old. The age that was most frequently indicated by respondents (50 percentage points) as the starting date for e-gaming was within 6–9 years old (median of 6–7 years old). An interesting finding is that after commutation of age ranges, 6–11 earns more than 65 percentage points of all gamers' population. Only 1.07% admits starting playing games as adults at of 20–25 years old (2.76% considering age of 16–25), which seems to be marginal to the research. The preceding results showed that early age of starting the e-gaming experience allows the games to be treated as alternative sources of entertainment in relation to other activities such as outdoor activities including games, films, TV, and art.

The authors were not able to interview children and young people from the declared starting age groups to confirm the results, which was the limitation of this research. Nevertheless, the results show the influence of early e-gaming exposition to use of e-games in later life.

Another question of the research was related to amounts of money e-gamers spend monthly at playing computer games.

The vast majority of respondents (79.14%) declared use of free applications installed on smartphones or PC games downloaded from the Internet. Another 18.18% respondents declared the possibility to pay up to 80 PLN monthly, and 2.67% were willing to pay a range of 81–300 PLN a month to play their favorite e-games. Respondents that can be considered as professional e-gamers (2.14%) declared to pay between 151 and 300 PLN; this group is constituted mainly of hobbyists, enthusiasts, and fanatics. The representatives of this group also indicated their interests in sport, which in this case is realized by means of various electronic tools (PC, smartphone or tablet, console, etc.).

The subsequent questions were used to indicate what kind of games the e-gamers played most frequently in the last year. The games were divided according to the typology indicated by the most frequent e-gamers:

- Arcade games (shooting, fighting) (e.g., *Counter Strike, Tom Clancy's Rainbow Six, Super Mario*)
- Action-adventure games (e.g., *Assassin's Creed, Half-Life*)
- Adventure games (e.g., *The Walking Dead, Wallace and Gromit*)
- Role-playing games (RPG) (e.g., *Diablo, Fallout*)
- Simulation games (e.g., *The Sims, FIFA 16, Need for Speed*)
- Strategic games (e.g., *StarCraft II, Civilization, Warhammer, Heroes of Might and Magic*)
- Survival horror games (e.g., *Resident Evil*)
- MMO games and their variants (e.g., *World of Warcraft, Lord of the Rings Online*)

Additionally respondents answered sets of questions related to the type of games they have recently played. Questions formulated in such a way might be influenced by the most recent games covering a period of a few months before the survey. Authors also consider that formulation of such questions allow for gathering more accurate responses than those related to frequency or type of played games.

The authors did not evaluate the most recent market trends, or most popular films and books that influence the gaming market. Results of the survey show the following figures: 80.75% indicated as players of the simplest simulation games; 57.75% of responses belongs to arcade games and 50.27% to action-adventure games. Additionally the number of positive responses exceeded 50%.

Games of greater complexity or duration seem to be less popular, with the survey showing smaller percentages of e-gamers admitting that they play a particular kind of game. Also games where the involvement of a gamer is at higher level are also less popular. Taking into consideration other external factors like the history of a game (e.g., "game was on a market since we remember"), popularity of a main character, or plot popularized in films or books contribute to a higher level of its presence in e-gamers' lives.

Another set of two questions was to investigate the predominance of a game over other kinds of entertainment. The aim of those questions was to address the aspect of choosing a game over other activities in two options: (1) ever or (2) within the period of the last year. Games were compared to alternative forms of entertainment, such as

- Going to the cinema
- Meeting friends
- Going on a date
- Going on a trip with friends
- Going to a party
- No such case

Responses showed that computer games are not as attractive as other activities to waive anything in the past (61.78%) or in the last year (77.40%). To indicate aspects that respondents are willing to resign from, the most frequently chosen answers were a meeting with friends (15.11%) and a party (9.78%). In case of giving up anything in the last year in favor of a game, the results were comparable. The respondents indicated a social meeting (8.17%) and a party (6.73%). In reality, the difference indicated in the percentage of people who are willing to give up other forms of entertainment amounts to 17.32 percentage points and decreases the actual numbers of indications in particular categories, the greatest with regard to social meetings (nearly 7 percentage points) and parties (over 3 percentage points).

Nearly 20% of respondents provided positive answers to the question concerning the greater prestige of a particular gamer among friends or acquaintances outside the game that he or she currently plays (Figure 33.1). It is not the score that would give evidence to the wide influence of this form of entertainment, or its particular importance for the circle of gamers' friends. In general, the present results are similar to the responses obtained in the case of the second question as regards the formation of a circle of friends made up of other gamers, who play the same games at a particular moment (Figure 33.2). A slight difference that amounts to about 6 percentage points probably results from the fact that, as reflected in the responses in the first part of the survey, many e-gamers treat games as the source of prestige or as "pure" individual entertainment (only one-third of e-gamers play multiplayer games). Thus, in general (56.15% of respondents), e-gamers do not contact each other with regard to matters that are not related to playing games (Figure 33.3). The possible examination of the reasons for the increase of prestige due to playing computer games is worth taking into consideration in the future. Considering the large variety of social networking, the level of responses indicating the fact of forming circles of friends based around gaming is relatively

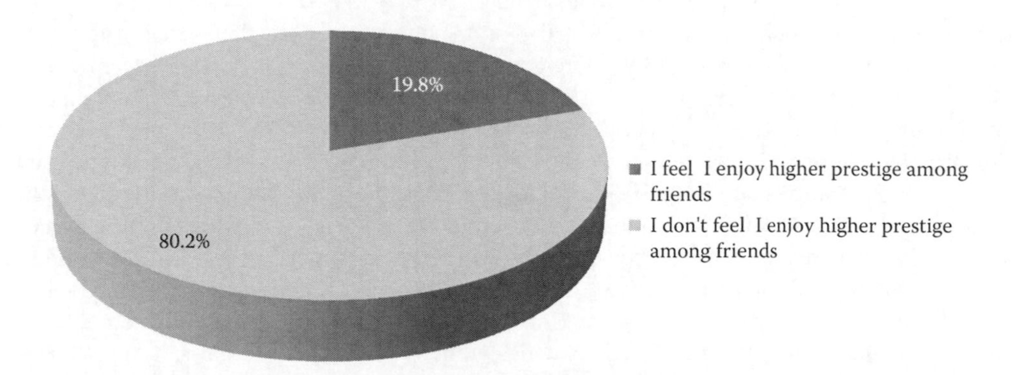

FIGURE 33.1 Playing and higher prestige among friends and acquaintances (n = 254).

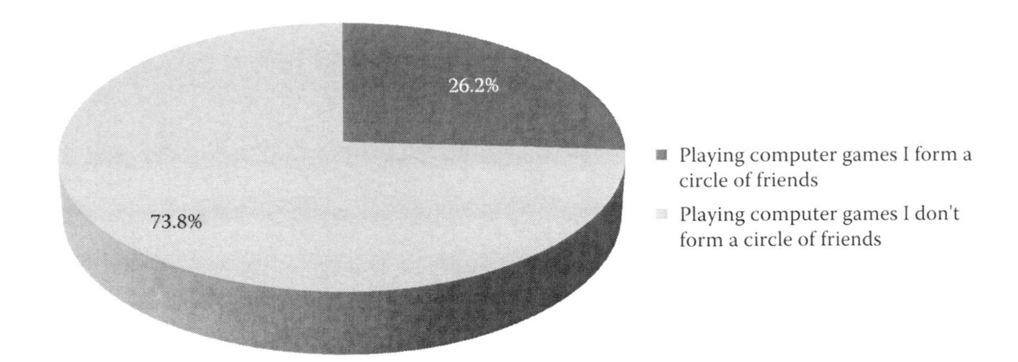

FIGURE 33.2 Forming circles of friends around playing computer games (n = 254).

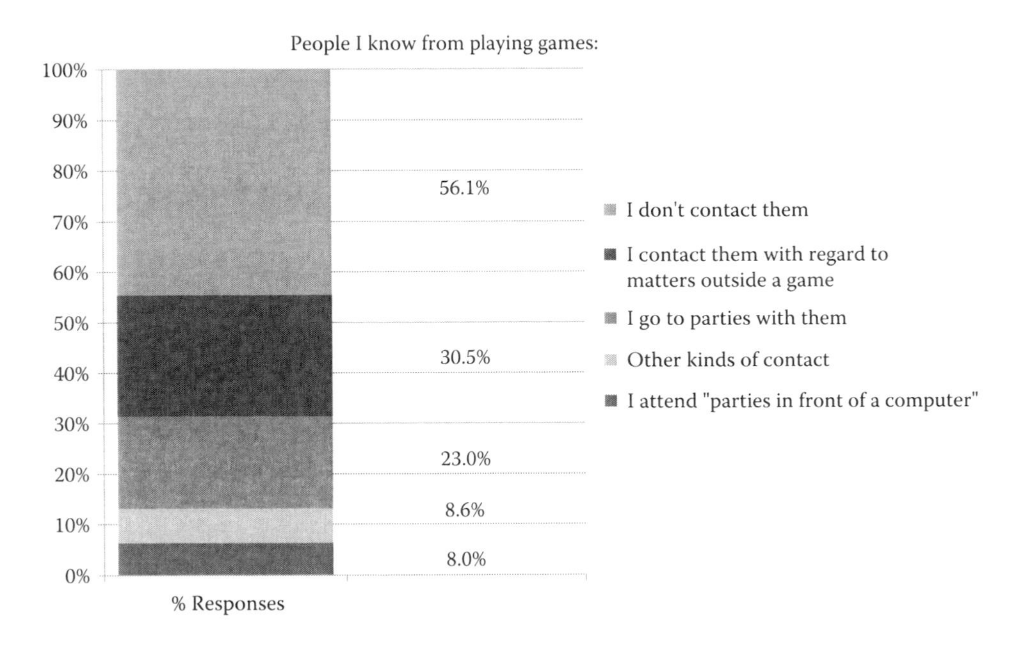

FIGURE 33.3 Social behavior toward other players (n = 254). Multiple answers were possible.

low. One of the reasons for it may be strong identification of gamers with the virtual world generated in the game and not with the real world that surrounds them.

However, on the other hand, more than 30% of respondents contact their friends also outside games. It amounts to 4 percentage points more than in the case of the circle of friends formed around the games the respondents play, which is a positive phenomenon, typical for this kind of entertainment. E-gamers indicated the fact that over 22% of them go to parties with the group of friends they share e-gaming interest with. Based on the preceding indication we may conclude that for e-gamers other e-gamers are attractive partners to maintain contact with, both in terms of virtual and direct contact. Slightly over 8% of gamers admit that they attend LAN parties ("parties in front of a computer"), which is a specific way of spending time with other e-gamers, consisting in chatting and usually drinking hard drinks. Similarly, around 8% of respondents declare that they spend their time meeting friends and being involved in other kinds of activities, for example, gaming conventions.

Subsequent questions concerned the e-gamer's emotions after playing a game in two situations: if a gamer won or if a gamer lost (Figure 33.4). The situation when the e-gamer wins a game affects his mood: 62.03% of respondents believe that they feel better after they play the game than before

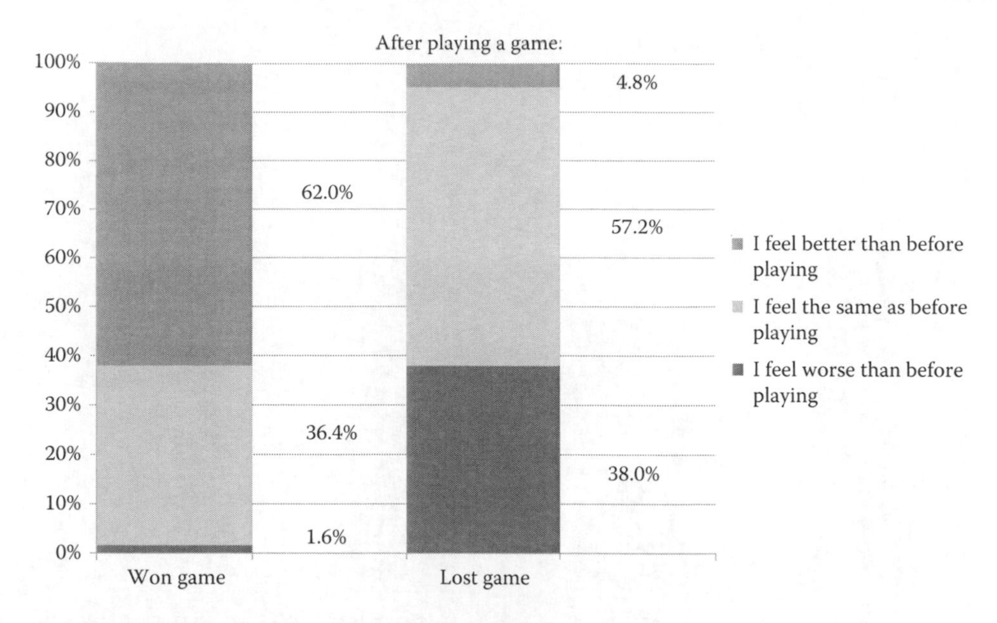

FIGURE 33.4 E-gamers' mood after playing a game (n = 254).

starting the game, and 36.36% of interviewees believe they feel the same. In the case of a game they lose, almost 38% of e-gamers feel worse than before starting the game, and over 57% say that they are indifferent to the situation. Low are the indications of good mood after losing a game (4.81%) or bad after winning the game (1.60%), which is understandable. It is quite natural considering the wide popularity of the casual games where winning a game is a relatively simple task, and the e-gamer needs a few minutes of practice to start playing the game. In total, the obtained findings indicate a deep involvement of e-gamers in playing the games, even though the ratio of over 50% indifference in the case of losing a game may be worrisome.

Interesting results have been obtained in the case of responses to the queries concerning the influence of factors related to studying or working, such as interpersonal, learning and memorizing, or management skills. Despite the fluctuating levels of negative responses (52%–68%) in this case (they did not change as a result of participating in games), the increasingly larger number of e-gamers indicate also positive effects of games on their skills. The tendencies indicated by respondents are as follows:

- 28% improvement of interpersonal skills
- 45% improvement of learning and memorizing skills
- 36% improvement of management skills

The participation of neutral skills (the skills that have not changed) is marginal (Figure 33.5). The score is close to the one recorded in a similar extended nationwide study carried out in recent years (Chmielarz and Szumski 2016). The idea of the in-depth research into the reasons of the perceived improvement of skills, as indicated by the gamers is worth considering. The next element is an assessment of the degree of subjectivity of such an impression.

In order to further analyze the problem, the respondents were also asked about the influence of games on their selected psychophysical skills such as divisibility of attention, reflex, the speed and the accuracy of decision-making, courage, and stress resistance. According to the respondents, the greatest changes were indicated with regard to reflex (59.89%). The scores above 50% were reached in the case of opinions on the positive influence of games on the speed of decision making (56%) and divisibility of attention (55%). The smallest group of people (21%) marked that they felt more

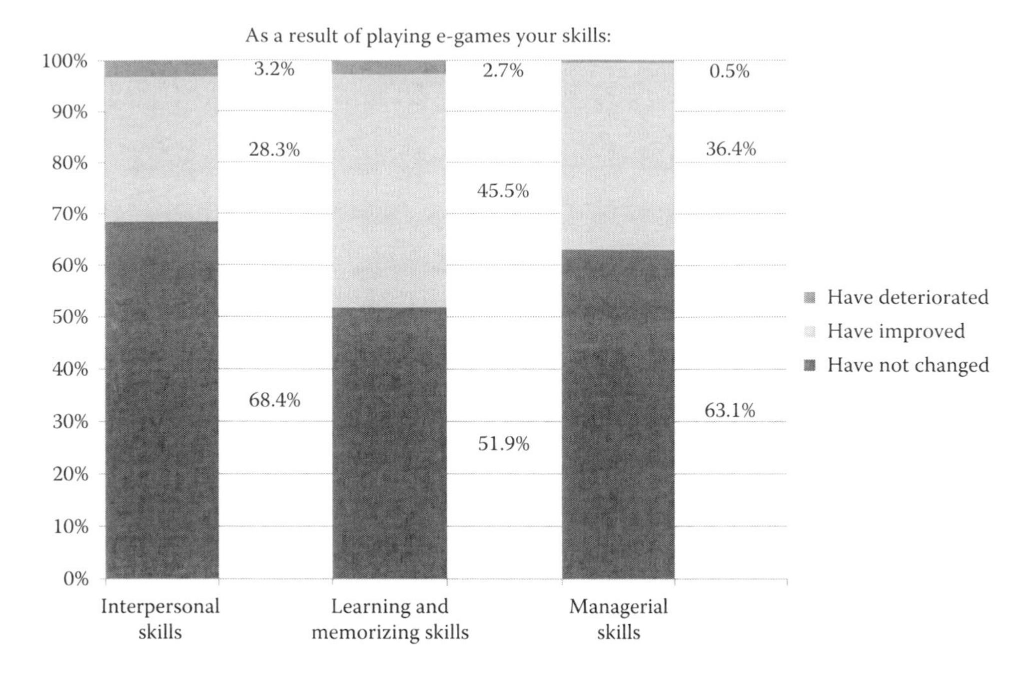

FIGURE 33.5 Effect of playing games on the e-gamers' skills (n = 254).

courageous as a result of playing computer games. The same number indicated the fact that, in their opinion, nothing has changed due to playing computer games, whereas an absolute majority of e-gamers (83%) claim that playing computer games does not cause any deterioration of their psychological and physical qualities. Among the remaining responses the greatest number indicated deterioration of divisibility of attention (6.95%) or stress resistance (4.28%). The deterioration of the remaining psychophysical features was of marginal importance. In the case of the speed of decision making and multitasking, they may be relatively easily verified. Another issue is the fact whether the positive influence is indicated also outside the game (Figure 33.6).

The question concerning the experience of discovering other people's cheating was of a slightly different character. The results are illustrated in Figure 33.7. Nearly half of e-gamers (47.59%) provided negative responses to the query; however, it should be noted that some of the e-players may have not been aware of this phenomenon, especially in the case of games with high-speed gameplay, among other first-person shooter (FPS). The remaining respondents noticed some of the most obvious cases of cheating. They most frequently pointed to cases of fraud such as map hack (37.43%) and the function of flying (36.36%) as well as removing limitations or disadvantages in a game (28.34%). Among the functions that were not included in the survey, the respondents indicated, for example, money hack and teleport hack, that were not foreseen in the game.

The authors also examined the economic aspects of participating in computer games in terms of the willingness to make money or earn real money by participating in computer games (Figure 33.8). We may notice a specific logical contradiction present in the findings. Namely, nearly 80% of e-gamers claim that they have not considered the possibility to earn money playing computer games, and simultaneously only 51% of the same group of e-gamers admit that they have not earned any money playing computer games. Thus, the responses would indicate that some e-gamers were earning money without being aware of such an opportunity. The second possibility is that it is merely a matter of declaration that the respondents have not considered earning money while playing computer games, and still they were doing it. An important element of the future in-depth studies should be the possible ways of making money on playing games, that is, whether they are mainly limited to selling virtual items or we are presented with a broader spectrum of possible options.

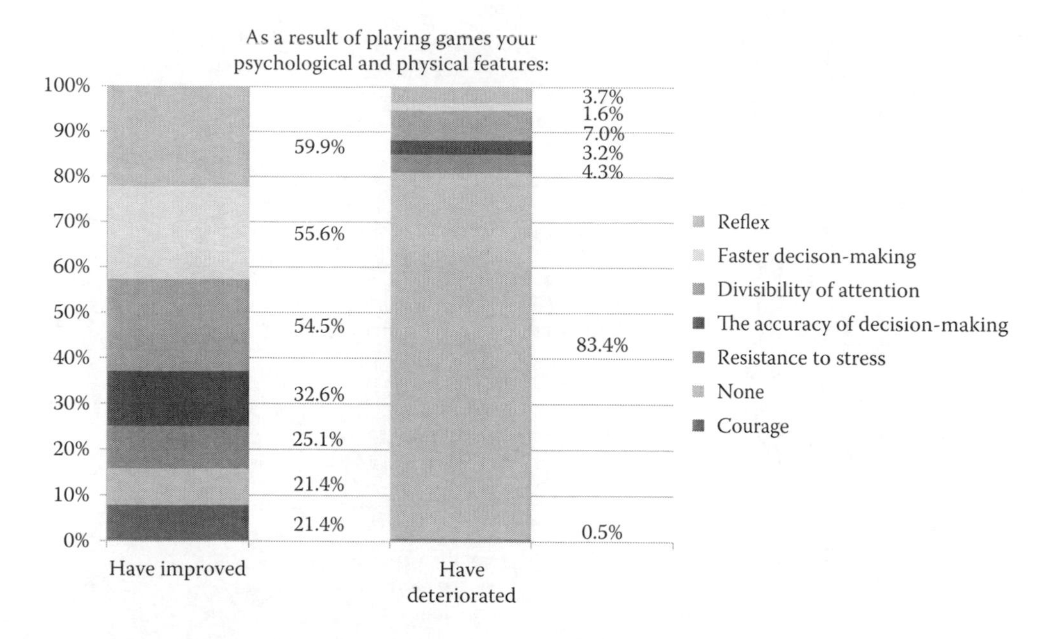

FIGURE 33.6 Influence of playing computer games on e-gamers' skills (n = 254). Multiple answers were possible.

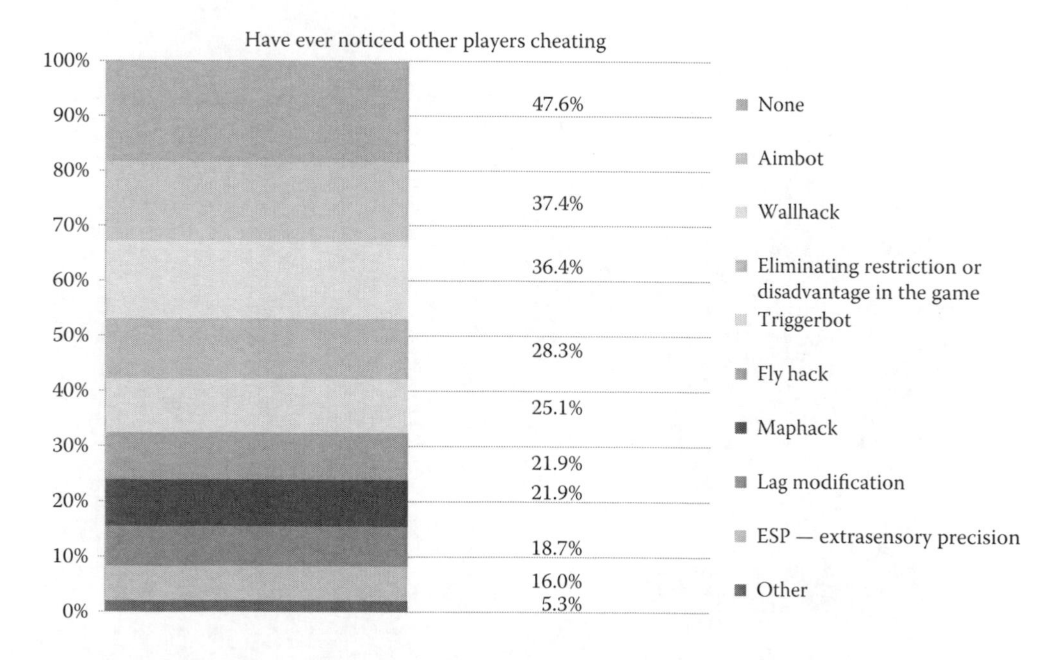

FIGURE 33.7 Cheating in computer games (n = 254). Multiple answers were possible.

Few people among the e-gamers know the names or pseudonyms of professional e-gamers (Figure 33.9). More than 82% claim that they do not know the most popular idols among players. Among the remaining group of respondents, the greatest number of them have heard about Brian Lewis (Astro) and Johnathan Wendel (Fatal1ty) (13.78% and 7.48%, respectively). The survey participants also indicated the names of other popular e-gamers not included in the list. The obtained

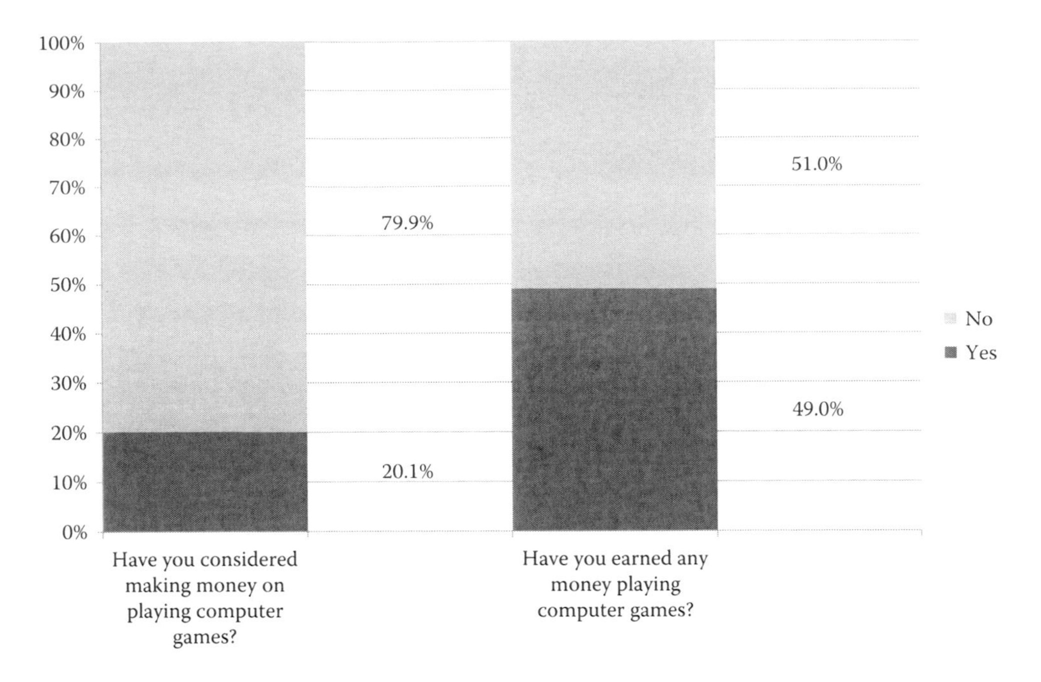

FIGURE 33.8 Earning money on participating in computer games (n = 254).

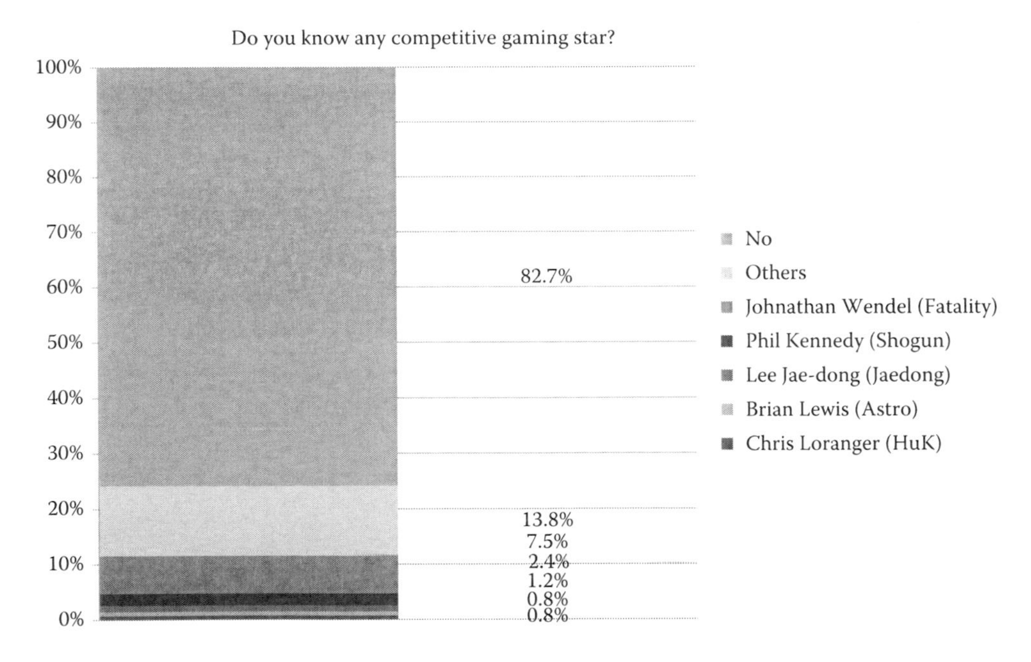

FIGURE 33.9 Level of e-gamers' knowledge concerning the most popular professional players of computer games (n = 254). Multiple answers were possible.

results confirm the earlier observations concerning switching to simple entertainment games, as a result of marked changes in e-games technology.

Further problems the survey participants had to comment on concerned the e-gamers' active and passive participation in Polish and international events related to the e-games industry (meetings, trade fairs/shows, games, championships, etc.). The obtained findings turn out to be surprising.

Among students of the Faculty of Management in the University of Warsaw and the students of BSc studies at Vistula University, around 85% to 98% of respondents never participate in such events. The highest score (nearly 8%) was indicated in the case of single or multiple times (almost 7%) when the respondents passively participated in the events connected with e-gaming in Poland. None of the members of the examined group admitted to regular, active participation in the events in Poland: 3% of respondents participated actively in the events a few times and 4% only once. The situation concerning international events is even worse: less than 3% participated in such events several times, and around 2% participated in such an event once; an active and repeated participation was indicated only by 0.39%, and 1% of respondents claim that they participated in such an event once. There remains one more problem to examine: What kind of environment do e-gamers who regularly participate in such events come from? Simultaneously there appears a chance to take advantage of the circumstances: a huge market gap that exists in the area creates an opportunity that should not be wasted (Figure 33.10).

The last group of survey questions concerned the e-gamers' interests connected with playing games and other pastimes. In the first case the questions concerned additional hobbies connected with games they participated in, for example, writing stories about games, recording and publishing game videos, and mixing gaming music or video clips from the games. And here e-gamers rather concentrate on the game itself and not on related activities connected with the games. Almost 94% of e-gamers are not interested in it at all. The marginal number, close to 4%, record and publish game videos, mainly on the Internet. This fits within the trend known from the Web 2.0 concept, according to which a few percent of people create content for the remaining 90%. The findings are presented in Figure 33.11.

A slightly higher level of interest is indicated in the case of passive reception of the content related to fan fiction (e.g., zines, game videos, game-related amateur films). However, 75% of e-gamers claim that they have never been interested in it, but 15.75% of respondents are interested in it at present, and over 9% are recipients of such content (Figure 33.12). The passive reception of

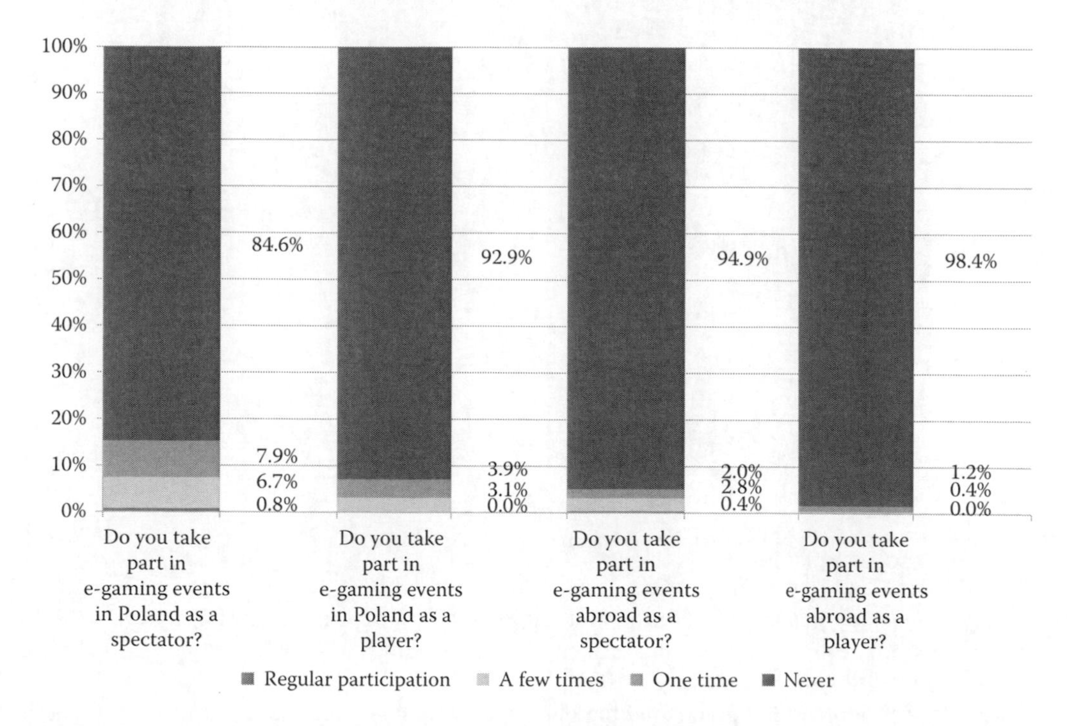

FIGURE 33.10 Passive and active participation in national and international events connected with e-gaming work (n = 254).

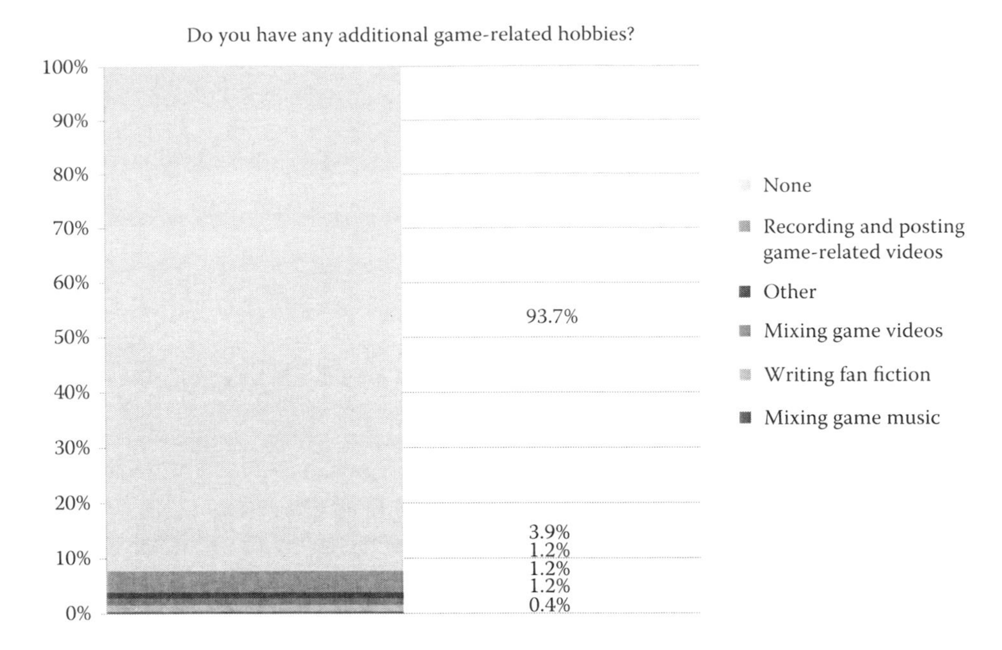

FIGURE 33.11 Additional hobbies connected with games (n = 254).

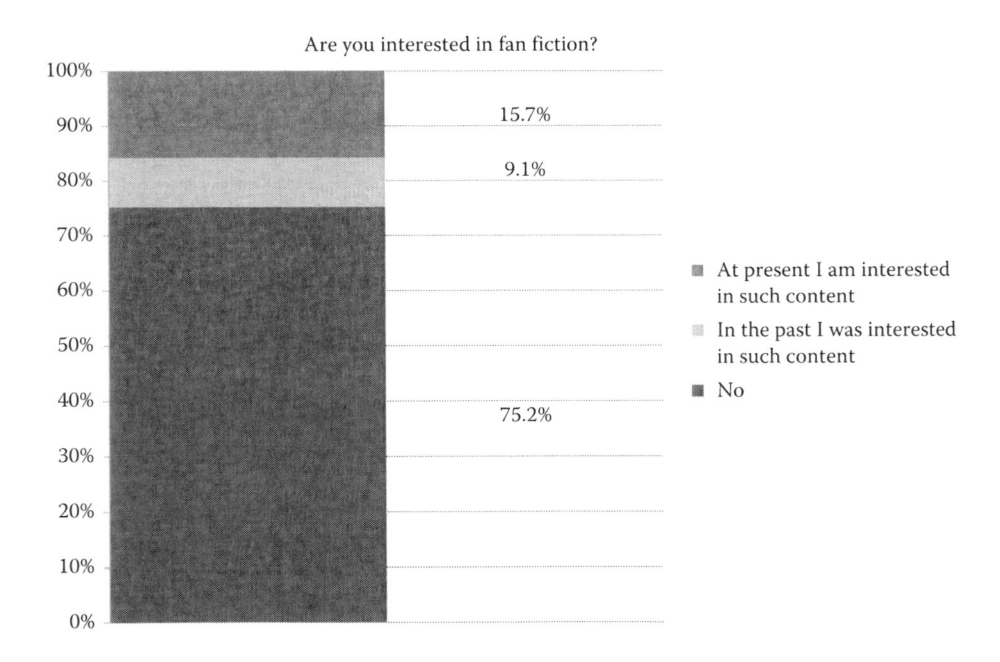

FIGURE 33.12 Reception of fan fiction content among the respondents (n = 254).

the content is usually connected with the cult games such as, for example, The *Legend of Zelda* series. Fan fiction presents itself as a niche phenomenon, and thus, we may conclude that 15.75% of respondents indicating their interest in such content is a relatively high score.

If we consider hobbies that go beyond the participation in computer games, it turns out that the e-gamers have very broad interests (Figure 33.13). The greatest number of people (72%–85% in other studies) are keen on physical activity (sport, leisure). Over 66% watch films; reading books

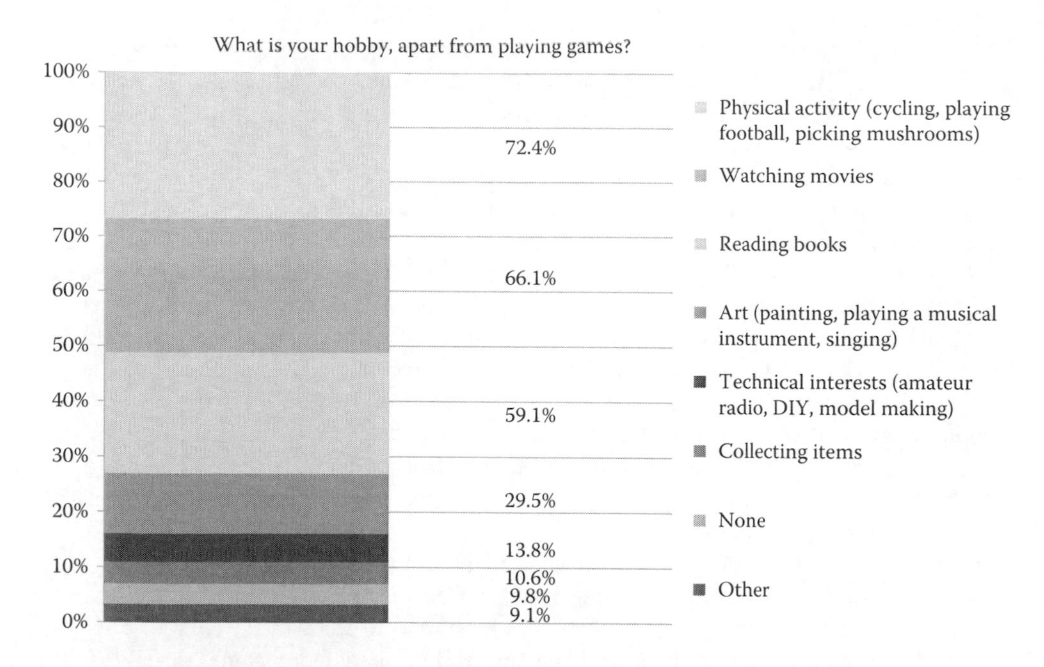

FIGURE 33.13 Additional hobbies of e-gamers (n = 254). Multiple responses were possible.

as a hobby was indicated by 60% of respondents (the score is 10 percentage points lower than in the case of other studies, e.g., #JestemGraczem 2014). Over 29% of survey participants are actively engaged in hobbies related to arts, such as, for example, painting or playing a musical instrument, and almost 14% of them are interested in technical hobbies (such as model building or DIY). Nearly 11% pursue collection hobbies. Less than 10% of e-gamers admitted that they do not have any other hobbies apart from playing computer games.

33.4 CONCLUSIONS

The conducted research and the presented findings point to the following conclusions:

- Almost all respondents (over 99% of the sample) in the current study were students, which was reflected in the obtained scores. The older the students, the weaker interest in completing the questionnaire or its findings. It is caused by the increasing number of tasks connected with studies as well as the heavy workload connected with regular or temporary work (nearly 36% of working students). The latter is confirmed in the scores of other surveys (#JestemGraczem 2014; Marketing przy Kawie 2014; Newzoo 2014; Żywiczyńska 2014b), despite the fact that, in total, from 16% to 25% students participated in the study (depending on study) (even though it was always the largest group of players).
- Among people who completed questionnaires there were markedly more women (almost 60%) than in other survey studies (around 43%–48%) (Żywiczyńska 2014a), conducted 2 or 3 years ago. Thus, we may conclude that there occurs a specific change with regard to the number of women playing computer games. Naturally, we should also be aware of the fact that the present study examined mainly the responses of students of economic faculties, and in this case the general number of female students in these faculties is greater than men. Still, the survey included the option "I don't play computer games," which the women could indicate.
- Frequency examined in this study covers the following options: every day and up to a few times a week. The study shows that 39% of respondents play computer games frequently.

This result differs from other studies and is of 20 percentage points less (39% here, as compared to 62%–63%). The rationale of such discrepancy can be guided by the chosen research sample, whereas this study was done with students as indicated in the initial part of this chapter, whereas other surveys were done also with pupils who are considered another large group of potential gamers (Marketing przy Kawie, 2014). The group of pupils has more free time than students, especially senior students.

- The research showed that the smartphone is the most popular device used for playing computer games (over 80%, mainly on the Android system due to a large number of free games). It does not exclude parallel use of other devices, mainly PC (over 65%) and a regular console (63%) or mobile (11%). Mobile solutions such as smartphones and tablets are constantly taking a role of a PC. There is noticeable change of trend of use of different gaming devices. Two or three years ago the proportions were more or less reversed; approximately 90% of respondents (#jestemgraczem, 2014; Żywiczyńska 2014a, b) used mainly personal computer, and only half of them a smartphone or a tablet. The reason for such situation is that mobile devices allow for occasional use of many kinds of generally simple games at any place or time.
- Dynamic development of smartphones and tablets in the last two years as well as wider access to mobile applications including games, affected the popularity of different game types that e-gamers are playing in their free time than it was 2 to 3 years ago. A good example is the loss of popularity of RPG games to simple simulation games (over 80%) and arcade games (58%) and action-adventure games (50%). Noticeable is that arcade games are again gaining an important share of the computer game market. Comparing the present research with the earlier studies related to this area (#jestemgraczem, 2014; Żywiczyńska 2014a, b) it seems that RPG games lost their popularity due to the increasing importance of mobile devices use (here 44%, but 65% in other studies).
- 97% of gamers use the games installed on a smartphone or a PC, with a surprisingly low percentage (10%) of people using Facebook games.
- 54% of the respondents, after completing one game, take a break before they start playing another game, and only 14% immediately start to play another game.
- Participation in computer games does not significantly raise an e-gamer's prestige; also, it does not influence the creation of circles of friends among e-gamers, and over a half of e-gamers admit that they do not contact each other outside the game.
- After winning a game, the positive emotions of e-gamers are significantly on the increase; losing the game, on the other hand, does not result in creating strong emotions. For example, 50% of the survey participants claim that they remain indifferent to the outcome of the game,
- E-gamers positively evaluate participation in games. Almost half of respondents feel that their learning and memorizing skills have improved.
- In the opinions of respondents, games have improved mainly such psychophysical skills as reflexes, the pace of decision-making, and divisibility of attention. Over 80% of people maintain that playing games does not deteriorate their psychophysical qualities.
- Over half of e-gamers have noticed cases of cheating when playing games and admit that various hacks may be applied by players.
- Computer games to a greater degree are, seen as the source of income (e.g., being involved as a player, or selling a game account, etc.) even though e-gamers are trying to prove otherwise.
- In total, e-gamers generally do not know the idols of e-sports circles (more than 80% of respondents). They do not take part, neither as players or spectators, in e-sports related events in Poland and abroad (over 90%). Also, they do not have any additional hobbies (such as writing fan fiction, recording and posting game videos, mixing gaming music, or mixing game video clips) connected with this kind of entertainment (this indication was reported in more than 90% of cases).

- Relatively more people (around 25%) are interested in the fan fiction content due to the fact that they do not require such intense player's engagement as the aforementioned activities.
- The surveyed e-gamers have broad interests outside the area of gaming. They are mainly active with regard to physical exercise (sport, leisure), they watch films, read books, or are interested in art.

The findings of the second stage are not as positive as those obtained in the preliminary stage of the research. They tend to support the thesis concerning the fact that games are treated as a form of entertainment rather than sport, and they belong to a domain of amateurs rather than people who would like to take up playing games professionally. In order to move this form of entertainment to the next level, we need to examine the reasons why the group comprising enthusiasts and people who are serious about e-games is so small, and further studies should include fan clubs and groups of players participating in the tournament games (both home and abroad).

REFERENCES

Akcjonariat Obywatelski. 2013. Polski rynek gier komputerowych na tle rynku światowego, http://akcjonariatobywatelski.pl/pl/centrum-edukacyjne/gospodarka/1033,Polski-rynek-gier-komputerowych-na-tle-rynku-swiatowego.html, accessed January 2016.

Chmielarz, W. 2015a. Porównanie wykorzystania sklepów internetowych z aplikacjami mobil-nymi w Polsce z punktu widzenia klienta indy-widualnego (Comparison of the Use of Mobile Applications Websites in Poland from the Point of View of Individual Client), Vol. II, Part IX Inżynieria ja-kości produkcji i usług, in: R. Knosala (ed.), *Innowacje w zarządzaniu i inżynierii produkcji*, Oficyna Wydawnicza Polskiego Towarzystwa Zarządzania Produk-cją, Opole, pp. 234–245.

Chmielarz, W. 2015b. Study of smartphones usage from the customer's point of view, *Procedia Computer Science*, 65, 2015, 1085–1094.

Chmielarz W. and Szumski O. 2016. Charakterystyka e-graczy i ich preferencji w grach komputerowych, Innowacje w zarządzaniu i inży-nierii produkcji'2016, Zakopane, 2016.

GRY-OnLine. 2004. Klasyfikacja gier, http://www.gry-online.pl/S018.asp?ID=208andSTR=2, accessed January 2016.

GRY-OnLine. 2014. 13,4 miliona graczy w Polsce i inne informacje o naszym rynku, http://www.gry-online.pl/S013.asp?ID=82806; accessed January 2016.

IT-Pomoc.pl. 2016. Czym jest gra kompute-rowa, http://it-pomoc.pl/komputer/gra-komputerowa; accessed January 2016.

#JestemGraczem. 2014. Badanie graczy w Polsce, http://www.jestemgraczem.com/wyniki, accessed January 2016.

KIPA. 2016. Definicje gier komputerowych, http://www.kipa.pl/index.php/promocja-filmu/gry-komputerowe/definicje-gier-komputerowych, accessed January 2016.

Marketing przy Kawie. 2014. Jacy są Polacy grający w gry komputerowe? http://www.marketing-news.pl/message.php?art=43734, accessed January 2016.

Mijal M. and Szumski O. 2013. Zastosowania gier FPS w organizacji, in: Chmielarz W., Ki-sielnicki J., Parys T. (eds.), *Informatyka @ przyszłości*, Wydawnictwo Naukowe WZ UW, Warsaw, 2013, pp. 165–176.

Newzoo. 2014. Global games market report, http://www.newzoo.com/product/global-games-market-report-premium/, accessed January 2016.

PTBG. 2010. Homo Ludens 1/(2), http://ptbg.org.pl/HomoLudens/vol/2/, accessed January 2016.

Wiedza i Edukacja. 2009. Analiza gier, http://wiedzaiedukacja.eu/archives/tag/analiza-gier, accessed January 2016.

Wikpedia. 2016. Gra komputerowea, https://pl.wikipedia.org/wiki/Gra_komputerowa, accessed January 2016.

Zając J. 2014. Jestem graczem w social media, http://blog.sotrender.com/pl/2014/12/jestem-graczem-w-social-media/, accessed January 2016.

Żywiczyńska E. 2014a. Co tak naprawdę wie-my o graczach, http://zgranarodzina.edu.pl/2014/10/12/co-tak-naprawde-wiemy-o-graczach/, accessed January 2016.

Żywiczyńska E. 2014b. Optymizm czy myślenie ży-czeniowe. Zaskakujące wyniki badania #jestemgra-czem, http://zgranarodzina.edu.pl/2014/12/20/optymizm-czy-myslenie-zyczeniowe-zaskakujace-wyniki-badania-jestemgraczem/, accessed January 2016.

34 ICT Drivers of Intelligent Enterprises

Monika Łobaziewicz

CONTENTS

34.1 Introduction ... 455
34.2 An Intelligent Enterprise: What Does It Mean? ... 456
34.3 ICT Drivers of Intelligent Organization Research ... 458
34.4 ICT Drivers Empowering the Intelligent Enterprise ... 462
34.4.1 Mobile Workforce Integration ... 462
34.4.2 Smart Virtual Workplace ... 462
34.4.3 E-Collaboration ... 462
34.4.4 Business Flexibility ... 463
34.4.5 Scalability and Customization ... 463
34.4.6 Business Continuity ... 463
34.4.7 Converting Data into Business Intelligence ... 463
34.5 Conclusions ... 464
References ... 464

34.1 INTRODUCTION

With the evolution of information technologies (IT), systems, applications, and professional IT tools to drill and to analyze data generated in organizations, a discussion about intelligent enterprise has been ongoing. The way which and how data and information are processed is vital for any significant improvement in any organization today. It can be noticed that the information and communications technology (ICT) solutions and advanced technologies supporting business or manufacturing processes take its place as a primary driver of profitability and market differentiation in every industry or branch. But without the knowledge and skill for using data or information generated through these technological solutions to create organizational value, it is difficult to compete in the marketplace. In recent years, the term *intelligent* has gotten special meaning, next to the term *innovation*. Thus, enterprise intelligence may be considered as a methodological structure institutionalizing the adaptation through the continuous adjustment and evaluation. This allows the company to deal with unknown market situations; to adapt to new approaches, strategic, and tactical concepts; and to develop key competences. It means an "engine of innovation" allowing for a company to constantly reposition itself for the desired competitiveness in the dynamically changing market (Thannhuber 2005).

The aim of this chapter is the identification of ICT drivers empowering the intelligent enterprise based on the research conducted in Poland and other countries and the scientific discussion about their role in the intelligent enterprise management. The amount of research dedicated to this subject is very small, thus it should be continued. The chapter consists of three parts. The first is dedicated to the concept of intelligent enterprise. Presented in the second section are final results of research conducted by Polish Agency for Enterprise Development and MIT Sloan Management Review partnered with the IBM Institute for Business Value concerning ICT drivers empowering intelligent organizations. The third part presents the results of surveys conducted by the author about ICT drivers.

34.2 AN INTELLIGENT ENTERPRISE: WHAT DOES IT MEAN?

The concept of an intelligent enterprise has its source in the science of management and is particularly correlated with the following issues: intelligent information systems, intelligent ICT solutions for enterprise management, business intelligence (BI), a learning organization, knowledge management, and intellectual capital management.

Also, the discussion about an intelligent enterprise is the result of rising digital complexity (Gartner 2014), unprecedented data volumes (Kurylko 2014), decreasing cost of digital data storage (Stockton 2014), virtually unlimited compute power (Dickey 2014), improvements in deep learning, and cognitive-computing technologies.

The wide scope of theoretical and practical basics causes leads to a number of approaches and definitions, the further consequence of which is the lack of a commonly accepted, universal definition. An intelligent organization is able to collect, process, interpret, and communicate information needed for decision-making processes (Wilensky 1967). Nonaka and Tekeuchi (1995) believe that an intelligent organization is not based on specialized research and development (R&D) departments, but is based on its members, the way in which they behave, and the culture they present, as part of which each is an employee and entrepreneur with his or her knowledge. Surely, intelligent organizations constantly learn and are open to knowledge. This process of learning consists of observations of the external and internal environments, development of the perception of understanding the environment, giving meaning through an interpretation and undertaking activities, and improvement of organization behavior (Hamel and Prahalad 1994; Muryjas and Wawer 2014).

An intelligent organization has a capacity to teach not to do things that are just old habits or routine. It is able to change those activities that bring no progress or are wrong (Christensen 1997).

As the *Accenture Technology Vision 2015* report reveals, an intelligent enterprise embeds software intelligence into every aspect of its business to drive new levels of operational efficiency, evolution, and innovation. An intelligent enterprise is based on huge data and smart systems (Accenture 2015).

An intelligent enterprise is able to turn raw data into useful organization knowledge using business intelligent solutions with in-depth analytical capabilities. In an intelligent enterprise, information systems are integrated knowledge gathering and analyzing tools for data analysis and dynamic end-user querying of a variety of enterprise data sources (Gupta and Sharma 2004).

Therefore, an intelligent enterprise has following abilities:

- High adaptability to a changing environment
- Ability to influence and shape the environment
- Ability to find new strategic domains (in the product-market system) in their external environment and the rapid reconfiguration of resources according to the new domain
- Ability to make a positive contribution to the development of their environment in the context of sustainable development (Schwaninger 2009)

The research conducted by IBM (LaValle 2009) have pointed the essential characteristics that describe an intelligent enterprise opposite to the traditionally managed enterprise (Table 34.1).

The most important features that distinguish the two types of models include

- The transition from focusing on improving production processes to focusing on innovation as an essential factor of value creation; human capital, a creative factor of the development of intelligent organization is its key value
- The transition from high to low states of working human capital by optimizing supply chain management systems and customer relationship management
- The transition from high to low states of physical capital, and recognition of intellectual capital as a fundamental value creation medium (Davenport et al. 2006)
- Using digital technology to weave businesses into the broader digital fabric that extends to customers, business partners, employees

TABLE 34.1
Traditional Enterprise and Intelligent Enterprise

Criteria	Traditional Enterprise	Intelligent Enterprise
Aware: Gathering, sensing, and using structured and unstructured information	1. Collects data from its own transactional systems and internally generated data 2. Processes some large databases in batches to create snapshots of the past 3. Keeps large just-in-case stores of information that are not interpretable, not understandable, and ultimately not usable	1. Collect and analyze data from anywhere, including external sources, new instrument data, and unstructured and societal data 2. Process incredible quantities of data at incredible new speeds as needed 3. Gain insights from previously unquantifiable and unusable data
Linked: Connecting internal and external functions	1. Uses professional expert knowledge when it is definitely required 2. Has expertise and accesses wisdom based upon who people know and who is close by 3. Has a distance to an open collaboration, treat business partners more as clients	1. Mobilize experts to work together both within the enterprise as well as collaborate with external entities for mutual advantage 2. Generate a new type of collective wisdom from larger and more sophisticated crowds of experts 3. Keep information more relevant and use it beyond its spot application, having implications both up and down the value chain
Precise: Using only the most relevant information to support timely decisions and actions closer to the point of impact and consequence	1. Uses content and structured information transnationally 2. Has users who must seek out information based on the immediate need 3. Does not give employees the information on time when they need it. 4. Delivers volumes of data separately and rarely in context of the situation or parceled together into actionable packages	1. Manage and analyze vast stores of content 2. Prepare data and automate analysis to ensure the quality and timeliness of information 3. Deliver information in ways that are useful to the context of the situation being handled 4. Deliver just the right amount of quantitative data, definitions, knowledge bases, unstructured data, and expert networks to meet the decision-maker's need at the right moment
Questioning: Challenges are the status quo while creating new opportunities	1. Focuses on getting today's tasks done 2. Views innovation as a discrete function of R&D or product managers 3. Makes its decisions and moves on with little interest in whether expectations are met	1. Get the job done and think about the improvement 2. Create knowledge workers and involve them in innovations and improvements 3. Evaluate outcomes relative to expectations, tracking, and understanding exceptions both good and bad
Empowering: Enabling and extending employees' memory, insight, and reach, as well as the authority to decide and act	1. Takes decisions "up to down" 2. Aligns incentives to how much someone works or what they produce 3. Hunts, searches, and compiles information	1. Automate and orchestrate routine tasks in order to focus people on new, unsolved issues, and opportunities 2. Delegate decision making to the best agents for the situation whether they are employees, workflows, bots, or customers, requiring less managerial and administrative oversight as employees solve issues immediately and locally 3. Align incentives to smart results while also considering how the results were achieved 4. Give people access to user-friendly, fact-based tools available over the channels and devices of their own choosing

(*Continued*)

TABLE 34.1 (*Continued*)
Traditional Enterprise and Intelligent Enterprise

Criteria	Traditional Enterprise	Intelligent Enterprise
Anticipating: Predicting and preparing for the future	1. Uses personal experience and informed guesswork to make decisions 2. Uses historical data for reporting and tracking 3. Manages performance and risk separately with all future variance and chance managed reactively.	1. Build simulations and models to understand future implications for alternatives based on facts 2. See opportunities and threats as they are happening and even beforehand 3. Track events in real time, applying sophisticated rules, enabling automation, and speed of response 4. Be informed about opportunity and risk, and know what to do tactically about events well before action is needed

These features of the business model of an intelligent organization cause a shift of attention from the improvement of existing processes as a development factor to an ongoing changes of existing processes through the mechanisms of a learning organization. In the intelligent organization attention is moved in the management of enterprises from material resources to intangible resources that require specific management competences for a continuous process of converting information into intelligence.

34.3 ICT DRIVERS OF INTELLIGENT ORGANIZATION RESEARCH

In Poland the only research concerning intelligent organizations was done by Polish Agency for Enterprise Development (PARP) in 2010 on a group of 300 enterprises. The purpose of the research was finding an answer to a question whether small- and medium-sized enterprises (SMEs) in Poland use the solutions dedicated for intelligent organizations and whether they are more effective than other enterprises. In the research it was assumed that an intelligent organization fulfills the following conditions:

- It has a strategy of development, which includes long-term goals to achieve and possible ways of their achievement.
- It has a well-developed personnel management policy.
- It has a company website and intranetwork, and uses specialized computer software.
- During the process of making a purchase or sale, it exchanges knowledge with the environment.

Studies have shown that for SMEs 26.5% of the participating companies had a strategy of development, 31.6% of them had a personnel management policy, 47% had developed computer software, and 38% exchanged knowledge with the environment during the process of making a purchase or sale. However, the results indicated that almost 63% of companies that belong to the sector of large enterprises have both a strategy of development policy and the personnel management policy.

Therefore, the results imply that bigger organizations meet the criteria of intelligent organization to a larger extent than SMEs. Further analysis of the ICT solutions used in Polish enterprises shows that innovative enterprises more often apply the solutions of intelligent organizations than companies that are not innovative. The probability that an innovative company would implement intelligent solutions is about double in comparison with a company that is not innovative, which means that the implementation of the innovation process triggers mechanisms leading to the usage of appropriate solutions applicable to intelligent organizations.

In Poland, intelligent organizations do not have a clear innovative profile yet established, as the research in this subject became possible after 2007–2013, which was the time period of obtaining resources under the Operational Programme Innovative Economy. Only after this period was it possible to find out what types of innovations Polish companies implemented and how many of them fell under this program. When the Operational Programme Intelligent Development 2014–2020 started, it was known that the type of innovation is not a factor differentiating companies in terms of their willingness to implement solutions typical for intelligent organizations. Most often implemented were process innovations (28%), slightly less more often than organizational innovations (24%) and product innovations (21%). The tendency to introduce into practice solutions appropriate for intelligent organizations increases with the size of company turnover, which is most often positively correlated with its largeness. As far as the sector of business activity is concerned, intelligent organizations have the biggest share among industrial companies (14%), as well as trade and service companies.

Intelligent organizations are more common among companies with greater turnover. These results are in line with expectations, as the meeting of the criteria of intelligent organizations requires financial investment, which at a low scale of operation is not always profitable.

The results indicate a stronger focus on technological development among intelligent organizations, and their better adaptation to the challenges of the knowledge-based economy, where the speed of access to knowledge and the possibility of its use is a key factor of competitiveness.

Intelligent organizations in Poland more often use ICT solutions to support management processes in comparison with other organizations. The most commonly introduced is the system for electronic circulation of documents (e-workflow) and databases and data warehouses (83%), as well as intranet (76%). Other solutions are being used much less frequently: every fourth intelligent organization uses customer relationship management (CRM) and solutions that support group work, every fifth intelligent organization practices of human resources management (having marginal use in remaining organizations), and every sixth intelligent organization practices business intelligence (three times more often than other organizations) (Figure 34.1).

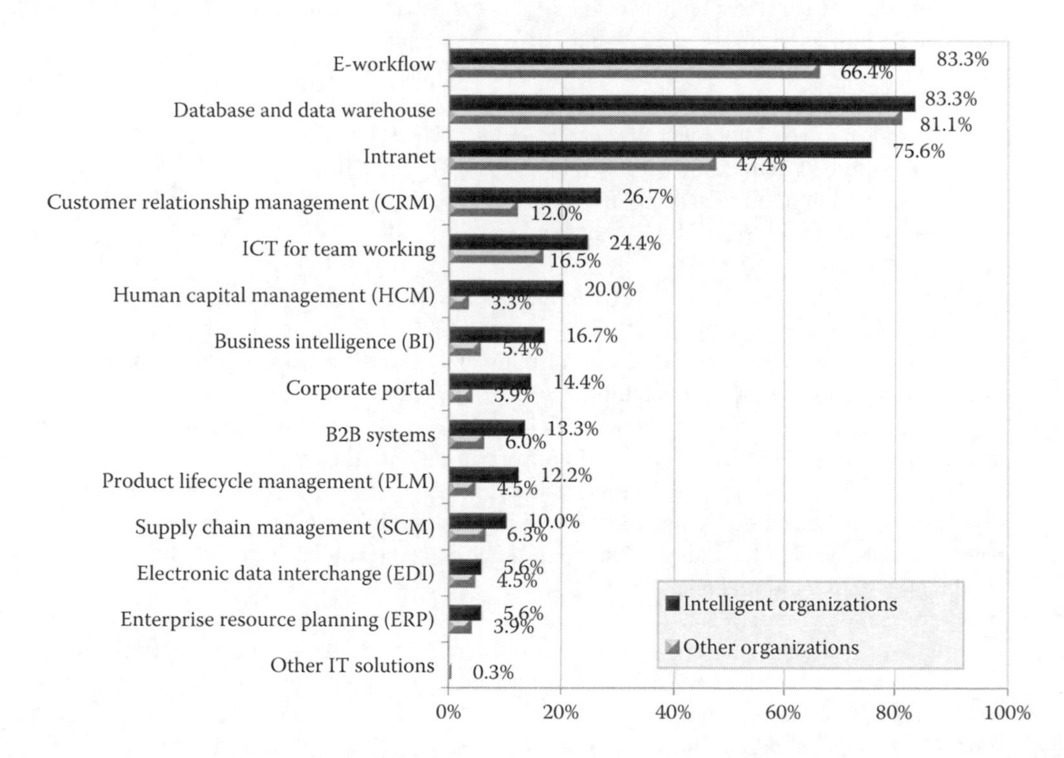

FIGURE 34.1 Adopted computer software solutions as tools that support knowledge management processes by intelligent organizations and other organizations.

Intelligent organizations much more often than other organizations are planning to implement ICT solutions to support management processes. Respondents believe that the vast majority of companies plan to implement electronic document flow (100% intelligent organizations and 69% of other organizations), databases and data warehouses (96% and 75%), and intranet (96% and 50%). One in three intelligent organization is planning in this period to implement more advanced solutions, namely, human capital management, customer relationship management, or business intelligence.

The last problem that concerns ICT solutions that support the management of intelligent organizations is an assessment of their effectiveness. Enterprises that have used various ICT tools generally highly evaluated their effectiveness. The few critical comments were focused on low efficiency of databases and data warehouses. The results achieved by the implementation of supply chain management (78%) and customer relationship management (70%) were very positively evaluated. As far as the effectiveness of various ICT tools by intelligent organizations is concerned, but taking into account the size of organizations, it is worth emphasizing that ICT tools are generally assessed as less effective by small businesses than by middle-sized and large businesses. This is due to the specific nature of these tools, which do not necessarily have to be effective in organizations with poorly developed organizational structure and not very complicated processes. The respondents representing small companies indeed pointed out that these tools are useful. However, in this group only a few people indicated very high usability of ICT tools (Kordel et al. 2010).

Finally, Polish companies are still learning how to create the intelligence. This is especially a challenge for SME companies. In Poland the situation of SME sector companies is changing dynamically as there appear opportunities for the implementation of computerization of companies, in particular in the field of e-business, implementation of ERP systems and dedicated systems, or advanced ICT applications. At the moment, there are no advanced studies that show how these solutions have contributed to the development of intelligent organizations and what is the level of growth of the intelligent enterprises between 2010 and 2015.

In 2010 MIT Sloan Management Review partnered with the IBM Institute for Business Value to conduct a survey among nearly 3,000 executives, managers, and analysts working across more than 30 industries, and involved intelligent organizations of various sizes in more than 100 countries. They also interviewed academic experts and subject matter experts from a number of industries and disciplines to understand the practical issues facing intelligent organizations (LaValle et al. 2010). As a result, the survey provided the following results:

- Intelligent enterprise is focused on the biggest and highest value opportunities, which is against traditional organizations.
- Intelligent enterprise uses each business opportunity, starting with questions, not data. Traditionally, organizations are tempted to start by gathering all available data before beginning their analysis. Too often, this leads to an all-encompassing focus on data management: collecting, cleansing, and converting data that leaves little time, energy, or resources to understand its potential uses. Intelligent organizations should first define the insights and questions needed to meet the big business objective and then identify those pieces of data needed for targets. They can target specific subject areas, and use readily available data in the initial analytic models.
- Intelligent enterprise drives actions and delivers value. New methods and tools to embed information into business processes using cases, ICT analytics solutions, optimization, workflows, and simulations are making insights more understandable and actionable.
- Intelligent enterprise develops existing capabilities adding new ones. To do this, it uses sophisticated modeling and visualization tools based on ICT. On the contrary, new tools should supplement earlier ones or continue to be used side by side as needed.
- Intelligent enterprise uses an information agenda to plan for the future.

Nowadays, big data is getting bigger. Information is coming from interconnected supply chains. Strategic information arrives through unstructured digital channels: social media, smartphone applications, and an ever-increasing stream of emerging Internet-based gadgets. The information agenda identifies foundational information practices and tools while aligning IT and business goals through enterprise information plans and financially justified deployment road maps. This agenda helps establish necessary links between those who drive the priorities of the organization by line of business and set the strategy, and those who manage data and information. A comprehensive agenda also enables managers to keep pace with changing business goals. It provides a vision and high-level road map for information that aligns business needs to growth (Hopkins et al. 2010).

The last research related to an annual view of the technology trends that will have a profound impact on enterprises for the next 3 to 5 years was carried out by Accenture Technology Labs in 2015. *Accenture Technology Vision 2015* (Accenture 2015) highlights five emerging trends that reflect the shifts being seen among the digital power brokers of tomorrow: "the internet of me," "outcome economy," "the platform (r)evolution," "the intelligent enterprise," and "workforce reimagined" (Figure 34.2). These trends represent the newest expression that every modern business is a digital-driven business. The Accenture research put a multiyear perspective on technology that should impact the strategies and operational priorities for organizations worldwide. As *Accenture Technology Vision 2015* provides, in the intelligent enterprise the level of operational excellence will emerge from the latest gains in software intelligence. Until now, increasingly capable software has been geared to help managers and employees make better and faster decisions, but the influx of big data, advanced data processing, and cognitive computing will force business leaders to rethink and to redesign their strategy and operating models. Managers of intelligent enterprises must now view software intelligence as an across-the-board functionality, one that will drive new levels of evolution and discovery, propelling innovation throughout the enterprise. With increased software intelligence, more than ever the enterprise must put its attention to data-driven outcomes as opportunities for the innovation. Now, this is the era of software intelligence, in which applications and tools based on ICT take on more human-like intelligence (Accenture 2015).

Surely, IT engineers have been working on intelligent systems for years, but these technologies have recently become viable. This is the consequence of today's mix of vast amounts of data and tremendously scalable computing. In the intelligent enterprise, software intelligence should now be

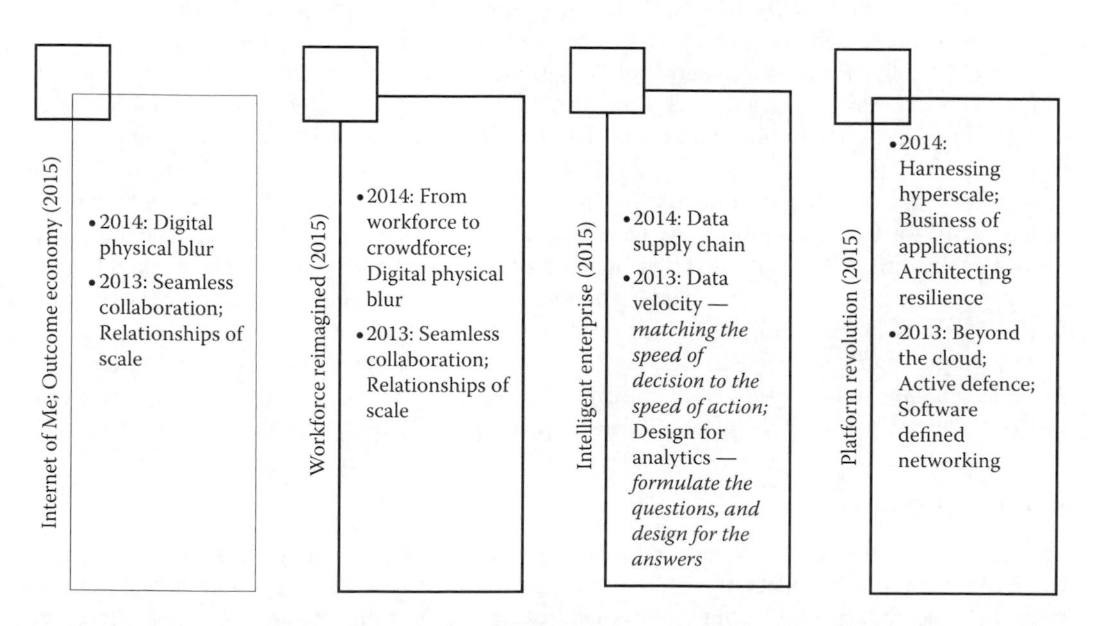

FIGURE 34.2 Emerging trends of the digital power brokers of tomorrow.

seen as one of the core capabilities, because it can not only support operational excellence throughout the organization but also help in innovation stimulation. Now, it is up to business leaders to ensure that the software is intelligent enough to consistently make the right decisions.

34.4 ICT DRIVERS EMPOWERING THE INTELLIGENT ENTERPRISE

The literature review and the results of the research conducted by PARP and MIT Sloan Management Review partnered with the IBM Institute for Business Value became the impetus for the author to conduct research in a group of 20 enterprises with which she collaborates.

34.4.1 Mobile Workforce Integration

In modern organizations access to specific knowledge is critical, and mobile connections to operating systems, applications, and platforms are important for intelligent enterprises that want to operate efficiently and effectively in the fast-paced business environment. Mobile technologies drive technical innovation to improve networks, and ensure employees remain fully integrated with their company and clients wherever they are. Thus, in the intelligent organization its coherency is determined by the intelligence of its network that becomes the organization with wireless tentacles spreading from it to embrace location-aware services, not just for the benefit of tracking, tracing, safety, and security, but also as a means to promote the organization itself through marketing and sales initiatives. Successful workforce integration focuses on clear rules and involves the whole system. To achieve effective mobile workforce integration, people need to acknowledge and overcome resistance to quick changes and need to accept a turbulent working environment. As workers and their enterprises become deskless, mobile technologies will continue driving technical innovation and new services to smartphones or other mobile devices. Communications, data, and business applications need to seamlessly extend to these mobile devices and include functions such as single-number reachability, and presence and easy routing through the company network.

34.4.2 Smart Virtual Workplace

As the approaches to virtualization of IT infrastructure, networks, and storage devices continue to mature, infrastructures become software-driven and IT management more efficient. This efficiency enables services to be provided dynamically according to enterprise, business partner, or customer requirements. The smart virtual workplace provides end-to-end desktop virtualization allowing employees to access applications and data safely over any network from the device of any choice. New trends show that businesses will increasingly turn to hybrid cloud solutions to enable scalable business processes. Many companies use public clouds for less sensitive applications, but they prefer private clouds for their vital processing tasks with allocation of these tasks as well as data storage for each application being controlled by cloud and edge terminals. Hybrid clouds can quickly scale to a company's needs, and services can be paid for as needed. They combine the best of two worlds, offering true benefits to intelligent enterprises aiming to stay ahead in their markets.

On the other hand, it should be emphasized that cloud-based solutions have been designed for SMEs, as a means to ensure increased agility at low cost. Large organizations still remain on the outskirts. The main problem is related to data security and big data processing.

34.4.3 E-Collaboration

E-collaboration is the de facto standard for business communication today, nearly eliminating the need for business travel to meet in person. While knowledge sharing increases, formal and informal groups become collaborative communities that provide coaching and create harmony to reach personal, group, and organizational goals. Intelligent enterprises continue to integrate these into their business processes and reinvent their customer engagement models. Unified ICT tools allow wide-spread teams to work together in real-time, enabling multiple individuals to interact as efficiently and effectively with coworkers, clients, and suppliers.

34.4.4 Business Flexibility

The intelligent enterprise must deal with both the complexity of strategy (strategy level) and daily competitive demands (operational level). As a result, supporting systems based on ICT must now be highly flexible and resilient in order to seamlessly communicate and interoperate with other disparate technologies and systems. The cloud architecture enables flexible deployment, therefore the IaaS (infrastructure as a service) model becomes very useful. It makes internetworking easier, and deploys servers and endpoints from multiple sources.

34.4.5 Scalability and Customization

Intelligent enterprises align their IT infrastructure capabilities with business requirements. Modularity of systems and applications allow companies to have only what is needed at present, trimming upfront costs and leaving open the possibility of expanding or incorporating new technologies in the future. With the increase of consolidation and intensive virtualization, the traditional data center environment will transform to the "hyperscale" data center. It requires a fundamentally different approach than that taken with typical enterprise IT systems. Rather than building "monolithic" platforms, distributed architecture design is implemented around distributed processing frameworks. That requires software and tools automating node deployment, recovering from failure (rerouting of workloads), and other management and monitoring tools.

34.4.6 Business Continuity

It is obvious that companies across the globe need to have access to their data 24 hours a day. Data digitalization and rapidity of their processing require more accurate, reliable, and sophisticated IT tools converting all data into intelligence for better business outcomes. On the other hand, managers need them to be not complicated in their use. For IT managers there is a challenge. Moreover, for a high level of operational uptime, infrastructure components must be fault tolerant with the ability to recover from complex failures and data storage must be secure. In this case, clustering provides the best solution when ensuring uninterrupted workflow on standby systems when failure strikes. This clustering can take the form of software or fault-tolerant server solutions, which deliver exceptional uptime through dual modular hardware redundancy. These servers will provide continuous availability for all components resulting in optimal data integrity.

34.4.7 Converting Data into Business Intelligence

Advanced ICT solutions enable extracting crucial information from huge amounts of data collected in real cyberspace. Intelligent enterprises are able to manage big data to drive better business intelligence, product development, and customer service. Important is the fact that they enable effective use of unstructured data captured from different systems, mobile devices, social media, log files, and e-mails to perform real-time context analytics. A contextually aware presence allows employees and managers to understand received information and its content to make right decisions at the right time.

Therefore, intelligent enterprises are not only the users of advanced tools based on ICT technologies to optimize business activity, drive workforce engagement, and create a competitive edge, but they are also able to leverage and to create value from the data and information generated by ICT solutions. With knowledge, a professional approach, and endless determination, it is impossible to create this value.

34.5 CONCLUSIONS

According to the Europe 2020 strategy for intelligent, sustainable, and inclusive growth, the EU puts forward three mutually reinforcing priorities (European Commission 2010). The first is smart growth—developing an economy based on knowledge and innovation. The second is sustainable growth—promoting a more resource efficient, greener, and more competitive economy. The third priority is inclusive growth—fostering a high-employment economy delivering social and territorial cohesion. The document includes targets for the whole European community to achieve. New goals and priorities concern many fields of science, economy, business management, and education, but above all they focus on the sector of research and development (Organisation for Economic Cooperation and Development [OECD] 2004; European Commission 2010). Therefore, the importance is the use of information technologies that are shaping almost every aspect of modern business. In this situation, the key factors are the resources and the effectiveness of the use of ICT solutions combined with the management of flow of information, knowledge, and financial resources.

The research conducted by the Polish Agency for Enterprise Development showed that small- and medium-sized enterprises in Poland have still many challenges to overcome. They must change a lot to deserve to be called intelligent organizations. In comparison with other EU countries, Polish companies use commonly available, popular ICT tools, as they do not have enough funds for investment in comparison to their competitors. The concept of the intelligent organization is relatively new in the context of modern enterprise management style, which was created in response to ever-growing competition, and the high pace of technology development and rapidly changing business conditions that require skillful knowledge combining what the organization has with and what the technological achievements offer.

Research conducted by MIT Sloan Management Review partnered with the IBM Institute for Business Value showed that the intelligent enterprise must be able to manage data and information effectively using appropriate analytical tools based on ICT. The way in which ICT tools are exploited depends on the knowledge and skills of managers and analysts who in this case play a significant role.

Research conducted by the author correlate with the aforementioned findings. They represent the first approach to the problem of an intelligent organization that in doing business uses a variety of ICT-based solutions. The author made an attempt to identify which of the ICT drivers have the biggest influence on shaping an intelligent organization. Advanced research will be continued in the near future.

REFERENCES

Accenture. 2015. *Accenture Technology Vision 2015: Digital Business Era—Stretch Your Boundaries*. https://www.accenture.com/us-en/it-technology-trends-2015 (accessed March 30, 2016).

Christensen, C. M. 1997. Making strategy: Learning by doing. *Harvard Business Review*, 4: 141–156.

Davenport, T. H., Leibold, M., and Voelpel, S. 2006. *Strategic Management in the Innovation Economy*. Erlangen, Germany: Wiley.

Dickey, M. R. 2014. How Times Reporter Got an Algorithm to Write Articles for Him. Business Insider. 17 March. http://www.businessinsider.com/quakebot-robot-la-times-2014-3.

European Commission. 2010. Europe 2020: Strategia na rzecz inteligentnego i zrównoważonego rozwoju sprzyjającego włączeniu społecznemu, Brussels.

Gartner. 2014. Gartner Reveals Top Predictions for IT Organizations and Users for 2015 and Beyond. October 7. http://www.gartner.com/newsroom/id/2866617.

Gupta, J. N. D., and Sharma, S. K. (eds.). 2004. *Intelligent Enterprises of the 21st Century*. Hershey, PA: Idea Group Publishing.

Hamel, G., and Prahalad, C. K. 1994. *Competing for the Future*. Boston: Harvard Business School Press.

Hopkins, M. S., Lavalle, S., and Balboni, F. 2010. 10 Insights: A first look at the new intelligent enterprise survey on winning with data. *MIT Sloan Management Review*, 52(1): 22–31.

Kordel, P., Kornecki, J., Kowalczyk, A. et al. 2010. *Inteligentne organizacje—zarządzanie wiedzą i kompetencjami pracowników.* Warsaw: Polish Agency for Enterprise and Development.
Kurylko, D. T. 2014. Mercedes-Benz's Autonomous Driving Features Dominate the Industry—and Will for Years. 2014. *Automotive News.* August 4. http://www.autonews.com/article/20140804/OEM06/308049979/mercedes-benzs-autonomousdriving-features-dominate-the-industry.
LaValle, S. 2009. IBM Global Business Services Executive Report: Business Analytics and Optimization for the Intelligent Enterprise. http://www-05.ibm.com/de/services/bao/pdf/gbe03211-usen-00.pdf.
LaValle, S., Hopkins, M. S., Lesser, E., Shockley, R., and Kruschwitz, N. 2010. Findings of the new intelligent enterprise study. In: *Big Idea: Data & Analytics.* October 24.
Muryjas, P., and Wawer, M. 2014. Business intelligence as a support in human resources strategies realization in contemporary organizations. *Actual Problems of Economics* 152(2): 183–190.
Nonaka, I., and Takeuchi, H. 1995. *The Knowledge Creating Company.* New York: Oxford University Press.
OECD. 2004. *Innovation in the Knowledge Economy: Implications for Education and Learning.* Organisation for Economic Cooperation and Development, Centre for Educational Research and Innovation. Paris: OECD Publishing, pp. 14–15.
Schwaninger, M. 2009. *Intelligent Organizations: Powerful Models for Systemic Management.* Berlin: Springer-Verlag.
Stockton, N. 2014. Using Cameras and Fancy Algorithms to Track Spinning Space Junk. *Wired.* September 11. https://www.wired.com/2014/09/algorithm-spinning-space-junk/.
Thannhuber, M. J. 2005. *The Intelligent Enterprise: Theoretical Concepts and Practical Implications.* New York: Physica-Verlag, Springer, p. 72.
Wilensky, M. L. 1967. *Organisational Intelligence.* London: Basic Books.

35 MVDR Beamformer Model for Array Response Vector Mismatch Reduction

Suhail Najm Shahab, Ayib Rosdi Zainun,
Essa Ibrahim Essa, Nurul Hazlina Noordin,
Izzeldin Ibrahim Mohamed, and Omar Khaldoon

CONTENTS

35.1 Introduction 467
35.2 MVDR Beamformer Design Model 468
35.3 Numerical Results and Analysis 471
35.3.1 Case 1 471
35.3.2 Case 2 478
35.4 Conclusion 479
Acknowledgments 479
References 480

35.1 INTRODUCTION

The growth in the number of wireless devices and applications has led to a crowding of the wireless spectrum and more stringent requirements for receiver designs. Radio frequency interference continues to be a persistent problem in many communication systems and will potentially exacerbate as the unused wireless spectrum continues to shrink. There are, in general, two types of interfering signals: (1) intentional jammers used in military applications, such as electronic warfare (EW) and (2) unintentional, yet harmful interference, primarily associated with wireless commercial systems (Van Trees, 2002). In recent years, there has been a rapid increase in the number of wireless devices for both commercial and defense applications. This has been adding strain on the spectrum utilization of wireless communication systems. Because there exists a limited amount of available frequency spectrum, interference is bound to occur as the spectrum saturates (Gross, 2011). Antenna arrays are used in wireless communications to focus electromagnetic energy on a signal of interest while simultaneously minimizing energy in jammer directions.

The minimum variance distortionless response (MVDR) or Capon beamformer (Capon, 1969) is one of the adaptive optimum statistical beamformers that ensures a distortionless response for a predefined steering direction (Godara, 1997; Van Trees, 2002; Gross, 2015). The basic idea of the MVDR technique is to estimate the beamforming excitation coefficients in an adaptive way by minimizing the variance of the residual interference and noise while enforcing a set of linear constraints to ensure that the real user signal is not distorted (Pan et al., 2014).

The most common MVDR problem is that the signal model must be quite accurate in order not to form unity gain in the user of interest (UOI) direction nulls in the direction of the user not of interest (UNOI). There are many ways to make the MVDR beamformer robust against this error such as diagonal loading (Lin et al., 2007; Gu et al., 2008), beamspace processing (Feldman and Griffiths, 1994), and spatial averaging (Van Trees, 2002). Another problem with the MVDR beamformer is the finite size of data snapshots (Wax and Anu, 1996a; Fertig, 2000; Van Trees, 2002; Mestre and Lagunas, 2006; Chen and Lee, 2012; Ghadian et al., 2015) and the array response vector uncertainty (Wax and Anu, 1996b; Besson and Vincent, 2005). When the size of data snapshots is small this will result in a poorly represented beampattern and degrade the MVDR performance.

Many attempts have been made to find the optimal weight vector (Lin et al., 2007; Gu et al., 2008). For example, the idea of diagonal loading (Lin et al., 2007; Gu et al., 2008) is to adapt a covariance matrix by adding a displacement value to the diagonal elements of the estimated covariance matrix. In the study carried out by Lu et al. (2013), the authors demonstrate the array response vector mismatch due to array calibration misadjustment based on MVDR technique. The method used an iterative algorithm to reconstruct the covariance matrix to overcome the array response pointing errors. More recently, a study carried out by Abdulrahman et al. (2015) for enhancing the MVDR performance against the array response error by repositioning the reference element in the linear antenna array to be in the middle. The results show in the enhanced model the minimum number of data snapshots required to produce satisfactory resolution is 30 snapshots.

However, the effects of finite snapshot size on the beampattern accuracy and the output Signal-to-Interference plus Noise Ratio (SINR) are still unknown from the expressions. The present work introduces a new method to estimate the array response vector by adding the null-forming linear constraint to the MVDR technique. Therefore, the weights coefficients are calculated to place null toward the UNOI direction accurately and the unity gain response toward the direction of user-of-interest. Simulation results confirm the accuracy of the theoretical results.

The remainder of this chapter is organized as follows. In Section 35.2, MVDR beamformer based on linear antenna array design method with the signal propagation model is described. The simulation results and performance evaluation are provided in Section 35.3. Finally, in Section 35.4, the chapter's conclusions and summary of MVDR performance are described.

35.2 MVDR BEAMFORMER DESIGN MODEL

The basic theory of the beamforming algorithm and the signal structure is presented in this section. The signal model considers L signals impinging on a uniform linear array (ULA) of M isotropic antenna elements, and the spacing between adjacent antennas is a half wavelength. Assume that the L signal coming from angles of θ_l and ϕ_l is incident upon an antenna array of M elements, shown in Figure 35.1. Here, the impinging angles of θ and ϕ are the azimuthal and elevation angles, respectively.

The received signal, $r_m(k) \in \mathbb{C}^{M\times K}$, at the mth antenna at the kth snapshot incident upon the antenna array can be written as:

$$r_m(k) = \sum_{d=1}^{D} x_d(k)v_d(\theta,\phi) + \sum_{i=1}^{I} x_i(k)v_i(\theta,\phi) + \sum_{m=1}^{M} n_m(k) \tag{35.1}$$

where $x_d(k)$, $x_i(k)$, and $n_m(k)$ denote the dth user-of-interest signals, ith interference signals, and additive background white Gaussian noise at the mth elements, respectively. Among those L incident signals, it is assumed that $x_d(k)$ is the desired user-of-interest and $x_i(k) + n_m(k)$ are the user-not-of-interest signals. The array response vector $v(\theta_l,\phi_l)$ is a $\in \mathbb{C}^{1\times M}$ of a ULA with M antenna elements where (θ_l,ϕ_l) are the direction of arrival (DOAs) of the lth signal component given as (Godara, 1997, 2004):

$$v(\theta,\phi) - [1, e^{-j\beta\delta\sin\theta\sin\phi}, \ldots, e^{-j(M-1)\beta\delta\sin\theta\sin\phi}]^* \tag{35.2}$$

where $\beta = 2\pi/\lambda$ is the free-space wavenumber, δ is the spacing between adjacent antenna elements, and λ is the free-space wavelength. The $\theta \in [-\pi/2, \pi/2]$, $\phi \in [0, \pi/2]$ and $(.)^*$ denote the complex conjugate. The $v(\theta_l,\phi_l)$ is a function of the incident angles, the location of the antenna, and the array

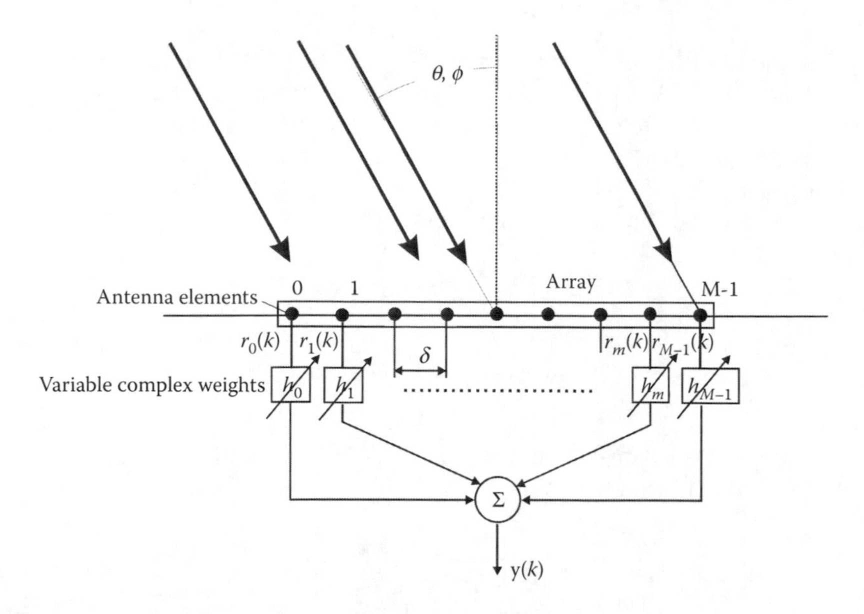

FIGURE 35.1 Uniform linear antenna array geometry.

geometry. It plays an important role in smart antenna systems, containing information of the impinging angles. The output of the beamformer at the kth snapshots, $y(k)$ after signal processing is defined as

$$y(k)=\sum_{m=1}^{M}h_m^{\dagger}r_m(k) \tag{35.3}$$

where h is a complex multiplicative weight vector given as $[h_0, h_1, \ldots, h_m, h_{M-1}]^T$ multiplied by the received signal at the mth antenna element, and $(.)^{\dagger}$ and $(.)^T$ denote, respectively, the complex conjugate transpose of a vector or matrix and transpose of a vector or matrix. The array cross-correlation (covariance) matrix $\Gamma_r \in \mathbb{C}^{M\times M}$ is defined as (Krim and Viberg, 1996)

$$\Gamma_r=\{r_m(k)r_m^{\dagger}(k)\} \tag{35.4}$$

The array covariance matrix Γ_r in Equation 35.4 is the statistical second-order property of the impinging signals. In real applications, Γ_r is estimated using the received array snapshots. The estimated array covariance matrix is given by (Gross, 2015)

$$\hat{\Gamma}_r \cong \Gamma_d+\Gamma_{i+n}\cong \frac{1}{K}\sum_{k=1}^{K}r_m(k)r_m^{\dagger}(k) \tag{35.5}$$

$$\Gamma_d=\sum_{d=1}^{D}\sigma_d^2 v_d(\theta,\phi)v_d^{\dagger}(\theta,\phi) \tag{35.6}$$

$$\Gamma_{i+n}=\sum_{i=1}^{I}\sigma_i^2\, v_i(\theta,\phi)v_i^{\dagger}(\theta,\phi)+\sigma_n^2\,\Lambda_m \tag{35.7}$$

where K is the number of available snapshots. Γ_d denotes the array correlation matrix corresponding to the desired user-of-interest, and Γ_{i+n} refers to the array correlation matrix corresponding to the undesired user-not-of-interest. The terms σ_d^2, σ_i^2, and σ_n^2 denote the real user, interference, and noise powers, respectively. $\Lambda_m \in \mathbb{R}^{M\times M}$ stands for the identity matrix. It is known from the literature that the optimization criterion for MVDR (Capon, 1969) forms weights in a way that will attempt to maintain unity gain of the beamformer in the beam angle direction while steering nulls in the direction of interference (Souden et al., 2010). The weights are calculated by solving the following minimization equations with unity gain restraint:

$$h_{MVDR} = \arg\min_{0\leq h\leq 1} E\{|y(k)|^2\} \tag{35.8}$$

$$\min_h h^\dagger \hat{\Gamma}_r h \ s.t.\ h^\dagger v_d(\theta,\phi) = 1 \tag{35.9}$$

The preceding equations are solved by using Lagrange multipliers, and the MVDR weight (h_{MVDR}) is given as (Balanis and Ioannides, 2007)

$$h_{MVDR} = \frac{v(\theta,\phi)\hat{\Gamma}_r^{-1}}{v^\dagger(\theta,\phi)\hat{\Gamma}_r^{-1} v(\theta,\phi)} \tag{35.10}$$

The zero-forcing (ZF) beamforming technique (Davies, 1967) has been used extensively for interference suppression purpose in wireless communication. The ZF method cancels several plane waves impinging from known directions and directs the mainbeam toward the desired user-of-interest source. Assume that the desired signal with steering vector $v_d(\theta, \phi)$ and I interference sources are impinging on the array. The null-forming weight vector is calculated based on the unity power reception in the direction of the desired signal, and null-forming is used to steer the nulls (zero or near-zero antenna power) reception in the interference sources directions with steering vectors $v_i(\theta, \phi)$ as given by (Friedlander and Porat, 1989; Qamar and Khan, 2009)

$$h^\dagger v_d(\theta,\phi) = 1 \tag{35.11}$$

$$h^\dagger v_i(\theta,\phi) = 0;\ i = 2,3,\ldots,M \tag{35.12}$$

However, the computational burden associated with this approach is quite high and makes it difficult for real-world applications. Based on the ZF idea, the proposed array response vector to overcome the limited number of data snapshots to obtain high resolution in terms of beampattern accuracy can be defined as

$$v^p(\theta,\phi) = [\Lambda_m - v_i(\theta,\phi)\{v_i^\dagger(\theta,\phi)v_i(\theta,\phi)\}v_i^\dagger(\theta,\phi)]v_d(\theta,\phi) \tag{35.13}$$

Then, the ZF algorithm and MVDR technique can be combined to achieve the (beam+null)-forming. The combined constraints can be rewritten as

$$\min_h h^\dagger \hat{\Gamma}_r h \ s.t.\ h^\dagger v_d(\theta,\phi) = 1; h^\dagger v_i(\theta,\phi) = 0 \tag{35.14}$$

Thus, the new complex weight vector calculation according to the proposed array response vector from Equation 35.13 is

$$h_{MVDR}^p = \frac{v^p(\theta,\phi)\hat{\Gamma}_r^{-1}}{v^{p\dagger}(\theta,\phi)\hat{\Gamma}_r^{-1} v^p(\theta,\phi)} \tag{35.15}$$

The null-forming beamformer can be formulated as

$$h^p_{MVDR}V^p(\theta,\phi)=h^p_{MVDR}[v^p_d(\theta,\phi),v^p_i(\theta,\phi),...,v^p_{M-1}(\theta,\phi)] \quad (35.16)$$

where $V^p(\theta,\phi)$ is the array response matrix that contains the desired array response vector and the interfering signals response vector.

Antenna radiation patterns are typically expressed in terms of radiated power. The output power is defined as (Godara, 1997)

$$P_y=E\{y(k)y^\dagger(k)\}=h^\dagger E\{y(k)y(k)\}h=h^\dagger\hat{\Gamma}_r h \quad (35.17)$$

Equation 35.16 can be rewritten as

$$P_y=h^\dagger\Gamma_d h+h^\dagger\Gamma_{i+n}h=P_d+P_{i+n} \quad (35.18)$$

$$P_d=\sum_{d=1}^{D}\sigma_d^2\left|h^\dagger\, v_d(\theta,\phi)\right| \quad (35.19)$$

$$P_{i+n}=\sum_{i=1}^{I}\sigma_i^2\left|h^\dagger\, v_i(\theta,\phi)\right|+\sigma_n^2 \quad (35.20)$$

where the P_d denotes the power of the desired signal and P_{i+n} refers to the power output in the direction of UNOI. Finally, the SINR is defined as the ratio of the average power of the desired signal divided by the average power of the undesired signal computed as (El Zooghby, 2005)

$$SINR\triangleq\frac{P_d}{P_{i+n}}\triangleq\frac{\sum_{d=1}^{D}\sigma_d^2\left|h^\dagger\, v_d(\theta,\varphi)\right|}{\sum_{i=1}^{I}\sigma_i^2\left|h^\dagger\, v_i(\theta,\varphi)\right|+\sigma_n^2} \quad (35.21)$$

35.3 NUMERICAL RESULTS AND ANALYSIS

In this section, the results of these two beamformers are discussed where the MATLAB platform has been used to model the performance results in terms of mathematical functions. To compare the MVDR algorithm and the proposed approach, we perform a simulation according to the parameters given in Table 35.1. It has been assumed that all users are stationary in a multipath fading environment and the performance of each algorithm is evaluated under the same noise and interference conditions.

35.3.1 Case 1

First, the performance of the conventional MVDR beamformer and the proposed method are investigated and the weight vector calculated based on Equations 35.10 and 35.15. The first case simulation shows the beampattern for a different number of snapshots of the intended user, and two interfering sources are introduced from two different azimuth angles 14.5° and 40.5° and the desired information signal arrives at the user from an azimuth angle of 0°. The antenna array is an ULA and the antenna elements are separated by 0.5 λ. The noise level and the interference level are fixed at 10 dB. The element spacing δ is set to 0.0577 m to satisfy the half of the wavelength separation

TABLE 35.1
Key System Parameters

Key System Parameters	Values
Array antenna configuration	LAA
Antenna type	Isotropic
Carrier frequency (Fc)	2.6 GHz
Beam scanning range	±90° (Azimuth)
Number of element (M)	8
Element spacing (δ)	$\lambda/2$
# UOI	1
# UNOIs	2
SNR (dB)	10
INR (dB)	10
Snapshots (K)	500, 100, 50, 10

between neighboring elements. Different numbers of data samples are used, and the results are shown in Figures 35.2 through 35.5 to illustrate a comparison between these two algorithms.

Figure 35.2a illustrates a comparison between the conventional MVDR ($MVDR_{con}$) and the proposed MVDR ($MVDR_{pro}$) algorithms for 500 snapshots. It is evident that the $MVDR_{pro}$ algorithm is more efficient. All UNOI sources are perfectly null with more precision, and, hence, the peaks of sild lobe levels (SLLs) are lower than $MVDR_{con}$ as it depicted in Figure 35.2b and c. It is clearly shown in Figure 35.2b that the SLL for the $MVDR_{pro}$ found to be –12.6 dB compare to −12.2 dB for the $MVDR_{con}$. It shows the SINR at the output of the $MVDR_{pro}$ beamformer is 21.6 dB, whereas the $MVDR_{pro}$ is 18.8 dB. It can observe that the beampatterns are fairly similar for both algorithms.

Second, the arriving signals are time varying, therefore, the calculations are based on time snapshots of the incoming signal. Figure 35.3a and c shows the beampattern for $K = 100$th. Figure 35.3a shows that the direction of the main lobe for both algorithms is steered toward the desired direction, 0°. It also shows that the height of the SLL from $MVDR_{con}$ is higher than that from $MVDR_{pro}$ by 0.8 dB; therefore, the interference suppression from $MVDR_{pro}$ is less than from $MVDR_{con}$ as shown in Figure 35.3b. The graph shows that the $MVDR_{pro}$ achieves null for $\theta_i = 14.5°$ and 40.5° of −52 dB and −41 dB comparing –33 dB and –36 dB achieved by $MVDR_{con}$. The SINR obtained for $MVDR_{con}$ and $MVDR_{pro}$ are 18.2 dB and 20.3 dB, respectively. That means $MVDR_{pro}$ achieves deep null even with the snapshots size of 100.

Third, the simulation that illustrates the 50 data snapshots is shown in Figure 35.4a through c. The beam angle was steered toward 0°, and $MVDR_{pro}$ clearly present null at the two interfering signals whereas the null position of $MVDR_{con}$ shifted by 3.5° at 40.5° as illustrated in Figure 35.4c. Figure 35.4c shows that the interference suppression of $MVDR_{pro}$ is lower than $MVDR_{con}$ due to the higher height of SLLs. As can be seen from Figure 35.4c, the UNOI signals that arrive from 14.5° and 40.5° are rejected by the $MVDR_{con}$, and the $MVDR_{pro}$ beamformer creates a null of −42 dB, −32 dB and −49 dB, −37 dB, respectively. To illustrate the SINR performance of the proposed approach, the SINR value is ≈17.5 dB; this represents a 9% improvement to the $MVDR_{con}$ of 16.1 dB.

Last, the graphical representation of a beampattern is created by plotting P_y versus all possible incident directions. The number of snapshots (K) creating $\hat{\Gamma}_r$ is 10 as demonstrated in Figure 35.5a through e. Figure 35.5a is an example of a normalized beampattern when an ULA with eight antennas is used, the spacing between adjacent antennas is half of the carrier frequency, the desired user-of-interest comes from 0°, and the two interfering signals come from 14.5° and 40.5°. It can be seen that there is a single dominant main lobe peak directed toward 0°. The $MVDR_{pro}$ mainbeams are steered toward the incident angle of the desired user-of-interest while the $MVDR_{con}$ provides the

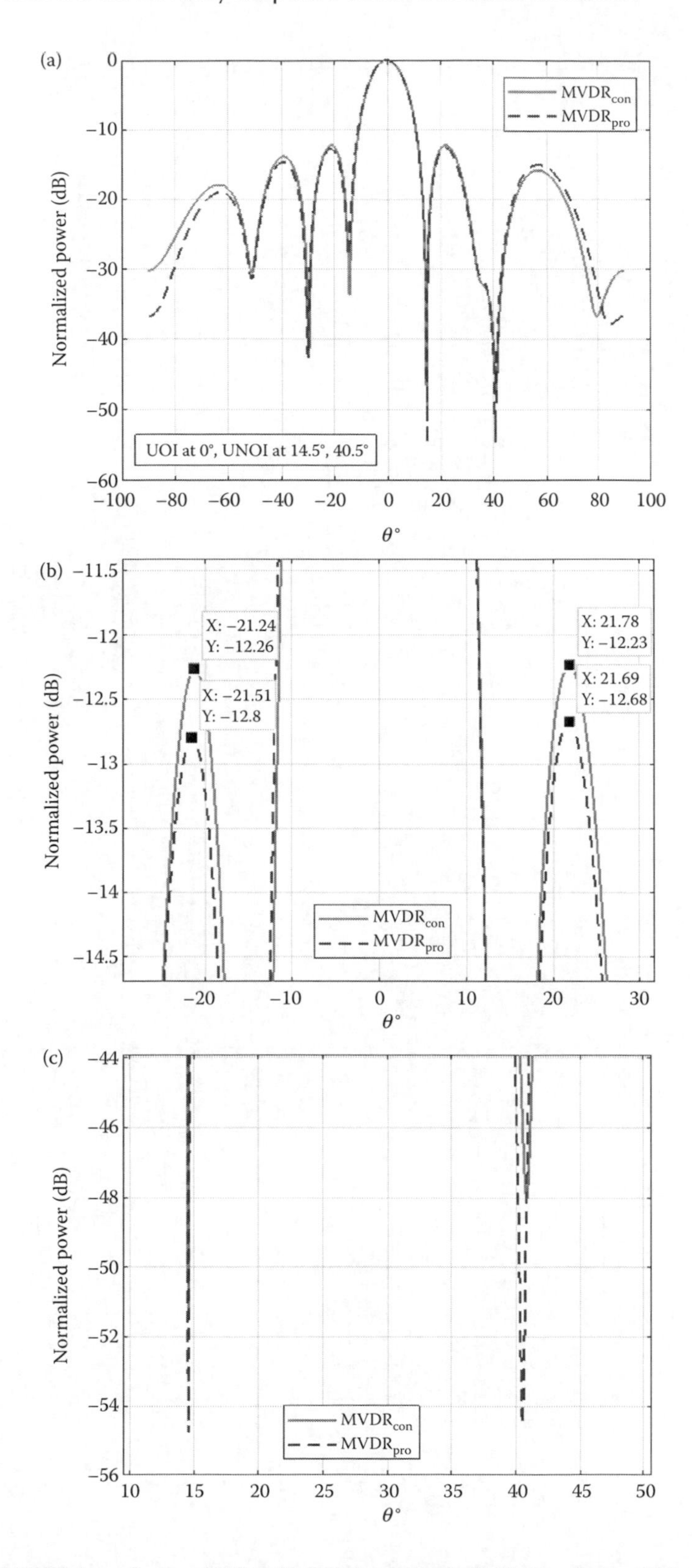

FIGURE 35.2 MVDR beamformer. UOI at 0°, UNOIs at 14.5°, 40.5°, $M = 8$, $K = 500$: (a) typical MVDR beampattern, (b) zoom in SLL pattern, (c) zoom in the null-forming pattern.

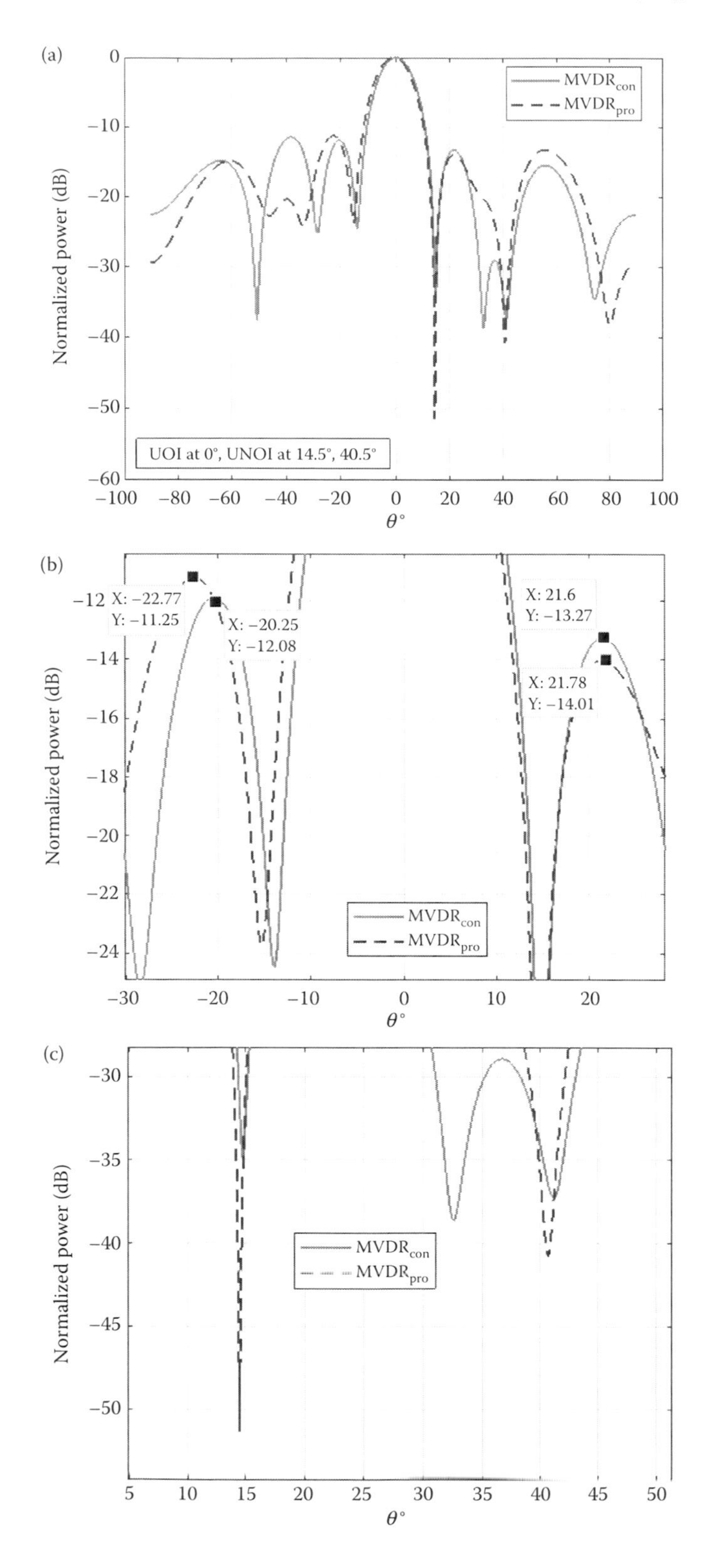

FIGURE 35.3 MVDR beamformer. UOI at 0°, UNOIs at 14.5°, 40.5°, $M = 8$, $K = 100$: (a) typical MVDR beampattern, (b) zoom in SLL pattern, (c) zoom in the null-forming pattern.

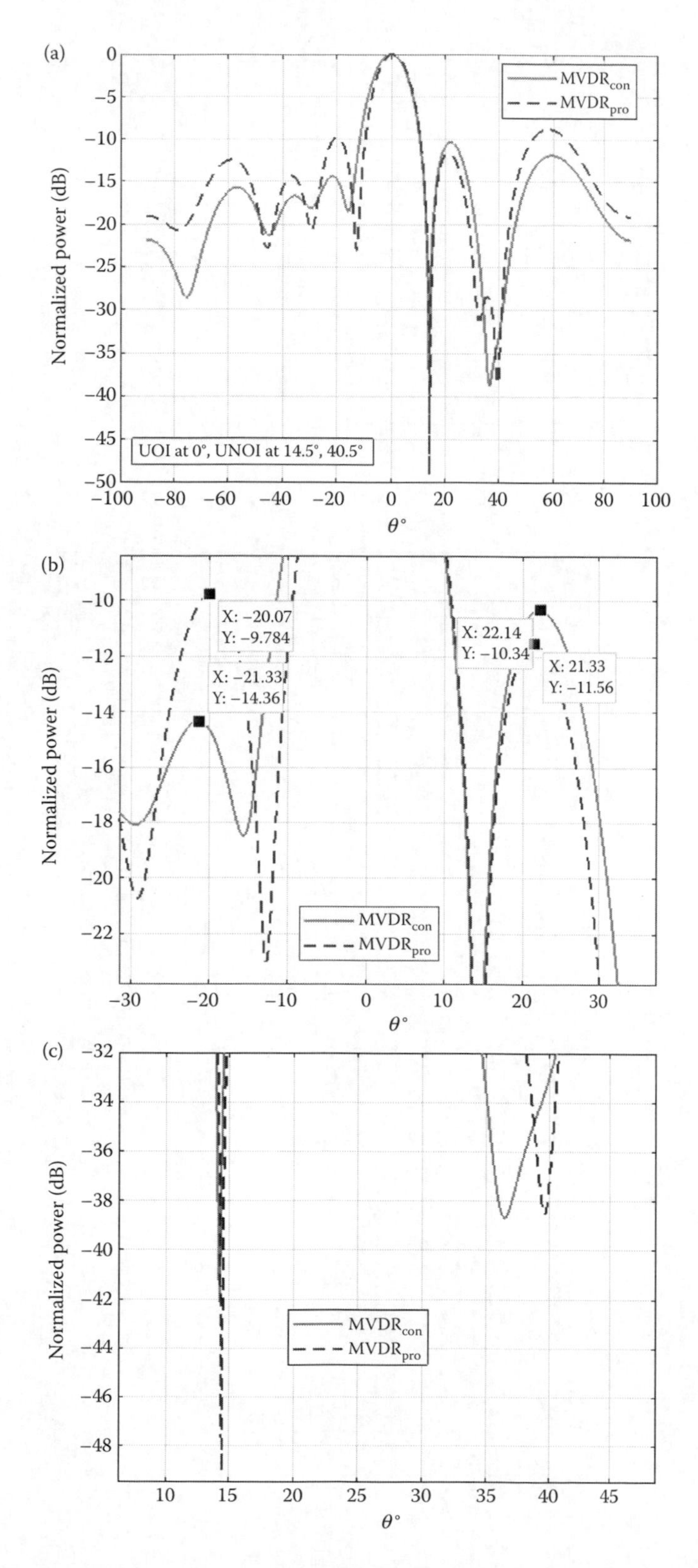

FIGURE 35.4 MVDR beamformer. UOI at 0°, UNOIs at 14.5°, 40.5°, $M = 8$, $K = 50$: (a) typical MVDR beampattern, (b) zoom in SLL pattern, (c) zoom in the null-forming pattern.

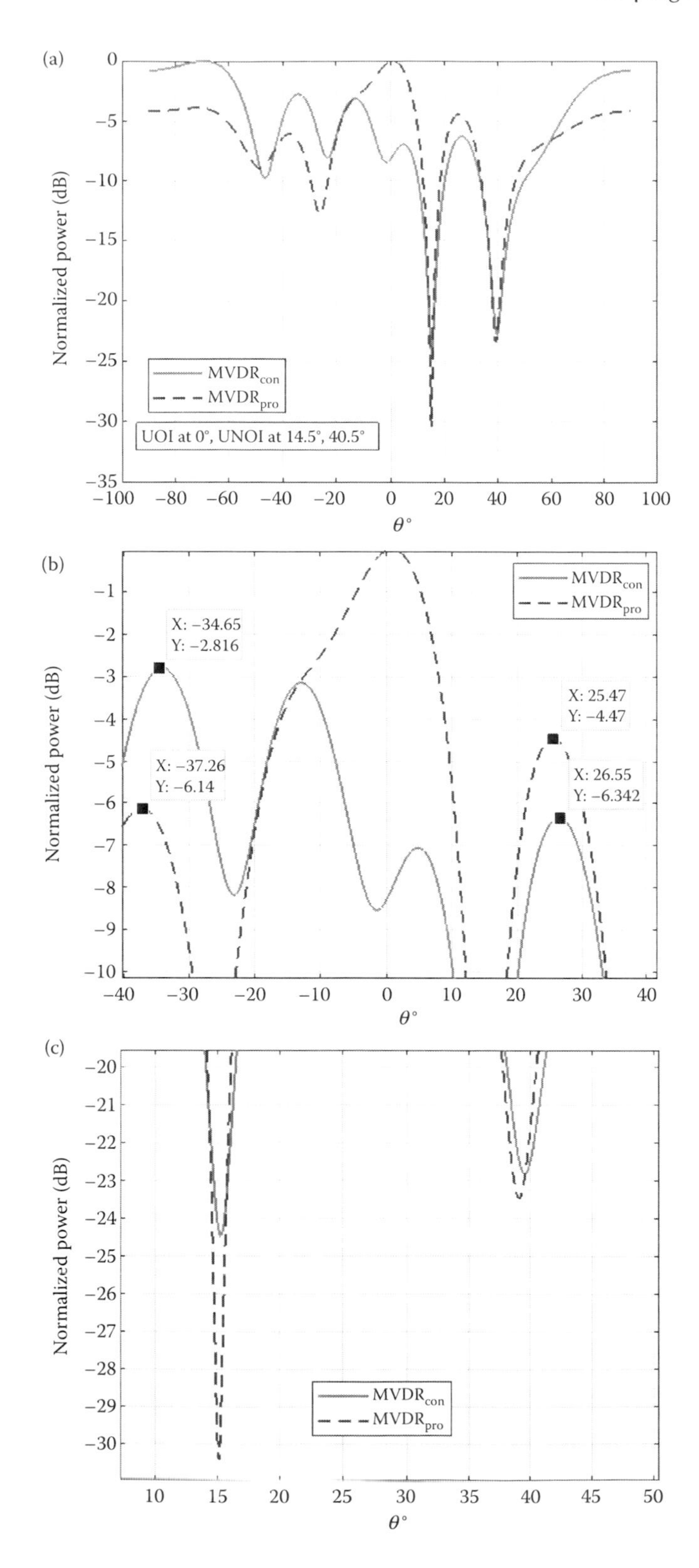

FIGURE 35.5 MVDR beamformer. UOI at 0°, UNOIs at 14.5°, 40.5°, $M = 8$, $K = 10$: (a) typical MVDR beampattern, (b) zoom in SLL pattern, (c) zoom in the null-forming pattern. *(Continued)*

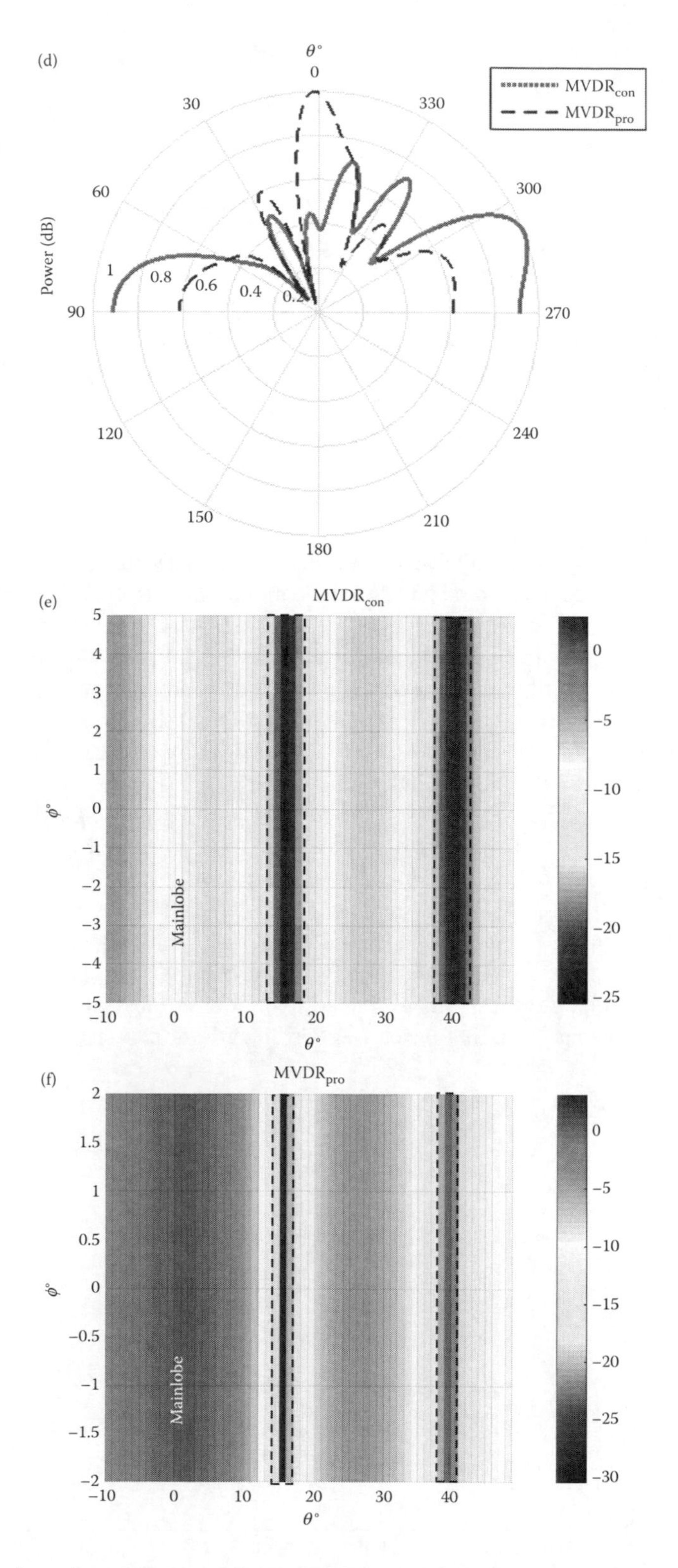

FIGURE 35.5 (*Continued*) MVDR beamformer. UOI at 0°, UNOIs at 14.5°, 40.5°, $M = 8$, $K = 10$: (d) polar power pattern, (e) and (f) 3D beampattern for azimuth and elevation angles.

mainbeam steered 10° off the desired target direction. From Figure 35.5b, the height of the SLL is found to be –2.8 dB and −4.4 dB for $MVDR_{pro}$ and $MVDR_{con}$, respectively. Figure 35.5c shows that the null-forming of the interfering signal from 14.5° and 40.5° is nullified by –33 dB and –23 dB, while in comparison with $MVDR_{con}$ the interfering signal from 14.5° is suppressed by −24 dB and −23 dB.

Figure 35.5e and f shows the 3D beampatterns for azimuthal and elevation scan angle plots of the $MVDR_{con}$ and $MVDR_{pro}$. The power is measured in decibels (dB) and a color bar is used for a sense of the relative scale of the power. The inner rectangle dashed black line represents the null width that encompasses the UNOI targets while the main lobe is represented by "Mainlobe" in both figures. Furthermore, it can be easily seen by comparing these two figures, the null width in the $\theta°$ by $MVDR_{pro}$ is narrower than $MVDR_{con}$ and the null-forming is deeper than the conventional MVDR. It is observed that the SINRs of the $MVDR_{con}$ is −11.4 dB, whereas the $MVDR_{pro}$ results show 15.1 dB giving a 33% improvement.

35.3.2 Case 2

The simulation in case 2 is divided into two scenarios. The first scenario is a comparison between the performance of the proposed method and the conventional MVDR. Figure 35.6 present the SINR obtained for the range of the 1st snapshot to the 100th snapshots. The solid line represents the conventional ($MVDR_{con}$) and the dashed line demonstrates the $MVDR_{pro}$ as illustrated in Figure 35.6.

The result shows that $MVDR_{pro}$ has better tracking ability to compute the final weight vector compared to $MVDR_{con}$ with the data snapshots less than 50. $MVDR_{pro}$ showed superior performance, for example, at $K = 30$. It is found that the SINR for $MVDR_{con}$ is equal to 12.1 dB compared to 16.7 dB obtained by $MVDR_{pro}$. The SINR for $MVDR_{con}$ increased gradually after the snapshots size reach more than 50. It is observed that the SINRs of the $MVDR_{con}$ and proposed algorithm increase with the increase of K. The output SINR for the proposed method gives a 56% improvement for data snapshots ≤100. The weight coefficients depend on the covariance matrix estimation and the array steering response. The proposed approach enhances the output SINR by placing accurate nulls in the direction of interferences. Therefore, the $MVDR_{pro}$ seems to have stable performance.

Figure 35.7 shows that the SINR versus SNR is varied from –20 to 20 for both beamformers. The performance of both algorithms was evaluated using eight linear antenna elements ($M = 8$). The spacing between antennas is half of the carrier wavelength for fixed training data size $K = 150$.

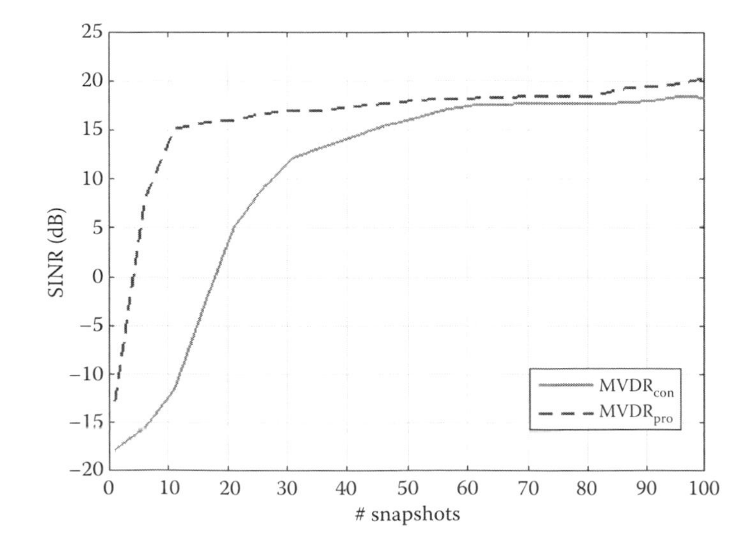

FIGURE 35.6 Output SINR versus K for array response mismatch.

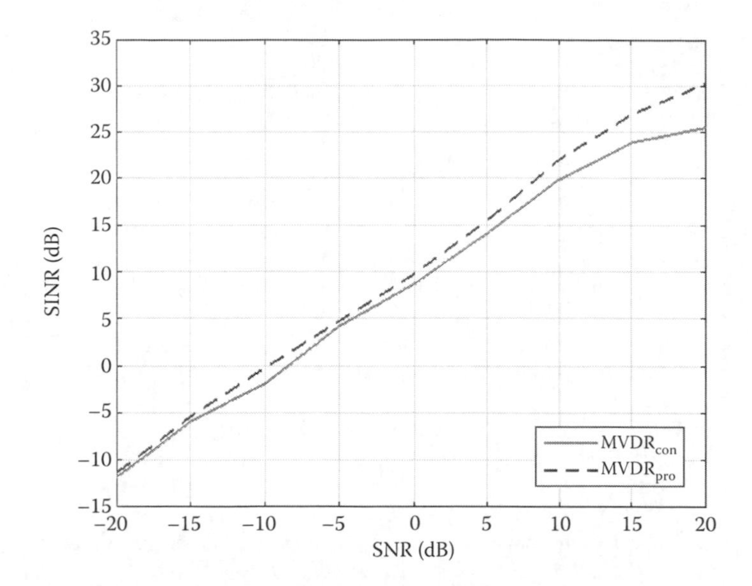

FIGURE 35.7 Output SINR versus SNR for array response mismatch.

There are four signals impinging from 10°, ±25°, and −60.7°. The signal arriving from 10° is the desired user-of-interest and all others are interfering signals. In Figure 35.7, the solid line represents the conventional $MVDR_{con}$ and the dashed line demonstrates the $MVDR_{pro}$. This increase in resolution of $MVDR_{pro}$ is due to the interference signal being perfectly nulled even at a range of SNR.

It clearly is observed in Figure 35.7 that the $MVDR_{con}$ beamformer does increase the resolution because the data did not actually estimate the location exactly. Furthermore, it can be seen that the proposed approach has accurate beampatterns toward the main lobe target and the null-forming toward the disturbance sources. Therefore, the $MVDR_{pro}$ algorithm seems to have stable and good performance regardless of the SNR values. In addition, the mean SINR achieved by the $MVDR_{pro}$ is always greater than the SINR achieved by the MVDR technique, and their difference increases with increasing SNR.

35.4 CONCLUSION

This chapter looked at the MVDR beamforming problem in a multiple access interference environment. The MVDR beamformer is sensitive to errors such as array response mismatch and the effect of finite snapshots. To keep the snapshot requirement to a minimum, the proposed method provides a robust solution for the MVDR array response vector mismatch problem and limited data snapshots. The new beamforming method is developed for uniform linear array configuration. This method has high resolution and is statistically efficient and consistent. The performances are evaluated and compared to the MVDR technique. The present method provides different performances in terms of beampattern accuracy and SINR improvement. From the results of the simulation, the conclusions drawn are that the proposed beamformer successfully increased the MVDR resolution and accurately null the user-not-of-interest signals. The main lobe steers to the user-of-interest direction even with very small data snapshots. In general, we can observe that for these two methods, the beampattern is almost the same for a large number of data snapshots.

ACKNOWLEDGMENTS

This research was supported by Universiti Malaysia Pahang, through the Fundamental Research Grant Scheme (FRGS) funded by the Ministry of Education (RDU 140129).

REFERENCES

Abdulrahman, O. K., Rahman, M. M., Hassnawi, L., and Ahmad, R. B. 2015. Modifying MVDR beamformer for reducing direction-of-arrival estimation mismatch. *Arabian Journal for Science and Engineering*, 41, 3321–3334.

Balanis, C. A. and Ioannides, P. I. 2007. *Introduction to Smart Antennas*. San Rafael, CA: Morgan & Claypool.

Besson, O. and Vincent, F. E. 2005. Performance analysis of beamformers using generalized loading of the covariance matrix in the presence of random steering vector errors. *IEEE Transactions on Signal Processing*, 53, 452–459.

Capon, J. 1969. High-resolution frequency-wavenumber spectrum analysis. *Proceedings of the IEEE*, 57, 1408–1418.

Chen, Y. L. and Lee, J.-H. 2012. Finite data performance analysis of LCMV antenna array beamformers with and without signal blocking. *Progress in Electromagnetics Research*, 130, 281–317.

Davies, D. 1967. Independent angular steering of each zero of the directional pattern for a linear array. *IEEE Transactions on Antennas and Propagation*, 15, 296–298.

El Zooghby, A. 2005. *Smart Antenna Engineering*. Norwood, MA: Artech House.

Feldman, D. D. and Griffiths, L. J. 1994. A projection approach for robust adaptive beamforming. *IEEE Transactions on Signal Processing*, 42, 867–876.

Fertig, L. B. 2000. Statistical performance of the MVDR beamformer in the presence of diagonal loading. *Proceedings of the 2000 IEEE Sensor Array and Multichannel Signal Processing Workshop, Cambridge, MA*, pp. 77–81.

Friedlander, B. and Porat, B. 1989. Performance analysis of a null-steering algorithm based on direction-of-arrival estimation. *IEEE Transactions on Acoustics, Speech and Signal Processing*, 37, 461–466.

Ghadian, M., Jabbarian-Jahromi, M., and Kahaei, M. 2015. Recursive sparsity-based MVDR algorithm for interference cancellation in sensor arrays. *IETE Journal of Research*, 62, 212–220.

Godara, L. C. 1997. Application of antenna arrays to mobile communications. II. Beam-forming and direction-of-arrival considerations. *Proceedings of the IEEE*, 85, 1195–1245.

Godara, L. C. 2004. *Smart Antennas*. Boca Raton, FL: CRC Press.

Gross, F. 2015. *Smart Antennas with MATLAB: Principles and Applications in Wireless Communication*. New York: McGraw-Hill Professional.

Gross, F. B. 2011. *Frontiers in Antennas: Next Generation Design & Engineering*. New York: McGraw-Hill.

Gu, Y. J., Shi, Z.-G., Chen, K. S., and Li, Y. 2008. Robust adaptive beamforming for steering vector uncertainties based on equivalent DOAs method. *Progress in Electromagnetics Research*, 79, 277–290.

Krim, H. and Viberg, M. 1996. Two decades of array signal processing research: The parametric approach. *Signal Processing Magazine, IEEE*, 13, 67–94.

Lin, J.-R., Peng, Q.-C., and Shao, H.-Z. 2007. On diagonal loading for robust adaptive beamforming based on worst-case performance optimization. *ETRI Journal*, 29, 50–58.

Lu, Z., Li, Y., Gao, M.-L., and Zhang, Y. 2013. Interference covariance matrix reconstruction via steering vectors estimation for robust adaptive beamforming. *Electronics Letters*, 49, 1373–1374.

Mestre, X. and Lagunas, M. Á. 2006. Finite sample size effect on minimum variance beamformers: Optimum diagonal loading factor for large arrays. *IEEE Transactions on Signal Processing*, 54, 69–82.

Pan, C., Chen, J., and Benesty, J. 2014. Performance study of the MVDR beamformer as a function of the source incidence angle. *IEEE/ACM Transactions on Audio, Speech, and Language Processing*, 22, 67–79.

Qamar, R. A. and Khan, N. M. 2009. Null steering, a comparative analysis. *IEEE 13th International Multitopic Conference (INMIC09)*, Islamabad, pp. 1–5.

Souden, M., Benesty, J., and Affes, S. 2010. A study of the LCMV and MVDR noise reduction filters. *IEEE Transactions on Signal Processing*, 58, 4925–4935.

Van Trees, H. L. 2002. *Optimum Array Processing: Part IV of Detection, Estimation, and Modulation Theory*. New York: Wiley.

Wax, M. and Anu, Y. 1996a. Performance analysis of the minimum variance beamformer. *IEEE Transactions on Signal Processing*, 44, 928–937.

Wax, M. and Anu, Y. 1996b. Performance analysis of the minimum variance beamformer in the presence of steering vector errors. *IEEE Transactions on Signal Processing*, 44, 938–947.

Index

A

Academic branches
  EA in, 54
  IT and computer, 52–54
Accenture Technology Vision (2015), 456, 461
Access categories (ACs), 102
Accessibility, 110, 111
Accuracy, 129
ACE, 143, 144
ACID properties, *see* Atomicity, consistency, isolation, durability properties
Acknowledgement messages (ACK), 99
ACM, disadvantage of, 275
Acoustic analysis of vocal emotion, 81
Acoustic parameters, 81–82
ACPSE, 21
ACs, *see* Access categories
Action, 21, 25
Action-adventure games, 442
Active learner, 114
ActiveRouter class, 246
Actor concept extraction, 223
Actuators, 95
Ad hoc on-demand distance vector (AODV), 100, 244
Adaptability, 110
Adaptation
  adaptation-oriented domain modeling, 116
  modeling, 116–117
  standardization of adaptation components and service levels, 117
Adaptive channel identification, 335
Adaptive cognitive radio detection system, 234–236
Adaptive filter, 333–334
  coefficients, 339
Adaptive intelligent-software radio (AI-SR), 231
Adaptive learning algorithms, 340
Adaptive Learning Environment (ALE), 116
Adaptive LMS algorithm, 338
Adaptive neural-fuzzy inference system (ANFIS), 259–260, 262
  architecture, 262
  feature extraction, 261
  layers of, 262–263
  methodology and analysis of simulation, 263–267
  MFSK and mathematical expressions, 260–261
  moments, 261–262
  structure, 265
  WT, 261
Adaptive wireless channel identification, 333
ADC, *see* Analog-to-digital converter; Average distance change
Additive white Gaussian noise (AWGN), 263–264, 283
AddWaitingSenders operation, 213
ADL, *see* Advanced Distributed Learning
Administrative functions, 364
Adult learners, 17, 32, 35, 37
  model, 32
Advanced Distributed Learning (ADL), 111
  Initiative, 121
Advanced Video Coding (AVC), 154
Adventure games, 442
Affective Computing, 309, 310–311
Affine projection algorithm (AFP algorithm), 337, 340
Affordability, 110
AFP algorithm, *see* Affine projection algorithm
AFSSET, 73, 74
AGC, *see* Automatic gain control
Agricultural sector, 403
"Agricultural Sector of Ukraine", 408–409
Agriculture, 408
"Agro-ecological center of the Poltava region", 409
Agro-industrial complex, 405
AICC, *see* Aviation Industry Computer-Based Training Committee
AI-SR, *see* Adaptive intelligent-software radio
ALE, *see* Adaptive Learning Environment
*Alloc*1, 210
Amazon AWS IoT, 191
Amazon DynamoDB database service, 191
Amazon Kinesis, 191
Amazon Machine Learning, 191
Amazon Web Services IoT (AWS IoT), 187, 189
  system structure, 191
AMI, *see* Automatic modulation identification
Analog-to-digital converter (ADC), 233
Analytical grounds for green logistics management, 415–418
Analytics-driven managers, 426
Anchor points, 381
Andragogic model, 32
Andragogical engineering, 19, 30
Android, 441
ANFIS, *see* Adaptive neural-fuzzy inference system
Animal marking, 67
ANN, *see* Artificial neural network
Anomaly detection, 139
Antenna array, 471–472
Anti-collision methods, 66
AODV, *see* Ad hoc on-demand distance vector
Apache Hadoop, 126
Apache HBase, 126, 127
Apache Hive, 126
Apache Mahout, 131
Apache Pig, 126
Apache Scalable Advanced Massive Online Analysis (SAMOA), 131
Apache Software Foundation (ASF), 131
Application architecture, 51
Apprenticeship, 38, 50
APTEEN, 144
Arab e-government programs, 178

Arcade games, 442
ArcCatalog, 365
ArcGIS, 365
Architecture, 50
ArcMap, 365
ArcObjects, 365
ARFF file, 79
Ariadne Foundation, 121
ARM Cortext-M4 microcontroller, 190–191
Arnold's cat map, 269
Array response vector, 468
Array steering response, 478
Artificial division, 28
Artificial intelligence; *see also* E-learning
  accuracy measures for classification algorithms, 89, 90
  ASTEMOI system, 80–81, 84–85
  classification, 79
  conventional e-learning framework, 77
  determining emotion from vocal analysis, 81–84
  emotions, 78
  Felder questionnaire, 79–80
  MAS, 79
  performance and evaluation, 89–90
  performance factors for classification algorithms, 89
  predictive model by exploring data, 85–89
  techniques, 166
  Weka, 79
Artificial neural network (ANN), 128, 259
AS, *see* Autonomous system
ASF, *see* Apache Software Foundation
Association Agreement between Ukraine and European Union, 405
Association rule, 128, 139
ASTEMOI system, 80
  architecture, 80–81
  ITS, 78
  motivation improvement through, 85
  and motivation of learner, 84
  problems treated by, 81
  self-determination theory, 84–85
Asymmetric cryptography, 321
Atomicity, consistency, isolation, durability properties (ACID properties), 127
Attributes (ATTs), 315
  data, 365
ATTs, *see* Attributes
Audio file format, 300
Auditory learning style, 113
Authentication, 326–327, 329
Automatic gain control (AGC), 233
Automatic modulation identification (AMI), 259, 261
Automatic recognition of emotions, 82–83
Autonomous system (AS), 156
AVC, *see* Advanced Video Coding
Average distance change (ADC), 270
Average latency, 251–253
Average packet service time (PSTavg), 102
Aviation Industry Computer-Based Training Committee (AICC), 112, 118, 120–121
AWGN, *see* Additive white Gaussian noise
AWS IoT, *see* Amazon Web Services IoT
Azure IoT Suite, 192
Azure Storage, 192
Azure Stream Analytics, 192

B

B2B applications, *see* Business-to-business applications
BA, *see* Business analytics
BARC algorithm, *see* Battery aware reliable clustering algorithm
Barrier coverage, 166
Base stations (BS), 137, 141
Baseband processing unit, 232
Baseband signal processing unit, 232
Battery aware reliable clustering algorithm (BARC algorithm), 146
Bayes theorem, 88
Bayesian networks, 128
BCDCP, 144
BC-PDM, *see* Big Cloud–Parallel Data Mining
Beamforming algorithm, 468
Beampattern accuracy, 470
Behavior, 32, 33
BER, *see* Bit error rate
Bernoulli distribution, 86
Best-fit algorithm, 210
BFSK, *see* Binary FSK
BI, *see* Business intelligence
Biannual HOPE convention, 74
Big Cloud–Parallel Data Mining (BC-PDM), 131
Big data, 123, 124, 463
  Big Data Research and Development Initiative, 124–125
  classification algorithms, 129–133
  technologies, 123–124, 126–127
Big data mining, 124, 127–128; *see also* Data mining
  classification, 128
  issues and challenges, 129
  knowledge discovery, 128–129
  tools, 131
Binary classification, 128
Binary coverage model, 167
Binary FSK (BFSK), 260
  signal of, 260
Binary semaphores, 206
Bit error rate (BER), 103, 300
Bitmaps (. bmp), 304
Black hole attack, 243–244, 249
  average latency, 251–252, 253
  choice of simulator, 245–246
  delivery probability for DTN protocol, 249
  design and implementation, 246
  DTN routing protocols, 244–245
  evaluation and discussion, 247
  evaluation criteria, 251
  experiment aims and setup, 248
  implementation of extended simulator, 247
  message delivery ratio, 251
  Message Listener, 247
  message statistics report for DTN routing protocols, 249
  methodology, 245
  overhead ratio for DTN routing protocols, 250
  protocol performance, 249–251
  replication-based routings, 245
  report, 247
  routing protocols in MANETs, 244

screenshot from GUI event log displaying black hole events, 248
simulation scenario, 246
Black holes
module, 247
nodes, 246
impact of number of black holes within all groups, 252–253
impact of percentage of black holes within certain groups, 253–254
total message drops and drops by, 254–255
Blended learning models, 19
Blind algorithm, 334
Blind channel identification methods, 334
Blind identification architecture, 340–341
Block set, 209
BlockAddr function, 209
Block-size function, 209
Brain Storm Optimization algorithm (BSO algorithm), 166
brainstorming process, 168, 169
coverage performance of, 172, 173
coverage problem in WSN, 166–167
formal algorithmic steps, 169
GAs, 169–170
problem statement, 167–168
running time for, 174
simulation experiments, 171–174
solution approach, 170
Brainstorming process, 168, 169
Brightness of gray scale image, 8
Broadcasts, 333
Broyden–Fletcher–Goldfarb–Shanno algorithm (L-BFGS algorithm), 316
BRs, *see* Building resources
BS, *see* Base stations
BSO algorithm, *see* Brain Storm Optimization algorithm
Building resources (BRs), 313
Business
applications, 462
calendar cycle, 55
continuity, 463
cycle, 55
events, 55
flexibility, 463
function, 55
information system, 55
organizations, 55
process architecture, 51
strategy architecture, 51
Business analytics (BA), 426–427, 429
Business intelligence (BI), 428, 456, 460
converting data into, 463
Business-to-business applications (B2B applications), 120

C

4C, *see* Connectivity-Cooperation-Communication and Creativity
C#.NET, 364, 365
C4.5 techniques, 128
Calculus of communicating systems (CCS), 211
Camera module, 9, 10
Camera nodes (CNs), 101
Camera serial interface (CSI), 9
CAMS algorithm, *see* Context Aware Multipath Selection algorithm
CamShift algorithm, 3, 4
account for search window size, 5–6
flow chart, 7
mean-shift algorithm, 4
in natural lighting, 12
object tracking using, 4–6
in very low light, 12
Canada Geographic Information System, 364
Capon beamformer, 467
Carbon dioxide emissions, 415
CART techniques, 128
Cartographic functions, 364
CASE tool, *see* Computer-aided software engineering tool
Case-based reasons, 117
Cassandra, 127
CC3200 chip, 190–191
CC4C Method, *see* Cooperative Coarse Sensing Scheme for Cognitive Radio Network Method
CCA, *see* Clear channel assessment
CCS, *see* Calculus of communicating systems; Concentric Clustering Scheme
Centralized PSO (PSO-C), 146
Centroid position of search window, 5
CF, *see* Collaborative filtering
CFAR, *see* Constant false alarm rate
CGMP, *see* Cisco Group Management Protocol
CH, *see* Cluster head; Cooccurrence histogram
Chance, 378–379
Change point (CP), 418
Channel output signal, 335
Channel selection filter, 233
Channel utilization "Utili", 102
Chaos theory in steganography, 300–304
Chaotic algorithm, 303
Chaotic flow sequences, 273, 274, 276
Chaotic maps, 301
using hybrid of steganography and cryptography keys, 304
used as cryptography key, 303–304
used as steganography key, 301, 303
Chaotic pixel value difference approach (C-PVD approach), 304
Chaotic pseudo random bit sequence (C-PRBG), 304
Chaotic sequences, 269
Chartered Institute for Personnel and Development emphasizes, 428
Chef information officers (CIOs), 426
Chosen message attack, 299
Chosen stego attack, 299
CIOs, *see* Chef information officers
Cipher text, 321
Cisco Group Management Protocol (CGMP), 154
Cisco IoT Reference Model, 189
City logistics, 414
Classical supply chains, 395
Classical value creation chains, 393, 395
Classification, 79, 139
algorithms based on MapReduce, 130–131, 132–133
of data, 128
method of nearest neighbor, 87
Clear channel assessment (CCA), 103
Climate change, 415

"Climate for Transport Change", 415
Clock overflow, 214
Cloud computing, 123, 189
Cloud-based solutions, 462
Cluster, 4, 141
  analysis, 4
  count, 142
  formation methodology, 142
Cluster head (CH), 101, 141
  mobility, 142
  selection, 142
Clustered WSNs, 142
  clustering protocols, 144–149
  data aggregation, 138–139
  data clustering, 141–143
  data mining, 139–141
  data streams, 143–144
  for data streams, 148–149
  sensor nodes, 137–138
Clustering, 4, 128, 138, 139; *see also* Data clustering
  algorithms schemes in WSNs, 144–147
  parameters, 142–143
  phenomenon, 141
  scheme, 142
Clustering on Demand (COD), 148
Clustering protocols, 142–144
  clustered WSNs for data streams, 148–149
  clustering routing protocols in WSNs, 144
  for data streams, 147–148
  proactive and reactive clustering, 144, 145
  taxonomy of, 149
CluStream algorithm, 147–148
CMA, *see* Constant modulus algorithm
CMOS, *see* Complementary metal-oxide semiconductor
CMRP, *see* Cross-Layer-Based Clustered Multipath Routing protocol
CMS, *see* Content management system
CMVT protocol, *see* Cross-Layer and Multipath-Based Video Transmission scheme protocol
CNRST, *see* National Center for Scientific and Technical Research
CNs, *see* Camera nodes
CO_2 emissions, 416, 418
CoAP, *see* Constrained application protocol
Coarse sensing stage, 286
Coarse–fine sensing method, 282
COD, *see* Clustering on Demand
Coded video transmission, 160–163
COE, *see* Council of Europe
Cognition, 230
Cognitive radio (CR), 229, 230–231, 260, 281; *see also* Two-stage sensing method
  adaptive CR detection system, 234–236
  advantages of software defined radio, 231–232
  architecture for generic CR, 232–233
  detection cycle of, 231
  frequency spectrum, 230
  physical architecture, 232
  simulation results, 236–240
  statistical features, 233–234
  two hypotheses, 233
Cognitive radio network (CRN), 322
  infrastructure-based CRN, 283
Cognitive radio users, 281
Cognitive users (CUs), 281
Collaborative filtering (CF), 131
Collision issues, 66
Color detection algorithm, 3, 4
  flow chart, 8
  in natural lighting, 11
  object tracking using, 6
  in very low light, 11
  visible spectrum, 8
Column stores, 127
Common energy model, 172
Communications, 462; *see also* Cognitive radio (CR)
  protocol, 98
  subsystem, 96
  technologies, 403
  unit, 138
Community specifications, 110
Comparative sentences, 317
Competences obtained during studies, 56–57
Complementary metal-oxide semiconductor (CMOS), 93
Complete interrupt event, 205, 206
Compound emotion, 316
Comprehensive modeling, 21, 22
Computational complexity, 341
Computer games on gamers
  additional hobbies connected with games, 450
  additional hobbies of e-gamers, 451
  analysis of results and relevant discussion, 441
  cheating in computer games, 447
  earning money on participating in computer games, 448
  e-gamer playing games, 440
  E-gamers' mood after playing game, 445
  forming circles of friends around playing computer games, 444
  free applications installed on smartphones, 442
  influence of playing computer games on e-gamers skills, 447
  level of e-gamers knowledge concerning popular professional players, 448
  passive and active participation in national and international events, 449
  playing and higher prestige among friends and acquaintances, 443
  effect of playing games on e-gamers skills, 446
  reception of fan fiction content among respondents, 450
  research methodology, 440–441
  social behavior toward players, 444
Computer science (CS), 52–53
Computer vision, 3
  camera module, 9
  histogram equalization, 8–9
  object tracking using CamShift algorithm, 4–6
  object tracking using color detection, 6–8
  Raspberry Pi Model B, 9
  results, 10
  system requirement, 9
Computer-aided software engineering tool (CASE tool), 224
"Computing world", 309
Concentric Clustering Scheme (CCS), 144
Concept modeling and models, 18

construct ion of setting of didactic intervention, 19–20
typology of models, 18–19
Conceptual setting construction for teaching, 29–30
Conceptual standards, 112
Connected k-coverage, 167
Connectivity-Cooperation-Communication and Creativity (4C), 20
Console, 439
Constant false alarm rate (CFAR), 237
Constant modulus algorithm (CMA), 334
Constrained application protocol (CoAP), 189
Constraint, 138
Contemporary enterprises in Eastern Poland
business analytics in today's organizations, 426–427
findings and discussion, 430
frequency of utilization of analytics in HR areas, 430–434
HR analytics approach, 427–430
level of benefits of HR analytics in enterprise management, 436
research methodology, 430, 431
types of analytics utilizing in particular HR areas, 434–435
Content communication, 112
Content management system (CMS), 117
Content packaging, 111
Context Aware Multipath Selection algorithm (CAMS algorithm), 100
Context(s), 112, 214
applications, 25
dependence, 317
function, 203
identification and modeling of learning styles, 26
organization and relevance of study, 26–27
population targets, 25
of research, 25
switch event(s), 203
switch flag, 204
switching, 203–205
Controller, 95
Conventional e-learning framework, 77
Conventional MVDR algorithm ($MVDR_{con}$ algorithm), 472, 478
Convergence rate, 341
Convolution equation, 337
Cooccurrence histogram (CH), 4
Cooperative Coarse Sensing Scheme for Cognitive Radio Network Method (CC4C Method), 282, 285–286
Cooperative scenario (CSS), 281–282, 290–294
Coordinate systems, 365
Coordination subsystem, 96
CouchDB, 127
Council of Europe (COE), 66
Covariance matrix estimation, 478
Cover medium, 298
Coverage problem, 165–168
CP, *see* Change point
C-PRBG, *see* Chaotic pseudo random bit sequence
C-PVD approach, *see* Chaotic pixel value difference approach
CQ schemes, *see* Custom queuing schemes
CR, *see* Cognitive radio
Craig's models of operating system kernels, comparison of proposed guidelines with, 212–213
Craig's work, 211–212
createNewMessage, method, 246
Credentials, 50
CRM, *see* Customer relationship management
CRN, *see* Cognitive radio network
Cross-disciplinary approach, 314
Cross-Layer and Multipath-Based Video Transmission scheme protocol (CMVT protocol), 101
Cross-layer architecture, 94, 97–98
Cross-layer protocols for WMSNs, 98
multichannel routing, 98–100
multipath routing, 100–101
single-path routing, 101–106
Cross-layer QoS architecture (QoSMOS), 102
Cross-Layer-Based Clustered Multipath Routing protocol (CMRP), 101
CR-WMSN, 102
Cryptography, 297, 321; *see also* Steganography
chaotic maps used as cryptography key, 303–304
chaotic maps using hybrid of steganography and cryptography keys, 304
CS, *see* Computer science
CSI, *see* Camera serial interface
CSS, *see* Cooperative scenario
Cumulants, 234
Current context, 203
CUs, *see* Cognitive users
Custom queuing schemes (CQ schemes), 156
Customer orientation, 392
Customer relationship management (CRM), 459, 460

D

Data, 462
aggregation, 138–139, 142
analysis, 45
arrival, 140
data/information architecture, 51
data-generating phenomenon, 140
management system, 64
preparation, 366–367
source, 141
transfer, 197
transformation, 140
transmission, 144
volume, 125
warehouses, 460
Data clustering, 141
clustering parameters, 142–143
designing clusters in WSNs, 142
hierarchical clustering structure, 141
Data collection
methodology, 43–44
procedure, 180–182
Data exploration and Weka, 86
Decision Tree Algorithm J48, 87
IBk algorithm, 87–88
logistic regression, 86–87
naïve Bayes classifier, 88–89
Data gathering, 365
protocols, 141

Data mining, 139; *see also* Big data mining
  application areas of WSN data mining, 140
  challenges, 140
  classes of, 139
  implementation of WSN data mining, 141
  managing and processing data, 139–140
  taxonomy, 140
  in WSNs, 139
Data streams, 138, 143–144
  algorithms of, 144
  characteristics, 143
  clustered WSN s for, 148–149
  clustering protocols for, 147–148
  mining, 138
  traditional data mining and data stream mining, 143
Database management system (DBMS), 55, 126
Database(s), 460
  domain, 55
  object class, 55
  object information system, 55
Dataset, 366
dB, *see* Decibels
DBMS, *see* Database management system
DCMI, *see* Dublin Core Metadata Initiative
DCT, *see* Discrete cosine transform
DD, *see* Directed Diffusion
DD protocols, *see* Direct Delivery protocols
DD-Stream, 148
De jure standards, 110
Decibels (dB), 478
Decision and prediction models, 30
Decision theoretic (DT), 259
Decision Tree Algorithm J48, 87
Decision tree approach, 87
Decision-making process, 33, 425
Decryption, 326
Delay operations, 214
Delay-tolerant networks (DTNs), 243–244
  average latency, 251–252, 253
  black hole attacks, 243, 247, 249, 251
  black hole module, 247
  choice of simulator, 245–246
  delivery probability for DTN protocol, 249
  design and implementation, 246
  evaluation and discussion, 247
  experiment aims and setup, 248
  implementation of extended simulator, 247
  message delivery ratio, 251
  Message Listener, 247
  message statistics report for DTN routing protocols, 249
  methodology, 245
  impact of number of black holes within all groups, 252–253
  overhead ratio for DTN routing protocols, 250
  impact of percentage of black holes within certain groups, 253–254
  protocol performance, 249–251
  replication-based routings, 245
  routing protocols, 244–245
  routing protocols in MANETs, 244
  screenshot from GUI event log displaying black hole events, 248
  simulation scenario, 246
  total message drops and drops by black holes, 254–255
Deloitte Corporation, 426
Deming circle, 391
Density-based clustering algorithms, 148
DenStream algorithm, 147–148
Department of Defense Architecture Framework (DoDAF), 51
Dependency modeling, 128
Deployment, 165
Dequeue operations, 213
Descrambling process, 271
Description–correlation–experimentation loop, 32
Design system model, 283, 284
Desktops, 439–440
DESs, *see* Discrete event simulations
Destination-sequenced distance vector (DSDV), 244
DFT, *see* Discrete Fourier transform
Dialogue systems, 78
Didactic intervention, construct ion of setting of, 19–20
Didactic preparation of intervention online, 17
Didactics, e-learning domains, 112
Diegetic model, 34–35
Differentiated Services Code Point (DSCP), 159
Differentiated services protocol (DiffServ protocol), 156
DiffServ protocol, *see* Differentiated services protocol
Digital learning, 41, 43; *see also* E-learning
Digital signal processing (DSP), 230
Digital Subscriber Line Access Multiplexer DSLAM (DSL Modem), 156
Digital television (DTV), 118
"Digital Universe Study", 123
Digital-driven business, 461
Direct Delivery protocols (DD protocols), 249
Directed Diffusion (DD), 144
Directional questionnaires, 32
Discrete cosine transform (DCT), 301
Discrete event simulations (DESs), 156
Discrete Fourier transform (DFT), 148
Discrete wavelet transform (DWT), 260
Distance function, 87
Distance Learning E-Course, 42, 44
Distance learning platforms, 19
Distance scrambling factor (DSF), 270
Distributed algorithm, 166–167
Distributed approaches, 140
Distributed protocols, 142–143
Distributed source coding (DSC), 103
Distributed Weight-Based Energy-Efficient Hierarchical Clustering protocol (DWEHC), 144
Divisive-Agglomerative Clustering (ODAC), 148
DocumentDB, 192
Document-level SA, 311
Document-oriented database, 127
DoDAF, *see* Department of Defense Architecture Framework
Domain model, 33
Domains of e-learning standards, 112–113
Doppler shift, 337
Double-byte bias algorithm to analyze RC4 cipher, 352
  double-byte biases, 354–357
  measuring distributions of keystream bytes, 353
DSC, *see* Distributed source coding
DSCLU, 148
DSCP, *see* Differentiated Services Code Point

DSDV, *see* Destination-sequenced distance vector
DSF, *see* Distance scrambling factor
DSL Modem, *see* Digital Subscriber Line Access Multiplexer DSLAM
DSP, *see* Digital signal processing
DSR, *see* Dynamic source routing
D-Stream algorithm, 147–148
DT, *see* Decision theoretic
DTNhost class, 246
DTNs, *see* Delay-tolerant networks
DTV, *see* Digital television
Dublin Core approach, 119
Dublin Core Metadata Initiative (DCMI), 121
Durability, 110, 111
DWEHC, *see* Distributed Weight-Based Energy-Efficient Hierarchical Clustering protocol
DWT, *see* Discrete wavelet transform
Dynamic end-user querying, 456
Dynamic network topology, 140
Dynamic source routing (DSR), 244

E

EA, *see* Enterprise architecture
EAM, *see* Enterprise Architecture Modelling
EAN–UCC, *see* European Article Numbering–Uniform Code Council
EAR, *see* Energy-Aware Routing
EB, *see* Exabytes
EC algorithm, *see* Elliptic curve algorithm
E-collaboration, 462–463
Economic growth, 405
Economic mechanisms, 403
EDCA, *see* Enhanced Distributed Channel Access
"Edge computing", 189
Education and Research Institute (ERI), 408
Educational program structure, place of course in, 57–59
  content of topics, 58–59
  duration of course and types of educational work, 58
EduCAUSE, 119
Educom, 119
EECS, *see* Energy Efficient Clustering Scheme
EEUC, 144
E-gamer, 439–440
EGE, *see* European Group on Ethics in Science and New Technologies
E-governance development, 405
E-government, *see* Electronic government
EGRs, *see* Emotion generation rules
EIVP, *see* Evaluation of impacts on private life
E-learning, 18
  adaptation standards, 116–117
  approach, method, and object of teaching–training modeling, 22–27
  concept modeling and models, 18–20
  contributions towards modeling of pedagogical scenarios, 27–35
  modeling notion in Moroccan academic curriculum, 20–21
  modeling of prescribed and effective scenarios, 21–22
  personalization standards, 114–116
  platforms, 44
  quality standards, 119–120
E-learning communities standards, 120–122
  AICC, 120–121
  Ariadne Foundation, 121–122
  DCMI, 121
  IMS, 120
E-learning management systems standards, 117
  e-learning quality standards, 119–120
  IMS standard, 119
  LMSs, 117–118
  LOM, 119
  metadata and interoperability standards, 118
  SCORM standard, 118
  T-SCORM standard, 118
E-learning standards, 110
  advantages of developing, 111
  categories, 111
  domains, 112–113
  for E-learning styles, 113–114
  entities, 113
  standards for e-learning styles, 113–114
  types, 112
Electromagnetic radio spectrum, 229
Electronic certificate, 67
Electronic government (E-government), 177; *see also* Internet of Things (IoT)
  appropriate training, 183
  Arab e-government programs, 178
  data collection procedure, 180–182
  "Frequently Asked Questions" service, 184
  ICT investment in Saudi Arabia, 179
  implications, 184
  portal, 180
  problems arising, E-services, 182
  research objectives and methodology, 180
  research problem statement, 179–180
  sampling, 180–182
  services, 181
Electronic Product Code (EPCglobal), 64
Electronic warfare (EW), 467
Elliptic curve algorithm (EC algorithm), 321
Emotion generation rules (EGRs), 315
Emotional agent, 81
Emotional artificial intelligence, 310
Emotional intelligence (EQ), 309
Emotions, 78, 309, 314
  detection, 312
  measuring techniques, 316
Emotive vocabulary, 315
Empirical sciences, 29
Encryption, 325–326
  algorithms, 347
  data, 303
End product, 51
Endogenous event, 16
End-to-end (ETE), 155
  delay, 100, 101, 154, 157–158
  desktop virtualization, 462
  IPTV system, 155
  latency, 96
  packet latency, 102
  path, 243, 244
Energy consumption, method to improving, 283
  CC4C Method, 285–286
  TSEE OB-CSS Schemes, 285

Energy detection, 281
  spectrum sensing, 283
Energy efficiency, 138
Energy Efficient Clustering Scheme (EECS), 143, 144
Energy-Aware Routing (EAR), 144
"Engine of innovation", 455
Engineering manuals and guidelines, 195
Enhanced Distributed Channel Access (EDCA), 102
Enqueue operations, 213
"Ensemble of classifiers", 316
Enterprise architecture (EA), 49, 51
  in academic branches, 54
  application area and normative references, 54–57
  competences obtained during studies, 56–57
  components, 55
  course parts and types of classes, 59
  different architectural views within framework, 51
  discipline, 50
  frameworks, 50, 51
  integration and security architectures, 51–52
  IT and computer academic branches, 52–54
  learning goals, 54, 56
  methodical recommendations for course, 59–60
  place of course in educational program structure, 57–59
  from professional certificates to academic credentials, 49, 50
Enterprise Architecture Modelling (EAM), 54
Enterprise resource planning systems (ERP systems), 426–427
Enterprises, 404, 460
  creation of opportunities in management of, 378–384
  level of benefits of HR analytics in enterprise management, 436
Entities
  of e-learning standards, 113
  entity–relationship data model, 18
Environmental degradation, 414
Environmental factor, 414
Environmental green logistics, rail transport in era of, 421–422
Environmental objectives, 413
Environmental Systems Research Institute (ESRI), 365
EPC UHF1 standard, 65
EPCglobal, *see* Electronic Product Code
EPCglobal Class 1 Gen2, *see* ISO 18000–6C
EQ, *see* Emotional intelligence
ERI, *see* Education and Research Institute
ERP systems, *see* Enterprise resource planning systems
ESRI, *see* Environmental Systems Research Institute
ETE, *see* End-to-end
ETL processes, *see* Extract, transform, and load processes
Euler's totient function, 323
European Article Numbering–Uniform Code Council (EAN–UCC), 64
European Commission, 68
European Commission's Transport White Paper, The, 415
European Environment Agency (2008), 415
European Group on Ethics in Science and New Technologies (EGE), 68
Evaluation method, 141
Evaluation of impacts on private life (EIVP), 66
Event, 22
Event-B model, 195, 198–200
EW, *see* Electronic warfare
Exabytes (EB), 123, 125
Exogenous event, 16
Explicit knowledge, 115
"Extend dependency" extraction, 223
Extended simulator implementation, 247
eXtensible Stylesheet Language Transformation (XSLT), 224
"Exteriority–interiority" modeling, 16
External regulation, 84–85
Extract, transform, and load processes (ETL processes), 131
Extroversion, 114

F

Facebook, 125, 126, 441, 442
Fan fiction, 450
Fast Fourier transform (FFT), 260
Favorable states, 82
FBSO, *see* Fuzzy brain storm optimization
FC, *see* First Contact; Fusion center
FCA, *see* Formal concept analysis
FCC, *see* Federal Communications Commission
FCM, *see* Fuzzy c-means
FEAF, *see* Federal Enterprise Architecture Framework
FEAPO, *see* Federation of Enterprise Architecture Professional Organizations
Federal Communications Commission (FCC), 230
Federal Enterprise Architecture Framework (FEAF), 50
Federal Information Processing Standard for Personal Identity Verification (FIPS PUB 201), 73
Federation of Enterprise Architecture Professional Organizations (FEAPO), 50
Feeling, 114
Felder questionnaire, 79–80
Felder-Silverman Index of Learning Styles, 114
FFT, *see* Fast Fourier transform
Field of view (FoV), 148
FIFO order, *see* First in, first out order
Filter coefficients, 344
Financial transactions, 67
Financing opportunity, 384
Fine sensing stage, 286
Fingerprint algorithms, 298–299
Finite impulse response (FIR), 333
FIPS PUB *201, see* Federal Information Processing Standard for Personal Identity Verification
FIR, *see* Finite impulse response
First Contact (FC), 250
First dimension, *see* Information processing dimension
First in, first out order (FIFO order), 208
First language (L1), 41
First-order moment, 5
First-person shooter (FPS), 46
FIS, *see* Fuzzy inference system
Flat network topology, 144
Flat networks, 138
FLC, *see* Fuzzy logic controller
Fleming VAK model, 113
FlockStream algorithm, 148
Flooding, 144
Flow of speech, 82
FoeBuD, 74

Formal concept analysis (FCA), 115
Formal standards, 110
Forward-backward spatial smoothing algorithm, 334
Fourth dimension, *see* Progression to understanding dimension
FoV, *see* Field of view
FPS, *see* First-person shooter; Frames per second
FR, *see* Fusion rule
Frames per second (FPS), 158
FreeRTOS, 197–198
  comparison of proposed guidelines with Craig's models, 212–213
  Craig's work, 211–212
  Event-B, 198
  interrupts, 205–206
  ISRs, 205–206
  memory management, 209–211, 215
  models, 213
  queue management, 206–209, 214–215
  scheduler states, 202–203
  scheduling and context switching, 203–205
  task management, 199–202, 213–214
Free-space wavenumber, 468–469
French legislation, 74
Frequency
  modulation, 260
  spectrum, 230
  of utilization of analytics in HR areas, 430–434
2FSK, surface graph of, 266
4FSK, surface graph of, 266
8FSK, surface graph of, 266
Fundamental frequency, 81
Fusion center (FC), 281–282
Fusion rule (FR), 281–282
Fuzzification layer, 262
Fuzzy brain storm optimization (FBSO), 166
Fuzzy clustering, 130
Fuzzy c-means (FCM), 147, 148
Fuzzy inference system (FIS), 262
Fuzzy logic controller (FLC), 155

G

Galois field of prime number (GF_p), 321
GAs, *see* Genetic algorithms
Gate, 219–220
GBR, *see* Gradient-Based Routing
Generalization, 220
Generalized System of Preferences (GSP), 408
Generic adaptive wireless channel architecture, 334
Genetic algorithms (GAs), 166, 169–170
  coverage performance of, 173
  running time for, 174
Geodatabase data model, 365
Geographic coordinate system, 365
Geographic information systems (GIS), 364
  ArcMap and ESRI, 365
  coordinate systems, 365
Geographic routing algorithms, 103
Georelational data model, 365
GF_p, *see* Galois field of prime number
GIS, *see* Geographic information systems
Global Information Technology Report, The, 406
Global learner, 114
Global Report on Development of Information Technologies, 406
GMDBSCAN, 148
Gold-adaptive learning resources mobilized model, 25
Google Trends, 123, 124
GOP, *see* Group of pictures
Gossiping, 144
GPS trajectories, 363
GPSR, *see* Greedy Perimeter Stateless Routing
Gradient-Based Routing (GBR), 144
Graph database, 127
Graphein, 321
Graphical representation, 224
GraphLab, 131
Grasha-Riechmann Student Learning Style Scale (GRSLSS), 114
Grasping, 113
Greedy Perimeter Stateless Routing (GPSR), 144
Green logistics, 413
  analytical grounds for green logistics management, 415–416, 417, 418
  areas of green logistics and transport, 413–415
  effective operation area for transportation, 416–421
  rail transport in era of environmental green logistics, 421–422
  trade-off analysis, 416
  transport sector, 413
Green supply chain management, 414
Greenhouse gas, 416
  emissions, 413, 415
GreenLogistics coefficient, 419
Grid schemes, 145–146
Grid-based clustering algorithm, 146
GROUP, clustering protocol, 143
Group algorithm, 146
Group modeling, 116
Group of pictures (GOP), 158
  size, 101
Grouping sensor nodes, 141
GRSLSS, *see* Grasha-Riechmann Student Learning Style Scale
GSP, *see* Generalized System of Preferences

H

H.264/AVC technology, 153
Haar-type DWT, 260
Hadoop distributed file system (HDFS), 126
Half wavelength, 468
Hamming distance, 329
Handle interrupt event, 205, 206
Hardware (HW), 231
  clock and timing properties, 214
  "hardware independent" layer, 198
HD LCD TV, *see* High-definition LCD TV
HDFS, *see* Hadoop distributed file system
HDM, *see* Hierarchical development methodology
HEED, *see* Hybrid Energy-Efficient Distributed Clustering
Henon chaotic map, 303–304
Heterogeneity, 129
Heuristic algorithms, 146
HEVC, *see* High Efficiency Video Coding
HGMR, 144
Hidden Markov models (HMMs), 316

Hidden message, 300
Hierarchical development methodology (HDM), 211–212
Hierarchical networks, 138
Hierarchical routing protocols, 141
Hierarchical schemes, 145
Hierarchical topology, 144
High-definition LCD TV (HD LCD TV), 153
High Efficiency Video Coding (HEVC), 154
High-level managers, 426, 433
High level of Internet piracy, 407–408
High level quality of voice, 81
High-order cumulants (HOCs), 260
High-order moments (HOMs), 260
High pass filters (HP filters), 261
HIPERLAN/2 standard-based MIMO-OFDM simulator, 334
Histogram equalization, 8–9, 13
History recording, 67–68
Hive query language (HQL), 126
HMMs, *see* Hidden Markov models
HOCs, *see* High-order cumulants
"Homomorphism", 21
HOMs, *see* High-order moments
HP filters, *see* High pass filters
HQL, *see* Hive query language
HR, *see* Human resource
HRM, *see* Human resource management
HSL color space, 6
HSV color system, *see* Hue saturation value color system
Hue saturation value color system (HSV color system), 4
Human bar code, *see* VeriChip chip
Human capital management, 460
Human learning, 21
Human marking, 67
Human resource (HR), 427
  analytics, 426, 427–430
  frequency of utilization of analytics in HR areas, 430–434
  function, 425
  level of benefits of HR analytics in enterprise management, 436
  types of analytics utilizing in particular HR areas, 434–435
Human resource management (HRM), 425
HW, *see* Hardware
Hybrid clouds, 462
Hybrid Energy-Efficient Distributed Clustering (HEED), 144, 145
Hybrid modeling, 22
Hybrid of steganography and cryptography keys, chaotic maps using, 304
HyperGraphDB, 127
Hypernym, 439–440
Hyperplane, 83, 84
"Hyperscale" data center, 463
Hypothesis, 436
  H_0, 283
  H_1, 283
  models, 30

I

IBk algorithm, 87–88
IBM, 456
ICT, *see* Information and communication technology
ID, *see* Identification number
IDC, *see* International Data Corporation
Identification number (ID), 65
Identify friend or foe (IFF), 64
iDTV, 118
IDWT, *see* Inverse discrete wavelet transform
IEC, *see* International Electrotechnical Commission
IEEE, *see* Institute of Electrical and Electronics Engineers
IEEE Internet of Things Initiative, 187
IEEE Learning Technology Standards Committee Standard (IEEE LTSC Standard), 119
IF, *see* Intermediate frequency
IFF, *see* Identify friend or foe
IGMP, *see* Internet Group Management Protocol
IIPA, *see* International Intellectual Property Alliance
ILP, *see* Integer linear programming
Image enhancement, 9
Image file format, 299
Image scrambling, 269
  parameters of chaotic flow systems in simulations, 273
  proposed scrambling algorithm, 270–273
  simulation results, 273–278
Implementation standards, 112
IMS Global Learning Consortium, Inc. (IMS), 111, 120
IMS Learner Information Package, 116
In situ verification and characterization of models, 30
"Include dependency" extraction, 222–223
Incremental DBSCAN, 148
Industrial specifications, 110
Information, 347
  and communication software, 403
  processing dimension, 80
  society development, 406–407
Information and communication technology (ICT), 68, 69, 177, 404, 455; *see also* Radio-frequency identification technology (RFID technology)
  drivers empowering intelligent enterprise, 462–463
  drivers of intelligent organization research, 458–462
  EGE, 68
  ethical aspects implanting in human body, 68
  intelligent enterprise, 456–458
  Opinion No 20, 68–73
Information security, 347–348
  steganography in, 300
Information systems (IS), 18, 53–54, 177
Information technology (IT), 19, 50, 53, 177, 425, 455
  and computer academic branches, 52–54
  CS, 52–53
  IS, 53–54
  managers, 463
Innovation, 455
Instance-based techniques, 128
Institute of Electrical and Electronics Engineers (IEEE), 110
Intangibility, 398
Integer linear programming (ILP), 166–167
Integrated transport system, 414
Integration architectures, 51–52
Integrity, 327, 329
Intellectual capital management, 456
Intelligent, 455
  ICT solutions for enterprise management, 456
  information systems, 456

intelligent organization research, ICT drivers of, 458–462
organization, 456, 462
Intelligent enterprise, 455–458
business continuity, 463
business flexibility, 463
converting data into business intelligence, 463
E-collaboration, 462–463
ICT drivers empowering, 462
mobile workforce integration, 462
scalability and customization, 463
smart virtual workplace, 462
Intelligent tutoring systems (ITS), 78
Interactiveness, 129
Inter-AS routing protocols, 156
Interdisciplinary, 313–314
Interjected regulation, 84–85
Intermediate frequency (IF), 232
Internal opportunities, 379
International Agro-Ecological Center, 409
International Data Corporation (IDC), 123, 426
International Electrotechnical Commission (IEC), 110
International Intellectual Property Alliance (IIPA), 408
International Organization for Standardization (ISO), 110
ISO 18000–6, 65
ISO 18000–6A, 65
ISO 18000–6B, 65
ISO 18000–6C, 65
ISO 9126 standards, 120
ISO/IEC 19796–1 standards, 120
International Telecommunication Union (ITU), 154
indoor channel model, 289
Internet, 177–178, 329, 442
internet-based gadgets, 461
Internet computer games, 440
Internet Group Management Protocol (IGMP), 156
Internet of Things (IoT), 123, 187; *see also* Electronic government (E-government)
Amazon AWS IoT, 191
applications and implementation, 187–188
development tools, 190
IEEE Internet of Things Initiative, 187
Microsoft Azure IoT Suite, 191–192
Texas Instruments' IoT kit, 190–191
TI CC3200 Internet-on-a-chip system on chip, 190
UW Tacoma IoT course, 188–190
Internet piracy, high level of, 407–408
"Internet pirates", 408
Internet protocol (IP), 153
compression, 156
multicasting, 156
Interoperability, 110, 111
standards, 118
Interrupt(s), 205–206
context, 206
data type, 205
handler function, 205
variable, 205
Interrupt service routines (ISRs), 205–206
Intersymbol Interference (ISI), 333–334
Intra-AS routing protocols, 156
Intracluster communication, 142
Intrinsic motivation, 84
Introjected regulation, 85
Introversion, 114
Intuition, 114
Intuitive learner, 114
Inverse discrete wavelet transform (IDWT), 304
IoT, *see* Internet of Things
IP, *see* Internet protocol
IPTV network, 153
designed and implemented IPTV network model, 156
evaluating network performance in case of coded video transmission, 160–163
evaluating with uncompressed video transmission, 159–160
hardware, 155
OPNET, 155–157
performance characterization, 157
protocols, 156
technology, 154
IS, *see* Information systems
ISDBSCAN, 148
ISI, *see* Intersymbol Interference
ISO, *see* International Organization for Standardization
ISRs, *see* Interrupt service routines
IT, *see* Information technology
ITS, *see* Intelligent tutoring systems
ITU, *see* International Telecommunication Union

J

J48 classifier, 87
Jaql, 126
JAVA Annotation Patterns Engine rules (JAPE rules), 218, 222
to extract "extend dependency", 223
to extract "include dependency", 223
to extract actor concept, 224
to extract objects, 222
to extract use case concepts, 222
JavaScript Object Notation (JSON), 126

K

k-coverage, 167
KD process, *see* Knowledge discovery process
Kernelized secure operating system (KSOS), 212
Key performance indicators (KPIs), 425
Key scheduling algorithm (KSA), 348
of RC4 algorithm, 349
Key–value stores, 127
Keyword selection, 317
Kinesthetic learning, 113
k-means
algorithm, 147–148
clustering, 170
k nearest neighbors method, 87, 128
KNIME, data mining tools, 128
KNN algorithm, 129
Knowledge-based economy, 459
Knowledge discovery process (KD process), 124, 127
Knowledge management, 456
Knowledge Tree framework, 117
Known cover attack, 299
Known message attack, 299
Known stego attack, 299
Kolb Learning Style Indicator, 113

KPIs, *see* Key performance indicators
Kryptos, 321
KSA, *see* Key scheduling algorithm
KSOS, *see* Kernelized secure operating system

L

Labeled emotion database, 317
Lagrange multipliers, 470
LAN parties, 444
Language, 41
Laptops, 439–440
Latvian as second language students (LSL students), 43
Layer-oriented functions, 364
L-BFGS algorithm, *see* Broyden–Fletcher–Goldfarb–Shanno algorithm
LCA, *see* Linked cluster algorithm
LD, *see* Learning design
LEACH, 142, 144, 145
Learner
  modeling, 116
  profile information, 111
  registration information, 111
Learner Information Package (LIP), 111
Learning
  approach, 385
  content, 112
  environment, 113
  goals, 54, 56
  organization, 456
  resources, 113
  styles scale, 114
  systems, 113
  technology, 112
Learning algorithms, 337
  AFP algorithm, 340
  LMS algorithm, 337–338
  NLMS algorithm, 338–339
  RLS algorithm, 339
Learning design (LD), 117
Learning management system (LMS), 117–118
Learning Object Management (LOM), 119
Learning object metadata (LOM), 118
Learning Technology Standards Committee (LTSC), 118
Least mean square algorithm (LMS algorithm), 337
Least significant bit (LSB), 301
  steganography technique, 303
Legend of Zelda series, 450
Lempel-Ziv-Welch technique (LZW technique), 303
Lesson plan model, 27–28
Level standards, 112
LFSRs, *see* Linear feedback shift registers
lib-SVM, 131
Life cycle of opportunity, 384
Linear feedback shift registers (LFSRs), 348
Linear programming techniques, 166–167
Linguistic studies, 317–315
Linguistically diverse first-year university students, solving problems of, 41
  aim, hypotheses, and research methods, 42
  data analyses and evaluation, 44–47
  data collection methodology, 43–44
  description of experiment, 44
  independent samples test, 46
  Latvian and linguistically diverse students' results, 45
  research methodology, 43–44
  students' evaluation of effectiveness of form, 45, 46
  theoretical background, 42–43
Linked cluster algorithm (LCA), 142, 146
LIP, *see* Learner Information Package
LMS, *see* Learning management system
LMS algorithm, *see* Least mean square algorithm
LNA, *see* Low noise amplifier
Local multidisciplinary, 314
LOCALSEARCH algorithm, 147–148
LOCK, *see* Logical coprocessing kernel
Lock mechanism, 214
Logged data, 44
Logic learning, 22
Logical coprocessing kernel (LOCK), 212
Logic-based techniques, 128
Logistic processes identification in organizations, 389–391
  in public services, 390
Logistic regression, 86–87
Logistics in corporate environmental strategies, 414
LOM, *see* Learning Object Management; Learning object metadata
Lossy compression, 153–154
Low noise amplifier (LNA), 232
Low pass filters (LP filters), 261
LP filters, *see* Low pass filters
LSB, *see* Least significant bit
LSL students, *see* Latvian as second language students
LTSC, *see* Learning Technology Standards Committee
LZW technique, *see* Lempel-Ziv-Welch technique

M

MAC protocols, *see* Media access control protocols
Machine translation technology, 42, 43
Machines learning, 83–84
MACM, *see* Modified android cat map
Macroscenaristic level, 28
Management level, 55
  analytics utilizing in particular HR areas depending on, 434–435
  Frequency of HR analytics utilization depending on, 431
Management through opportunities, 377
MANETs, *see* Mobile ad hoc networks
Map function, 126
MapPlaces, 366
MapReduce, 126, 129
  classification algorithms based on, 130–131
MapReduce based parallel backpropagation neural network (MRBPNN), 131
MapReduce-based k-nearest neighbor approach (MR-KNN), 131
Marking of objects, 66–67
M-ary frequency shift keying (MFSK), 259
M-ary FSK system, 263
MAS, *see* Multiagent system
Massive online open courses (MOOC), 18
Massively multiplayer online (MMO), 439–440
  games and variants, 442
Master map (M-map), 115
Mathematical model of wireless channel, 335

time-invariant channel, 335–336
time-variant channel, 336–337
MATLAB
program, 265
simulation program, 236
simulations, 234
MaxDist normalization, 167, 170
Mbps, *see* Megabits per second
MBTI, *see* Myers Briggs Type Indicator
McDonald's, 380
MDA, *see* Model-driven architecture
Mean square error (MSE), 300, 341
Meaning domain, 112
Mean-shift algorithm, 4
Mean-squared error (MSE), 335
Mechanism model, 30
Media access control protocols (MAC protocols), 98, 99
Megabits per second (Mbps), 153
Member nodes (MNs), 144
Membership functions (MFs), 262
Memory, 95, 96
blocks and addresses, 215
management, 209–211, 215
management in Craig's model, 213
MEMS, *see* Microelectromechanical systems
Merriam-Webster Dictionary, 378
Mesoscenaristic level, 28
Message delivery ratio, 251
Message Listener, 247
Message queuing telemetry transport (MQTT), 189
MessageStatsReport, 246
Metadata, 111, 118
Metamodel, 18
conceptual setting construction for teaching, 29–30
general approach of, 30–31
modeling in nonexact "human sciences", 29
of teaching process–learning online training, 27
theoretical setting of modeling approach and perspective, 29–30
MEVI, 101
MFs, *see* Membership functions
MFSK, *see* M-ary frequency shift keying; Multiple frequency shift keying signal
MHDM routing algorithm, *see* Minimum Hop Disjoint Multipath routing algorithm
MHDMwTS, *see* Minimum Hop Disjoint Multipath routing algorithm with Time Slice
Microcells in human body, 66
Microcomputers, 439–440
Microelectromechanical systems (MEMS), 93
Microscenaristic level, 28
Microsoft, 380–381
Azure IoT Suite, 191–192
Power BI, 192
Windows, 110
Middle managers, 434
Middle-level managers, 426
Military applications, 259
MIME, 118
"Mimicry", 21
MIMO-OFDM, *see* Multiple output-orthogonal frequency division multiplexing
MinDist normalization, 167, 170
Minimum enclosing ball clustering, 130
Minimum Hop Disjoint Multipath routing algorithm (MHDM routing algorithm), 100
Minimum Hop Disjoint Multipath routing algorithm with Time Slice (MHDMwTS), 100
protocol, 101
Minimum variance distortionless response (MVDR), 467
beamformer design model, 468–471
numerical results and analysis, 471–479
Missions, 56
Mixer, 232
M-map, *see* Master map
MME model, *see* Multi-label maximum entropy model
MMO, *see* Massively multiplayer online
MNs, *see* Member nodes
Mobile
devices, 363
technologies, 462
workforce integration, 462
Mobile ad hoc networks (MANETs), 244
Mobility/actuation unit, 96
Modal shift, 415
Model
of activities and learning tasks, 33
of context, 33
of educational scenario, 27
process product, 31
transformation, 220
Model-driven architecture (MDA), 217, 220
Modeled learning objects, 22
Modeling
conceptual setting construction for teaching by, 29–30
of theoretical foundations of virtual environment, 34
Modified android cat map (MACM), 304
Moments, 261–262
MongoDB, 127
Monte Carlo sampling, 171
MOOC, *see* Massive online open courses
Moodle, e-learning platform, 44
Moroccan academic curriculum, modeling notion in, 20–21
Moroccan context, 33
Motivation, 42, 84
MPMP transmission, *see* Multipath Multipriority transmission
MQTT, *see* Message queuing telemetry transport
MRADG, *see* Multi-Reaction Automata Direct Graph
MRBPNN, *see* MapReduce based parallel backpropagation neural network
MR-KNN, *see* MapReduce-based k-nearest neighbor approach
MR-Stream algorithm, 147
MSE, *see* Mean square error; Mean-squared error
MSOE-ODA, 17
Multiagent system (MAS), 79
Multichannel cross-layer architecture, 100
Multichannel routing, 98–100
Multiclass classification, 128
Multidisciplinary in TED, 313–315
Multifactor knowledge, 23
Multilabel classification, 128
Multi-label maximum entropy model (MME model), 316
Multilevel cluster hierarchy, 142

Multimedia
data, 94
environment, 19
sensors, 148
Multipath Multipriority transmission (MPMP transmission), 100
Multipath propagation, 333
Multipath routing, 100–101
Multiple access communication, 66
Multiple frequency shift keying signal (MFSK), 260
and mathematical expressions, 260–261
performance of MFSK system, 267
signal, 260
Multiple opinions in sentence, 317
Multiple output-orthogonal frequency division multiplexing (MIMO-OFDM), 334
Multi-Reaction Automata Direct Graph (MRADG), 322
embedding, 324–325
example, 326, 327, 328
MRADG with transition function, 325–326
number of values in design, 322–323
transition function, 323–324
MUSIC algorithm, 334
Mutexes, 214
MV, 153–154, 158, 160–161
MVDR, *see* Minimum variance distortionless response
Myers Briggs Type Indicator (MBTI), 113–114

N

Naive Bayes classifier (NBC), 88, 128, 130
Bayes theorem, 88
example, 88–89
National Center for Scientific and Technical Research (CNRST), 25
National Informatization Program, 405
National University of Life and Environmental Sciences of Ukraine (NULES), 408
Natural language processing (NLP), 312
NBC, *see* Naive Bayes classifier
NCC, *see* Normalized cross-correlation
Need for autonomy, 84
Need for competence, 84
Need for social belonging, 84
Negation handling, 317
Negative biases, 353
Neighbor filter classification (NFC), 129
Neo4j, 127
Network layer protocols, 98
Networked Readiness Index, 406
New Out-Of-Box Software (NOOBS), 9
NFC, *see* Neighbor filter classification
NIMBLE, 131
NLMS algorithm, *see* Normalized least mean square algorithm
NLP, *see* Natural language processing
Node mobility, 142–143
Nodes, 142
Nokia, 380–381
Noncooperative scenario, 289–290
Nonexact "human sciences", modeling in, 29
Nonmodeling, conceptual setting construction for teaching by, 29–30
Non-Popperian space of natural reasoning, 29
NOOBS, *see* New Out-Of-Box Software
Normalization methods, 167
Normalized cross-correlation (NCC), 300
Normalized least mean square algorithm (NLMS algorithm), 335, 337, 338–339
Not only SQL (NoSQL), 126–127
Noticing opportunity, 383–384
NP-complete problem, 166
NULES, *see* National University of Life and Environmental Sciences of Ukraine
Null process, 201
Null-forming beamformer, 471
Numerical results and analysis, 471
beampattern for different number of snapshots, 471–472
comparison between performance of proposed method and conventional MVDR, 478
key system parameters, 472
MVDR beamformer, 473, 474, 475, 476–477
output SINR *vs.* SNR for array response mismatch, 479
Numerical/economical sovereignty, 73

O

Object Management Group (OMG), 220
Object modeling technology (OMT), 18
Object tracking, 3, 10
using CamShift algorithm, 4–6
using color detection, 6–8
Objects, 36, 79
Observation, data collection, modeling, interpretation, objective, learning situations, evaluation (ODMIOSE), 17
Observatory education research and university teaching (ORDIPU teaching), 19
ODAC, *see* Divisive-Agglomerative Clustering
ODMIOSE, *see* Observation, data collection, modeling, interpretation, objective, learning situations, evaluation
OM, *see* Opinion mining
OMG, *see* Object Management Group
Omni-ID company, 67
OMT, *see* Object modeling technology
"On Agricultural Advisory Activities", 409
1D logistic map, 304
ONE simulator, 246, 247
ONE simulator, *see* Opportunistic Network Environment simulator
Online mining, 140
Online training, 16
Ontology for Selection of Personalization Strategy standard (OSPS standard), 115
Ontology semantics, 131
Open Shortest Path First (OSPF), 156
OpenCV libraries, 4, 10
Operating system (OS), 195
Operational Programme Innovative Economy, 459
Operational Programme Intelligent Development, 459
Opinion mining (OM), 311
Opinion No 20, 68, 71–72
applications and research, 69
classification of ICT implants, 69

ethical aspects of evolution of information society, 72–73
ethical background, 70–71
future research directions, 69
legal background, 70
other considerations, 73
Opinion spam, 317
OPNET, *see* Optimized Network Engineering Tools
Opportunistic Network Environment simulator (ONE simulator), 243–245
Opportunities, 378
approaches to opportunity, 383
creation in management of enterprise, 378
identification in context of strategic management, 384–385
identification of opportunities, 379
Nokia, 380–381
noticing opportunity, 383–384
Poland, 380
theoretical differences in meanings of *chance* and *opportunity*, 378
typology of opportunities, 382
ways of creating opportunities, 381
ways of generating opportunities, 382
Optimization objective, 141
Optimized Network Engineering Tools (OPNET), 154, 156–157
capabilities, 155
IPTV network configuration over, 155–156
Orange, data mining tools, 128
ORDIPU teaching, *see* Observatory education research and university teaching
Organizational specifications, 110
Organizations accomplishing functions, 56
Organizations performing missions, 56
OS, *see* Operating system
OSPF, *see* Open Shortest Path First
OSPS standard, *see* Ontology for Selection of Personalization Strategy standard
Overenhancement, 9
Overgeneration metric, 225

P

Packet delay variation (PDV), 154, 157
Packet jitter, 158
Palmtops, 439–440
PANEL, *see* Position-Based Aggregator Node Election protocol
PAP, *see* Personal Action Plan
Parallel SVM based on MapReduce algorithm (PSMR algorithm), 131
PARP, *see* Polish Agency for Enterprise Development
Particle Swarm Optimization (PSO), 144
Partnership-based relations, 392
Passenger transport, 421
Passive sensors, 310
Path acknowledgment, 100–101
Path build up, 100–101
Pattern recognition (PR), 259
Payload length, 103
PDV, *see* Packet delay variation
Peak signal-to-noise ratio (PSNR), 270, 300
Pedagogical model, 33–34
Pedagogical online training, 19
Pedagogical scenarios, contributions towards modeling of, 27
commentary, 33–35
directional and semidirectional questionnaires, 32
general approach of metamodel, 30–31
grid of analysis of approaches, 34
lesson plan model, 27–28
modeling levels, 28
theoretical setting of modeling approach, 29–30
PEGASIS, *see* Power-efficient gathering in sensor information systems
PEGASUS, *see* Peta-Scale Graph Mining System
Perception of information dimension, 80
Perceptron-based techniques, 128
Personal Action Plan (PAP), 18
Personalization, 114
Personalized e-learning
elements, 114–115
goals, 115
methods, 115–116
Personalized learning, 114
Personalized Learning System (PLS), 117
Peta-Scale Graph Mining System (PEGASUS), 131
Phase locked loop (PLL), 233
Physical channels, 336–337
PIA, *see* Privacy Impact Assessment
Piecewise linear chaotic map (PWLCM), 301
PigLatin, 126
PIM-SIM, 156
PIM-SM, *see* Protocol Independent Multicast-Sparse Mode
Pitch, *see* Fundamental frequency
Plant marking, 67
PLL, *see* Phase locked loop
PLS, *see* Personalized Learning System
POIs, *see* Positions of interest
Polarity reversers, 317
Polish Agency for Enterprise Development (PARP), 458
Polish language, 378
Polish Security Printing Works, 395
"Polluter pays", 415
Poltava Regional Agricultural Advisory Service (PRAAS), 409
Poltava Regional Public Organization Official Agricultural Advisory Service (TRPO OAAS), 409
Poltava State Agrarian Academy (PSAA), 409
Population targets, 25
"Portable" layer, 198
POS tagging, 224
Position-Based Aggregator Node Election protocol (PANEL), 144
Positions, 56
performing missions, 56
Positions of interest (POIs), 366–367
Positive biases, 353
Power-efficient gathering in sensor information systems (PEGASIS), 144, 145
Power source, 95
PQ schemes, *see* Priority queuing schemes
PR, *see* Pattern recognition
PRAAS, *see* Poltava Regional Agricultural Advisory Service
Praat software, 82

Practice, e-learning entities, 113
Precision, 225
Predictive model
in ASTEMOI system, 85
by exploring data, 85
exploring data and Weka, 86–89
Preemptive priority scheduling mechanism, 203
Prescribed and effective scenarios modeling, 21
hybrid modeling, 22
problem and difficulty of systemic modeling of effective courses, 21–22
PRGA, *see* Pseudo-random generation algorithm
PriGet (Priority get), 200
Primary user (PU), 229, 230
Primary user receiver (PU RX), 283
Primary user signals (PU signals), 281
Primary user transmitter (PU TX), 283
Principle
of data minimization, 70
of dignity, 70
of integrity and inviolability of body, 70
of proportionality, 70
Priority get, *see PriGet*
Priority inheritance protocol, 200
Priority inversion, 200
Priority queuing schemes (PQ schemes), 156
PriSet (Priority set), 200
Privacy crisis, 129
Privacy Impact Assessment (PIA), 66
PRNG, *see* Pseudorandom number generator
Proactive clustering algorithms, 144, 145
Probability (Probl), 170
Process, 199
attributes, 199
creation event, 199
management in Craig's models, 212–213
model, 31
orientation, 389, 397, 398
priority, 199–200
set, 199
table, 199
variable, 32
Process states, 200–201, 212–213
event(s), 201
variables, 200
Processing unit, 96, 138
Pro-environmental policy in railway sector, 421
Progression to understanding dimension, 80
Projected coordinate system, 365
PRoPHET, 247, 249–251
Proposed method, 364, 478
Proposed MVDR algorithm ($MVDR_{pro}$ algorithm), 472
Proposed system, 270, 271, 273, 364, 369
flowchart, 237
Proposed system database
data gathering, 365
preparing data, 366–367
Proposed technique
algorithm for building rating matrices, 370
creating rating matrix for implicit feedback, 369
rating matrix implementation, 370
splitting tracks, 367–369
Protocol Independent Multicast-Sparse Mode (PIM-SM), 156
Protocol steganography, 300
Provably secure operating system (PSOS), 211–212
PSAA, *see* Poltava State Agrarian Academy
Pseudo-random generation algorithm (PRGA), 348, 349
Pseudorandom number generator (PRNG), 303
PSMR algorithm, *see* Parallel SVM based on MapReduce algorithm
PSNR, *see* Peak signal-to-noise ratio
PSO, *see* Particle Swarm Optimization
PSO-based scheme, 146
PSO-C, *see* Centralized PSO
PSOS, *see* Provably secure operating system
PSTavg, *see* Average packet service time
PU, *see* Primary user
PU RX, *see* Primary user receiver
PU signals, *see* Primary user signals
PU TX, *see* Primary user transmitter
Public key steganography method, 298
Public key techniques, 326
Public logistics in supply chain, 391–393
Public sector
abstract activities, 399
classical supply chains, 395
classical value creation chains, 393
concept of recipient, 394
decision-making levels, 397
determinants of multilevel character of networks in, 391–393
effects-oriented control of public supply chain, 398
environment of process of providing public services, 396
exchange relations in, 393
flows occurring within supply chain, 395
public supply chain, 395, 396, 397
supply chain in, 393
Public services, 395–396
Public supply chain, 391, 392, 395–397
effects-oriented control of, 398
Public-key ciphers, 321
Public-key techniques, 321
Pure steganography method, 298
PWLCM, *see* Piecewise linear chaotic map

Q

QFD, *see* Quality function deployment
QoE, *see* Quality of experience
QOFL, *see* Quebec Board of rench Language
QoS, *see* Quality of service
QoSMOS, *see* Cross-layer QoS architecture
Quality, 112, 119
of assurance, 317
attributes, 120
Quality function deployment (QFD), 391
Quality of experience (QoE), 153, 157
Quality of service (QoS), 94, 142, 153, 156
Quantitative–qualitative analysis, 16
Quantum-based support vector machine algorithm, 130
Quebec Board of rench Language (QOFL), 38
Queue management, 206, 214–215
waiting messages, 208–209
Queue types, 214

R

R&D departments, *see* Research and development departments
RADG, *see* Reaction Automata Direct Graph
Radio frequency (RF), 232
  filter, 232
  interference, 467
  resources, 407
  signals, 64
Radio spectrum, 230
Radio-frequency identification technology (RFID technology), 64
  advantages, 65
  applications, 66–68
  badges, 67, 68
  environmental impact, 73
  ethics and life privacy, 66
  financial transactions, 67
  function, 65
  fundamentals, 64
  history recording, 67–68
  humans, animals, and plants marking, 67
  limitations, 65–66
  marking of objects, 66–67
  raised issues, 68
  safety, 68
  security issues in, 73–74
  standardization, 64–65
  stock regulation and inventory, 67
Rail transport in era of environmental green logistics, 421–422
Railway "open-air museum", 422
Random index, 349
Random normalization, 167
RapidMiner, data mining tools, 128
Raspberry Pi, 4
  camera module, 10
  Model B, 9
Rating matrix for implicit feedback, 369
RC4 algorithm, 347
  description, 348
  KSA, 348–349
  PRGA, 349
  weaknesses, 349–350
RC4 cipher; *see also* Steganography
  double-byte bias algorithm, 352–357
  encryption algorithms, 347
  RC4 algorithm description, 348–349
  RC4 algorithm weaknesses, 349–350
  RC4 keystream, 348
  RC4 stream cipher, 347
  single-byte bias algorithm, 350–352, 353
  single-byte bias attack algorithm, 354–355, 358–360
RCCP, *see* Receiver Congestion Control Protocol
RDF languages, 121
Reaction Automata Direct Graph (RADG), 322
  analysis, 329–330
  authentication, 326–327, 329
  cryptography, 321
  integrity, 327, 329
  MRADG, 322–326, 327, 328
Reactive clustering algorithms, 144, 145
Reader–reader collision, 66
Reader–tag communication, 65
Real-time operating systems (RTOSs), 196–197, 209
Real-time operation, 142
Reasoning dimension, 80
Recall, 224–225
Receiver Congestion Control Protocol (RCCP), 103
Receiver operating characteristics (ROC), 89, 233
Recognition task, 310
Recursive least square algorithm (RLS algorithm), 337, 339
Recursive mutex, 214
Reduce function, 126
Reference setting of research, 23
Reflective learner, 114
Regression, 128, 139
Regulation, 84–85
Rendezvous point (RP), 156, 157
Repair mechanisms, 142
Repair negative acknowledgment (RNACK), 103
Replication-based routings, 245
Request message (REQ message), 99
Research and development departments (R&D departments), 456
Research Institute for Information Technologies in Environmental Management, 408
Resources, 56
  constraint, 140
  life cycle analysis node, 56
Reusability, 110, 111
Reverse logistics, 414
RF, *see* Radio frequency
RFID technology, *see* Radio-frequency identification technology
Riga Technical University (RTU), 44
RIP, *see* Routing Information Protocol
RIPPER, 128
"Risk factor" level for students, 47
Rivest, Shamir, Adleman encryption (RSA encryption), 304
  transition function, 322
RLS algorithm, *see* Recursive least square algorithm
RNACK, *see* Repair negative acknowledgment
ROC, *see* Receiver operating characteristics
Role-playing games (RPG), 442
Roles entity, 113
Routing Information Protocol (RIP), 156
Routing protocols, 156, 244
RP, *see* Rendezvous point
RPG, *see* Role-playing games
RSA encryption, *see* Rivest, Shamir, Adleman encryption
RSAllocateFromHole, 213
RSL algorithm, 342
R-TOOL, 221
RTOSs, *see* Real-time operating systems
RTU, *see* Riga Technical University
"Rugs" (shock absorbers), 422
Run indexing, 246
RunToReady event, 201

S

SA, *see* Sentiment analysis
Safety, 68
SAM programs, *see* Software asset management programs

SAMOA, *see* Apache Scalable Advanced Massive Online Analysis
Sampling, 180–182
SAR, *see* Sequential Assignment Routing
Saudi portal, 180
SAX, *see* Simple Api for XML
Scalability, 129
Scalable Vector Graphics file (SVG file), 224
SCAP, *see* Source Congestion Avoidance Protocol
Scenarios, 27, 36, 37
Scenaristicweft, 39
Scenarization, 39
Scheduler states, 202–203, 214
Scheduling, 203–205
SCORM, *see* Sharable Content Object Reference Model
Scrambling process, 270
Screenwriting process analysis, 23
SDR, *see* Software defined radio; Software-defined radio
SDT, *see* Self-determination theory
Search window size, account for, 5–6
Second dimension, *see* Reasoning dimension
Secondary user (SU), 229, 283
Second-language learning process, 43
Secret key
  ciphers, 321
  steganography method, 298
Secret message, 297–299, 303, 304
Secure and Intelligent Internet of Things, 187–188
Secure Socket Layer protocol (SSL protocol), 347
Security architectures, 51–52
Self-determination theory (SDT), 78, 84–85
Self-government administration, 395–396
Self-organization, 165
Semantic labels (SLs), 315
Semaphores, 196, 197, 206, 212, 214
Semidirectional questionnaires, 32
Semistructured data, 125
Sensing, 114
  energy, 288
  judging and perceiving, 113
  learner, 114
  ratio, 288
  unit, 138
Sensor networks, 140, 143
  layers, 97–98
  wildlife tracking, 243
Sensor nodes (SNs), 94, 101, 137–138, 141
Sensors, 95, 165
  in object tracking, 140
  residual energy, 167
  unit, 96
Sentence splitting, 224
Sentence-level SA, 311
Sentiment analysis (SA), 310, 311–313, 316
Sequence analysis, 128
Sequencing, 118
Sequential Assignment Routing (SAR), 144
Sequential learner, 114
Sequential minimal optimization algorithm (SMO algorithm), 131
Serial number, 65
Service-oriented architecture (SOA), 59
SERVQUAL method, 391
SetProcessStatusToReady, 212–213
SetProcessStatusToRunning, 212–213
SetProcessStatusToWaiting, 212–213
Set-top boxes (STBs), 156
Sharable Content Object Reference Model (SCORM), 111–112, 118
Short-time Fourier transform (STFT), 261
Shrinkage, 427
Signal
  of BFSK, 260
  models, 334
  processing, 469
Signalto-Interference plus Noise Ratio (SINR), 468
Signal-to-noise ratio (SNR), 259–260, 282, 300
Simple Api for XML (SAX), 224
Simple rule-based engines, 117
Simulation
  cooperative scenario, 290–294
  evaluating network IPTV with uncompressed video transmission, 159–160
  evaluating network performance in case of coded video transmission, 160–163
  experiments, 171–174
  games, 442
  methodology and analysis of, 263–267
  noncooperative scenario, 289–290
  results and discussion, 158, 289
  scenario, 246
  simulation parameters, 290
  video coding traffic generation, 158
Simulator choice, 245–246
Single-byte bias algorithm
  to analyzing RC4 cipher, 350
  distribution of keystream bytes, 351–352
  measuring distributions of bytes of keystream, 350
  measuring distributions of RC4 keystream, 353
  on RC4, 354–355
  recovery rate for first 32 positions, 359–360
  single-byte bias attack, 358
Single-path routing, 101–106
SINR, *see* Signalto-Interference plus Noise Ratio
Sleep time function, 202
Slepian-Wolf DSC, 103
SLs, *see* Semantic labels
Small-and medium-sized enterprises (SMEs), 458
Smart objects, 189–190, 191
Smart virtual workplace, 462
Smartphones, 380, 439, 440, 441, 462
  applications, 461
  dynamic development, 452
  free applications installed on, 442
SME, *see* Subject matter expert
SMEs, *see* Small-and medium-sized enterprises
SMO algorithm, *see* Sequential minimal optimization algorithm
SNR, *see* Signal-to-noise ratio
SNs, *see* Sensor nodes
SOA, *see* Service-oriented architecture
SoC, *see* System on chip
Social factor, 414
Social media, 129, 461
Social services, 389–391
Software asset management programs (SAM programs), 407
Software defined radio (SDR), 259

advantages of, 231–232
communication systems, 231
Software-defined radio (SDR), 230
Solution approach, BSO algorithm, 170
Source Congestion Avoidance Protocol (SCAP), 103
Spatial
data, 364
domain techniques, 299
Speaking, 81, 82
SPECIfication and Assertion Language (SPECIAL), 211–212
"Specificity–generality" modeling, 16
Spectrum sensing process, 233, 286
Speed, 129, 329, 459
SPIN, 144
Spirit of time, 29
Spray and Wait protocols (SW protocols), 249, 250
Square soot-NLMS algorithm (SR-NLMS algorithm), 335
SSL protocol, *see* Secure Socket Layer protocol
Stability, 142
of functioning and development of ICT, 405
stability–generality, 16
Standardization process, 110
State diagram, 200
Statistical
analysis, 430
features, 233–234
learning techniques, 128
moments, 233, 261–262
proposed adaptive cognitive radio detection system, 234–236
second-order property, 469–470
STBs, *see* Set-top boxes
Steganography, 297; *see also* RC4 cipher
advantages and disadvantages of steganography, 300
algorithms, 299–300
chaos in, 300–304
techniques, 297–300
Stego only attack, 299
Stego tool algorithm, 299
STFT, *see* Short-time Fourier transform
Stock regulation and inventory, 67
"Store–carry–forward", 244–245
Storm, 126
Strategic games, 442
Strategic management theory
creation of opportunities in management of enterprise, 378–384
identification of opportunities in context of strategic management, 384–385
management through opportunities, 377
STREAM algorithm, 147–148
Structured data, 125
Style agent, 80
SU, *see* Secondary user
Subject matter expert (SME), 115
Subtractive fuzzy cluster means (SUBFCM), 148
Sufficient transport mechanisms, 154
Summarization, 128, 139
Supervised learning, 128
Supply chain management, 389, 414, 460
determinants of multilevel character of networks in public sector, 391–393
identification of logistic processes in organizations, 389–391
premises of public logistics in, 391–393
in public sector, 393–399
public value creation chain, 392
Support vector machine method (SVM method), 82, 124, 130–131
Sustainability, 415
Sustainable
development, 416
transport system, 422
SVG file, *see* Scalable Vector Graphics file
SVM method, *see* Support vector machine method
SW protocols, *see* Spray and Wait protocols
Symmetric
cryptography, 321
encryption, 347
Synchronization, 142
System
concepts extraction, 222
objectives, 367
and process approach, 391
System on chip (SoC), 190
Systemic computerized learning modeling, 23

T

Tablets, 439, 440, 452
Tacit-model (T-model), 115
Tag–reader
collision, 66
communication, 65–66
Tags, 64, 65
Task management, 199, 213–214
null process, 201
process priority, 199–200
process states, 200–201
process table, 199
timing behavior, 201–202
Task states, 214
TB, *see* Terabytes
TBF, *see* Trajectory-Based Forwarding
TCP/IP model, 94, 98, 190
TE, *see* Text emotion
TEA, *see* Text emotion analysis
Teacher's activity designer, 23
Teaching process, 27, 30
Teaching—learning model, 18, 30
Teaching–training modeling
approach, method, and object of, 22
construction of a conceptual framework for, 24
context of research, 25–27
multifactor knowledge, 23
reference setting of research, 23
training professionalizing in virtual environment, 25
variables of research modeling, 23–24
TEAF, *see* Treasury Enterprise Architecture Framework
Technology architecture, 51, 58
TED, *see* Text emotion detection
TEEN, *see* Threshold Sensitive Energy Efficient Sensor Network protocol
Terabytes (TB), 125
Texas Instruments' IoT kit, 190–191
Text emotion (TE), 309

Text emotion analysis (TEA), 310
Text emotion detection (TED), 309–317
  affective computing, 310–311
  challenges of, 316–317
  importance of, 310
  multidisciplinary in, 313–315
  SA, 311–313
  status of, 314–316
Text(s), 310
  file format, 299
  length, 317
  preprocessing, 223–224
  text-based approaches, 310
Thinking, 35, 43, 114, 169, 230
Third dimension, *see* Perception of information dimension
Three-dimension (3D), 153
  chaotic map equation, 303
  video, 153–154
3 V, *see* Volume, variety, velocity
Threshold Sensitive Energy Efficient Sensor Network protocol (TEEN), 144
TI CC3200 Internet-on-a-chip system on chip, 190
Tick length constant, 201
Tick variable, 201
Time saving energy efficient one-bit based-cooperative spectrum sensing (TSEEOB-CSSS), 282, 285
Time slice, 100–101
Time textual expressions, 309
Time-invariant channel, 335–336
  adaptive filtering architecture for wireless, 340
  general impulse response of, 336
  impulse response of, 342
  simulation, 342–344
Timer interrupt, 206
TimeToWake function, 202
Time-variant channel, 336–337
  simulation, 344
  wireless time-variant channel identification architecture, 341
Timing behavior, 201–202
TL-LEACH, *see* Two-Level Hierarchy LEACH
TLS protocol, *see* Transport Layer Security protocol
TME model, *see* Topic-level maximum entropy model
T-model, *see* Tacit-model
Tokenization, 224
Topic-level maximum entropy model (TME model), 315
ToS, *see* Type of service
Total message drops and by black holes, 254–255
TP, *see* True positive
TPGF, *see* Two-Phase Geographic Greedy Forwarding
Track segmentation algorithm, 368
TrackCollection, 366
Tracks, 366
Traditional classification algorithms, improving, 129–130
Traditional data mining, 143
  algorithms, 129
  techniques, 140
Traffic Generation, 158
Trainer model, 33
Training
  analysis tools, 29
  professionalizing in virtual environment, 25
  set, 128
Trajectory-Based Forwarding (TBF), 144
Transceiver, 95, 96, 138
Transdisciplinary approaches, 313–314
Transfer learning, 313
Transformation rule execution, 224
Transformation rules construction, 221
  extracting "extend dependency", 223
  extracting "include dependency", 222–223
  extracting actor concept, 223
  extracting system concepts, 222
  extracting use case concept, 222
Transmission data, type of, 94
Transponders, *see* Tags
Transport
  areas of green logistics and, 413–415
  sector, 413, 415
  services, 414
Transport area, analytical grounds for green logistics management in, 415–416, 417, 418
Transport Layer Security protocol (TLS protocol), 347
Transportation, green logistics effective operation area for, 416–421
Traveling, 363
Treasury Enterprise Architecture Framework (TEAF), 50
Trial-based learning, 87
TRPO OAAS, *see* Poltava Regional Public Organization Official Agricultural Advisory Service
True positive (TP), 89
Trust, 129
TSEEOB-CSSS, *see* Time saving energy efficient one-bit based-cooperative spectrum sensing
TTDD, *see* Two-Tier Data Dissemination
TTSDM, *see* Two-tiered service differentiation mechanism
Tumor, 144
Tutor agent, 80
TV content, 153
2D Arnold's cat map, 304
Two hypotheses, 233
Two-Level Hierarchy LEACH (TL-LEACH), 142, 144
Two-Phase Geographic Greedy Forwarding (TPGF), 100
Two-stage sensing method; *see also* Cognitive radio (CR); Communications
  design system model, 283, 284
  energy detection-based spectrum sensing, 283
  to improving energy consumption, 283–286
  proposed method, 286–289
  simulation results and discussion, 289–294
Two-Tier Data Dissemination (TTDD), 144
Two-tiered service differentiation mechanism (TTSDM), 102
Type attribute, 218
Type of service (ToS), 159
Typology of models, 18–19

U

UCgen, *see* Use Case Generator
UCLA, *see* University of California at Los Angeles
UCS, *see* Unequal Clustering Size
UID, *see* Unique identifier
Ukraine
  analysis of research and publications, 403–404

current state of information and communication provision, 403
economic growth, 405
European Union law and international law, 407
formulation of problem, 403
ICT development in Ukraine, 406
IIPA, 408
material of research, 404
process of creating advisory services, 409
ULA, *see* Uniform linear array
UML use case diagram, 220
evaluation, 224–225
execution of transformation rules, 224
generation from user requirements, 221
graphical representation, 224
proposed approach, 221
text preprocessing, 223–224
transformation rules construction, 221–223
UML use case diagram, *see* Unified Modeling Language use case diagram
UN Global E-Government Development Index, 406
Uncompressed video transmission, evaluating network IPTV with, 159–160
Unequal Clustering Size (UCS), 144
Unfavorable states, 82
Unified Modeling Language use case diagram (UML use case diagram), 217, 220, 221
Uniform linear array (ULA), 468, 469
Unique identifier (UID), 64
Unit impulse function, 336–337
Universal Transverse Mercator (UTM), 365
University of California at Los Angeles (UCLA), 212
University of Washington, Seattle (UW Seattle), 188
University of Washington, Tacoma (UW Tacoma), 188
IoT course, 188–190
UNOI, *see* User not of interest
Unstructured data, 125, 129, 463
Unsupervised learning, 79, 128
UOI, *see* User of interest
URS, *see* User Requirement Specifications
Use case
concept extraction, 222
diagram metamodel, 219
Use Case Generator (UCgen), 224
tool, 225
User not of interest (UNOI), 467–468
User of interest (UOI), 467–468
User Requirement Specifications (URS), 218
metamodel, 219
Users' GPS tracks dataset
creating rating matrix from POIs matrices, 373
GIS, 364–365
mobile devices, 363
proposed system database, 364–367
proposed technique, 367–369, 370
results, 369
sample of filling POIs matrix from track scanning, 372
system objectives, 367
user path on map, 371
UTM, *see* Universal Transverse Mercator

V

Validation variable, 22
Variable(s)
omen, 32
presage, 32
product, 32
program, 32
of research modeling, 23–24
Variety, 125, 129
VCO, *see* Voltage-controlled oscillator
Vector space of class's inverted position, 22
Velocity of data, 125, 129
Verbal learner, 114
VeriChip chip, 66, 67
Video
coding traffic generation, 158
compression techniques, 154
conferencing, 159
content material, 153
file format, 300
video-aware MMtransmission, 101
Video Codecs assessment
designed and implemented IPTV network model, 156–158
IPTV networks, 155–156
literature review, 154–155
lossy compression, 153–154
simulation results, 158–163
Vision-based systems, 3
Visual learner, 114
Visual learning style, 113
Vocal analysis, determining emotion from, 81
acoustic parameters, 81–82
automatic recognition of emotions, 82–83
hyperplane separating data belonging to two classes, 83
information conveyed in speech, 81
machines learning, 83–84
Praat software, 82
principle of emotion recognition system, 83
VOGC, *see* Voting-on-grid clustering
Voltage-controlled oscillator (VCO), 232
Volume, variety, velocity (3 V), 123, 125–126
Voting-on-grid clustering (VOGC), 146

W

W3C, *see* World Wide Web Consortium
Waiting
events, 214
messages, 208–209
to receive queue, 208
to send queue, 208
Warehouse management (WMS), 65
Watermarking algorithms, 299
Wavelet transform (WT), 261
WCA, *see* Weighted clustering algorithm
WCCP, *see* WMSN Congestion Control Protocol
Weak signals, 385
Web-based e-learning platforms, 44
WebTV, 153
WEF, *see* World Economic Forum
Weighted clustering algorithm (WCA), 146

Weighted high-order HMMs, 316
Weighted schemes, 146
Weka, 79
data exploration and, 86
data mining tools, 128
Decision Tree Algorithm J48, 87
IBk algorithm, 87–88
logistic regression, 86–87
naïve Bayes classifier, 88–89
WEP, *see* Wired equivalent privacy
Wi-Fi
Internet-on-a-chip, 190–191
network, 190–191
WIMAX broadband access technology, 154
Wired equivalent privacy (WEP), 347
Wireless
channel, 333–334
communication system, 335
environment, 103
Wireless blind channel identification
adaptive filter, 333–334
architecture, 340–341
existing work, 334–335
learning algorithms, 337–340
mathematical model of wireless channel, 335–337
results and discussions, 341
time-invariant channel simulation, 342–344
time-variant channel simulation, 344
Wireless multimedia sensor networks (WMSN), 93, 94, 95–97, 148
advanced technology in low-power circuits, 94
applications, 97
architecture, 96–97
cross-layer architecture, 97–98
cross-layer protocols for, 98–106
multichannel routing, 98–100
multipath routing, 100–101
sensor network layers, 97–98
single-path routing, 101–106
WSN, 94–95
Wireless sensor network (WSN), 93, 94, 137–138, 165, 166
applications, 95, 140–141
challenges, 140
clustering algorithms schemes in, 144–147
clustering routing protocols in, 144
coverage problem in, 166–167
data mining in, 139
designing clusters in, 142
implementation of WSN data mining, 141
managing and processing data, 139–140
taxonomy, 140
WMS, *see* Warehouse management
WMSN, *see* Wireless multimedia sensor networks
WMSN Congestion Control Protocol (WCCP), 103
World Economic Forum (WEF), 406
World Economic Forum and Accenture (2009), 413
World Wide Web Consortium (W3C), 110
World Wide Web Foundation on Internet development ranking, 406
WSN, *see* Wireless sensor network
WT, *see* Wavelet transform

X

xAPI, 111
XL-WMSN protocols, 102
XML languages, 117, 121
XSLT, *see* eXtensible Stylesheet Language Transformation

Y

"Yesser" e-government program, 180

Z

Zero-forcing beamforming technique (ZF beamforming technique), 470
Zero-order moment, 5
Z-MAC, 147
Zone of proximal development (ZPD), 25, 32, 33, 37